W9-AET-486

System Identification

Theory for the User

Second Edition

Prentice Hall Information and System Sciences Series
Thomas Kailath, Editor

Åström & Wittenmark	Computer-Controlled Systems: Theory and Design, 3rd ed.
Friedland	Advanced Control System Design
Haykin	Adaptive Filter Theory, 3rd ed.
Kailath	Linear Systems
Kung, Mak & Lin	Biometric Authentication: A Machine Learning Approach
Ljung	System Identification: Theory for the User, 2nd ed.
Ljung & Glad	Modeling of Dynamic Systems
Macovski	Medical Imaging Systems
Rugh	Linear System Theory, 2nd ed.
Soliman & Srinath	Continuous and Discrete Signals and Systems, 2nd ed.
Solo & Kong	Adaptive Signal Processing Algorithms: Stability & Performance
Srinath, Rajasekaran & Viswanathan	Introduction to Statistical Signal Processing with Applications
Wells	Applied Coding and Information Theory for Engineers

System Identification

Theory for the User

Second Edition

Lennart Ljung
Linköping University
Sweden

PRENTICE HALL PTR
Upper Saddle River, NJ 07458
http://www.phptr.com

Library of Congress Cataloging-in-Publication Data

Editorial/production supervision: *Jim Gwyn*
Cover design director: *Jerry Votta*
Cover design: *Anthony Gemmellaro*
Manufacturing manager: *Alexis R. Heydt*
Acquisitions editor: *Bernard Goodwin*
Marketing manager: *Kaylie Smith*
Composition: *PreTEX*

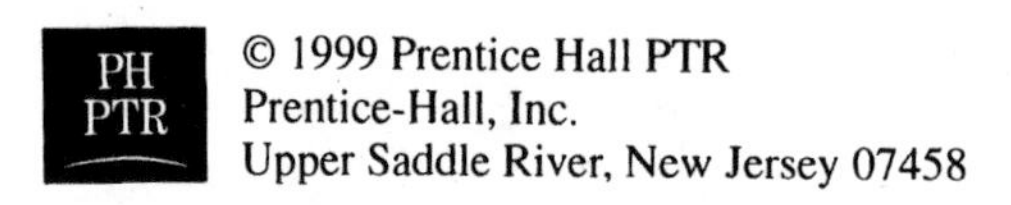

Prentice-Hall, Inc.
Upper Saddle River, New Jersey 07458

Prentice Hall books are widely used by corporations and government agencies for training, marketing, and resale. The publisher offers discounts on this book when ordered in bulk quantities. For more information, contact:

Corporate Sales Department
Phone: 800-382-3419
Fax: 201-236-7141
E-mail: corpsales@prenhall.com

Or write:

Prentice Hall PTR
Corp. Sales Department
One Lake Street
Upper Saddle River, NJ 07458

Text printed in the United States at Integrated Book Technology in Troy, New York.
10th Printing December 2007

ISBN 0-13-656695-2

Prentice-Hall International (UK) Limited, *London*
Prentice-Hall of Australia Pty. Limited, *Sydney*
Prentice-Hall of Canada, Inc., *Toronto*
Prentice-Hall Hispanoamericana, S. A., *Mexico*
Prentice-Hall of India Private Limited, *New Delhi*
Prentice-Hall of Japan, Inc., *Tokyo*
Prentice-Hall Asia Pte. Ltd., *Singapore*
Editora Prentice-Hall do Brasil, Ltda., *Rio de Janeiro*

To Ann-Kristin, Johan, and Arvid

Contents

part ii: methods

part iii: user's choices

Preface to the First Edition

System identification is a diverse field that can be presented in many different ways. The subtitle, *Theory for the User*, reflects the attitude of the present treatment. Yes, the book is about theory, but the focus is on theory that has direct consequences for the understanding and practical use of available techniques. My goal has been to give the reader a firm grip on basic principles so that he or she can confidently approach a practical problem, as well as the rich and sometimes confusing literature on the subject.

Stressing the utilitarian aspect of theory should not, I believe, be taken as an excuse for sloppy mathematics. Therefore, I have tried to develop the theory without cheating. The more technical parts have, however, been placed in appendixes or in asterisk-marked sections, so that the reluctant reader does not have to stumble through them. In fact, it is a redeeming feature of life that we are able to use many things without understanding every detail of them. This is true also of the theory of system identification. The practitioner who is looking for some quick advice should thus be able to proceed rapidly to Part III (User's Choices) by hopping through the summary sections of the earlier chapters.

The core material of the book should be suitable for a graduate-level course in system identification. As a prerequisite for such a course, it is natural, although not absolutely necessary, to require that the student should be somewhat familiar with dynamical systems and stochastic signals. The manuscript has been used as a text for system identification courses at Stanford University, the Massachusetts Institute of Technology, Yale University, the Australian National University and the Universities of Lund and Linköping. Course outlines, as well as a solutions manual for the problems, are available from the publisher.

The existing literature on system identification is indeed extensive and virtually impossible to cover in a bibliography. In this book I have tried to concentrate on recent and easily available references that I think are suitable for further study, as well as on some earlier works that reflect the roots of various techniques and results. Clearly, many other relevant references have been omitted.

Some portions of the book contain material that is directed more toward the serious student of identification theory than to the user. These portions are put either in appendixes or in sections and subsections marked with an asterisk (*). While occasional references to this material may be encountered, it is safe to regard it as optional reading; the continuity will not be impaired if it is skipped.

The problem sections for each chapter have been organized into four groups of different problem types:

- *G problems:* These could be of General interest and it may be worthwhile to browse through them, even without intending to solve them.
- *E problems:* These are regular pencil-and-paper Exercises to check the basic techniques of the chapter.
- *T problems:* These are Theoretically oriented problems and typically more difficult than the E problems.
- *D problems:* In these problems the reader is asked to fill in technical Details that were glossed over in the text.

Acknowledgments

Any author of a technical book is indebted to the people who taught him the subject and to the people who made the writing possible. My interest in system identification goes back to my years as a graduate student at the Automatic Control Department in Lund. Professor Karl Johan Åström introduced me to the subject, and his serious attitude to research has always been a reference model for me. Since then I have worked with many other people who added to my knowledge of the subject. I thank, therefore, my previous coauthors (in alphabetical order) Anders Ahlén, Peter Caines, David Falconer, Farhat Fnaiech, Ben Friedlander, Michel Gevers, Keith Glover, Ivar Gustavsson, Tom Kailath, Stefan Ljung, Martin Morf, Ton van Overbeek, Jorma Rissanen, Torsten Söderström, Göte Solbrand, Eva Trulsson, Bo Wahlberg, Don Wiberg, and Zhen-Dong Yuan.

The book has developed from numerous seminars and several short courses that I have given on the subject world-wide. Comments from the seminar participants have been instrumental in my search for a suitable structure and framework for presenting the topic.

Several persons have read and used the manuscript in its various versions and given me new insights. First, I would like to mention: Michel Gevers, who taught from an early version and gave me invaluable help in revising the text; Robert Kosut and Arye Nehorai, who taught from the manuscript at Stanford and Yale, respectively; and Jan Holst, who lead a discussion group with it at Denmark's Technical University, and also gathered helpful remarks. I co-taught the course at MIT with Fred Schweppe, and his lectures as well as his comments, led to many clarifying changes in the manuscript. Students in various courses also provided many useful comments. I mention in particular George Hart, Juan Lavalle, Ivan Mareels, Brett Ridgely, and Bo Wahlberg. Several colleagues were also kind enough to critique the manuscript. I am especially grateful to Hiro Akaike, Chris Byrnes, Peter Falb, Meir Feder, Gene Franklin, Claes Källström, David Ruppert, Torsten Söderström, Petre Stoica, and Peter Whittle.

Svante Gunnarsson and Sten Granath made the experiments described in Section 17.2, Bo Wahlberg contributed to the frequency-domain interpretations, and Alf Isaksson prepared Figure 13.9.

The preparation of the manuscript's many versions was impeccably coordinated and, to a large extent, also carried out by Ingegerd Stenlund. She had useful help from Ulla Salaneck and Karin Lönn. Marianne Anse-Lundberg expertly prepared all the illustrations. I deeply appreciate all their efforts.

Writing a book takes time, and I probably would not have been able to finish this one had I not had the privilege of sabbatical semesters. The first outline of this book was written during a sabbatical leave at Stanford University in 1980–1981. 1 wrote a first version of what turned out to be the last edition during a minisabbatical visit to the Australian National University in Canberra in 1984. The writing was completed during 1985–1986, the year I spent at MIT. I thank Tom Kailath, Brian Anderson and Sanjoy Mitter (and the U.S. Army Research Office under contract DAAG–29–84-K–005) for making these visits possible and for providing inspiring working conditions. My support from the Swedish National Board for Technical Development (STUF) has also been important.

Sabbatical or not, it was unavoidable that a lot of the writing (not to mention the thinking!) of the book had to be done on overtime. I thank my family, Ann-Kristin, Johan, and Arvid, for letting me use their time.

Lennart Ljung
Linköping, Sweden

Preface to the Second Edition

During the 10 years since the first edition appeared, System Identification has developed considerably. I have added new material in the second edition that reflects my own subjective view of this development. The new sections deal with *Subspace methods* for multivariable state-space models, with *Nonlinear Black-box models*, such as neural networks and fuzzy models, with *Frequency-domain methods*, with *input design, in particular periodic inputs*, and with more comprehensive discussions of *identification in closed loop* and of *data preprocessing*, including missing data and merging of data sets. In addition, the book has been updated and several chapters have been rewritten.

The new methods I have learnt directly from persons who have developed them, and they have also been kind enough to review my presentation: Urban Forssell, Michel Gevers, Alf Isaksson, Magnus Jansson, Peter Lindskog, Tomas McKelvey, Rik Pintelon, Johan Schoukens, Paul van den Hof, Peter Van Overschee, Michel Verhaegen, and Mats Viberg. The nonlinear black box material I learnt from and together with Albert Benveniste, Bernard Delyon, Pierre-Yves Glorennec, Håkan Hjalmarsson, Anatoli Juditsky, Jonas Sjöberg, and Qinghua Zhang as we prepared a tutorial survey for AUTOMATICA. The presentation in Chapter 5 is built upon this survey. Håkan Hjalmarsson and Erik Weyer have pointed out important mistakes in the first edition and helped me correct them. Pierre Carrette and Fredrik Gustafsson have read the manuscript and provided me with many useful comments. Ulla Salaneck has helped me keeping the project together and she also typed portions of the new material. I thank all these people for their invaluable help in my work.

I am grateful to SAAB AB and to Stora Skutskär Pulp Factory for letting me use data from their processes. My continued support from the Swedish Research Council for Engineering Sciences has also been instrumental for my work.

Software support is necessary for any User of System Identification. The examples in this book have been computed using MathWork's SYSTEM IDENTIFICATION TOOLBOX for use with MATLAB. Code and data to generate the examples can be found at the book's home page

http://www.control.isy.liu.se/~ljung/sysid

Readers are invited to this home page also for other material of interest to users of System Identification.

Operators and Notational Conventions

$\arg(z)$ = argument of the complex number z

$\arg\min f(x)$ = value of x that minimizes $f(x)$

$x_N \in AsF(n, m)$: sequence of random variables x_N converges in distribution to the F-distribution with n and m degrees of freedom

$x_N \in AsN(m, P)$: sequence of random variables x_N converges in distribution to the normal distribution with mean m and covariance matrix P; see (I.17)

$x_N \in As\chi^2(n)$: sequence of random variables x_N converges in distribution to the χ^2 distribution with n degrees of freedom

$\text{Cov}\, x$ = covariance matrix of the random vector x; see (I.4)

$\det A$ = determinant of the matrix A

$\dim \theta$ = dimension (number of rows) of the column vector θ

Ex = mathematical expectation of the random vector x; see (I.3)

$\overline{E}x(t) = \lim_{N\to\infty} \frac{1}{N} \sum_{t=1}^{N} Ex(t)$; see (2.60)

$\mathbf{O}(x)$ = ordo x: function tending to zero at the same rate as x

$\mathbf{o}(x)$ = small ordo x: function tending to zero faster than x

$x \in N(m, P)$: random variable x is normally distributed with mean m and covariance matrix P; see (I.6)

$\text{Re}\, z$ = real part of the complex number z

$\mathcal{R}(f)$ = range of the function f = the set of values that $f(x)$ may assume

$\mathbf{R}^d$ = Euclidian d-dimensional space

$x = \text{sol}\{f(x) = 0\}$: x is the solution (or set of solutions) to the equation $f(x) = 0$

$\text{tr}(A)$ = trace (the sum of the diagonal elements) of the matrix A

$\text{Var}\, x$ = variance of the random variable x

A^{-1} = inverse of the matrix A

A^T = transpose of the matrix A

A^{-T} = transpose of the inverse of the matrix A

$\overline{z}$ = complex conjugate of the complex number z

(superscript * is not used to denote transpose and complex conjugate: it is used only as a distinguishing superscript)

$y_s^t = \{y(s), y(s+1), \cdots, y(t)\}$

$y^t = \{y(1), y(2), \cdots y(t)\}$
$U_N(\omega)$ = Fourier transform of u^N; see (2.37)
$R_v(\tau) = \overline{E}v(t)v^T(t-\tau)$; see (2.61)
$R_{sw}(\tau) = \overline{E}s(t)w^T(t-\tau)$; see (2.62)
$\Phi_v(\omega)$ = spectrum of v = Fourier transform of $R_v(\tau)$; see (2.63)
$\Phi_{sw}(\omega)$ = cross spectrum between s and w = Fourier transform of $R_{sw}(\tau)$; see (2.64)
$\hat{R}_s^N(\tau) = \frac{1}{N}\sum_{t=1}^N s(t)s^T(t-\tau)$; see (6.10)
$\hat{\Phi}_u^N(\omega)$ = estimate of the spectrum of u based on u^N; see (6.48)
$\hat{v}(t|t-1)$ = prediction of $v(t)$ based on v^{t-1}
$\frac{d}{d\theta}V(\theta)$ = gradient of $V(\theta)$ with respect to θ: a column vector of dimension $\dim\theta$ if V is scalar valued
$V'(\theta)$ = gradient of V with respect to its argument
$\ell'_\varepsilon(\varepsilon, \theta)$ = partial derivative of ℓ with respect to ε
δ_{ij} = Kronecker's delta: zero unless $i = j$
$\delta(k) = \delta_{k0}$
$\mathcal{B}(\theta_0, \varepsilon)$ = ε neighborhood of θ_0 : $\{\theta \,|\, |\theta - \theta_0| < \varepsilon\}$
$\triangleq$ = the left side is defined by the right side
$|\cdot|$ = (Euclidian) norm of a vector
$||\cdot||$ = (Frobenius) norm of a matrix (see 2.91)

SYMBOLS USED IN TEXT

This list contains symbols that have some global use. Some of the symbols may have another local meaning.

$D_{\mathcal{M}}$ = set of values over which θ ranges in a model structure. See (4.122)
D_c = set into which the θ-estimate converges. See (8.23)
$e(t)$ = disturbance at time t; usually $\{e(t), t = 1, 2, \ldots\}$ is white noise (a sequence of independent random variables with zero mean values and variance λ)
$e_0(t)$ = "true" driving disturbance acting on a given system S; see (8.2)
$f_e(x)$, $f_e(x, \theta)$ = probability density function of the random variable e; see (I.2) and (4.4)
$G(q)$ = transfer function from u to y; see (2.20)
$G(q, \theta)$ = transfer function in a model structure, corresponding to the parameter value θ; see (4.4)
$G_0(q)$ = "true" transfer function from u to y for a given system; see (8.7)
$\hat{G}_N(q)$ = estimate of $G(q)$ based on Z^N
$G^*(q)$ = limiting estimate of $G(q)$; see (8.71)

$\tilde{G}_N(q)$ = difference $\hat{G}_N(q) - G_0(q)$; see (8.15)
$\mathcal{G}$ = set of transfer functions obtained in a given structure; see (8.44)
$H(q)$, $H(q,\theta)$, $H_0(q)$, $\hat{H}_N(q)$, $H^*(q)$, $\tilde{H}_N(q)$, $\mathcal{H}$: analogous to G but for the transfer function from e to y
$L(q)$ = prefilter for the prediction errors; see (7.10)
$\ell(\varepsilon)$, $\ell(\varepsilon,\theta)$, $\ell(\varepsilon,t,\theta)$ = norm for the prediction errors used in the criterion; see (7.11), (7.16), (7.18)
$\mathcal{M}$ = model structure (a mapping from a parameter space to a set of models); see (4.122)
$\mathcal{M}(\theta)$ = particular model corresponding to the parameter value θ; see (4.122)
$\mathcal{M}^*$ = set of models (usually generated as the range of a model structure); see (4.118).
P_θ = asymptotic covariance matrix of θ; see (9.11)
q, q^{-1} = forward and backward shift operators; see (2.15)
S = "the true system;" see (8.7)
$T(q) = [G(q)\; H(q)]$; see (4.109)
$T(q,\theta)$, $T_0(q)$, $\hat{T}_N(q)$, $\tilde{T}_N(q)$ = analogous to G and H
$u(t)$ = input variable at time t
$V_N(\theta, Z^N)$ = criterion function to be minimized; see (7.11)
$\overline{V}(\theta)$ = limit of criterion function; see (8.28)
$v(t)$ = disturbance variable at time t
$w(t)$ = usually a disturbance variable at time t; the precise meaning varies with the local context
$x(t)$ = state vector at time t; dimension = n
$y(t)$ = output variable at time t
$\hat{y}(t|\theta)$ = predicted output at time t using a model $\mathcal{M}(\theta)$ and based on Z^{t-1}; see (4.6)
$z(t) = [\,y(t) \quad u(t)\,]^T$; see (4.113)
$Z^N = \{u(0), y(0), \ldots, u(N), y(N)\}$
$\varepsilon(t,\theta)$ = prediction error $y(t) - \hat{y}(t|\theta)$
λ = used to denote variance; also, in Chapter 11, the forgetting factor; see (11.6), (11.63)
θ = vector used to parametrize models; dimension = d; see (4.4), (4.5), (5.66)
$\hat{\theta}_N$, θ_0, θ^*, $\tilde{\theta}_N$ = analogous to G
$\varphi(t)$ = regression vector at time t; see (4.11) and (5.67)
$\chi_0(t) = [\,u(t) \quad e_0(t)\,]^T$; see (8.14)
$\psi(t,\theta)$ = gradient of $\hat{y}(t|\theta)$ with respect to θ; a d-dimensional column vector; see (4.121c)
$\zeta(t)$, $\zeta(t,\theta)$ = "the correlation vector" (instruments); see (7.110)
$\mathbf{T}'(q,\theta)$ = gradient of $T(q,\theta)$ with respect to θ (a $d \times 2$ matrix); see (4.125)

ABBREVIATIONS AND ACRONYMS

ARARX: See Table 4.1
ARMA: AutoRegressive Moving Average (see Table 4.1)
ARMAX: AutoRegressive Moving Average with eXternal input (see Table 4.1)
ARX: AutoRegressive with eXternal input (see Table 4.1)
BJ: Box-Jenkins model structure (see Table 4.1)
ETFE: Empirical Transfer Function Estimate; see (6.24)
FIR: Finite Impulse Response model (see Table 4.1)
IV: Instrumental variables (see Section 7.6)
LS: Least Squares (see Section 7.3)
ML: Maximum Likelihood (see Section 7.4)
MSE: Mean Square Error
OE: Output error model structure (see Table 4.1)
PDF: Probability Density Function
PEM: Prediction-Error Method (see Section 7.2)
PLR: PseudoLinear Regression (see Section 7.5)
RIV: Recursive IV (see Section 11.3)
RLS: Recursive LS (see Section 11.2)
RPEM: Recursive PEM (see Section 11.4)
RPLR: Recursive PLR (see Section 11.5)
SISO: Single Input Single Output
w.p.: with probability
w.p. 1: with probability one; see (I.15)
w.r.t.: with respect to

1

INTRODUCTION

Inferring models from observations and studying their properties is really what science is about. The models ("hypotheses," "laws of nature," "paradigms," etc.) may be of more or less formal character, but they have the basic feature that they attempt to link observations together into some pattern. System identification deals with the problem of building mathematical models of dynamical systems based on observed data from the system. The subject is thus part of basic scientific methodology, and since dynamical systems are abundant in our environment, the techniques of system identification have a wide application area. This book aims at giving an understanding of available system identification methods, their rationale, properties, and use.

1.1 DYNAMIC SYSTEMS

In loose terms a *system* is an object in which variables of different kinds interact and produce observable signals. The observable signals that are of interest to us are usually called *outputs*. The system is also affected by external stimuli. External signals that can be manipulated by the observer are called *inputs*. Others are called *disturbances* and can be divided into those that are directly measured and those that are only observed through their influence on the output. The distinction between inputs and measured disturbances is often less important for the modeling process. See Figure 1.1. Clearly the notion of a system is a broad concept, and it is not surprising that it plays an important role in modern science. Many problems in various fields are solved in a system-oriented framework. Instead of attempting a formal definition of the system concept, we shall illustrate it by a few examples.

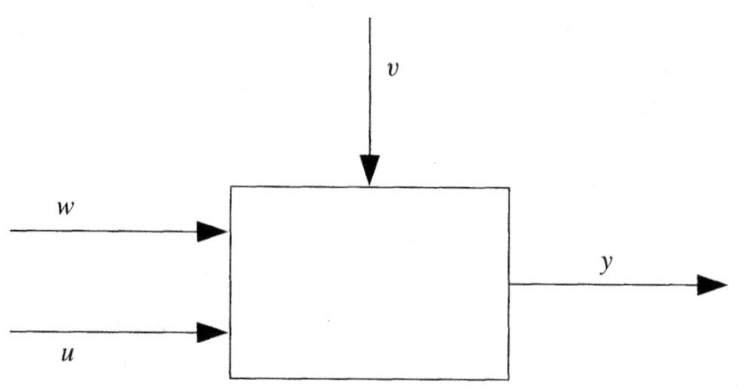

Figure 1.1 A system with output y, input u, measured disturbance w, and unmeasured disturbance v.

Example 1.1 A Solar-Heated House

Consider the solar-heated house depicted in Figure 1.2. The system operates in such a way that the sun heats the air in the solar panel. The air is then pumped into a heat storage, which is a box filled with pebbles. The stored energy can later be transferred to the house. We are interested in how solar radiation and pump velocity affect the temperature in the heat storage. This system is symbolically depicted in Figure 1.3. Figure 1.4 shows a record of observed data over a 50-hour period. The variables were sampled every 10 minutes. □

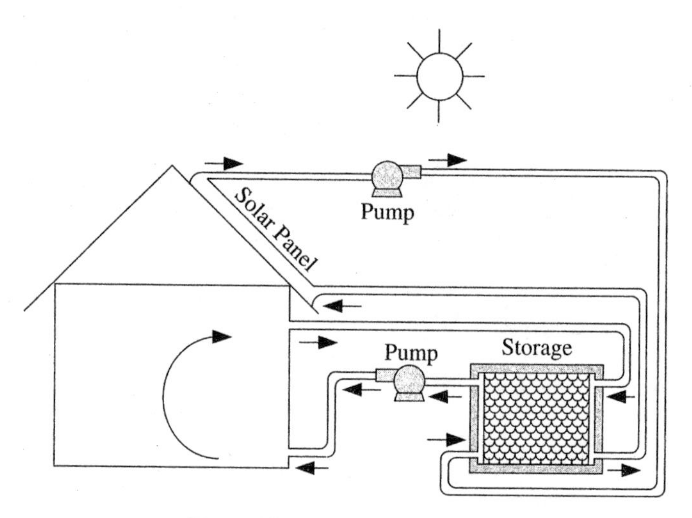

Figure 1.2 A solar-heated house.

Example 1.2 A Military Aircraft

For the development of an aircraft, a substantial amount of work is allocated to construct a mathematical model of its dynamic behavior. This is required both for the simulators, for the synthesis of autopilots, and for the analysis of its properties. Substantial physical insight is utilized, as well as wind tunnel experiments, in the course of this work, and a most important source of information comes from the test flights.

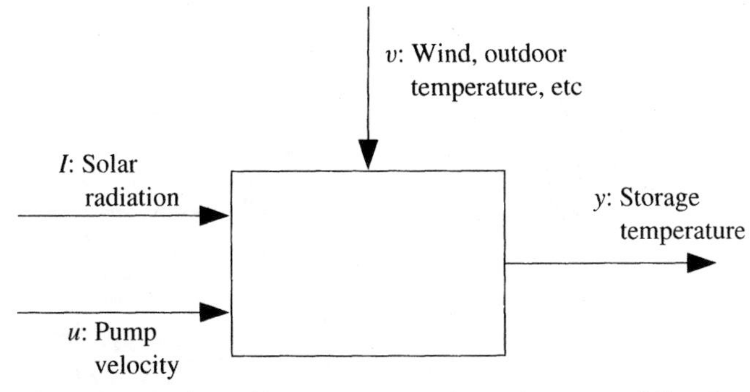

Figure 1.3 The solar-heated house system: u: input; I: measured disturbance; y: output; v: unmeasured disturbances.

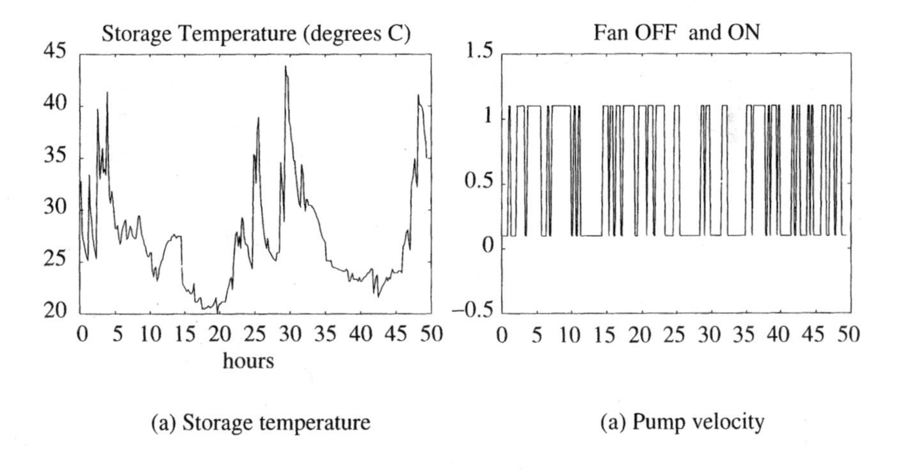

(a) Storage temperature

(a) Pump velocity

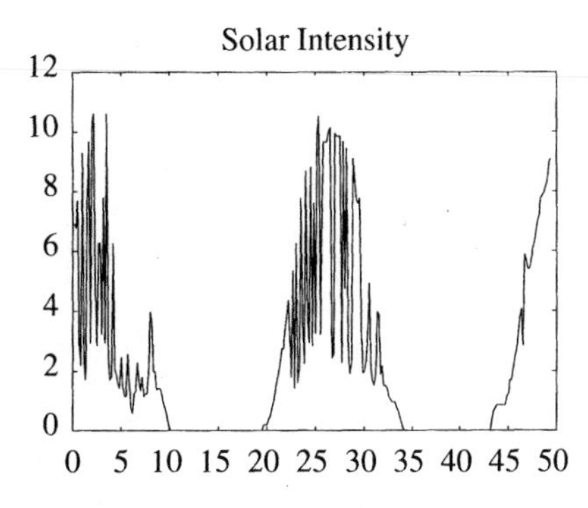

(a) Solar intensity

Figure 1.4 Storage temperature y, pump velocity u, and solar intensity I over a 50-hour period. Sampling interval: 10 minutes.

Figure 1.5 The Swedish fighter aircraft JAS-Gripen.

Figure 1.5 shows the Swedish aircraft JAS-Gripen, developed by SAAB AB, Sweden, and Figure 1.6 shows some results from test flights. Such data can be used to build a model of the pitch channel, i.e., how the pitch rate is affected by the three control signals: elevator, canard, and leading edge flap. The elevator in this case corresponds to aileron combinations at the back of the wings, while separate action is achieved from the ailerons at the leading edge (the front of the wings). The canards are a separate set of rudders at the front of the wings.

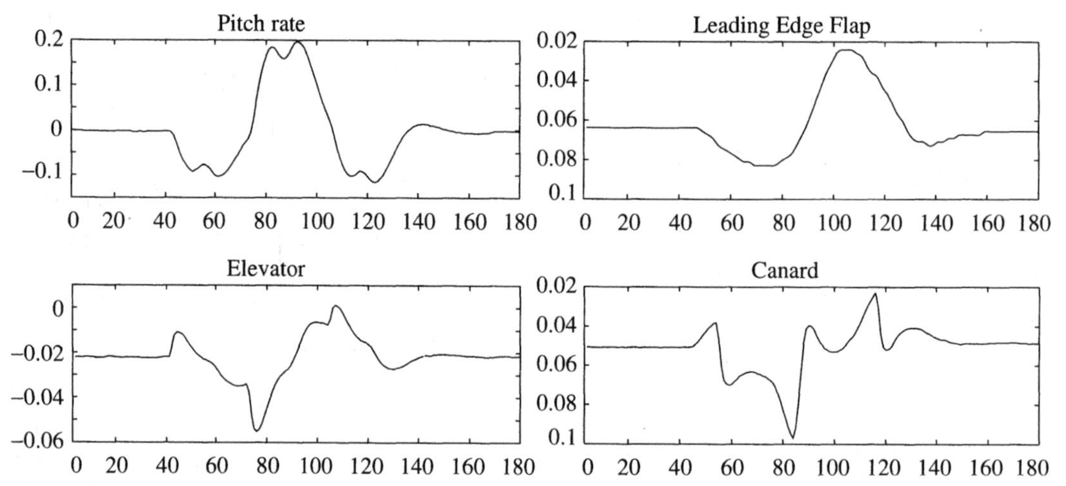

Figure 1.6 Results from test flights of the Swedish aircraft JAS-Gripen, developed by SAAB AB, Sweden. The pitch rate and the elevator, leading edge flap, and canard angles are shown.

The aircraft is unstable in the pitch channel at this flight condition, so clearly the experiment was carried out under closed loop control. In Section 17.3 we will return to this example and identify models based on the measured data. □

Example 1.3 Speech

The sound of the human voice is generated by the vibration of the vocal chords or, in the case of unvoiced sounds, the air stream from the throat, and formed by the shape of the vocal tract. See Figure 1.7. The output of this system is sound vibration (i.e., the air pressure), but the external stimuli are not measurable. See Figure 1.8. Data from this system are shown in Figure 1.9. □

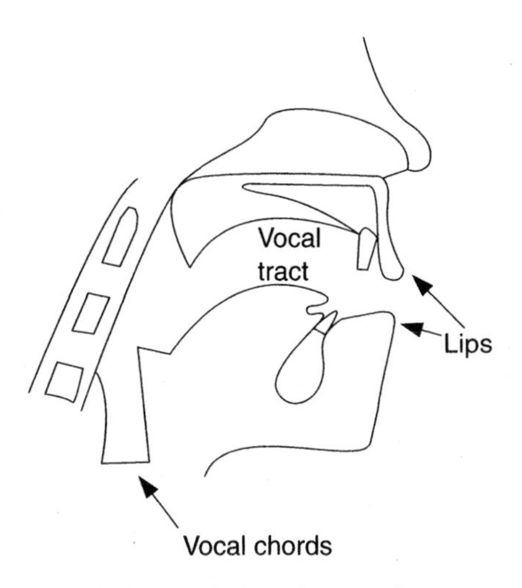

Figure 1.7 Speech generation.

The systems in these examples are all *dynamic*, which means that the current output value depends not only on the current external stimuli but also on their earlier values. Outputs of dynamical systems whose external stimuli are not observed (such as in Example 1.3) are often called *time series*. This term is especially common in economic applications. Clearly, the list of examples of dynamical systems can be very long, encompassing many fields of science.

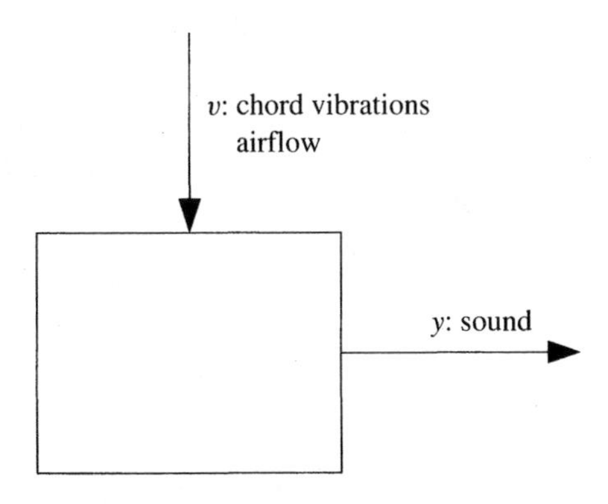

Figure 1.8 The speech system: y: output; v: unmeasured disturbance.

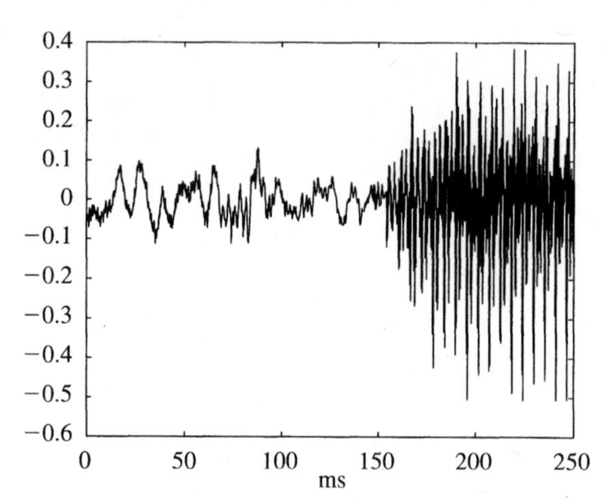

Figure 1.9 The speech signal (air pressure). Data sampled every 0.125 ms. (8 kHz sampling rate).

1.2 MODELS

Model Types and Their Use

When we interact with a system, we need some concept of how its variables relate to each other. With a broad definition, we shall call such an assumed relationship among observed signals a *model* of the system. Clearly, models may come in various shapes and be phrased with varying degrees of mathematical formalism. The intended use will determine the degree of sophistication that is required to make the model purposeful.

No doubt, in daily life many systems are dealt with using *mental models*, which do not involve any mathematical formalization at all. To drive a car, for example, requires the knowledge that turning the steering wheel to the left induces a left turn, together with subtle information built up in the muscle memory. The importance and degree of sophistication of the latter should of course not be underestimated.

For certain systems it is appropriate to describe their properties using numerical tables and/or plots. We shall call such descriptions *graphical models*. Linear systems, for example, can be uniquely described by their impulse or step responses or by their frequency functions. Graphical representation of these are widely used for various design purposes. The nonlinear characteristics of, say, a valve are also well suited to be described by a graphical model.

For more advanced applications, it may be necessary to use models that describe the relationships among the system variables in terms of mathematical expressions like difference or differential equations. We shall call such models *mathematical (or analytical) models*. Mathematical models may be further characterized by a number of adjectives (time continuous or time discrete, lumped or distributed, deterministic or stochastic, linear or nonlinear, etc.) signifying the type of difference or differential equation used. The use of mathematical models is inherent in all fields of engineering and physics. In fact, a major part of the engineering field deals with how to make good designs based on mathematical models. They are also instrumental for simulation and forecasting (prediction), which is extensively used in all fields, including nontechnical areas like economy, ecology and biology.

The model used in a computer simulation of a system is a program. For complex systems, this program may be built up by many interconnected subroutines and lookup tables, and it may not be feasible to summarize it analytically as a mathematical model. We use the term *software model* for such computerized descriptions. They have come to play an increasingly important role in decision making for complicated systems.

Building Models

Basically, a model has to be constructed from observed data. The mental model of car-steering dynamics, for example, is developed through driving experience. Graphical models are made up from certain measurements. Mathematical models may be developed along two routes (or a combination of them). One route is to split up the system, figuratively speaking, into subsystems, whose properties are well understood from previous experience. This basically means that we rely on earlier empirical work. These subsystems are then joined mathematically and a model of the whole system is obtained. This route is known as *modeling* and does not necessarily involve any experimentation on the actual system. The procedure of modeling is quite application dependent and often has its roots in tradition and specific techniques in the application area in question. Basic techniques typically involve structuring of the process into block diagrams with blocks consisting of simple elements. The reconstruction of the system from these simple blocks is now increasingly being done by computer, resulting in a software model rather than a mathematical model.

The other route to mathematical as well as graphical models is directly based on experimentation. Input and output signals from the system, such as those in Figures 1.4, 1.6, and 1.9, are recorded and subjected to data analysis in order to infer a model. This route is *system identification*.

The Fiction of a True System

The real-life actual system is an object of a different kind than our mathematical models. In a sense, there is an impenetrable but transparent screen between our world of mathematical descriptions and the real world. We can look through this window and compare certain aspects of the physical system with its mathematical description, but we can never establish any exact connection between them. The question of nature's susceptibility to mathematical description has some deep philosophical aspects, and in practical terms we have to take a more pragmatic view of models. Our acceptance of models should thus be guided by "usefulness" rather than "truth." Nevertheless, we shall occasionally use a concept of "the true system," defined in terms of a mathematical description. Such a fiction is helpful for devising identification methods and understanding their properties. In such contexts we assume that the observed data have been generated according to some well-defined mathematical rules, which of course is an idealization.

1.3 AN ARCHETYPICAL PROBLEM—ARX MODELS AND THE LINEAR LEAST SQUARES METHOD

In this section we shall consider a specific estimation problem that contains most of the central issues that this book deals with. The section will thus be a preview of the book. In the following section we shall comment on the general nature of the issues raised here and how they relate to the organization of the book.

The Model

We shall generally denote the system's input and output at time t by $u(t)$ and $y(t)$, respectively. Perhaps the most basic relationship between the input and output is the *linear difference equation*:

$$y(t) + a_1 y(t-1) + \ldots + a_n y(t-n) = b_1 u(t-1) + \ldots + b_m u(t-m) \quad (1.1)$$

We have chosen to represent the system in *discrete time*, primarily since observed data are always collected by sampling. It is thus more straightforward to relate observed data to discrete time models. In (1.1) we assume the *sampling interval* to be one time unit. This is not essential, but makes notation easier.

A pragmatic and useful way to see (1.1) is to view it as a way of *determining the next output value* given previous observations:

$$y(t) = -a_1 y(t-1) - \ldots - a_n y(t-n) + b_1 u(t-1) + \ldots + b_m u(t-m) \quad (1.2)$$

For more compact notation we introduce the vectors:

$$\theta = \begin{bmatrix} a_1 & \ldots & a_n & b_1 & \ldots & b_m \end{bmatrix}^T \quad (1.3)$$

$$\varphi(t) = \begin{bmatrix} -y(t-1) & \ldots & -y(t-n) & u(t-1) & \ldots & u(t-m) \end{bmatrix}^T \quad (1.4)$$

With these, (1.2) can be rewritten as

$$y(t) = \varphi^T(t)\theta$$

To emphasize that the calculation of $y(t)$ from past data (1.2) indeed depends on the parameters in θ, we shall rather call this calculated value $\hat{y}(t|\theta)$ and write

$$\hat{y}(t|\theta) = \varphi^T(t)\theta \quad (1.5)$$

The Least Squares Method

Now suppose for a given system that we do not know the values of the parameters in θ, but that we have recorded inputs and outputs over a time interval $1 \le t \le N$:

$$Z^N = \{u(1), y(1), \ldots, u(N), y(N)\} \quad (1.6)$$

An obvious approach is then to select θ in (1.1) through (1.5) so as to fit the calculated values $\hat{y}(t|\theta)$ as well as possible to the measured outputs by the least squares method:

$$\min_{\theta} V_N(\theta, Z^N) \quad (1.7)$$

where

$$V_N(\theta, Z^N) = \frac{1}{N}\sum_{t=1}^{N}(y(t) - \hat{y}(t|\theta))^2 = \frac{1}{N}\sum_{t=1}^{N}(y(t) - \varphi^T(t)\theta)^2 \tag{1.8}$$

We shall denote the value of θ that minimizes (1.7) by $\hat{\theta}_N$:

$$\hat{\theta}_N = \arg\min_{\theta} V_N(\theta, Z^N) \tag{1.9}$$

("arg min" means the minimizing argument, i.e., that value of θ which minimizes V_N.)

Since V_N is quadratic in θ, we can find the minimum value easily by setting the derivative to zero:

$$0 = \frac{d}{d\theta}V_N(\theta, Z^N) = \frac{2}{N}\sum_{t=1}^{N}\varphi(t)(y(t) - \varphi^T(t)\theta)$$

which gives

$$\sum_{t=1}^{N}\varphi(t)y(t) = \sum_{t=1}^{N}\varphi(t)\varphi^T(t)\theta \tag{1.10}$$

or

$$\hat{\theta}_N = \left[\sum_{t=1}^{N}\varphi(t)\varphi^T(t)\right]^{-1}\sum_{t=1}^{N}\varphi(t)y(t) \tag{1.11}$$

Once the vectors $\varphi(t)$ are defined, the solution can easily be found by modern numerical software, such as MATLAB.

Example 1.4 First Order Difference Equation

Consider the simple model:

$$y(t) + ay(t-1) = bu(t-1).$$

This gives us the estimate according to (1.4), (1.3) and (1.11):

$$\begin{bmatrix}\hat{a}_N \\ \hat{b}_N\end{bmatrix} = \begin{bmatrix}\sum y^2(t-1) & -\sum y(t-1)u(t-1) \\ -\sum y(t-1)u(t-1) & \sum u^2(t-1)\end{bmatrix}^{-1} \times \begin{bmatrix}-\sum y(t)y(t-1) \\ \sum y(t)u(t-1)\end{bmatrix}$$

All sums are from $t = 1$ to $t = N$. A typical convention is to take values outside the measured range to be zero. In this case we would thus take $y(0) = 0$. □

The simple model (1.1) and the well known least squares method (1.11) form the archetype of System Identification. Not only that—they also give the most commonly used parametric identification method and are much more versatile than perhaps perceived at first sight. In particular one should realize that (1.1) can directly be extended to several different inputs (this just calls for a redefinition of $\varphi(t)$ in (1.4)) and that the inputs and outputs do not have to be the raw measurements. On the contrary—it is often most important to think over the physics of the application and come up with suitable inputs and outputs for (1.1), formed from the actual measurements.

Example 1.5 An Immersion Heater

Consider a process consisting of an immersion heater immersed in a cooling liquid. We measure:

- $v(t)$: The voltage applied to the heater
- $r(t)$: The temperature of the liquid
- $y(t)$: The temperature of the heater coil surface

Suppose we need a model for how $y(t)$ depends on $r(t)$ and $v(t)$. Some simple considerations based on common sense and high school physics ("Semi-physical modeling") reveal the following:

- The change in temperature of the heater coil over one sample is proportional to the electrical power in it (the inflow power) minus the heat loss to the liquid
- The electrical power is proportional to $v^2(t)$
- The heat loss is proportional to $y(t) - r(t)$

This suggests the model:

$$y(t) = y(t-1) + \alpha v^2(t-1) - \beta(y(t-1) - r(t-1))$$

which fits into the form

$$y(t) + \theta_1 y(t-1) = \theta_2 v^2(t-1) + \theta_3 r(t-1)$$

This is a two input (v^2 and r) and one output model, and corresponds to choosing

$$\varphi(t) = \begin{bmatrix} -y(t-1) & v^2(t-1) & r(t-1) \end{bmatrix}^T$$

in (1.5). We could also enforce the suggestion from the physics that $\theta_1 + \theta_3 = -1$ by another choice of variables. □

Linear Regressions

Model structures such as (1.5) that are linear in θ are known in statistics as *linear regressions*. The vector $\varphi(t)$ is called the *regression vector*, and its components are the *regressors*. "Regress" here alludes to the fact that we try to calculate (or describe) $y(t)$ by "going back" to $\varphi(t)$. Models such as (1.1) where the regression vector—$\varphi(t)$—contains old values of the variable to be explained—$y(t)$—are then partly *auto-regressions*. For that reason the model structure (1.1) has the standard name ARX-model: Auto-Regression with eXtra inputs (or eXogeneous variables).

There is a rich statistical literature on the properties of the estimate $\hat{\theta}_N$ under varying assumptions. We shall deal with such questions in a much more general setting in Chapters 7 to 9, and the following section can be seen as a preview of the issues dealt with there. See also Appendix II.

Model Quality and Experiment Design

Let us consider the simplest special case, that of a Finite Impulse Response (FIR) model. That is obtained from (1.1) by taking $n = 0$:

$$y(t) = b_1 u(t-1) + \ldots + b_m u(t-m) \tag{1.12}$$

Suppose that the observed data really have been generated by a similar mechanism

$$y(t) = b_1^0 u(t-1) + \ldots + b_m^0 u(t-m) + e(t) \tag{1.13}$$

where $e(t)$ is a white noise sequence with variance λ, but otherwise unknown. (That is, $e(t)$ can be described as a sequence of independent random variables with zero mean values and variances λ.) Analogous to (1.5), we can write this as

$$y(t) = \varphi^T(t)\theta_0 + e(t) \tag{1.14}$$

The input sequence $u(t), t = 1, 2, \ldots$ is taken as a given, deterministic, sequence of numbers. We can now replace $y(t)$ in (1.12) by the above expression, and obtain

$$\hat{\theta}_N = R(N)^{-1}\left[\sum_{t=1}^{N}\varphi(t)\varphi^T(t)\theta_0 + \sum_{t=1}^{N}\varphi(t)e(t)\right]$$

$$R(N) = \sum_{t=1}^{N}\varphi(t)\varphi^T(t)$$

or

$$\tilde{\theta}_N = \hat{\theta}_N - \theta_0 = R(N)^{-1}\sum_{t=1}^{N}\varphi(t)e(t) \tag{1.15}$$

Since u and hence φ are given, deterministic variables, $R(N)$ is a deterministic matrix. If E denotes *mathematical expectation*, we therefore have

$$E\tilde{\theta}_N = E\left[R(N)^{-1}\sum_{t=1}^{N}\varphi(t)e(t)\right] = R(N)^{-1}\sum_{t=1}^{N}\varphi(t)Ee(t) = 0$$

since $e(t)$ has zero mean. The estimate is consequently *unbiased*.

We can also form the expectation of $\tilde{\theta}_N \tilde{\theta}_N^T$, i.e., the covariance matrix of the parameter error

$$
\begin{aligned}
P_N &= E\tilde{\theta}_N \tilde{\theta}_N^T = ER(N)^{-1} \sum_{t,s=1}^{N} \varphi(t)e(t)e(s)\varphi^T(s)R(N)^{-1} \\
&= R(N)^{-1} \sum_{t,s=1}^{N} \varphi(t)\varphi^T(s)R(N)^{-1}Ee(t)e(s) \\
&= R(N)^{-1} \sum_{t,s=1}^{N} \varphi(t)\varphi^T(s)R(N)^{-1}\lambda\delta(t-s) \\
&= R(N)^{-1}\lambda \sum_{t=1}^{N} \varphi(t)\varphi^T(t)R(N)^{-1} = \lambda R(N)^{-1}
\end{aligned}
\tag{1.16}
$$

where we used the fact that e is a sequence of independent variables so that $Ee(t)e(s) = \lambda\delta(t-s)$, with $\delta(0) = 1$ and $\delta(\tau) = 0$ if $\tau \neq 0$.

We have thus computed the covariance matrix of the estimate $\hat{\theta}_N$. It is determined entirely by the input properties $R(N)$ and the noise level λ. Moreover, define

$$
\overline{R} = \lim_{N\to\infty} \frac{1}{N} R(N) \tag{1.17}
$$

This will correspond to the *covariance matrix* of the input, i.e., the $i-j$-element of $\overline{R}$ is

$$
\lim_{N\to\infty} \frac{1}{N} \sum_{t=1}^{N} u(t-i)u(t-j)
$$

If the matrix $\overline{R}$ is non-singular, we find that the covariance matrix of the parameter estimate is approximately given by

$$
P_N = \frac{\lambda}{N}\overline{R}^{-1} \tag{1.18}
$$

and the approximation improves as $N \to \infty$. A number of things follow from this.

- The covariance decays like $1/N$, so the parameters approach the limiting value at the rate $1/\sqrt{N}$.
- The covariance is proportional to the Noise-To-Signal ratio. That is, it is proportional to the noise variance λ and inversely proportional to the input power.
- The covariance does not depend on the input's or noise's signal shapes, only on their variance/covariance properties.
- Experiment design, i.e., the selection of the input u, aims at making the matrix $\overline{R}^{-1}$ "as small as possible." Note that the same $\overline{R}$ can be obtained for many different signals u.

1.4 THE SYSTEM IDENTIFICATION PROCEDURE

Three Basic Entities

As we saw in the previous section, the construction of a model from data involves three basic entities:

1. A data set, like Z^N in (1.6).
2. A set of candidate models; a *Model Structure*, like the set (1.1) or (1.5).
3. A rule by which candidate models can be assessed using the data, like the Least Squares selection rule (1.9).

Let us comment on each of these:

1. *The data record.* The input-output data are sometimes recorded during a specifically designed identification experiment, where the user may determine which signals to measure and when to measure them and may also choose the input signals. The objective with *experiment design* is thus to make these choices so that the data become maximally informative, subject to constraints that may be at hand. Making the matrix $\overline{R}^{-1}$ in (1.18) small is a typical example of this, and we shall in Chapter 13 treat this question in more detail. In other cases the user may not have the possibility to affect the experiment, but must use data from the normal operation of the system.
2. *The set of models or the model structure.* A set of candidate models is obtained by specifying within which collection of models we are going to look for a suitable one. This is no doubt the most important and, at the same time, the most difficult choice of the system identification procedure. It is here that *a priori* knowledge and engineering intuition and insight have to be combined with formal properties of models. Sometimes the model set is obtained after careful *modeling*. Then a model with some unknown physical parameters is constructed from basic physical laws and other well-established relationships. In other cases standard linear models may be employed, without reference to the physical background. Such a model set, whose parameters are basically viewed as vehicles for adjusting the fit to the data and do not reflect physical considerations in the system, is called a *black box*. Model sets with adjustable parameters with physical interpretation may, accordingly, be called *gray boxes*. Generally speaking, a model structure is a parameterized mapping from past inputs and outputs Z^{t-1} (cf (1.6)) to the space of the model outputs:
$$\hat{y}(t|\theta) = g(\theta, Z^{t-1}) \tag{1.19}$$
Here θ is the finite dimensional vector used to parameterize the mapping. Chapters 4 and 5 describe common model structures.
3. *Determining the "best" model in the set, guided by the data.* This is the *identification method*. The assessment of model quality is typically based on how the models perform when they attempt to reproduce the measured data. The basic approaches to this will be dealt with independently of the model structure used. Chapter 7 treats this problem.

Model Validation

After having settled on the preceding three choices, we have, at least implicitly, arrived at a particular model: the one in the set that best describes the data according to the chosen criterion. It then remains to test whether this model is "good enough," that is, whether it is valid for its purpose. Such tests are known as *model validation*. They involve various procedures to assess how the model relates to observed data, to prior knowledge, and to its intended use. Deficient model behavior in these respects make us reject the model, while good performance will develop a certain confidence in it. A model can never be accepted as a final and true description of the system. Rather, it can at best be regarded as a good enough description of certain aspects that are of particular interest to us. Chapter 16 contains a discussion of model validation.

The System Identification Loop

The system identification procedure has a natural logical flow: first collect data, then choose a model set, then pick the "best" model in this set. It is quite likely, though, that the model first obtained will not pass the model validation tests. We must then go back and revise the various steps of the procedure.

The model may be deficient for a variety of reasons:

- The numerical procedure failed to find the best model according to our criterion.
- The criterion was not well chosen.
- The model set was not appropriate, in that it did not contain any "good enough" description of the system.
- The data set was not informative enough to provide guidance in selecting good models.

The major part of an identification application in fact consists of addressing these problems, in particular the third one, in an iterative manner, guided by prior information and the outcomes of previous attempts. See Figure 1.10. Interactive software obviously is an important tool for handling the iterative character of this problem.

1.5 ORGANIZATION OF THE BOOK

To master the loop of Figure 1.10, the user has to be familiar with a number of things.

1. Available techniques of identification and their rationale, as well as typical choices of model sets.
2. The properties of the identified model and their dependence on the basic items: data, model set, and identification criterion.
3. Numerical schemes for computing the estimate.
4. How to make intelligent choices of experiment design, model set, and identification criterion, guided by prior information as well as by observed data.

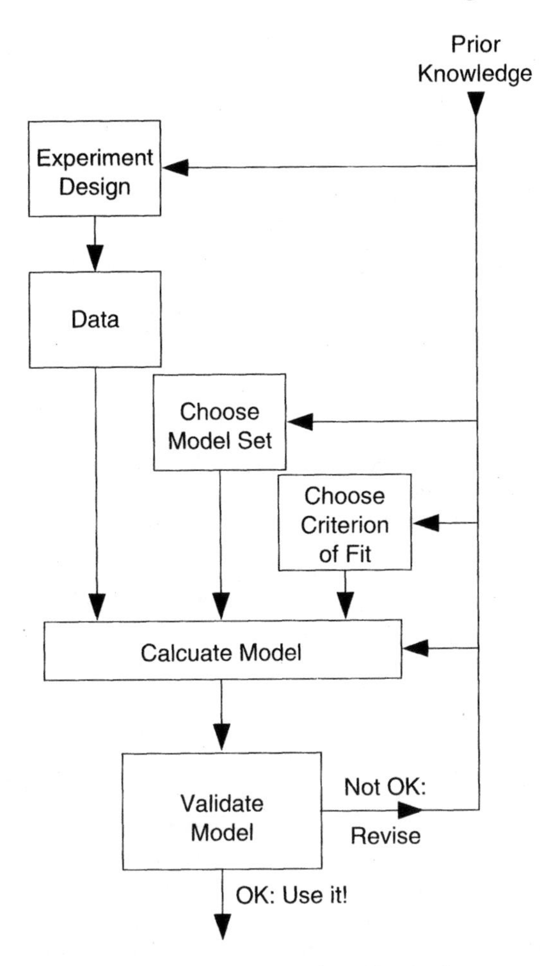

Figure 1.10 The system identification loop.

In fact, a user of system identification may find that he or she is primarily a user of an interactive software package. Items 1 and 3 are then part of the package and the important thing is to have a good understanding of item 2 so that task 4 can be successfully completed. This is what we mean by "Theory for the User" and this is where the present book has its focus.

The idea behind the book's organization is to present the list of common and useful model sets in Chapters 4 and 5. Available techniques are presented in Chapters 6 and 7, and the analysis follows in Chapters 8 and 9. Numerical techniques for off-line and on-line applications are described in Chapters 10 and 11. Task 4, the user's choices, is discussed primarily in Chapters 13 through 16, after some preliminaries in Chapter 12. In addition, Chapters 2 and 3 give the formal setup of the book, and Chapter 17 describes and assesses system identification as a tool for practical problems.

Figure 1.11 illustrates the book's structure in relation to the loop of system identification.

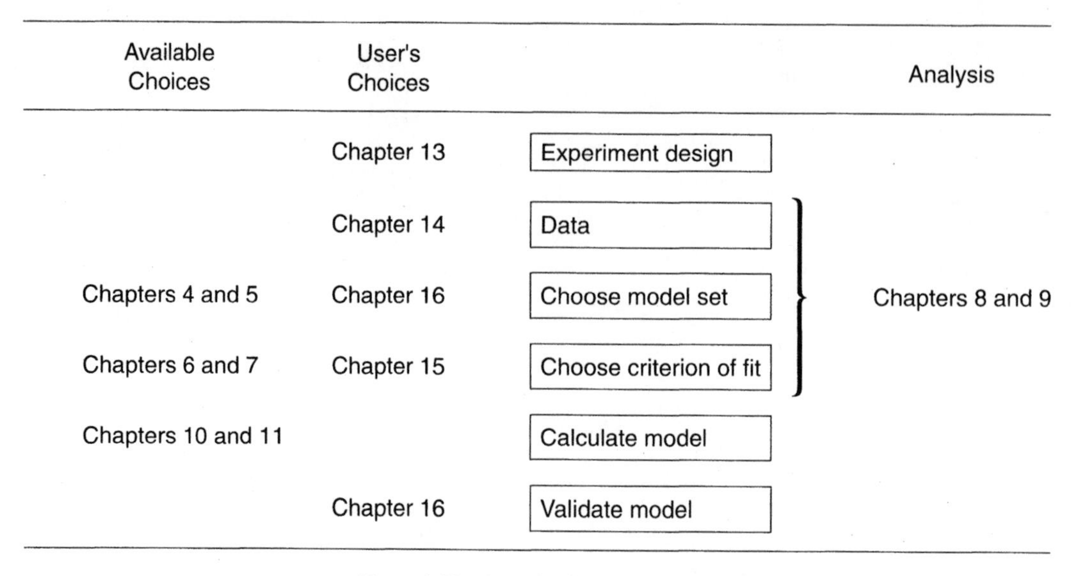

Figure 1.11 Organization of the book.

About the Framework

The system identification framework we set up here is fairly general. It does not confine us to linear models or quadratic criteria or to assuming that the system itself can be described within the model set. Indeed, this is one of the points that should be stressed about our framework. Nevertheless, we often give proofs and explicit expressions only for certain special cases, like single-input, single-output systems and quadratic criteria. The purpose is of course to enhance the underlying basic ideas and not conceal them behind technical details. References are usually provided for more general treatments.

Parameter estimation and identification are usually described within a probabilistic framework. Here we basically employ such a framework. However, we also try to retain a pragmatic viewpoint that is independent of probabilistic interpretations. That is, the methods we describe and the recommendations we put forward should make sense even without the probabilistic framework that may motivate them as "optimal solutions." The probabilistic and statistical environments of the book are described in Appendices I and II, respectively. These appendices may be read prior to the other chapters or consulted occasionally when required. In any case, the book does not lean heavily on the background provided there.

1.6 BIBLIOGRAPHY

The literature on the system identification problem and its ramifications is extensive. Among general textbooks on the subject we may mention Box and Jenkins (1970), Eykhoff (1974), Spriet and Vansteenkiste (1982), Ljung and Glad (1994a), and Johansson (1993)for treatments covering several practical issues, while Goodwin and Payne (1977), Davis and Vinter (1985), Hannan and Deistler (1988), Caines

(1988), Chen and Guo (1991), and Söderström and Stoica (1989)give more theoretically oriented presentations. Kashyap and Rao (1976), Rissanen (1989)and Bohlin (1991)emphasize the role of model validation and model selection in their treatment of system identification, while Söderström and Stoica (1983)focuses on instrumental-variable methods. A treatment based on frequency domain data is given in Schoukens and Pintelon (1991), and the so-called subspace approach is thoroughly discussed in Van Overschee and DeMoor (1996). Texts that concentrate on recursive identification techniques include Ljung and Söderström (1983), Solo and Kong (1995), Haykin (1986), Widrow and Stearns (1985), and Young (1984). Spectral analysis is closely related, and treated in many books like Marple (1987), Kay (1988)and Stoica and Moses (1997). Statistical treatments of time-series modeling such as Anderson (1971), Hannan (1970), Brillinger (1981), and Wei (1990)are most relevant also for the system identification problem. The co-called behavioral approach to modeling is introduced in Willems (1987).

Among edited collections of articles, we may refer to Mehra and Lainiotis (1976), Eykhoff (1981), Hannan, Krishnaiah, and Rao (1985), and Leondes (1987), as well as to the special journal issues Kailath, Mehra, and Mayne (1974), Isermann (1981), Eykhoff and Parks (1990), Kosut, Goodwin, and Polis (1992)and Söderström and Åström (1995). The proceedings from the IFAC (International Federation of Automatic Control) Symposia on Identification and System Parameter Estimation contain many articles on all aspects of the system identification problem. These symposia are held every three years, starting in Prague 1967.

Philosophical aspects on mathematical models of real-life objects are discussed, for example, in Popper (1934). Modeling from basic physical laws, rather than from data, is discussed in many books; see, for example, Wellstead (1979), Ljung and Glad (1994a), Frederick and Close (1978)and Cellier (1990)for engineering applications. Such treatments are important complements to the model set selection (see Section 1.3 and Chapter 16).

Many books discuss modeling and identification in various application areas. See, for example, Granger and Newbold (1977)or Malinvaud (1980)(econometrics), Godfrey (1983)(biology), Robinson and Treitel (1980), Mendel (1983)(geoscience), Dudley (1983)(electromagnetic wave theory), Markel and Gray (1976)(speech signals) and Beck and Van Straten (1983)(environmental systems). Rajbman (1976, 1981)has surveyed the Soviet literature.

I systems and models

2

TIME-INVARIANT LINEAR SYSTEMS

Time-invariant linear systems no doubt form the most important class of dynamical systems considered in practice and in the literature. It is true that they represent idealizations of the processes encountered in real life. But, even so, the approximations involved are often justified, and design considerations based on linear theory lead to good results in many cases.

A treatise of linear systems theory is a standard ingredient in basic engineering education, and the reader has no doubt some knowledge of this topic. Anyway, in this chapter we shall provide a refresher on some basic concepts that will be instrumental for the further development in this book. In Section 2.1 we shall discuss the impulse response and various ways of describing and understanding disturbances, as well as introduce the transfer-function concept. In Section 2.2 we study frequency-domain interpretations and also introduce the periodogram. Section 2.3 gives a unified setup of spectra of deterministic and stochastic signals that will be used in the remainder of this book. In Section 2.4 a basic ergodicity result is proved. The development in these sections is for systems with a scalar input and a scalar output. Section 2.5 contains the corresponding expressions for multivariable systems.

2.1 IMPULSE RESPONSES, DISTURBANCES, AND TRANSFER FUNCTIONS

Impulse Response

Consider a system with a scalar input signal $u(t)$ and a scalar output signal $y(t)$ (Figure 2.1). The system is said to be *time invariant* if its response to a certain input signal does not depend on absolute time. It is said to be *linear* if its output response to a linear combination of inputs is the same linear combination of the output responses of the individual inputs. Furthermore, it is said to be *causal* if the output at a certain time depends on the input up to that time only.

Figure 2.1 The system.

It is well known that a linear, time-invariant, causal system can be described by its *impulse response* (or *weighting function*) $g(\tau)$ as follows:

$$y(t) = \int_{\tau=0}^{\infty} g(\tau)u(t - \tau)d\tau \tag{2.1}$$

Knowing $\{g(\tau)\}_{\tau=0}^{\infty}$ and knowing $u(s)$ for $s \leq t$, we can consequently compute the corresponding output $y(s)$, $s \leq t$ for any input. *The impulse response is thus a complete characterization of the system.*

Sampling

In this book we shall almost exclusively deal with observations of inputs and outputs in discrete time, since this is the typical data-acquisition mode. We thus assume $y(t)$ to be observed at the *sampling instants* $t_k = kT$, $k = 1, 2, \ldots$:

$$y(kT) = \int_{\tau=0}^{\infty} g(\tau)u(kT - \tau)d\tau \tag{2.2}$$

The interval T will be called the *sampling interval.* It is, of course, also possible to consider the situation where the sampling instants are not equally spread.

Most often, in computer control applications, the input signal $u(t)$ is kept constant between the sampling instants:

$$u(t) = u_k, \qquad kT \leq t < (k+1)T \tag{2.3}$$

This is mostly done for practical implementation reasons, but it will also greatly simplify the analysis of the system. Inserting (2.3) into (2.2) gives

$$\begin{aligned} y(kT) &= \int_{\tau=0}^{\infty} g(\tau)u(kT - \tau)d\tau = \sum_{\ell=1}^{\infty} \int_{\tau=(\ell-1)T}^{\ell T} g(\tau)u(kT - \tau)d\tau \\ &= \sum_{\ell=1}^{\infty} \left[\int_{\tau=(\ell-1)T}^{\ell T} g(\tau)d\tau \right] u_{k-\ell} = \sum_{\ell=1}^{\infty} g_T(\ell)u_{k-\ell} \end{aligned} \tag{2.4}$$

where we defined

$$g_T(\ell) = \int_{\tau=(\ell-1)T}^{\ell T} g(\tau)d\tau \tag{2.5}$$

The expression (2.4) tells us what the output will be at the sampling instants. Note that no approximation is involved if the input is subject to (2.3) and that it is sufficient to know the sequence $\{g_T(\ell)\}_{\ell=1}^{\infty}$ in order to compute the response to the input. The relationship (2.4) describes a *sampled-data system*, and we shall call the sequence $\{g_T(\ell)\}_{\ell=1}^{\infty}$ the impulse response of that system.

Even if the input is not piecewise constant and subject to (2.3), the representation (2.4) might still be a reasonable approximation, provided $u(t)$ does not change too much during a sampling interval. See also the following expressions (2.21) to (2.26). Intersample behavior is further discussed in Section 13.3.

We shall stick to the notation (2.3) to (2.5) when the choice and size of T are essential to the discussion. For most of the time, however, we shall for ease of notation assume that T is one time unit and use t to enumerate the sampling instants. We thus write for (2.4)

$$y(t) = \sum_{k=1}^{\infty} g(k)u(t-k), \quad t = 0, 1, 2, \dots \tag{2.6}$$

For sequences, we shall also use the notation

$$y_s^t = (y(s), y(s+1), \dots, y(t)) \tag{2.7}$$

and for simplicity

$$y_1^t = y^t$$

Disturbances

According to the relationship (2.6), the output can be exactly calculated once the input is known. In most cases this is unrealistic. There are always signals beyond our control that also affect the system. Within our linear framework we assume that such effects can be lumped into an additive term $v(t)$ at the output (see Figure 2.2):

$$y(t) = \sum_{k=1}^{\infty} g(k)u(t-k) + v(t) \tag{2.8}$$

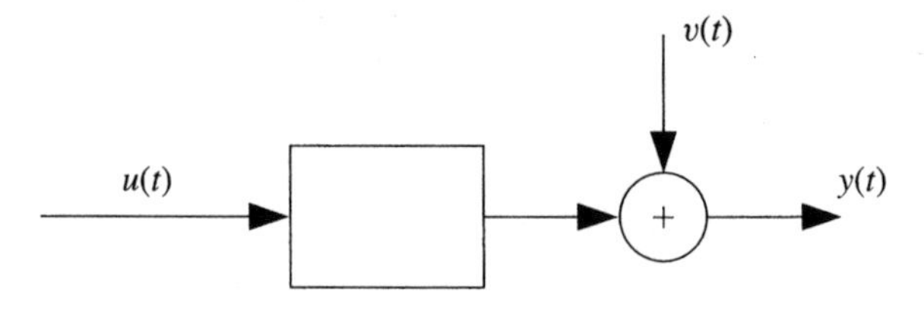

Figure 2.2 System with disturbance.

There are many sources and causes for such a disturbance term. We could list:

- Measurement noise: The sensors that measure the signals are subject to noise and drift.
- Uncontrollable inputs: The system is subject to signals that have the character of inputs, but are not controllable by the user. Think of an airplane, whose movements are affected by the inputs of rudder and aileron deflections, but also by wind gusts and turbulence. Another example could be a room, where the temperature is determined by radiators, whose effect we control, but also by people ($\approx$ 100 W per person) who may move in and out in an unpredictable manner.

The character of the disturbances could also vary within wide ranges. Classical ways of describing disturbances in control have been to study steps, pulses, and sinusoids, while in stochastic control the disturbances are modeled as realizations of stochastic processes. See Figures 2.3 and 2.4 for some typical, but mutually quite different, disturbance characteristics. The disturbances may in some cases be separately measurable, but in the typical situation they are noticeable only via their effect on the output. If the impulse response of the system is known, then of course the actual value of the disturbance $v(t)$ can be calculated from (2.8) at time t.

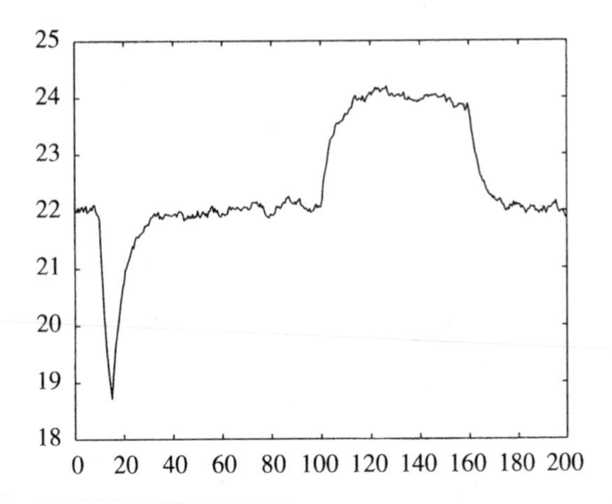

Figure 2.3 Room temperature.

The assumption of Figure 2.2 that the noise enters additively to the output implies some restrictions. Sometimes the measurements of the inputs to the system may also be noise corrupted ("error-in-variable" descriptions). In such cases we take a pragmatic approach and regard the measured input values as the actual inputs $u(t)$ to the process, and their deviations from the true stimuli will be propagated through the system and lumped into the disturbance $v(t)$ of Figure 2.2.

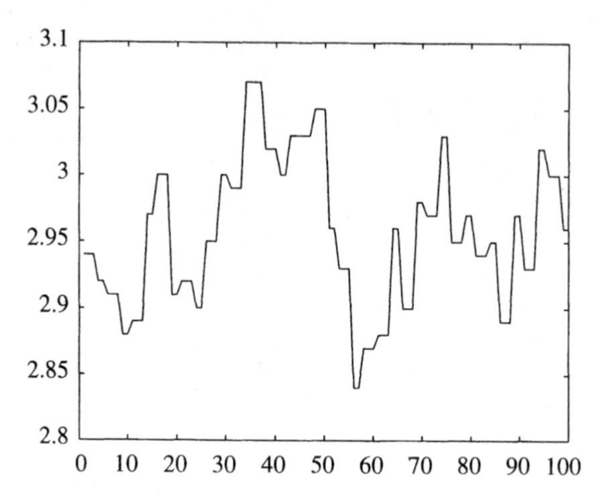

Figure 2.4 Moisture content in paper during paper-making.

Characterization of Disturbances

The most characteristic feature of a disturbance is that *its value is not known beforehand*. Information about past disturbances could, however, be important for making qualified guesses about future values. It is thus natural to employ a probabilistic framework to describe future disturbances. We then put ourselves at time t and would like to make a statement about disturbances at times $t + k$, $k \geq 1$. A complete characterization would be to describe the conditional joint probability density function for $\{v(t + k), k \geq 1\}$, given $\{v(s), s \leq t\}$. This would, however, in most cases be too laborious, and we shall instead use a simpler approach.

Let $v(t)$ be given as

$$v(t) = \sum_{k=0}^{\infty} h(k)e(t - k) \tag{2.9}$$

where $\{e(t)\}$ is *white noise*, i.e., a sequence of independent (identically distributed) random variables with a certain probability density function. Although this description does not allow completely general characterizations of all possible probabilistic disturbances, it is versatile enough for most practical purposes. In Section 3.2 we shall show how the description (2.9) allows predictions and probabilistic statements about future disturbances. For normalization reasons, we shall usually assume that $h(0) = 1$, which is no loss of generality since the variance of e can be adjusted.

It should be made clear that the specification of different probability density functions (PDF) for $\{e(t)\}$ may result in very different characteristic features of the disturbance. For example, the PDF

$$\begin{aligned} e(t) &= 0, \quad \text{with probability } 1 - \mu \\ e(t) &= r, \quad \text{with probability } \mu \end{aligned} \tag{2.10}$$

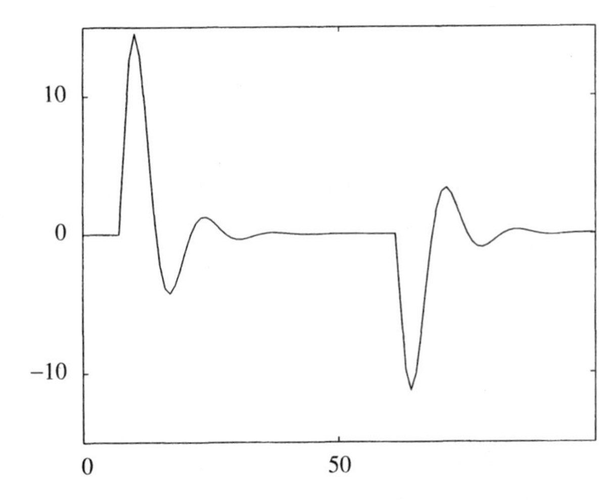

Figure 2.5 A realization of the process (2.9) with e subject to (2.10).

where r is a normally distributed random variable: $r \in N(0, \gamma)$ leads to, if μ is a small number, disturbance sequences with characteristic and "deterministic" profiles occurring at random instants. See Figure 2.5. This could be suitable to describe "classical" disturbance patterns, steps, pulses, sinusoids, and ramps (cf. Figure 2.3!). On the other hand, the PDF

$$e(t) \in N(0, \lambda) \tag{2.11}$$

gives a totally different picture. See Figure 2.6. Such a pattern is more suited to describe measurements noises and irregular and frequent disturbance actions.

Often we only specify the *second-order properties* of the sequence $\{e(t)\}$, that is, the mean and the variances. Note that (2.10) and (2.11) can both be described as "a sequence of independent random variables with zero mean values and variances λ" [$\lambda = \mu\gamma$ for (2.10)], despite the difference in appearance.

Remark. Notice that $\{e(t)\}$ and $\{v(t)\}$ as defined previously are *stochastic processes* (i.e., sequences of random variables). The disturbances that we observe and that are added to the system output as in Figure 2.2 are thus *realizations* of the *stochastic process* $\{v(t)\}$. Strictly speaking, one should distinguish in notation between the process and its realization, but the meaning is usually clear from the context, and we do not here adopt this extra notational burden. Often one has occasion to study signals that are mixtures of deterministic and stochastic components. A framework for this will be discussed in Section 2.3.

Covariance Function

In the sequel, we shall assume that $e(t)$ has zero mean and variance λ. With the description (2.9) of $v(t)$, we can compute the mean as

$$Ev(t) = \sum_{k=0}^{\infty} h(k)Ee(t-k) = 0 \tag{2.12}$$

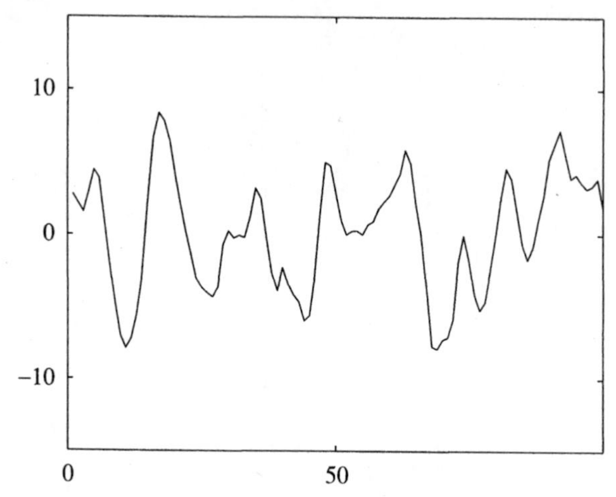

Figure 2.6 A realization of the same process (2.9) as in Figure 2.5, but with e subject to (2.11).

and the covariance as

$$\begin{aligned}
Ev(t)v(t-\tau) &= \sum_{k=0}^{\infty}\sum_{s=0}^{\infty} h(k)h(s)Ee(t-k)e(t-\tau-s) \\
&= \sum_{k=0}^{\infty}\sum_{s=0}^{\infty} h(k)h(s)\delta(k-\tau-s)\lambda \qquad (2.13) \\
&= \lambda\sum_{k=0}^{\infty} h(k)h(k-\tau)
\end{aligned}$$

Here $h(r) = 0$ if $r < 0$. We note that this covariance is independent of t and call

$$R_v(\tau) = Ev(t)v(t-\tau) \qquad (2.14)$$

the *covariance function of the process* v. This function, together with the mean, specifies the *second-order properties of* v. These are consequently uniquely defined by the sequence $\{h(k)\}$ and the variance λ of e. Since (2.14) and $Ev(t)$ do not depend on t, the process is said to be *stationary*.

Transfer Functions

It will be convenient to introduce a shorthand notation for sums like (2.8) and (2.9), which will occur frequently in this book. We introduce the *forward shift operator* q by

$$qu(t) = u(t+1)$$

and the *backward shift operator* q^{-1}:

$$q^{-1}u(t) = u(t-1)$$

We can then write for (2.6)

$$\begin{aligned} y(t) &= \sum_{k=1}^{\infty} g(k)u(t-k) = \sum_{k=1}^{\infty} g(k)\left(q^{-k}u(t)\right) \\ &= \left[\sum_{k=1}^{\infty} g(k)q^{-k}\right]u(t) = G(q)u(t) \end{aligned} \tag{2.15}$$

where we introduced the notation

$$G(q) = \sum_{k=1}^{\infty} g(k)q^{-k} \tag{2.16}$$

We shall call $G(q)$ the *transfer operator* or the *transfer function* of the linear system (2.6). Notice that (2.15) thus describes a relation between the sequences u^t and y^t.

Remark. We choose q as argument of G rather than q^{-1} (which perhaps would be more natural in view of the right side) in order to be in formal agreement with z-transform and Fourier-transform expressions. Strictly speaking, the term transfer *function* should be reserved for the z-transform of $\{g(k)\}_1^{\infty}$, that is,

$$G(z) = \sum_{k=1}^{\infty} g(k)z^{-k} \tag{2.17}$$

but we shall sometimes not observe that point. □

Similarly with

$$H(q) = \sum_{k=0}^{\infty} h(k)q^{-k} \tag{2.18}$$

we can write

$$v(t) = H(q)e(t) \tag{2.19}$$

for (2.9). Our basic description for a linear system with additive disturbance will thus be

$$\boxed{y(t) = G(q)u(t) + H(q)e(t)} \tag{2.20}$$

with $\{e(t)\}$ as a sequence of independent random variables with zero mean values and variances λ.

Continuous-time Representation and Sampling Transfer Functions (∗)

For many physical systems it is natural to work with a continuous-time representation (2.1), since most basic relationships are expressed in terms of differential equations. With $G_c(s)$ denoting the Laplace transform of the impulse response function $\{g(\tau)\}$ in (2.1), we then have the relationship

$$Y(s) = G_c(s)U(s) \tag{2.21}$$

between $Y(s)$ and $U(s)$, the Laplace transforms of the output and input, respectively. Introducing p as the differentiation operator, we could then write

$$y(t) = G_c(p)u(t) \tag{2.22}$$

as a shorthand operator form of (2.1) or its underlying differential equation. Now, (2.1) or (2.22) describes the output at all values of the continuous time variable t. If $\{u(t)\}$ is a known function (piecewise constant or not), then (2.22) will of course also serve as a description of the output of the sampling instants. We shall therefore occasionally use (2.22) also as a system description for the sampled output values, keeping in mind that the computation of these values will involve numerical solution of a differential equation. In fact, we could still use a discrete-time model (2.9) for the disturbances that influence our discrete-time measurements, writing this as

$$y(t) = G_c(p)u(t) + H(q)e(t), \qquad t = 1, 2, \ldots \tag{2.23}$$

Often, however, we shall go from the continuous-time representation (2.22) to the standard discrete-time one (2.15) by transforming the transfer function

$$G_c(p) \rightarrow G_T(q) \tag{2.24}$$

T here denotes the sampling interval. When the input is piecewise constant over the sampling interval, this can be done without approximation, in view of (2.4). See Problem 2G.4 for a direct transfer-function expression, and equations (4.67) to (4.71), for numerically more favorable expressions. One can also apply approximate formulas that correspond to replacing the differentiation operator p by a difference approximation. We thus have the Euler approximation

$$G_T(q) \approx G_c\left(\frac{q-1}{T}\right) \tag{2.25}$$

and Tustin's formula

$$G_T(q) \approx G_c\left(\frac{2}{T}\frac{q-1}{1+q}\right) \tag{2.26}$$

See Åström and Wittenmark (1984)for a further discussion.

(∗)Denotes sections and subsections that are optional reading; they can be omitted without serious loss of continuity. See Preface.

Some Terminology

The function $G(z)$ in (2.17) is a complex-valued function of the complex variable z. Values β_i, such that $G(\beta_i) = 0$, are called *zeros* of the transfer function (or of the system), while values α_i for which $G(z)$ tends to infinity are called *poles*. This coincides with the terminology for analytic functions (see, e.g., Ahlfors, 1979). If $G(z)$ is a rational function of z, the poles will be the zeros of the denominator polynomial.

We shall say that the transfer function $G(q)$ (or "the system G" or "the filter G") is *stable* if

$$G(q) = \sum_{k=1}^{\infty} g(k)q^{-k}, \qquad \sum_{k=1}^{\infty} |g(k)| < \infty \tag{2.27}$$

The definition (2.27) coincides with the system theoretic definition of bounded-input, bounded-output (BIBO) stability (e.g. Brockett, 1970): If an input $\{u(t)\}$ to $G(q)$ is subject to $|u(t)| \leq C$, then the corresponding output $z(t) = G(q)u(t)$ will also be bounded, $|z(t)| \leq C'$, provided (2.27) holds. Notice also that (2.27) assures that the (Laurent) expansion

$$G(z) = \sum_{k=1}^{\infty} g(k)z^{-k}$$

is convergent for all $|z| \geq 1$. This means that the function $G(z)$ *is analytic on and outside the unit circle*. In particular, it then has no poles in that area.

We shall often have occasion to consider families of filters $G_\alpha(q)$, $\alpha \in \mathcal{A}$:

$$G_\alpha(q) = \sum_{k=1}^{\infty} g_\alpha(k)q^{-k}, \qquad \alpha \in \mathcal{A} \tag{2.28}$$

We shall then say that such *a family is uniformly stable* if

$$|g_\alpha(k)| \leq g(k), \forall \alpha \in \mathcal{A}, \sum_{k=1}^{\infty} g(k) < \infty \tag{2.29}$$

Sometimes a slightly stronger condition than (2.27) will be required. We shall say that $G(q)$ is *strictly stable* if

$$\sum_{k=1}^{\infty} k|g(k)| < \infty \tag{2.30}$$

Notice that, for a transfer function that is rational in q, stability implies strict stability (and, of course, vice versa). See Problem 2T.3.

Finally, we shall say that a filter $H(q)$ is *monic* if its zeroth coefficient is 1 (or the unit matrix):

$$H(q) = \sum_{k=0}^{\infty} h(k)q^{-k}, \qquad h(0) = 1 \tag{2.31}$$

2.2 FREQUENCY-DOMAIN EXPRESSIONS

Sinusoid Response and the Frequency Function

Suppose that the input to the system (2.6) is a sinusoid:

$$u(t) = \cos \omega t \tag{2.32}$$

It will be convenient to rewrite this as

$$u(t) = \operatorname{Re} e^{i\omega t}$$

with Re denoting "real part." According to (2.6), the corresponding output will be

$$\begin{aligned} y(t) &= \sum_{k=1}^{\infty} g(k) \operatorname{Re} e^{i\omega(t-k)} = \operatorname{Re} \sum_{k=1}^{\infty} g(k) e^{i\omega(t-k)} \\ &= \operatorname{Re} \left\{ e^{i\omega t} \cdot \sum_{k=1}^{\infty} g(k) e^{-i\omega k} \right\} = \operatorname{Re} \left\{ e^{i\omega t} \cdot G(e^{i\omega}) \right\} \\ &= |G(e^{i\omega})| \cos(\omega t + \varphi) \end{aligned} \tag{2.33}$$

where

$$\varphi = \arg G(e^{i\omega}) \tag{2.34}$$

Here, the second equality follows since the $g(k)$ are real and the fourth equality from the definition (2.16) or (2.17). The fifth equality follows straightforward rules for complex numbers.

In (2.33) we assumed that the input was a cosine since time minus infinity. If $u(t) = 0, t < 0$, we obtain an additional term

$$-\operatorname{Re} \left\{ e^{i\omega t} \sum_{k=t}^{\infty} g(k) e^{-i\omega k} \right\}$$

in (2.33). This term is dominated by

$$\sum_{k=t}^{\infty} |g(k)|$$

and therefore is of transient nature (tends to zero as t tends to infinity), provided that $G(q)$ is stable.

In any case, (2.33) tells us that the output to (2.32) will also be a cosine of the same frequency, but with an amplitude magnified by $|G(e^{i\omega})|$ and a phase shift of $\arg G(e^{i\omega})$ radians. The complex number

$$G(e^{i\omega}) \tag{2.35}$$

which is the transfer function evaluated at the point $z = e^{i\omega}$, therefore gives full information as to what will happen in stationarity, when the input is a sinusoid of frequency ω. For that reason, the complex-valued function

$$G(e^{i\omega}), \qquad -\pi \le \omega \le \pi \tag{2.36}$$

is called the *frequency function* of the system (2.6). It is customary to graphically display this function as $\log |G(e^{i\omega})|$ and $\arg G(e^{i\omega})$ plotted against $\log \omega$ in a *Bode plot*. The plot of (2.36) in the complex plane is called the *Nyquist plot*. These concepts are probably better known in the continuous-time case, but all their basic properties carry over to the sampled-data case.

Periodograms of Signals over Finite Intervals

Consider the finite sequence of inputs $u(t)$, $t = 1, 2, \ldots, N$. Let us define the function $U_N(\omega)$ by

$$U_N(\omega) = \frac{1}{\sqrt{N}} \sum_{t=1}^{N} u(t) e^{-i\omega t} \tag{2.37}$$

The values obtained for $\omega = 2\pi k/N$, $k = 1, \ldots, N$, form the familiar discrete Fourier transform (DFT) of the sequence u_1^N. We can then represent $u(t)$ by the inverse DFT as

$$u(t) = \frac{1}{\sqrt{N}} \sum_{k=1}^{N} U_N(2\pi k/N) e^{i2\pi kt/N} \tag{2.38}$$

To prove this, we insert (2.37) into the right side of (2.38), giving

$$\begin{aligned} &\frac{1}{N} \sum_{k=1}^{N} \sum_{s=1}^{N} u(s) \exp\left(-\frac{i2\pi ks}{N}\right) \cdot \exp\left(\frac{i2\pi kt}{N}\right) \\ &= \frac{1}{N} \sum_{s=1}^{N} u(s) \sum_{k=1}^{N} \exp\left(\frac{2\pi ik(t-s)}{N}\right) \\ &= \frac{1}{N} \sum_{s=1}^{N} u(s) N \cdot \delta(t-s) = u(t) \end{aligned}$$

Here we used the relationship

$$\frac{1}{N} \sum_{k=1}^{N} e^{2\pi irk/N} = \begin{cases} 1, & r = 0 \\ 0, & 1 \le r < N \end{cases} \tag{2.39}$$

From (2.37) we note that $U_N(\omega)$ is periodic with period 2π:

$$U_N(\omega + 2\pi) = U_N(\omega) \tag{2.40}$$

Also, since $u(t)$ is real,

$$U_N(-\omega) = \overline{U_N(\omega)} \tag{2.41}$$

where the overbar denotes the complex conjugate. The function $U_N(\omega)$ is therefore uniquely defined by its values over the interval $[0, \pi]$. It is, however, customary to consider $U_N(\omega)$ for $-\pi \le \omega \le \pi$, and in accordance with this, (2.38) is often written

$$u(t) = \frac{1}{\sqrt{N}} \sum_{k=-N/2+1}^{N/2} U_N(2\pi k/N) e^{i2\pi kt/N} \tag{2.42}$$

making use of (2.40) and the periodicity of $e^{i\omega}$. In (2.42) and elsewhere we assume N to be even; for odd N analogous summation boundaries apply.

In (2.42) we represent the signal $u(t)$ as a linear combination of $e^{i\omega t}$ for N different frequencies ω. As is further elaborated in Problem 2D.1, this can also be rewritten as sums of $\cos \omega t$ and $\sin \omega t$ for the same frequencies, thus avoiding complex numbers.

The number $U_N(2\pi k/N)$ tells us the "weight" that the frequency $\omega = 2\pi k/N$ carries in the decomposition of $\{u(t)\}_{t=1}^{N}$. Its absolute square value $|U_N(2\pi k/N)|^2$ is therefore a measure of the contribution of this frequency to the "signal power." This value

$$|U_N(\omega)|^2 \tag{2.43}$$

is known as the *periodogram* of the signal $u(t)$, $t = 1, 2, \ldots, N$.

Parseval's relationship,

$$\sum_{k=1}^{N} |U_N(2\pi k/N)|^2 = \sum_{t=1}^{N} u^2(t) \tag{2.44}$$

reinforces the interpretation that the energy of the signal can be decomposed into energy contributions from different frequencies. Think of the analog decomposition of light into its spectral components!

Example 2.1 Periodogram of a Sinusoid

Suppose that

$$u(t) = A \cos \omega_0 t \tag{2.45}$$

where $\omega_0 = 2\pi/N_0$ for some integer $N_0 > 1$. Consider the interval $t = 1, 2, \ldots, N$, where N is a multiple of $N_0 : N = s \cdot N_0$. Writing

$$\cos \omega_0 t = \frac{1}{2} \left[e^{i\omega_0 t} + e^{-i\omega_0 t} \right]$$

gives

$$U_N(\omega) = \frac{1}{\sqrt{N}} \sum_{t=1}^{N} \frac{A}{2} \left[e^{i(\omega_0 - \omega)t} + e^{-i(\omega_0 + \omega)t} \right]$$

Using (2.39), we find that

$$|U_N(\omega)|^2 = \begin{cases} N \cdot \frac{A^2}{4}, & \text{if } \omega = \pm\omega_0 = \frac{2\pi}{N_0} = \frac{2\pi s}{N} \\ 0, & \text{if } \omega = \frac{2\pi k}{N}, \quad k \neq s \end{cases} \tag{2.46}$$

The periodogram thus has two spikes in the interval $[-\pi, \pi]$. □

Example 2.2 Periodogram of a Periodic Signal

Suppose $u(t) = u(t + N_0)$ and we consider the signal over the interval $[1, N]$, $N = s \cdot N_0$. According to (2.42), the signal over the interval $[1, N_0]$ can be written

$$u(t) = \frac{1}{\sqrt{N_0}} \sum_{r=-N_0/2+1}^{N_0/2} A_r e^{2\pi i t r/N_0} \tag{2.47}$$

with

$$A_r = \frac{1}{\sqrt{N_0}} \sum_{t=1}^{N_0} u(t) e^{-2\pi i t r/N_0} \tag{2.48}$$

Since u is periodic, (2.47) applies over the whole interval $[1, N]$. It is thus a sum of N_0 sinusoids, and the results of the previous example (or straightforward calculations) show that

$$|U_N(\omega)|^2 = \begin{cases} s \cdot |A_r|^2, & \text{if } \omega = \frac{2\pi r}{N_0}, \quad r = 0, \pm 1 \pm \cdots \pm \frac{N_0}{2} \\ 0, & \text{if } \omega = \frac{2\pi k}{N}, \quad k \neq r \cdot s \end{cases} \tag{2.49}$$

□

The periodograms of Examples 2.1 and 2.2 turned out to be well behaved. For signals that are realizations of stochastic processes, the periodogram is typically a very erratic function of frequency. See Figure 2.8 and Lemma 6.2.

Transformation of Periodograms(*)

As a signal is filtered through a linear system, its periodogram changes. We show next how a signal's Fourier transform is affected by linear filtering. Results for the transformation of periodograms are then immediate.

Theorem 2.1. Let $\{s(t)\}$ and $\{w(t)\}$ be related by the strictly stable system $G(q)$:

$$s(t) = G(q)w(t) \tag{2.50}$$

The input $w(t)$ for $t \leq 0$ is unknown, but obeys $|w(t)| \leq C_w$ for all t. Let

$$S_N(\omega) = \frac{1}{\sqrt{N}} \sum_{t=1}^{N} s(t) e^{-i\omega t} \tag{2.51}$$

$$W_N(\omega) = \frac{1}{\sqrt{N}} \sum_{t=1}^{N} w(t) e^{-i\omega t} \tag{2.52}$$

Then

$$S_N(\omega) = G(e^{i\omega}) W_N(\omega) + R_N(\omega) \tag{2.53}$$

where

$$|R_N(\omega)| \leq 2C_w \cdot \frac{C_G}{\sqrt{N}} \tag{2.54}$$

with

$$C_G = \sum_{k=1}^{\infty} k |g(k)| \tag{2.55}$$

Proof. We have by definition

$$\begin{aligned} S_N(\omega) &= \frac{1}{\sqrt{N}} \sum_{t=1}^{N} s(t) e^{-it\omega} = \frac{1}{\sqrt{N}} \sum_{k=1}^{\infty} \sum_{t=1}^{N} g(k) w(t-k) e^{-it\omega} \\ &= [\text{change variables: } t - k = \tau] \\ &= \frac{1}{\sqrt{N}} \sum_{k=1}^{\infty} g(k) e^{-ik\omega} \cdot \sum_{\tau=1-k}^{N-k} w(\tau) e^{-i\tau\omega} \end{aligned}$$

Now

$$\begin{aligned} &\left| W_N(\omega) - \frac{1}{\sqrt{N}} \sum_{\tau=1-k}^{N-k} w(\tau) e^{-i\tau\omega} \right| \\ &\leq \left| \frac{1}{\sqrt{N}} \sum_{\tau=1-k}^{0} w(\tau) e^{-i\tau\omega} \right| + \left| \frac{1}{\sqrt{N}} \sum_{N-k+1}^{N} w(\tau) e^{-i\tau\omega} \right| \\ &\leq \frac{2}{\sqrt{N}} \cdot k \cdot C_w \end{aligned} \tag{2.56}$$

Hence

$$\left|S_N(\omega) - G(e^{i\omega})W_N(\omega)\right| = \left|\sum_{k=1}^{\infty} g(k)e^{-ik\omega}\left[\frac{1}{\sqrt{N}}\sum_{\tau=1-k}^{N-k} w(\tau)e^{i\tau\omega} - W_N(\omega)\right]\right|$$

$$\leq \frac{2}{\sqrt{N}}\sum_{k=1}^{\infty}\left|kg(k)C_w e^{-ik\omega}\right| \leq \frac{2C_w \cdot C_G}{\sqrt{N}}$$

and (2.53) to (2.55) follow. □

Corollary. Suppose $\{w(t)\}$ is periodic with period N. Then $R_N(\omega)$ in (2.53) is zero for $\omega = 2\pi k/N$.

Proof. The left side of (2.56) is zero for a periodic $w(\tau)$ at $\omega = 2\pi k/N$. □

2.3 SIGNAL SPECTRA

The periodogram defines, in a sense, the frequency contents of a signal over a finite time interval. This information may, however, be fairly hidden due to the typically erratic behavior of a periodogram as a function of ω. We now seek a definition of a similar concept for signals over the interval $t \in [1, \infty)$. Preferably, such a concept should more clearly demonstrate the different frequency contributions to the signal.

A definition for our framework is, however, not immediate. It would perhaps be natural to define the spectrum of a signal s as

$$\lim_{N\to\infty} |S_N(\omega)|^2 \tag{2.57}$$

but this limit fails to exist for many signals of practical interest. Another possibility would be to use the concept of the spectrum, or spectral density, of a stationary stochastic process as the Fourier transform of its covariance function. However, the processes that we consider here are frequently not stationary, for reasons that are described later. We shall therefore develop a framework for describing signals and their spectra that is applicable to deterministic as well as stochastic signals.

A Common Framework for Deterministic and Stochastic Signals

In this book we shall frequently work with signals that are described as stochastic processes with deterministic components. The reason is, basically, that we prefer to consider the input sequence as deterministic, or at least partly deterministic, while disturbances on the system most conveniently are described by random variables. In this way the system output becomes a stochastic process with deterministic components. For (2.20) we find that

$$Ey(t) = G(q)u(t)$$

so $\{y(t)\}$ is not a stationary process.

To deal with this problem, we introduce the following definition.

Definition 2.1. *Quasi-Stationary Signals* A signal $\{s(t)\}$ is said to be quasi-stationary if it is subject to

$$\text{(i)} \qquad Es(t) = m_s(t), \qquad |m_s(t)| \leq C, \qquad \forall t$$

$$\text{(ii)} \qquad Es(t)s(r) = R_s(t,r), \qquad |R_s(t,r)| \leq C \tag{2.58}$$

$$\lim_{N\to\infty} \frac{1}{N} \sum_{t=1}^{N} R_s(t, t-\tau) = R_s(\tau), \qquad \forall \tau \tag{2.59}$$

Here expectation E is with respect to the "stochastic components" of $s(t)$. If $\{s(t)\}$ itself is a deterministic sequence, the expectation is without effect and quasi-stationarity then means that $\{s(t)\}$ is a bounded sequence such that the limits

$$R_s(\tau) = \lim_{N\to\infty} \frac{1}{N} \sum_{t=1}^{N} s(t)s(t-\tau)$$

exist. If $\{s(t)\}$ is a stationary stochastic process, (2.58) and (2.59) are trivially satisfied, since then $Es(t)s(t-\tau) \stackrel{\Delta}{=} R_s(\tau)$ does not depend on t.

For easy notation we introduce the symbol $\overline{E}$ by

$$\overline{E}f(t) = \lim_{N\to\infty} \frac{1}{N} \sum_{t=1}^{N} Ef(t) \tag{2.60}$$

with an implied assumption that the limit exists when the symbol is used. Assumption (2.59), which simultaneously is a definition of $R_s(\tau)$, then reads

$$\overline{E}s(t)s(t-\tau) = R_s(\tau) \tag{2.61}$$

Sometimes, with some abuse of notation, we shall call $R_s(\tau)$ the *covariance function* of s, keeping in mind that this is a correct term only if $\{s(t)\}$ is a stationary stochastic process with mean value zero.

Similarly, we say that two signals $\{s(t)\}$ and $\{w(t)\}$ are *jointly quasi-stationary* if they both are quasi-stationary and if, in addition, the *cross-covariance* function

$$R_{sw}(\tau) = \overline{E}s(t)w(t-\tau) \tag{2.62}$$

exists. We shall say that jointly quasi-stationary signals are *uncorrelated* if their cross-covariance function is identically zero.

Definition of Spectra

When limits like (2.61) and (2.62) hold, we define the (power) *spectrum* of $\{s(t)\}$ as

$$\Phi_s(\omega) = \sum_{\tau=-\infty}^{\infty} R_s(\tau)e^{-i\tau\omega} \tag{2.63}$$

and the *cross spectrum* between $\{s(t)\}$ and $\{w(t)\}$ as

$$\Phi_{sw}(\omega) = \sum_{\tau=-\infty}^{\infty} R_{sw}(\tau)e^{-i\tau\omega} \tag{2.64}$$

provided the infinite sums exist. In the sequel, as we talk of a signal's "spectrum," we always implicitly assume that the signal has all the properties involved in the definition of spectrum.

While $\Phi_s(\omega)$ always is real, $\Phi_{sw}(\omega)$ is in general a complex-valued function of ω. Its real part is known as the *cospectrum* and its imaginary part as the *quadrature spectrum*. The argument $\arg \Phi_{sw}(\omega)$ is called the *phase spectrum*, while $|\Phi_{sw}(\omega)|$ is the *amplitude spectrum*.

Note that, by definition of the inverse Fourier transform, we have

$$\overline{E}s^2(t) = R_s(0) = \frac{1}{2\pi}\int_{-\pi}^{\pi} \Phi_s(\omega)d\omega \tag{2.65}$$

Example 2.3 Periodic Signals

Consider a deterministic, periodic signal with period M, i.e., $s(t) = s(t + M)$. We then have $Es(t) = s(t)$ and $Es(t)s(r) = s(t)s(r)$ and:

$$\frac{1}{N}\sum_{t=1}^{N} s(t)s(t-\tau) = \frac{MK}{N}\cdot\frac{1}{K}\sum_{\ell=0}^{K-1}\left\{\frac{1}{M}\sum_{t=1}^{M} s(t+\ell M)s(t-\tau+\ell M)\right\}$$

$$+\frac{1}{N}\sum_{t=KM+1}^{N} s(t)s(t-\tau)$$

where K is chosen as the maximum number of full periods, i.e., $N - MK < M$. Due to the periodicity, the sum within braces in fact does not depend on ℓ. Since there are at most $M-1$ terms in the last sum, this means that the limit as $N \to \infty$ will exist with

$$\overline{E}s(t)s(t-\tau) = R_s(\tau) = \frac{1}{M}\sum_{t=1}^{M} s(t)s(t-\tau)$$

A periodic, deterministic signal is thus quasi-stationary. We clearly also have $R_s(\tau + kM) = R_s(\tau)$.

For the spectrum we then have:

$$\Phi_s(\omega) = \sum_{\tau=-\infty}^{\infty} R_s(\tau)e^{-i\omega\tau} = \sum_{\ell=-\infty}^{\infty}\sum_{\tau=0}^{M-1} R_s(\tau+\ell M)e^{-i\omega\tau}e^{-i\omega\ell M}$$

$$= \Phi_s^p(\omega)\sum_{\ell=-\infty}^{\infty} e^{-i\ell M\omega} = \Phi_s^p(\omega)F(\omega, M)$$

where

$$\Phi_s^p(\omega) = \sum_{\tau=0}^{M-1} R_s(\tau)e^{i\omega\tau}, \quad F(\omega, M) = \sum_{\ell=-\infty}^{\infty} e^{-i\ell M\omega}$$

The function F is not well defined in the usual sense, but using the Dirac delta-function, it is well known (and well in line with (2.39)) that:

$$F(\omega, M) = \frac{2\pi}{M}\sum_{k=1}^{M} \delta(\omega - 2\pi k/M), \quad 0 \le \omega < 2\pi \tag{2.66}$$

This means that we can write:

$$\Phi_s(\omega) = \frac{2\pi}{M}\sum_{k=0}^{M-1} \Phi_s^p(2\pi k/M)\delta(\omega - 2\pi k/M), \quad 0 \le \omega < 2\pi \tag{2.67}$$

In $\Phi_s^p(2\pi k/M)$ we recognize the k:th Fourier coefficient of the periodic signal $R_s(\tau)$. Recall that the spectrum is periodic with period 2π, so different representations of (2.67) can be given.

The spectrum of a signal that is periodic with period M has thus (at most) M delta spikes, at the frequencies $2\pi k/M$, and is zero elsewhere. Although the limit (2.63) does not exist in a formal sense in this case, it is useful to extend the spectrum concept for signals with periodic and constant components in this way. □

Example 2.4 Spectrum of a Sinusoid

Consider again the signal (2.45), now extended to the interval $[1, \infty)$. We have

$$\frac{1}{N}\sum_{k=1}^{N} Eu(k)u(k - \tau) = \frac{1}{N}\sum_{k=1}^{N} A^2\cos(\omega_0 k)\cos\left(\omega_0(k - \tau)\right), \tag{2.68}$$

(Expectation is of no consequence since u is deterministic.) Now

$$\cos(\omega_0 k)\cos\left(\omega_0(k - \tau)\right) = \frac{1}{2}\left(\cos(2\omega_0 k - \omega_0\tau) + \cos\omega_0\tau\right)$$

which shows that

$$\bar{E}u(t)u(t - \tau) = \frac{A^2}{2}\cos\omega_0\tau = R_u(\tau)$$

The spectrum now is

$$\Phi_u(\omega) = \sum_{\tau=-\infty}^{\infty} \frac{A^2}{2}\cos(\omega_0\tau)e^{-i\omega\tau} = \frac{A^2}{4}\left(\delta(\omega - \omega_0) + \delta(\omega + \omega_0)\right)\cdot 2\pi \tag{2.69}$$

This result fits well with the finite interval expression (2.46) and the general expression (2.67) for periodic signals. □

Example 2.5 Stationary Stochastic Processes

Let $\{v(t)\}$ be a stationary stochastic process with covariance function (2.14). Since (2.59) then equals (2.14), our definition of spectrum coincides with the conventional one. Suppose now that the process v is given as (2.9). Its covariance function is then given by (2.13). The spectrum is

$$\begin{aligned}\Phi_v(\omega) &= \sum_{\tau=-\infty}^{\infty} \lambda e^{-i\tau\omega} \sum_{k=\max(0,\tau)}^{\infty} h(k)h(k-\tau) \\ &= \lambda \sum_{\tau=-\infty}^{\infty} \sum_{k=\max(0,\tau)}^{\infty} h(k)e^{-ik\omega}h(k-\tau)e^{i(k-\tau)\omega} \\ &= [k-\tau=s] = \lambda \sum_{s=0}^{\infty} h(s)e^{is\omega} \sum_{k=0}^{\infty} h(k)e^{-ik\omega} = \lambda \left|H(e^{i\omega})\right|^2\end{aligned}$$

using (2.18). This result is very important for our future use:

> The stochastic process described by $v(t) = H(q)e(t)$, where $\{e(t)\}$ is a sequence of independent random variables with zero mean values and covariances λ, has the spectrum
>
> $$\Phi_v(\omega) = \lambda \left|H(e^{i\omega})\right|^2 \tag{2.70}$$

This result, which was easy to prove for the special case of a stationary stochastic process, will be proved in the general case as Theorem 2.2 later in this section. Figure 2.7 shows the spectrum of the process of Figures 2.5 and 2.6, while the periodogram of the realization of Figure 2.6 is shown in Figure 2.8. □

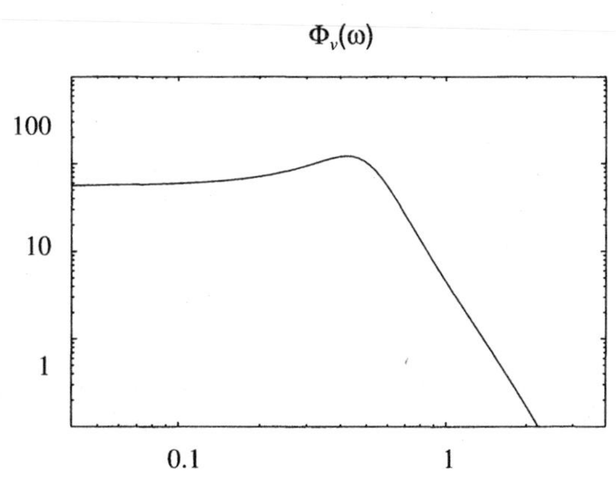

Figure 2.7 The spectrum of the process $v(t) - 1.5v(t-1) + 0.7v(t-2) = e(t) + 0.5e(t-1)$, $\{e(t)\}$ being white noise.

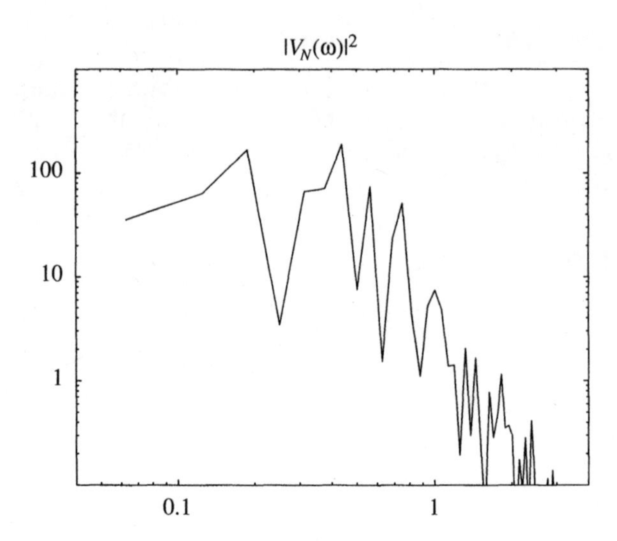

Figure 2.8 The periodogram of the realization of Figure 2.6.

Example 2.6 Spectrum of a Mixed Deterministic and Stochastic Signal

Consider now a signal

$$s(t) = u(t) + v(t) \tag{2.71}$$

where $\{u(t)\}$ is a deterministic signal with spectrum $\Phi_u(\omega)$ and $\{v(t)\}$ is a stationary stochastic process with zero mean value and spectrum $\Phi_v(\omega)$. Then

$$\begin{aligned} \overline{E}s(t)s(t-\tau) &= \overline{E}u(t)u(t-\tau) + \overline{E}u(t)v(t-\tau) \\ &\quad + \overline{E}v(t)u(t-\tau) + \overline{E}v(t)v(t-\tau) \\ &= R_u(\tau) + R_v(\tau) \end{aligned} \tag{2.72}$$

since $\overline{E}v(t)u(t-\tau) = 0$. Hence

$$\Phi_s(\omega) = \Phi_u(\omega) + \Phi_v(\omega) \tag{2.73}$$

□

Connections to the Periodogram

While the original idea (2.57) does not hold, a conceptually related result can be proved; that is, the expected value of the periodogram *converges weakly* to the spectrum:

$$E\,|S_N(\omega)|^2 \underset{w}{\rightarrow} \Phi_s(\omega) \tag{2.74}$$

By this is meant that

$$\lim_{N\to\infty} \int_{-\pi}^{\pi} E\,|S_N(\omega)|^2\,\Psi(\omega)\,d\omega = \int_{-\pi}^{\pi} \Phi_s(\omega)\Psi(\omega)\,d\omega \tag{2.75}$$

for all sufficiently smooth functions $\Psi(\omega)$. We have

Lemma 2.1. Suppose that $\{s(t)\}$ is quasi-stationary with spectrum $\Phi_s(\omega)$. Let

$$S_N(\omega) = \frac{1}{\sqrt{N}} \sum_{t=1}^{N} s(t)e^{-it\omega}$$

and let $\Psi(\omega)$ be an arbitrary function for $|\omega| \le \pi$ with Fourier coefficients a_τ, such that

$$\sum_{\tau=-\infty}^{\infty} |a_\tau| < \infty$$

Then (2.75) holds.

Proof.

$$\begin{aligned} E\,|S_N(\omega)|^2 &= \frac{1}{N} \sum_{k=1}^{N} \sum_{\ell=1}^{N} Es(k)s(\ell)e^{i\omega(k-\ell)} \\ &= [\ell - k = \tau] = \sum_{\tau=-(N-1)}^{N-1} R_N(\tau)e^{-i\omega\tau} \end{aligned} \tag{2.76}$$

where

$$R_N(\tau) = \frac{1}{N} \sum_{k=1}^{N} Es(k)s(k-\tau) \tag{2.77}$$

with the convention that $s(k)$ is taken as zero outside the interval $[1, N]$. Multiplying (2.76) by $\Psi(\omega)$ and integrating over $[-\pi, \pi]$ gives

$$\int_{-\pi}^{\pi} E\,|S_N(\omega)|^2\,\Psi(\omega)\,d\omega = \sum_{\tau=-(N-1)}^{N-1} R_N(\tau)a_\tau$$

by the definition of a_τ. Similarly, allowing interchange of summation and integration, we have

$$\int_{-\pi}^{\pi} \Phi_s(\omega)\Psi(\omega)\,d\omega = \sum_{\tau=-\infty}^{\infty} R_s(\tau)a_\tau$$

Hence

$$\int_{-\pi}^{\pi} E\,|S_N(\omega)|^2\,\Psi(\omega)\,d\omega - \int_{-\pi}^{\pi} \Phi_s(\omega)\Psi(\omega)\,d\omega$$

$$= \sum_{\tau=-(N-1)}^{N} a_\tau\,[R_N(\tau) - R_s(\tau)] + \sum_{|\tau|\geq N} a_\tau R_s(\tau)$$

Problem 2D.5 now completes the proof. □

Notice that for stationary stochastic processes the result (2.74) can be strengthened to "ordinary" convergence (see Problem 2D.3). Notice also that, in our framework, results like (2.74) can be applied also to realizations of stochastic processes simply by ignoring the expectation operator. We then view the realization in question as a given "deterministic" sequence, and will then, of course, have to require that the conditions (2.58) and (2.59) hold for this particular realization [disregard "E" also in (2.58) and (2.59)].

Transformation of Spectra by Linear Systems

As signals are filtered through linear systems, their properties will change. We saw how the periodogram was transformed in Theorem 2.1 and how white noise created stationary stochastic processes in (2.70). For spectra we have the following general result.

Theorem 2.2. Let $\{w(t)\}$ be a quasi-stationary signal with spectrum $\Phi_w(\omega)$, and let $G(q)$ be a stable transfer function. Let

$$s(t) = G(q)w(t) \tag{2.78}$$

Then $\{s(t)\}$ is also quasi-stationary and

$$\Phi_s(\omega) = \left|G(e^{i\omega})\right|^2 \Phi_w(\omega) \tag{2.79}$$

$$\Phi_{sw}(\omega) = G(e^{i\omega})\Phi_w(\omega) \tag{2.80}$$

Proof. The proof is given in Appendix 2A. □

Corollary. Let $\{y(t)\}$ be given by

$$y(t) = G(q)u(t) + H(q)e(t) \tag{2.81}$$

where $\{u(t)\}$ is a quasi-stationary, deterministic signal with spectrum $\Phi_u(\omega)$, and $\{e(t)\}$ is white noise with variance λ. Let G and H be stable filters. Then $\{y(t)\}$ is quasi-stationary and

$$\Phi_y(\omega) = \left|G(e^{i\omega})\right|^2 \Phi_u(\omega) + \lambda \left|H(e^{i\omega})\right|^2 \tag{2.82}$$

$$\Phi_{yu}(\omega) = G(e^{i\omega})\Phi_u(\omega) \tag{2.83}$$

Proof. The corollary follows from the theorem using Examples 2.5 and 2.6. □

Spectral Factorization

Typically, the transfer functions $G(q)$ and $H(q)$ used here are rational functions of q. Then results like (2.70) and Theorem 2.2 describe spectra as real-valued rational functions of $e^{i\omega}$ (which means that they also are rational functions of $\cos\omega$).

In practice, the converse of such results is of major interest: Given a spectrum $\Phi_v(\omega)$, can we then find a transfer function $H(q)$ such that the process $v(t) = H(q)e(t)$ has this spectrum with $\{e(t)\}$ being white noise? It is quite clear that this is not possible for all positive functions $\Phi_v(\omega)$. For example, if the spectrum is zero on an interval, then the function $H(z)$ must be zero on a portion of the unit circle. But since by necessity $H(z)$ should be analytic outside and on the unit circle for the expansion (2.18) to make sense, this implies that $H(z)$ is zero everywhere and cannot match the chosen spectrum.

The exact conditions under which our question has a positive answer are discussed in texts on stationary processes, such as Wiener (1949)and Rozanov (1967). For our purposes it is sufficient to quote a simpler result, dealing only with spectral densities $\Phi_v(\omega)$ that are rational in the variable $e^{i\omega}$ (or $\cos\omega$).

Spectral factorization: Suppose that $\Phi_v(\omega) > 0$ is a rational function of cos ω (or $e^{i\omega}$). Then there exists a monic rational function of z, $R(z)$, with no poles and no zeros on or outside the unit circle such that

$$\Phi_v(\omega) = \lambda \left|R(e^{i\omega})\right|^2$$

The proof of this result consists of a straightforward construction of R, and it can be found in standard texts on stochastic processes or stochastic control (e.g., Rozanov, 1967; Åström, 1970).

Example 2.7 ARMA Processes

If a stationary process $\{v(t)\}$ has rational spectrum $\Phi_v(\omega)$, we can represent it as

$$v(t) = R(q)e(t) \tag{2.84}$$

where $\{e(t)\}$ is white noise with variance λ. Here $R(q)$ is a rational function

$$R(q) = \frac{C(q)}{A(q)}$$

$$C(q) = 1 + c_1 q^{-1} + \cdots + c_{n_c} q^{-n_c}$$

$$A(q) = 1 + a_1 q^{-1} + \cdots + a_{n_a} q^{-n_a}$$

so that we may write

$$\begin{aligned} v(t) + a_1 v(t-1) + \cdots + a_{n_a} v(t-n_a) \\ = e(t) + c_1 e(t-1) + \cdots + c_{n_c} e(t-n_c) \end{aligned} \tag{2.85}$$

for (2.84). Such a representation of a stochastic process is known as an *ARMA model.* If $n_c = 0$, we have an autoregressive (AR) model:

$$v(t) + a_1 v(t-1) + \cdots + a_{n_a} v(t-n_a) = e(t) \tag{2.86}$$

And if $n_a = 0$, we have a moving average (MA) model:

$$v(t) = e(t) + c_1 e(t-1) + \cdots + c_{n_c} e(t-n_c) \tag{2.87}$$

□

The spectral factorization concept is important since it provides a way of representing the disturbance in the standard form $v = H(q)e$ from information about its spectrum only. The spectrum is usually a sound engineering way of describing properties of signals: "The disturbances are concentrated around 50 Hz" or "We are having low-frequency disturbances with little power over 1 rad/s." Rational functions are able to approximate functions of rather versatile shapes. Hence the spectral factorization result will provide a good modeling framework for disturbances.

Second-order Properties

The signal spectra, as defined here, describe the *second-order properties* of the signals (for stochastic processes, their second-order statistics, i.e., first and second moments). Recall from Section 2.1 that stochastic processes may have very different looking realizations even if they have the same covariance function (see Figures 2.5 and 2.6)! The spectrum thus describes only certain aspects of a signal. Nevertheless, it will turn out that many properties related to identification depend only on the spectra of the involved signals. This motivates our detailed interest in the second-order properties.

2.4 SINGLE REALIZATION BEHAVIOR AND ERGODICITY RESULTS (*)

All the results of the previous section are also valid, as we pointed out, for the special case of a given deterministic signal $\{s(t)\}$. Definitions of spectra, their transformations (Theorem 2.2) and their relationship with the periodogram (Lemma 2.1) hold unchanged; we may just disregard the expectation E and interpret $\overline{E} f(t)$ as

$$\lim_{N\to\infty} \frac{1}{N} \sum_{t=1}^{N} f(t)$$

There is a certain charm with results like these that do not rely on a probabilistic framework: we anyway observe just one realization, so why should we embed this observation in a stochastic process and describe its average properties taken over an ensemble of potential observations? There are two answers to this question. One is that such a framework facilitates certain calculations. Another is that it allows us to deal with the question of what would happen if the experiment were repeated.

Nevertheless, it is a valid question to ask whether the spectrum of the signal $\{s(t)\}$, as defined in a probabilistic framework, differs from the spectrum of the actually observed, single realization were it to be considered as a given, deterministic signal. This is the problem of *ergodic* theory, and for our setup we have the following fairly general result.

Theorem 2.3. Let $\{s(t)\}$ be a quasi-stationary signal. Let $Es(t) = m(t)$. Assume that

$$s(t) - m(t) = v(t) = \sum_{k=0}^{\infty} h_t(k)e(t-k) = H_t(q)e(t) \tag{2.88}$$

where $\{e(t)\}$ is a sequence of independent random variables with zero mean values, $Ee^2(t) = \lambda_t$, and bounded fourth moments, and where $\{H_t(q), t = 1, 2, \dots, \}$ is a uniformly stable family of filters. Then, with probability 1 as N tends to infinity,

$$\frac{1}{N}\sum_{t=1}^{N} s(t)s(t-\tau) \to \bar{E}s(t)s(t-\tau) = R_s(\tau) \tag{2.89a}$$

$$\frac{1}{N}\sum_{t=1}^{N} [s(t)m(t-\tau) - Es(t)m(t-\tau)] \to 0 \tag{2.89b}$$

$$\frac{1}{N}\sum_{t=1}^{N} [s(t)v(t-\tau) - Es(t)v(t-\tau)] \to 0 \tag{2.89c}$$

The proof is given in Appendix 2B.

The theorem is quite important. It says that, provided the stochastic part of the signal can be described as filtered white noise as in (2.88), then

> *the spectrum of an observed single realization of $\{s(t)\}$, computed as for a deterministic signal, coincides, with probability 1, with that of the process $\{s(t)\}$, defined by ensemble averages (E) as in (2.61).*

This de-emphasizes the distinction between deterministic and stochastic signals when we consider second-order properties only. A signal $\{s(t)\}$ whose spectrum is $\Phi_s(\omega) \equiv \lambda$ may, for all purposes related to second-order properties, be regarded as a realization of white noise with variance λ.

The theorem also gives an answer to the question of whether our "theoretical" spectrum, defined in (2.63) using the physically unrealizable concepts of E and lim, relates to the actually observed periodogram (2.43). According to Theorem 2.3 and Lemma 2.1, "smoothed" versions of $|S_N(\omega)|^2$ will look like $\Phi_s(\omega)$ for large N. Compare Figures 2.7 and 2.8. This link between our theoretical concepts and the real data is of course of fundamental importance. See Section 6.3.

2.5 MULTIVARIABLE SYSTEMS (*)

So far, we have worked with systems having a scalar input and a scalar output. In this section we shall consider the case where the output signal has p components and the input signal has m components. Such systems are called *multivariable*. The extra work involved in dealing with models of multivariable systems can be split up into two parts:

1. The easy part: mostly notational changes, keeping track of transposes, and noting that certain scalars become matrices and might not commute.
2. The difficult part: multioutput models have a much richer internal structure, which has the consequence that their parametrization is nontrivial. See Appendix 4A. (Multiple-input, single-output, MISO, models do not expose these problems.)

Let us collect the p components of the output signal into a p-dimensional column vector $y(t)$ and similarly construct an m-dimensional input vector $u(t)$. Let the disturbance $e(t)$ also be a p-dimensional column vector. The basic system description then looks just like (2.20):

$$y(t) = G(q)u(t) + H(q)e(t) \tag{2.90}$$

where now $G(q)$ is a transfer function matrix of dimension $p \times m$ and $H(q)$ has dimension $p \times p$. This means that the i, j entry of $G(q)$, denoted by $G_{ij}(q)$, is the scalar transfer function from input number j to output number i. The sequence $\{e(t)\}$ is a sequence of independent random p-dimensional vectors with zero mean values and covariance matrices $Ee(t)e^T(t) = \Lambda$.

Now, all the development in this chapter goes through with proper interpretation of matrix dimensions. Note in particular the following:

- The impulse responses $g(k)$ and $h(k)$ will be $p \times m$ and $p \times p$ matrices, respectively, with norms

$$\|g(k)\| = \left(\sum_{i,j} |g_{ij}|^2 \right)^{1/2} \tag{2.91}$$

  replacing absolute values in the definitions of stability.
- The definitions of covariances become [cf. (2.59)]

$$\overline{E}s(t)s^T(t - \tau) = R_s(\tau) \tag{2.92}$$

$$\overline{E}s(t)w^T(t - \tau) = R_{sw}(\tau) \tag{2.93}$$

  These are now matrices, with norms as in (2.91).
- Definitions of spectra remain unchanged, while the counterpart of Theorem 2.2 reads

$$\Phi_s(\omega) = G(e^{i\omega})\Phi_w(\omega)G^T(e^{-i\omega}) \tag{2.94}$$

This result will then also handle how cross spectra transform. If we have

$$y(t) = G(q)u(t) + H(q)w(t) = [\,G(q) \quad H(q)\,]\begin{bmatrix} u(t) \\ w(t) \end{bmatrix}$$

where u and w are jointly quasi-stationary, scalar sequences, we have:

$$\begin{aligned}\Phi_y(\omega) &= \begin{bmatrix} G(e^{i\omega}) & H(e^{i\omega}) \end{bmatrix}\begin{bmatrix} \Phi_u(\omega) & \Phi_{uw}(\omega) \\ \Phi_{uw}(-\omega) & \Phi_w(\omega) \end{bmatrix}\begin{bmatrix} G(e^{-i\omega}) \\ H(e^{-i\omega}) \end{bmatrix} \\ &= \left|G(e^{i\omega})\right|^2 \Phi_u(\omega) + \left|H(e^{i\omega})\right|^2 \Phi_w(\omega) \\ &\quad + G(e^{i\omega})\Phi_{uw}(\omega)H(e^{-i\omega}) + H(e^{i\omega})\Phi_{uw}(-\omega)G(e^{-i\omega}) \\ &= \left|G(e^{i\omega})\right|^2 \Phi_u(\omega) + \left|H(e^{i\omega})\right|^2 \Phi_w(\omega) \\ &\quad + 2\mathrm{Re}\left(G(e^{i\omega})\Phi_{uw}(\omega)H(e^{-i\omega})\right) \end{aligned} \tag{2.95}$$

where we used that $G(e^{i\omega})$ and $G(e^{-i\omega})$ are complex conjugates as well as $\Phi_{uw}(\omega)$ and $\Phi_{uw}(-\omega)$. The counterpart of the corollary to Theorem 2.2 for multivariable systems reads

$$\Phi_y(\omega) = G(e^{i\omega})\Phi_u(\omega)G^T(e^{-i\omega}) + H(e^{i\omega})\Lambda H^T(e^{-i\omega}) \tag{2.96}$$

- The spectral factorization result now reads: Suppose that $\Phi_v(\omega)$ is a $p \times p$ matrix that is positive definite for all ω and whose entries are rational functions of $\cos\omega$ (or $e^{i\omega}$). Then there exists a $p \times p$ monic matrix function $H(z)$ whose entries are rational functions of z (or z^{-1}) such that the (rational) function $\det H(z)$ has no poles and no zeros on or outside the unit circle. (For a proof, see Theorem 10.1 in Rozanov, 1967).
- The formulation of Theorem 2.3 carries over without changes. (In fact, the proof in Appendix 2B is carried out for the multivariable case).

2.6 SUMMARY

We have established the representation

$$y(t) = G(q)u(t) + H(q)e(t) \tag{2.97}$$

as the basic description of a linear system subject to additive random disturbances. Here $\{e(t)\}$ is a sequence of independent random variables with zero mean values and variances λ (in the multioutput case, covariance matrices Λ). Also,

$$G(q) = \sum_{k=1}^{\infty} g(k)q^{-k}$$

$$H(q) = 1 + \sum_{k=1}^{\infty} h(k)q^{-k}$$

The filter $G(q)$ is *stable* if

$$\sum_{k=1}^{\infty} |g(k)| < \infty$$

As the reader no doubt is aware of, other particular ways of representing linear systems, such as state-space models and difference equations, are quite common in practice. These can, however, be viewed as particular ways of representing the sequences $\{g(k)\}$ and $\{h(k)\}$, and they will be dealt with in some detail in Chapter 4.

We have also discussed the frequency function $G(e^{i\omega})$, bearing information about how an input sinusoid of frequency ω is transformed by the system. Frequency-domain concepts in terms of the frequency contents of input and output signals were also treated. The Fourier transform of a finite-interval signal was defined as

$$U_N(\omega) = \frac{1}{\sqrt{N}} \sum_{t=1}^{N} u(t) e^{-i\omega t} \tag{2.98}$$

A signal $s(t)$ such that the limits

$$\overline{E} s(t) s(t-\tau) = R_s(\tau)$$

exist is said to be *quasi-stationary*. Here

$$\overline{E} f(t) = \lim_{N \to \infty} \frac{1}{N} \sum_{1}^{N} E f(t)$$

Then the *spectrum* of $s(t)$ is defined as

$$\Phi_s(\omega) = \sum_{\tau=-\infty}^{\infty} R_s(\tau) e^{-i\omega\tau} \tag{2.99}$$

For y generated as in (2.97) with $\{u(t)\}$ and $\{e(t)\}$ independent, we then have

$$\Phi_y(\omega) = \left|G(e^{i\omega})\right|^2 \Phi_u(\omega) + \lambda \left|H(e^{i\omega})\right|^2$$

2.7 BIBLIOGRAPHY

The material of this chapter is covered in many textbooks on systems and signals. For a thorough elementary treatment, see Oppenheim and Willsky (1983). A discussion oriented more toward signals as time series is given in Brillinger (1981), which also contains several results of the same character as our Theorems 2.1 and 2.2.

A detailed discussion of the sampling procedure and connections between the physical time-continuous system and the sampled-data description (2.6) is given in Chapter 4 of Åström and Wittenmark (1984). Chapter 6 of that book also contains an illuminating discussion of disturbances and how to describe them mathematically. The idea of describing stochastic disturbances as linearly filtered white noise goes back to Wold (1938).

Fourier techniques in the analysis and description of signals go back to the Babylonians. See Oppenheim and Willsky (1983), Section 4.0, for a brief historical account. The periodogram was evidently introduced by Schuster (1894)to study periodic phenomena without having to consider relative phases. The statistical properties of the periodogram were first studied by Slutsky (1929). See also Brillinger (1983). Concepts of spectra are intimately connected to the harmonic analysis of time series, as developed by Wiener (1930), Wold (1938), Kolmogorov (1941), and others. Useful textbooks on these concepts (and their estimation) include Jenkins and Watts (1968) and Brillinger (1981). Our definition of the Fourier transform (2.37) with summation from 1 to N and a normalization with $1/\sqrt{N}$ suits our purposes, but is not standard. The placement of 2π in the definition of the spectrum or in the inverse transform, as we have it in (2.65), varies in the literature. Our choice is based on the wish to let white noise have a constant spectrum whose value equals the variance of the noise. The particular framework chosen here to accommodate mixtures of stochastic processes and deterministic signals is apparently novel, but has a close relationship to the classical concepts.

The result of Theorem 2.2 is standard when applied to stationary stochastic processes. See, for example, James, Nichols, and Phillips (1947)or Åström (1970). The extension to quasi-stationary signals appears to be new.

Spectral factorization turned out to be a key issue in the prediction of time series. It was formulated and solved by Wiener (1949)and Paley and Wiener (1934). The multivariable version is treated in Youla (1961). The concept is now standard in textbooks on stationary processes (see, e.g., Rozanov, 1967).

The topic of single realization behavior is a standard problem in probability theory. See, for example, Ibragimov and Linnik (1971), Billingsley (1965), or Chung (1974)for general treatments of such problems.

2.8 PROBLEMS[†]

2G.1 Let $s(t)$ be a p-dimensional signal. Show that

$$\overline{E}\,|s(t)|^2 = \frac{1}{2\pi}\int_{-\pi}^{\pi} \operatorname{tr}\left(\Phi_s(\omega)\right)\, d\omega$$

2G.2 Let $\Phi_s(\omega)$ be the (power) spectrum of a scalar signal defined as in (2.63). Show that

- **i.** $\Phi_s(\omega)$ is real.
- **ii.** $\Phi_s(\omega) \geq 0\,\forall\omega$.
- **iii.** $\Phi_s(-\omega) = \Phi_s(\omega)$.

2G.3 Let $s(t) = \begin{bmatrix} y(t) \\ u(t) \end{bmatrix}$ and let its spectrum be

$$\Phi_s(\omega) = \begin{bmatrix} \Phi_y(\omega) & \Phi_{yu}(\omega) \\ \Phi_{uy}(\omega) & \Phi_u(\omega) \end{bmatrix}$$

[†] See the Preface for an explanation of the numbering system.

Show that $\Phi_s(\omega)$ is a Hermitian matrix: that is,

$$\Phi_s(\omega) = \Phi_s^*(\omega)$$

where * denotes transpose and complex conjugate. What does this imply about the relationships between the cross spectra $\Phi_{yu}(\omega)$, $\Phi_{uy}(\omega)$, and $\Phi_{yu}(-\omega)$?

2G.4 Let a continuous time system representation be given by

$$y(t) = G_c(p)u(t)$$

The input is constant over the sampling interval T. Show that the sampled input-output data are related by

$$y(t) = G_T(q)u(t)$$

where

$$G_T(q) = \int_{s=-i\infty}^{i\infty} G_c(s)\frac{e^{sT}-1}{s}\frac{1}{q-e^{sT}}\,ds$$

Hint: Use (2.5).

2E.1 A stationary stochastic process has the spectrum

$$\Phi_v(\omega) = \frac{1.25 + \cos\omega}{1.64 + 1.6\cos\omega}$$

Describe $\{v(t)\}$ as an ARMA process.

2E.2 Suppose that $\{\eta(t)\}$ and $\{\xi(t)\}$ are two mutually independent sequences of independent random variables with

$$E\eta(t) = E\xi(t) = 0, \qquad E\eta^2(t) = \lambda_\eta, \qquad E\xi^2(t) = \lambda_\xi$$

Consider

$$w(t) = \eta(t) + \xi(t) + \gamma\xi(t-1)$$

Determine a MA(1) process

$$v(t) = e(t) + ce(t-1)$$

where $\{e(t)\}$ is white noise with

$$Ee(t) = 0, \qquad Ee^2(t) = \lambda_e$$

such that $\{w(t)\}$ and $\{v(t)\}$ have the same spectra; that is, find c and λ_e so that $\Phi_v(\omega) \equiv \Phi_w(\omega)$.

2E.3 **(a)** In Problem 2E.2 assume that $\{\eta(t)\}$ and $\{\xi(t)\}$ are jointly Gaussian. Show that if $\{e(t)\}$ also is chosen as Gaussian then the joint probability distribution of the process $\{w(t)\}$ [i. e., the joint PDFs of $w(t_1), w(t_2), \ldots, w(t_p)$ for any collection of time instances t_i] coincides with that of the process $\{v(t)\}$. Then, for all practical purposes, *the processes* $\{v(t)\}$ *and* $\{w(t)\}$ *are indistinguishable.*

(b) Assume now that $\eta(t) \in N(0, \lambda_\eta)$, while

$$\xi(t) = \begin{cases} 1, & \text{w.p.} \quad \dfrac{\lambda_\xi}{2} \\ -1, & \text{w.p.} \quad \dfrac{\lambda_\xi}{2} \\ 0, & \text{w.p.} \quad 1 - \lambda_\xi \end{cases}$$

Show that, although v and w have the same spectra, we cannot find a distribution for $e(t)$ so that they have the same joint PDFS. *Consequently the process $w(t)$ cannot be represented as an MA(1) process, although it has a second-order equivalent representation of that form.*

2E.4 Consider the "state-space description"

$$x(t+1) = fx(t) + w(t)$$

$$y(t) = hx(t) + \nu(t)$$

where x, f, h, w, and ν are scalars. $\{w(t)\}$ and $\{\nu(t)\}$ are mutually independent white Gaussian noises with variances R_1 and R_2, respectively. Show that $y(t)$ can be represented as an ARMA process:

$$y(t) + a_1 y(t-1) + \cdots + a_n y(t-n) = e(t) + c_1 e(t-1) + \cdots + c_n e(t-n)$$

Determine n, a_i, c_i, and the variance of $e(t)$ in terms of f, h, R_1, and R_2. What is the relationship between $e(t)$, $w(t)$, and $\nu(t)$?

2E.5 Consider the system

$$y(t) = G(q)u(t) + v(t)$$

controlled by the regulator

$$u(t) = -F_2(q)y(t) + F_1(q)r(t)$$

where $\{r(t)\}$ is a quasi-stationary reference signal with spectrum $\Phi_r(\omega)$. The disturbance $\{v(t)\}$ has spectrum $\Phi_v(\omega)$. Assume that $\{r(t)\}$ and $\{v(t)\}$ are uncorrelated and that the resulting closed-loop system is stable. Determine the spectra $\Phi_y(\omega)$, $\Phi_u(\omega)$, and $\Phi_{yu}(\omega)$.

2E.6 Consider the system

$$\frac{d}{dt}y(t) + ay(t) = u(t) \tag{2.100}$$

Suppose that the input $u(t)$ is piecewise constant over the sampling interval

$$u(t) = u_k, \qquad kT \le t < (k+1)T$$

(a) Derive a sampled-data system description for u_k, $y(kT)$.

(b) Assume that there is a time delay of T seconds so that $u(t)$ in (2.100) is replaced by $u(t-T)$. Derive a sampled-data system description for this case.

(c) Assume that the time delay is $1.5T$ so that $u(t)$ is replaced by $u(t-1.5T)$. Then give the sampled-data description.

2E.7 Consider a system given by

$$y(t) + ay(t-1) = bu(t-1) + e(t) + ce(t-1)$$

where $\{u(t)\}$ and $\{e(t)\}$ are independent white noises, with variances μ and λ, respectively. Follow the procedure suggested in Appendix 2C to multiply the system description by $e(t)$, $e(t-1)$, $u(t)$, $u(t-1)$, $y(t)$, and $y(t-1)$, respectively, and take expectation to show that

$$R_{ye}(0) = \lambda, \qquad R_{ye}(1) = (c-a)\lambda$$

$$R_{yu}(0) = 0, \qquad R_{yu}(1) = b\mu$$

$$R_y(0) = \frac{b^2\mu + \lambda + c^2\lambda - 2ac\lambda}{1-a^2},$$

$$R_y(1) = \frac{\lambda(c - a + a^2c - ac^2) - ab^2\mu}{1-a^2}$$

2T.1 Consider a continuous time system (2.1):

$$y(t) = \int_{\tau=0}^{\infty} g(\tau)u(t-\tau)\,d\tau$$

Let $g_T(\ell)$ be defined by (2.5), and assume that $u(t)$ is not piecewise constant, but that

$$\left|\frac{d}{dt}u(t)\right| \le C_1$$

Let $u_k = u\left((k+\frac{1}{2})T\right)$ and show that

$$y(kT) = \sum_{\ell=1}^{\infty} g_T(\ell)u_{k-\ell} + r_k$$

where

$$|r_k| \le C_2 \cdot T^2$$

Give a bound for C_2.

2T.2 If the filters $R_1(q)$ and $R_2(q)$ are (strictly) stable, then show that $R_1(q)R_2(q)$ is also (strictly) stable (see also Problem 3D.1).

2T.3 Let $G(q)$ be a rational transfer function; that is,

$$G(q) = \frac{b_1 q^{n-1} + \cdots + b_n}{q^n + a_1 q^{n-1} + \cdots + a_n}$$

Show that if $G(q)$ is stable, then it is also strictly stable.

2T.4 Consider the time-varying system

$$x(t+1) = F(t)x(t) + G(t)u(t)$$

$$y(t) = H(t)x(t)$$

Write

$$y(t) = \sum_{k=1}^{t} g_t(k)u(t-k)$$

[take $g_t(k) = 0$ for $k > t$]. Assume that

$$F(t) \to \overline{F}, \qquad \text{as } t \to \infty$$

where $\overline{F}$ has all eigenvalues inside the unit circle. Show that the family of filters $\{g_t(k), t = 1, 2, \ldots\}$ is uniformly stable.

2D.1 Consider $U_N(\omega)$ defined by (2.37). Show that $U_N(2\pi - \omega) = U_N(\omega) = \overline{U}_N(\omega)$ and rewrite (2.38) in terms of real-valued quantities only.

2D.2 Establish (2.39).

2D.3 Let $\{u(t)\}$ be a stationary stochastic process with $R_u(\tau) = Eu(t)u(t-\tau)$, and let $\Phi_u(\omega)$ be its spectrum. Assume that

$$\sum_{1}^{\infty} |\tau R_u(\tau)| < \infty$$

Let $U_N(\omega)$ be defined by (2.37). Prove that

$$E\,|U_N(\omega)|^2 \to \Phi_u(\omega), \qquad \text{as } N \to \infty$$

This is a strengthening of Lemma 2.1 for stationary processes.

2D.4 Let $G(q)$ be a stable system. Prove that

$$\lim_{N\to\infty} \frac{1}{N} \sum_{k=1}^{N} k\,|g(k)| = 0$$

Hint: Use Kronecker's lemma: Let a_k, b_k be sequences such that a_k is positive and decreasing to zero. Then $\Sigma a_k b_k < \infty$ implies

$$\lim_{N\to\infty} a_N \sum_{1}^{N} b_k = 0$$

(see, e.g., Chung, 1974, for a proof of Kronecker's lemma).

2D.5 Let $b_N(\tau)$ be a doubly indexed sequence such that, $\forall\tau$,

$$b_N(\tau) \to b(\tau), \qquad \text{as } N \to \infty$$

(but not necessarily uniformly in τ). Let a_τ be an infinite sequence, and assume that

$$\sum_{1}^{\infty} |a_\tau| < \infty, \qquad |b(\tau)| \le C \qquad \forall\tau$$

Show that

$$\lim_{N\to\infty} \left[\sum_{\tau=-N}^{N} a_\tau \left(b_N(\tau) - b(\tau)\right) + \sum_{|\tau|>N} a_\tau b(\tau) \right] = 0$$

Hint: Study Appendix 2A.

APPENDIX 2A: PROOF OF THEOREM 2.2

We carry out the proof for the multivariate case. Let $w(s) = 0$ for $s \leq 0$, and consider

$$\begin{aligned} R_s^N(\tau) &= \frac{1}{N}\sum_{t=1}^{N} Es(t)s^T(t-\tau) \\ &= \frac{1}{N}\sum_{t=1}^{N}\sum_{k=0}^{t}\sum_{\ell=0}^{t-\tau} g(k)Ew(t-k)w^T(t-\tau-\ell)g^T(\ell) \end{aligned} \tag{2A.1}$$

With the convention that $w(s) = 0$ if $s \notin [0, N]$, we can write

$$R_s^N(\tau) = \sum_{k=0}^{N}\sum_{\ell=0}^{N} g(k)\frac{1}{N}\sum_{t=1}^{N} Ew(t-k)w^T(t-\tau-\ell)g^T(\ell) \tag{2A.2}$$

If $w(s) \neq 0$, $s \leq 0$, then $s(t)$ gets the negligible contribution

$$\bar{s}(t) = \sum_{k=t}^{\infty} g(k)w(t-k)$$

Let

$$R_w^N(\tau) = \frac{1}{N}\sum_{t=1}^{N} Ew(t)w^T(t-\tau)$$

We see that $R_w^N(\tau + \ell - k)$ and the inner sum in (2A.2) differ by at most $\max(k, |\tau + \ell|)$ summands, each of which are bounded by C according to (2.58). Thus

$$\begin{aligned} &\left| R_w^N(\tau + \ell - k) - \frac{1}{N}\sum_{t=1}^{N} Ew(t-k)w^T(t-\tau-\ell) \right| \\ &\leq C\frac{\max(k, |\tau+\ell|)}{N} \leq \frac{C}{N}(k + |\tau + \ell|) \end{aligned} \tag{2A.3}$$

Let us define

$$R_s(\tau) = \sum_{k=0}^{\infty}\sum_{\ell=0}^{\infty} g(k)R_w(\tau + \ell - k)g^T(\ell) \tag{2A.4}$$

Then

$$R_s(\tau) - R_s^N(\tau) \le \sum_{k=N+1}^{\infty} \sum_{\ell=N+1}^{\infty} |g(k)||g(\ell)||R_w(\tau + \ell - k)|$$

$$+ \sum_{k=0}^{N} \sum_{\ell=0}^{N} |g(k)||g(\ell)||R_w(\tau + \ell - k) - R_w^N(\tau + \ell - k)|$$

$$+ \frac{C}{N} \sum_{k=0}^{N} k\,|g(k)| \cdot \sum_{\ell=0}^{N} |g(\ell)|$$

$$+ \frac{C}{N} \sum_{\ell=0}^{N} |\tau + \ell||g(\ell)| \cdot \sum_{k=0}^{N} |g(k)| \tag{2A.5}$$

The first sum tends to zero as $N \to \infty$ since $|R_w(\tau)| \le C$ and $G(q)$ is stable. It follows from the stability of $G(q)$ that

$$\frac{1}{N} \sum_{k=0}^{N} k\,|g(k)| \to 0, \qquad \text{as } N \to \infty \tag{2A.6}$$

(see Problem 2D.4). Hence the last two sums of (2A.5) tend to zero as $N \to \infty$. Consider now the second sum of (2A.5). Select an arbitrary $\varepsilon > 0$, and choose $N = N_\varepsilon$ such that

$$\sum_{k=N_\varepsilon+1}^{\infty} |g(k)| < \frac{\varepsilon}{[C \cdot C_1]} \quad \text{where} \quad C_1 = \sum_{k=0}^{\infty} |g(k)| \tag{2A.7}$$

This is possible since G is stable. Then select N'_ε such that

$$\max_{\substack{1 \le \ell \le N_\varepsilon \\ 1 \le k \le N_\varepsilon}} \left| R_w(\tau + \ell - k) - R_w^N(\tau + \ell - k) \right| < \frac{\varepsilon}{C_1^2}$$

for $N > N'_\varepsilon$. This is possible since

$$R_w^N(\tau) \to R_w(\tau), \qquad \text{as } N \to \infty \tag{2A.8}$$

(w is quasi-stationary) and since only a finite number of $R_w(s)$'s are involved (no uniform convergence of (2A.8) is necessary). Then, for $N > N'_\varepsilon$, we have that the second sum of (2A.5) is bounded by

$$\sum_{k=0}^{N_\varepsilon}\sum_{\ell=0}^{N_\varepsilon} |g(k)||g(\ell)| \cdot \frac{\varepsilon}{C_1^2} + \sum_{k=N_\varepsilon+1}^{\infty}\sum_{\ell=0}^{\infty} |g(k)||g(\ell)| \cdot 2C$$
$$+ \sum_{k=0}^{N_\varepsilon}\sum_{\ell=N_\varepsilon+1}^{\infty} |g(k)||g(\ell)| \cdot 2C$$

which is less than 5ε, according to (2A.7). Hence, also, the second sum of (2A.5) tends to zero as $N \to \infty$, and we have proved that the limit of (2A.5) is zero and hence that $s(t)$ is quasi-stationary.

The proof that $\bar{E}s(t)w^T(t-\tau)$ exists is analogous and simpler.

For $\Phi_s(\omega)$ we now find that

$$\Phi_s(\omega) = \sum_{\tau=-\infty}^{\infty}\left(\sum_{k=0}^{\infty}\sum_{\ell=0}^{\infty} g(k)R_w(\tau+\ell-k)g^T(\ell)\right)e^{-i\tau\omega}$$
$$= \sum_{\tau=-\infty}^{\infty}\sum_{k=0}^{\infty} g(k)e^{-ik\omega}\sum_{\ell=0}^{\infty} R_w(\tau+\ell-k)e^{-i(\tau+\ell-k)\omega}g^T(\ell)e^{i\ell\omega}$$
$$= [\tau+\ell-k = s]$$
$$= \sum_{k=0}^{\infty} g(k)e^{-ik\omega} \cdot \sum_{s=-\infty}^{\infty} R_w(s)e^{-is\omega} \cdot \sum_{\ell=0}^{\infty} g^T(\ell)e^{i\ell\omega}$$
$$= G(e^{i\omega})\Phi_s(\omega)G^T(e^{-i\omega})$$

Hence (2.79) is proved. The result (2.80) is analogous and simpler.

For families of linear filters we have the following result:

Corollary. Let $\{G_\theta(q),\ \theta \in D\}$ be a uniformly stable family of linear filters, and let $\{w(t)\}$ be a quasi-stationary sequence. Let $s_\theta(t) = G_\theta(q)w(t)$ and $R_s(\tau,\theta) = \bar{E}s_\theta(t)s_\theta^T(t-\tau)$. Then:

$$\sup_{\theta\in D}\left\|\frac{1}{N}\sum_{t=1}^{N} s_\theta(t)s_\theta^T(t-\tau) - R_s(\tau,\theta)\right\| \to 0 \quad \text{as } N \to \infty$$

Proof. We only have to establish that the convergence in (2A.5) is uniform in $\theta \in D$. In the first step all the $g(k)$-terms carry an index θ : $g_\theta(k)$. Interpreting

$$g(k) = \sup_{\theta \in D} |g_\theta(k)|$$

(2A.5) will of course still hold. Since the family $G_\theta(q)$ is uniformly stable, the sum over $g(k)$ will be convergent, which was the only property used to prove that (2A.5) tends to zero. This completes the proof of the corollary. □

APPENDIX 2B: PROOF OF THEOREM 2.3

In this appendix we shall show a more general variant of Theorem 2.3, which will be of value for the convergence analysis of Chapter 8. We also treat the multivariable case.

Theorem 2B.1. Let $\{G_\theta(q), \theta \in D_\theta\}$ and $\{M_\theta(q), \theta \in D_\theta\}$ be uniformly stable families of filters, and assume that the deterministic signal $\{w(t)\}$, $t = 1, 2, \ldots,$ is subject to

$$|w(t)| \le C_w, \qquad \forall t \tag{2B.1}$$

Let the signal $s_\theta(t)$ be defined, for each $\theta \in D_\theta$, by

$$s_\theta(t) = G_\theta(q)v(t) + M_\theta(q)w(t) \tag{2B.2}$$

where $\{v(t)\}$ is subject to the conditions of Theorem 2.3 (see (2.88) and let $Ee(t)e^T(t) = \Lambda_t$). Then

$$\sup_{\theta \in D_\theta} \left\| \frac{1}{N} \sum_{t=1}^{N} \left[s_\theta(t)s_\theta^T(t) - Es_\theta(t)s_\theta^T(t) \right] \right\| \to 0 \text{ w.p. } 1, \text{ as } N \to \infty \tag{2B.3}$$

Remark. We note that with $\dim s = 1$, $D_\theta = \{\theta^*\}$ (only one element), $G_\theta^*(q) = 1$, $M_\theta^*(q) = 1$ and $w(t) = m(t)$, then (2B.3) implies (2.89). With

$$s_\theta(t) = \begin{bmatrix} s(t) \\ m(t) \\ v(t) \end{bmatrix} = \begin{bmatrix} 1 \\ 0 \\ 1 \end{bmatrix} v(t) + \begin{bmatrix} w(t) \\ w(t) \\ 0 \end{bmatrix} \tag{2B.4}$$

the different cross products in (2B.3) imply all the results (2.89).

To prove Theorem 2B.1, we first establish two lemmas.

Lemma 2B.1. Let $\{v(t)\}$ obey the conditions of Theorem 2.3 and let

$$C_H = \sum_{k=1}^{\infty} \sup_t |h_t(k)|, \qquad C_l = \sup_t E\,|e(t)|^4, \qquad C_w = \sup_t |w(t)|$$

Then, for all r, N, k, and l,

$$E\left\| \sum_{t=r}^{N} [v(t-k)v^T(t-\ell) - Ev(t-k)v^T(t-\ell)] \right\|^2 \leq 4 \cdot C_e \cdot C_H^4 \cdot (N-r) \tag{2B.5}$$

$$E\left\| \sum_{t=r}^{N} v(t-k)w^T(t-l) \right\|^2 \leq 4 \cdot C_e \cdot C_H^2 \cdot C_w^2 \cdot (N-r) \tag{2B.6}$$

Proof of Lemma 2B.1. With no loss of generality, we may take $k = l = 0$. We then have

$$S_r^N \triangleq \sum_{t=r}^{N} v(t)v^T(t) - Ev(t)v^T(t) = \sum_{t=r}^{N}\sum_{k=0}^{\infty}\sum_{\ell=0}^{\infty} h_t(k)\alpha(t,k,\ell)h_t^T(\ell) \tag{2B.7}$$

where

$$\alpha(t,k,\ell) = e(t-k)e^T(t-\ell) - \Lambda_{t-\ell}\delta_{k\ell} \tag{2B.8}$$

For the square of the i, j entry of the matrix (2B.7), we have

$$(S_r^N(i,\,j))^2 = \sum_{t=r}^{N}\sum_{s=r}^{N}\sum_{k_1=0}^{\infty}\sum_{\ell_l=0}^{\infty}\sum_{k_2=0}^{\infty}\sum_{\ell_2=0}^{\infty} \gamma(t,s,k_1,k_2,\ell_1,\ell_2)$$

with

$$\gamma(t,s,k_1,k_2,\ell_1,\ell_2) = h_t^{(i)}(k_1)\alpha(t,k_1,\ell_1)\left[h_t^{(j)}(\ell_1)\right]^T h_s^{(i)}(k_2)\alpha(s,k_2,\ell_2)\left[h_s^{(j)}(\ell_2)\right]^T$$

Superscript (i) indicates the ith row vector. Since $\{e(t)\}$ is a sequence of independent variables, the expectation of γ is zero, unless at least some of the time indexes involved in $\alpha(t,k_1,\ell_1)$ and $\alpha(s,k_2,\ell_2)$ coincide, that is, unless

$$t-k_1 = s-k_2 \quad \text{or} \quad t-k_1 = s-\ell_2 \quad \text{or} \quad t-\ell_1 = s-k_2 \quad \text{or} \quad t-\ell_1 = s-\ell_2$$

For given values of $t,\ k_1,\ k_2,\ \ell_1$, and ℓ_2 this may happen for at most four values of s. For these we also have

$$E\gamma(t, s, k_1, k_2, \ell_1, \ell_2) \leq C_e \cdot |h(k_1)| \cdot |h(k_2)| \cdot |h(\ell_1)| \cdot |h(\ell_2)|$$

Hence

$$E\left(S_r^N(i, j)\right)^2 \leq \sum_{k_1=0}^{\infty} |h(k_1)| \cdot \sum_{k_2=0}^{\infty} |h(k_2)| \cdot \sum_{\ell_1=0}^{\infty} |h(\ell_1)|$$

$$\cdot \sum_{\ell_2=0}^{\infty} |h(\ell_2)| \cdot \sum_{t=r}^{N} 4 \cdot C_e \leq 4 \cdot C_e C_H^4 (N - r)$$

which proves (2B.5) of the lemma. The proof of (2B.6) is analogous and simpler. □

Corollary to Lemma 2B.1. Let

$$w(t) = \sum_{k=0}^{\infty} \alpha_t(k) e(t - k), \quad v(t) = \sum_{k=0}^{\infty} \beta_t(k) e(t - k)$$

Then

$$E\left\| \sum_{t=r}^{N} w(t)v(t) - Ew(t)v(t) \right\|^2 \leq C \cdot C_w^2 \cdot C_v^2 \cdot (N - r)$$

$$C_w = \sum_{k=0}^{\infty} \sup_t |\alpha_t(k)|, \quad C_v = \sum_{k=0}^{\infty} \sup_t |\beta_t(k)|$$

Lemma 2B.2. Let

$$R_r^N = \sup_{\theta \in D_\theta} \left\| \sum_{t=r}^{N} s_\theta(t) s_\theta^T(t) - E s_\theta(t) s_\theta^T(t) \right\| \tag{2B.9}$$

Then

$$E(R_r^N)^2 \leq C(N - r) \tag{2B.10}$$

Proof of Lemma 2B.2. First note the following fact: If

$$\varphi = \sum_{k=0}^{\infty} a(k) z(k) \tag{2B.11}$$

where $\{a(k)\}$ is a sequence of deterministic matrices, such that

$$\sum_{k=0}^{\infty} \|a(k)\| \leq C_a$$

and $\{z(k)\}$ is a sequence of random vectors such that $E\,|z(k)|^2 \le C_z$ then:

$$
\begin{aligned}
E|\varphi|^2 &= \sum_{k=0}^{\infty}\sum_{\ell=0}^{\infty} \operatorname{tr}\left[a(k)Ez(k)z^T(\ell)a^T(\ell)\right] \\
&\le \sum_{k=0}^{\infty}\sum_{\ell=0}^{\infty} \|a(k)\| \cdot \left[E\,|z(k)|^2\right]^{1/2} \cdot \left[E\,|z(\ell)|^2\right]^{1/2} \cdot \|a(\ell)\| \quad \text{(2B.12)} \\
&\le C_z \cdot \left[\sum_{k=0}^{\infty} \|a(k)\|\right]^2 \le C_z \cdot C_a^2
\end{aligned}
$$

Here the first inequality is Schwarz's inequality. We now have

$$
\begin{aligned}
R_\theta(N,r) &= \sum_{t=r}^{N} \left[s_\theta(t)s^T\theta(t) - Es_\theta(t)s_\theta^T(t)\right] \\
&= \sum_{t=r}^{N}\sum_{k=0}^{\infty}\sum_{\ell=0}^{\infty} g_\theta(k)\left[v(t-k)v^T(t-\ell) - Ev(t-k)v^T(t-\ell)\right]g_\theta^T(\ell) \\
&\quad + \sum_{t=r}^{N}\sum_{k=0}^{\infty}\sum_{\ell=0}^{\infty} g_\theta(k)v(t-k)w^T(t-\ell)m_\theta^T(\ell) \\
&\quad + \sum_{t=r}^{N}\sum_{k=0}^{\infty}\sum_{\ell=0}^{\infty} m_\theta(k)w(t-k)v^T(t-\ell)g_\theta^T(\ell) \qquad \text{(2B.13)}
\end{aligned}
$$

This gives

$$
\begin{aligned}
\sup_\theta \|R_\theta(N,r)\| &\le \sum_{k=0}^{\infty}\sum_{\ell=0}^{\infty} \sup_\theta \|g_\theta(k)\| \cdot \sup_\theta \|g_\theta(\ell)\| \cdot \left\|S_r^N(k,\ell)\right\| \\
&\quad + 2\sum_{k=0}^{\infty}\sum_{\ell=0}^{\infty} \sup_\theta \|g_\theta(k)\| \sup_\theta \|m_\theta(\ell)\| \cdot \left\|\tilde{S}_r^N(k,\ell)\right\| \qquad \text{(2B.14)}
\end{aligned}
$$

with S_r^N and $\tilde{S}_r^N$ defined by

$$
\begin{aligned}
S_r^N(k,\ell) &= \sum_{t=r}^{N} \left[v(t-k)v^T(t-\ell) - Ev(t-k)v^T(t-\ell)\right] \\
\tilde{S}_r^N(k,\ell) &= \sum_{t=r}^{N} v(t-k)w^T(t-\ell)
\end{aligned}
$$

Since $G_\theta(q)$ and $M_\theta(q)$ are uniformly stable families of filters,

$$\sup_\theta \|g_\theta(k)\| \le \bar{g}(k), \quad \sum_{k=1}^{\infty} \bar{g}(k) = C_G < \infty,$$

$$\sup_\theta \|m_\theta(k)\| \le \bar{m}(k), \quad \sum_{k=1}^{\infty} \bar{m}(k) = C_M < \infty$$

Applying (2B.11) and (2B.12) together with Lemma 2B.1 to (2B.14) gives

$$E\left[\sup_\theta \|R_\theta(N,r)\|\right]^2 \le 2 \cdot C_G^4 \cdot 4 \cdot C_e \cdot C_H^4 \cdot (N-r)$$
$$+\ 8 \cdot C_G^2 \cdot C_M^2 \cdot 4 \cdot C_w^2 \cdot C_H^2 \cdot (N-r) \le C \cdot (N-r)$$

which proves Lemma 2B.2. □

We now turn to the proof of Theorem 2B.1. Denote

$$r(t,\theta) = s_\theta(t)s_\theta^T(t) - Es_\theta(t)s_\theta^T(t) \tag{2B.15}$$

and let

$$R_r^N = \sup_{\theta \in D_\theta} \|R_\theta(N,r)\| \tag{2B.16}$$

with $R_\theta(N,r)$ defined by (2B.13). According to Lemma 2B.2

$$E\left(\frac{1}{N^2}R_1^{N^2}\right)^2 \le \left(\frac{1}{N^2}\right)^2 \cdot C \cdot N^2 \le \frac{C}{N^2}$$

Chebyshev's inequality (I.19) gives

$$P\left(\frac{1}{N^2}R_1^{N^2} > \varepsilon\right) \le \frac{1}{\varepsilon^2}E\left(\frac{R_1^{N^2}}{N^2}\right)^2$$

Hence:

$$\sum_{k=1}^{\infty} P\left(\frac{1}{k^2}R_1^{k^2} > \varepsilon\right) \le \frac{C}{\varepsilon^2} \cdot \sum_{k=1}^{\infty} \frac{1}{k^2} < \infty$$

which, via Borel-Cantelli's lemma [see (I.18)], implies that

$$\frac{1}{k^2}R_1^{k^2} \to 0, \qquad \text{w.p. 1} \qquad \text{as } k \to \infty \tag{2B.17}$$

Now suppose that

$$\sup_{N^2 \le k \le (N+1)^2} \frac{1}{k}R_1^k$$

is obtained for $k = k_N$ and $\theta = \theta_N$. Hence

$$\begin{aligned}
\sup_{N^2 \le k \le (N+1)^2} \frac{1}{k} R_1^k &= \frac{1}{k_N} \left| \sum_{t=1}^{k_N} r(t, \theta_N) \right| \\
&\le \frac{1}{k_N} \left| \sum_{t=1}^{N^2} r(t, \theta_N) \right| + \frac{1}{k_N} \left| \sum_{t=N^2+1}^{k_N} r(t, \theta_N) \right| \qquad \text{(2B.18)} \\
&\le \frac{1}{k_N} \cdot R_1^{N^2} + \frac{1}{k_N} \cdot R_{N^2+1}^{k_N}
\end{aligned}$$

Since $k_N \ge N^2$, the first term on the right side of (2B.18) tends to zero w.p.1 in view of (2B.17). For the second one we have, using Lemma 2B.2,

$$\begin{aligned}
E\left| \frac{1}{k_N} R_{N^2+1}^{k_N} \right|^2 &\le \frac{1}{N^4} \cdot E \max_{N^2 \le k \le (N+1)^2} \left| R_{N^2+1}^k \right|^2 \\
&\le \frac{1}{N^4} \sum_{k=N^2+1}^{(N+1)^2} E\left| R_{N^2+1}^k \right|^2 \le \frac{1}{N^4} \sum_{k=N^2+1}^{(N+1)^2} C \cdot (k - N^2) \le \frac{C}{N^2}
\end{aligned}$$

which using Chebyshev's inequality (I.19) and the Borel-Cantelli lemma as before, shows that also the second term of (2B.18) tends to zero w.p.1. Hence

$$\sup_{N^2 \le k \le (N+1)^2} \frac{1}{k} R_1^k \to 0, \quad \text{w.p.1}, \quad \text{as } N \to \infty \qquad \text{(2B.19)}$$

which proves the theorem.

Corollary to Theorem 2B.1. Suppose that the conditions of the theorems hold, but that (2.88) is weakened to

$$E\left[e(t) | e(t-1), \ldots, e(0)\right] = 0, \quad E\left[e^2(t) | e(t-1), \ldots, e(0)\right] = \lambda$$

$$E\left|e(t)\right|^4 \le C$$

Then the theorem still holds. [That is: $\{e(t)\}$ need not be white noise, it is sufficient that it is a martingale difference.]

Proof. Independence was used only in Lemma 2B.1. It is easy to see that this lemma holds also under the weaker conditions. □

APPENDIX 2C: COVARIANCE FORMULAS

For several calculations we need expressions for variances and covariances of signals in ARMA descriptions. These are basically given by the inverse formulas

$$R_s(\tau) = \frac{1}{2\pi}\int_{-\pi}^{\pi} \Phi_s(\omega)e^{i\omega\tau}\, d\omega \tag{2C.1a}$$

$$R_{sv}(\tau) = \frac{1}{2\pi}\int_{-\pi}^{\pi} \Phi_{sv}(\omega)e^{i\omega\tau}\, d\omega \tag{2C.1b}$$

With the expressions according to Theorem 2.2 for the spectra, (2C.1) takes the form

$$\begin{aligned} R_s(\tau) &= \frac{\lambda}{2\pi}\int_{-\pi}^{\pi} \left|\frac{C(e^{i\omega})}{A(e^{i\omega})}\right|^2 e^{i\omega\tau}\, d\omega = [z = e^{i\omega}] \\ &= \frac{\lambda}{2\pi}\oint \frac{C(z)C(1/z)}{A(z)A(1/z)} z^{\tau-1}\, dz \end{aligned} \tag{2C.2}$$

for an ARMA process. The last integral is a complex integral around the unit circle, which could be evaluated using residue calculus. Åström, Jury, and Agniel (1970)(see also Åström, 1970, Ch. 5) have derived an efficient algorithm for computing (2C.2) for $\tau = 0$. It has the following form:

$$A(z) = a_0 z^n + a_1 z^{n-1} + \cdots + a_n, \qquad C(z) = c_0 z^n + c_1 z^{n-1} + \cdots + c_n$$

Let $a_i^n = a_i$ and $c_i^n = c_i$ and define a_i^k, c_i^k recursively by

$$a_i^{n-k} = \frac{a_0^{n-k+1}a_i^{n-k+1} - a_{n-k+1}^{n-k+1}a_{n-k+1-i}^{n-k+1}}{a_0^{n-k+1}}$$

$$c_i^{n-k} = \frac{a_0^{n-k+1}c_i^{n-k+1} - c_{n-k+1}^{n-k+1}a_{n-k+1-i}^{n-k+1}}{a_0^{n-k+1}}$$

$$i = 0, 1, \ldots, n-k, \qquad k = 1, 2, \ldots, n$$

Then for (2C.2)

$$R_s(0) = \frac{1}{a_0}\sum_{k=0}^{n} \frac{(c_k^k)^2}{a_0^k} \tag{2C.3}$$

An explicit expression for the variance of a second-order ARMA process

$$\begin{aligned} y(t) + a_1 y(t-1) + a_2 y(t-2) &= e(t) + c_1 e(t-1) + c_2 e(t-2) \\ Ee^2(t) &= 1 \end{aligned} \tag{2C.4}$$

is

$$\operatorname{Var} y(t) = \frac{(1+a_2)\left(1+(c_1)^2+(c_2)^2\right) - 2a_1c_1(1+c_2) - 2c_2\left(a_2-(a_1)^2+(a_2)^2\right)}{(1-a_2)(1-a_1+a_2)(1+a_1+a_2)} \tag{2C.5}$$

To find $R_y(\tau)$ and the cross covariances $R_{ye}(\tau)$ by hand calculations in simple examples, the easiest approach is to multiply (2C.4) by $e(t)$, $e(t-1)$, $e(t-2)$, $y(t)$, $y(t-1)$, and $y(t-2)$ and take expectation. This gives six equations for the six variables $R_{ye}(\tau)$, $R_y(\tau)$, or $\tau = 0, 1, 2$. Note that $R_{ye}(\tau) = 0$ for $\tau < 0$.

For numerical computation in MATLAB it is easiest to represent the ARMA process in state-space form, with

$$[\, y(t-1) \quad \ldots \quad y(t-n) \quad e(t-1) \quad \ldots \quad e(t-n)\,]^T$$

as states, and then use **`dlyap`** to compute the state covariance matrix. This will contain all variances and covariances of interest.

3

SIMULATION AND PREDICTION

The system descriptions given in Chapter 2 can be used for a variety of design problems related to the true system. In this chapter we shall discuss some such uses. The purpose of this is twofold. First, the idea of how to predict future output values will turn out to be most essential for the development of identification methods. The expressions provided in Section 3.2 will therefore be instrumental for the further discussion in this book. Second, by illustrating different uses of system descriptions, we will provide some insights into what is required for such descriptions to be adequate for their intended uses. A leading idea of our framework for identification will be that the effort spent in developing a model of a system must be related to the application it is going to be used for. Throughout the chapter we assume that the system description is given in the form (2.97):

$$y(t) = G(q)u(t) + H(q)e(t) \tag{3.1}$$

3.1 SIMULATION

The most basic use of a system description is to simulate the system's response to various input scenarios. This simply means that an input sequence $u^*(t)$, $t = 1, 2, \ldots, N$, chosen by the user is applied to (3.1) to compute the undisturbed output

$$y^*(t) = G(q)u^*(t), \qquad t = 1, 2, \ldots, N \tag{3.2}$$

This is the output that the system would produce had there been no disturbances, according to the description (3.1). To evaluate the disturbance influence, a random-number generator (in the computer) is used to produce a sequence of numbers $e^*(t)$, $t = 1, 2, \ldots, N$, that can be considered as a realization of a white-noise stochastic process with variance λ. Then the disturbance is calculated as

$$v^*(t) = H(q)e^*(t) \tag{3.3}$$

By suitably presenting $y^*(t)$ and $v^*(t)$ to the user, an idea of the system's response to $\{u^*(t)\}$ can be formed.

This way of experimenting on the model (3.1) rather than on the actual, physical process to evaluate its behavior under various conditions has become widely used in engineering practice of all fields and no doubt reflects the most common use of mathematical descriptions. To be true, models used in, say, flight simulators or nuclear power station training simulators are of course far more complex than (3.1), but they still follow the same general idea (see also Chapter 5).

3.2 PREDICTION

We shall start by discussing how future values of $v(t)$ can be predicted in case it is described by

$$v(t) = H(q)e(t) = \sum_{k=0}^{\infty} h(k)e(t-k) \tag{3.4}$$

For (3.4) to be meaningful, we assume that H is stable; that is,

$$\sum_{k=0}^{\infty} |h(k)| < \infty \tag{3.5}$$

Invertibility of the Noise Model

A crucial property of (3.4), which we will impose, is that it should be *invertible*; that is, if $v(s)$, $s \leq t$, are known, then we shall be able to compute $e(t)$ as

$$e(t) = \tilde{H}(q)v(t) = \sum_{k=0}^{\infty} \tilde{h}(k)v(t-k) \tag{3.6}$$

with

$$\sum_{k=0}^{\infty} \left|\tilde{h}(k)\right| < \infty$$

How can we determine the filter $\tilde{H}(q)$ from $H(q)$? The following lemma gives the answer.

Lemma 3.1. Consider $\{v(t)\}$ defined by (3.4) and assume that the filter H is stable. Let

$$H(z) = \sum_{k=0}^{\infty} h(k)z^{-k} \tag{3.7}$$

and assume that the function $1/H(z)$ is analytic in $|z| \geq 1$:

$$\frac{1}{H(z)} = \sum_{k=0}^{\infty} \tilde{h}(k)z^{-k} \tag{3.8}$$

Define the filter $H^{-1}(q)$ by

$$H^{-1}(q) = \sum_{k=0}^{\infty} \tilde{h}(k) q^{-k} \tag{3.9}$$

Then $\tilde{H}(q) = H^{-1}(q)$ satisfies (3.6).

Remark. That (3.8) exists for $|z| \geq 1$ also means that the filter $H^{-1}(q)$ is stable. For convenience, we shall then say that $H(q)$ is an *inversely stable* filter.

Proof. From (3.7) and (3.8) it follows that

$$1 = \sum_{k=0}^{\infty} \sum_{s=0}^{\infty} h(k) \tilde{h}(s) z^{-(k+s)} = [k + s = \ell] = \sum_{\ell=0}^{\infty} \sum_{k=0}^{\ell} h(k) \tilde{h}(\ell - k) z^{-\ell}$$

which implies that

$$\sum_{k=0}^{\ell} h(k) \tilde{h}(\ell - k) = \begin{cases} 1, & \text{if } \ell = 0 \\ 0, & \text{if } \ell \neq 0 \end{cases} \tag{3.10}$$

Now let $\{v(t)\}$ be defined by (3.4) and consider

$$\begin{aligned} \sum_{k=0}^{\infty} \tilde{h}(k) v(t - k) &= \sum_{k=0}^{\infty} \tilde{h}(k) \sum_{s=0}^{\infty} h(s) e(t - k - s) \\ &= \sum_{k=0}^{\infty} \sum_{s=0}^{\infty} \tilde{h}(k) h(s) e(t - k - s) = [k + s = \ell] \\ &= \sum_{\ell=0}^{\infty} \left[\sum_{k=0}^{\ell} \tilde{h}(k) h(\ell - k) \right] e(t - \ell) = e(t) \end{aligned}$$

according to (3.10), which proves the lemma. □

Note: The lemma shows that the properties of *the filter* $H(q)$ are quite analogous to those of *the function* $H(z)$. It is not a triviality that the inverse filter $H^{-1}(q)$ can be drived by inverting the function $H(z)$; hence the formulation of the result as a lemma. However, all similar relationships between $H(q)$ and $H(z)$ will also hold, and from a practical point of view it will be useful to switch freely between the filter and its z-transform. See also Problem 3D.1.

The lemma shows that the inverse filter (3.6) in a natural way relates to the original filter (3.4). In view of its definition, we shall also write

$$H^{-1}(q) = \frac{1}{H(q)} \tag{3.11}$$

for this filter. All that is needed is that the function $1/H(z)$ be analytic in $|z| \geq 1$; that is, it has no poles on or outside the unit circle. We could also phrase the condition as $H(z)$ must have no zeros on or outside the unit circle. This ties in very nicely with the spectral factorization result (see Section 2.3) according to which, for rational strictly positive spectra, we can always find a representation $H(q)$ with these properties.

Example 3.1 A Moving Average Process

Suppose that

$$v(t) = e(t) + ce(t-1) \tag{3.12}$$

That is,

$$H(q) = 1 + cq^{-1}$$

According to (2.87), this process is a *moving average of order 1*, MA(1). Then

$$H(z) = 1 + cz^{-1} = \frac{z+c}{z}$$

has a pole in $z = 0$ and a zero in $z = -c$, which is inside the unit circle if $|c| < 1$. If so, the inverse filter is determined as

$$H^{-1}(z) = \frac{1}{H(z)} = \frac{1}{1 + cz^{-1}} = \sum_{k=0}^{\infty} (-c)^k z^{-k}$$

and $e(t)$ is recovered from (3.12) as

$$e(t) = \sum_{k=0}^{\infty} (-c)^k v(t-k)$$

□

One-step-ahead Prediction of v

Suppose now that we have observed $v(s)$ for $s \leq t-1$ and that we want to predict the value of $v(t)$ based on these observations. We have, since H is monic,

$$v(t) = \sum_{k=0}^{\infty} h(k)e(t-k) = e(t) + \sum_{k=1}^{\infty} h(k)e(t-k) \tag{3.13}$$

Now, the knowledge of $v(s)$, $s \leq t-1$ implies the knowledge of $e(s)$, $s \leq t-1$, in view of (3.6). The second term of (3.13) is therefore known at time $t-1$. Let us denote it, provisionally, by $m(t-1)$:

$$m(t-1) = \sum_{k=1}^{\infty} h(k)e(t-k)$$

Suppose that $\{e(t)\}$ are identically distributed, and let the probability distribution of $e(t)$ be denoted by $f_e(x)$:

$$P(x \leq e(t) \leq x + \Delta x) \approx f_e(x)\Delta x$$

This distribution is independent of the other values of $e(s)$, $s \neq t$, since $\{e(t)\}$ is a sequence of independent random variables. What we can say about $v(t)$ at time $t-1$ is consequently that the probability that $v(t)$ assumes a value between $m(t-1)+x$ and $m(t-1)+x+\Delta x$ is $f_e(x)\Delta x$. This could also be phrased as

> *the (posterior) probability density function of $v(t)$, given observations up to time $t-1$, is $f_v(x) = f_e(x - m(t-1))$.*

Formally, these calculations can be written as

$$
\begin{aligned}
f_v(x)\Delta x &\approx P\left(x \le v(t) \le x + \Delta x | v_{-\infty}^{t-1}\right) \\
&= P\left(x \le m(t-1) + e(t) \le x + \Delta x\right) \\
&= P\left(x - m(t-1) \le e(t) \le x + \Delta x - m(t-1)\right) \\
&\approx f_e\left(x - m(t-1)\right)\Delta x
\end{aligned}
$$

Here $P(A|v_{-\infty}^{t-1})$ means the conditional probability of the event A, given $v_{-\infty}^{t-1}$.

This is the most complete statement that can be made about $v(t)$ at time $t-1$. Often we just give one value that characterizes this probability distribution and hence serves as a *prediction* of $v(t)$. This could be chosen as the value for which the PDF $f_e(x - m(t-1))$ has its maximum, the most probable value of $v(t)$, which also is called the *maximum a posteriori (MAP)* prediction. We shall, however, mostly work with the mean value of the distribution in question, the *conditional expectation* of $v(t)$ denoted by $\hat{v}(t|t-1)$. Since the variable $e(t)$ has zero mean, we have

$$
\hat{v}(t|t-1) = m(t-1) = \sum_{k=1}^{\infty} h(k)e(t-k) \tag{3.14}
$$

It is easy to establish that the conditional expectation also minimizes the mean-square error of the prediction error:

$$
\min_{\hat{v}(t)} E\left(v(t) - \hat{v}(t)\right)^2 \Rightarrow \hat{v}(t) = \hat{v}(t|t-1)
$$

where the minimization is carried out over all functions $\hat{v}(t)$ of $v_{-\infty}^{t-1}$. See Problem 3D.3.

Let us find a more convenient expression for (3.14). We have, using (3.6) and (3.11),

$$
\begin{aligned}
\hat{v}(t|t-1) &= \left[\sum_{k=1}^{\infty} h(k)q^{-k}\right] e(t) = [H(q) - 1]\, e(t) \\
&= \frac{H(q)-1}{H(q)} v(t) = \left[1 - H^{-1}(q)\right] v(t) = \sum_{k=1}^{\infty} -\tilde{h}(k)v(t-k)
\end{aligned} \tag{3.15}
$$

Applying $H(q)$ to both sides gives the alternative expression

$$H(q)\hat{v}(t|t-1) = [H(q)-1]\,v(t) = \sum_{k=1}^{\infty} h(k)v(t-k) \tag{3.16}$$

Example 3.2 A Moving Average Process

Consider the process (3.12). Then (3.16) shows that the predictor is calculated as

$$\hat{v}(t|t-1) + c\hat{v}(t-1|t-2) = cv(t-1) \tag{3.17}$$

Alternatively we can determine $H^{-1}(q)$ from Example 3.1 and use (3.15):

$$\hat{v}(t|t-1) = -\sum_{k=1}^{\infty}(-c)^k v(t-k)$$

□

Example 3.3 An Autoregressive Process

Consider a process

$$v(t) = \sum_{k=0}^{\infty} a^k e(t-k), \qquad |a| < 1$$

Then

$$H(z) = \sum_{k=0}^{\infty} a^k z^{-k} = \frac{1}{1-az^{-1}}$$

which gives

$$H^{-1}(z) = 1 - az^{-1}$$

and the predictor, according to (3.15),

$$\hat{v}(t|t-1) = av(t-1) \tag{3.18}$$

□

One-step-ahead Prediction of y

Consider the description (3.1), and assume that $y(s)$ and $u(s)$ are known for $s \le t-1$. Since

$$v(s) = y(s) - G(q)u(s) \tag{3.19}$$

this means that also $v(s)$ are known for $s \le t-1$. We would like to predict the value

$$y(t) = G(q)u(t) + v(t)$$

based on this information. Clearly, the conditional expectation of $y(t)$, given the information in question, is

$$\begin{aligned}\hat{y}(t|t-1) &= G(q)u(t) + \hat{v}(t|t-1) \\ &= G(q)u(t) + \left[1 - H^{-1}(q)\right]v(t) \\ &= G(q)u(t) + \left[1 - H^{-1}(q)\right][y(t) - G(q)u(t)]\end{aligned}$$

using (3.15) and (3.19), respectively. Collecting the terms gives

$$\hat{y}(t|t-1) = H^{-1}(q)G(q)u(t) + \left[1 - H^{-1}(q)\right]y(t) \tag{3.20}$$

or

$$H(q)\hat{y}(t|t-1) = G(q)u(t) + [H(q) - 1]\,y(t) \tag{3.21}$$

Remember that these expressions are shorthand notation for expansions. For example, let $\{\ell(k)\}$ be defined by

$$\frac{G(z)}{H(z)} = \sum_{k=1}^{\infty} \ell(k)z^{-k} \tag{3.22}$$

[This expansion exists for $|z| \geq 1$ if $H(z)$ has no zeros and $G(z)$ no poles in $|z| \geq 1$.] Then (3.20) means that

$$\hat{y}(t|t-1) = \sum_{k=1}^{\infty} \ell(k)u(t-k) + \sum_{k=1}^{\infty} -\tilde{h}(k)y(t-k) \tag{3.23}$$

Unknown Initial Conditions

In the reasoning so far we have made use of the assumption that the whole data record from time minus infinity to $t-1$ is available. Indeed, in the expression (3.20) as in (3.23) all these data appear explicitly. In practice, however, it is usually the case that only data over the interval $[0, t-1]$ are known. The simplest thing would then be to replace the unknown data by zero (say) in (3.23):

$$\hat{y}(t|t-1) \approx \sum_{k=1}^{t} \ell(k)u(t-k) + \sum_{k=1}^{t} -\tilde{h}(k)y(t-k) \tag{3.24}$$

One should realize that this is now only an approximation of the actual conditional expectation of $y(t)$, given data over $[0, t-1]$. The exact prediction involves time-varying filter coefficients and can be computed using the Kalman filter [see (4.94)]. For most practical purposes, (3.24) will, however, give a satisfactory solution. The reason is that the coefficients $\{\ell(k)\}$ and $\{\tilde{h}(k)\}$ typically decay exponentially with k (see Problem 3G.1).

The Prediction Error

From (3.20) and (3.1), we find that the prediction error $y(t) - \hat{y}(t|t-1)$ is given by

$$y(t) - \hat{y}(t|t-1) = -H^{-1}(q)G(q)u(t) + H^{-1}(q)y(t) = e(t) \tag{3.25}$$

The variable $e(t)$ thus represents that part of the output $y(t)$ that cannot be predicted from past data. For this reason it is also called the *innovation* at time t.

k-step-ahead Prediction of y (*)

Having treated the problem of one-step-ahead prediction in some detail, it is easy to generalize to the following problem: Suppose that we have observed $v(s)$ for $s \leq t$ and that we want to predict the value $v(t+k)$. We have

$$\begin{aligned} v(t+k) &= \sum_{\ell=0}^{\infty} h(\ell)e(t+k-\ell) \\ &= \sum_{\ell=0}^{k-1} h(\ell)e(t+k-\ell) + \sum_{\ell=k}^{\infty} h(\ell)e(t+k-\ell) \end{aligned} \tag{3.26}$$

Let us define

$$\overline{H}_k(q) = \sum_{\ell=0}^{k-1} h(\ell)q^{-\ell}, \qquad \tilde{H}_k(q) = \sum_{\ell=k}^{\infty} h(\ell)q^{-\ell+k} \tag{3.27}$$

The second sum of (3.26) is known at time time t, while the first sum is independent of what has happened up to time t and has zero mean. The conditional mean of $v(t+k)$, given $v_{-\infty}^{t}$ is thus given by

$$\hat{v}(t+k|t) = \sum_{\ell=k}^{\infty} h(\ell)e(t+k-\ell) = \tilde{H}_k(q)e(t) = \tilde{H}_k(q) \cdot H^{-1}(q)v(t)$$

This expression is the k-step-ahead predictor of v.

Now suppose that we have measured $y_{-\infty}^{t}$ and know $u_{-\infty}^{t+k-1}$ and would like to predict $y(t+k)$. We have, as before

$$y(t+k) = G(q)u(t+k) + v(t+k)$$

which gives

$$\begin{aligned} \hat{y}(t+k|y_{-\infty}^{t}, u_{-\infty}^{t+k-1}) \triangleq \hat{y}(t+k|t) &= G(q)u(t+k) + \hat{v}(t+k|t) \\ &= G(q)u(t+k) + \tilde{H}_k(q)H^{-1}(q)v(t) \qquad (3.28) \\ &= G(q)u(t+k) + \tilde{H}_k(q)H^{-1}(q)\left[y(t) - G(q)u(t)\right] \end{aligned}$$

Introduce

$$\begin{aligned} W_k(q) &\triangleq 1 - q^{-k}\tilde{H}_k(q)H^{-1}(q) = \left[H(q) - q^{-k}\tilde{H}_k(q)\right]H^{-1}(q) \\ &= \overline{H}_k(q)H^{-1}(q) \end{aligned} \qquad (3.29)$$

Then simple manipulation on (3.28) gives

$$\hat{y}(t+k|t) = W_k(q)G(q)u(t+k) + \tilde{H}_k(q)H^{-1}(q)y(t) \qquad (3.30)$$

or, using the first equality in (3.29),

$$\hat{y}(t|t-k) = W_k(q)G(q)u(t) + \left[1 - W_k(q)\right]y(t) \qquad (3.31)$$

This expression, together with (3.27) and (3.29), defines the k-step-ahead predictor for y. Notice that this predictor can also be viewed as a one-step-ahead predictor associated with the model

$$y(t) = G(q)u(t) + W_k^{-1}(q)e(t) \qquad (3.32)$$

The prediction error is obtained from (3.30) as

$$\begin{aligned} e_k(t+k) &\triangleq y(t+k) - \hat{y}(t+k|t) = -W_k(q)G(q)u(t+k) \\ &\quad + \left[q^k - \tilde{H}_k(q)H^{-1}(q)\right]y(t) \\ &= W_k(q)\left[y(t+k) - G(q)u(t+k)\right] = W_k(q)H(q)e(t+k) \\ &= \overline{H}_k(q)e(t+k) \end{aligned} \qquad (3.33)$$

Here we used (3.29) in the second and fourth equalities. According to (3.27), $\overline{H}_k(q)$ is a polynomial in q^{-1} of order $k-1$. Hence the prediction error is a moving average of $e(t+k), \ldots, e(t+1)$.

The Multivariable Case (*)

For a multivariable system description (3.1) (or 2.90), we define the $p \times p$ matrix filter $H^{-1}(q)$ as

$$H^{-1}(q) = \sum_{k=0}^{\infty} \tilde{h}(k) q^{-k}$$

Here $\tilde{h}(k)$ are the $p \times p$ matrices defined by the expansion of the matrix function

$$[H(z)]^{-1} = \sum_{k=0}^{\infty} \tilde{h}(k) z^{-k} \tag{3.34}$$

This expansion can be interpreted entrywise in the matrix $[H(z)]^{-1}$ (formed by standard manipulations for matrix inversion). It exists for $|z| \geq 1$ provided the function det $H(z)$ has no zeros in $|z| \geq 1$. With $H^{-1}(q)$ thus defined, all calculations and formulas given previously are valid also for the multivariable case.

3.3 OBSERVERS

In many cases in systems and control theory, one does not work with a full description of the properties of disturbances as in (3.1). Instead a noise-free or "*deterministic*" model is used:

$$y(t) = G(q)u(t) \tag{3.35}$$

In this case one probably keeps in the back of one's mind, though, that (3.35) is not really the full story about the input-output properties.

The description (3.35) can of course also be used for "computing," "guessing," or "predicting" future values of the output. The lack of noise model, however, leaves several possibilities for how this can best be done. The concept of *observers* is a key issue for these calculations. This concept is normally discussed in terms of state-space representations of (3.35) (see Section 4.3); see, for example, Luenberger (1971)or Åström and Wittenmark (1984), but it can equally well be introduced for the input-output form (3.35).

An Example

Let

$$G(z) = b \sum_{k=1}^{\infty} (a)^{k-1} z^{-k} = \frac{bz^{-1}}{1 - az^{-1}} \tag{3.36}$$

This means that the input-output relationship can be represented either as

$$y(t) = b \sum_{k=1}^{\infty} (a)^{k-1} u(t-k) \tag{3.37}$$

that is

$$y(t) = \frac{bq^{-1}}{1 - aq^{-1}} u(t)$$

or as

$$(1 - aq^{-1})y(t) = bq^{-1}u(t)$$

i.e.

$$y(t) - ay(t - 1) = bu(t - 1) \tag{3.38}$$

Now, if we are given the description (3.35) and (3.36) together with data $y(s)$, $u(s)$, $s \leq t-1$, and are asked to produce a "guess" or to "calculate" what $y(t)$ might be, we could use either

$$\hat{y}(t|t - 1) = b\sum_{k=1}^{\infty}(a)^{k-1}u(t - k) \tag{3.39}$$

or

$$\hat{y}(t|t - 1) = ay(t - 1) + bu(t - 1) \tag{3.40}$$

As long as the data and the system description are correct, there would also be no difference between (3.39) and (3.40); they are both "observers" (in our setting "predictors" would be a more appropriate term) for the system. The choice between them would be carried out by the designer in terms of how vulnerable they are to imperfections in data and descriptions. For example, if input-output data are lacking prior to time $s = 0$, then (3.39) suffers from an error that decays like a^t (effect of wrong initial conditions), whereas (3.40) is still correct for $t \geq 1$. On the other hand, (3.39) is unaffected by measurement errors in the output, whereas such errors are directly transferred to the prediction in (3.40). From the discussion of Section 3.2, it should be clear that, if (3.35) is complemented with a noise model as in (3.1), then the choice of predictor becomes unique (cf. Problem 3E.3). This follows since the conditional mean of the output, computed according to the assumed noise model, is a uniquely defined quantity.

A Family of Predictors for (3.35)

The example (3.36) showed that the choice of predictor could be seen as a trade-off between sensitivity with respect to output measurement errors and rapidly decaying effects of erroneous initial conditions. To introduce design variables for this trade-off, choose a filter $W(q)$ such that

$$W(q) = 1 + \sum_{\ell=k}^{\infty} w_\ell q^{-\ell} \tag{3.41}$$

Apply it to both sides of (3.35):

$$W(q)y(t) = W(q)G(q)u(t)$$

which means that

$$y(t) = [1 - W(q)]\, y(t) + W(q)G(q)u(t)$$

In view of (3.41), the right side of this expression depends only on $y(s)$, $s \leq t-k$, and $u(s)$, $s \leq t-1$. Based on that information, we could thus produce a "guess" or prediction of $y(t)$ as

$$\hat{y}(t|t-k) = [1 - W(q)]\, y(t) + W(q)G(q)u(t) \tag{3.42}$$

The trade-off considerations for the choice of W would then be:

1. Select $W(q)$ so that both W and WG have rapidly decaying filter coefficients in order to minimize the influence of erroneous initial conditions.
2. Select $W(q)$ so that measurement imperfections in $y(t)$ are maximally attenuated. (3.43)

The later issue can be illuminated in the frequency domain: Suppose that $y(t) = y_M(t)+v(t)$, where $y_M(t) = G(q)u(t)$ is the useful signal and $v(t)$ is a measurement error. Then the prediction error according to (3.42) is

$$\varepsilon(t) = y(t) - \hat{y}(t|t-k) = W(q)v(t) \tag{3.44}$$

The spectrum of this error is, according to Theorem 2.2,

$$\Phi_\varepsilon(\omega) = \left|W(e^{i\omega})\right|^2 \Phi_v(\omega) \tag{3.45}$$

where $\Phi_v(\omega)$ is the spectrum of v. The problem is thus to select W, subject to (3.41), such that the error spectrum (3.45) has an acceptable size and suitable shape.

A comparison with the k-step prediction case of Section 3.2 shows that the expression (3.42) is identical to (3.31) with $W(q) = W_k(q)$. It is clear that the qualification of a complete noise model in (3.1) allows us to analytically compute the filter W in accordance with aspect 2. This was indeed what we did in Section 3.2. However, aspect 1 was neglected there, since we assumed all past data to be available. Normally, as we pointed out, this aspect is also less important.

Fundamental Role of the Predictor Filter

It turns out that for most uses of system descriptions it is the predictor form (3.20), or as in (3.31) and (3.42), that is more important than the description (3.1) or (3.35) itself. We use (3.31) and (3.42) to predict, or "guess," future outputs; we use it for control design to regulate the predicted output, and so on. Now, (3.31) and (3.42) are just linear filters into which sequences $\{u(t)\}$ and $\{y(t)\}$ are fed, and that produce $\hat{y}(t|t-k)$ as output. The thoughts that the designer had when he or she selected this filter are immaterial once it is put to use: *The filter is the same whether $W = W_k$ was chosen as a trade-off (3.43) or computed from H as in (3.27) and (3.29).* The noise model H in (3.1) is from this point of view just an alibi for determining the predictor. This is the viewpoint we are going to adopt. The predictor filter is the fundamental system description (Figure 3.1). Our rationale for arriving at the filter is secondary. This also means that the difference between a "stochastic system" (3.1) and a "deterministic" one (3.35) is not fundamental. Nevertheless, we find it convenient to use the description (3.1) as the basic system description. It is in a one-to-one correspondence with the one-step-ahead predictor (3.20) (see Problem 3D.2) and relates more immediately to traditional system descriptions.

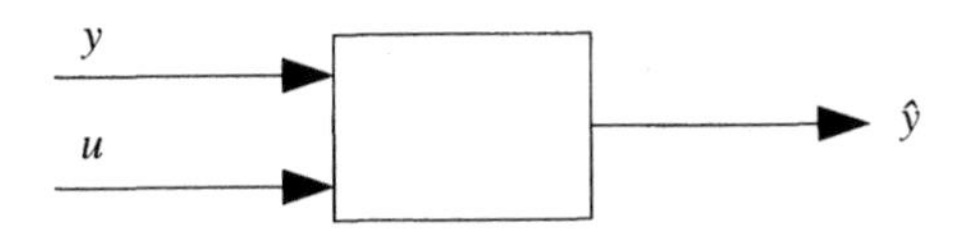

Figure 3.1 The predictor filter.

3.4 SUMMARY

Starting from the representation

$$y(t) = G(q)u(t) + H(q)e(t)$$

we have derived an expression for the one-step-ahead prediction of $y(t)$ [i.e., the best "guess" of $y(t)$ given $u(s)$ and $y(s)$, $s \leq t-1$]. This expression is given by

$$\hat{y}(t|t-1) = H^{-1}(q)G(q)u(t) + \left[1 - H^{-1}(q)\right]y(t) \tag{3.46}$$

We also derived a corresponding k-step-ahead predictor (3.31). We pointed out that one can arrive at such predictors also through deterministic observer considerations, not relying on a noise model H. We have stressed that the bottom line in most uses of a system description is how these predictions actually are computed; the underlying noise assumptions are merely vehicles for arriving at the predictors. The discussion of Chapters 2 and 3 can thus be viewed as a methodology for "guessing" future system outputs.

It should be noted that calculations such as (3.46) involved in determining the predictors and regulators are typically performed with greater computational efficiency once they are applied to transfer functions G and H with more specific structures. This will be illustrated in the next chapter.

3.5 BIBLIOGRAPHY

Prediction and control are standard textbook topics. Accounts of the k-step-ahead predictor and associated control problems can be found in Åström (1970)and Åström and Wittenmark (1984). Prediction is treated in detail in, for example, Anderson and Moore (1979)and Box and Jenkins (1970). An early account of this theory is Whittle (1963).

Prediction theory was developed by Kolmogorov (1941), Wiener (1949), Kalman (1960), and Kalman and Bucy (1961). The hard part in these problems is indeed to find a suitable representation of the disturbance. Once we arrive at (3.1) via spectral factorization, or at its time-varying counterpart via the Riccati equation [see (4.95) and Problem 4G.3], the calculation of a reasonable predictor is, as demonstrated here, easy. Note, however (as pointed out in Problem 2E.3), that for non-Gaussian processes normally only the second-order properties can be adequately described by (3.1), which consequently is too simple a representation to accommodate more complex noise structures. The calculations carried out in Section 3.2 are given in Åström (1970)for the case where G and H are rational with the same denominators. Rissanen and Barbosa (1969)have given expressions for the prediction in input-output models of this kind when the lack of knowledge of the infinite past is treated properly [i.e., when the ad hoc solution (3.24) is not accepted]. The result is, of course, a time-varying predictor.

3.6 PROBLEMS

3G.1 Suppose that the transfer function $G(z)$ is rational and that its poles are all inside $|z| < \mu$, where $\mu < 1$. Show that

$$|g(k)| \leq c \cdot \mu^k$$

where $g(k)$ is defined as in (2.16).

3G.2 Let $A(q)$ and $B(q)$ be two monic stable and inversely stable filters. Show that

$$\frac{1}{2\pi} \int_{-\pi}^{\pi} \left|A(e^{i\omega})\right|^2 \cdot \left|B(e^{i\omega})\right|^2 \, d\omega \geq 1$$

with equality only if $A(q) = 1/B(q)$.

3E.1 Let

$$H(q) = 1 - 1.1q^{-1} + 0.3q^{-2}$$

Compute $H^{-1}(q)$ as an explicit infinite expansion.

3E.2 Determine the 3-step-ahead predictors for

$$y(t) = \frac{1}{1 - aq^{-1}} e(t)$$

and

$$y(t) = (1 + cq^{-1})e(t)$$

respectively. What are the variances of the associated prediction errors?

3E.3 Show that if (3.35) and (3.36) are complemented with the noise model $H(q) = 1$ then (3.39) is the natural predictor, whereas the noise model

$$H(q) = \sum_{k=0}^{\infty} (a^k) q^{-k}$$

leads to the predictor (3.40).

3E.4 Let $e(t)$ have the distribution

$$e(t) = \begin{cases} 1, & \text{w.p. } 0.5 \\ -0.5, & \text{w.p. } 0.25 \\ -1.5, & \text{w.p. } 0.25 \end{cases}$$

Let

$$v(t) = H(q)e(t)$$

and let $\hat{v}(t|t-1)$ be defined as in the text. What is the most probable value (MAP) of $v(t)$ given the information $\hat{v}(t|t-1)$? What is the probability that $v(t)$ will assume a value between $\hat{v}(t|t-1) - \frac{1}{4}$ and $\hat{v}(t|t-1) + \frac{1}{4}$?

3T.1 Suppose that $A(q)$ is inversely stable and monic. Show that $A^{-1}(q)$ is monic.

3T.2 Suppose the measurement error spectrum of v in (3.44) and (3.45) is given by

$$\Phi_v(\omega) = \lambda \left|R(e^{i\omega})\right|^2$$

for some monic stable and inversely stable filter $R(q)$. Find the filter W, subject to (3.41) with $k = 1$, that minimizes

$$\overline{E}\varepsilon^2(t)$$

Hint: Use Problem 3G.2.

3T.3 Consider the system description of Problem 2E.4:

$$x(t+1) = fx(t) + w(t)$$

$$y(t) = hx(t) + \nu(t)$$

(x scalar). Assume that $\{\nu(t)\}$ is white Gaussian noise with variance R_2 and that $\{w(t)\}$ is a sequence of independent variables with

$$w(t) = \begin{cases} 1, & \text{w.p. } 0.05 \\ -1, & \text{w.p. } 0.05 \\ 0, & \text{w.p. } 0.9 \end{cases}$$

Determine a monic filter $W(q)$ such that the predictor

$$\hat{y}(t) = (1 - W(q))\, y(t)$$

minimizes

$$\overline{E}\left(y(t) - \hat{y}(t)\right)^2$$

What can be said about

$$E\left(y(t) | y_{\infty}^{t-1}\right)?$$

3T.4 Consider the noise description

$$v(t) = e(t) + ce(t-1), \quad |c| > 1, \quad Ee^2(t) = \lambda \tag{3.47}$$

Show that $e(t)$ cannot be reconstructed from v^t by a causal, stable filter. However, show that $e(t)$ can be computed from v_{t+1}^{∞} by an anticausal, stable filter. Thus construct a stable, anticausal predictor for $v(t)$ given $v(s)$, $s \geq t+1$.

Determine a noise $\overline{v}(t)$ with the same second-order properties as $v(t)$, such that

$$\overline{v}(t) = \overline{e}(t) + c^*\overline{e}(t-1), \quad |c^*| < 1, \quad E\overline{e}^2(t) = \lambda^* \tag{3.48}$$

Show that $\overline{v}(t)$ can be predicted from $\overline{v}^{t-1}$ by a stable, causal predictor. [Measuring just second-order properties of the noise, we cannot distinguish between (3.47) and (3.48). However, when $e(t)$ in (3.47) is a physically well defined quantity (although not measured by us), we may be interested in which one of (3.47) and (3.48) has generated the noise. See Benveniste, Goursat, and Ruget (1980).]

3D.1 In the chapter we have freely multiplied, added, subtracted, and divided by transfer-function operators $G(q)$ and $H(q)$. Division was formalized and justified by Lemma 3.1 and (3.11). Justify similarly addition and multiplication.

3D.2 Suppose a one-step-ahead predictor is given as

$$\hat{y}(t|t-1) = L_1(q)u(t-1) + L_2(q)y(t-1)$$

Calculate the system description (3.1) from which this predictor was derived.

3D.3 Consider a stochastic process $\{v(t)\}$ and let

$$\hat{v}(t) = E\left(v(t)|v^{t-1}\right)$$

Define

$$e(t) = v(t) - \hat{v}(t)$$

Let $\tilde{v}(t)$ be an arbitrary function of v^{t-1}. Show that

$$E\left(v(t) - \tilde{v}(t)\right)^2 \geq Ee^2(t)$$

Hint: Use $Ex^2 = E_z E(x^2|z)$.

4

MODELS OF LINEAR TIME-INVARIANT SYSTEMS

A model of a system is a description of (some of) its properties, suitable for a certain purpose. The model need not be a true and accurate description of the system, nor need the user have to believe so, in order to serve its purpose.

System identification is the subject of constructing or selecting models of dynamical systems to serve certain purposes. As we noted in Chapter 1, a first step is to determine a class of models within which the search for the most suitable model is to be conducted. In this chapter we shall discuss such classes of models for linear time-invariant systems.

4.1 LINEAR MODELS AND SETS OF LINEAR MODELS

A linear time-invariant model is specified, as we saw in Chapter 2, by the impulse response $\{g(k)\}_1^\infty$, the spectrum $\Phi_v(\omega) = \lambda \left|H(e^{i\omega})\right|^2$ of the additive disturbance, and, possibly, the probability density function (PDF) of the disturbance $e(t)$. A complete model is thus given by

$$\begin{aligned} y(t) &= G(q)u(t) + H(q)e(t) \\ &f_e(\cdot), \text{ the PDF of } e \end{aligned} \tag{4.1}$$

with

$$G(q) = \sum_{k=1}^{\infty} g(k)q^{-k}, \qquad H(q) = 1 + \sum_{k=1}^{\infty} h(k)q^{-k} \tag{4.2}$$

A particular model thus corresponds to specification of the three functions G, H, and f_e. It is in most cases impractical to make this specification by enumerating the infinite sequences $\{g(k)\}$, $\{h(k)\}$ together with the function $f_e(x)$. Instead one chooses to work with structures that permit the specification of G and H in terms of a finite number of numerical values. Rational transfer functions and finite-dimensional state-space descriptions are typical examples of this. Also, most often the PDF f_e is not specified as a function, but described in terms of a few numerical characteristics, typically the first and second moments:

$$\begin{aligned} Ee(t) &= \int x f_e(x)\, dx = 0 \\ Ee^2(t) &= \int x^2 f_e(x)\, dx = \lambda \end{aligned} \tag{4.3}$$

It is also common to assume that $e(t)$ is Gaussian, in which case the PDF is entirely specified by (4.3). The specification of (4.1) in terms of a finite number of numerical values, or coefficients, has another and most important consequence for the purposes of system identification. Quite often it is not possible to determine these coefficients a priori from knowledge of the physical mechanisms that govern the system's behavior. Instead the determination of all or some of them must be left to estimation procedures. This means that the coefficients in question enter the model (4.1) as *parameters to be determined.* We shall generally denote such parameters by the vector θ, and thus have a model description

$$y(t) = G(q,\theta)u(t) + H(q,\theta)e(t) \tag{4.4a}$$

$$f_e(x,\theta), \text{ the PDF of } e(t); \ \{e(t)\} \text{ white noise} \tag{4.4b}$$

The parameter vector θ then ranges over a subset of $\mathbf{R}^d$, where d is the dimension of θ:

$$\theta \in D_{\mathcal{M}} \subset \mathbf{R}^d \tag{4.5}$$

Notice that (4.4) to (4.5) no longer is *a* model; it is a *set of models*, and it is for the estimation procedure to select that member in the set that appears to be most suitable for the purpose in question. [One may sometimes loosely talk about "the model (4.4)," but this is abuse of notation from a formal point of view.] Using (3.20), we can compute the one-step-ahead prediction for (4.4). Let it be denoted by $\hat{y}(t|\theta)$ to emphasize its dependence on θ. We thus have

$$\hat{y}(t|\theta) = H^{-1}(q,\theta)G(q,\theta)u(t) + \left[1 - H^{-1}(q,\theta)\right]y(t) \tag{4.6}$$

This predictor form does not depend on $f_e(x,\theta)$. In fact, as we stressed in Section 3.3, we could very well arrive at (4.6) by considerations that are not probabilistic. Then the specification (4.4) does not apply. We shall use the term *predictor models* for models that just specify G and H as in (4.4) or in the form (4.6). Similarly, *probabilistic models* will signify descriptions (4.4) that give a complete characterization of the probabilistic properties of the system. A parametrized set of models like (4.6) will be called a *model structure* and will be denoted by $\mathcal{M}$. The particular model associated with the parameter value θ will be denoted by $\mathcal{M}(\theta)$. (A formal definition is given in Section 4.5.)

In the following three sections, different ways of describing (4.4) in terms of θ (i.e., different ways of parametrizing the model set) will be discussed. A formalization of the concepts of model sets, parametrizations, model structures, and uniqueness of parametrization will then be given in Section 4.5, while questions of identifiability are discussed in Section 4.6.

4.2 A FAMILY OF TRANSFER-FUNCTION MODELS

Perhaps the most immediate way of parametrizing G and H is to represent them as rational functions and let the parameters be the numerator and denominator coefficients. In this section we shall describe various ways of carrying out such parametrizations. Such model structures are also known as black-box models.

Equation Error Model Structure

Probably the most simple input-output relationship is obtained by describing it as a linear difference equation:

$$\begin{aligned} y(t) + a_1 y(t-1) + \cdots + a_{n_a} y(t-n_a) \\ = b_1 u(t-1) + \cdots + b_{n_b} u(t-n_b) + e(t) \end{aligned} \tag{4.7}$$

Since the white-noise term $e(t)$ here enters as a direct error in the difference equation, the model (4.7) is often called an *equation error model* (structure). The adjustable parameters are in this case

$$\theta = [a_1 \quad a_2 \dots a_{n_a} \quad b_1 \dots b_{n_b}]^T \tag{4.8}$$

If we introduce

$$A(q) = 1 + a_1 q^{-1} + \cdots + a_{n_a} q^{-n_a}$$

and

$$B(q) = b_1 q^{-1} + \cdots + b_{n_b} q^{-n_b}$$

we see that (4.7) corresponds to (4.4) with

$$G(q,\theta) = \frac{B(q)}{A(q)}, \qquad H(q,\theta) = \frac{1}{A(q)} \tag{4.9}$$

Remark. It may seem annoying to use q as an argument of $A(q)$, being a polynomial in q^{-1}. The reason for this is, however, simply to be consistent with the conventional definition of the z-transform; see (2.17).

We shall also call the model (4.7) an ARX model, where AR refers to the autoregressive part $A(q)y(t)$ and X to the extra input $B(q)u(t)$ (called the exogeneous variable in econometrics). In the special case where $n_a = 0$, $y(t)$ is modeled as a finite impulse response (FIR). Such model sets are particularly common in signal-processing applications.

The signal flow can be depicted as in Figure 4.1. From that picture we see that the model (4.7) is perhaps not the most natural one from a physical point of view: the white noise is assumed to go through the denominator dynamics of the system before being added to the output. Nevertheless, the equation error model set has a very important property that makes it a prime choice in many applications: The predictor defines a linear regression.

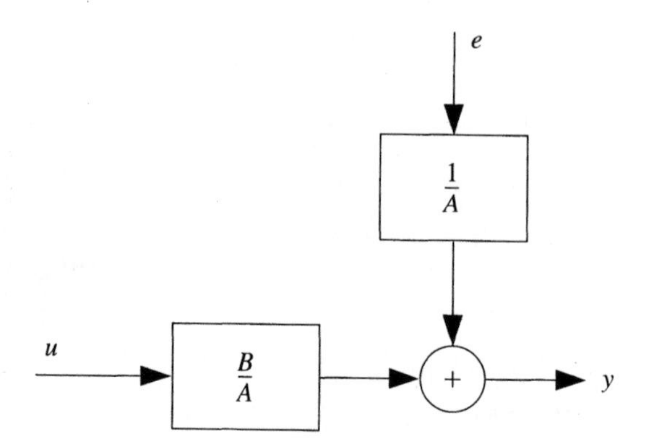

Figure 4.1 The ARX model structure.

Linear Regressions

Let us compute the predictor for (4.7). Inserting (4.9) into (4.6) gives

$$\hat{y}(t|\theta) = B(q)u(t) + [1 - A(q)]\, y(t) \tag{4.10}$$

Clearly, this expression could have more easily been derived directly from (4.7). Let us reiterate the view expressed in Section 3.3: Without a stochastic framework, the predictor (4.10) is a natural choice if the term $e(t)$ in (4.7) is considered to be "insignificant" or "difficult to guess." It is thus perfectly natural to work with the expression (4.10) also for "deterministic" models.

Now introduce the vector

$$\varphi(t) = [-y(t-1)\ldots - y(t-n_a) \quad u(t-1)\ldots u(t-n_b)]^T \tag{4.11}$$

Then (4.10) can be rewritten as

$$\hat{y}(t|\theta) = \theta^T\varphi(t) = \varphi^T(t)\theta \tag{4.12}$$

This is the important property of (4.7) that we alluded to previously. The predictor is a scalar product between a known data vector $\varphi(t)$ and the parameter vector θ. Such a model is called a *linear regression* in statistics, and the vector $\varphi(t)$ is known as the *regression vector*. It is of importance since powerful and simple estimation methods can be applied for the determination of θ.

In case some coefficients of the polynomials A and B are known, we arrive at linear regressions of the form

$$\hat{y}(t|\theta) = \varphi^T(t)\theta + \mu(t) \tag{4.13}$$

where $\mu(t)$ is a known term. See Problem 4E.1 and also (5.67). The estimation of θ in linear regressions will be treated in Section 7.3. See also Appendix II.

ARMAX Model Structure

The basic disadvantage with the simple model (4.7) is the lack of adequate freedom in describing the properties of the disturbance term. We could add flexibility to that by describing the equation error as a moving average of white noise. This gives the model

$$\begin{aligned} y(t) + a_1 y(t-1) + \cdots + a_{n_a} y(t-n_a) &= b_1 u(t-1) + \cdots \\ + b_{n_b} u(t-n_b) + e(t) + c_1 e(t-1) + \cdots &+ c_{n_c} e(t-n_c) \end{aligned} \tag{4.14}$$

With

$$C(q) = 1 + c_1 q^{-1} + \cdots + c_{n_c} q^{-n_c}$$

it can be rewritten

$$A(q)y(t) = B(q)u(t) + C(q)e(t) \tag{4.15}$$

and clearly corresponds to (4.4) with

$$G(q,\theta) = \frac{B(q)}{A(q)}, \qquad H(q,\theta) = \frac{C(q)}{A(q)} \tag{4.16}$$

where now

$$\theta = \begin{bmatrix} a_1 \ldots a_{n_a} \, b_1 \ldots b_{n_b} \, c_1 \ldots c_{n_c} \end{bmatrix}^T \tag{4.17}$$

In view of the moving average (MA) part $C(q)e(t)$, the model (4.15) will be called ARMAX. The ARMAX model has become a standard tool in control and econometrics for both system description and control design. A version with an enforced integration in the noise description is the ARIMA(X) model (I for integration, with or without the X-variable u), which is useful to describe systems with slow disturbances; see Box and Jenkins (1970). It is obtained by replacing $y(t)$ and $u(t)$ in (4.15) by their differences $\Delta y(t) = y(t) - y(t-1)$ and is further discussed in Section 14.1.

Pseudolinear Regressions

The predictor for (4.15) is obtained by inserting (4.16) into (4.6). This gives

$$\hat{y}(t|\theta) = \frac{B(q)}{C(q)}u(t) + \left[1 - \frac{A(q)}{C(q)}\right] y(t)$$

or

$$C(q)\hat{y}(t|\theta) = B(q)u(t) + [C(q) - A(q)]\, y(t) \tag{4.18}$$

This means that the prediction is obtained by filtering u and y through a filter with denominator dynamics determined by $C(q)$. To start it up at time $t = 0$ requires knowledge of

$$\begin{aligned} &\hat{y}(0|\theta) \ldots \hat{y}(-n_c + 1|\theta) \\ &y(0) \ldots y(-n^* + 1), \qquad n^* = \max(n_c, n_a) \\ &u(0) \ldots u(-n_b + 1) \end{aligned}$$

If these are not available, they can be taken as zero, in which case the prediction differs from the true one with an error that decays as $c \cdot \mu^t$, where μ is the maximum magnitude of the zeros of $C(z)$. It is also possible to start the recursion at time $\max(n^*, n_b)$ and include the unknown initial conditions $\hat{y}(k|\theta)$, $k = 1, \ldots, n_c$, in the vector θ.

The predictor (4.18) can be rewritten in formal analogy with (4.12) as follows. Adding $[1 - C(q)]\, \hat{y}(t|\theta)$ to both sides of (4.18) gives

$$\hat{y}(t|\theta) = B(q)u(t) + [1 - A(q)]\, y(t) + [C(q) - 1]\,[y(t) - \hat{y}(t|\theta)] \tag{4.19}$$

Introduce the prediction error

$$\varepsilon(t, \theta) = y(t) - \hat{y}(t|\theta)$$

and the vector

$$\begin{aligned} \varphi(t, \theta) = [-y(t-1) \ldots &- y(t - n_a) \quad u(t-1) \ldots \\ &u(t - n_b) \quad \varepsilon(t - 1, \theta) \ldots \varepsilon(t - n_c, \theta)]^T \end{aligned} \tag{4.20}$$

Then (4.19) can be rewritten as

$$\hat{y}(t|\theta) = \varphi^T(t, \theta)\theta \tag{4.21}$$

Notice the similarity with the linear regression (4.12). The equation (4.21) itself is, however, no linear regression, due to the nonlinear effect of θ in the vector $\varphi(t, \theta)$. To stress the kinship to (4.12), we shall call it a *pseudolinear regression*.

Other Equation-Error-Type Model Structures

Instead of modeling the equation error in (4.7) as a moving average, as we did in (4.14), it can of course be described as an autoregression. This gives a model set

$$A(q)y(t) = B(q)u(t) + \frac{1}{D(q)}e(t) \tag{4.22}$$

with

$$D(q) = 1 + d_1 q^{-1} + \cdots + d_{n_d} q^{-n_d}$$

which, analogously to the previous terminology, could be called ARARX. More generally, we could use an ARMA description of the equation error, leading to an "ARARMAX" structure

$$A(q)y(t) = B(q)u(t) + \frac{C(q)}{D(q)}e(t) \tag{4.23}$$

which of course contains (4.7), (4.15), and (4.22) as special cases. This would thus form the family of equation-error-related model sets, and is depicted in Figure 4.2. The relationship to (4.4) as well as expressions for the predictions are straightforward.

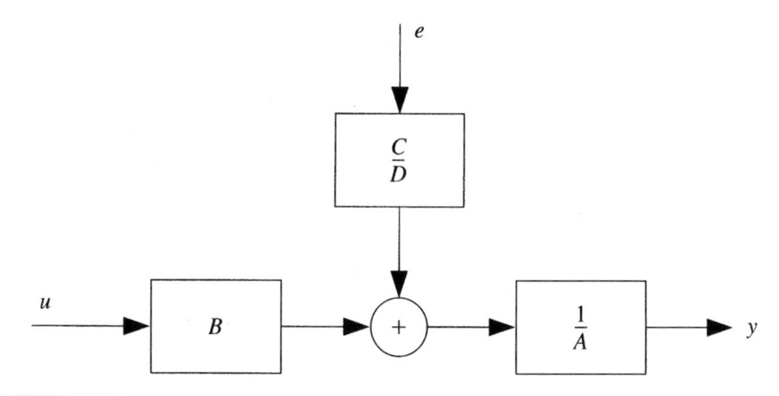

Figure 4.2 The equation error model family: The model structure (4.23).

Output Error Model Structure

The equation error model structures all correspond to descriptions where the transfer functions G and H have the polynomial A as a common factor in the denominators. See Figure 4.2. From a physical point of view it may seem more natural to parametrize these transfer functions independently.

If we suppose that the relation between input and undisturbed output w can be written as a linear difference equation, and that the disturbances consist of white measurement noise, then we obtain the following description:

$$\begin{aligned} w(t) + f_1 w(t-1) + \cdots + f_{n_f} w(t-n_f) \\ = b_1 u(t-1) + \cdots + b_{n_b} u(t-n_b) \end{aligned} \tag{4.24a}$$

$$y(t) = w(t) + e(t) \tag{4.24b}$$

With

$$F(q) = 1 + f_1 q^{-1} + \cdots + f_{n_f} q^{-n_f}$$

we can write the model as

$$y(t) = \frac{B(q)}{F(q)} u(t) + e(t) \tag{4.25}$$

The signal flow of this model is shown in Figure 4.3.

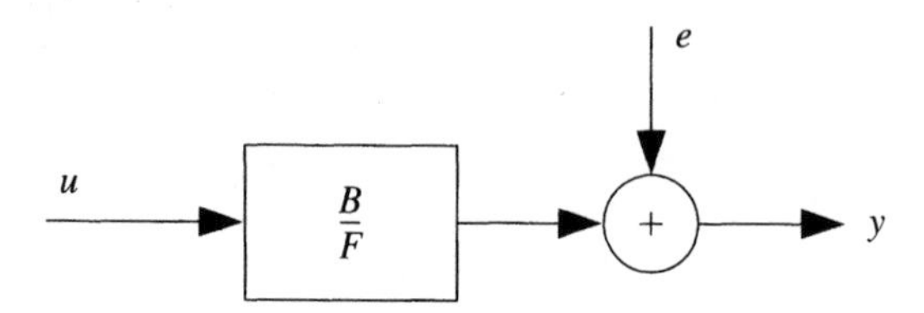

Figure 4.3 The output error model structure.

We call (4.25) an *output error (OE) model* (structure). The parameter vector to be determined is

$$\theta = [b_1 \quad b_2 \ldots b_{n_b} \; f_1 \quad f_2 \ldots f_{n_f}]^T \tag{4.26}$$

Since $w(t)$ in (4.24) is never observed, it should rightly carry an index θ, since it is constructed from u using (4.24a). That is,

$$\begin{aligned} w(t, \theta) + f_1 w(t-1, \theta) + \cdots + f_{n_f} w(t - n_f, \theta) \\ = b_1 u(t-1) + \cdots + b_{n_b} u(t - n_b) \end{aligned} \tag{4.27}$$

Comparing with (4.4), we find that $H(q, \theta) = 1$, which gives the natural predictor

$$\hat{y}(t|\theta) = \frac{B(q)}{F(q)} u(t) = w(t, \theta) \tag{4.28}$$

Note that $\hat{y}(t|\theta)$ is constructed from past inputs only. With the aid of the vector

$$\varphi(t, \theta) = [u(t-1) \ldots u(t - n_b) \quad -w(t-1, \theta) \ldots -w(t - n_f, \theta)]^T \tag{4.29}$$

this can be rewritten as

$$\hat{y}(t|\theta) = \varphi^T(t, \theta)\theta \tag{4.30}$$

which is in formal agreement with the ARMAX-model predictor (4.21). Note that in (4.29) the $w(t-1, \theta)$ are not observed, but, using (4.28), they can be computed: $w(t-k, \theta) = \hat{y}(t-k|\theta)$, $k = 1, 2, \ldots, n_f$.

Box-Jenkins Model Structure

A natural development of the output error model (4.25) is to further model the properties of the output error. Describing this as an ARMA model gives

$$y(t) = \frac{B(q)}{F(q)}u(t) + \frac{C(q)}{D(q)}e(t) \tag{4.31}$$

In a sense, this is the most natural finite-dimensional parametrization, starting from the description (4.4): the transfer functions G and H are independently parametrized as rational functions. The model set (4.31) was suggested and treated in Box and Jenkins (1970). This model also gives us the family of output-error-related models. See Figure 4.4 and compare with Figure 4.2. According to (4.6), the predictor for (4.31) is

$$\hat{y}(t|\theta) = \frac{D(q)B(q)}{C(q)F(q)}u(t) + \frac{C(q) - D(q)}{C(q)}y(t) \tag{4.32}$$

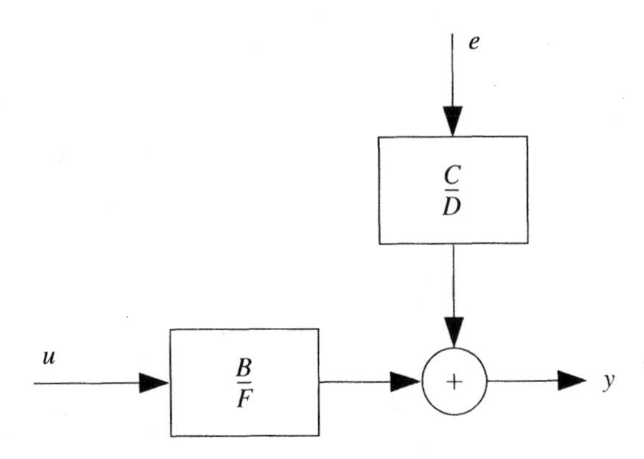

Figure 4.4 The BJ-model structure (4.31).

A General Family of Model Structures

The structures we have discussed in this section actually may give rise to 32 different model sets, depending on which of the five polynomials A, B, C, D, and F are used. (We have, however, only explicitly displayed six of these possibilities here.) Several of these model sets belong to the most commonly used ones in practice, and we have therefore reason to return to them both for explicit algorithms and for analytic results. For convenience, we shall therefore use a generalized model structure

$$A(q)y(t) = \frac{B(q)}{F(q)}u(t) + \frac{C(q)}{D(q)}e(t) \tag{4.33}$$

Sometimes the dynamics from u to y contains a delay of n_k samples, so some leading coefficients of B are zero; that is,

$$B(q) = b_{n_k}q^{-n_k} + b_{n_k+1}q^{-n_k-1} + \cdots + b_{n_k+n_b-1}q^{-n_k-n_b+1} = q^{-n_k}\overline{B}(q), \quad b_{n_k} \neq 0$$

It may then be a good idea to explicitly display this delay by

$$A(q)y(t) = q^{-n_k}\frac{\overline{B}(q)}{F(q)}u(t) + \frac{C(q)}{D(q)}e(t) \tag{4.34}$$

For easier notation we shall, however, here mostly use $n_k = 1$ and (4.33). From expressions for (4.33) we can always derive the corresponding ones for (4.34) by replacing $u(t)$ by $u(t - n_k + 1)$.

The structure (4.33) is too general for most practical purposes. One or several of the five polynomials would be fixed to unity in applications. However, by developing algorithms and results for (4.33), we also cover all the special cases corresponding to more realistic model sets.

From (4.6) we know that the predictor for (4.33) is

$$\hat{y}(t|\theta) = \frac{D(q)B(q)}{C(q)F(q)}u(t) + \left[1 - \frac{D(q)A(q)}{C(q)}\right]y(t) \tag{4.35}$$

The common special cases of (4.33) are summarized in Table 4.1.

TABLE 4.1 Some Common Black-box SISO Models as Special Cases of (4.33)

Polynomials Used in (4.33)	Name of Model Structure
B	FIR (finite impulse response)
AB	ARX
ABC	ARMAX
AC	ARMA
ABD	ARARX
$ABCD$	ARARMAX
BF	OE (output error)
$BFCD$	BJ (Box-Jenkins)

A Pseudolinear Form for (4.35) (*)

The expression (4.35) can also be written as a recursion:

$$C(q)F(q)\hat{y}(t|\theta) = F(q)\left[C(q) - D(q)A(q)\right]y(t) + D(q)B(q)u(t) \tag{4.36}$$

From (4.36) we find that the prediction error

$$\varepsilon(t, \theta) = y(t) - \hat{y}(t|\theta)$$

can be written

$$\varepsilon(t, \theta) = \frac{D(q)}{C(q)}\left[A(q)y(t) - \frac{B(q)}{F(q)}u(t)\right] \tag{4.37}$$

It is convenient to introduce the auxiliary variables

$$w(t, \theta) = \frac{B(q)}{F(q)}u(t) \tag{4.38a}$$

and

$$v(t, \theta) = A(q)y(t) - w(t, \theta) \tag{4.38b}$$

Then

$$\varepsilon(t, \theta) = y(t) - \hat{y}(t|\theta) = \frac{D(q)}{C(q)}v(t, \theta) \tag{4.39}$$

Let us also introduce the "state vector"

$$\begin{aligned}\varphi(t, \theta) = [&-y(t-1), \ldots, -y(t-n_a), u(t-1), \ldots, u(t-n_b),\\ &- w(t-1, \theta), \ldots, -w(t-n_f, \theta), \varepsilon(t-1, \theta), \ldots, \varepsilon(t-n_c, \theta),\\ &-v(t-1, \theta), \ldots, -v(t-n_d, \theta)]^T\end{aligned} \tag{4.40}$$

With the parameter vector

$$\theta = \left[a_1 \ldots a_{n_a}\, b_1 \ldots b_{n_b}\, f_1 \ldots f_{n_f}\, c_1 \ldots c_{n_c}\, d_1 \ldots d_{n_d}\right]^T \tag{4.41}$$

and (4.40) we can give a convenient expression for the prediction. To find this, we proceed as follows: From (4.38a) and (4.39) we obtain

$$\begin{aligned}w(t, \theta) = &\ b_1 u(t-1) + \cdots + b_{n_b} u(t-n_b)\\ &- f_1 w(t-1, \theta) - \cdots - f_{n_f} w(t-n_f, \theta)\end{aligned} \tag{4.42}$$

and

$$\begin{aligned}\varepsilon(t, \theta) = &\ v(t, \theta) + d_1 v(t-1, \theta) + \ldots + d_{n_d} v(t-n_d, \theta)\\ &- c_1 \varepsilon(t-1, \theta) - \ldots - c_{n_c} \varepsilon(t-n_c, \theta)\end{aligned} \tag{4.43}$$

Now inserting

$$v(t, \theta) = y(t) + a_1 y(t-1) + \cdots + a_{n_a} y(t-n_a) - w(t, \theta)$$

into (4.43) and substituting $w(t, \theta)$ with the expression (4.42), we find that

$$\varepsilon(t, \theta) = y(t) - \theta^T \varphi(t, \theta) \tag{4.44}$$

Hence

$$\hat{y}(t|\theta) = \theta^T \varphi(t, \theta) = \varphi^T(t, \theta)\theta \tag{4.45}$$

The two expressions, (4.36) and (4.45) can both be used for the calculation of the prediction. It should be noticed that the expressions simplify considerably in the special cases of the general model (4.33) that have been discussed in this section.

Other Model Expansions

The FIR model structure

$$G(q, \theta) = \sum_{k=1}^{n} b_k q^{-k} \tag{4.46}$$

has two important advantages: it is a linear regression (being a special case of ARX) and it is an output error model (being a special case of OE). This means, as we shall see later, that the model can be efficiently estimated and that it is robust against noise. The basic disadvantage is that many parameters may be needed. If the system has a pole close to the unit circle, the impulse response decays slowly, so n has then to be large to approximate the system well. This leads to the question whether it would be possible to retain the linear regression and output error features, while offering better possibilities to treat slowly decaying impulse responses. Generally speaking, such models would look like

$$G(q, \theta) = \sum_{k=1}^{n} \theta_k L_k(q, \alpha) \tag{4.47}$$

where $L_k(q, \alpha)$ represents a function expansion in the delay operator, which may contain a user-chosen parameter α. This parameter would be treated as fixed in the model structure, in order to make (4.47) a linear regression. A simple choice would be

$$L_k(q, \alpha) = \frac{q^{-k}}{q - \alpha}$$

where α is an estimate of the system pole closest to the unit circle. More sophisticated choices in terms of orthonormal basis expansions, see, e.g., Van den Hof, Heuberger, and Bokor (1995), have attracted wide interest. In particular, *Laguerre polynomials* have been used in this context, (Wahlberg, 1991):

$$L_k(q, \alpha) = \frac{1}{q - \alpha} \left(\frac{1 - \alpha q}{q - \alpha} \right)^{k-1} \tag{4.48}$$

where, again, it is natural to let α be an estimate of the dominating pole (time constant).

Continuous-time Black-box Models (*)

The linear system description could also be parameterized in terms of the continuous-time transfer function (2.22):

$$y(t) = G_c(p, \theta)u(t) \tag{4.49}$$

Adjustments to observed, sampled data could then be achieved either by solving the underlying differential equations or by applying an exact or approximate sampling procedure (2.24). The model (4.49) could also be fitted in the frequency domain, to Fourier transformed band-limited input-output data, as described in Section 7.7.

In addition to obvious counterparts of the structures already discussed, two specific model sets should be mentioned. The first-order system model with a time delay

$$G_c(s, \theta) = \frac{Ke^{-s\tau_c}}{(s\tau + 1)}, \qquad \theta = [K, \tau_c, \tau]^T \tag{4.50}$$

has been much used in process industry applications. Orthonormal function series expansions

$$G_c(s, \theta) = \sum_{k=0}^{d-1} a_k f_k(s), \qquad \theta = [a_0, \cdots, a_{d-1}]^T \tag{4.51}$$

have been discussed in the early literature, and also, e.g., by Belanger (1985). Like for discrete-time models, Laguerre polynomials appear to be a good choice:

$$f_k(s) = \sqrt{2\alpha}\frac{(s - \alpha)^k}{(s + \alpha)^{k+1}}$$

α being a time-scaling factor. Clearly, the model (4.49) can then be complemented with a model for the disturbance effects at the sampling instants as in (2.23).

Multivariable Case: Matrix Fraction Descriptions (*)

Let us now consider the case where the input $u(t)$ is an m-dimensional vector and the output $y(t)$ is a p-dimensional vector. Most of the ideas that we have described in this section have straightforward multivariable counterparts. The simplest case is the generalization of the equation error model set (4.7). We obtain

$$\begin{aligned} y(t) &+ A_1 y(t-1) + \cdots + A_{n_a} y(t - n_a) \\ &= B_1 u(t-1) + \cdots + B_{n_b} u(t - n_b) + e(t) \end{aligned} \tag{4.52}$$

where the A_i are $p \times p$ matrices and the B_i are $p \times m$ matrices.

Analogous to (4.9), we may introduce the polynomials

$$\begin{aligned} A(q) &= I + A_1 q^{-1} + \cdots + A_{n_a} q^{-n_a} \\ B(q) &= B_1 q^{-1} + \cdots + B_{n_b} q^{-n_b} \end{aligned} \tag{4.53}$$

These are now *matrix polynomials* in q^{-1} meaning that $A(q)$ is a matrix whose entries are polynomials in q^{-1}. We note that the system is still given by

$$y(t) = G(q,\theta)u(t) + H(q,\theta)e(t) \tag{4.54}$$

with

$$G(q,\theta) = A^{-1}(q)B(q), \qquad H(q,\theta) = A^{-1}(q) \tag{4.55}$$

The inverse $A^{-1}(q)$ of the matrix polynomial is interpreted and calculated in a straightforward way as discussed in connection with (3.34). Clearly, $G(q,\theta)$ will be a $p \times m$ matrix whose entries are rational functions of q^{-1} (or q). The factorization in terms of two matrix polynomials is also called a (left) *matrix fraction description* (MFD). A thorough treatment of such descriptions is given in Chapter 6 of Kailath (1980).

We have not yet discussed the *parametrization* of (4.52) (i.e., which elements of the matrices should be included in the parameter vector θ). This is a fairly subtle issue, which will be further discussed in Appendix 4A. An immediate analog of (4.8) could, however, be noted: Suppose all matrix entries in (4.52) (a total of $n_a \cdot p^2 + n_b \cdot p \cdot m$) are included in $\boldsymbol{\theta}$. We may then define the $[n_a \cdot p + n_b \cdot m] \times p$ matrix

$$\boldsymbol{\theta} = \left[A_1 A_2 \cdots A_{n_a}\ B_1 \cdots B_{n_b}\right]^T \tag{4.56}$$

and the $[n_a \cdot p + n_b \cdot m]$-dimensional column vector

$$\varphi(t) = \begin{bmatrix} -y(t-1) \\ \vdots \\ -y(t-n_a) \\ u(t-1) \\ \vdots \\ u(t-n_b) \end{bmatrix} \tag{4.57}$$

to rewrite (4.52) as

$$y(t) = \boldsymbol{\theta}^T \varphi(t) + e(t) \tag{4.58}$$

in obvious analogy with the linear regression (4.12). This can be seen as p different linear regressions, written on top of each other, all with the same regression vector.

When additional structure is imposed on the parametrization, it is normally no longer possible to use (4.58), since the different output components will not employ identical regression vectors. Then a d-dimensional column vector θ and a $p \times d$ matrix $\boldsymbol{\varphi}^T(t)$ has to be formed so as to represent (4.52) as

$$y(t) = \boldsymbol{\varphi}^T(t)\theta + e(t) \tag{4.59}$$

See Problems 4G.6 and 4E.12 for some more aspects on (4.58) and (4.59).

In light of the different possibilities for SISO systems, it is easy to visualize a number of variants for the MIMO case, like the vector difference equation (VDE)

$$\begin{aligned} y(t) + A_1 y(t-1) + \cdots + A_{n_a} y(t-n_a) \\ = B_1 u(t-1) + \cdots + B_{n_b} u(t-n_b) \\ + e(t) + C_1 e(t-1) + \cdots + C_{n_c} e(t-n_c) \end{aligned} \tag{4.60a}$$

or

$$G(q,\theta) = A^{-1}(q)B(q), \qquad H(q,\theta) = A^{-1}(q)C(q) \tag{4.60b}$$

which is the natural extension of the ARMAX model. A multivariable Box-Jenkins model takes the form

$$G(q,\theta) = F^{-1}(q)B(q), \qquad H(q,\theta) = D^{-1}(q)C(q) \tag{4.61}$$

and so on. The parametrizations of these MFD-descriptions are discussed in Appendix 4A.

4.3 STATE-SPACE MODELS

In the state-space form the relationship between the input, noise, and output signals is written as a system of first-order differential or difference equations using an auxiliary state vector $x(t)$. This description of linear dynamical systems became an increasingly dominating approach after Kalman's (1960) work on prediction and linear quadratic control. For our purposes it is especially useful in that insights into physical mechanisms of the system can usually more easily be incorporated into state-space models than into the models described in Section 4.2.

Continuous-time Models Based on Physical Insight

For most physical systems it is easier to construct models with physical insight in continuous time than in discrete time, simply because most laws of physics (Newton's law of motion, relationships in electrical circuits, etc.) are expressed in continuous time. This means that modeling normally leads to a representation

$$\dot{x}(t) = F(\theta)x(t) + G(\theta)u(t) \tag{4.62}$$

Here F and G are matrices of appropriate dimensions ($n \times n$ and $n \times m$, respectively, for an n-dimensional state and an m-dimensional input). The overdot denotes differentiation with respect to (w.r.t) time t. Moreover, θ is a vector of parameters that typically correspond to unknown values of physical coefficients, material constants, and the like. The modeling is usually carried out in terms of state variables x that have physical significance (positions, velocities, etc.), and then the measured outputs will be known combinations of the states. Let $\eta(t)$ be the measurements that would be obtained with ideal, noise-free sensors:

$$\eta(t) = Hx(t) \tag{4.63}$$

Using p for the differentiation operator, (4.62) can be written

$$[pI - F(\theta)]x(t) = G(\theta)u(t)$$

which means that the transfer operator from u to η in (4.63) is

$$\eta(t) = G_c(p,\theta)u(t)$$

$$G_c(p,\theta) = H\,[pI - F(\theta)]^{-1}\,G(\theta) \tag{4.64}$$

We have thus obtained a continuous-time transfer-function model of the system, as in (2.22), that is parametrized in terms of physical coefficients.

In reality, of course, some noise-corrupted version of $\eta(t)$ is obtained, resulting from both measurement imperfections and disturbances acting on (4.62). There are several different possibilities to describe these noise and disturbance effects. Here we first take the simplest approach. Other cases are discussed in (4.84) and (4.96) to (4.99), in Problem 4G.7, and in Section 13.7. Let the measurements be sampled at time instants $t = kT$, $k = 1, 2, \cdots$, and the disturbance effects at those time instants be $v_T(kT)$. Hence the measured output is

$$y(kT) = Hx(kT) + v_T(kT) = G_c(p,\theta)u(t) + v_T(kT) \tag{4.65}$$

Sampling the Transfer Function

As we discussed in Section 2.1, there are several ways of transporting $G_c(p,\theta)$ to a representation that is explicitly discrete time. Suppose that the input is constant over the sampling interval T as in (2.3):

$$u(t) = u_k = u(kT), \qquad kT \le t < (k+1)T \tag{4.66}$$

Then the differential equation (4.62) can easily be solved from $t = kT$ to $t = kT+T$, yielding

$$x(kT + T) = A_T(\theta)x(kT) + B_T(\theta)u(kT) \tag{4.67}$$

where

$$A_T(\theta) = e^{F(\theta)T} \tag{4.68a}$$

$$B_T(\theta) = \int_{\tau=0}^{T} e^{F(\theta)\tau}G(\theta)\,d\tau \tag{4.68b}$$

(See, e.g., Åström and Wittenmark, 1984.)

Introducing q for the forward shift of T time units, we can rewrite (4.67) as

$$[qI - A_T(\theta)]x(kT) = B_T(\theta)u(kT) \tag{4.69}$$

or

$$\eta(kT) = G_T(q,\theta)u(kT) \tag{4.70}$$

$$G_T(q,\theta) = H\,[qI - A_T(\theta)]^{-1}\,B_T(\theta) \tag{4.71}$$

Hence (4.65) can equivalently be given in the sampled-data form

$$y(t) = G_T(q, \theta)u(t) + v_T(t), \qquad t = T, 2T, 3T, \dots \tag{4.72}$$

When (4.66) holds, no approximation is involved in this representation. Note, however, that in view of (4.68) $G_T(q, \theta)$ could be quite a complicated function of θ.

Example 4.1 DC Servomotor

In this example we shall study a physical process, where we have some insight into the dynamic properties. Consider the dc motor depicted in Figure 4.5 with a block diagram in Figure 4.6. The input to this system is assumed to be the applied voltage, u, and the output the angle of the motor shaft, η. The relationship between applied voltage u and the resulting current i in the rotor circuit is given by the well-known relationship

$$u(t) = R_a i(t) + L_a \frac{di(t)}{dt} + s(t) \tag{4.73}$$

where $s(t)$ is the back electromotive force, due to the rotation of the armature circuit in the magnetic field:

$$s(t) = k_v \frac{d}{dt}\eta(t)$$

The current i gives a turning torque of

$$T_a(t) = k_a \cdot i(t)$$

on the motor shaft, which is also affected by a torque $T_\ell(t)$ from the load. Newton's law then gives

$$J\frac{d^2}{dt^2}\eta(t) = T_a(t) - T_\ell(t) - f\frac{d}{dt}\eta(t) \tag{4.74}$$

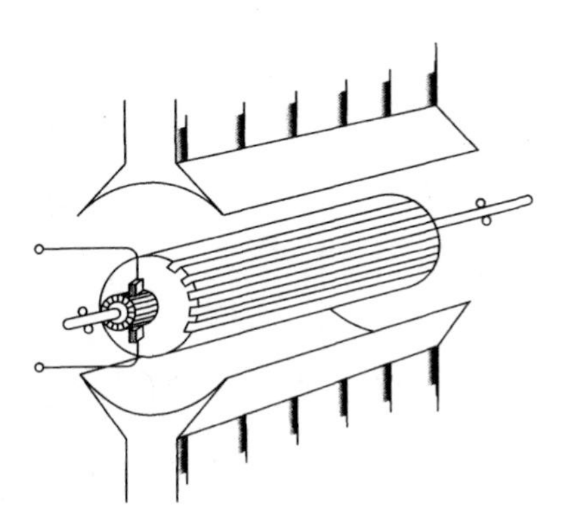

Figure 4.5 The dc motor.

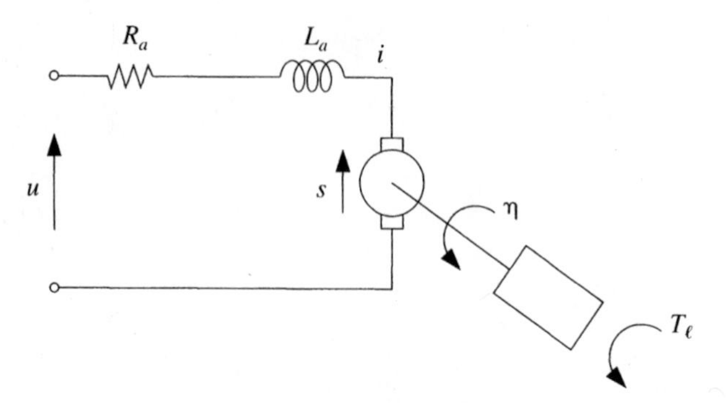

Figure 4.6 Block diagram of the dc motor.

where J is the moment of inertia of the rotor plus load and f represents viscous friction. Assuming that the inductance of the armature circuit can be neglected, $L_a \approx 0$, the preceding equations can be summarized in state-space form as

$$\frac{d}{dt}x(t) = \begin{bmatrix} 0 & 1 \\ 0 & -1/\tau \end{bmatrix} x(t) + \begin{bmatrix} 0 \\ \beta/\tau \end{bmatrix} u(t) + \begin{bmatrix} 0 \\ \gamma'/\tau \end{bmatrix} T_\ell(t) \tag{4.75}$$

$$\eta(t) = \begin{bmatrix} 1 & 0 \end{bmatrix} x(t)$$

with

$$x(t) = \begin{bmatrix} \eta(t) \\ \frac{d}{dt}\eta(t) \end{bmatrix}$$

$$\tau = \frac{JR_a}{fR_a + k_a k_v}, \quad \beta = \frac{k_a}{fR_a + k_a k_v}, \quad \gamma' = -\frac{R_a}{fR_a + k_a k_v}$$

Assume now that the torque T_ℓ is identically zero. To determine the dynamics of the motor, we now apply a piecewise constant input and sample the output with the sampling interval T. The state equation (4.75) can then be described by

$$x(t+T) = A_T(\theta)x(t) + B_T(\theta)u(t) \tag{4.76}$$

where

$$\theta = \begin{bmatrix} \tau \\ \beta \end{bmatrix}$$

and, according to (4.68),

$$A_T(\theta) = \begin{bmatrix} 1 & \tau(1 - e^{-T/\tau}) \\ 0 & e^{-T/\tau} \end{bmatrix}, \quad B_T(\theta) = \begin{bmatrix} \beta(\tau e^{-T/\tau} - \tau + T) \\ \beta(1 - e^{-T/\tau}) \end{bmatrix} \tag{4.77}$$

Also assume that $y(t)$, the actual measurement of the angle $\eta(t)$, is made with a certain error $v(t)$:

$$y(t) = \eta(t) + v(t) \tag{4.78}$$

This error is mainly caused by limited accuracy (e.g., due to the winding of a potentiometer) and can be described as a sequence of independent random variables with zero mean and known variance R_2 (computed from the truncation error in the measurement), provided the measurements are not too frequent. We thus have a model

$$y(t) = G_T(q, \theta)u(t) + v(t)$$

with $v(t)$ being white noise. The natural predictor is thus

$$\hat{y}(t|\theta) = G_T(q, \theta)u(t) = [1 \quad 0][qI - A_T(\theta)]^{-1} B_T(\theta)u(t) \tag{4.79}$$

This predictor is parametrized using only two parameters β and τ. Notice that if we used our physical insight to conclude only that the system is of second order we would use, say, a second-order ARX or OE model containing *four* adjustable parameters. As we shall see, using fewer parameters has some positive effects on the estimation procedure: the variance of the parameter estimates will decrease. The price is, however, not insignificant. The predictor (4.79) is a far more complicated function of its two parameters than the corresponding ARX or OE model of its four parameters. □

Equations (4.67) and (4.65) constitute a standard discrete-time state-space model. For simplicity we henceforth take $T = 1$ and drop the corresponding index. We also introduce an arbitrary parametrization of the matrix that relates x to η: $H = C(\theta)$. We thus have

$$x(t + 1) = A(\theta)x(t) + B(\theta)u(t) \tag{4.80a}$$

$$y(t) = C(\theta)x(t) + v(t) \tag{4.80b}$$

corresponding to

$$y(t) = G(q, \theta)u(t) + v(t) \tag{4.81}$$

$$G(q, \theta) = C(\theta)[qI - A(\theta)]^{-1} B(\theta) \tag{4.82}$$

Although sampling a time-continuous description is a natural way to obtain the model (4.80), it could also for certain applications be posed directly in discrete time, with the matrices A, B, and C directly parametrized in terms of θ, rather than indirectly via (4.68).

Noise Representation and the Time-invariant Kalman Filter

In the representation (4.80) and (4.81) we could further model the properties of the noise term $\{v(t)\}$. A straightforward but entirely valid approach would be to postulate a noise model of the kind

$$v(t) = H(q, \theta)e(t) \tag{4.83}$$

with $\{e(t)\}$ being white noise with variance λ. The θ-parameters in $H(q, \theta)$ could be partly in common with those in $G(q, \theta)$ or be extra additional noise model parameters.

For state-space descriptions, it is, however, more common to split the lumped noise term $v(t)$ into contributions from *measurement noise* $\nu(t)$ and *process noise* $w(t)$ acting on the states, so that (4.80) is written

$$\begin{aligned} x(t+1) &= A(\theta)x(t) + B(\theta)u(t) + w(t) \\ y(t) &= C(\theta)x(t) + \nu(t) \end{aligned} \tag{4.84}$$

Here $\{w(t)\}$ and $\{\nu(t)\}$ are assumed to be sequences of independent random variables with zero mean values and covariances

$$\begin{aligned} Ew(t)w^T(t) &= R_1(\theta) \\ E\nu(t)\nu^T(t) &= R_2(\theta) \\ Ew(t)\nu^T(t) &= R_{12}(\theta) \end{aligned} \tag{4.85}$$

The disturbances $w(t)$ and $\nu(t)$ may often be signals whose physical origins are known. In Example 4.1 the load variation $T_\ell(t)$ was a "process noise," while the inaccuracy in the potentiometer angular sensor $v(t)$ was the "measurement noise." In such cases it may of course not always be realistic to assume that these signals are white noises. To arrive at (4.84) and (4.85) will then require extra modeling and extension of the state vector. See Problem 4G.2.

Let us now turn to the problem of predicting $y(t)$ in (4.84). This state-space description is one to which the celebrated Kalman filter applies (see, e.g., Anderson and Moore, 1979, for a thorough treatment). The conditional expectation of $y(t)$, given data $y(s), u(s), s \leq 1$ (i.e., from the infinite past up to time $t-1$), is, provided ν and w are Gaussian processes, given by

$$\begin{aligned} \hat{x}(t+1,\theta) &= A(\theta)\hat{x}(t,\theta) + B(\theta)u(t) + K(\theta)\left[y(t) - C(\theta)\hat{x}(t,\theta)\right] \\ \hat{y}(t|\theta) &= C(\theta)\hat{x}(t,\theta) \end{aligned} \tag{4.86}$$

Here $K(\theta)$ is given as

$$K(\theta) = \left[A(\theta)\overline{P}(\theta)C^T(\theta) + R_{12}(\theta)\right]\left[C(\theta)\overline{P}(\theta)C^T(\theta) + R_2(\theta)\right]^{-1} \tag{4.87a}$$

where $\overline{P}(\theta)$ is obtained as the positive semidefinite solution of the stationary Riccati equation:

$$\begin{aligned} \overline{P}(\theta) &= A(\theta)\overline{P}(\theta)A^T(\theta) + R_1(\theta) - \left[A(\theta)\overline{P}(\theta)C^T(\theta) + R_{12}(\theta)\right] \\ &\quad \times \left[C(\theta)\overline{P}(\theta)C^T(\theta) + R_2(\theta)\right]^{-1}\left[A(\theta)\overline{P}(\theta)C^T(\theta) + R_{12}(\theta)\right]^T \end{aligned} \tag{4.87b}$$

The predictor filter can thus be written as

$$\hat{y}(t|\theta) = C(\theta)\left[qI - A(\theta) + K(\theta)C(\theta)\right]^{-1} B(\theta)u(t)$$
$$+ C(\theta)\left[qI - A(\theta) + K(\theta)C(\theta)\right]^{-1} K(\theta)y(t) \tag{4.88}$$

The matrix $\overline{P}(\theta)$ is the covariance matrix of the state estimate error:

$$\overline{P}(\theta) = \overline{E}\left[x(t) - \hat{x}(t,\theta)\right]\left[x(t) - \hat{x}(t,\theta)\right]^T \tag{4.89}$$

Innovations Representation

The prediction error

$$y(t) - C(\theta)\hat{x}(t,\theta) = C(\theta)\left[x(t) - \hat{x}(t,\theta)\right] + v(t) \tag{4.90}$$

in (4.86) amounts to that part of $y(t)$ that cannot be predicted from past data: "the innovation." Denoting this quantity by $e(t)$ as in (3.25), we find that (4.86) can be rewritten as

$$\begin{aligned} \hat{x}(t+1,\theta) &= A(\theta)\hat{x}(t,\theta) + B(\theta)u(t) + K(\theta)e(t) \\ y(t) &= C(\theta)\hat{x}(t,\theta) + e(t) \end{aligned} \tag{4.91a}$$

The covariance of $e(t)$ can be determined from (4.90) and (4.89):

$$Ee(t)e^T(t) = \Lambda(\theta) = C(\theta)\overline{P}(\theta)C^T(\theta) + R_2(\theta) \tag{4.91b}$$

Since $e(t)$ appears explicitly, this representation is known as the *innovations form* of the state-space description. Using the shift operator q, we can clearly rearrange it as

$$y(t) = G(q,\theta)u(t) + H(q,\theta)e(t) \tag{4.92a}$$

$$\begin{aligned} G(q,\theta) &= C(\theta)\left[qI - A(\theta)\right]^{-1} B(\theta) \\ H(q,\theta) &= C(\theta)\left[qI - A(\theta)\right]^{-1} K(\theta) + I \end{aligned} \tag{4.92b}$$

showing its relationship to the general model (4.4) and to a direct modeling of $v(t)$ as in (4.83). See also Problem 4G.3.

Directly Parametrized Innovations Form

In (4.91) the Kalman gain $K(\theta)$ is computed from $A(\theta)$, $C(\theta)$, $R_1(\theta)$, $R_{12}(\theta)$, and $R_2(\theta)$ in the fairly complicated manner given by (4.87). It is an attractive idea to sidestep (4.87) and the parametrization of the R-matrices by directly parametrizing $K(\theta)$ in terms of θ. This has the important advantage that the predictor (4.88) becomes a much simpler function of θ. Such a model structure we call a *directly parametrized innovations form*.

The R-matrices describing the noise properties contain $\frac{1}{2}n(n+1)+np+\frac{1}{2}p(p+1)$ matrix elements (discounting symmetric ones), while the Kalman gain K contains np elements ($p=\dim y$, $n=\dim x$). If we have no prior knowledge about the R-matrices and thus would need many parameters to describe them, it would therefore be a better alternative to parametrize $K(\theta)$, also from the point of view of keeping $\dim\theta$ small. On the other hand, physical insight into (4.84) may entail knowing, for example, that the process noise affects only one state and is independent of the measurement noise, which might have a known variance. Then the parametrization of $K(\theta)$ via (4.85) and (4.87) may be done using less parameters than would be required in a direct parametrization of $K(\theta)$.

Remark. The parametrization in terms of (4.85) also gives a parametrization of the $p(p+1)/2$ elements of $\Lambda(\theta)$ in (4.91). A direct parametrization of (4.91) would involve extra parameters for Λ, which, however, would not affect the predictor. (Compare also Problems 7E.4 and 8E.2.)

Directly parametrized innovations forms also contain black-box models that are in close relationship to those discussed in Section 4.2.

Example 4.2 Companion Form Parametrizations

In (4.91) let

$$\theta^T = [a_1 \quad a_2 \quad a_3 \quad b_1 \quad b_2 \quad b_3 \quad k_1 \quad k_2 \quad k_3]$$

and

$$A(\theta) = \begin{bmatrix} -a_1 & 1 & 0 \\ -a_2 & 0 & 1 \\ -a_3 & 0 & 0 \end{bmatrix}$$

$$B(\theta) = \begin{bmatrix} b_1 \\ b_2 \\ b_3 \end{bmatrix} \qquad K(\theta) = \begin{bmatrix} k_1 \\ k_2 \\ k_3 \end{bmatrix}$$

$$C(\theta) = \begin{bmatrix} 1 & 0 & 0 \end{bmatrix}$$

These matrices are said to be in *companion form* or in observer canonical form (see, e.g., Kailath, 1980). It is easy to verify that with these matrices

$$C(\theta)\,[qI - A(\theta)]^{-1}\,B(\theta) = \frac{b_1q^{-1} + b_2q^{-2} + b_3q^{-3}}{1 + a_1q^{-1} + a_2q^{-2} + a_3q^{-3}}$$

and

$$C(\theta)\,[qI - A(\theta)]^{-1}\,K(\theta) = \frac{k_1q^{-1} + k_2q^{-2} + k_3q^{-3}}{1 + a_1q^{-1} + a_2q^{-2} + a_3q^{-3}}$$

so that

$$1 + C(\theta)[qI - A(\theta)]^{-1} K(\theta) = \frac{1 + c_1 q^{-1} + c_2 q^{-2} + c_3 q^{-3}}{1 + a_1 q^{-1} + a_2 q^{-2} + a_3 q^{-3}}$$

with

$$c_i \triangleq a_i + k_i, \qquad i = 1, 2, 3$$

With this we have consequently obtained a parametrization of the ARMAX model set (4.15) and (4.16) for $n_a = n_b = n_c = 3$. □

The corresponding parametrization of a multioutput model is more involved and is described in Appendix 4A.

Time-varying Predictors (∗)

For the predictor filter (4.86) and (4.87) we assumed all previous data from time minus infinity to be available. If data prior to time $t = 0$ are lacking, we could replace them by zero, thus starting the recursion (4.86) at $t = 0$ with $\hat{x}(0) = 0$, and take the penalty of a suboptimal estimate. This was also our philosophy in Section 3.2.

An advantage with the state-space formulation is that a correct treatment of incomplete information about $t < 0$ can be given at the price of a slightly more complex predictor. If the information about the history of the system prior to $t = 0$ is given in terms of an initial state estimate $x_0(\theta) = \hat{x}(0, \theta)$ and associated uncertainty

$$\Pi_0(\theta) = E[x(0) - x_0(\theta)][x(0) - x_0(\theta)]^T \tag{4.93}$$

then the Kalman filter tells us that the one-step-ahead prediction is given by, (see, e.g., Anderson and Moore, 1979),

$$\hat{x}(t+1, \theta) = A(\theta)\hat{x}(t, \theta) + B(\theta)u(t) + K(t, \theta)[y(t) - C(\theta)\hat{x}(t, \theta)] \tag{4.94}$$

$$\hat{y}(t|\theta) = C(\theta)\hat{x}(t, \theta), \qquad \hat{x}(0, \theta) = x_0(\theta)$$

$$\begin{aligned} K(t, \theta) &= \left[A(\theta)P(t, \theta)C^T(\theta) + R_{12}(\theta)\right] \\ &\quad \times \left[C(\theta)P(t, \theta)C^T(\theta) + R_2(\theta)\right]^{-1} \end{aligned} \tag{4.95}$$

$$\begin{aligned} P(t+1, \theta) &= A(\theta)P(t, \theta)A^T(\theta) + R_1(\theta) - K(t, \theta) \\ &\quad \times \left[C(\theta)P(t, \theta)C^T(\theta) + R_2(\theta)\right] K^T(t, \theta), \quad P(0, \theta) = \Pi_0(\theta) \end{aligned}$$

Now $K(t, \theta)$ determined by (4.95) converges, under general conditions, fairly rapidly to $K(\theta)$ given by (4.87) (see, e.g., Anderson and Moore, 1979). For many problems it is thus reasonable to apply the limit form (4.86) with (4.87) directly to simplify calculations. For short data records, though, the solution (4.93) to (4.95) gives a useful possibility to deal with the transient properties in a correct way, including possibly a parametrization of the unknown initial conditions $x_0(\theta)$ and $\Pi_0(\theta)$. Clearly, the steady-state approach (4.86) with (4.87) is a special case of (4.94) to (4.95), corresponding to $x_0(\theta) = 0$, $\Pi_0(\theta) = \overline{P}(\theta)$.

Sampling Continuous-time Process Noise (*)

Just as for the systems dynamics, we may have more insight into the nature of the process noise in continuous time. We could then pose a disturbed state-space model

$$\dot{x}(t) = F(\theta)x(t) + G(\theta)u(t) + \overline{w}(t) \tag{4.96}$$

where $\overline{w}(t)$ is formal white noise with covariance function

$$E\overline{w}(t)\overline{w}^T(s) = \overline{R}_1(\theta)\delta(t - s) \tag{4.97}$$

where δ is Dirac's delta function. When the input is piecewise constant as in (4.66), the corresponding discrete-time state equation becomes

$$x(kT + T) = A_T(\theta)x(kT) + B_T(\theta)u(kT) + w_T(kT) \tag{4.98}$$

where A_T and B_T are given by (4.68) and $w_T(kT)$, $k = 1, 2, \cdots$ is a sequence of independent random vectors with zero means and covariance matrix

$$Ew_T(kT)w_T^T(kT) = R_1(\theta) = \int_0^T e^{F(\theta)\tau}\overline{R}_1(\theta)e^{F^T(\theta)\tau}\,d\tau \tag{4.99}$$

See Åström (1970)for a derivation.

State-space Models

In summary, we have found that state-space models provide us with a spectrum of modeling possibilities: We may use physical modeling in continuous time with or without a corresponding time-continuous noise description to obtain structures with physical parameters θ. We can use physical parametrization of the dynamics part combined with a black-box parametrization of the noise properties, such as in the directly parametrized innovations form (4.91), or we can arrive at a noise model that is also physically parametrized via (4.96) to (4.99). Finally, we can use black-box state-space structures, such as the one of Example 4.2. These have the advantage over the input-output black box that the flexibility in choice of representation can secure better numerical properties of the parametrization (Problem 16E.1).

4.4 DISTRIBUTED PARAMETER MODELS (*)

Models that involve partial differential equations (PDE), directly or indirectly, when relating the input signal to the output signal are usually called *distributed parameter models*. "Distributed" then refers to the state vector, which in general belongs to a function space, rather than $\mathbf{R}^n$. There are basically two ways to deal with such models. One is to replace the space variable derivative by a difference expression or to truncate a function series expansion so as to approximate the PDE by an ordinary differential equation. Then a "lumped" finite-dimensional model, of the kind we discussed in Section 4.3, is obtained. ("Lumped" refers to the fact that the distributed states are lumped together into a finite collection.) The other approach is to stick to the original PDE for the calculations, and only at the final, numerical, stage introduce approximations to facilitate the computations. It should be noted that this second approach also remains within the general model structure (4.4), provided the underlying PDE is linear and time invariant. This is best illustrated by an example.

Example 4.3 Heating Dynamics

Consider the physical system schematically depicted in Figure 4.7. It consists of a well-insulated metal rod, which is heated at one end. The heating power at time t is the input $u(t)$, while the temperature measured at the other end is the output $y(t)$. This output is sampled at $t = 1, 2, \ldots$.

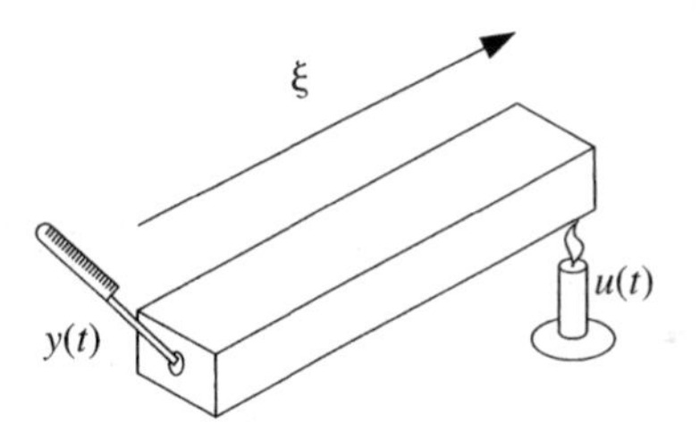

Figure 4.7 The heat-rod system.

Under ideal conditions, this system is described by the heat-diffusion equation. If $x(t, \xi)$ denotes the temperature at time t, ξ length units from one end of the rod, then

$$\frac{\partial x(t, \xi)}{\partial t} = \kappa \frac{\partial^2 x(t, \xi)}{\partial \xi^2} \tag{4.100}$$

where κ is the coefficient of thermal conductivity. The heating at the far end means that

$$\left. \frac{\partial x(t, \xi)}{\partial \xi} \right|_{\xi = L} = K \cdot u(t) \tag{4.101}$$

where K is a heat-transfer coefficient. The near end is insulated so that

$$\left. \frac{\partial x(t, \xi)}{\partial \xi} \right|_{\xi = 0} = 0 \tag{4.102}$$

The measurements are

$$y(t) = x(t, 0) + v(t), \qquad t = 1, 2, \ldots \tag{4.103}$$

where $\{v(t)\}$ accounts for the measurement noise. The unknown parameters are

$$\theta = \begin{bmatrix} \kappa \\ K \end{bmatrix} \tag{4.104}$$

Approximating

$$\frac{\partial^2 x(t, \xi)}{\partial \xi^2} = \frac{x(t, \xi + \Delta L) - 2x(t, \xi) + x(t, \xi - \Delta L)}{(\Delta L)^2}, \qquad \xi = k \cdot \Delta L$$

transfers (4.100) to a state-space model of order $n = L/\Delta L$, where the state variables $x(t, k \cdot \Delta L)$ are lumped representatives for $x(t, \xi)$, $k \cdot \Delta L \leq \xi < (k+1) \cdot \Delta L$. This often gives a reasonable approximation of the heat-diffusion equation.

Here we instead retain the PDE (4.100) by Laplace transforming it. Thus let $X(s, \xi)$ be the Laplace transform of $x(t, \xi)$ with respect to t for fixed ξ. Then (4.100) to (4.102) take the form

$$\begin{aligned} sX(s, \xi) &= \kappa X''(s, \xi) \\ X'(s, L) &= K \cdot U(s) \\ X'(s, 0) &= 0 \end{aligned} \tag{4.105}$$

Prime and double prime here denote differentiation with respect to ξ, and $U(s)$ is the Laplace transform of $u(t)$. Solving (4.105) for fixed s gives

$$X(s, \xi) = A(s)e^{-\xi\sqrt{s/\kappa}} + B(s)e^{\xi\sqrt{s/\kappa}}$$

where the constants $A(s)$ and $B(s)$ are determined from the boundary values

$$\begin{aligned} X'(s, 0) &= 0 \\ X'(s, L) &= K \cdot U(s) \end{aligned}$$

which gives

$$A(s) = B(s) = \frac{K \cdot U(s)}{\sqrt{s/\kappa}\,(e^{L\sqrt{s/\kappa}} - e^{-L\sqrt{s/\kappa}})} \tag{4.106}$$

Inserting this into (4.103) gives

$$Y(s) = X(s, 0) + V(s) = G_c(s, \theta)U(s) + V(s) \tag{4.107}$$

$$G_c(s, \theta) = \frac{2K}{\sqrt{s/\kappa}\,(e^{L\sqrt{s/\kappa}} - e^{-L\sqrt{s/\kappa}})} \tag{4.108}$$

where $V(s)$ is the Laplace transform of the noise $\{v(t)\}$. We have thus arrived at a model parametrization of the kind (4.49). With some sampling procedure and a model for the measurement noise sequence, it can be carried further to the form (4.4). Note that $G_c(s, \theta)$ is an analytic function of s although not rational. All our concepts of poles, zeros, stability, and so on, can still be applied. □

We can thus include distributed parameter models in our treatment of system identification methods. There is a substantial literature on this subject. See, for example, Banks, Crowley, and Kunisch (1983)and Kubrusly (1977). Not surprisingly, computational issues, choice of basis functions, and the like, play an important role in this literature.

4.5 MODEL SETS, MODEL STRUCTURES, AND IDENTIFIABILITY: SOME FORMAL ASPECTS (*)

In this chapter we have dealt with models of linear systems, as well as with parametrized sets of such models. When it comes to analysis of identification methods, it turns out that certain properties will have to be required from these models and model sets. In this section we shall discuss such formal aspects. To keep notation simple, we treat explicitly only SISO models.

Some Notation

For the expressions we shall deal with in this section, it is convenient to introduce some more compact notation. With

$$T(q) = [\,G(q) \quad H(q)\,] \text{ and } \chi(t) = \begin{bmatrix} u(t) \\ e(t) \end{bmatrix} \tag{4.109}$$

we can rewrite (4.1) as

$$y(t) = T(q)\chi(t) \tag{4.110}$$

The model structure (4.4) can similarly be written

$$y(t) = T(q,\theta)\chi(t), \qquad T(q,\theta) = [\,G(q,\theta) \quad H(q,\theta)\,] \tag{4.111}$$

Given the model (4.110), we can determine the one-step-ahead predictor (3.46), which we can rewrite as

$$\hat{y}(t|t-1) = W(q)z(t) \tag{4.112}$$

with

$$W(q) = \begin{bmatrix} W_u(q) & W_y(q) \end{bmatrix} \qquad z(t) = \begin{bmatrix} u(t) \\ y(t) \end{bmatrix} \tag{4.113}$$

$$W_u(q) = H^{-1}(q)G(q), \qquad W_y(q) = \begin{bmatrix} 1 - H^{-1}(q) \end{bmatrix} \tag{4.114}$$

Clearly, (4.114) defines a one-to-one relationship between $T(q)$ and $W(q)$:

$$T(q) \leftrightarrow W(q) \tag{4.115}$$

Remark. Based on (4.110), we may prefer to work with the k-step-ahead predictor (3.31). To keep the link (4.115), we can view (3.31) as the one-step-ahead predictor for the model (3.32).

Models

We noted already in (4.1) that a model of a linear system consists of specified transfer functions $G(z)$ and $H(z)$, possibly complemented with a specification of the prediction error variance λ, or the PDF $f_e(x)$ of the prediction error e. In Sections 3.2 and 3.3, we made the point that what matters in the end is by which expression future outputs are predicted. The one-step-ahead predictor based on the model (4.1) is given by (4.112).

While the predictor (4.112) via (4.115) is in a one-to-one relationship with (4.110), it is useful to relax the link (4.115) and regard (4.112) as the basic model. This will, among other things, allow a direct extension to nonlinear and time-varying models, as shown in Section 5.7. We may thus formally define what we mean by a model:

Definition 4.1. A *predictor model* of a linear, time-invariant system is a stable filter $W(q)$, defining a predictor (4.112) as in (4.113).

Stability, which was defined in (2.27) (applying to both components of $W(q)$) is necessary to make the right side of (4.112) well defined. While predictor models are meaningful also in a deterministic framework without a stochastic alibi, as discussed in Section 3.3, it is useful also to consider models that specify properties of the associated prediction errors (innovations).

Definition 4.2. A *complete probabilistic model* of a linear, time-invariant system is a pair $(W(q), f_e(x))$ of a predictor model $W(q)$ and the PDF $f_e(x)$ of the associated prediction errors.

Clearly, we can also have models where the PDFs are only partially specified (e.g., by the variance of e).

In this section we shall henceforth only deal with predictor models and therefore drop this adjective. The concepts for probabilistic models are quite analogous.

We shall say that two models $W_1(q)$ and $W_2(q)$ are *equal* if

$$W_1(e^{i\omega}) = W_2(e^{i\omega}), \qquad \text{almost all } \omega \tag{4.116}$$

A model

$$W(q) = [\, W_u(q) \quad W_y(q) \,]$$

will be called a *k-step-ahead predictor model* if

$$W_y(q) = \sum_{\ell=k}^{\infty} w_y(\ell) q^{-\ell}, \qquad \text{with } w_y(k) \neq 0 \tag{4.117}$$

and an *output error model* (or a *simulation model*) if $W_y(q) \equiv 0$.

Note that the definition requires the predictors to be stable. This does not necessarily mean that the system dynamics is stable.

Example 4.4 Unstable System

Suppose that

$$G(q) = \frac{bq^{-1}}{1 + aq^{-1}}, \qquad \text{with } |a| > 1$$

and

$$H(q) = \frac{1}{1 + aq^{-1}}$$

This means that the model is described by

$$y(t) + ay(t-1) = bu(t-1) + e(t)$$

and the dynamics from u to y is unstable. The predictor functions are, however:

$$W_y(q) = -aq^{-1}, \qquad W_u(q) = bq^{-1}$$

implying that

$$\hat{y}(t|t-1) = -ay(t-1) + bu(t-1)$$

which clearly satisfies the condition of Definition 4.1. □

Model Sets

Definition 4.1 describes one given model of a linear system. The identification problem is to determine such a model. The search for a suitable model will typically be conducted over a set of candidate models. Quite naturally, we define a *model set* $\mathcal{M}^*$ as

$$\mathcal{M}^* = \{W_\alpha(q)|\alpha \in \mathcal{A}\} \tag{4.118}$$

This is just a collection of models, each subject to Definition 4.1, here "enumerated" with an index a covering an index set $\mathcal{A}$.

Typical model sets could be

$$\mathcal{M}^* = \mathcal{L}^* = \{\text{all linear models}\}$$

that is, all models that are subject to Definition 4.1, or

$$\mathcal{M}_n^* = \{\text{all models such that } W_y(q) \text{ and } W_u(q) \text{ are polynomials of } q^{-1} \text{ of degree at most } n\} \tag{4.119}$$

or a finite model set

$$\mathcal{M}^* = \{W_1(q), W_2(q), W_3(q)\} \tag{4.120}$$

We say that *two model sets are equal,* $\mathcal{M}_1^* = \mathcal{M}_2^*$, if for any W_1 in $\mathcal{M}_1^*$ there exists a W_2 in $\mathcal{M}_2^*$ such that $W_1 = W_2$ [defined by (4.116)], and vice versa.

Model Structures: Parametrization of Model Sets

Most often a model set of interest is noncountable. Since we have to conduct a search over it for "the best model," it is then interesting how the indexation is chosen. The basic idea is to parametrize (index) the set "smoothly" over a "nice" area and perform the search over the parameter set (the index set). To put this formally, we let the model be indexed by a d-dimensional vector θ:

$$W(q, \theta)$$

To formalize "smoothly," we require that for any given z, $|z| \geq 1$, the complex-valued function $W(z, \theta)$ of θ be differentiable:

$$\Psi(z, \theta) = \frac{d}{d\theta} W(z, \theta) \tag{4.121a}$$

Here

$$\begin{aligned} \Psi(z, \theta) &= \begin{bmatrix} \Psi_u(z, \theta) & \Psi_y(z, \theta) \end{bmatrix} \\ &= \begin{bmatrix} \frac{d}{d\theta} W_u(z, \theta) & \frac{d}{d\theta} W_y(z, \theta) \end{bmatrix} \end{aligned} \tag{4.121b}$$

is a $d \times 2$ matrix. Thus the gradient of the prediction $\hat{y}(t|\theta)$ is given by

$$\psi(t, \theta) = \frac{d}{d\theta} \hat{y}(t|\theta) = \Psi(q, \theta) z(t) \tag{4.121c}$$

Since the filters Ψ will have to be computed and used when the search is carried out, we also require them to be stable. We thus have the following definition:

Definition 4.3. *A model structure* $\mathcal{M}$ is a differentiable mapping from a connected, open subset $D_{\mathcal{M}}$ of $\mathbf{R}^d$ to a model set $\mathcal{M}^*$, such that the gradients of the predictor functions are stable.

To put this definition in mathematical notation we have

$$\mathcal{M} : D_{\mathcal{M}} \ni \theta \rightarrow \mathcal{M}(\theta) = W(q, \theta) \in \mathcal{M}^* \tag{4.122}$$

such that the filter Ψ in (4.121) exists and is stable for $\theta \in D_{\mathcal{M}}$. We will thus use $\mathcal{M}(\theta)$ to denote the particular model corresponding to θ and reserve $\mathcal{M}$ for the mapping itself.

Remark. The requirement that $D_{\mathcal{M}}$ should be open is in order for the derivatives in (4.121) to be unambiguously well defined. When using model structures, we may prefer to work with compact sets $D_{\mathcal{M}}$. Clearly, as long as $D_{\mathcal{M}}$ is contained in an open set where (4.121) are defined, no problems will occur. Differentiability can also be defined over more complicated subsets of $\mathbf{R}^d$ than open ones, that is, *differentiable manifolds* (see, e.g., Boothby, 1975). See the chapter bibliography for further comments.

Example 4.5 An ARX Structure

Consider the ARX model

$$y(t) + ay(t-1) = b_1u(t-1) + b_2u(t-2) + e(t)$$

The predictor is given by (4.10), which means that

$$W(q, \theta) = [b_1q^{-1} + b_2q^{-2} \quad -aq^{-1}], \qquad \theta = [a \quad b_1 \quad b_2]^T$$

and

$$\Psi(q, \theta) = \begin{bmatrix} 0 & -q^{-1} \\ q^{-1} & 0 \\ q^{-2} & 0 \end{bmatrix}$$

□

The parametrized model sets that we have explicitly studied in this chapter have been in terms of (4.4), that is,

$$y(t) = G(q, \theta)u(t) + H(q, \theta)e(t), \qquad \theta \in D_{\mathcal{M}} \tag{4.123}$$

or using (4.111)

$$y(t) = T(q, \theta)\chi(t)$$

It is immediate to verify that, in view of (4.114),

$$\Psi(q, \theta) = \frac{1}{(H(q, \theta))^2}\mathbf{T}'(q, \theta)\begin{bmatrix} H(q, \theta) & 0 \\ -G(q, \theta) & 1 \end{bmatrix} \tag{4.124}$$

where $\mathbf{T}'(q, \theta)$ is the $d \times 2$ matrix

$$\mathbf{T}'(q, \theta) = \frac{d}{d\theta}T(q, \theta) = \left[\frac{d}{d\theta}G(q, \theta) \quad \frac{d}{d\theta}H(q, \theta)\right] \tag{4.125}$$

Differentiability of W is thus assured by differentiability of T.

It should be clear that all parametrizations we have considered in this chapter indeed are model structures in the sense of Definition 4.3. We have, for example:

Lemma 4.1. The parametrization (4.35) together with (4.41) with θ confined to $D_{\mathcal{M}} = \{\theta | F(z) \cdot C(z)$ has no zeros on or outside the unit circle$\}$ is a model structure.

Proof. We need only verify that the gradients of

$$W_u(z, \theta) = \frac{B(z)D(z)}{C(z)F(z)}$$

and

$$W_y(z, \theta) = 1 - \frac{D(z)A(z)}{C(z)}$$

with respect to θ are analytical in $|z| \geq 1$ for $\theta \in D_{\mathcal{M}}$. But this is immediate since, for example

$$\frac{\partial}{\partial c_k} W_u(z, \theta) = -\frac{B(z)D(z)z^{-k}}{[C(z)]^2 \, F(z)}$$ □

Lemma 4.2. Consider the state-space parametrization (4.91). Assume that the entries of the matrices $A(\theta)$, $B(\theta)$, $K(\theta)$, and $C(\theta)$ are differentiable with respect to θ. Suppose that $\theta \in D_{\mathcal{M}}$, with

$$D_{\mathcal{M}} = \{\theta | \text{all eigenvalues of } A(\theta) - K(\theta)C(\theta) \text{ are inside the unit circle}\}$$

Then the parametrization of the corresponding predictor is a model structure.

Proof. See Problem 4D.1. □

Notice that when $K(\theta)$ is obtained as the solution of (4.87), then by a standard Kalman filter property (see Anderson and Moore, 1979),

$$D_{\mathcal{M}} = \{\theta | [A(\theta), R_1(\theta)] \text{ stabilizable and } [A(\theta), C(\theta)] \text{ detectable}\} \tag{4.126}$$

When relating different model structures, we shall use the following concept.

Definition 4.4. A model structure $\mathcal{M}_1$ is said to be *contained in* $\mathcal{M}_2$,

$$\mathcal{M}_1 \subset \mathcal{M}_2 \tag{4.127}$$

if $D_{\mathcal{M}_1} \subset D_{\mathcal{M}_2}$ and the mapping $\mathcal{M}_1$ is obtained by restricting $\mathcal{M}_2$ to $\theta \in D_{\mathcal{M}_1}$. The archetypical situation for (4.127) is when $\mathcal{M}_2$ defines nth-order models and $\mathcal{M}_1$ defines mth-order models, $m < n$. One could think of $\mathcal{M}_1$ as obtained from $\mathcal{M}_2$ by fixing some parameters (typically to zero).

The following property of a model structure is sometimes useful:

Definition 4.5. A model structure $\mathcal{M}$ is said to have an *independently parametrized transfer function and noise model* if

$$\theta = \begin{bmatrix} \rho \\ \eta \end{bmatrix}, \quad D_{\mathcal{M}} = D_\rho \times D_\eta, \quad \rho \in D_\rho, \quad \eta \in D_\eta$$
$$T(q, \theta) = \begin{bmatrix} G(q, \rho) & H(q, \eta) \end{bmatrix} \tag{4.128}$$

We note that in the family (4.33) the special cases with $A(q) \equiv 1$ correspond to independent parametrizations of G and H.

Remark On "Finite Model Structures": Sometimes the set of candidate models is finite as in (4.120). It may still be desirable to index it using a parameter vector θ, now ranging over a finite set of points. Although such a construction does not qualify as a "model structure" according to Definition 4.3, it should be noted that the estimation procedures of Sections 7.1 to 7.4, as well as the convergence analysis of Sections 8.1 to 8.5, still make sense in this case.

Model Set as a Range of a Model Structure

A model structure will clearly define a model set by its range:

$$\mathcal{M}^* = \mathcal{R}(\mathcal{M}) = \text{Range } \mathcal{M} = \{\mathcal{M}(\theta) | \theta \in D_{\mathcal{M}}\}$$

An important problem for system identification is to find a model structure whose range equals a given model set. This may sometimes be an easy problem and sometimes highly nontrivial.

Example 4.6 Parametrizing $\mathcal{M}_3^*$

Consider the set $\mathcal{M}_n^*$ defined by (4.119) with $n = 3$. If we take

$$\theta = \begin{bmatrix} a_1 & a_2 & a_3 & b_1 & b_2 & b_3 \end{bmatrix}^T, \qquad d = 6$$

$$D_{\mathcal{M}} = \mathbf{R}^6$$

and

$$W_y(q,\theta) = -a_1 q^{-1} - a_2 q^{-2} - a_3 q^{-3}$$

$$W_u(q,\theta) = b_1 q^{-1} + b_2 q^{-2} + b_3 q^{-3}$$

we have obviously constructed a model structure whose range equals $\mathcal{M}_3^*$ □

A given model set can typically be described as the range of several different model structures (see Problems 4E.6 and 4E.9).

Model Set as a Union of Ranges of Model Structures

In the preceding example it was possible to describe the desired model set as the range of a model structure. We shall later encounter model sets for which this is not possible, at least not with model structures with desired identifiability properties. The remedy for these problems is to describe the model set as a union of ranges of different model structures:

$$\mathcal{M}^* = \bigcup_{i=1}^{\beta} \mathcal{R}(\mathcal{M}_i) \tag{4.129}$$

This idea has been pursued in particular for representing linear multioutput systems. We shall give the details of this procedure in Appendix 4A. Let us here only remark that model sets described by (4.129) are useful also for working with models of different orders, and that they are often used, at least implicitly, when the order of a suitable model is unknown and is to be determined.

Identifiability Properties

Identifiability is a concept that is central in identification problems. Loosely speaking, the problem is whether the identification procedure will yield a unique value of the parameter θ, and/or whether the resulting model is equal to the true system. We shall deal with the subject in more detail in the analysis chapter (see Sections 8.2 and 8.3). The issue involves aspects on whether the data set (the experimental conditions) is informative enough to distinguish between different models as well as properties of the model structure itself: If the data are informative enough to distinguish between nonequal models, then the question is whether different values of θ can give equal models. With our terminology, the latter problem concerns the *invertibility of the model structure* $\mathcal{M}$ (i.e., whether $\mathcal{M}$ is injective). We shall now discuss some concepts related to such invertibility properties. Remember that these are only one leg of the identifiability concept. They are to be complemented in Sections 8.2 and 8.3.

Definition 4.6. A model structure $\mathcal{M}$ is *globally identifiable at* θ^* if

$$\mathcal{M}(\theta) = \mathcal{M}(\theta^*), \qquad \theta \in D_{\mathcal{M}} \Rightarrow \theta = \theta^* \tag{4.130}$$

Recall that model equality was defined in (4.116), requiring the predictor transfer functions to coincide. According to (4.115), this means that the underlying transfer functions G and H coincide.

Once identifiability at a point is defined, we proceed to properties of the whole set.

Definition 4.7. A model structure $\mathcal{M}$ is *strictly globally identifiable* if it is globally identifiable at all $\theta^* \in D_{\mathcal{M}}$.

This definition is quite demanding. As we shall see, it is difficult to construct model structures that are strictly globally identifiable. The difficulty for linear systems, for example, is that global identifiability may be lost at points on hyper-surfaces corresponding to lower-order systems. Therefore, we introduce a weaker and more realistic property:

Definition 4.8. A model structure $\mathcal{M}$ is *globally identifiable* if it is globally identifiable at almost all $\theta^* \in D_{\mathcal{M}}$.

Remark. This means that $\mathcal{M}$ is globally identifiable at all $\theta^* \in \tilde{D}_{\mathcal{M}} \subset D_{\mathcal{M}}$, where

$$\delta D_{\mathcal{M}} = \left\{\theta | \theta \in D_{\mathcal{M}}; \theta \notin \tilde{D}_{\mathcal{M}}\right\}$$

is a set of Lebesgue measure zero in $\mathbf{R}^d$ (recall that $D_{\mathcal{M}}$ and hence $\delta D_{\mathcal{M}}$ is a subset of $\mathbf{R}^d$).

For corresponding local properties, the most natural definition of local identifiability of $\mathcal{M}$ at θ^* would be to require that there exists an ε such that

$$\mathcal{M}(\theta) = \mathcal{M}(\theta^*), \quad \theta \in \mathcal{B}(\theta^*, \varepsilon) \Rightarrow \theta = \theta^* \tag{4.131}$$

where $\mathcal{B}(\theta^*, \varepsilon)$ denotes an ε-neighborhood of θ^*.

(Strict) local identifiability of a model structure can then be defined analogously to Definitions 4.7 and 4.8. See also Problem 4G.4.

Use of the Identifiability Concept

The identifiability concept concerns the unique representation of a given system description in a model structure. Let

$$S : y(t) = G_0(q)u(t) + H_0(q)e(t) \tag{4.132}$$

be such a description. We could think of it as a "true" or "ideal" description of the actual system, but such an interpretation is immaterial for the moment. Let $\mathcal{M}$ be a model structure based on one-step-ahead predictors for

$$y(t) = G(q, \theta)u(t) + H(q, \theta)e(t) \tag{4.133}$$

Then define the set $D_T(S, \mathcal{M})$ as those θ-values in $D_{\mathcal{M}}$ for which $S = \mathcal{M}(\theta)$. We can write this as

$$D_T(S, \mathcal{M}) = \left\{\theta \in D_{\mathcal{M}} | G_0(z) = G(z, \theta), H_0(z) = H(z, \theta) \text{ almost all } z\right\} \tag{4.134}$$

This set is empty in case $S \notin \mathcal{M}$. (Here, with abuse of notation, $\mathcal{M}$ also denotes the range of the mapping $\mathcal{M}$.)

Now suppose that $S \in \mathcal{M}$ so that $S = \mathcal{M}(\theta_0)$ for some value θ_0. Furthermore, suppose that *$\mathcal{M}$ is globally identifiable at θ_0*. Then

$$D_T(S, \mathcal{M}) = \{\theta_0\} \tag{4.135}$$

One aspect of the choice of a good model structure is to select $\mathcal{M}$ so that (4.135) holds for the given description S. Since S is unknown to the user, this will typically involve tests of several different structures $\mathcal{M}$. The identifiability concepts will then provide useful guidance in finding an $\mathcal{M}$ such that (4.135) holds.

4.6 IDENTIFIABILITY OF SOME MODEL STRUCTURES

Definition 4.6 and (4.116) together imply that a model structure is globally identifiable at θ^* if and only if

$$G(z,\theta) \equiv G(z,\theta^*) \text{ and } H(z,\theta) \equiv H(z,\theta^*) \quad \text{for almost all } z \Rightarrow \theta = \theta^* \tag{4.136}$$

For local identifiability, we consider only θ confined to a sufficiently small neighborhood of θ^*. A general approach to test local identifiability is given by the criterion in Problem 4G.4.

Global identifiability is more difficult to deal with in general terms. In this section we shall only briefly discuss identifiability of physical parameters and give some results for general black-box SISO models. Black-box multivariable systems are dealt with in Appendix 4A.

Parametrizations in Terms of Physical Parameters

Modeling physical processes typically leads to a continuous-time state-space model (4.62) to (4.63), summarized as (4.65) ($T = 1$):

$$y(t) = G_c(p,\theta)u(t) + v(t) \tag{4.137}$$

For proper handling we should sample G_c, and include a noise model H so that (4.136) can be applied for identifiability tests. A simpler test to apply is

$$G_c(s,\theta) = G_c(s,\theta^*) \text{ almost all } s \Rightarrow \theta = \theta^*? \tag{4.138}$$

It is true that this is not identical to (4.136): When sampling G_c, ambiguities may occur; two different G_c can give the same G_T [cf. (2.24)]. Equation (4.138) is thus not sufficient for (4.136) to hold. However, with a carefully selected sampling interval, this ambiguity should not cause any problems. Also, a θ-parametrized noise model may help in resolving (4.138). This condition is thus not necessary for (4.136) to hold. However, in most applications the noise characteristics are not so significant that they indeed bear information about the physical parameters. All this means that (4.138) is a reasonable test for global identifiability of the corresponding model structure at θ^*.

Now, (4.138) is a difficult enough problem. Except for special structures there are no general techniques available other than brute-force solution of the equations underlying (4.138). See Problems 4E.5 and 4E.6 for some examples. A comprehensive treatment of (4.138) for state-space models is given by Walter (1982), and Godfrey (1983)discusses the same problem for compartmental models. See also Godfrey and Distefano (1985). A general approach based on differential algebra is described in Ljung and Glad (1994b).

SISO Transfer-function Model Structures

We shall now aim at an analysis of the general black-box SISO model structure (4.33) together with (4.41). Let us first illustrate the character of the analysis with two simple special cases.

Consider the ARX model structure (4.7) together with (4.9):

$$G(z, \theta) = \frac{B(z)}{A(z)}, \qquad H(z, \theta) = \frac{1}{A(z)}$$
$$\theta = \left[a_1 \dots a_{n_a} \, b_1 \dots b_{n_b}\right]^T \tag{4.139}$$

Equality for H in (4.136) implies that the A-polynomials coincide, which in turn implies that the B-polynomials must coincide for the G to be equal. It is thus immediate to verify that (4.136) holds for all θ^* in the model structure (4.139). Consequently, *the structure (4.139) is strictly globally identifiable.*

Let us now turn to the OE model structure (4.25) with orders n_b and n_f. At $\theta = \theta^*$ we have

$$G(z, \theta^*) = \frac{B^*(z)}{F^*(z)} = \frac{b_1^* z^{-1} + \cdots + b_{nb}^* z^{-nb}}{1 + f_1^* z^{-1} + \cdots + f_{nf}^* z^{-nf}}$$
$$= z^{nf-nb} \frac{b_1^* z^{nb-1} + \cdots + b_{nb}^*}{z^{nf} + f_1^* z^{nf-1} + \cdots + f_{nf}^*} = z^{nf-nb} \frac{z^{nb} B^*(z)}{z^{nf} F^*(z)} \tag{4.140}$$

We shall work with the polynomial $\tilde{F}^*(z) = z^{nf} F^*(z)$ in the variable z, rather than with $F^*(z)$, which is a polynomial in z^{-1}. The reason is that $z^{nf} F^*(z)$ always has degree n_f regardless of whether f_{nf}^* is zero. Let $\tilde{B}^*(z) = z^{nb} B^*(z)$, and let θ be an arbitrary parameter value. We can then write (4.136),

$$G(z, \theta^*) = G(z, \theta) = \frac{B(z)}{F(z)} = z^{nf-nb} \frac{\tilde{B}(z)}{\tilde{F}(z)}$$

as

$$\tilde{F}(z)\tilde{B}^*(z) - \tilde{F}^*(z)\tilde{B}(z) \equiv 0 \tag{4.141}$$

Since $\tilde{F}^*(z)$ is a polynomial of degree n_f, it has n_f zeros:

$$\tilde{F}^*(\alpha_i) = 0, \qquad i = 1, \dots, n_f$$

Suppose that $\tilde{B}^*(\alpha_i) \neq 0$, $i = 1, \dots, n_f$; that is, $\tilde{B}^*(z)$ and $\tilde{F}^*(z)$ are *coprime* (have no common factors). Then (4.141) implies that

$$\tilde{F}(\alpha_i) = 0, \qquad i = 1, \dots, n_f$$

[if a zero α_i has multiplicity n_i, then differentiate (4.141) $n_i - 1$ times to conclude that it is a zero of the same multiplicity to $\tilde{F}(z)$]. Consequently, we have $\tilde{F}(z) \equiv \tilde{F}^*(z)$, which in turn implies that $\tilde{B}(z) = \tilde{B}^*(z)$ so that $\theta = \theta^*$. If, on the other hand, $\tilde{F}^*$ and $\tilde{B}^*$ do have a common factor so that

$$\tilde{F}^*(z) = \gamma(z)\tilde{F}_1^*(z), \qquad \tilde{B}^*(z) = \gamma(z)\tilde{B}_1^*(z)$$

then all θ, such that

$$\tilde{F}(z) = \beta(z)\tilde{F}_1^*(z), \qquad \tilde{B}(z) = \beta(z)\tilde{B}_1^*(z)$$

for arbitrary $\beta(z)$ will yield equality in (4.141). Hence the model structure is neither globally nor locally identifiable at θ^* [$\beta(z)$ can be chosen arbitrarily close to $\gamma(z)$]. We thus find that *the OE structure (4.25) is globally and locally identifiable at* θ^* *if and only if the corresponding numerator and denominator polynomials* $z^{n_f}F^*(z)$ *and* $z^{n_b}B^*(z)$ *are coprime.*

The generalization to the black-box SISO structure (4.33) is now straightforward:

Theorem 4.1. Consider the model structure $\mathcal{M}$ corresponding to

$$A(q)y(t) = \frac{B(q)}{F(q)}u(t) + \frac{C(q)}{D(q)}e(t) \tag{4.142}$$

with θ, given by (4.41), being the coefficients of the polynomials involved. The degrees of the polynomials are n_a, n_b, and so on. This model structure is globally identifiable at θ^* if and only if all of (i) to (vi) hold:

i. There is no common factor to all $z^{n_a}A^*(z)$, $z^{n_b}B^*(z)$, and $z^{n_c}C^*(z)$.

ii. There is no common factor to $z^{n_b}B^*(z)$ and $z^{n_f}F^*(z)$.

iii. There is no common factor to $z^{n_c}C^*(z)$ and $z^{n_d}D^*(z)$.

iv. If $n_a \geq 1$, then there must be no common factor to $z^{n_f}F^*(z)$ and $z^{n_d}D^*(z)$.

v. If $n_d \geq 1$, then there must be no common factor to $z^{n_a}A^*(z)$ and $z^{n_b}B^*(z)$.

vi. If $n_f \geq 1$, then there must be no common factor to $z^{n_a}A^*(z)$ and $z^{n_c}C^*(z)$.

The starred polynomials correspond to θ^*.

Notice that several of the conditions (i) to (vi) will be automatically satisfied in the common special cases of (4.142). Notice also that any of the conditions (i) to (vi) can be violated only for "special" θ^*, placed on hyper-surfaces in $\mathbf{R}^d$. We thus have the following corollary:

Corollary. The model structure given by (4.142) is globally identifiable.

Looking for a "True" System Within Identifiable Structures

We shall now illustrate the usefulness of Theorem 4.1 by applying it to the problem of finding an $\mathcal{M}$ such that (4.135) holds for a given S. Suppose that S is given by

$$S: \; G_0(q) = \frac{B_0(q)}{A_0(q)F_0(q)}, \qquad H_0(q) = \frac{C_0(q)}{A_0(q)D_0(q)} \tag{4.143}$$

with orders n_a^0, n_b^0, and so on (after all possible cancellations of common factors). This system belongs to the model structure $\mathcal{M}$ in (4.142) provided all the model orders are at least as large as the true ones:

$$n_a \geq n_a^0, \qquad n_b \geq n_b^0, \text{ etc.} \tag{4.144}$$

When (4.144) holds, let θ_0 be a value that gives the description (4.143):

$$S = \mathcal{M}(\theta_0) \tag{4.145}$$

Now, clearly, $\mathcal{M}$ will be globally identifiable at θ_0 and (4.135) will hold if we have equality in all of (4.144). The true orders $n_a^0, \ldots,$ are, however, typically not known, and it would be quite laborious to search for all combinations of model orders until equalities in (4.144) were obtained. The point of Theorem 4.1 is that such a search is not necessary; the structure $\mathcal{M}$ is globally identifiable at θ_0 under weaker conditions.

We have the following reformulation of Theorem 4.1:

Theorem 4.2. Consider the system description S in (4.143) with true polynomial orders n_a^0, n_b^0, and so on, as defined in the text. Consider model structure $\mathcal{M}$ of Theorem 4.1. Then $S \in \mathcal{M}$ and corresponds to a globally identifiable θ-value if and only if

i. $\min(n_a - n_a^0, n_b - n_b^0, n_c - n_c^0) = 0$.

ii. $\min(n_b - n_b^0, n_f - n_f^0) = 0$.

iii. $\min(n_c - n_c^0, n_d - n_d^0) = 0$.

iv. If $n_a \geq 1$, then also $\min(n_f - n_f^0, n_d - n_d^0) = 0$.

v. If $n_d \geq 1$, then also $\min(n_a - n_a^0, n_b - n_b^0) = 0$.

vi. If $n_f \geq 1$, then also $\min(n_a - n_a^0, n_c - n_c^0) = 0$.

With Theorem 4.2, the search for a true system within identifiable model structures is simplified. If, for example, S can be described in ARMAX form with finite orders n_a^0, n_b^0 and n_c^0, then we may take $n_a = n_b = n_c = n (n_f = n_d = 0)$ in $\mathcal{M}$, giving a model structure, say, $\mathcal{M}_n$. By increasing n one unit at a time, we will sooner or later strike a structure where (i) holds and thus S can be uniquely represented.

SISO State-space Models

Consider now a state-space model structure (4.91). It is quite clear that the matrices $A(\theta)$, $B(\theta)$, $C(\theta)$, and $K(\theta)$ cannot be "filled" with parameters, since the corresponding input-output description (4.92) is defined by $3n$ parameters only ($n = \dim x$). To obtain identifiable structures, it is thus natural to seek parametrizations of the matrices that involve $3n$ parameters; the coefficients of the two $(n-1)$th order numerator polynomials and the coefficients of the common, monic nth order denominator polynomial or some transformation of these coefficients. One such parametrization is the observer canonical form of Example 4.2, which we can write in symbolic form as

$$\begin{aligned} x(t+1,\theta) &= A(\theta)x(t,\theta) + B(\theta)u(t) + K(\theta)e(t) \\ y(t) &= C(\theta)x(t,\theta) + e(t) \end{aligned} \tag{4.146a}$$

$$A(\theta) = \left[\begin{array}{c|c} \times & \\ \times & I_{n-1} \\ \vdots & \\ \hline \times & 0\dots0 \end{array}\right], \quad B(\theta) = \begin{bmatrix} \times \\ \times \\ \vdots \\ \times \end{bmatrix}, \quad K(\theta) = \begin{bmatrix} \times \\ \times \\ \vdots \\ \times \end{bmatrix} \tag{4.146b}$$

$$C(\theta) = \begin{bmatrix} 1 & 0\dots0 \end{bmatrix}$$

Here I_{n-1} is the $(n-1)\times(n-1)$ unit matrix, while $\times$ marks an adjustable parameter. This representation is observable by construction.

According to Example 4.2, this structure is in one-to-one correspondence with an ARMAX structure with $n_a = n_b = n_c = n$. From Theorem 4.1 we know that this is identifiable at θ^*, provided the corresponding polynomials do not all have a common factor, meaning that the model could be represented using a smaller value of n. It is well known that for state-space models this can only happen if the model is uncontrollable and/or unobservable. Since (4.146) is observable by construction, we thus conclude that *this structure is globally and locally identifiable at* θ^* *if and only if the two-input system* $\{A(\theta^*), [\,B(\theta^*) \quad K(\theta^*)\,]\}$ *is controllable.* Note that this result applies to the particular state-space structure (4.146) only.

4.7 SUMMARY

In this chapter we have studied sets of predictors of the type

$$\hat{y}(t|\theta) = W_u(q,\theta)u(t) + W_y(q,\theta)y(t), \qquad \theta \in D_{\mathcal{M}} \subset \mathbf{R}^d \tag{4.147}$$

These are in one-to-one correspondence with model descriptions

$$y(t) = G(q,\theta)u(t) + H(q,\theta)e(t), \qquad \theta \in D_{\mathcal{M}} \tag{4.148}$$

with $\{e(t)\}$ as white noise, via

$$W_u(q,\theta) = H^{-1}(q,\theta)G(q,\theta)$$
$$W_y(q,\theta) = \left[1 - H^{-1}(q,\theta)\right]$$

When choosing models it is usually most convenient to go via (4.148), even if (4.147) is the "operational" version.

We have denoted parametrized model sets, or *model structures* by $\mathcal{M}$, while a particular model corresponding to the parameter value θ is denoted by $\mathcal{M}(\theta)$. Such a parametrization is instrumental in conducting a search for "best models." Two different philosophies may guide the choice of parametrized model sets:

1. *Black-box model structures:* The prime idea is to obtain flexible model sets that can accommodate a variety of systems, without looking into their internal structures. The input-output model structures of Section 4.2, as well as canonically parametrized state-space models (see Example 4.2), are of this character.
2. *Model structures with physical parameters:* The idea is to incorporate physical insight into the model set so as to bring the number of adjustable parameters down to what is actually unknown about the system. Continuous-time state-space models are typical representatives for this approach.

We have also in this chapter introduced formal requirements on the predictor filters $W_u(q,\theta)$ and $W_y(q,\theta)$ (Definition 4.3) and discussed concepts of parameter identifiability (i.e., whether the parameter θ can be uniquely determined from the predictor filters). These properties were investigated for the most typical black-box model structures in Section 4.6 and Appendix 4A. The bottom line of these results is that identifiability can be secured, provided certain orders are chosen properly. The number of such orders to be chosen typically equals the number of outputs.

4.8 BIBLIOGRAPHY

The selection of a parameterized set of models is, as we have noted, vital for the identification problem. This is the link between system identification and parameter estimation techniques. Most articles and books on system identification thus contain material on model structures, even if not presented in as explicit terms as here.

The simple equation error model (4.7) has been widely studied in many contexts. See, for example, Åström (1968), Hsia (1977), Mendel (1973), and Unbehauen, Göhring, and Bauer (1974)for discussions related to identification. Linear models like (4.12) are prime objects of study in statistics; see, for example, Rao (1973)or Draper and Smith (1981). The ARMAX model was introduced into system identification in Åström and Bohlin (1965)and is since then a basic model. The ARARX model structure was introduced into the control literature by Clarke (1967), but was apparently first used in a statistical framework by Cochrane and Orcutt (1949). The term pseudo-linear regression for the representation (4.21) was introduced by

Solo (1978). Output error models are treated, for example, in Dugard and Landau (1980)and Kabaila and Goodwin (1980). The general family (4.33) was first discussed in Ljung (1979). It was used in Ljung and Söderström (1983). Multivariable MFDs are discussed in Kailath (1980). When no input is present, the corresponding model structures reduce to AR, MA, and ARMA descriptions. These are discussed in many textbooks on time series, e.g., Box and Jenkins (1970); Hannan (1970); and Brillinger (1981).

Black-box continuous transfer function models of the type (4.50) have been used in many cases oriented toward control applications. Ziegler and Nichols (1942)determine parameters in such models from step responses and self-oscillatory modes (see Section 6.1).

State-space models in innovations forms as well as the general forms are treated in standard textbooks on control (e.g., Åström and Wittenmark, 1984). The use of continuous-time representations for estimation using discrete data has been discussed, for example, in Mehra and Tyler (1973)and Åström and Källström (1976). The continuous-time model structure is usually arrived at after an initial modeling step. See, for example, Wellstead (1979), Nicholson (1981), Ljung and Glad (1994a)and Cellier (1990)for general modeling techniques and examples. Direct identification of continuous-time systems is discussed in Unbehauen and Rao (1987).

Distributed parameter models and their estimation are treated in, for example, Banks, Crowley, and Kunisch (1983), Kubrusly (1977), Qureshi, Ng, and Goodwin (1980) and Polis and Goodson (1976). Example 4.3 is studied experimentally in Leden, Hamza, and Sheirah (1976).

The prediction aspect of models was emphasized in Ljung (1974)and Ljung (1978). Identifiability is discussed in many contexts. A survey is given in Nguyen and Wood (1982). Often identifiability is related to convergence of the parameter estimates. Such definitions are given in Åström and Bohlin (1965), Staley and Yue (1970), and Tse and Anton (1972). Identifiability definitions in terms of the model structure only was introduced by Bellman and Åström (1970), who called it "structural identifiability." Identifiability definitions in terms of the set $D_T(S, \mathcal{M})$ [defined by (4.134)] were given in Gustavsson, Ljung, and Söderström (1977). The particular definitions of the concept of model structure and identifiability given in Section 4.5 are novel. In Ljung and Glad (1994b)identifiability is treated from an algebraic perspective. It is shown that any globally identifiable structure can be rearranged as a linear regression.

A more general model structure concept than Definition 4.3 would be to let $D_{\mathcal{M}}$ be a differentiable manifold (see, e.g., Byrnes, 1976). However, in our treatment that possibility is captured by letting a model set be described as a union of (overlapping) ranges of model structures as in (4.129). This manifold structure for linear systems was first described by Kalman (1974), Hazewinkel and Kalman (1976) and Clark (1976).

The identifiability of multivariable model structures has been dealt with in numerous articles. See, for example, Kailath (1980), Luenberger (1967), Glover and Willems (1974), Rissanen (1974), Ljung and Rissanen (1976), Guidorzi (1981), Gevers and Wertz (1984), Van Overbeek and Ljung (1982), and Correa and Glover (1984).

In addition to the parameterizations described in the appendix, approaches based on balanced realizations are described in Maciejowski (1985), Ober (1987), and Hanzon and Ober (1997). Parameterizations that are not identifiable, but may still have numerical advantages, are discussed by McKelvey (1994)and McKelvey and Helmersson (1996).

4.9 PROBLEMS

4G.1 Consider the predictor (4.18). Show that the effect from an erroneous initial condition in $\hat{y}(s|\theta)$, $s \leq 0$, is bounded by $c \cdot \mu^t$, where μ is the maximum magnitude of the zeros of $C(z)$.

4G.2 *Colored measurement noise:* Suppose that a state-space representation is given as

$$\begin{aligned} x(t+1) &= A_1(\theta)x(t) + B_1(\theta)u(t) + w_1(t) \\ y(t) &= C_1(\theta)x(t) + v(t) \end{aligned} \tag{4.149}$$

where $\{w_1(t)\}$ is white with variance $\overline{R}_1(\theta)$, but the measurement noise $\{v(t)\}$ is not white. A model for $v(t)$ can, however, be given as

$$v(t) = H(q,\theta)\nu(t) \tag{4.150}$$

with $\{\nu(t)\}$ being white noise with variance $R_2(\theta)$ and $H(q,\theta)$ monic. Introduce a state-space representation for (4.150):

$$\begin{aligned} \xi(t+1) &= A_2(\theta)\xi(t) + K(\theta)\nu(t) \\ v(t) &= C_2(\theta)\xi(t) + \nu(t) \end{aligned} \tag{4.151}$$

Combine (4.149) and (4.150) into a single representation that complies with the structure (4.84) to (4.85). Determine $R_1(\theta)$, $R_{12}(\theta)$, and $R_2(\theta)$. Note that if $w_1(t)$ is zero then the new representation will be directly in the innovations form (4.91).

4G.3 *Verification of the Steady-State Kalman Filter:* The state-space model (4.84) can be written (suppressing the argument θ and assuming dim $y = 1$)

$$y(t) = G(q)u(t) + v_1(t)$$

where

$$\begin{aligned} G(q) &= C(qI - A)^{-1}B \\ v_1(t) &= C(qI - A)^{-1}w(t) + \nu(t) \end{aligned}$$

Let $R_{12} = 0$. The spectrum of $\{v_1(t)\}$ then is

$$\Phi_1(\omega) = C(e^{i\omega} \cdot I - A)^{-1}R_1(e^{-i\omega} \cdot I - A^T)^{-1}C^T + R_2$$

using Theorem 2.2. The innovations model (4.91) can be written

$$\begin{aligned} y(t) &= G(q)u(t) + v_2(t) \\ v_2(t) &= H(q)e(t), \qquad H(q) = C(qI - A)^{-1}K + 1 \end{aligned}$$

The spectrum of $\{v_2(t)\}$ thus is

$$\Phi_2(\omega) = \lambda\left[C(e^{i\omega} \cdot I - A)^{-1}K + 1\right]\left[C(e^{-i\omega} \cdot I - A)^{-1}K + 1\right]^T$$

where λ is the variance of $e(t)$.

(a) Show by direct calculation that

$$\Phi_1(\omega) - \Phi_2(\omega) \equiv 0$$

utilizing the expressions (4.87) and (4.91b). The two representations thus have the same second-order properties, and if the noises are Gaussian, they are indistinguishable in practice (see Problem 2E.3).

(b) Show by direct calculation that

$$\begin{aligned} 1 - H^{-1}(q) &= 1 - \left[1 + C(qI - A)^{-1}K\right]^{-1} \\ &= C(qI - A + KC)^{-1}K \end{aligned}$$

and

$$\begin{aligned} H^{-1}(q)G(q) &= \left[1 + C(qI - A)^{-1}K\right]^{-1} C(qI - A)^{-1}B \\ &= C(qI - A + KC)^{-1}B \end{aligned}$$

(c) Note that the predictor (4.86) can be written as (4.88):

$$\hat{y}(t|\theta) = C(qI - A + KC)^{-1}Bu(t) + C(qI - A + KC)^{-1}Ky(t)$$

and thus that (a) and (b) together with (3.20) constitute a derivation of the steady-state Kalman filter.

4G.4 Consider a model structure $\mathcal{M}$, with predictor function gradient $\Psi(z, \theta)$ defined in (4.121). Define the $d \times d$ matrix

$$\Gamma_1(\theta) = \int_{-\pi}^{\pi} \Psi(e^{i\omega}, \theta)\Psi^T(e^{-i\omega}, \theta)\, d\omega$$

(a) Show that $\mathcal{M}$ is locally identifiable at θ if $\Gamma_1(\theta)$ is nonsingular.

(b) Let $\mathbf{T}'(z, \theta)$ be defined by (4.125), and let

$$\Gamma_2(\theta) = \int_{-\pi}^{\pi} \mathbf{T}'(e^{i\omega}, \theta)\left[\mathbf{T}'(e^{-i\omega}, \theta)\right]^T d\omega$$

Use (4.124) to show that $\Gamma_2(\theta)$ is nonsingular if and only if $\Gamma_1(\theta)$ is. [Note that by assumption $H(q)$ has no zeros on the unit circle.] $\Gamma_2(\theta)$ can thus be used to test local identifiability.

4G.5 Consider an output error structure with several inputs

$$y(t) = \frac{B_1(q)}{F(q)}u_1(t) + \cdots + \frac{B_m(q)}{F(q)}u_m(t) + e(t)$$

Show that this structure is globally identifiable at a value θ^* if and only if there is no common factor to all of the $m + 1$ polynomials

$$z^{n_f} F^*(z), \quad z^{n_b} B_i^*(z), \quad i = 1, \ldots, m$$

$$n_f = \text{degree } F^*(z), \quad n_b = \max \text{degree } B_i^*(z)$$

θ^* here corresponds to the starred polynomials.

4G.6 The Kronecker product of an $m \times n$ matrix $A = (a_{ij})$ and a $p \times r$ matrix $B = (b_{ij})$ is defined as (see, e.g., Barnett, 1975)

$$A \otimes B = \begin{bmatrix} a_{11}B & a_{12}B \dots a_{1n}B \\ a_{21}B & a_{22}B \dots a_{2n}B \\ \vdots & \vdots \quad \vdots \\ a_{m1}B & a_{m2}B \dots a_{mn}B \end{bmatrix}$$

This is an $mp \times nr$ matrix. Define the operator "col" as the operation to form a column vector out of a matrix by stacking its columns on top of each other:

$$\text{col } B = \begin{bmatrix} B^1 \\ B^2 \\ \vdots \\ B^r \end{bmatrix}, \qquad (rp \times 1 \text{ vector})$$

where B^j is the jth column of B.

Consider (4.56) to (4.59). Show that (4.58) can be transformed into (4.59) with

$$\theta = \text{col } \boldsymbol{\theta}^T$$

$$\boldsymbol{\varphi}(t) = \varphi(t) \otimes I_p$$

where I_p is the $p \times p$ unit matrix. Are other variants of θ and $\boldsymbol{\varphi}$ also possible?

4G.7 Consider the continuous-time state-space model (4.96) to (4.97). Assume that the measurements are made in wideband noise with high variance, idealized as

$$\overline{y}(t) = Hx(t) + \overline{v}(t)$$

where $\overline{v}(t)$ is formal continuous-time white noise with covariance function

$$E\overline{v}(t)\overline{v}^T(s) = \overline{R}_2(\theta)\delta(t - s)$$

Assume that $\overline{v}(t)$ is independent of $\overline{w}(t)$. Let the output be defined as

$$y((k+1)T) = y_{k+1} = \frac{1}{T}\int_{t=kT}^{(k+1)T} \overline{y}(t)\, dt$$

Show that the sampled-data system can be represented as (4.98) and (4.99) but with

$$y(kT) = C_T(\theta)x(kT) + D_T(\theta)u(kT - T) + v_T(kT)$$

$$C_T(\theta) = \frac{1}{T}H\Phi_T(\theta)$$

$$Ew_T(kT)v_T^T(kT) = R_{12}(\theta) = \frac{1}{T}\int_0^T e^{F(\theta)\tau}\overline{R}_1(\theta)\Phi_{T-\tau}^T(\theta)H^T\, d\tau$$

$$Ev_T(kT)v_T^T(kT) = R_2(\theta) = \frac{1}{T}\overline{R}_2(\theta) + \frac{1}{T^2}\int_0^T H\Phi_{T-\tau}(\theta)\overline{R}_1(\theta)\Phi_{T-\tau}^T(\theta)H^T\, d\tau$$

$$\Phi_T(\theta) = \int_0^T e^{F(\theta)\tau}\, d\tau; \qquad D_T(\theta) = -\frac{1}{T}\int_0^T H\Phi_\tau(\theta)\, d\tau$$

4G.8 Consider the ARX model (4.7). Introduce the δ-operator

$$\delta = 1 - q^{-1}$$

and reparametrize the models in terms of coefficients of powers of δ. Work out the details of a second-order example. Such a parametrization has the advantage of being less sensitive to numerical errors when the sampling interval is short, Middleton and Goodwin (1990).

4E.1 Consider the ARX model structure

$$y(t) + a_1 y(t-1) + \cdots + a_{n_a} y(t - n_a)$$
$$= b_1 u(t-1) + \cdots + b_{n_b} u(t - n_b) + e(t)$$

where b_1 is known to be 0.5. Write the corresponding predictor in the linear regression form (4.13).

4E.2 Consider the continuous-time model (4.75) of the dc servo with $T_\ell(t) \equiv 0$. Apply the Euler approximation (2.25) to obtain an approximate discrete-time transfer function that is a simpler function of θ.

4E.3 Consider the small network of tanks in Figure 4.8. Each tank holds 10 volume units of fluid. Through the pipes A and E flows 1 volume unit per second, through the pipe B, α units, and through C and D, $1 - \alpha$ units per second. The concentration of a certain substance in the fluid is u in pipe A (the input) and y in pipe E (the output). Write down a structured state-space model for this system. Assume that each tank is perfectly mixed (i.e., the substance has the same concentration throughout the tank). (Models of this character are known as *compartmental models* and are very common in chemical and biological applications; see Godfrey, 1983.)

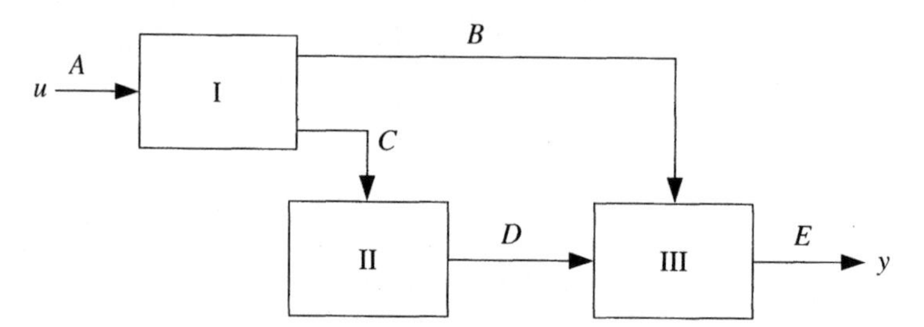

Figure 4.8 A network of tanks.

4E.4 Consider the RLC circuit in Figure 4.9 with ideal voltage source $u_v(t)$ and ideal current source $u_i(t)$. View this circuit as a linear time-invariant system with two inputs

$$u(t) = \begin{bmatrix} u_v(t) \\ u_i(t) \end{bmatrix}$$

and one output: the voltage $y(t)$. R, L, and C are unknown constants. Discuss several model set parametrizations that could be feasible for this system and describe their advantages and disadvantages.

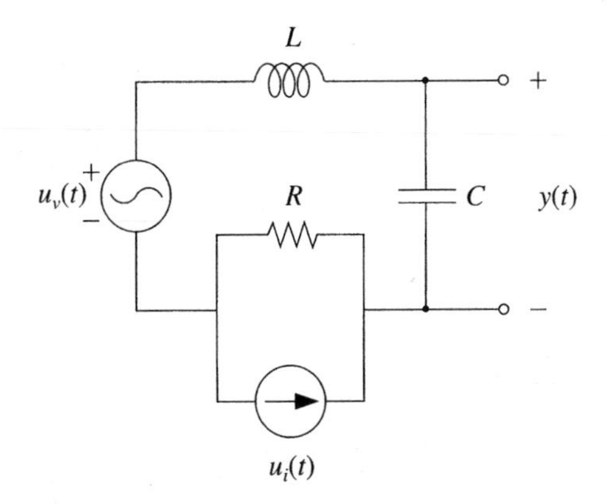

Figure 4.9 A simple circuit.

Hint: The basic equations for this circuit are

$$u_v(t) = L\frac{di(t)}{dt} + y(t) + R\,[i(t) + u_i(t)]$$

$$y(t) = \frac{1}{C}\int_0^t i(\tau)\,d\tau$$

4E.5 A state-space model of ship-steering dynamics can be given as follows:

$$\frac{d}{dt}\begin{bmatrix} v(t) \\ r(t) \\ h(t) \end{bmatrix} = \begin{bmatrix} a_{11} & a_{12} & 0 \\ a_{21} & a_{22} & 0 \\ 0 & 1 & 0 \end{bmatrix}\begin{bmatrix} v(t) \\ r(t) \\ h(t) \end{bmatrix} + \begin{bmatrix} b_{11} \\ b_{21} \\ 0 \end{bmatrix} u(t)$$

where $u(t)$ is the rudder angle, $v(t)$ the sway velocity, $r(t)$ the turning rate, and $h(t)$ the heading angle.

(a) Suppose only $u(t)$ and $y(t) = h(t)$ are measured. Show that the six parameters a_{ij}, b_{ij} are not identifiable.

(b) Try also to show that if $u(t)$ and $y(t) = \begin{bmatrix} v(t) \\ h(t) \end{bmatrix}$ are measured then all six parameters are globally identifiable at values such that the model is controllable. If you cannot complete the calculations, indicate how you would approach the problem (reference: Godfrey and DiStefano, 1985).

4E.6 Consider the model structure (4.91) with

$$A(\theta) = \begin{bmatrix} -a_1 & 1 \\ -a_2 & 0 \end{bmatrix}, \qquad B(\theta) = \begin{bmatrix} b_1 \\ b_2 \end{bmatrix},$$

$$C(\theta) = \begin{bmatrix} 1 & 0 \end{bmatrix}, \qquad K(\theta) = \begin{bmatrix} k_1 \\ k_2 \end{bmatrix}$$

$$\theta = \begin{bmatrix} a_1 & a_2 & b_1 & b_2 & k_1 & k_2 \end{bmatrix}^T, \qquad \theta \in D_1 \subset \mathbf{R}^6$$

and another structure

$$A(\eta) = \begin{bmatrix} \lambda_1 & 0 \\ 0 & \lambda_2 \end{bmatrix}, \qquad B(\eta) = \begin{bmatrix} \mu_1 \\ \mu_2 \end{bmatrix}$$

$$C(\eta) = \begin{bmatrix} \gamma_1 & \gamma_2 \end{bmatrix}, \qquad K(\eta) = \begin{bmatrix} \kappa_1 \\ \kappa_2 \end{bmatrix}$$

$$\eta = \begin{bmatrix} \lambda_1 & \lambda_2 & \mu_1 & \mu_2 & \gamma_1 & \gamma_2 & \kappa_1 & \kappa_2 \end{bmatrix}^T, \qquad \eta \in D_2 \subset \mathbf{R}^8$$

Determine D_1 and D_2 so that the two model structures determine the same model set. What about identifiability properties?

4E.7 Consider the heated metal rod of Example 4.3. Introduce a five-state lumped approximation and write down the state-space model explicitly.

4E.8 Consider the OE model structure with $n_b = 2, n_f = 1$, and b_1 fixed to unity:

$$y(t) = \frac{q^{-1} + b_2 q^{-2}}{1 + f_1 q^{-1}} u(t) + e(t), \qquad \theta = [b_2 \quad f_1]^T$$

Determine $\Gamma_2(\theta)$ of Problem 4G.4 explicitly. When is it singular?

4E.9 Consider the model structures

$$\mathcal{M}_1 : \; y(t) = -ay(t-1) + bu(t-1)$$

$$\theta = \begin{bmatrix} a \\ b \end{bmatrix}, \qquad D_{\mathcal{M}_1} = \{|a| \leq 1, b > 0\}$$

and

$$\mathcal{M}_2 : \; y(t) = -(\cos\alpha)y(t-1) + e^{\beta}u(t-1)$$

$$\eta = \begin{bmatrix} \alpha \\ \beta \end{bmatrix}, \qquad D_{\mathcal{M}_2} = \{0 \leq \alpha \leq \pi, -\infty < \beta < \infty\}$$

Show that $\mathcal{R}(\mathcal{M}_1) = \mathcal{R}(\mathcal{M}_2)$. Discuss possible advantages and disadvantages with the two structures.

4E.10 Consider the dc-motor model (4.75). Assume that the torque T_ℓ can be seen as a white-noise zero mean disturbance with variance σ^2 (i.e., the variations in T_ℓ are random and fast compared to the dynamics of the motor). Apply (4.97) to (4.99) to determine $R_1(\theta)$ and $R_{12}(\theta)$ in a sampled model (4.84) and (4.85) of the motor, with $A(\theta)$ and $B(\theta)$ given by (4.77) and

$$\theta = \begin{bmatrix} \tau \\ \beta \\ \gamma \end{bmatrix}, \qquad \gamma = \gamma' \cdot \sigma$$

As an alternative, we could use a directly parametrized innovations form (4.91) with $A(\theta)$ and $B(\theta)$ again given by (4.77), but

$$K(\theta) = \begin{bmatrix} k_1 \\ k_2 \end{bmatrix} \text{ and } \theta = [\tau \quad \beta \quad k_1 \quad k_2]^T$$

Discuss the advantages and disadvantages of these two parametrizations.

4E.11 Consider the system description

$$x(t+1) = ax(t) + bu(t) + \xi(t)$$
$$y(t) = x(t) + e(t)$$

where $e(t)$ is white Gaussian noise and $\xi(t)$ has the distribution

$$\xi(t) = 0, \qquad w.p.\ 1-\lambda$$
$$\xi(t) = +1, \qquad w.p.\ \lambda/2$$
$$\xi(t) = -1, \qquad w.p.\ \lambda/2$$

The coefficients a, b, and λ are adjustable parameters. Can this description be cast into the form (4.4)? If so, at the expense of what approximations?

4E.12 Consider a multivariable ARX model set

$$y(t) + A_1 y(t-1) + A_2 y(t-2) = B_1 u(t-1) + e(t)$$

where $\dim y = p = 2$, $\dim u = m = 1$, and where the matrices are parametrized as

$$A_1 = \begin{bmatrix} \times & \times \\ \alpha & \times \end{bmatrix}, \quad A_2 = \begin{bmatrix} \times & \times \\ 0 & \times \end{bmatrix}, \quad B_1 = \begin{bmatrix} \beta \\ \times \end{bmatrix}$$

where α and β are known values and $\times$ indicates a parameter to be estimated. Write the predictor in the form

$$\hat{y}(t|\theta) = \varphi^T(t)\theta + \mu(t)$$

with $\mu(t)$ as a known term and give explicit expressions for φ and θ. Can this predictor be written in the form (4.58)?

4T.1 Determine the k-step-ahead predictor for the ARMAX model (4.15).

4T.2 Give an expression for the k-step-ahead predictor for (4.91).

4T.3 Suppose that $W_u(q)$ and $W_y(q)$ are given functions, known to be determined as k-step-ahead predictors for the system description

$$y(t) = G(q)u(t) + H(q)e(t)$$

Can $G(e^{i\omega})$ and $H(e^{i\omega})$ be uniquely computed from $W_u(e^{i\omega})$ and $W_y(e^{i\omega})$? What if G and H are known to be of the ARMAX structure

$$G(q) = \frac{B(q)}{A(q)}, \qquad H(q) = \frac{C(q)}{A(q)}$$

where A, B, and C have known (and suitable) orders?

4D.1 Prove Lemma 4.2.

APPENDIX 4A: IDENTIFIABILITY OF BLACK-BOX MULTIVARIABLE MODEL STRUCTURES

The topic of multivariable model structures and canonical forms for multivariable systems is often regarded as difficult, and there is an extensive literature in the field. We shall here give a no-frills account of the problem, and the reader is referred to the literature for more insights and deeper results. See the bibliography.

The issue still is whether (4.136) holds at a given θ. Our development parallels the one in Section 4.6. We start by discussing polynomial parametrizations or MFDS, such as (4.55) to (4.61), and then turn to state-space models. Throughout the section, p denotes the number of outputs and m the number of inputs.

Matrix Fraction Descriptions (MFD)

Consider first the simple multivariable ARX structure (4.52) or (4.56). This uses

$$G(z,\theta) = A^{-1}(z)B(z), \qquad H(z,\theta) = A^{-1}(z) \tag{4A.1}$$

with θ comprising all the coefficients of the matrix polynomials (in $1/z$) $A(z)$ and $B(z)$. These could be of arbitrary orders. Just as for the SISO case (4.139), it is immediate to verify that (4.136) holds for all θ^*. Hence the model structure given by the MFD (4A.1) is *strictly globally identifiable.*

Let us now turn to the output error model structure

$$G(z,\theta) = F^{-1}(z)B(z), \qquad H(z,\theta) = I \tag{4A.2}$$

It should be noted that the analysis of (4A.2) contains also the analysis of the multivariable ARMAX structure and multivariable Box-Jenkins models. See the corollary to Theorem 4A.1, which follows.

The matrix polynomial $F(z)$ is here a $p \times p$ matrix

$$F(z) = \begin{bmatrix} F_{11}(z) & F_{12}(z)\dots F_{1p}(z) \\ F_{21}(z) & F_{22}(z)\dots F_{2p}(z) \\ \vdots & \vdots \quad \vdots \\ F_{p1}(z) & F_{p2}(z)\dots F_{pp}(z) \end{bmatrix} = F^{(0)} + F^{(1)}z^{-1} + \cdots + F^{(\nu)}z^{-\nu} \tag{4A.3}$$

whose entries are polynomials in z^{-1}:

$$F_{ij}(z) = f_{ij}^{(0)} + f_{ij}^{(1)}z^{-1} + \cdots + f_{ij}^{(\nu_{ij})}z^{-\nu_{ij}} \tag{4A.4}$$

The degree of the F_{ij} polynomial will thus be denoted by ν_{ij} and $\nu = \max \nu_{ij}$. Similarly, $B(z)$ is a $p \times m$ matrix polynomial. Let the degrees of its entries be denoted by μ_{ij}.

The structure issue is really to select the orders v_{ij} and μ_{ij} [i.e., $p(p+m)$ integers]. This will give a staggering amount of possible model structures. Some special cases discussed in the literature are

$$1.\ v_{ij} = n,\ \mu_{ij} = r \tag{4A.5}$$

$$2.\ v_{ij} = 0,\ i \neq j; \qquad v_{ii} = n_i,\ \mu_{ij} = r_i \tag{4A.6}$$

$$3.\ v_{ij} = n_j,\ \text{all } i; \qquad \mu_{ij} = r_j,\ \text{all } i \tag{4A.7}$$

In all these cases we fix the leading matrix to be a unit matrix:

$$F^{(0)} = I; \qquad \text{i.e., } f_{ij}^{(0)} = \delta_{ij} \tag{4A.8}$$

The form (4A.5) is called the "full polynomial form" in Söderström and Stoica (1983). It clearly is a special case of (4A.7). It is used and discussed in Hannan (1969, 1976), Kashyap and Rao (1976), Jakeman and Young (1979), and elsewhere.

The form (4A.6) gives a diagonal F-matrix and has been used, for example, in Kashyap and Nasburg (1974), Sinha and Caines (1977), and Gauthier and Landau (1978).

The structure (4A.7) where the different columns are given different orders is discussed, for example, in Guidorzi (1975), Gauthier and Landau (1978), and Gevers and Wertz (1984).

Remark. In the literature, especially the one discussing canonical forms rather than identification applications, often the polynomials

$$\overline{F}(z) = z^v F(z) = F^{(0)} z^v + F^{(1)} z^{v-1} + \cdots + F^{(v)} \tag{4A.9}$$

in the variable z are considered instead of $F(z)$ (just as we did the SISO case). Canonical representations of $F(z)$ [such as the "Hermite form"; see Dickinson, Kailath, and Morf, 1974; Hannan, 1971a; or Kailath, 1980] will then typically involve singular matrices $F^{(0)}$. Such representations are not suitable for our purposes since $y(t)$ cannot be solved for explicitly in terms of past data.

The identifiability properties of the diagonal form (4A.6) can be analyzed by SISO arguments. For the others we need some theory for matrix polynomials.

Some Terminology for Matrix Polynomials

Kailath (1980), Chapter 6, gives a detailed account of various concepts and properties of matrix polynomials. We shall here need just a few:

A $p \times p$ matrix polynomial $P(x)$ is said to be *unimodular* if $\det P(x) =$ constant. Then $P^{-1}(x)$ is also a matrix polynomial. Two polynomials $P(x)$ and $Q(x)$ with the same number of rows have a *common left divisor* if there exists a matrix polynomial $L(x)$ such that

$$P(x) = L(x)\tilde{P}(x)$$

$$Q(x) = L(x)\tilde{Q}(x)$$

for some matrix polynomials $\tilde{P}(x)$ and $\tilde{Q}(x)$.

$P(x)$ and $Q(x)$ are said to be *left coprime* if all common left divisors are unimodular. This is a direct extension of the corresponding concept for scalar polynomials. A basic theorem says that if $P(x)$ and $Q(x)$ are left coprime then there exist matrix polynomials $A(x)$ and $B(x)$ such that

$$P(x)A(x) + Q(x)B(x) = I \text{ (identity matrix)} \tag{4A.10}$$

Loss of Identifiability in Multivariable MFD Structures

We can now state the basic identifiability result.

Theorem 4A.1. Consider the output error MFD model structure (4A.2) with the polynomial degrees chosen according to the scheme (4A.7). Let θ comprise all the coefficients in the resulting matrix polynomials, and let $F_*(z)$ and $B_*(z)$ be the polynomials in $1/z$ that correspond to the value θ^*. Let

$$\begin{aligned} D_p(z) &= \text{diag}(z^{n_1}, \ldots, z^{n_p}) \\ D_m(z) &= \text{diag}(z^{r_1}, \ldots, z^{r_m}) \end{aligned}$$

be diagonal matrices, with n_i and r_i defined in (4A.7), and define $\tilde{F}_*(z) = F_*(z)D_p(z)$, $\tilde{B}_*(z) = B_*(z)D_m(z)$ as polynomials in z. Then the model structure in question is globally and locally identifiable at θ^* if and only if

$$\tilde{F}_*(z) \text{ and } \tilde{B}_*(z) \text{ are left coprime} \tag{4A.11}$$

Proof. Let θ correspond to $F(z)$ and $B(z)$, and assume that

$$G(z, \theta) = G(z, \theta^*) = F^{-1}(z)B(z) = F_*^{-1}(z)B_*(z)$$

This can also be written as

$$D_p(z)\tilde{F}^{-1}(z)\tilde{B}(z)D_m^{-1}(z) = D_p(z)\tilde{F}_*^{-1}(z)\tilde{B}_*(z)D_m^{-1}(z)$$

where $\tilde{F}$ and $\tilde{B}$ are defined analogously to $\tilde{F}_*$ and $\tilde{B}_*$. This gives

$$\tilde{B}_*(z) = \tilde{F}_*(z)\tilde{F}^{-1}(z)\tilde{B}(z) \tag{4A.12}$$

When $\tilde{B}_*$ and $\tilde{F}_*$ are left coprime there exist, according to (4A.10), matrix polynomials $X(z)$ and $Y(z)$ such that

$$\tilde{F}_*(z)X(z) + \tilde{B}_*(z)Y(z) = I$$

Inserting (4A.12) into this expression gives

$$\tilde{F}_*(z)\tilde{F}^{-1}(z)\left[\tilde{F}(z)X(z) + \tilde{B}(z)Y(z)\right] = I$$

or

$$\tilde{F}(z)X(z) + \tilde{B}(z)Y(z) = \tilde{F}(z)\tilde{F}_*^{-1}(z) \triangleq U(z)$$

Since the left side is a matrix polynomial in z, so is $U(z)$. We have

$$\tilde{F}(z) = U(z)\tilde{F}_*(z) \tag{4A.13}$$

Note that, by (4A.8),

$$I = \lim_{z\to\infty} F(z) = \lim_{z\to\infty} \tilde{F}(z)D_p^{-1}(z) = \lim_{z\to\infty} \tilde{F}_*(z)D_p^{-1}(z)$$

Hence, multiplying (4A.13) by $D_p^{-1}(z)$ gives

$$I = \lim_{z\to\infty} U(z)$$

which since $U(z)$ is a polynomial in z, shows that $U(z) \equiv I$, and hence $F(z) \equiv F_*(z)$, which in turn implies that $B(z) = B_*(z)$, and the if-part of the theorem has been proved. If (4A.11) does not hold, a common, nonunimodular, left factor $U_*(z)$ can be pulled out from $F_*(z)$ and $B_*(z)$ and be replaced by an arbitrary matrix with the same orders as $U_*(z)$ [subject to the constraint (4A.8)]. This proves the only-if-part of the theorem. □

The theorem can immediately be extended to a model structure

$$G(z,\theta) = F^{-1}(z)B(z), \qquad H(z,\theta) = D^{-1}(z)C(z) \tag{4A.14}$$

with F and D subject to the degree structure (4A.7). It can also be extended to the multivariable ARMAX structure:

$$G(z,\theta) = A^{-1}(z)B(z), \qquad H(z,\theta) = A^{-1}(z)C(z) \tag{4A.15}$$

Corollary 4A.1. Consider the ARMAX model structure (4A.15) with the degrees of the polynomial $A(z)$ subject to (4A.7). Let $\tilde{A}_*(z)$ and $\tilde{\beta}_*(z) = [\,\tilde{B}_*(z) \quad \tilde{C}_*(z)\,]$, a $p \times (m+p)$ matrix polynomial, be the polynomials that correspond to θ^*, as described in the theorem. Then the structure is identifiable at θ^* if and only if

$$\tilde{A}_*(z) \text{ and } \tilde{\beta}_*(z) \text{ are left coprime}$$

The usefulness of these identifiability results lies in the fact that only p orders (the column degrees) have to be chosen with care to find a suitable identifiable structure, despite the fact that $p \cdot m$ [or even $p \cdot (m + p)$ in the ARMAX case] different transfer functions are involved.

State-space Model Structures

For a multivariable state-space model (4.146), we introduce a parametric structure, analogous to (4.146):

$$A(\theta) = \begin{bmatrix} 0 & 1 & 0 & 0 & 0 & 0 & 0 & 0 & 0 \\ 0 & 0 & 1 & 0 & 0 & 0 & 0 & 0 & 0 \\ \times & \times & \times & \times & \times & \times & \times & \times & \times \\ 0 & 0 & 0 & 0 & 1 & 0 & 0 & 0 & 0 \\ \times & \times & \times & \times & \times & \times & \times & \times & \times \\ 0 & 0 & 0 & 0 & 0 & 0 & 1 & 0 & 0 \\ 0 & 0 & 0 & 0 & 0 & 0 & 0 & 1 & 0 \\ 0 & 0 & 0 & 0 & 0 & 0 & 0 & 0 & 1 \\ \times & \times & \times & \times & \times & \times & \times & \times & \times \end{bmatrix}, \quad B(\theta) = \begin{bmatrix} \times & \times \\ \times & \times \\ \times & \times \\ \times & \times \\ \times & \times \\ \times & \times \\ \times & \times \\ \times & \times \\ \times & \times \end{bmatrix}$$

$$K(\theta) = \begin{bmatrix} \times & \times & \times \\ \times & \times & \times \\ \times & \times & \times \\ \times & \times & \times \\ \times & \times & \times \\ \times & \times & \times \\ \times & \times & \times \\ \times & \times & \times \\ \times & \times & \times \end{bmatrix}, \quad C(\theta) = \begin{bmatrix} 1 & 0 & 0 & 0 & 0 & 0 & 0 & 0 & 0 \\ 0 & 0 & 0 & 1 & 0 & 0 & 0 & 0 & 0 \\ 0 & 0 & 0 & 0 & 0 & 1 & 0 & 0 & 0 \end{bmatrix} \quad (4A.16)$$

The number of rows with $\times$'s in $A(\theta)$ equals the number of outputs. We have thus illustrated the structure for $n = 9$, $p = 3$, $m = 2$. In words, the general structure can be defined as:

> *Let $A(\theta)$ initially be a matrix filled with zeros and with ones along the superdiagonal. Let then row numbers $r_1, r_2, \ldots, r_p$, where $r_p = n$, be filled with parameters. Take $r_0 = 0$ and let $C(\theta)$ be filled with zeros, and then let row i have a one in column $r_{i-1}+1$. Let $B(\theta)$ and $K(\theta)$ be filled with parameters.* (4A.17)

The parametrization is uniquely characterized by the p numbers r_i that are to be chosen by the user. We shall also use

$$\nu_i = r_i - r_{i-1}$$

and call

$$\overline{\nu}_n = \{\nu_1, \dots, \nu_p\} \tag{4A.18}$$

the *multiindex* associated with (4A.17). Clearly,

$$n = \sum_{i=1}^{p} \nu_i \tag{4A.19}$$

By a *multiindex* $\overline{\nu}_n$ we henceforth understand a collection of p numbers $\nu_i \geq 1$ subject to (4A.19). For given n and p, there exist $\binom{n-1}{p-1}$ different multi-indexes. Notice that the structure (4A.17) contains $2np + mn$ parameters regardless of $\overline{\nu}_n$.

The key property of a "canonical" parametrization like (4A.16) is that the corresponding state vector $x(t, \theta)$ can be interpreted in a pure input-output context. This can be seen as follows. Fix time t, and assume that $u(s) = e(s) \equiv 0$ for $s \geq t$. Denote the corresponding outputs that are generated by the model for $s \geq t$ by $\hat{y}_\theta(s|t-1)$. We could think of them as projected outputs for future times s as calculated at time $t-1$. The state-space equations give directly

$$\begin{aligned}
\hat{y}_\theta(t|t-1) &= C(\theta)x(t,\theta) \\
\hat{y}_\theta(t+1|t-1) &= C(\theta)A(\theta)x(t,\theta) \\
&\vdots \\
\hat{y}_\theta(t+n-1|t-1) &= C(\theta)A^{n-1}(\theta)x(t,\theta)
\end{aligned} \tag{4A.20}$$

With

$$O(\theta) = \begin{bmatrix} C(\theta) \\ C(\theta)A(\theta) \\ \vdots \\ C(\theta)A^{n-1}(\theta) \end{bmatrix} \tag{4A.21}$$

(the $np \times n$ observability matrix) and

$$\hat{Y}_n^\theta(t) = \begin{bmatrix} \hat{y}_\theta(t|t-1) \\ \vdots \\ \hat{y}_\theta(t+n-1|t-1) \end{bmatrix}$$

We can write (4A.20) as

$$\hat{Y}_n^\theta(t) = O_n(\theta)x(t,\theta) \tag{4A.22}$$

It is straightforward to verify that (4A.17) has a fundamental property: The $np \times n$ observability matrix $O_n(\theta)$ will have n rows that together constitute the unit matrix, regardless of θ. The reader is invited to verify that row number $kp + i$ of O_n will be

$$[0 \quad 0 \dots 0 \quad 1 \quad 0 \dots 0]$$

with 1 in position $r_{i-1}+k+1$. This holds for $1 \le i \le p, 0 \le k < \nu_i$. Thus (4A.22) implies that the state variables corresponding to the structure (4A.17) are

$$x_{r_{i-1}+k+1}(t,\theta) = \hat{y}_\theta^{(i)}(t+k|t-1), \qquad i = 1,\dots,p, \qquad 0 \le k < \nu_i \quad \text{(4A.23)}$$

Here superscript (i) denotes the ith component of y. This interpretation of state variables as predictors is discussed in detail in Akaike (1974b)and Rissanen (1974). By the relation (4A.23), n rows are picked out from the np vector $\hat{Y}_n^\theta(t)$ in (4A.22). The indexes of these rows are uniquely determined by the multiindex $\overline{\nu}_n$. Let them be denoted by

$$I_{\overline{\nu}_n} = \{(k-1)p + i; \quad 1 \le k \le \nu_i; \quad 1 \le i \le p\} \qquad \text{(4A.24)}$$

The key relationship is (4A.23). It shows that the state variables depend only on the input-output properties of the associated model.

Consider now two values θ^* and θ that give the same input-output properties of (4A.17). Then $\hat{y}_\theta(t+k|t-1) = \hat{y}_{\theta^*}(t+k|t-1)$, since these are computed from input-output properties only. Thus $x(t,\theta) = x(t,\theta^*)$. Now, if θ^* corresponds to a minimal realization, so must θ, and Theorem 6.2.4 of Kailath (1980)gives that there exists an invertible matrix T such that

$$\begin{aligned} A(\theta^*) &= TA(\theta)T^{-1}, \qquad & B(\theta^*) &= TB(\theta) \\ K(\theta^*) &= TK(\theta), \qquad & C(\theta^*) &= C(\theta)T^{-1} \end{aligned} \qquad \text{(4A.25)}$$

corresponding to the change of basis

$$x(t,\theta^*) = Tx(t,\theta) \qquad \text{(4A.26)}$$

But (4A.26) together with our earlier observation that $x(t,\theta^*) = x(t,\theta)$ shows that $T = I_n$ and hence that $\theta^* = \theta$.

We have now proved the major part of the following theorem:

Theorem 4A.2. Consider the state-space model structure (4A.17). This structure is globally and locally identifiable at θ^* if and only if $\{A(\theta^*), [\, B(\theta^*) \quad K(\theta^*)\,]\}$ is controllable.

Proof. The if-part was proved previously. To show the only-if-part, we find that if θ^* does not give a controllable system then its input-output properties can be described by a lower-dimensional model with an additional, arbitrary, noncontrollable model. This can be accomplished by infinitely many different θ's. □

It follows from the theorem that the parametrization (4A.17) is globally identifiable, and as such is a good candidate to describe systems of order n. What is not clear yet is whether *any* nth-order linear system can be represented in the form (4A.17) for *an arbitrary choice* of multiindex $\overline{\nu}_n$. That is the question we now turn to.

Hankel-Matrix Interpretation

Consider a multivariable system description

$$y(t) = G_0(q)u(t) + H_0(q)e(t) = T_0(q)\chi(t) \tag{4A.27}$$

with

$$T_0(q) = [\, G_0(q) \quad H_0(q)\,], \qquad \chi(t) = \begin{bmatrix} u(t) \\ e(t) \end{bmatrix}$$

Assume that $T_0(q)$ has full row rank [i.e., $LT_0(q)$ is not identically zero for any nonzero $1 \times p$ vector L]. Let

$$T_0(q) = [\,0 \quad I\,] + \sum_{k=1}^{\infty} H_k q^{-k} \tag{4A.28}$$

be the impulse response of the system. The matrices H_k are here $p \times (p + m)$. Define the matrix

$$\mathcal{H}_{r,s} = \begin{bmatrix} H_1 & H_2 & \dots & H_s \\ H_2 & H_3 & \dots & H_{s+1} \\ H_3 & H_4 & \dots & H_{s+2} \\ \vdots & \vdots & & \vdots \\ H_r & H_{r+1} & \dots & H_{r+s-1} \end{bmatrix} \tag{4A.29}$$

This structure with the same block elements along antidiagonals is known as a *block Hankel matrix*. Consider the semifinite matrix $\mathcal{H}_r = \mathcal{H}_{r,\infty}$. For this matrix we have the following two fundamental results.

Lemma 4A.1. Suppose that the n rows $I_{\overline{\nu}_n}$ [see (4A.24)] of $\mathcal{H}_n$ span all the rows of $\mathcal{H}_{n+1}$. Then the system (4A.27) can be represented in the state-space form (4A.17) corresponding to the multiindex $\overline{\nu}_n$.

The proof consists of an explicit construction and is given at the end of this appendix.

Lemma 4A.2. Suppose that

$$\text{rank}\ \mathcal{H}_{n+1} \le n \tag{4A.30}$$

Then there exists a multiindex $\overline{\nu}_n$ such that the n rows $I_{\overline{\nu}_n}$ span $\mathcal{H}_{n+1}$. The proof of this lemma is also given at the end of this appendix.

It follows from the two lemmas that (4A.30) is a sufficient condition for (4A.27) to be an n-dimensional linear system (i.e., to admit a state-space representation of order n). It is, however, well known that this is also a necessary condition. ($\mathcal{H}$ is obtained as the product of the observability and controllability matrices.) We thus conclude:

Any linear system that can be represented in state-space form of order n can also be represented in the particular form (4A.17) for some multiindex $\overline{\nu}_n$ *(4A.31)*

When (4A.30) holds, we thus find that the np rows of $\mathcal{H}_n$ span an n-dimensional (or less) linear space. The *generic* situation is then that the same space is spanned by *any* subset of n rows of $\mathcal{H}_n$. (By this term we mean if we randomly pick from a uniform distribution np row vectors to span an n-dimensional space the probability is 1 that *any* subset of n vectors will span the same space.) We thus conclude:

A state-space representation in the form (4A.17) for a particular multiindex $\overline{\nu}_n$ *is capable of describing almost all n-dimensional linear systems. (4A.32)*

Overlapping Parametrizations

Let $\mathcal{M}_{\overline{\nu}_n}$ denote the model structure (4A.17) corresponding to $\overline{\nu}_n$. The result (4A.31) then implies that the model set

$$\overline{\mathcal{M}} = \bigcup_{\overline{\nu}_n} \mathcal{R}(\mathcal{M}_{\overline{\nu}_n}) \tag{4A.33}$$

(union over all possible multiindices $\overline{\nu}_n$) covers all linear n-dimensional systems. We have thus been able to describe the set of all linear n-dimensional systems as the union of ranges of identifiable structures [cf. (4.129)]. From (4A.32), it follows that the ranges of $\mathcal{M}_{\overline{\nu}_n}$ overlap considerably. This is no disadvantage for identification; on the contrary, one may then change from one structure to another without losing information. The practical use of such overlapping parametrizations for identification is discussed in van Overbeek and Ljung (1982). Using a topological argument, Delchamps and Byrnes (1982)give estimates on the number of overlapping structures needed in (4A.33). See also Hannan and Kavalieris (1984).

Connections Between Matrix Fraction and State-Space Descriptions

In the SISO case the connection between a state-space model in observability form and the corresponding ARMAX model is simple and explicit (see Example 4.2). Unfortunately, the situation is much more complex in the multivariable case. We refer to Gevers and Wertz (1984), Guidorzi (1981), and Beghelli and Guidorzi (1983)for detailed discussions.

We may note, though, the close connection between the indexes ν_i used in (4A.17) and the column degrees n_i in (4A.7). Both determine the number of time shifts of the ith component of y that are explicitly present in the representations. The shifts are, however, forward for the state space and backward for the MFD. The relationship between the ν_i and the observability indexes is sorted out in the proof of Lemma 4A.2.

A practical difference between the two representations is that the state-space representation naturally employs the state $x(t)$ (n variables) as a memory vector for simulation and other purposes. When (4A.2) is simulated in a straightforward fashion, the different delayed components of y and u are stored, a total number of $np + m \cdot \sum r_i$ variables. This is of course not necessary, but an efficient organization of the variables to be stored amounts to a state-space representation. There are consequently several advantages associated with state-space representations for multivariable systems.

Proofs of Lemmas 4A.1 and 4A.2

It now remains only to prove Lemmas 4A.1 and 4A.2.

Proof of Lemma 4A.1. Let

$$S(t) = \begin{bmatrix} \chi(t-1) \\ \chi(t-2) \\ \vdots \end{bmatrix}$$

Let [cf. (4A.20) to (4A.22)]

$$\hat{y}_0(t|t-k) = \sum_{\ell=k}^{\infty} H_\ell \chi(t-\ell) \tag{4A.34}$$

and

$$\hat{Y}_N(t) = \begin{bmatrix} \hat{y}_0(t|t-1) \\ \vdots \\ \hat{y}_0(t+N-1|t-1) \end{bmatrix}$$

Then, from (4A.28) and (4A.29),

$$\hat{Y}_N(t) = \mathcal{H}_N S(t) \tag{4A.35}$$

Now enumerate the row indexes i_r of $I_{\bar{\nu}_n}$ in (4A.24) as follows:

$$\begin{aligned}
i_1 &= 1, & i_2 &= p+1, \ldots, i_{\nu_1} = (\nu_1 - 1)\cdot p + 1 \\
i_{\nu_1+1} &= 2, & i_{\nu_1+2} &= p+2, \ldots, i_{r_2} = (\nu_2 - 1)\cdot p + 2 \\
&\vdots \\
&\vdots \\
i_{r_{p-1}+1} &= p, & i_{r_{p-1}+2} &= p+p, \ldots, i_{r_p} = (\nu_p - 1)\cdot p + p
\end{aligned} \tag{4A.36}$$

Recall that

$$r_k = \sum_{1}^{k} \nu_j$$

Now construct the n-vector $x(t)$ by taking its rth component to be the i_rth component of $\hat{Y}_N(t)$. Let us now focus on the components $i_1 + p, i_2 + p, \ldots, i_n + p$ of

(4A.35). Collect these components into a vector $\xi(t+1)$. They all correspond to rows of $\mathcal{H}_{n+1}$. But this matrix is spanned by $x(t)$ by the assumption of the lemma. Hence

$$\xi(t+1) = Fx(t) \tag{4A.37}$$

for some matrix F. Now several of the components of $\xi(t+1)$ will also belong to $x(t)$, as shown in (4A.36). The corresponding rows of F will then be zeros everywhere except for a 1 in one position. A moment's reflection on (4A.36) shows that the matrix F will in fact have the structure (4A.17). Also, with H given by (4A.17),

$$y(t) = Hx(t) + e(t) \tag{4A.38}$$

Let us now return to (4A.37). Consider component r of $x(t+1)$, which by definition equals row i_r of $\hat{Y}_N(t+1)$. This row is given as $\hat{y}_0^{(j)}(t+k|t)$ for some values j and k that depend on i_r. But, according to (4A.34), we have

$$\hat{y}_0(t+k|t) = \hat{y}_0(t+k|t-1) + H_k\chi(t) \tag{4A.39}$$

Hence

$$x_r(t+1) = \hat{y}_0^{(j)}(t+k|t) = \hat{y}_0^{(j)}(t-1+(k+1)|t-1) + [H_k\chi(t)]_j$$

But the first term of the right side equals component number $i_r + p$ of $\hat{Y}_N(t)$ [i.e., $\xi_r(t+1)$]. Hence

$$x(t+1) = \xi(t+1) + M\chi(t) \tag{4A.40}$$

for some matrix M. Equations (4A.37), (4A.38), and (4A.40) now form a state-space representation of (4A.27) within the structure (4A.17) and the lemma is proved. □

Proof of Lemma 4A.2. The defining property of the Hankel matrix $\mathcal{H}_N$ in (4A.29) means that the same matrix is obtained by either deleting the first block column (and the last block row) or by deleting the first block row. This implies that, if row i of block row k [i.e., row $(k-1)p+i$] lies in the linear span of all rows above it, then so must row i of block $k+1$.

Now suppose that

$$\text{rank } \mathcal{H}_{n+1} = n$$

and let us search the rows from above for a set of linearly independent ones. A row that is not linearly dependent on the ones above it is thus included in the basis; the others are rejected. When the search is finished, we have selected n rows from $\mathcal{H}_{n+1}$. The observation mentioned previously implies that, if row $kp+i$ is included in this basis for $k \geq 1$, then so is row $(k-1)p+i$. Hence the row indexes will obey the structure

$$\begin{array}{llll}
1, & p+1, & 2p+1, \ldots, (\sigma_1 - 1)p + 1 \\
2, & p+2, & 2p+2, \ldots, (\sigma_2 - 1)p + 2 \\
\vdots & & & \\
p, & p+p, & 2p+p, \ldots, (\sigma_p - 1)p + p
\end{array}$$

for some numbers $\{\sigma_i\}$ that are known as the *observability indexes* of the system. Since the total number of selected rows is n, we have

$$\sum_{1}^{p} \sigma_i = n$$

The rows thus correspond to the multiindex $\overline{\sigma}_n$ as in (4A.24) and the lemma is proved. Notice that several other multiindexes may give a spanning set of rows; one does not have to look for the first linearly independent rows. □

5

MODELS FOR TIME-VARYING AND NONLINEAR SYSTEMS

While linear, time-invariant models no doubt form the most common way of describing a dynamical system, it is also quite often useful or necessary to employ other descriptions. In this chapter we shall first, in Section 5.1, discuss linear, time-varying models. In Section 5.2 we deal with models with nonlinearities in form of static, non-linear elements at the input and/or the output. We also describe how to handle nonlinearities that can be introduced by suitable nonlinear transformations of the raw measured data. In Section 5.3 we describe nonlinear models in state-space form. So far the development concerns model structures, where the nonlinearities are brought in based on some physical insight. In Section 5.4 we turn to general nonlinear models of black-box type, and describe the general features of such models. Particular instances of these, like artificial neural networks, wavelet models, etc. are then dealt with in Section 5.5, while fuzzy models are discussed in Section 5.6. Finally, in Section 5.7 we give a formal account of what we mean by a model in general, thus complementing the discussion in Section 4.5 on general linear models.

5.1 LINEAR TIME-VARYING MODELS

Weighting Function

In Chapter 2 we defined a linear system as one where a linear combination of inputs leads to the same linear combination of the corresponding outputs. A general linear system can then be described by

$$y(t) = \sum_{k=1}^{\infty} g_t(k)u(t-k) + v(t) \tag{5.1}$$

If we write

$$g_t(k) = \tilde{g}(t, t-k)$$

we find that

$$y(t) = \sum_{s=-\infty}^{t-1} \tilde{g}(t,s)u(s) + v(t) \tag{5.2}$$

where $\tilde{g}(t,s), t = s, s+1, \dots,$ is the response at time t to a unit input pulse at time s. The function $\tilde{g}(t,s)$ is also known as the *weighting function*, since it describes the weight that the input at time s has in the output at time t.

The description (5.1) is quite analogous to the time-invariant model (2.8), except that the sequence $g_t(k)$ carries the time index t. In general, we could introduce a time-varying transfer function by

$$G_t(q) = \sum_{k=1}^{\infty} g_t(k)q^{-k} \tag{5.3}$$

and repeat most of the discussion in Section 4.2 for time-varying transfer functions. In practice, though, it is easier to deal with time variation in state-space forms.

Time-Varying State-Space Model

Time variation in state-space models (4.91) is simply obtained by letting the matrices be time varying:

$$\begin{aligned} x(t+1,\theta) &= A_t(\theta)x(t,\theta) + B_t(\theta)u(t) + K_t(\theta)e(t) \\ y(t) &= C_t(\theta)x(t,\theta) + e(t) \end{aligned} \tag{5.4}$$

The predictor corresponding to (4.86) then becomes

$$\begin{aligned} \hat{x}(t+1,\theta) &= [A_t(\theta) - K_t(\theta)C_t(\theta)]\hat{x}(t,\theta) + B_t(\theta)u(t) + K_t(\theta)y(t) \quad (5.5) \\ \hat{y}(t|\theta) &= C_t(\theta)\hat{x}(t,\theta) \end{aligned}$$

Notice that this can be written

$$\hat{y}(t|\theta) = \sum_{k=1}^{\infty} w_t^u(k,\theta)u(t-k) + \sum_{k=1}^{\infty} w_t^y(k,\theta)y(t-k) \tag{5.6}$$

with

$$\begin{aligned} w_t^u(k,\theta) &= C_t(\theta)\prod_{j=t-k}^{t-1}\left[A_j(\theta) - K_j(\theta)C_j(\theta)\right]B_{t-k}(\theta) \\ w_t^y(k,\theta) &= C_t(\theta)\prod_{j=t-k}^{t-1}\left[A_j(\theta) - K_j(\theta)C_j(\theta)\right]K_{t-k}(\theta) \end{aligned} \tag{5.7}$$

Similarly, we could start with a time-varying model like (4.84) and (4.85), where the matrices A, B, C, R_1, R_{12}, and R_2 are functions of t. The corresponding predictor will then be given by (4.94) and (4.95).

Two common problems associated with time-invariant systems in fact lead to time-varying descriptions: *nonequal sampling intervals and linearization.* If the system (4.62) and (4.63) is sampled at time instants $t = t_k, k = 1, 2, \ldots$, we can still apply the sampling formulas (4.66) to (4.68) to go from t_k to t_{k+1}, using $T_k = t_{k+1} - t_k$. If this sampling interval is not constant, (4.67) will be a time-varying system. A related case is multirate systems, i.e., when *different variables are sampled at different rates.* Then the $C_t(\theta)$ matrix in (5.4) will be time varying in order to pick out the states that are sampled at instant t. Missing output data can also be seen as non-uniform sampling. See Section 14.2.

Linearization of Nonlinear Systems

Perhaps the most common use of time-varying linear systems is related to linearization of a nonlinear system around a certain trajectory. Suppose that a nonlinear system is described by

$$\begin{aligned} x(t+1) &= f(x(t), u(t)) + r(x(t), u(t))\, w(t) \\ y(t) &= h(x(t)) + m(x(t), u(t))\, v(t) \end{aligned} \tag{5.8}$$

Suppose also that the disturbance terms $\{w(t)\}$ and $\{v(t)\}$ are white and small, and that the nominal, disturbance-free ($w(t) \equiv 0$; $v(t) \equiv 0$) behavior of the system corresponds to an input sequence $u^*(t)$ and corresponding trajectory $x^*(t)$. Neglecting nonlinear terms, the differences

$$\begin{aligned} \Delta x(t) &= x(t) - x^*(t) \\ \Delta y(t) &= y(t) - h\left(x^*(t)\right) \\ \Delta u(t) &= u(t) - u^*(t) \end{aligned}$$

are then subject to

$$\begin{aligned} \Delta x(t+1) &= F(t)\Delta x(t) + G(t)\Delta u(t) + \overline{w}(t) \\ \Delta y(t) &= H(t)\Delta x(t) + \overline{v}(t) \end{aligned} \tag{5.9}$$

where

$$F(t) = \frac{\partial}{\partial x} f(x, u)\bigg|_{x^*(t), u^*(t)}, \qquad G(t) = \frac{\partial}{\partial u} f(x, u)\bigg|_{x^*(t), u^*(t)}$$

$$H(t) = \frac{\partial}{\partial x} h(x)\bigg|_{x^*(t)}$$

Here we have neglected cross terms with the disturbance term (like $\Delta x \cdot \nu$), in view of our assumption of small disturbances. In (5.9), $\overline{w}(t)$ and $\overline{v}(t)$ are white disturbances with the following covariance properties:

$$\begin{aligned} R_1(t) &= E\overline{w}(t)\overline{w}^T(t) = r\left(x^*(t), u^*(t)\right) Ew(t)w^T(t)r^T\left(x^*(t), u^*(t)\right) \\ R_2(t) &= E\overline{v}(t)\overline{v}^T(t) = m\left(x^*(t), u^*(t)\right) E\nu(t)\nu^T(t)m^T\left(x^*(t), u^*(t)\right) \quad (5.10) \\ R_{12}(t) &= r\left(x^*(t), u^*(t)\right) Ew(t)\nu^T(t)m^T\left(x^*(t), u^*(t)\right) \end{aligned}$$

This model is now a linear, time-varying, approximate description of (5.8) in a vicinity of the nominal trajectory.

5.2 MODELS WITH NONLINEARITIES

A nonlinear relationship between the input sequence and the output sequence as in (5.8) clearly gives much richer possibilities to describe systems. At the same time, the situation is very flexible and it is quite demanding to infer general nonlinear structures from finite data records. Even a first order model (5.8) (dim $x=1$) without disturbances is specified only up to members in a general infinite-dimensional function space (functions $f(\cdot, \cdot)$ and $h(\cdot)$), while the corresponding linear model is characterized in terms of two real numbers. Various possibilities to parameterize general mappings from past data to the predictor will be discussed in Section 5.4. It is however always wise first to try and utilize physical insight into the character of possible nonlinearities and construct suitable model structures that way. In this section we shall deal with what might be called *semi-physical modeling*, by which we mean using simple physical understanding to come up with the essential nonlinearities and incorporate those in the models.

Wiener and Hammerstein Models

It is a quite common situation that while the dynamics itself can be well described by a linear system, there are static nonlinearities at the input and/or at the output. This will be the case if the actuators are nonlinear, e.g., due to saturation, or if the sensors have nonlinear characteristics. A model with a static nonlinearity at the input is called a *Hammerstein model*, while we talk about a *Wiener model* if the nonlinearity is at the output. See Figure 5.1. The combination will then be a *Wiener-Hammerstein* model. The parameterization of such models is rather straightforward. Consider the Hammerstein case. The static nonlinear function $f(\cdot)$ can be parameterized either in terms of physical parameters, like saturation point and saturation level, or in black-box terms like spline-function coefficients. This gives $f(\cdot, \eta)$. If the linear model is $G(q, \theta)$, the predicted output model will be

$$\hat{y}(t|\theta, \eta) = G(q, \theta) f(u(t), \eta) \quad (5.11)$$

which is not much more complicated than the models of the previous chapter.

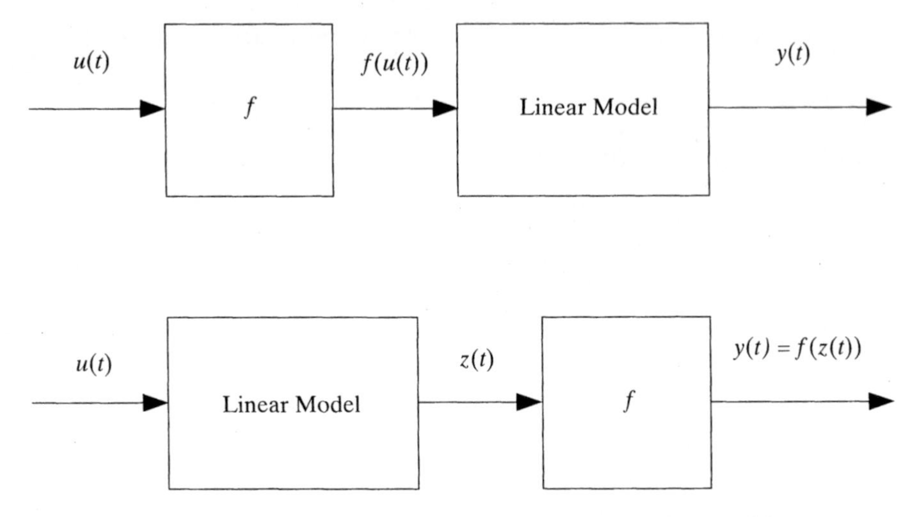

Figure 5.1 Above: A Hammerstein model. Below: A Wiener model.

Linear Regression Models

In (4.12) we defined a linear regression model structure where the prediction is linear in the parameters:

$$\hat{y}(t|\theta) = \varphi^T(t)\theta \tag{5.12}$$

To describe a linear difference equation, the components of the vector $\varphi(t)$ (i.e., the regressors), were chosen as lagged input and output values; see (4.11). When using (5.12) it is, however, immaterial how $\varphi(t)$ is formed; what matters is that it is a known quantity at time t. We can thus let it contain arbitrary transformations of measured data. Let, as usual, y^t and u^t denote the input and output sequences up to time t. Then we could write

$$\hat{y}(t|\theta) = \theta_1\varphi_1(u^t, y^{t-1}) + \ldots + \theta_d\varphi_d(u^t, y^{t-1}) = \varphi^T(t)\theta \tag{5.13}$$

with arbitrary functions φ_i of past data. The structure (5.13) could be regarded as a finite-dimensional parameterization of a general, unknown, nonlinear predictor. The key is how to choose the functions $\varphi_i(u^t, y^{t-1})$. There are a few possibilities:

- *Try black box expansions.* We could construct the regressors as typical (polynomial) combinations of past inputs and outputs and see if the model is able to describe the data. This will be contained in the treatment of Section 5.4 and also corresponds to the so-called GMDH-approach (see Ivakhnenko, 1968 and Farlow, 1984). It normally gives a large number of possible regressors. It is somewhat simpler for the Hammerstein model, where we may approximate the static nonlinearity by a polynomial expansion:

$$f(u) = \alpha_1 u + \alpha_2 u^2 + \ldots + \alpha_m u^m \tag{5.14}$$

Each power of u could then pass a different numerator dynamics:

$$A(q)y(t) = B_1(q)u(t) + B_2(q)u^2(t) + \ldots + B_m(q)u^m(t) \tag{5.15}$$

This clearly is a linear regression model structure.

- *Use physical insight.* A few moments reflection, using high school physics, often may reveal which are the essential nonlinearities in a system. This will suggest which regressors to try in (5.13). See Example 1.5. We call this *semi-physical modeling*. It could be as simple as taking the product of voltage and current measurements to create a power signal. See Example 5.1 below for another simple case.

Example 5.1 A Solar-Heated House

Consider the problem of identifying the dynamics of a solar-heated house, described in Example 1.1. We need a model of how the storage temperature $y(t)$ is affected by pump velocity and solar intensity. A straightforward linear model of the type (4.7) would be

$$\begin{aligned} y(t) &+ a_1 y(t-1) + a_2 y(t-2) \\ &= b_1 u(t-1) + b_2 u(t-2) + c_1 I(t-1) + c_2 I(t-2) \end{aligned} \tag{5.16}$$

With this we have not used any physical insight into the heating process, but introduced the black-box model (5.16) in an ad hoc manner. A moment's reflection reveals that a linear model is not very realistic. Clearly, the effects of solar intensity and pump velocity are not additive. When the pump is off, the sun does not at all affect the storage temperature.

Let us go through what happens in the heating system. Introduce $x(t)$ for the temperature of the solar panel collector at time t. With some simplifications, the physics can be described as follows in discrete time: The heating of the air in the collector $[= x(t+1) - x(t)]$ is equal to heat supplied by the sun $[= d_2 \cdot I(t)]$ minus loss of heat to the environment $[= d_3 \cdot x(t)]$ minus the heat transported to storage $[= d_0 \cdot x(t) \cdot u(t)]$; that is,

$$x(t+1) - x(t) = d_2 I(t) - d_3 x(t) - d_0 x(t) \cdot u(t) \tag{5.17}$$

In the same way, the increase of storage temperature $[= y(t+1) - y(t)]$ is equal to supplied heat $[= d_0 x(t) \cdot u(t)]$ minus losses to the environment $[= d_1 y(t)]$; that is,

$$y(t+1) - y(t) = d_0 x(t) u(t) - d_1 y(t) \tag{5.18}$$

In equations (5.17) and (5.18) the coefficients d_k are unknown constants, whose numerical values are to be determined. The temperature $x(t)$ is not, however, measured, so we first eliminate $x(t)$ from (5.17) and (5.18). This gives

$$\begin{aligned} y(t) &= (1-d_1)y(t-1) + (1-d_3)\frac{y(t-1)u(t-1)}{u(t-2)} \\ &+ (d_3-1)(1-d_1)\frac{y(t-2)u(t-1)}{u(t-2)} + d_0 d_2 u(t-1) I(t-2) \\ &- d_0 u(t-1) y(t-1) + d_0(1-d_1)u(t-1)y(t-2) \end{aligned} \tag{5.19}$$

The relationship between the measured quantities y, u, and I and the parameters d_i is now more complicated. It can be simplified by reparametrization:

$$
\begin{aligned}
\theta_1 &= (1 - d_1) & \varphi_1(t) &= y(t-1) \\
\theta_2 &= (1 - d_3) & \varphi_2(t) &= \frac{y(t-1)u(t-1)}{u(t-2)} \\
\theta_3 &= (d_3 - 1)(1 - d_1) & \varphi_3(t) &= \frac{y(t-2)u(t-1)}{u(t-2)} \\
\theta_4 &= d_0 d_2 & \varphi_4(t) &= u(t-1)I(t-2) \\
\theta_5 &= -d_0 & \varphi_5(t) &= u(t-1)y(t-1) \\
\theta_6 &= d_0(1 - d_1) & \varphi_6(t) &= u(t-1)y(t-2) \\
\theta^T &= \begin{bmatrix} \theta_1 & \theta_2 \dots \theta_6 \end{bmatrix} & \varphi^T(t) &= \begin{bmatrix} \varphi_1(t) & \varphi_2(t) \dots \varphi_6(t) \end{bmatrix}
\end{aligned}
\tag{5.20}
$$

Then (5.19) can be rewritten as a true linear regression,

$$
y(t) = \hat{y}(t|\theta) = \varphi^T(t)\theta \tag{5.21}
$$

where we have a linear relationship between the new parameters θ and the constructed measurements $\varphi(t)$. (Notice that φ does not depend on θ.) The price for this is that the knowledge of algebraic relationships between the θ_i, according to (5.20), has been lost. The simple structure (5.21) turns out to give a reasonably good model, Ljung and Glad (1994a). □

5.3 NONLINEAR STATE-SPACE MODELS

A General Model Set

The most general description of a finite-dimensional system is

$$
\begin{aligned}
x(t+1) &= f\left(t, x(t), u(t), w(t); \theta\right) \\
y(t) &= h\left(t, x(t), u(t), v(t); \theta\right)
\end{aligned}
\tag{5.22}
$$

Here $w(t)$ and $v(t)$ are sequences of independent random variables and θ denotes a vector of unknown parameters. The problem to determine a predictor based on (5.22) and on formal probabilistic grounds is substantial. In fact, this nonlinear prediction problem is known to have no finite-dimensional solution except in some isolated special cases.

Nevertheless, predictors for (5.22) can of course be constructed, either with *ad hoc* approaches or by some approximation of the unrealizable optimal solution. For the latter problem, there is abundant literature (see, e.g., Jazwinski, 1970, or Anderson and Moore, 1979). In either case the resulting predictor takes the form

$$
\hat{y}(t|\theta) = g(t, Z^{t-1}; \theta) \tag{5.23}
$$

Here, for easier notation, we introduced

$$
Z^t = (y^t, u^t) = (y(1), u(1) \dots y(t), u(t))
$$

to denote the input-output measurements available at time t. This is the form in which the model is put to use for identification purposes. We may thus view (5.23) as the *basic model*, and disregard the route that took us from the underlying description [like (5.22)] to the form (5.23). This is consistent with the view of *models as predictors* that we took in Chapter 4; the only difference is that (5.23) is a nonlinear function of past data, rather than a linear one. Just as in Chapter 4, the model (5.23) can be complemented with assumptions about the associated prediction error

$$\varepsilon(t, \theta) = y(t) - g(t, Z^{t-1}; \theta) \tag{5.24}$$

such as its covariance matrix, $\Lambda(t; \theta)$ or its PDF $f_e(x, t; \theta)$.

Nonlinear Simulation Model

A particularly simple way of deriving a predictor from (5.22) is to disregard the process noise $w(t)$ and take

$$\begin{aligned} x(t+1, \theta) &= f\,(t, x(t, \theta), u(t), 0; \theta) \\ \hat{y}(t|\theta) &= h\,(t, x(t, \theta), u(t), 0; \theta) \end{aligned} \tag{5.25}$$

We call such a predictor a *simulation model*, since $\hat{y}(t|\theta)$ is constructed by simulating a noise-free model (5.22) using the actual input. Clearly, a simulation model is almost as easy to use starting from a continuous-time representation:

$$\begin{aligned} \frac{d}{dt}x(t, \theta) &= f\,(t, x(t, \theta), u(t), 0; \theta) \\ \hat{y}(t|\theta) &= h\,(t, x(t, \theta), u(t), 0; \theta) \end{aligned} \tag{5.26}$$

Example 5.2 Delignification

Consider the problem of reducing the lignin content of wood chips in a chemical mixture. This is, basically, the process of cellulose cooking for obtaining pulp for paper making.

Introduce the following notation:

$x(t)$: lignin concentration at time t

$u_1(t)$: absolute temperature at time t

$u_2(t)$: concentration of hydrogen sulfite, $[HSO_3^-]$

$u_3(t)$: concentration of hydrogen, $[H^+]$

Then basic chemical laws tell us that

$$\frac{d}{dt}x(t) = -k_1 e^{-E_L/u_1(t)}\,[x(t)]^m \cdot [u_2(t)]^\alpha \cdot [u_3(t)]^\beta \cdot \tag{5.27}$$

Here E_L is the Arrhenius constant and k_1, m, α, and β are other constants associated with the reaction. Simulating (5.27) with the measured values of $\{u_i(t), i = 1, 2, 3\}$ for given values of $\theta^T = [E_L, k_1, m, \alpha, \beta]$ gives a sequence of corresponding lignin concentrations $\{x(t, \theta)\}$. In this case the system output is also the lignin concentration, so $\hat{y}(t|\theta) = x(t, \theta)$. These predicted, or simulated, values can then be compared with the actually measured values so that the errors associated with a particular value of θ can be evaluated. Such an application is described in detail in Hagberg and Schöön (1974). □

5.4 NONLINEAR BLACK-BOX MODELS: BASIC PRINCIPLES

As we have seen before, a model is a mapping from past data to the space of the output. In the nonlinear case, this mapping has the general structure (5.23):

$$\hat{y}(t|\theta) = g(Z^{t-1}, \theta) \tag{5.28}$$

We here omit the explicit time dependence. If we have no particular insight into the system's properties, we should seek parameterizations of g that are flexible and cover "all kinds of reasonable system behavior." This would give us a *nonlinear black-box* model structure—a nonlinear counterpart of the linear black models discussed in Section 4.2. How to achieve such parameterizations is the topic of the present and following sections.

A Structure for the General Mapping: Regressors

Now, the model structure family (5.28) is really too general, and it is useful to write g as a concatenation of two mappings: one that takes the increasing number of past observations Z^t and maps them into a finite dimensional vector $\varphi(t)$ of fixed dimension, and one that takes this vector to the space of the outputs

$$g(Z^{t-1}, \theta) = g(\varphi(t), \theta) \tag{5.29}$$

where

$$\varphi(t) = \varphi(Z^{t-1}) \tag{5.30}$$

We shall as before call this vector φ the *regression vector* and its components will be referred to as *regressors*. We may allow also the more general case that the formation of the regressors is itself parameterized

$$\varphi(t) = \varphi(Z^{t-1}, \theta) \tag{5.31}$$

which we for short write $\varphi(t, \theta)$, see, e.g., (4.40). For simplicity, the extra argument θ will however be used explicitly only when essential for the discussion.

The choice of the nonlinear mapping in (5.28) has thus been decomposed into two partial problems for dynamical systems:

1. How to choose the regression vector $\varphi(t)$ from past inputs and outputs.
2. How to choose the nonlinear mapping $g(\varphi, \theta)$ from the regressor space to the output space.

The choice of regressors is of course application dependent. Typically the regressors are chosen in the same way as for linear models: past measurements and possibly past model outputs and past prediction errors, as in (4.11), (4.20), and (4.40). We would then, analogously to Table 4.1 talk about NARX, NARMAX, NOE model structures, etc. For the nonlinear black-box case, it is most common to use only measured (not estimated) quantities in the regressors, i.e., NFIR and NARX.

Basic Features of Function Expansions and Basis Functions

Now let us turn to the nonlinear mapping $g(\varphi, \theta)$ which for any given θ maps $\mathbf{R}^d$ to $\mathbf{R}^p$. In the sequel, for simplicity we shall take $p = 1$, i.e. we deal only with scalar outputs. At this point it does not matter how the regression vector $\varphi = [\,\varphi_1 \quad \dots \quad \varphi_d\,]^T$ is constructed. It is just a vector that lives in $\mathbf{R}^d$.

It is natural to think of the parameterized function family as function expansions:

$$g(\varphi, \theta) = \sum_{k=1}^{n} \alpha_k g_k(\varphi), \quad \theta = [\,\alpha_1 \quad \dots \quad \alpha_n\,]^T \tag{5.32}$$

We refer to g_k as *basis functions*, since the role they play in (5.32) is similar to that of a functional space basis. We are going to show that the expansion (5.32), with different basis functions, plays the role of a unified framework for investigating most known nonlinear black-box model structures.

Now, the key question is: How to choose the basis functions g_k? An immediate choice might be to try Taylor expansion

$$g_k(\varphi) = \varphi^k \tag{5.33}$$

where for $d > 1$ we would have to interpret φ^k as the variables that cover all products of the components φ_j with summed exponents being k. Such expansions are called *Volterra expansions* in the modeling application, and have long been tried.

The expansions that have attracted most of the recent interest are however of a different kind, and the following facts are essential to understand the connections between most known nonlinear black-box model structures:

- All the g_k are formed from one "mother basis function" that we generically denote by $\kappa(x)$.
- This function $\kappa(x)$ is a function of a scalar variable x.
- Typically g_k are dilated (scaled) and translated versions of κ. For the scalar case $d = 1$ we may write

  $$g_k(\varphi) = g_k(\varphi, \beta_k, \gamma_k) = \kappa(\beta_k(\varphi - \gamma_k)) \tag{5.34}$$

  We thus use β_k to denote the dilation parameters and γ_k to denote translation parameters.

A scalar example: Fourier series. Take $\kappa(x) = \cos(x)$. Then (5.32), (5.34) will be the Fourier series expansion, with β_k as the frequencies and γ_k as the phases.

Another scalar example: Piece-wise constant functions. Take κ as the unit interval indicator function

$$\kappa(x) = \begin{cases} 1 & \text{for } 0 \leq x < 1 \\ 0 & \text{else} \end{cases} \tag{5.35}$$

and take, for example, $\gamma_k = k\Delta$, $\beta_k = 1/\Delta$ and $\alpha_k = f(k\Delta)$. Then (5.32), (5.34) gives a piece-wise constant approximation of any function f over intervals of length Δ. Clearly we would have obtained a quite similar result by a smooth version of the indicator function, e.g., the Gaussian bell:

$$\kappa(x) = \frac{1}{\sqrt{2\pi}} e^{-x^2/2} \tag{5.36}$$

A variant of the piece-wise constant case. Take κ to be the unit step function

$$\kappa(x) = \begin{cases} 0 & \text{for } x < 0 \\ 1 & \text{for } x \geq 0 \end{cases} \tag{5.37}$$

We then just have a variant of (5.35), since the indicator function can be obtained as the difference of two steps. A smooth version of the step, like the *sigmoid* function

$$\kappa(x) = \sigma(x) = \frac{1}{1 + e^{-x}} \tag{5.38}$$

will of course give quite similar results.

Classification of Single-Variable Basis Functions

Two classes of single-variable basis functions can be distinguished depending on their nature:

- *Local Basis Functions* are functions, where the significant variation takes place in local environment.
- *Global Basis Functions* are functions that have significant variation over the whole real axis.

Clearly the Fourier series and the Volterra expansions are examples of a global basis function, while (5.35)–(5.38) are all local functions.

Construction of Multi-Variable Basis Functions

In the multi-dimensional case ($d > 1$), g_k is a function of several variables. In most nonlinear black-box models, it is constructed from the single-variable function κ in some simple manner. There are three common methods for constructing multi-variable basis functions from single-variable basis functions.

1. **Tensor product.** The tensor product is obtained by taking the product of the single-variable function, applied to each component of φ. This means that the basis functions are constructed from the scalar function κ as

$$g_k(\varphi) = g_k(\varphi, \beta_k, \gamma_k) = \prod_{j=1}^{d} \kappa(\beta_k^j(\varphi_j - \gamma_k^j)) \tag{5.39}$$

2. **Radial construction.** The idea is to let the value of the function depend only on φ's distance from a given center point

$$g_k(\varphi) = g_k(\varphi, \beta_k, \gamma_k) = \kappa\big(\|\varphi - \gamma_k\|_{\beta_k}\big) \tag{5.40}$$

where $\|\cdot\|_{\beta_k}$ denotes any chosen norm on the space of the regression vector φ. The norm could typically be a quadratic norm

$$\|\varphi\|^2_{\beta_k} = \varphi^T \beta_k \varphi \tag{5.41}$$

with β_k as a positive definite matrix of dilation (scale) parameters. In simple cases β_k may be just scaled versions of the identity matrix.

3. **Ridge construction.** The idea is to let the function value depend only on φ's distance from a given hyperplane:

$$g_k(\varphi) = g_k(\varphi, \beta_k, \gamma_k) = \kappa(\beta_k^T \varphi + \gamma_k), \ \varphi \in \mathbf{R}^d \tag{5.42}$$

The ridge function is thus constant for all φ in the hyperplane $\{\varphi \in \mathbf{R}^d : \beta_k^T \varphi = \text{ constant}\}$. As a consequence, even if the mother basis function κ has local support, the basis functions g_k will have unbounded support in this subspace.

Approximation Issues

For any of the described choices the resulting model becomes

$$g(\varphi, \theta) = \sum_{k=1}^{n} \alpha_k \kappa(\beta_k(\varphi - \gamma_k)) \tag{5.43}$$

with the different interpretations of the argument $\beta_k(\varphi - \gamma_k)$ just discussed. The expansion is entirely determined by:

- The scalar valued function $\kappa(x)$ of a scalar variable x.
- The way the basis functions are expanded to depend on a vector φ.

The parameterization in terms of θ can be characterized by three types of parameters:

- The *coordinates* α
- The *scale* or *dilation* parameters β
- The *location* parameters γ

These three parameter types affect the model in quite different ways. The coordinates enter linearly, which means that (5.43) is a *linear regression* for fixed scale and location parameters.

A key issue is how well the function expansion is capable of approximating any possible "true system" $g_0(\varphi)$. There is rather extensive literature on this subject. For an identification oriented survey, see, e.g., Juditsky et.al. (1995).

The bottom line is easy: *For almost any choice of $\kappa(x)$—except being a polynomial—the expansion (5.43) can approximate any "reasonable" function $g_0(\varphi)$ arbitrarily well for sufficiently large n.*

It is not difficult to understand this. It is sufficient to check that the delta function—or the indicator function for arbitrarily small areas— can be arbitrarily well approximated within the expansion. Then clearly all reasonable functions can also be approximated. For a local κ with radial construction this is immediate: It is itself a delta function approximator. For the ridge construction this is somewhat more tricky and has been proved by Cybenko (1989). Intuitively we may think of a fixed $\gamma_k = \gamma$ and "many" hyperplanes β_k that all intersect at γ and are such that κ is nonzero only "close" to the hyperplane (norm of β is large). Then most of the support of the resulting function will be at γ.

The question of how *efficient* the expansion is, i.e., how large n is required to achieve a certain degree of approximation, is more difficult, and has no general answer. See, e.g. Barron (1993). We may point to the following aspects:

i. If the scale and location parameters β and γ are allowed to depend on the function g_0 to be approximated, then the number of terms n required for a certain degree of approximation is much less than if $\beta_k, \gamma_k, \ k = 1, \ldots$ is an *a priori* fixed sequence. To realize this, consider the following simple example:

Example 5.3 Constant Functions

Suppose we use piece-wise constant functions to approximate any scalar valued function of a scalar variable φ:

$$g_0(\varphi), \quad 0 \leq \varphi \leq B \tag{5.44}$$

$$\hat{g}_n(\varphi) = \sum_{k=1}^{n} \alpha_k \kappa(\beta_k(\varphi - \gamma_k)) \tag{5.45}$$

where κ is the unit indicator function (5.35). Let us suppose that the requirement is $[\sup_\varphi |g_0(\varphi) - \hat{g}_n(\varphi)| \leq \epsilon]$ and we know a bound on the derivative of g_0:$\sup |g_0'(\varphi)| \leq C$. For (5.45) to be able to deliver such a good approximation for *any* such function g_0 we need to take $\gamma_k = k\Delta$, $\beta_k = 1/\Delta$ with $\Delta \leq 2\epsilon/C$, i.e., we need $n \geq CB/(2\epsilon)$. That is, we need a fine grid Δ that is prepared for the worst case $|g_0'(\varphi)| = C$ at any point.

If the actual function to be approximated turns out to be much smoother, and has a large derivative only in a small interval, we can adapt the choice of $\beta_k = 1/\Delta_k$ so that $\Delta_k \approx 2\epsilon/|g_0'(\varphi^*)|$ for the interval around $\gamma_k = k\Delta_k \approx \varphi^*$ which may give the desired degree of approximation with much fewer terms in the expansion. □

ii. For the local, radial approach the number of terms required to achieve a certain degree of approximation δ of a s times differentiable function is proportional to

$$n \sim \frac{1}{\delta^{(d/s)}}, \qquad \delta \ll 1 \tag{5.46}$$

See Problem 5G.2. It thus increases exponentially with the number of regressors. This is often referred to as *the curse of dimensionality*.

Network Questions for Nonlinear Black-Box Structures (*)

So far we have viewed the model structures as *basis function expansions*, where the basis functions may contain adjustable parameters. Such structures are often referred to as *networks*, primarily since typically one "mother basic function" κ is repeated a large number of times in the expansion. Graphical illustrations of the structure therefore look like networks.

Multi-Layer Networks. The network aspect of the function expansion is even more pronounced if the basic mappings are convolved with each other in the following manner: Let the outputs of the basis functions be denoted by

$$\varphi_k^{(2)}(t) = g_k(\varphi(t)) = \kappa(\varphi(t), \beta_k, \gamma_k)$$

and collect them into a vector:

$$\varphi^{(2)}(t) = \left[\varphi_1^{(2)}(t), \ldots, \varphi_n^{(2)}(t)\right].$$

Now, instead of taking a linear combination of these $\varphi_k^{(2)}(t)$ as the output of the model (as in (5.32)), we could treat them as new regressors and insert them into another "layer" of basis functions forming a second expansion

$$g(\varphi, \theta) = \sum_l \alpha_l^{(2)} \kappa(\varphi^{(2)}, \beta_l^{(2)}, \gamma_l^{(2)}) \tag{5.47}$$

where θ denotes the whole collection of involved parameters: $\beta_k, \gamma_k, \alpha_l^{(2)}, \beta_l^{(2)}, \gamma_l^{(2)}$. Within neural network terminology, (5.47) is called a *two-hidden layer* network. The basis functions $\kappa(\varphi(t), \beta_k, \gamma_k)$ then constitute the first hidden layer, while $\kappa(\varphi^{(2)}, \beta_l^{(2)}, \gamma_l^{(2)})$ give the second one. The layers are "hidden" because they do not show up explicitly in the output $g(\varphi, \theta)$ in (5.47), but they are of course available to the user. See Figure 5.2 for an illustration. Clearly we can repeat the procedure an arbitrary number of times to produce *multi-layer* networks. This term is primarily used for sigmoid neural networks, but applies as such to any basis function expansion (5.32).

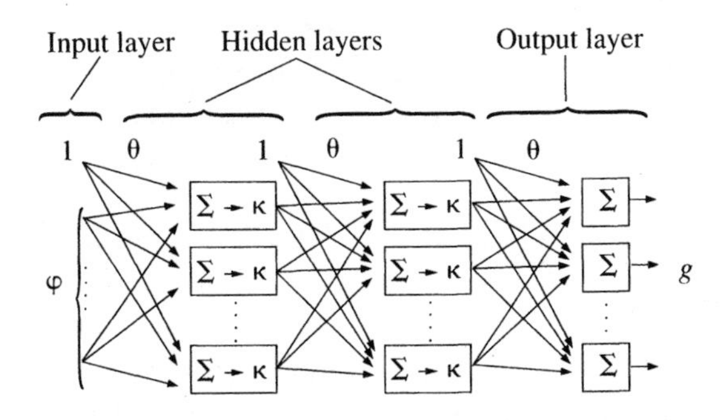

Figure 5.2 Feedforward network with two hidden layers.

The question of how many layers to use however is not easy. In principle, with many basis functions, one hidden layer is sufficient for modeling most practically reasonable systems. Sontag (1993)contains many useful and interesting insights into the importance of second hidden layers in the nonlinear structure.

Recurrent Networks. Another very important concept for applications to dynamical systems is that of *recurrent networks*. This refers to the situation that some of the regressors used at time t are outputs from the model structure at previous time instants:

$$\varphi_k(t) = g(\varphi(t-k), \theta)$$

See the illustration in Figure 5.3. It can also be the case that some component $\varphi_j(t)$ of the regressor at time t is obtained as a value from some interior node (not just at the output layer) at a previous time instant. Such model dependent regressors make the structure considerably more complex, but offer at the same time quite useful flexibility. The regression vectors (4.20) and (4.40) are examples where previous model outputs and other internal signals are used as regressors.

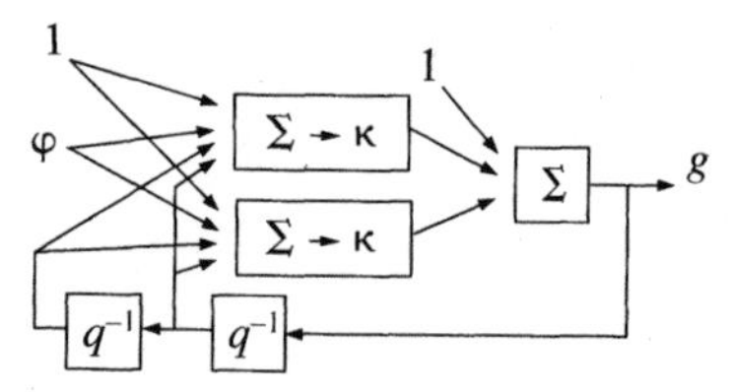

Figure 5.3 Example of a recurrent network. q^{-1} delays the signal by one time sample.

One might distinguish between *input/output based networks* and *state-space based networks*, although the difference is less distinct in the nonlinear case. The former would be using only past outputs from the network as recurrent regressors, while the latter may feed back any interior point in the network to the input layer as a recurrent regressor. Experience with state-space based networks is quite favorable, e.g., Nerrand et.al. (1993) .

Estimation Aspects

An important reason for dealing with these nonlinear black-box models in parallel with other models is that the estimation theory, the asymptotic properties and the basic algorithms are the same as for the other model structures discussed in this book. We shall return to special features in algorithms and methods that are particular to nonlinear black boxes, but most of the discussion in Chapters 7 and onwards applies also to these structures.

5.5 NONLINEAR BLACK-BOX MODELS: NEURAL NETWORKS, WAVELETS AND CLASSICAL MODELS

In this section we shall briefly review some popular model structures. They all have the general form of function expansions (5.43), and are composed of basis functions g_k obtained by parameterizing some particular "*mother basis function*" κ as described in the previous section.

Neural Networks

Neural networks have become a very popular choice of model structure in recent years. The name refers to certain structural similarities with the neural synapse system in animals. From our perspective these models correspond to certain choices in the general function expansion.

Sigmoid Neural Networks. The combination of the model expansion (5.32), with a ridge basis function (5.42) and the sigmoid choice (5.38) for mother function, gives the celebrated *one hidden layer feedforward sigmoid neural net.*

Wavelet and Radial Basis Networks. The combination of the Gaussian bell type mother function (5.36) and the radial construction (5.40) is found in both wavelet networks, Zhang and Benveniste (1992), and radial basis neural networks, Poggio and Girosi (1990).

Wavelets

Wavelet decomposition is a typical example for the use of local basis functions. Loosely speaking, the "mother basis function" (usually referred to as *mother wavelet* in the wavelet literature, and there denoted by ψ rather than κ) is dilated and translated to form a wavelet basis. In this context it is common to let the expansion (5.32) be doubly indexed according to scale and location, and use the specific choices (for one dimensional case) $\beta_j = 2^j$ and $\gamma_k = 2^{-j}k$. This gives, in our notation,

$$g_{j,k}(\varphi) = \kappa(2^j \varphi - k), \quad j, k \text{ positive integers} \tag{5.48}$$

Compared to the simple example of a piece-wise constant function approximation in (5.35), we have here *multi-resolution* capabilities, so that the intervals are multiply covered using basis functions of different resolutions (i.e. different scale parameters). With suitably chosen mother wavelet and appropriate translation and dilation parameters, the wavelet basis can be made orthonormal, which makes it easy to compute the coordinates $\alpha_{j,k}$ in (5.32).

The multi-variable wavelet functions can be constructed by tensor products of scalar wavelet functions, but other constructions are also possible. However, the burden in computation and storage of wavelet basis functions rapidly increases with the regressor dimension d, so the use of these orthonormal expansions are in practice limited to the case of few regressors ($d \lesssim 3$).

"Classical" Nonlinear Black-Box Models

Kernel Estimators. Another well known example for use of local basis functions is *Kernel estimators*, Nadaraya (1964), Watson (1969). A kernel function $\kappa(\cdot)$ is typically a bell-shaped function, and the kernel estimator has the form

$$g(\varphi) = \sum_{k=1}^{n} \alpha_k \; \kappa\left(\left\|\frac{\varphi - \gamma_k}{h}\right\|\right) \tag{5.49}$$

where h is a small positive number, γ_k are given points in the space of regression vector φ. This clearly is a special case of (5.32), (5.34), with one fixed scale parameter $h = 1/\beta$ for all the basis functions. This scale parameter is typically tuned to the problem, though. A common choice of κ in this case is the *Epanechnikov kernel*:

$$\kappa(x) = \begin{cases} 1 - x^2 & \text{for } |x| < 1 \\ 0 & \text{for } |x| \geq 1 \end{cases} \tag{5.50}$$

Nearest Neighbors or Interpolation. The nearest neighbor approach to identification has a strong intuitive appeal: When encountered with a new value φ^* we look into the past data Z^N to find that regression vector $\varphi(t) = \varphi^\dagger$ that is closest to φ^*. We then associate the regressor φ^* with the measurement $y(t)$ corresponding to $\varphi(t) = \varphi^\dagger$.

This corresponds to using (5.43) with κ as the indicator function (5.35), expanded to a hypercube by the radial approach (5.40). The location and scale parameters in (5.40) are chosen such that the cubes $\kappa(\|\varphi - \gamma_k\|_{\beta_k})$ are tightly laid such that exactly each data point $\varphi(k) \in Z^N$ falls at the center of one cube ($\gamma_k = \varphi(k)$). The corresponding coordinate α_k will be the value of $y(k)$ at this data point. The number of terms in the expansion will then equal the number of data points in Z^N. The value of $g(\varphi^*, \theta)$ will now equal $\alpha_t = y(t)$ for that t for which $\kappa(\beta_t(\varphi^* - \gamma_t)) = 1$, i.e. for that $\varphi(t)$ which is closest to φ^*. That gives the nearest neighbor approach.

B-Splines. B-splines are local basis functions which are piece-wise polynomials. The connections of the pieces of polynomials have continuous derivatives up to a certain order, depending on the degree of the polynomials, De Boor (1978), Schumaker (1981). Splines are very nice functions, since they are computationally very simple and can be made as smooth as desired. For these reasons, they have been widely used in classic interpolation problems.

5.6 FUZZY MODELS

For complex systems it may be difficult to set up precise mathematical models. This has led to so-called *fuzzy models*, which are based on verbal and imprecise descriptions on the relationships between the measured signals in a system, Zadeh (1975). The fuzzy models typically consist of so-called rule bases, but can be cast exactly into the framework of model structures of the class (5.32). In this case, the basis functions g_k are constructed from the fuzzy set membership functions and inference rules of combining fuzzy rules and how to "defuzzify" the results. When the fuzzy models contain parameters to be adjusted, they are also called *neuro-fuzzy* models, Jang and Sun (1995). In this section, we shall give a simplified description of how this works.

Fuzzy Rule Bases as Models

Fuzzy Rules. A *fuzzy rule base* is a collection of statements about some relationship between measured variables, like

- `If it is cold outdoors and the heater has low power, then the room will be cold`
- `If it is cold outdoors and the heater has high power, then the room temperature will be normal`
- `If it is hot outdoors and the heater has high power, then the room will be hot`
- ...

We can think of this as a verbal model relating the "regressors," φ_1=outdoor temperature and φ_2=heater power, to the variable to be explained, y=room temperature. The question then is, what does this tell us when we know that the outdoor temperature is $7°C$ and the heater is set to 100 W? To handle that, we need to:

1. Quantify what "cold outdoors" means, related to a particular temperature reading, and similarly for the heater power.
2. Decide which rules are applicable, and how to combine their conclusions to come up with a statement about the room temperature.

We shall now discuss each of these questions.

Fuzzy Sets and Membership Functions. Each of the regressors that is used in the rule base is associated with a number of *attributes*, like for the outdoor temperature, "cold," "nice," "hot". These attributes are seen as sets, to which the temperature can belong "to a certain degree." This is the fuzzy nature of the sets. Each attribute—each fuzzy set—is associated with a *membership function* that, for a particular value of the variable, describes the "degree of membership" to this set. If the attribute for the variable φ is denoted by A, the membership function is usually denoted by $\mu_A(\varphi)$, and will be a function assuming values between 0 and 1. Typical membership functions are built up by piece-wise linear functions like a "soft step"

$$\kappa_1(x) = \begin{cases} 1 & \text{for } x < -1 \\ 1 - (x+1)/2 & \text{for } -1 \le x < 1 \\ 0 & \text{for } 1 \le x \end{cases} \tag{5.51}$$

or a triangle

$$\kappa_2(x) = \begin{cases} 0 & \text{for } |x| > 1 \\ 1 - |x| & \text{for } |x| \le 1 \end{cases} \tag{5.52}$$

Clearly by scale and location parameters, the step and the peak can be placed anywhere and be of arbitrary width:

$$\mu_A(\varphi) = \kappa_i(\beta(\varphi - \gamma)) \tag{5.53}$$

We shall here only deal with attributes and membership functions, such that the degree of memberships to the different attributes associated with a given variable always sum up to one. That is, if φ has r attributes $A_i,\ i = 1, \dots, r$, then the membership functions should be subject to

$$\sum_{i=1}^{r} \mu_{A_i}(\varphi) = 1, \quad \forall \varphi \tag{5.54}$$

Such a collection of sets and membership functions is called *a strong fuzzy partition.*

For the outdoor temperature, e.g., we could associate the attribute "cold" with the membership function $\mu_{\text{cold}}(\varphi) = \kappa_1(0.2(\varphi - 10))$, that is, any temperature below $5°C$ is definitely "cold," and any temperature above $15°C$ is definitely "not cold," while temperatures in between are cold to varying degrees. Similarly, we could have $\mu_{\text{nice}}(\varphi) = \kappa_2(0.1(\varphi - 15))$ and $\mu_{\text{hot}}(\varphi) = \kappa_1(-0.2(\varphi - 20))$. This is illustrated in Figure 5.4. It is clearly a strong fuzzy partition.

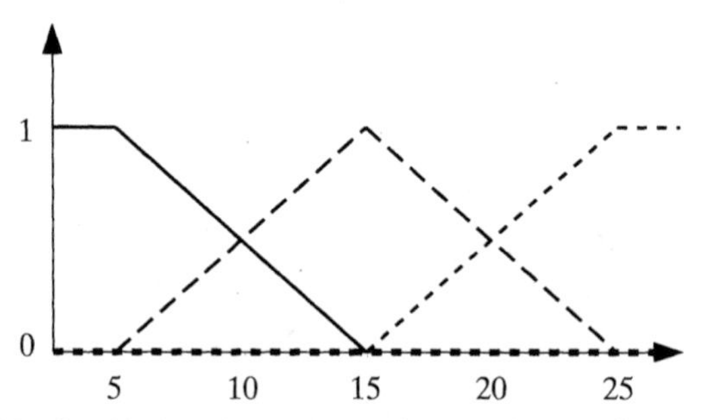

Figure 5.4 Membership functions for **cold** (solid line), **nice** (dashed line), and **hot** (dotted line).

Combining Rules. We can write down a fuzzy rule base, relating a regression vector φ to an output variable y, more formally as

$$\begin{array}{c} \text{if } (\varphi_1 \text{ is } A_{1,1}) \dots \text{ and } (\varphi_d \text{ is } A_{1,d}) \text{ then } (y \text{ is } B_1) \\ \dots\dots\dots\dots \\ \text{if } (\varphi_1 \text{ is } A_{p,1}) \dots \text{ and } (\varphi_d \text{ is } A_{p,d}) \text{ then } (y \text{ is } B_p) \end{array} \tag{5.55}$$

where the fuzzy sets $A_{j,i}$ are doubly indexed: i is the index of the regressor variable (measurement), and j is the index of the rule. We denote the membership functions by $\mu_{A_{j,i}}(\varphi_i)$ and $\mu_{B_j}(y)$, respectively.

We shall assume that the rule base is such that there is a rule that covers each possible combination of attributes for the different regressors. See Example 5.4 below. This means in particular that the number of rules, p, must be equal to the product of the number of attributes associated with each regressor. Such a rule base is called *complete*. This means in particular that

$$\sum_{j=1}^{p}\prod_{i=1}^{d}\mu_{A_{j,i}}(\varphi_i) \equiv 1. \tag{5.56}$$

(see Problem 5E.3.) The joint membership function for all the conditions of rule number j could rather naturally be taken as the product of the individual ones:

$$\mu_{A_{j,*}}(\varphi) = \prod_{i=1}^{d}\mu_{A_{j,i}}(\varphi_i) \tag{5.57}$$

This number could be seen as a "measure of the applicability" of rule number j for the given value of φ. With this rule we could associate a numerical value of y associated with the conclusion B_j of rule j. This number—let it be denoted by α_j—could be the center of mass of the membership function μ_{B_j} or the value for which the membership function has its peak. In any case it is natural to associate the value

$$y = \sum_{j=1}^{p}\alpha_j\mu_{A_{j,*}}(\varphi) \tag{5.58}$$

with the regressors φ. In view of (5.56) and (5.57), this is a weighted mean of the "average conclusion" of each of the rules, weighted by the applicability of the rule.

Example 5.4 A DC-Motor

Consider an electric motor with input voltage u and output angular velocity y. We would like to explain how the angular velocity at time t, i.e. $y(t)$, depends on the applied voltage $u(t-1)$ and the velocity at the previous time sample. That is, we are using the regressors $\varphi(t) = [\varphi_1(t), \varphi_2(t)]^T$, where $\varphi_1(t) = u(t-1)$ and $\varphi_2(t) = y(t-1)$. Let us now device a rule base of the kind (5.55), where we choose $A_{1,1}$ and $A_{2,1}$ to be "low voltage," and $A_{3,1}$ and $A_{4,1}$ to be "high voltage." We choose $A_{1,2}$ and $A_{3,2}$ to be "slow speed," while $A_{2,2}$ and $A_{4,2}$ are "fast speed." The membership function for "low voltage" is taken as $\mu_{A_{1,1}}(\varphi_1) = \kappa_1(0.5(\varphi_1 - 3))$, with κ_1 defined by (5.51). The membership function for "high voltage" is taken as $\mu_{A_{3,1}} = 1 - \mu_{A_{1,1}} = \kappa_1(-0.5(\varphi_1 - 3))$. The membership functions for slow and fast speed are chosen analogously, with breaking points 8 and 15 rad/sec. See Figure 5.5. The statements B_i about the output are chosen to be "slow," "medium," and "fast," with membership functions that are triangular shaped like κ_2 in (5.52) with peaks at 5, 10, and 20 rad/sec. We thus obtain a rule base:

If $\varphi_1(t)$ **is low and** $\varphi_2(t)$ **is slow then** $y(t)$ **is slow.**
If $\varphi_1(t)$ **is low and** $\varphi_2(t)$ **is fast then** $y(t)$ **is medium.**
If $\varphi_1(t)$ **is high and** $\varphi_2(t)$ **is slow then** $y(t)$ **is medium.**
If $\varphi_1(t)$ **is high and** $\varphi_2(t)$ **is fast then** $y(t)$ **is fast.**

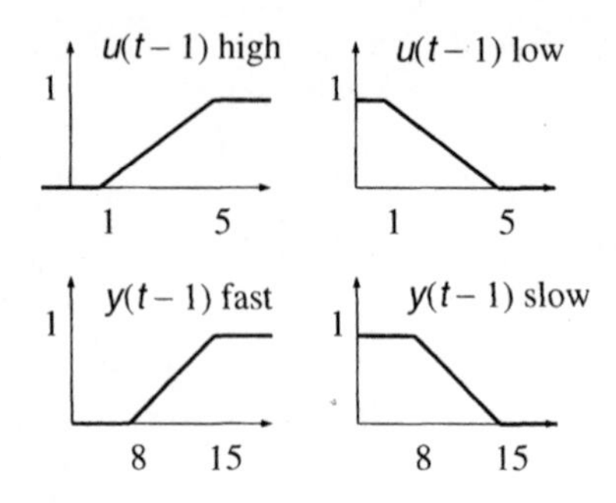

Figure 5.5 Membership functions for the DC motor.

Let us see what prediction of $y(t)$ the rule base gives, in case $\varphi(t) = [4, 17]^T$. This means that the voltage is "low" with a membership degree of 0.25, while it is "high" with a degree of 0.75. The past velocity $\varphi_2(t) = y(t-1)$ is "fast" to a degree of 1 and "slow" to a degree of 0. The four rules in the rule base then are weighted together according to (5.58) as

$$y(t) = 5 \cdot (0.25 \cdot 0) + 10 \cdot (0.25 \cdot 1) + 10 \cdot (0.75 \cdot 0) + 20 \cdot (0.75 \cdot 1) = 17.5$$

□

Back to the General Black-Box Formulation

Now, if some or all of the rules in the rule base need "tuning," we may introduce parameters to be tuned. The parameters could, e.g., be some or all of the (defuzzified) "conclusions" α_k of the rules. They could also be parameters associated with the membership functions $\mu_{A_{j,i}}(\varphi_i)$ for the regressors, typically scale and location parameters as in (5.53).

This means that (5.58) takes the form

$$y = g(\varphi, \theta) = \sum_{k=1}^{n} \alpha_k g_k(\varphi, \beta, \gamma) \tag{5.59}$$

where the "basis functions" g_k are obtained from the parameterized membership functions as

$$g_k(\varphi, \beta, \gamma) = \prod_{j=1}^{d} \kappa_j(\beta_k^j(\varphi_j - \gamma_k^j)) \tag{5.60}$$

We are thus back to the basic situation of (5.32) and (5.34), where the expansion into the d-dimensional regressor space is obtained by the tensor product construction (5.39).

Normally, not all of the parameters α_k, β_k and γ_k should be freely adjustable. For example, the requirement (5.56) imposes certain constraints on the scale and location parameters. If the fuzzy partition is fixed and not adjustable, i.e., β and γ fixed, then we get a particular case of the kernel estimate (5.49), which is also a linear regression model.

Thus *fuzzy models are just particular instances of the general model structure* (5.32).

One of the major potential advantages is that the fuzzy rule basis may give functions g_j that reflect (verbal) physical insight about the system. This may be useful also to come up with reasonable initial values of the dilation and location parameters to be estimated. One should realize, though, that the knowledge encoded in a fuzzy rule base may be nullified if many parameters are let loose. For example, in the DC-motor rule base, if all the numerical values for the motor speed (5, 10, 10, and 20) are replaced by free parameters, which are then estimated, our knowledge about what causes high speed has not been utilized by the resulting model structure.

5.7 FORMAL CHARACTERIZATION OF MODELS (*)

In this section we shall give a counterpart of the discussion of Section 4.5 for general, possibly time-varying, and nonlinear models. We assume that the output is p-dimensional and that the input is m-dimensional. Z^t denotes, as before, the input-output data up to and including time t.

Models

A model $\mathcal{m}$ *of a dynamical system* is a sequence of functions $g_{\mathcal{m}}(t, Z^{t-1})$, $t = 1, 2, \ldots$, from $\mathbf{R} \times \mathbf{R}^{p(t-1)} \times \mathbf{R}^{m(t-1)}$ to $\mathbf{R}^p$, representing a way of guessing or predicting the output $y(t)$ from past data:

$$\hat{y}(t|t-1) = g_{\mathcal{m}}(t, Z^{t-1}) \tag{5.61}$$

A model that defines only the predictor function is called a *predictor model.* When (5.61) is complemented with the conditional (given Z^{t-1}) probability density function (CPDF) of the associated prediction errors

$$f_e(x, t, Z^{t-1}) : \text{ CPDF of } y(t) - \hat{y}(t|t-1), \qquad \text{given } Z^{t-1} \tag{5.62}$$

we call the model a *complete probabilistic model.* A typical model assumption is that the prediction errors are independent. Then f_e does not depend on Z^{t-1}:

$$f_e(x, t) : \text{ PDF of } y(t) - \hat{y}(t|t-1), \text{ these errors independent} \tag{5.63}$$

Sometimes one may prefer not to specify the complete PDF but only its second moment (the covariance matrix):

$$\begin{aligned} \Lambda_{\mathcal{m}}(t) : \; & \text{covariance matrix of } y(t) - \hat{y}(t|t-1) \\ & \text{these errors independent} \end{aligned} \tag{5.64}$$

A model (5.61) together with (5.64) could be called a *partial probabilistic model.*

A model can further be classified according to the following properties.

1. The model m is said to be *linear* if $g_m(t, Z^{t-1})$ is linear in y^{t-1} and u^{t-1}:
$$g_m(t, Z^{t-1}) = W_t^y(q)y(t) + W_t^u(q)u(t) \tag{5.65}$$
2. A model m is said to be *time invariant* if $g_m(t, Z^{t-1})$ is invariant under a shift of absolute time. If Λ_m or f_e is specified, it is further required that they be independent of t.
3. A model m is said to be a *k-step-ahead predictor* if $g_m(t, Z^{t-1})$ is a function of y^{t-k}, u^{t-1} only.
4. A model m is said to be a *simulation model* or an *output error model* if $g_m(t, Z^{t-1})$ is a function of u^{t-1} only.

Analogously to the linear case, we could define the stability of the predictor function and equality between different models [see (4.116)]. We refrain, however, from elaborating on these points here.

Model Sets and Model Structures

Sets of models $\mathcal{M}^*$ as well as model structures $\mathcal{M}$ as differentiable mappings

$$\mathcal{M} : \theta \rightarrow g(t, Z^{t-1}; \theta) \in \mathcal{M}^*; \theta \in D_{\mathcal{M}} \subset \mathbf{R}^d \tag{5.66}$$

[and $\Lambda(t; \theta)$ or $f_e(x, t; \theta)$ if applicable] from subsets of $\mathbf{R}^d$ to model sets can be defined analogously to Definition 4.3. Once equality between models has been defined, identifiability concepts can be developed as in Section 4.5.

We shall say that a model structure $\mathcal{M}$ is a *linear regression* if $D_{\mathcal{M}} = \mathbf{R}^d$ and the predictor function is a linear (or affine) function of θ:

$$g(t, Z^{t-1}; \theta) = \varphi^T(t, Z^{t-1})\theta + \mu(t, Z^{t-1}) \tag{5.67}$$

Another View of Models (*)

The definition of models as predictors is a rather pragmatic way of approaching the model concept. A more abstract line of thought can be developed as follows.

As users, we communicate with the system only through the input-output data sequences $Z^t = (y^t, u^t)$. Therefore, any assumption about the properties of the system will be an assumption about Z^t. We could thus say that

$$\text{A model of a system is an assumed relationship for } Z^t, t = 1, 2, \ldots. \tag{5.68}$$

Often, experiments on a system are not exactly reproducible. For a given input sequence u^N, we may obtain different output sequences y^N at different experiments due to the presence of various disturbances. In such cases it is natural to regard y^t as a random variable of which we observe different realizations. A model of the system would then be a description of the probabilistic properties of Z^t (or, perhaps, of y^t, given u^t). This model m could be formulated in terms of a probability measure P_m or the probability density function (PDF) for Z^t:

$$\overline{f}_m(t, Z^t) \tag{5.69}$$

That is,

$$P_m(Z^t \in B) = \int_{x^t \in B} \overline{f}_m(t, x^t)dx^t \tag{5.70}$$

Sometimes it is preferable to consider the input u^t as a given deterministic sequence and focus attention on the conditional PDF of y^t, given u^t:

$$\overline{f}_m(t, y^t|u^t) \tag{5.71}$$

A model (5.69) or (5.71) would normally be quite awkward to construct and work with, and other, indirect ways of forming $\overline{f}_m$ will be preferred. Indeed, the stochastic models of Sections 4.2 and 4.3 are implicit descriptions of the probability density function for the measured signals. The introduction of unmeasurable, stochastic disturbances $\{w(t)\}$, $\{e(t)\}$, and so on, is a convenient way of describing the probabilistic properties of the observed signal, and also often corresponds to an intuitive feeling for how the output is generated. It is, however, worth noting that the effect of these unmeasurable disturbances in the model is just to define the PDF for the observed signals.

The assumed PDF $\overline{f}_m$ in (5.69) is in a sense the most general model that can be applied for an observed data record y^t, u^t. It includes deterministic models as a special case. It also corresponds to a general statistical problem: how to describe the properties of an observed data vector. For our current purposes, it is, however, not a suitably structured model. The natural direction of time flow in the data record, as well as the notions of causality, is not present in (5.69).

Given $\overline{f}_m(t, Z^t)$ in (5.69), it is, at least conceptually, possible to compute the conditional mean of $y(t)$ given y^{t-1}, u^{t-1}; that is,

$$\hat{y}(t|t-1) = E_m\left[y(t)|y^{t-1}, u^{t-1}\right] = g_m(t, Z^{t-1}) \tag{5.72}$$

and the distribution of $y(t) - g_m(t, Z^{t-1})$, say $f_e(x, t, Z^{t-1})$. From (5.69) we can thus compute a model (5.61) along with a CPDF f_e in (5.62). Conversely, given the predictor function $g_m(t, Z^{t-1})$ and an assumed PDF $f_e(x, t)$ for the associated prediction errors, we can calculate the joint PDF for the data y^t, u^t as in (5.69). This follows from the following lemma:

Lemma 5.1. Suppose that u^t is a given, deterministic sequence, and assume that the generation of y^t is described by the model

$$y(t) = g_m(t, Z^{t-1}) + \varepsilon_m(t) \tag{5.73}$$

where the conditional PDF of $\varepsilon_m(t)$ (given y^{t-1}, u^{t-1}) is $f_e(x, t)$. Then the joint probability density function for y^t, given u^t, is

$$\overline{f}_m(t, y^t|u^t) = \prod_{k=1}^{t} f_e\big(y(k) - g_m(k, Z^{k-1}), k\big) \tag{5.74}$$

Here we have, for convenience, denoted the dummy variable x_k for the distribution of $y(k)$ by $y(k)$ itself.

Proof. The output $y(t)$ is generated by (5.73). Hence the CPDF of $y(t)$, given Z^{t-1}, is

$$p(x_t|Z^{t-1}) = f_e(x_t - g_m(t, Z^{t-1}), t) \tag{5.75}$$

Using Bayes's rule (I.10), the joint CPDF of $y(t)$ and $y(t-1)$, given Z^{t-2} can be expressed as

$$\begin{aligned} p\left(x_t, x_{t-1}|Z^{t-2}\right) &= p\left(x_t|y(t-1) = x_{t-1}, Z^{t-2}\right) \cdot p\left(x_{t-1}|Z^{t-2}\right) \\ &= f_e\left(x_t - g_m(t, Z^{t-1}), t\right) \cdot f_e\left(x_{t-1} - g_m(t-1, Z^{t-2}), t-1\right) \end{aligned}$$

where $y(t-1)$ in $g_m(t, Z^{t-1})$ should be replaced by x_{t-1}. Here we have assumed u^t to be a given deterministic sequence. Iterating the preceding expression to $t = 1$ gives the joint probability density function of $y(t), y(t-1), \ldots, y(1)$, given u^t, that is, the function $\overline{f}_m$ in (5.74). □

The important conclusion from this discussion is that the predictor model (5.61), complemented with an assumed PDF for the associated prediction errors, is *no more and no less general* than the general, unstructured joint PDF model (5.69).

Remark. Notice the slight difference, though, in the conditional PDF for the prediction errors. The general form (5.69) may in general lead to a conditional PDF that in fact depends on Z^{t-1}; $f_e(x, t, Z^{t-1})$ as in (5.62). This means that the prediction errors are not necessarily independent, while they do form a martingale difference sequence:

$$E\left[\varepsilon_m(t)|\varepsilon_m(t-1), \ldots, \varepsilon_m(1)\right] = 0 \tag{5.76}$$

In the predictor formulation (5.73), we assumed the CPDF $f_e(x, t)$ not to depend on Z^{t-1}, which is an implied assumption of independence of $\varepsilon_m(t)$ on previous data. Clearly, though, we could have relaxed that assumption with obvious modifications in (5.74) as a result.

5.8 SUMMARY

The development of models for nonlinear systems is much like that of linear systems. The basic difference from a formal point of view is that the predictor becomes a nonlinear function of past observations. The important difference from a practical point of view is that the potential richness of possibilities makes unstructured black-box-type models much more demanding than in the linear case. It is much more important that knowledge about the character of the nonlinearities is built into the model structures. This can be done in different ways. An ambitious physical modeling attempt may lead to a well structured state-space model with unknown physical parameters as in (5.26). More leisurely "semi-physical" modeling may lead to valuable insights into how the regressors in (5.13) should be formed. Note that the physical insight need not be in analytic form. The nonlinearities could very well be defined in look-up tables, and the model parameters could be entries in these tables.

When physical insight is lacking, it remains to resort to black-box structures, as described in Sections 5.4 to 5.6. This contains approaches such as artificial neural networks, kernel methods, and fuzzy models.

We have also given a short summary, in Section 5.7, of formal aspects of dynamical systems. We have stressed that a model in the first place is a predictor function from past observations to the future output. The predictor function may possibly be complemented with a model assumption of properties of the associated prediction errors, such as its variance or its PDF.

5.9 BIBLIOGRAPHY

Models for identification of nonlinear systems are discussed in the surveys by Mehra (1979), Billings (1980), and Haber and Unbehauen (1990). Leontaritis and Billings (1985) give a thorough treatment of various parametric models, such as NARMAX (nonlinear ARMAX; cf. Problem 5G.1). The Hammerstein model was apparently first discussed in Narendra and Gallman (1966). Identification of the Wiener model is described, e.g., in Wigren (1993)and Kalafatis, Wang, and Cluett (1997). A general treatment of Wiener and Volterra models is given in Rugh (1981), and the direct estimation of the kernels is discussed in Korenberg (1991). Tools for semiphysical modelling are further discussed in Lindskog and Ljung (1995).

The nonlinear simulation model, (5.25) and (5.26), is frequently used in application areas where considerable prior information is available. See, for example, Grübel (1985)for an application to aircraft dynamics.

Applications of various parametric estimation techniques to nonlinear parametric structures are discussed in, for example, Billings and Voon (1984), Gabr and Subba Roa (1984), and Stoica and Söderström (1982b). Recursive techniques are treated by Fnaiech and Ljung (1986).

Our treatment of nonlinear black box models is based on Sjöberg et.al. (1995). There is a very extensive literature on this topic. See, e.g., the books Brown and Harris (1994), Haykin (1994), and Kung (1993). Identification with fuzzy models is treated in, e.g., Takagi and Sugeno (1985), Jang and Sun (1995), and Hellendorn and Driankov (1997). A related approach with so-called "operating regime based identification" is treated in Johansen and Foss (1995). See also Skeppstedt, Ljung, and Millnert (1992).

A general discussion of the model concept is given by Willems (1985).

5.10 PROBLEMS

5G.1 Consider the following nonlinear structure:

$$x(t) = f\,(x(t-1), \ldots, x(t-n), u(t-1), \ldots, u(t-n); \theta) \tag{5.77a}$$

$$y(t) = x(t) + v(t) \tag{5.77b}$$

$$v(t) = H(q, \theta)e(t) \tag{5.77c}$$

Here (5.77a) describes the nonlinear noise-free dynamics, parametrized by θ, while (5.77b) describes the measurements as the noise-free output, corrupted by the noise $\{v(t)\}$, which is modeled in the general way (2.19). Show that the natural predictor for (5.77) is given by

$$\hat{y}(t|\theta) = \left[1 - H^{-1}(q,\theta)\right] y(t) + H^{-1}(q,\theta)x(t,\theta)$$

where $x(t,\theta)$ is defined by

$$x(t,\theta) = f\,(x(t-1,\theta),\ldots,x(t-n,\theta),u(t-1),\ldots,u(t-n);\theta)$$

5G.2 To investigate how the smoothness of a function $f(x)$ and the number of arguments, $\dim x$, affect how many basis functions are required to approximate it, reason as follows: If $f(x)$ is p times differentiable with a bounded pth derivative, we can expand it in Taylor series

$$f(x) = f(x_0) + (x - x_0)f'(x_0) + \ldots + (x - x_0)^p f^{(p)}(\xi), \quad \left|f^{(p)}(\xi)\right| \le C$$

where the kth derivative $f^{(k)}$ is a tensor with d^k elements, and the multiplication with $(x - x_0)$ has to be interpreted accordingly. Suppose that we seek an approximation of f over the unit sphere $|x| \le 1$ with an error less than δ. This is to be accomplished by local approximations around centerpoints x_i of the radial basis type. Show that the necessary number of such center points is given by (5.46), and that the number of parameters associated with each centerpoint is $\sim d^p$.

5E.1 Consider the bilinear model structure described by

$$x(t) + a_1x(t-1) + a_2x(t-2) = b_1u(t-1) + b_2u(t-2) + c_1x(t-1)u(t-1)$$
$$y(t) = x(t) + v(t)$$

where

$$\theta = \begin{bmatrix} a_1 & a_2 & b_1 & b_2 & c_1 \end{bmatrix}^T$$

(a) Assume $\{v(t)\}$ to be white noise and compute the predictor $\hat{y}(t|\theta)$ and give an expression for it in the pseudolinear regression form

$$\hat{y}(t|\theta) = \varphi^T(t,\theta)\theta$$

with a suitable vector $\varphi(t,\theta)$.

(b) Now suppose that $\{v(t)\}$ is not white, but can be modeled as an (unknown) first-order ARMA process. Then suggest a suitable predictor for the system.

5E.2 Consider the system in Figure 5.1 (upper plot), where the nonlinearity is saturation with an unknown gain:

$$f\,(u(t)) = \begin{cases} \theta_1 \cdot \theta_2, & \text{if } u(t) > \theta_2 \\ \theta_1 u(t), & \text{if } |u(t)| \le \theta_2 \\ -\theta_1 \cdot \theta_2, & \text{if } u(t) < -\theta_2 \end{cases}$$

Suppose that the linear system can be described by a second-order ARX model. Write down, explicitly, the predictor for this model, parametrized in θ_1, θ_2 and the ARX parameters.

5E.3 Show (5.56) by induction as follows: Suppose first that there is just one regressor, associated with k attributes. Then there must be k rules in the rule base, for it to be complete, covering all the attributes. Thus (5.56) follows from the assumption (5.54). Now suppose that (5.56) holds for d regressors and that there are K rules. Here $K = k_1 \cdot k_2 \cdots k_d$, with k_j as the number of attributes for regressor j. Now, add another regressor φ_{d+1} with k_{d+1} attributes, subject to (5.54). For the rule base to remain complete, it must now be complemented with $K \cdot k_{d+1}$ new rules, covering the combinations of the previous cases with each attribute of the new regressor. Show the induction step, that (5.56) holds also when φ_{d+1} has been added.

5T.1 Time-continuous bilinear system descriptions are common in many fields (see Mohler, 1973). A model can be written

$$\dot{x}(t) = A(\theta)x(t) + B(\theta)u(t) + G(\theta)x(t)u(t) + w(t) \tag{5.78a}$$

where $x(t)$ is the state vector, $w(t)$ is white Gaussian noise with variance matrix R_1, and $u(t)$ is a scalar input. The output of the system is sampled as

$$y(t) = C(\theta)x(t) + e(t), \qquad \text{for } t = kT \tag{5.78b}$$

where $e(t)$ is white Gaussian measurement noise with variance R_2. The input is piecewise constant:

$$u(t) = u_k, \qquad kT \leq t < (k+1)T$$

Derive an expression for the prediction of $y\,((k+1)T)$, given u_r and $y(rT)$ for $r \leq k$, based on the model (5.78).

5T.2 Consider the Monod growth model structure

$$\dot{x}_1 = \frac{\theta_1 \cdot x_2}{\theta_2 + x_2} \cdot x_1 - \alpha_1 x_1$$

$$\dot{x}_2 = \frac{-1}{\theta_3} \cdot \frac{\theta_1 \cdot x_2}{\theta_2 + x_2} \cdot x_1 - \alpha_1(x_2 - \alpha_2)$$

$y = [x_1 \quad x_2]^T$ is measured and α_1 and α_2 are known constants. Discuss whether the parameters θ_1, θ_2 and θ_3 are identifiable.

Remark: Although we did not give any formal definition of identifiability for nonlinear model structures, they are quite analogous to the definitions in Sections 4.5 and 4.6. Thus, test whether two different parameter values can give the same input-output behavior of the model.

[See Holmberg and Ranta (1982). x_1 here is the concentration of the biomass that is growing, while x_2 is the concentration of the growth limiting substrate. θ_1 is the maximum growth rate, θ_2 is the Michaelis Menten constant, and θ_3 is the yield coefficient.]

II methods

6

NONPARAMETRIC TIME- AND FREQUENCY-DOMAIN METHODS

A linear time-invariant model can be described by its transfer functions or by the corresponding impulse responses, as we found in Chapter 4. In this chapter we shall discuss methods that aim at determining these functions by direct techniques without first selecting a confined set of possible models. Such methods are often also called *nonparametric* since they do not (explicitly) employ a finite-dimensional parameter vector in the search for a best description. We shall discuss the determination of the transfer function $G(q)$ from input to output. Section 6.1 deals with time-domain methods for this, and Sections 6.2 to 6.4 describe frequency-domain techniques of various degrees of sophistication. The determination of $H(q)$ or the disturbance spectrum is discussed in Section 6.5.

It should be noted that throughout this chapter we assume the system to operate in open loop [i.e., $\{u(t)\}$ and $\{v(t)\}$ are independent]. Closed-loop configurations will typically lead to problems for nonparametric methods, as outlined in some of the problems. These issues are discussed in more detail in Chapter 13.

6.1 TRANSIENT-RESPONSE ANALYSIS AND CORRELATION ANALYSIS

Impulse-Response Analysis

If a system that is described by (2.8)

$$y(t) = G_0(q)u(t) + v(t) \tag{6.1}$$

is subjected to a pulse input

$$u(t) = \begin{cases} \alpha, & t = 0 \\ 0, & t \neq 0 \end{cases} \tag{6.2}$$

then the output will be

$$y(t) = \alpha g_0(t) + v(t) \tag{6.3}$$

by definition of G_0 and the impulse response $\{g_0(t)\}$. If the noise level is low, it is thus possible to determine the impulse-response coefficients $\{g_0(t)\}$ from an experiment with a pulse input. The estimates will be

$$\hat{g}(t) = \frac{y(t)}{\alpha} \tag{6.4}$$

and the errors $v(t)/\alpha$. This simple idea is *impulse-response analysis*. Its basic weakness is that many physical processes do not allow pulse inputs of such an amplitude that the error $v(t)/\alpha$ is insignificant compared to the impulse-response coefficients. Moreover, such an input could make the system exhibit nonlinear effects that would disturb the linearized behavior we have set out to model.

Step-Response Analysis

Similarly, a step

$$u(t) = \begin{cases} \alpha, & t \geq 0 \\ 0, & t < 0 \end{cases}$$

applied to (6.1) gives the output

$$y(t) = \alpha \sum_{k=1}^{t} g_0(k) + v(t) \tag{6.5}$$

From this, estimates of $g_0(k)$ could be obtained as

$$\hat{g}(t) = \frac{y(t) - y(t-1)}{\alpha} \tag{6.6}$$

which has an error $[v(t) - v(t-1)]/\alpha$. If we really aim at determining the impulse-response coefficients using (6.6), we would suffer from large errors in most practical applications. However, if the goal is to determine some basic control-related characteristics, such as delay time, static gain, and dominating time constants [i.e., the model (4.50)], step responses (6.5) can very well furnish that information to a sufficient degree of accuracy. In fact, well-known rules for tuning simple regulators such as the Ziegler-Nichols rule (Ziegler and Nichols, 1942) are based on model information reached in step responses.

Based on plots of the step response, some characteristic numbers can be graphically constructed, which in turn can be used to determine parameters in a model of given order. We refer to Rake (1980)for a discussion of such characteristics.

Correlation Analysis

Consider the model description (6.1):

$$y(t) = \sum_{k=1}^{\infty} g_0(k)u(t-k) + v(t) \tag{6.7}$$

If the input is a quasi-stationary sequence [see (2.59)] with

$$\overline{E}u(t)u(t-\tau) = R_u(\tau)$$

and

$$\overline{E}u(t)v(t-\tau) \equiv 0 \qquad \text{(open-loop operation)}$$

then according to Theorem 2.2 (expressed in the time domain)

$$\overline{E}y(t)u(t-\tau) = R_{yu}(\tau) = \sum_{k=1}^{\infty} g_0(k)R_u(k-\tau) \tag{6.8}$$

If the input is chosen as white noise so that

$$R_u(\tau) = \alpha\delta_{\tau 0}$$

then

$$g_0(\tau) = \frac{R_{yu}(\tau)}{\alpha}$$

An estimate of the impulse response is thus obtained from an estimate of $R_{yu}(\tau)$; for example,

$$\hat{R}^N_{yu}(\tau) = \frac{1}{N}\sum_{t=\tau}^{N} y(t)u(t-\tau) \tag{6.9}$$

If the input is not white noise, we may estimate

$$\hat{R}^N_{u}(\tau) = \frac{1}{N}\sum_{t=\tau}^{N} u(t)u(t-\tau) \tag{6.10}$$

and solve

$$\hat{R}^N_{yu}(\tau) = \sum_{k=1}^{M} \hat{g}(k)\hat{R}^N_u(k-\tau) \tag{6.11}$$

for $\hat{g}(k)$. If the input is open for manipulation, it is of course desirable to choose it so that (6.10) and (6.11) become easy to solve. Equipment for generating such signals and solving for $\hat{g}(k)$ is commercially available. See Godfrey (1980)for a more detailed treatment.

In fact, the most natural way to estimate $g(k)$ when the input is not "exactly white" is to truncate (6.7) at n, and treat it as an n:th order FIR model (4.46) with the parametric (least-squares) methods of Chapter 7. Another way is to filter both inputs and outputs by a prefilter that makes the input as white as possible ("input prewhitening") and then compute the correlation function (6.9) for these filtered sequences.

6.2 FREQUENCY-RESPONSE ANALYSIS

Sine-wave Testing

The fundamental physical interpretation of the transfer function $G(z)$ is that the complex number $G(e^{i\omega})$ bears information about what happens to an input sinusoid [see (2.32) to (2.34)]. We thus have for (6.1) that with

$$u(t) = \alpha\cos\omega t, \qquad t = 0, 1, 2, \ldots \tag{6.12}$$

then

$$y(t) = \alpha |G_0(e^{i\omega})| \cos(\omega t + \varphi) + v(t) + \text{transient} \tag{6.13}$$

where

$$\varphi = \arg G_0(e^{i\omega}) \tag{6.14}$$

This property also gives a clue to a simple way of determining $G_0(e^{i\omega})$:

> With the input (6.12), determine the amplitude and the phase shift of the resulting output cosine signal, and calculate an estimate $\hat{G}_N(e^{i\omega})$ based on that information. Repeat for a number of frequencies in the interesting frequency band.

This is known as *frequency analysis* and is a simple method for obtaining detailed information about a linear system.

Frequency Analysis by the Correlation Method

With the noise component $v(t)$ present in (6.13), it may be cumbersome to determine $|G_0(e^{i\omega})|$ and φ accurately by graphical methods. Since the interesting component of $y(t)$ is a cosine function of known frequency, it is possible to correlate it out from the noise in the following way. Form the sums

$$I_c(N) = \frac{1}{N}\sum_{t=1}^{N} y(t)\cos\omega t, \qquad I_s(N) = \frac{1}{N}\sum_{t=1}^{N} y(t)\sin\omega t \tag{6.15}$$

Inserting (6.13) into (6.15), ignoring the transient term, gives

$$\begin{aligned} I_c(N) &= \frac{1}{N}\sum_{t=1}^{N} \alpha |G_0(e^{i\omega})| \cos(\omega t + \varphi)\cos\omega t + \frac{1}{N}\sum_{t=1}^{N} v(t)\cos\omega t \\ &= \alpha |G_0(e^{i\omega})| \frac{1}{2}\frac{1}{N}\sum_{t=1}^{N} [\cos\varphi + \cos(2\omega t + \varphi)] \\ &\quad + \frac{1}{N}\sum_{t=1}^{N} v(t)\cos\omega t \qquad (6.16) \\ &= \frac{\alpha}{2}|G_0(e^{i\omega})|\cos\varphi + \alpha |G_0(e^{i\omega})| \frac{1}{2}\frac{1}{N}\sum_{t=1}^{N}\cos(2\omega t + \varphi) \\ &\quad + \frac{1}{N}\sum_{t=1}^{N} v(t)\cos\omega t \end{aligned}$$

The second term tends to zero as N tends to infinity, and so does the third term if $v(t)$ does not contain a pure periodic component of frequency ω. If $\{v(t)\}$ is a stationary stochastic process such that

$$\sum_{0}^{\infty} \tau |R_v(\tau)| < \infty$$

then the variance of the third term of (6.16) decays like $1/N$ (Problem 6T.2). Similarly,

$$I_s(N) = -\frac{\alpha}{2}|G_0(e^{i\omega})| \sin\varphi + \alpha|G_0(e^{i\omega})|\frac{1}{2}\frac{1}{N}\sum_{t=1}^{N}\sin(2\omega t + \varphi)$$
$$+ \frac{1}{N}\sum_{t=1}^{N} v(t) \sin \omega t \tag{6.17}$$

These two expressions suggest the following estimates of $|G_0(e^{i\omega})|$ and φ:

$$|\hat{G}_N(e^{i\omega})| = \frac{\sqrt{I_c^2(N) + I_s^2(N)}}{\alpha/2} \tag{6.18a}$$

$$\hat{\varphi}_N = \arg \hat{G}_N(e^{i\omega}) = -\arctan\frac{I_s(N)}{I_c(N)} \tag{6.18b}$$

Rake (1980)gives a more detailed account of this method. By repeating the procedure for a number of frequencies, a good picture of $G_0(e^{i\omega})$ over the frequency domain of interest can be obtained. Equipment that performs such *frequency analysis by the correlation method* is commercially available.

An advantage with this method is that a Bode plot of the system can be obtained easily and that one may concentrate the effort to the interesting frequency ranges. The main disadvantage is that many industrial processes do not admit sinusodial inputs in normal operation. The experiment must also be repeated for a number of frequencies which may lead to long experimentation periods.

Relationship to Fourier Analysis

Comparing (6.15) to the definition (2.37),

$$Y_N(\omega) = \frac{1}{\sqrt{N}}\sum_{t=1}^{N} y(t)e^{-i\omega t} \tag{6.19}$$

shows that

$$I_c(N) - i I_s(N) = \frac{1}{\sqrt{N}} Y_N(\omega) \tag{6.20}$$

As in (2.46) we find that, for (6.12),

$$U_N(\omega) = \frac{\sqrt{N}\alpha}{2}, \qquad \text{if } \omega = \frac{2\pi r}{N} \text{ for some integer } r \tag{6.21}$$

It is straightforward to rearrange (6.18) as

$$\hat{G}_N(e^{i\omega}) = \frac{\sqrt{N}Y_N(\omega)}{N\alpha/2} \tag{6.22}$$

which, using (6.21), means that

$$\hat{G}_N(e^{i\omega}) = \frac{Y_N(\omega)}{U_N(\omega)} \tag{6.23}$$

Here ω is precisely the frequency of the input signal. Comparing with (2.53), we also find (6.23) a most reasonable estimate (especially since $R_N(\omega)$ in (2.53) is zero for periodic inputs, according to the corollary of Theorem 2.1).

6.3 FOURIER ANALYSIS

Empirical Transfer-function Estimate

We found the expression (6.23) to correspond to frequency analysis with a single sinusoid of frequency ω as input. In a linear system, different frequencies pass through the system independently of each other. It is therefore quite natural to extend the frequency analysis estimate (6.23) also to the case of multifrequency inputs. That is, we introduce the following estimate of the transfer function:

$$\hat{\hat{G}}_N(e^{i\omega}) = \frac{Y_N(\omega)}{U_N(\omega)} \tag{6.24}$$

with Y_N and U_N defined by (6.19), also for the case where the input is not a single sinusoid. This estimate is also quite natural in view of Theorem 2.1.

We shall call $\hat{\hat{G}}_N(e^{i\omega})$ the *empirical transfer-function estimate* (ETFE), for reasons that we shall discuss shortly. In (6.24) we assume of course that $U_N(\omega) \neq 0$. If this does not hold for some frequencies, we simply regard the ETFE as undefined at those frequencies. We call this estimate empirical, since no other assumptions have been imposed than linearity of the system. In the case of multifrequency inputs, the

ETFE consists of $N/2$ essential points. [Recall that estimates at frequencies intermediate to the grid $\omega = 2\pi k/N$, $k = 0, 1, \ldots, N-1$, are obtained by trigonometrical interpolation in (2.37)]. Also, since y and u are real, we have

$$\hat{\hat{G}}_N(e^{2\pi ik/N}) = \overline{\hat{\hat{G}}_N(e^{2\pi i(N-k)/N})} \tag{6.25}$$

[compare (2.40) and (2.41)].

The original data sequence consisting of $2N$ numbers $y(t)$, $u(t)$, $t = 1, 2, \ldots, N$, has thus been condensed into the N numbers

$$\mathrm{Re}\hat{\hat{G}}_N(e^{2\pi ik/N}), \qquad \mathrm{Im}\hat{\hat{G}}_N(e^{2\pi ik/N}), \qquad k = 0, 1, \ldots, \frac{N}{2} - 1$$

This is quite a modest data reduction, revealing that most of the information contained in the original data y, u still is quite "raw."

In addition to an extension of frequency analysis, the ETFE can be interpreted as a way of (approximately) solving the set of convolution equations

$$y(t) = \sum_{k=1}^{N} g_0(k)u(t-k), \qquad t = 1, 2, \ldots, N \tag{6.26}$$

for $g_0(k)$, $k = 1, 2, \ldots, N$, using Fourier techniques.

Properties of the ETFE

Assume that the system is subject to (6.1). Introducing

$$V_N(\omega) = \frac{1}{\sqrt{N}} \sum_{t=1}^{N} v(t)e^{-i\omega t} \tag{6.27}$$

for the disturbance term, we find from Theorem 2.1 that

$$\hat{\hat{G}}_N(e^{i\omega}) = G_0(e^{i\omega}) + \frac{R_N(\omega)}{U_N(\omega)} + \frac{V_N(\omega)}{U_N(\omega)} \tag{6.28}$$

where the term $R_N(\omega)$ is subject to (2.54) and decays as $1/\sqrt{N}$.

Let us now investigate the influence of the term $V_N(\omega)$ on $\hat{\hat{G}}_N(e^{i\omega})$. Since $v(t)$ is assumed to have zero mean value,

$$EV_N(\omega) = 0, \qquad \forall\omega$$

so that

$$E\hat{\hat{G}}_N(e^{i\omega}) = G_0(e^{i\omega}) + \frac{R_N(\omega)}{U_N(\omega)} \tag{6.29}$$

Here expectation is with respect to $\{v(t)\}$, assuming $\{u(t)\}$ to be a given sequence of numbers.

Let the covariance function $R_v(\tau)$ and the spectrum $\Phi_v(\omega)$ of the process $\{v(t)\}$ be defined by (2.14) and (2.63). Then evaluate

$$EV_N(\omega)V_N(-\xi) = \frac{1}{N}\sum_{r=1}^{N}\sum_{s=1}^{N} Ev(r)e^{-i\omega r}v(s)e^{+i\xi s}$$

$$= \frac{1}{N}\sum_{r=1}^{N}\sum_{s=1}^{N} e^{i(\xi s-\omega r)}R_v(r-s) = [r-s=\tau]$$

$$= \frac{1}{N}\sum_{r=1}^{N} e^{i(\xi-\omega)r}\cdot \sum_{\tau=r-N}^{r-1} R_v(\tau)e^{-i\xi\tau}$$

Now

$$\sum_{\tau=r-N}^{r-1} R_v(\tau)e^{-i\xi\tau} = \Phi_v(\xi) - \sum_{\tau=-\infty}^{r-N-1} e^{-i\xi\tau}R_v(\tau) - \sum_{\tau=r}^{\infty} e^{-i\xi\tau}R_v(\tau)$$

and

$$\frac{1}{N}\sum_{r=1}^{N} e^{i(\xi-\omega)r} = \begin{cases} 1, & \text{if } \xi = \omega \\ 0, & \text{if } (\xi-\omega) = \dfrac{k2\pi}{N}, \quad k = \pm1, \pm2, \dots, \pm(N-1) \end{cases}$$

Consider

$$\left|\frac{1}{N}\sum_{r=1}^{N} e^{i(\xi-\omega)r}\sum_{\tau=-\infty}^{r-N-1} e^{-i\xi\tau}R_v(\tau)\right| \le \frac{1}{N}\sum_{r=1}^{N}\sum_{\tau=-\infty}^{r-N-1}|R_v(\tau)|$$

$$\le \text{[change order of summation]}$$

$$\le \frac{1}{N}\sum_{\tau=-\infty}^{-1}|\tau|\cdot|R_v(\tau)| \le \frac{C}{N}$$

provided

$$\sum_{-\infty}^{\infty}|\tau\cdot R_v(\tau)| < \infty \tag{6.30}$$

Similarly,

$$\left|\frac{1}{N}\sum_{r=1}^{N} e^{i(\xi-\omega)r}\cdot\sum_{\tau=r}^{\infty} e^{-i\xi\tau}R_v(\tau)\right| \le \frac{1}{N}\sum_{\tau=1}^{\infty}\tau\cdot|R_v(\tau)| \le \frac{C}{N}$$

Combining these expressions, we find that

$$E V_N(\omega) V_N(-\xi) = \begin{cases} \Phi_v(\omega) + \rho_2(N), & \text{if } \xi = \omega \\ \rho_2(N), & \text{if } |\xi - \omega| = \dfrac{k2\pi}{N}, \quad k = 1, 2, \dots N - 1 \end{cases} \tag{6.31}$$

with $|\rho_2(N)| \leq 2C/N$. These calculations can be summarized as the following result.

Lemma 6.1. Consider a strictly stable system

$$y(t) = G_0(q)u(t) + v(t) \tag{6.32}$$

with a disturbance $\{v(t)\}$ being a stationary stochastic process with spectrum $\Phi_v(\omega)$ and covariance function $R_v(\tau)$, subject to (6.30). Let $\{u(t)\}$ be independent of $\{v(t)\}$ assume that $|u(t)| \leq C$ for all t. Then with $\hat{\hat{G}}_N(e^{i\omega})$ defined by (6.24), we have

$$E\hat{\hat{G}}_N(e^{i\omega}) = G_0(e^{i\omega}) + \frac{\rho_1(N)}{U_N(\omega)} \tag{6.33a}$$

where

$$|\rho_1(N)| \leq \frac{C_1}{\sqrt{N}} \tag{6.33b}$$

and

$$E[\hat{\hat{G}}_N(e^{i\omega}) - G_0(e^{i\omega})][\hat{\hat{G}}_N(e^{-i\xi}) - G_0(e^{-i\omega})]$$

$$= \begin{cases} \frac{1}{|U_N(\omega)|^2}[\Phi_v(\omega) + \rho_2(N)], & \text{if } \xi = \omega \\ \frac{\rho_2(N)}{U_N(\omega)U_N(-\xi)}, & \text{if } |\xi - \omega| = \dfrac{2\pi k}{N}, \quad k = 1, 2, \dots, N - 1 \end{cases} \tag{6.34a}$$

where

$$|\rho_2(N)| \leq \frac{C_2}{N} \tag{6.34b}$$

Here U_N is defined by (2.37), and we restrict ourselves to frequencies for which $\hat{\hat{G}}_N$ is defined. According to Theorem 2.1 and (6.30), the constants can be taken as

$$C_1 = \left(2\sum_{k=1}^{\infty} |kg_0(k)|\right) \cdot \max|u(t)| \tag{6.35a}$$

$$C_2 = C_1^2 + \sum_{k=-\infty}^{\infty} |\tau R_v(\tau)| \tag{6.35b}$$

If $\{u(t)\}$ is periodic, then according to the Corollary of Theorem 2.1 $\rho_1(N) = 0$ at $\omega = 2\pi k/N$, so we can take $C_1 = 0$.

Remark. Note that the input is regarded as a given sequence. Probabilistic quantities, such as E, "bias," and "variance" refer to the probability space of $\{v(t)\}$. This does not, of course, exclude that the input may be generated as a realization of a stochastic process independent of $\{v(t)\}$.

The properties of the ETFE are closely related to those of periodogram estimates of spectra. See (2.43) and (2.74). We have the following result.

Lemma 6.2. Let $v(t)$ be given by

$$v(t) = H(q)e(t)$$

where $\{e(t)\}$ is a white-noise sequence with variance λ and fourth moment μ^2, and H is a strictly stable filter. Let $V_N(\omega)$ be defined by (6.27), and let $\Phi_v(\omega)$ be the spectrum of $v(t)$. Then

$$E|V_N(\omega)|^2 = \Phi_v(\omega) + \rho_3(N) \tag{6.36}$$

$$E(|V_N(\omega)|^2 - \Phi_v(\omega))(|V_N(\xi)|^2 - \Phi_v(\xi))$$
$$= \begin{cases} [\Phi_v(\omega)]^2 + \rho_4(N), & \text{if } \xi = \omega \quad \omega \neq 0, \pi \\ \rho_4(N), & \text{if } |\xi - \omega| = \dfrac{2\pi k}{N}, \quad k = 1, 2, \ldots N-1 \end{cases} \tag{6.37}$$

where

$$|\rho_3(N)| \leq \frac{C}{N}, \qquad |\rho_4(N)| \leq \frac{C}{N}$$

Proof. Equation (6.36) is a restatement of (6.31). A simple proof of (6.37) is outlined in Problem 6D.2 under somewhat more restrictive conditions. A full proof can be given by direct evaluation of (6.37). See, for example, Brillinger (1981), Theorem 5.2.4, for that. See Problem 6G.5 for ideas on how the bias term can be improved by the use of data tapering. □

These lemmas, together with the results of Section 2.3, tell us the following:

Case 1. *The input is periodic.* When the input is periodic and N is a multiple of the period, we know from Example 2.2 that $|U_N(\omega)|^2$ increases like const $\cdot$ N for some ω and is zero for others [see (2.49)]. The number of frequencies $\omega = 2\pi k/N$ for which $|U_N(\omega)|^2$ is nonzero, and hence for which the ETFE is defined, is fixed and no more than the period length of the signal. We thus find that

- The ETFE $\hat{\hat{G}}_N(e^{i\omega})$ is defined only for a fixed number of frequencies.
- At these frequencies the ETFE is unbiased and its variance decays like $1/N$.

We note that the results (6.16) on frequency analysis by the correlation method are obtained as a special case.

Case 2. *The input is a realization of a stochastic process.* Lemma 6.2 shows that the periodogram $|U_N(\omega)|^2$ is an erratic function of ω, which fluctuates around $\Phi_u(\omega)$ which we assume to be bounded. Lemma 6.1 thus tells us that

- The ETFE is an asymptotically unbiased estimate of the transfer function at increasingly (with N) many frequencies.
- The variance of the ETFE does not decrease as N increases, and it is given as the noise-to-signal ratio at the frequency in question.
- The estimates at different frequencies are asymptotically uncorrelated.

It follows from this discussion that, in the case of a periodic input signal, the ETFE will be of increasingly good quality at the frequencies that are present in the input. However, when the input is not periodic, the variance does not decay with N, but remains equal to the noise-to-signal ratio at the corresponding frequency. This latter property makes the empirical estimate a very crude estimate in most cases in practice.

It is easy to understand the reason why the variance does not decrease with N. We determine as many independent estimates as we have data points. In other words, we have no feature of data and information compression. This in turn is due to the fact that we have only assumed linearity about the true system. Consequently, the system's properties at different frequencies may be totally unrelated. From this it also follows that the only possibility to increase the information per estimated parameter is to assume that the system's behavior at one frequency is related to that at another. In the subsequent section, we shall discuss one approach to how this can be done.

6.4 SPECTRAL ANALYSIS

Spectral analysis for determining transfer functions of linear systems was developed from statistical methods for spectral estimation. Good accounts of this method are given in Chapter 10 in Jenkins and Watts (1968) and in Chapter 6 in Brillinger (1981), and the method is widely discussed in many other textbooks on time series analysis. In this section we shall adopt a slightly non-standard approach to the subject by deriving the standard techniques as a smoothed version of the ETFE.

Smoothing the ETFE

We mentioned at the end of the previous section that the only way to improve on the poor variance properties of the ETFE is to assume that the values of the true transfer function at different frequencies are related. We shall now introduce the rather reasonable prejudice that

$$\text{The true transfer function } G_0(e^{i\omega}) \text{ is a smooth function of } \omega. \tag{6.38}$$

If the frequency distance $2\pi/N$ is small compared to how quickly $G_0(e^{i\omega})$ changes, then

$$\hat{\hat{G}}_N(e^{2\pi ik/N}), \qquad k \text{ integer}, \qquad 2\pi k/N \approx \omega \tag{6.39}$$

are uncorrelated, unbiased estimates of roughly the same constant $G_0(e^{i\omega})$, each with a variance of

$$\frac{\Phi_v(2\pi k/N)}{|U_N(2\pi k/N)|^2}$$

according to Lemma 6.1. Here we neglected terms that tend to zero as N tends to infinity.

If we assume $G_0(e^{i\omega})$ to be constant over the interval

$$\frac{2\pi k_1}{N} = \omega_0 - \Delta\omega < \omega < \omega_0 + \Delta\omega = \frac{2\pi k_2}{N} \tag{6.40}$$

then it is well known that the best (in a minimum variance sense) way to estimate this constant is to form a weighted average of the "measurements" (6.39) for the frequencies (6.40), each measurement weighted according to its inverse variance [compare Problem 6E.3, and Lemma II.2, (II.65), in Appendix II:

$$\hat{G}_N(e^{i\omega_0}) = \frac{\displaystyle\sum_{k=k_1}^{k_2} \alpha_k \hat{\hat{G}}_N(e^{2\pi ik/N})}{\displaystyle\sum_{k=k_1}^{k_2} \alpha_k} \tag{6.41a}$$

$$\alpha_k = \frac{|U_N(2\pi k/N)|^2}{\Phi_v(2\pi k/N)} \tag{6.41b}$$

For large N we could with good approximation work with the integrals that correspond to the (Riemann) sums in (6.41):

$$\hat{G}_N(e^{i\omega_0}) = \frac{\int_{\xi=\omega_0-\Delta\omega}^{\omega_0+\Delta\omega} \alpha(\xi)\hat{\hat{G}}_N(e^{i\xi})d\xi}{\int_{\xi=\omega_0-\Delta\omega}^{\omega_0+\Delta\omega} \alpha(\xi)d\xi}, \quad \alpha(\xi) = \frac{|U_N(\xi)|^2}{\Phi_v(\xi)} \tag{6.42}$$

If the transfer function G_0 is not constant over the interval (6.40) it is reasonable to use an additional weighting that pays more attention to frequencies close to ω_0:

$$\hat{G}_N(e^{i\omega_0}) = \frac{\int_{-\pi}^{\pi} W_\gamma(\xi - \omega_0)\alpha(\xi)\hat{\hat{G}}_N(e^{i\xi})d\xi}{\int_{-\pi}^{\pi} W_\gamma(\xi - \omega_0)\alpha(\xi)d\xi} \tag{6.43}$$

Here $W_\gamma(\xi)$ is a function centered around $\xi = 0$ and γ is a "shape parameter," which we shall discuss shortly.

Clearly, (6.42) corresponds to

$$W_\gamma(\xi) = \begin{cases} 1, & |\xi| < \Delta\omega \\ 0, & |\xi| > \Delta\omega \end{cases} \tag{6.44}$$

Now, if the noise spectrum $\Phi_v(\omega)$ is known, the estimate (6.43) can be realized as written. If $\Phi_v(\omega)$ is not known we could argue as follows: Suppose that the noise spectrum does not change very much over frequency intervals corresponding to the "width" of the weighting function $W_\gamma(\xi)$:

$$\int_{-\pi}^{\pi} W_\gamma(\xi - \omega_0) \left| \frac{1}{\Phi_v(\xi)} - \frac{1}{\Phi_v(\omega_0)} \right| d\xi = \text{"small"} \tag{6.45}$$

Then $\alpha(\xi)$ in (6.42) can be replaced by $\alpha(\xi) = |U_N(\xi)|^2/\Phi_v(\omega_0)$, which means that the constant $\Phi_v(\omega_0)$ cancels when (6.43) is formed. Under (6.45) the estimate

$$\hat{G}_N(e^{i\omega_0}) = \frac{\int_{-\pi}^{\pi} W_\gamma(\xi - \omega_0)|U_N(\xi)|^2 \hat{\hat{G}}_N(e^{i\xi})d\xi}{\int_{-\pi}^{\pi} W_\gamma(\xi - \omega_0)|U_N(\xi)|^2 d\xi} \tag{6.46}$$

is thus a good approximation of (6.42) and (6.43).

We may remark that, if (6.45) does not hold, it might be better to include a procedure where $\Phi_v(\omega)$ is estimated and use that estimate in (6.43).

Connection with the Blackman-Tukey Procedure (*)

Consider the denominator of (6.46). It is a weighted average of the periodogram $|U_N(\xi)|^2$. Using the result (2.74), we find that, as $N \to \infty$,

$$\int_{-\pi}^{\pi} W_\gamma(\xi - \omega_0)|U_N(\xi)|^2 d\xi \to \int_{-\pi}^{\pi} W_\gamma(\xi - \omega_0)\Phi_u(\xi)d\xi \tag{6.47}$$

where $\Phi_u(\omega)$ is the spectrum of $\{u(t)\}$, as defined by (2.61) to (2.63). If, moreover,

$$\int_{-\pi}^{\pi} W_\gamma(\xi)d\xi = 1$$

and the weighting function $W_\gamma(\xi)$ is concentrated around $\xi = 0$ with a width over which $\Phi_u(\omega)$ does not change much, then the right side of (6.47) is close to $\Phi_u(\omega_0)$. We may thus interpret the left side as an estimate of this quantity:

$$\hat{\Phi}_{yu}^N(\omega_0) = \int_{-\pi}^{\pi} W_\gamma(\xi - \omega_0)|U_N(\xi)|^2 d\xi \tag{6.48}$$

Similarly, since

$$|U_N(\xi)|^2 \hat{\hat{G}}_N(e^{i\xi}) = |U_N(\xi)|^2 \frac{Y_N(\xi)}{U_N(\xi)} = Y_N(\xi)\overline{U}_N(\xi) \tag{6.49}$$

we have that the numerator of (6.46)

$$\hat{\Phi}_{yu}^N(\omega_0) = \int_{-\pi}^{\pi} W_\gamma(\xi - \omega_0) Y_N(\xi) \overline{U}_N(\xi) d\xi \tag{6.50}$$

is an estimate of the cross spectrum between output and input. The transfer function estimate (6.46) is thus the ratio of two spectral estimates:

$$\hat{G}_N(e^{i\omega_0}) = \frac{\hat{\Phi}_{yu}^N(\omega_0)}{\hat{\Phi}_u^N(\omega_0)} \tag{6.51}$$

which makes sense, in view of (2.80). The spectral estimates (6.48) and (6.50) are the standard estimates, suggested in the literature, for spectra and cross spectra as smoothed periodograms. See Blackman and Tukey (1958), Jenkins and Watts (1968), or Brillinger (1981).

An alternative way of expressing these estimates is common. The Fourier coefficients for the periodogram $|U_N(\omega)|^2$ are

$$\hat{R}_u^N(\tau) = \frac{1}{2\pi} \int_{-\pi}^{\pi} |U_N(\omega)|^2 e^{i\tau\omega} d\omega = \frac{1}{N} \sum_{t=1}^{N} u(t) u(t - \tau) \tag{6.52}$$

[For this expression to hold exactly, the values $u(s)$ outside the interval $1 \leq s \leq N$ have to be interpreted by periodic continuation; i.e., $u(s) = u(s - N)$ if $s > N$; see Problem 6D.1.]

Similarly, let the Fourier coefficients of the function $2\pi W_\gamma(\xi)$ be

$$w_\gamma(\tau) = \int_{-\pi}^{\pi} W_\gamma(\xi) e^{i\xi\tau} d\xi \tag{6.53}$$

Since the integral (6.48) is a convolution, its Fourier coefficients will be the product of (6.52) and (6.53), so a Fourier expansion of (6.48) gives

$$\hat{\Phi}_u^N(\omega) = \sum_{\tau=-\infty}^{\infty} w_\gamma(\tau) \hat{R}_u^N(\tau) e^{-i\tau\omega} \tag{6.54}$$

The idea is now that the nice, smooth function $W_\gamma(\xi)$ is chosen so that its Fourier coefficients vanish for $|\tau| > \delta_\gamma$, where typically $\delta_\gamma \ll N$. It is consequently sufficient to form (6.52) (using the rightmost expression) for $|\tau| \leq \delta_\gamma$, and then take

$$\hat{\Phi}_u^N(\omega) = \sum_{\tau=-\delta_\gamma}^{\delta_\gamma} w_\gamma(\tau) \hat{R}_u^N(\tau) e^{-i\tau\omega} \tag{6.55}$$

This is perhaps the most convenient way of forming the spectral estimate. The expressions for $\hat{\Phi}_{yu}^N(\omega)$ are of course analogous.

Weighting Function $W_\gamma(\xi)$: The Frequency Window

Let us now discuss the weighting function $W_\gamma(\xi)$. In spectral analysis, it is often called the *frequency window*. [Similarly, $w_\gamma(\tau)$ is called the *lag window*.] If this window is "wide," then many different frequencies will be weighted together in (6.40). This should lead to a small variance of $\hat{G}_N(e^{i\omega_0})$. At the same time, a wide window will involve frequency estimates farther away from ω_0, with expected values that may differ considerably from $G_0(e^{i\omega_0})$. This will cause large bias. The width of the window will thus control the trade-off between bias and variance. To make this trade-off a bit more formal, we shall use the scalar γ to describe the width, so a large value of γ corresponds to a narrow window.

We shall characterize the window by the following numbers

$$\int_{-\pi}^{\pi} W_\gamma(\xi)d\xi = 1, \quad \int_{-\pi}^{\pi} \xi W_\gamma(\xi)d\xi = 0, \quad \int_{-\pi}^{\pi} \xi^2 W_\gamma(\xi)d\xi = M(\gamma) \quad (6.56a)$$

$$\int_{-\pi}^{\pi} |\xi|^3 W_\gamma(\xi)d\xi = C_3(\gamma), \quad \int_{-\pi}^{\pi} W_\gamma^2(\xi)d\xi = \frac{1}{2\pi}\overline{W}(\gamma) \quad (6.56b)$$

As γ increases (and the frequency window gets more narrow), the number $M(\gamma)$ decreases, while $\overline{W}(\gamma)$ increases.

Some typical windows are given in Table 6.1. [See, also, Table 3.3.1 in Brillinger (1981)for a more complete collection of windows.] Notice that the scaling quantity

TABLE 6.1 Some Windows for Spectral Analysis

	$2\pi W_\gamma(\omega)$	$w_\gamma(\tau), \quad 0 \le \lvert\tau\rvert \le \gamma$
Bartlett	$\frac{1}{\gamma}\left(\frac{\sin \gamma\omega/2}{\sin \omega/2}\right)^2$	$1 - \frac{\lvert\tau\rvert}{\gamma}$
Parzen	$\frac{4(2+\cos\omega)}{\gamma^3}\left(\frac{\sin \gamma\omega/4}{\sin \omega/2}\right)^4$	$\begin{cases} 1 - \frac{6\tau^2}{\gamma^2}\left(1 - \frac{\lvert\tau\rvert}{\gamma}\right), & 0 \le \lvert\tau\rvert \le \frac{\gamma}{2} \\ 2\left(1 - \frac{\lvert\tau\rvert}{\gamma}\right)^3, & \frac{\gamma}{2} \le \lvert\tau\rvert \le \gamma \end{cases}$
Hamming	$\frac{1}{2}D_\gamma(\omega) + \frac{1}{4}D_\gamma(\omega - \pi/\gamma)$ $+\frac{1}{4}D_\gamma(\omega + \pi/\gamma)$, where $D_\gamma(\omega) = \frac{\sin(\gamma + \frac{1}{2})\omega}{\sin \omega/2}$	$\frac{1}{2}\left(1 + \cos\frac{\pi\tau}{\gamma}\right)$

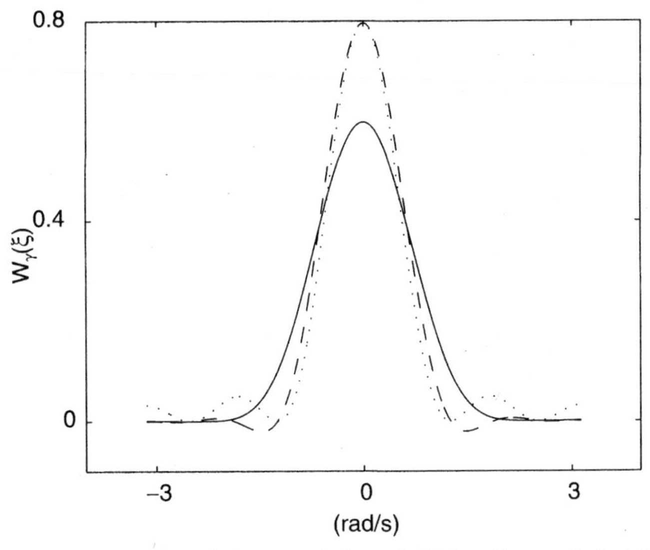

Figure 6.1 Some common frequency windows. Solid line: Parzen; dashed line: Hamming; dotted line: Bartlett, $\gamma = 5$.

γ has been chosen so that $\delta_\gamma = \gamma$ in (6.55). The frequency windows are shown graphically in Figure 6.1. For these windows, we have

$$\begin{aligned}
&\text{Bartlett:} \quad M(\gamma) = \frac{2.78}{\gamma} \qquad \overline{W}(\gamma) \approx 0.67\gamma \\
&\text{Parzen:} \quad M(\gamma) = \frac{12}{\gamma^2} \qquad \overline{W}(\gamma) \approx 0.54\gamma \\
&\text{Hamming:} \quad M(\gamma) = \frac{\pi^2}{2\gamma^2} \qquad \overline{W}(\gamma) \approx 0.75\gamma
\end{aligned} \tag{6.57}$$

The expressions are asymptotic for large γ but are good approximations for $\gamma \gtrsim 5$. See also Problem 6T.1 for a further discussion of how to scale windows.

Asymptotic Properties of the Smoothed Estimate

The estimate (6.46) has been studied in several treatments of spectral analysis. Results that are asymptotic in both N and γ can be derived as follows (see Appendix 6A). Consider the estimate (6.46), and suppose that the true system obeys the assumptions of Lemma 6.1. We then have

Bias

$$E\hat{G}_N(e^{i\omega}) - G_0(e^{i\omega}) = M(\gamma) \cdot \left[\frac{1}{2}G_0''(e^{i\omega}) + G_0'(e^{i\omega})\frac{\Phi_u'(\omega)}{\Phi_u(\omega)}\right] + \underset{\gamma\to\infty}{\mathbf{O}(C_3(\gamma))} + \underset{N\to\infty}{\mathbf{O}(1/\sqrt{N})} \tag{6.58}$$

Here $\mathbf{O}(x)$ is ordo x. See notational conventions at the beginning of the book. Prime and double prime denote differentiation with respect to ω, one and twice respectively.

Variance

$$E|\hat{G}_N(e^{i\omega}) - E\hat{G}_N(e^{i\omega})|^2 = \frac{1}{N} \cdot \overline{W}(\gamma) \cdot \frac{\Phi_v(\omega)}{\Phi_u(\omega)} + \mathbf{o}(\overline{W}(\gamma)/N) \quad \begin{matrix} \gamma\to\infty \\ N\to\infty, \gamma/N\to 0 \end{matrix} \tag{6.59}$$

We repeat that expectation here is with respect to the noise sequence $\{v(t)\}$ and that the input is supposed to be a deterministic quasi-stationary signal.

Let us use the asymptotic expressions to evaluate the mean-square error (MSE):

$$E|\hat{G}_N(e^{i\omega}) - G_0(e^{i\omega})|^2 \sim M^2(\gamma)|R(\omega)|^2 + \frac{1}{N}\overline{W}(\gamma)\frac{\Phi_v(\omega)}{\Phi_u(\omega)} \tag{6.60}$$

Here

$$R(\omega) = \frac{1}{2}G_0''(e^{i\omega}) + G_0'(e^{i\omega})\frac{\Phi_u'(\omega)}{\Phi_u(\omega)} \tag{6.61}$$

Some additional results can also be shown (see Brillinger, 1981, Chapter 6, and Problems 6D.3 and 6D.4).

- The estimates $\text{Re}\hat{G}_N(e^{i\omega})$ and $\text{Im}\hat{G}_N(e^{i\omega})$ are asymptotically uncorrelated and each have a variance equal to half that in (6.59). (6.62)
- The estimates $\hat{G}_N(e^{i\omega})$ at different frequencies are asymptotically uncorrelated. (6.63)
- The estimates $\text{Re}\hat{G}_N(e^{i\omega_k}), \text{Im}\hat{G}_N(e^{i\omega_k}), k = 1, 2, \ldots, M$, at an arbitrary collection of frequencies are asymptotically jointly normal distributed with means and covariances given by (6.58) to (6.63). (6.64)
- For a translation to properties of $|\hat{G}_N(e^{i\omega})|$, arg $\hat{G}_N(e^{i\omega})$, see Problem 9G.1.

From (6.60) we see that a desired property of the window is that both M and $\overline{W}$ should be small. We may also calculate the value of the width parameter γ that minimizes the MSE. Suppose that both γ and N tend to infinity and γ/N tends to zero, so that the asymptotic expressions are applicable. Suppose also that (6.57) holds with $M(\gamma) = M/\gamma^2$ and $\overline{W}(\gamma) = \gamma \cdot \overline{W}$ Then (6.60) gives

$$\gamma_{\text{opt}} = \left(\frac{4M^2|R(\omega)|^2\Phi_u(\omega)}{\overline{W}\Phi_v(\omega)}\right)^{1/5} \cdot N^{1/5} \tag{6.65}$$

This value can of course not be realized by the user, since the constant contains several unknown quantities. We note, however, that in any case it increases like $N^{1/5}$, and it should, in principle, be allowed to be frequency dependent. The frequency window consequently should get more narrow when more data are available, which is a very natural result.

The optimal choice of γ leads to a mean-square error that decays like

$$\text{MSE} \sim C \cdot N^{-4/5} \tag{6.66}$$

In practical use the trade-off (6.65) and (6.66) cannot be reached in formal terms. Instead, a typical procedure would be to start by taking $\gamma = N/20$ (see Table 6.1) and then compute and plot the corresponding estimates $\hat{G}_N(e^{i\omega})$ for various values of γ. As γ is increased, more and more details of the estimate will appear. These will be due to decreased bias (true resonance peaks appearing more clearly and the like), as well as to increased variance (spurious, random peaks). The procedure will be stopped when the user feels that the emerging details are predominately spurious.

Actually, as we noted, (6.65) points to the fact that the optimal window size should be frequency dependent. This can easily be implemented in (6.46), but not in (6.55), and most procedures do not utilize this feature.

Example 6.1 A Simulated System

The system

$$y(t) - 1.5y(t-1) + 0.7y(t-2) = u(t-1) + 0.5u(t-2) + e(t) \tag{6.67}$$

where $\{e(t)\}$ is white noise with variance 1 was simulated with the input as a PRBS signal (see Section 13.3) over 1000 samples. Part of the resulting data record is shown in Figure 6.2. The corresponding ETFE is shown in Figure 6.3a. An estimate $\hat{G}_N(e^{i\omega})$ was formed using (6.46), with $W_\gamma(\xi)$ being a Parzen window with various values of γ. Figure 6.3bcd shows the results for $\gamma = 10, 50$, and 200. Here $\gamma = 50$ appears to be a reasonable choice of window size. □

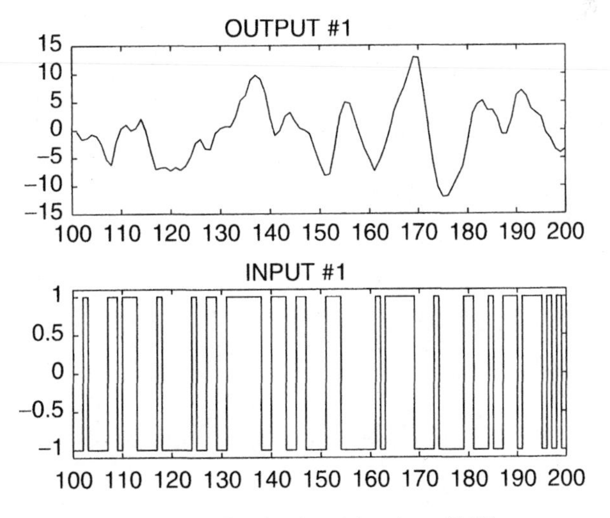

Figure 6.2 The simulated data from (6.67).

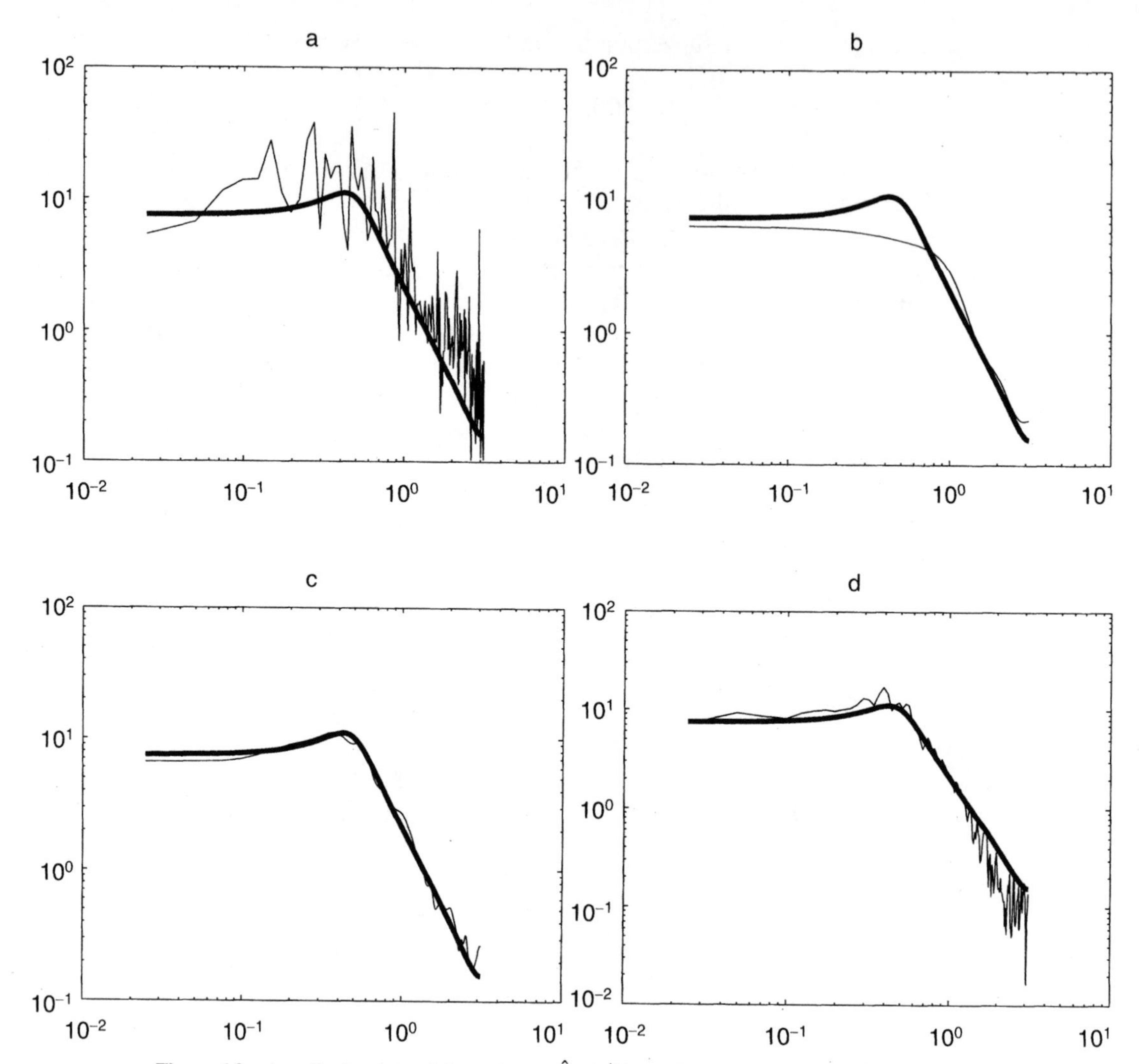

Figure 6.3 Amplitude plots of the estimate $\hat{G}_N(e^{i\omega})$. a: ETFE, b: $\gamma = 10$, c: $\gamma = 50$, d: $\gamma = 200$. Thick lines: true system; thin lines: estimate.

Another Way of Smoothing the ETFE (*)

The guiding idea behind the estimate (6.46) is that the ETFEs at neighboring frequencies are asymptotically uncorrelated, and that hence the variance could be reduced by averaging over these. The ETFEs obtained over different data sets will also provide uncorrelated estimates, and another approach would be to form averages over these. Thus, split the data set Z^N into M batches, each containing R data ($N = R \cdot M$). Then form the ETFE corresponding to the kth batch:

$$\hat{\hat{G}}_R^{(k)}(e^{i\omega}), \qquad k = 1, 2, \ldots, M \tag{6.68}$$

The estimate can then be formed as a direct average

$$\hat{G}_N(e^{i\omega}) = \frac{1}{M}\sum_{k=1}^{M} \hat{\hat{G}}_R^{(k)}(e^{i\omega}) \tag{6.69}$$

or one that is weighted according to the inverse variances:

$$\hat{G}_N(e^{i\omega}) = \frac{\sum_{k=1}^{M} \beta_R^{(k)}(\omega) \cdot \hat{\hat{G}}_R^{(k)}(e^{i\omega})}{\sum_{k=1}^{M} \beta_R^{(k)}(\omega)} \tag{6.70}$$

with

$$\beta_R^{(k)}(\omega) = |U_R^{(k)}(\omega)|^2 \tag{6.71}$$

being the periodogram of the kth subbatch. The inverse variance of $\hat{\hat{G}}_R^{(k)}(e^{i\omega})$ is $\beta_R^{(k)}(\omega)/\Phi_v(\omega)$, but the factor $\Phi_v(\omega)$ cancels when (6.70) is formed.

An advantage with the estimate (6.70) is that the fast Fourier transform (FFT) can be efficiently used when Z^N can be decomposed so that R is a power of 2. Compare Problem 6G.4. The method is known as Welch's method, Welch (1967).

6.5 ESTIMATING THE DISTURBANCE SPECTRUM (*)

Estimating Spectra

So far we have described how to estimate G_0 in a relationship (6.1):

$$y(t) = G_0(q)u(t) + v(t) \tag{6.72}$$

We shall now turn to the problem of estimating the spectrum of $\{v(t)\}$, $\Phi_v(\omega)$. Had the disturbances $v(t)$ been available for direct measurement, we could have used (6.48):

$$\hat{\Phi}_v^N(\omega) = \int_{-\pi}^{\pi} W_\gamma(\xi - \omega)|V_N(\xi)|^2 d\xi \tag{6.73}$$

Here $W_\gamma(\cdot)$ is a frequency window of the kind described earlier.

It is entirely analogous to the analysis of the previous section to calculate the properties of (6.73). We have:

Bias:

$$E\hat{\Phi}_v^N(\omega) - \Phi_v(\omega) = \tfrac{1}{2}M(\gamma) \cdot \Phi_v''(\omega) + \underset{\gamma\to\infty}{\mathbf{O}(C_1(\gamma))} + \underset{N\to\infty}{\mathbf{O}(1/\sqrt{N})} \tag{6.74}$$

Variance:

$$\operatorname{Var}\hat{\Phi}_v^N(\omega) = \frac{\overline{W}(\gamma)}{N} \cdot \Phi_v^2(\omega) + \mathop{\mathbf{o}(1/N)}_{N\to\infty}, \qquad \omega \neq 0, \pm\pi \tag{6.75}$$

Moreover, estimates at different frequencies are asymptotically uncorrelated.

The Residual Spectrum

Now the $v(t)$ in (6.72) are not directly measurable. However, given an estimate $\hat{G}_N$ of the transfer function, we may replace v in the preceding expression by

$$\hat{v}(t) = y(t) - \hat{G}_N(q)u(t) \tag{6.76}$$

which gives the estimate

$$\hat{\Phi}_v^N(\omega) = \int_{-\pi}^{\pi} W_\gamma(\xi - \omega)|Y_N(\xi) - \hat{G}_N(e^{i\xi})U_N(\xi)|^2 d\xi \tag{6.77}$$

If $\hat{G}_N(e^{i\xi})$ is formed using (6.46) with the same window $W_\gamma(\cdot)$, this expression can be rearranged as follows [using (6.48) to (6.51)]:

$$\begin{aligned}
&\int_{-\pi}^{\pi} W_\gamma(\xi - \omega)|Y_N(\xi)|^2 d\xi + \int_{-\pi}^{\pi} W_\gamma(\xi - \omega)|U_N(\xi)|^2|\hat{G}_N(e^{i\xi})|^2 d\xi \\
&\quad - 2\operatorname{Re}\int_{-\pi}^{\pi} W_\gamma(\xi - \omega)\hat{G}_N(e^{i\xi})U_N(\xi)\overline{Y_N(\xi)}d\xi \\
&\approx \int_{-\pi}^{\pi} W_\gamma(\xi - \omega)|Y_N(\xi)|^2 d\xi + |\hat{G}_N(e^{i\omega})|^2 \int_{-\pi}^{\pi} W_\gamma(\xi - \omega)|U_N(\xi)|^2 d\xi \\
&\quad - 2\operatorname{Re}\hat{G}_N(e^{i\omega}) \int_{-\pi}^{\pi} W_\gamma(\xi - \omega)U_N(\xi)\overline{Y_N(\xi)}d\xi \\
&= \hat{\Phi}_y^N(\omega) + \frac{|\hat{\Phi}_{yu}^N(\omega)|^2}{(\hat{\Phi}_u^N(\omega))^2} \cdot \hat{\Phi}_u^N(\omega) - 2\operatorname{Re}\frac{\hat{\Phi}_{yu}^N(\omega)}{\hat{\Phi}_u^N(\omega)} \cdot \overline{\hat{\Phi}_{yu}^N(\omega)}
\end{aligned}$$

Here the approximate equality follows from replacing the smooth function $\hat{G}_N(e^{i\xi})$ over the small interval around $\xi = \omega$ with its value at ω. Hence we have

$$\hat{\Phi}_v^N(\omega) = \hat{\Phi}_y^N(\omega) - \frac{|\hat{\Phi}_{yu}^N(\omega)|^2}{\hat{\Phi}_u^N(\omega)} \tag{6.78}$$

Asymptotically, as $N \to \infty$ and $\gamma \to \infty$, so that $\hat{G}_N(e^{i\omega}) \to G_0(e^{i\omega})$ according to (6.60), we find that the estimate (6.77) tends to (6.73). The asymptotic properties (6.74) and (6.75) will also hold for (6.77) and (6.78). In addition to the properties already listed, we may note that the estimates $\hat{\Phi}_v^N(\omega)$ are asymptotically uncorrelated with $\hat{G}_N(e^{i\omega})$. Moreover $\hat{\Phi}_v^N(\omega_k)$, $\hat{G}_N(e^{i\omega_k})$, $k = 1, 2, \ldots, r$, are asymptotically jointly normal random variables with mean and covariances given by (6.58) to (6.64) and (6.74) to (6.75). A detailed account of the asymptotic theory is given in Chapter 6 of Brillinger (1981).

Coherency Spectrum

Denote

$$\hat{\kappa}_{yu}^N(\omega) = \sqrt{\frac{|\hat{\Phi}_{yu}^N(\omega)|^2}{\hat{\Phi}_y^N(\omega)\hat{\Phi}_u^N(\omega)}} \tag{6.79}$$

Then

$$\hat{\Phi}_v^N(\omega) = \hat{\Phi}_y^N(\omega)[1 - (\hat{\kappa}_{yu}^N(\omega))^2] \tag{6.80}$$

The function $\kappa_{yu}(\omega)$ is called the *coherency spectrum* (between y and u) and can be viewed as the (frequency dependent) correlation coefficient between the input and output sequences. If this coefficient is 1 at a certain frequency, then there is perfect correlation between input and output at that frequency. There is consequently no noise interfering at that frequency, which is confirmed by (6.80).

6.6 SUMMARY

In this chapter we have shown how simple techniques of transient and frequency response can give valuable insight into the properties of linear systems. We have introduced the empirical transfer-function estimate (ETFE)

$$\hat{\hat{G}}_N(e^{i\omega}) = \frac{Y_N(\omega)}{U_N(\omega)} \tag{6.81}$$

based on data over the interval $1 \le t \le N$. Here

$$Y_N(\omega) = \frac{1}{\sqrt{N}}\sum_{t=1}^{N} y(t)e^{-it\omega}, \qquad U_N(\omega) = \frac{1}{\sqrt{N}}\sum_{t=1}^{N} u(t)e^{-it\omega},$$

The ETFE has the property (see Lemma 6.1) that it is asymptotically unbiased, but has a variance of $\Phi_v(\omega)/|U_N(\omega)|^2$.

We showed how smoothing the ETFE leads to the spectral analysis estimate

$$\hat{G}_N(e^{i\omega}) = \frac{\int_{-\pi}^{\pi} W_\gamma(\xi - \omega)|U_N(\xi)|^2\hat{\hat{G}}_N(e^{i\xi})d\xi}{\int_{-\pi}^{\pi} W_\gamma(\xi - \omega)|U_N(\xi)|^2 d\xi} \tag{6.82}$$

A corresponding estimate of the noise spectrum is

$$\hat{\Phi}_v^N(\omega) = \int_{-\pi}^{\pi} W_\gamma(\xi - \omega)|Y_N(\xi) - \hat{G}_N(e^{i\xi})U_N(\xi)|^2 d\xi \tag{6.83}$$

The properties of these estimates were summarized in (6.58) to (6.64) and (6.74) to (6.75).

These properties depend on the parameter γ, which describes the width of the associated frequency window W_γ. A narrow such window (large γ) gives small bias but high variance for the estimate, while the converse is true for wide windows.

6.7 BIBLIOGRAPHY

Section 6.1: Wellstead (1981)gives a general survey of nonparametric methods for system identification. A survey of transient response methods is given in Rake (1980). Several ways of determining numerical characteristics from step responses are discussed in Schwarze (1964). Correlation techniques are surveyed in Godfrey (1980).

Section 6.2: Frequency analysis is a classical identification method that is described in many textbooks on control. For detailed treatments, see Rake (1980)which also contains several interesting examples.

Section 6.3: General Fourier techniques are also discussed in Rake (1980). The term "empirical transfer function estimate" for $\hat{\hat{G}}$ is introduced in this chapter, but the estimate as such is well known.

Sections 6.4 and 6.5: Spectral analysis is a standard subject in textbooks on time series. See, for example, Grenander and Rosenblatt (1957)(Chapters 4 to 6), Anderson (1971)(Chapter 9), and Hannan (1970)(Chapter V). Among books devoted entirely to spectral analysis, we could point to Kay (1988), Marple (1987), and Stoica and Moses (1997). These texts deal primarily with estimation of power (auto-) spectra. Among specific treatments of frequency-domain techniques, including estimation of transfer functions, we note Brillinger (1981)for a thorough analytic study, Jenkins and Watts (1968)for a more leisurely discussion of both statistical properties and application aspects, and Bendat and Piersol (1980)for an application-oriented approach. Another extensive treatment is Priestley (1981). Overviews of different frequency-domain techniques are given in Brillinger and Krishnaiah (1983), and a control-oriented survey is given by Godfrey (1980). The treatment given here is based on Ljung (1985a). The first reference to the idea of smoothing the periodogram to obtain a better spectral estimate appears to be Daniell (1946). A comparative discussion of windows for spectral analysis is given in Gecklini and Yavuz (1978)and Papoulis (1973).

In addition to direct frequency-domain methods for estimating spectra, many efficient methods are based on parametric fit, such as those to be discussed in the following chapter. So called maximum entropy methods (MEM) have found wide use in signal-processing applications. See Burg (1967)for the first idea and Marple (1987)for a comparative survey of different approaches.

6.8 PROBLEMS

6G.1 Consider the system

$$y(t) = G_0(q)u(t) + v(t)$$

controlled by the regulator

$$u(t) = -F(q)y(t) + r(t)$$

where $r(t)$ is an external reference signal. r and v are independent and their spectra are $\Phi_r(\omega)$ and $\Phi_v(\omega)$, respectively. The usual spectral analysis estimate of G_0 is given in (6.51) as well as (6.46). Show that as N and γ tend to infinity then $\hat{G}_N(e^{i\omega})$ will converge to

$$G_*(e^{i\omega}) = \frac{G_0(e^{i\omega})\Phi_r(\omega) - F(e^{-i\omega})\Phi_v(\omega)}{\Phi_r(\omega) + |F(e^{i\omega})|^2\Phi_v(\omega)}$$

What happens in the two special cases $\Phi_r \equiv 0$ and $F \equiv 0$, respectively? *Hint:* Compare Problem 2E.5.

6G.2 *Prefiltering.* Prefilter inputs and outputs:

$$u_F(t) = L_u(q)u(t), \qquad y_F(t) = L_y(q)y(t)$$

If (6.32) holds, then the filtered variables obey

$$y_F(t) = G_0^F(q)u_F(t) + v_F(t)$$

$$G_0^F(q) = \frac{L_y(q)}{L_u(q)}G_0(q), \qquad v_F(t) = L_y(q)v(t)$$

Apply spectral analysis to u_F, y_F, thus forming an estimate $\hat{G}_N^F(e^{i\omega})$. The estimate of the original transfer function then is

$$\hat{G}_N(e^{i\omega}) = \frac{L_u(e^{i\omega})}{L_y(e^{i\omega})}\hat{G}_N^F(e^{i\omega})$$

Determine the asymptotic properties of $\hat{G}_N(e^{i\omega})$ and discuss how L_u and L_y can be chosen for smallest MSE (cf. Ljung, 1985a).

6G.3 In Figure 6.3 the amplitude of the ETFE appears to be systematically larger than the true amplitude, despite the fact that the ETFE is unbiased according to Lemma 6.1. However, $\hat{\hat{G}}$ being an unbiased estimate of G_0 does not imply that $|\hat{\hat{G}}|$ is an unbiased estimate of $|G_0|$. In fact, prove that

$$E|\hat{\hat{G}}_N(e^{i\omega})|^2 = |G_0(e^{i\omega})|^2 + \frac{\Phi_v(\omega)}{|U_N(\omega)|^2}$$

asymptotically for large N, under the assumptions of Lemma 6.1.

6G.4 The Cooley-Tukey spectral estimate for a process $\{v(t)\}$ is defined as

$$\hat{\Phi}_v^N(\omega) = \frac{1}{M}\sum_{k=1}^{M} |V_R^{(k)}(\omega)|^2$$

where $|V_R^{(k)}(\omega)|^2$ is the periodogram estimate of the kth subbatch of data:

$$V_R^{(k)}(\omega) = \frac{1}{\sqrt{R}}\sum_{\ell=1}^{R} v((k-1)R+\ell)e^{-i\omega\ell})$$

See Cooley and Tukey (1965)or Hannan (1970), Chapter V. The cross-spectral estimate is defined analogously. This estimate has the advantage that the FFT (fast Fourier transform) can be applied (most efficiently if R is a power of 2). Show that the estimate (6.70) is the ratio of two appropriate Cooley-Tukey spectral estimates.

6G.5 "*Tapers*" or "*faders.*" The bias term $\rho_3(N)$ in (6.36) can be reduced if tapering is introduced: Let $V_N^{(T)}(\omega)$ be defined by

$$V_N^{(T)}(\omega) = \sum_{t=1}^{N} h_t v(t) e^{-it\omega}$$

where $\{h_t\}_1^N$ is a sequence of numbers (*a tapering function*) such that

$$\sum_{t=1}^{N} h_t^2 = 1$$

Let

$$H_N(\omega) = \sum_{t=1}^{N} h_t e^{-it\omega}$$

Show that, under the conditions of Lemma 6.2,

$$E|V_N^{(T)}(\omega)|^2 = \int_{-\pi}^{\pi} |H_N(\omega-\xi)|^2 \Phi_v(\xi)d\xi$$

Show that our standard periodogram estimate, which uses $h_t \equiv 1/\sqrt{N}$, gives

$$|H_N(\omega)|^2 = \frac{1}{N}\left[\frac{\sin N\omega/2}{\sin \omega/2}\right]^2$$

Other tapering coefficients (or "faders" or "convergence factors") may give functions $|H_N(\omega)|^2$ that are more "δ-function like" than in the preceding equation (see, e.g., Table 6.1). The tapered periodogram can of course also be used to obtain smoothed spectra. They will typically lead to decreased bias and (slightly) increased variance (Brillinger, 1981, Theorem 5.2.3 and Section 5.8).

6E.1 Determine an estimate for $G_0(e^{i\omega})$ based on the impulse-response estimates (6.4). Show that this estimate coincides with the ETFE (6.24).

6E.2 Consider the system

$$y(t) = G_0(q)u(t) + v(t)$$

This system is controlled by output proportional feedback

$$u(t) = -Ky(t)$$

Let the ETFE $\hat{\hat{G}}_N(e^{i\omega})$ be computed in the straightforward way (6.24). What will this estimate be? Compare with Lemma 6.1.

6E.3 Let w_k, $k = 1, \dots, M$, be independent random variables, all with mean values 1 and variances $E(w_k - 1)^2 = \lambda_k$. Consider

$$w = \sum_{k=1}^{M} \alpha_k w_k$$

Determine α_k, $k = 1, \dots, M$, so that

(a) $Ew = 1$.

(b) $E(w - 1)^2$ is minimized.

6T.1 A general approach to treat the relationships between the scaling parameter γ and the lag and frequency windows $w_\gamma(\tau)$, and $W_\gamma(\omega)$ [see (6.53)] can be given as follows. Choose an even function $w(x)$ such that $w(0) = 1$ and $w(x) = 0$, $|x| > 1$, with Fourier transform

$$W(\lambda) = \int_{-\infty}^{\infty} w(x)e^{-ix\lambda}dx$$

Let

$$\overline{W} = 2\pi \int_{-\infty}^{\infty} W^2(\lambda)d\lambda, \quad = \int_{-\infty}^{\infty} \lambda^2 W(\lambda)d\lambda$$

Then define the lag window

$$w_\gamma(\tau) = w(\tau/\gamma)$$

This gives a frequency window

$$W_\gamma(\omega) = \sum_{\tau=-\gamma}^{\gamma} w_\gamma(\tau)e^{-i\tau\omega}$$

Show that, for large γ,

(a) $W_\gamma(\omega) \approx \gamma \cdot W(\gamma\omega)$

(b) $M(\gamma) \approx M/\gamma^2$

(c) $\overline{W}(\gamma) \approx \overline{W} \cdot \gamma$

where $M(\gamma)$ and $\overline{W}(\gamma)$ are defined by (6.56). Moreover, compute and compare $W_\gamma(\omega)$ and $\gamma \cdot W(\gamma \cdot \omega)$ for $w(x) = 1 - |x|$, $|x| \le 1$ (the Bartlett window). [Compare (6.57). See also Hannan (1970), Section V.4.]

6T.2 Let $\{v(t)\}$ be a stationary stochastic process with zero mean value and covariance function $R_v(\tau)$, such that

$$\sum_{-\infty}^{\infty} |\tau R_v(\tau)| < \infty$$

Let

$$S_N = \frac{1}{N}\sum_{t=1}^{N} \alpha_t v(t), \qquad |\alpha_t| \le C_1$$

Show that

$$E|S_N|^2 \le \frac{C_2}{N}$$

for some constant C_2.

6D.1 Prove (6.52) with the proper interpretation of values outside the interval $1 \le t \le N$.

6D.2 Prove a relaxed version of Lemma 6.2 with $|\rho_k(N)| \le C/\sqrt{N}$ by a direct application of Theorem 2.1 and the properties of periodograms of white noise.

6D.3 Prove (6.63) by using expressions analogous to (6A.3) and (6A.4).

6D.4 Prove (6.62) by using (6.63) and

$$\text{Re}\hat{G}(e^{i\omega}) = \frac{\hat{G}(e^{i\omega}) + \hat{G}(e^{-i\omega})}{2}$$

$$\text{Im}\hat{G}(e^{i\omega}) = \frac{\hat{G}(e^{i\omega}) - \hat{G}(e^{-i\omega})}{2i}$$

APPENDIX 6A: DERIVATION OF THE ASYMPTOTIC PROPERTIES OF THE SPECTRAL ANALYSIS ESTIMATE

Consider the transfer function estimate (6.46). In this appendix we shall derive the asymptotic properties (6.58) and (6.59). In order not to get too technical, some elements of the derivation will be kept heuristic. Recall that $\{u(t)\}$ here is regarded as a deterministic quasi-stationary sequence, and, hence, such that (6.47) holds.

We then have

$$E\hat{G}_N(e^{i\omega_0}) = \frac{\int_{-\pi}^{\pi} W_\gamma(\xi - \omega_0)|U_N(\xi)|^2\left[G_0(e^{i\xi}) + \rho_1(N)/U_N(\xi)\right]d\xi}{\int_{-\pi}^{\pi} W_\gamma(\xi - \omega_0)|U_N(\xi)|^2 d\xi}$$

$$\approx \frac{\int_{-\pi}^{\pi} W_\gamma(\xi - \omega_0)\Phi_u(\xi)G_0(e^{i\xi})d\xi}{\int_{-\pi}^{\pi} W_\gamma(\xi - \omega_0)\Phi_u(\xi)d\xi} \tag{6A.1}$$

using first Lemma 6.1 and then (6.47), neglecting the decaying term $\rho_1(N)$.

Now, expanding in Taylor series (prime denoting differentiation with respect to ω),

$$G_0(e^{i\xi}) \approx G_0(e^{i\omega_0}) + (\xi - \omega_0)G_0'(e^{i\omega_0}) + \frac{1}{2}(\xi - \omega_0)^2 G_0''(e^{i\omega_0})$$

$$\Phi_0(\xi) \approx \Phi_u(\omega_0) + (\xi - \omega_0)\Phi_u'(\omega_0) + \frac{1}{2}(\xi - \omega_0)^2 \Phi_u''(\omega_0)$$

and noting that, according to (6.56),

$$\int_{-\pi}^{\pi} (\xi - \omega_0) W_\gamma(\xi - \omega_0) d\xi = 0$$

$$\int_{-\pi}^{\pi} (\xi - \omega_0)^2 W_\gamma(\xi - \omega_0) d\xi = M(\gamma)$$

we find that the numerator of (6A.1) is approximately

$$G_0(e^{i\omega_0})\Phi_u(\omega_0) + M(\gamma)\left[\frac{1}{2}\Phi_u'' G_0 + \frac{1}{2}G_0''\Phi_u + \Phi_u' G_0'\right]$$

and the denominator

$$\Phi_u(\omega_0) + \frac{1}{2}M(\gamma)[\Phi_u'']$$

where we neglect effects that are of order $C_3(\gamma)$ [an order of magnitude smaller than $M(\gamma)$ as $\gamma \to \infty$; see (6.56)]. Equation (6A.1) thus gives

$$E\hat{G}_N(e^{i\omega_0}) \approx G_0(e^{i\omega_0}) + M(\gamma)\left[\frac{1}{2}G_0''(e^{i\omega_0}) + G_0'(e^{i\omega_0})\frac{\Phi_u'(\omega_0)}{\Phi_u(\omega_0)}\right]$$

which is (6.58).

For the variance expression, we first have from (6.28) and (6.46) that

$$\hat{G}_N(e^{i\omega_0}) - E\hat{G}_N(e^{i\omega_0}) \approx \frac{\int_{-\pi}^{\pi} W_\gamma(\xi - \omega_0)|U_N(\xi)|^2[V_N(\xi)/U_N(\xi)]d\xi}{\int_{-\pi}^{\pi} W_\gamma(\xi - \omega_0)|U_N(\xi)|^2 d\xi} \tag{6A.2}$$

Let us study the numerator of this expression. We write this, approximately, as a Riemann sum [see (6.41); we could have kept it discrete all along]:

$$\int_{-\pi}^{\pi} W_\gamma(\xi - \omega_0)\overline{U}_N(\xi)V_N(\xi)d\xi \approx A_N$$

$$\triangleq \frac{2\pi}{N}\sum_{k=-(N/2)+1}^{N/2} W_\gamma\left(\frac{2\pi k}{N} - \omega_0\right)\overline{U}_N\left(\frac{2\pi k}{N}\right)V_N\left(\frac{2\pi k}{N}\right) \tag{6A.3}$$

We have, with summation from $1 - N/2$ to $N/2$,

$$
\begin{aligned}
E A_N \overline{A}_N &= \frac{4\pi^2}{N^2} \sum_k \sum_\ell W_\gamma\left(\frac{2\pi k}{N} - \omega_0\right) W_\gamma\left(\frac{2\pi \ell}{N} - \omega_0\right) \overline{U}_N\left(\frac{2\pi k}{N}\right) \\
&\quad \times \overline{U}_N\left(-\frac{2\pi \ell}{N}\right) E V_N\left(\frac{2\pi k}{N}\right) V_N\left(-\frac{2\pi \ell}{N}\right) \qquad \text{(6A.4)} \\
&\approx \frac{4\pi^2}{N^2} \sum \left[W_\gamma\left(\frac{2\pi k}{N} - \omega_0\right)\right]^2 \cdot \left|U_N\left(\frac{2\pi k}{N}\right)\right|^2 \Phi_v\left(\frac{2\pi k}{N}\right)
\end{aligned}
$$

using (6.31) and neglecting the term $\rho_2(N)$.

Returning to the integral form, we thus have, using (6.47)

$$
E A_N \overline{A}_N \approx \frac{2\pi}{N} \int_{-\pi}^{\pi} W_\gamma^2(\xi - \omega_0) \Phi_u(\xi) \Phi_v(\xi) d\xi \approx \frac{1}{N} \overline{W}(\gamma) \Phi_u(\omega_0) \Phi_v(\omega_0)
$$

using (6.56) and the fact that, for large γ, $W_\gamma(\xi)$ is concentrated around $\xi = 0$.

The denominator of (6A.2) approximately equals $\Phi_u(\omega_0)$ for the same reason. We thus find that

$$
\operatorname{Var}[\hat{G}_N(e^{i\omega_0})] \approx \frac{(1/N)\overline{W}(\gamma)\Phi_u(\omega_0)\Phi_v(\omega_0)}{[\Phi_u(\omega_0)]^2}
$$

and (6.59) has been established.

7

PARAMETER ESTIMATION METHODS

Suppose a set of candidate models has been selected, and it is parametrized as a model structure (see Sections 4.5 and 5.7), using a parameter vector θ. The search for the best model within the set then becomes a problem of determining or estimating θ. There are many different ways of organizing such a search and also different views on what one should search for. In the present chapter we shall concentrate on the latter aspect: what should be meant by a "good model"? Computational issues (i.e., how to organize the actual search) will be dealt with in Chapters 10 and 11. The evaluation of the properties of the models that result under various conditions and using different methods is carried out in Chapters 8 and 9. In Chapter 15 we return to the estimation methods, and give a more user-oriented summary of recommended procedures.

7.1 GUIDING PRINCIPLES BEHIND PARAMETER ESTIMATION METHODS

Parameter Estimation Methods

We are now in the situation that we have selected a certain model structure $\mathcal{M}$, with particular models $\mathcal{M}(\theta)$ parametrized using the parameter vector $\theta \in D_{\mathcal{M}} \subset \mathbf{R}^d$. The set of models thus defined is

$$\mathcal{M}^* = \{\mathcal{M}(\theta) | \theta \in D_{\mathcal{M}}\} \tag{7.1}$$

Recall that each model represents a way of predicting future outputs. The predictor could be a linear filter, as discussed in Chapter 4:

$$\mathcal{M}(\theta) : \hat{y}(t|\theta) = W_y(q,\theta)y(t) + W_u(q,\theta)u(t) \tag{7.2}$$

This could correspond to one-step-ahead prediction for an underlying system description

$$y(t) = G(q,\theta)u(t) + H(q,\theta)e(t) \tag{7.3}$$

in which case

$$W_y(q,\theta) = \left[1 - H^{-1}(q,\theta)\right], \qquad W_u(q,\theta) = H^{-1}(q,\theta)G(q,\theta) \tag{7.4}$$

but it could also be arrived at from other considerations.

The predictor could also be a nonlinear filter, as discussed in Chapter 5, in which case we write it as a general function of past data Z^{t-1}:

$$\mathcal{M}(\theta) : \hat{y}(t|\theta) = g(t, Z^{t-1}; \theta) \tag{7.5}$$

The model $\mathcal{M}(\theta)$ may also contain (model) assumptions about the character of the associated prediction errors, such as their variances ($\lambda(\theta)$) or their probability distribution (PDF $f_e(x, \theta)$).

We are also in the situation that we have collected, or are about to collect, a batch of data from the system:

$$Z^N = [y(1), u(1), y(2), u(2), \ldots, y(N), u(N)] \tag{7.6}$$

The problem we are faced with is to decide upon how to use the information contained in Z^N to select a proper value $\hat{\theta}_N$ of the parameter vector, and hence a proper member $\mathcal{M}(\hat{\theta}_N)$ in the set $\mathcal{M}^*$. Formally speaking, we have to determine a mapping from the data Z^N to the set $D_{\mathcal{M}}$:

$$Z^N \to \hat{\theta}_N \in D_{\mathcal{M}} \tag{7.7}$$

Such a mapping is a *parameter estimation method*.

Evaluating the Candidate Models

We are looking for a test by which the different models' ability to "describe" the observed data can be evaluated. We have stressed that the essence of a model is its prediction aspect, and we shall also judge its performance in this respect. Thus let the prediction error given by a certain model $\mathcal{M}(\theta_*)$ be given by

$$\varepsilon(t, \theta_*) = y(t) - \hat{y}(t|\theta_*) \tag{7.8}$$

When the data set Z^N is known, these errors can be computed for $t = 1, 2, \ldots, N$.

A "good" model, we say, is one that is good at predicting, that is, one that produces small prediction errors when applied to the observed data. Note that there is considerable flexibility in selecting various predictor functions, and this gives a corresponding freedom in defining "good" models in terms of prediction performance. A guiding principle for parameter estimation thus is:

> *Based on Z^t we can compute the prediction error $\varepsilon(t, \theta)$ using (7.8). At time $t = N$, select $\hat{\theta}_N$ so that the prediction errors $\varepsilon(t, \hat{\theta}_N)$, $t = 1, 2, \ldots, N$, become as small as possible.* *(7.9)*

The question is how to qualify what "small" should mean. In this chapter we shall describe two such approaches. One is to form a scalar-valued norm or criterion function that measures the size of ε. This approach is dealt with in Sections 7.2 to 7.4. Another approach is to demand that $\varepsilon(t, \hat{\theta}_N)$ be uncorrelated with a given data sequence. This corresponds to requiring that certain "projections" of $\varepsilon(t, \hat{\theta}_N)$ are zero and is further discussed in Sections 7.5 and 7.6.

7.2 MINIMIZING PREDICTION ERRORS

The prediction-error sequence in (7.8) can be seen as a vector in $\mathbf{R}^N$. The "size" of this vector could be measured using any norm in $\mathbf{R}^N$, quadratic or nonquadratic. This leaves a substantial amount of choices. We shall restrict the freedom somewhat by only considering the following way of evaluating "how large" the prediction-error sequence is: Let the prediction-error sequence be filtered through a stable linear filter $L(q)$:

$$\varepsilon_F(t, \theta) = L(q)\varepsilon(t, \theta), \qquad 1 \le t \le N \tag{7.10}$$

Then use the following norm:

$$V_N(\theta, Z^N) = \frac{1}{N} \sum_{t=1}^{N} \ell\left(\varepsilon_F(t, \theta)\right) \tag{7.11}$$

where $\ell(\cdot)$ is a scalar-valued (typically positive) function.

The function $V_N(\theta, Z^N)$ is, for given Z^N, a well-defined scalar-valued function of the model parameter θ. It is a natural measure of the validity of the model $\mathcal{M}(\theta)$. The estimate $\hat{\theta}_N$ is then defined by minimization of (7.11):

$$\hat{\theta}_N = \hat{\theta}_N(Z^N) = \arg\min_{\theta \in D_{\mathcal{M}}} V_N(\theta, Z^N) \tag{7.12}$$

Here arg min means "the minimizing argument of the function." If the minimum is not unique, we let arg min denote the set of minimizing arguments. The mapping (7.7) is thus defined implicitly by (7.12).

This way of estimating θ contains many well-known and much used procedures. We shall use the general term *prediction-error identification methods* (PEM) for the family of approaches that corresponds to (7.12). Particular methods, with specific "names" attached to themselves, are obtained as special cases of (7.12), depending on the choice of $\ell(\cdot)$, the choice of prefilter $L(\cdot)$, the choice of model structure, and, in some cases, the choice of method by which the minimization is realized. We shall give particular attention to two especially well known members in the family (7.12) in the subsequent two sections. First, however, let us discuss some aspects on the choices of $L(q)$ and $\ell(\cdot)$ in (7.10) and (7.11). See also Section 15.2.

Choice of L

The effect of the filter L is to allow extra freedom in dealing with non-momentary properties of the prediction errors. Clearly, if the predictor is linear and time invariant, and y and u are scalars, then the result of filtering ε, is the same as first filtering the input-output data and then applying the predictors.

The effect of L is best understood in a frequency-domain interpretation and a full discussion will be postponed to Section 14.4. It is clear, however, that by the use of L, effects of high-frequency disturbances, not essential to the modeling problem, or slow drift terms and the like, can be removed. It also seems reasonable that certain properties of the models may be enhanced or suppressed by a properly selected L. L thus acts like *frequency weighting*.

The following particular aspect of the filtering (7.10) should be noted. If a model (7.3) is used, the filtered error $\varepsilon_F(t, \theta)$ is given by

$$\varepsilon_F(t,\theta) = L(q)\varepsilon(t,\theta) = \left[L^{-1}(q)H(q,\theta)\right]^{-1}[y(t) - G(q,\theta)u(t)] \tag{7.13}$$

The effect of prefiltering is thus identical to changing the noise model from $H(q, \theta)$ to

$$\overline{H}_L(q,\theta) = L^{-1}(q)H(q,\theta) \tag{7.14}$$

When we describe and analyze methods that employ general noise models in linear systems, we shall usually confine ourselves to $L(q) \equiv 1$, since the option of prefiltering is taken care of by the freedom in selecting $H(q, \theta)$. A discussion of the use and effects of $L(q)$ in practical terms will be given in Section 14.4.

Choice of ℓ

For the choice of $\ell(\cdot)$, a first candidate would be a quadratic norm:

$$\ell(\varepsilon) = \tfrac{1}{2}\varepsilon^2 \tag{7.15}$$

and this is indeed a standard choice, which is convenient both for computation and analysis. Questions of robustness against bad data may, however, warrant other norms, which we shall discuss in some detail in Section 15.2. One may also conceive situations where the "best" norm is not known beforehand so that it is reasonable to parametrize the norm itself:

$$\ell(\varepsilon, \theta) \tag{7.16}$$

Often the parametrization of the norm is independent of the model parametrization:

$$\theta = \begin{bmatrix} \theta' \\ \alpha \end{bmatrix} : \; \ell\left(\varepsilon(t,\theta),\theta\right) = \ell\left(\varepsilon(t,\theta'),\alpha\right) \tag{7.17}$$

An exception to this case is given in Problem 7E.4.

Time-varying Norms

It may happen that measurements at different time instants are considered to be of varying reliability. The reason may be that the degree of noise corruption changes or that certain measurements are less representative for the system's properties. In such cases we are motivated to let the norm ℓ be time varying:

$$V_N(\theta, Z^N) = \frac{1}{N}\sum_{t=1}^{N} \ell\left(\varepsilon(t,\theta),\theta,t\right) \tag{7.18}$$

In this way less reliable measurements can be associated with less weight in the criterion.

We shall frequently work with a criterion where the weighting is made explicitly by a weighting function $\beta(N, t)$:

$$V_N(\theta, Z^N) = \sum_{t=1}^{N} \beta(N,t)\ell\left(\varepsilon(t,\theta),\theta\right) \tag{7.19}$$

For fixed N, the N-dependence of $\beta(N, t)$ is of course immaterial. However, when estimates $\hat{\theta}_N$ for different N are compared, as for example in recursive identification (see Chapter 11), it becomes interesting to discuss how $\beta(N, t)$ varies with N. We shall return to this issue in Section 11.2.

Frequency-domain Interpretation of Quadratic Prediction-error Criteria for Linear Time-invariant Models

Let us consider the quadratic criterion error (7.12) and (7.15) for the standard linear model (7.3)

$$\begin{aligned} V_N(\theta, Z^N) &= \frac{1}{N}\sum_{t=1}^{N} \tfrac{1}{2}\varepsilon^2(t, \theta) \\ \varepsilon(t, \theta) &= H^{-1}(q, \theta)\,[y(t) - G(q, \theta)u(t)] \end{aligned} \tag{7.20}$$

Let $E_N(2\pi k/N, \theta), k = 0, 1, \ldots, N-1$, be the DFT of $\varepsilon(t, \theta), t = 1, 2, \ldots, N$:

$$E_N(2\pi k/N, \theta) = \frac{1}{\sqrt{N}}\sum_{t=1}^{N} \varepsilon(t, \theta)e^{-2\pi ikt/N}$$

Then, by Parseval's relation (2.44),

$$V_N(\theta, Z^N) = \frac{1}{N}\frac{1}{2}\sum_{k=1}^{N-1} |E_N(2\pi k/N, \theta)|^2 \tag{7.21}$$

Now let

$$w(t, \theta) = G(q, \theta)u(t)$$

Then the DFT of $w(t, \theta)$ is, according to Theorem 2.1,

$$W_N(\omega, \theta) = G(e^{i\omega}, \theta)U_N(\omega) + R_N(\omega)$$

with

$$|R_N(\omega)| \le \frac{C}{\sqrt{N}}$$

The DFT of $s(t, \theta) = y(t) - w(t, \theta)$ then is

$$S_N(\omega, \theta) = Y_N(\omega) - G(e^{i\omega}, \theta)U_N(\omega) - R_N(\omega)$$

Finally,

$$\varepsilon(t, \theta) = H^{-1}(q, \theta)s(t, \theta)$$

has the DFT, again using Theorem 2.1,

$$E_N(\omega) = H^{-1}(e^{i\omega}, \theta)S_N(\omega, \theta) + \tilde{R}_N(\omega)$$

with

$$\left|\tilde{R}_N(\omega)\right| \leq \frac{C}{\sqrt{N}}$$

Inserting this into (7.21) gives

$$V_N(\theta, Z^N) = \frac{1}{N}\sum_{k=0}^{N-1}\frac{1}{2}\left|H(e^{2\pi ik/N}, \theta)\right|^{-2}$$
$$\times \left|Y_N(2\pi k/N) - G(e^{2\pi ik/N}, \theta)U_N(2\pi k/N)\right|^2 + \overline{R}_N$$

with $|\overline{R}_N| \leq C/\sqrt{N}$, or, using the definition of the ETFE $\hat{\hat{G}}_N$ in (6.24),

$$V_N(\theta, Z^N) = \frac{1}{N}\sum_{k=0}^{N-1}\left\{\frac{1}{2}\left|\hat{\hat{G}}_N(e^{2\pi ik/N}) - G(e^{2\pi ik/N}, \theta)\right|^2\right.$$
$$\left.\times\; Q_N(2\pi k/N, \theta) + \overline{R}_N\right\} \tag{7.22}$$

with

$$Q_N(\omega, \theta) = \frac{|U_N(\omega)|^2}{\left|H(e^{i\omega}, \theta)\right|^2} \tag{7.23}$$

First notice that, apart from the remainder term $\overline{R}_N$, the expression (7.22) coincides with the weighted least-squares criterion for a model:

$$\hat{\hat{G}}_N(e^{2\pi ik/N}) = G(e^{2\pi ik/N}, \theta) + v(k) \tag{7.24}$$

Compare with (II.96) and (II.97). According to Lemma 6.1, the variance of $v(k)$ is, asymptotically, $\Phi_v(2\pi k/N)/\left|U_N(2\pi k/N)\right|^2$, so the weighting coefficient $Q_N(\omega, \theta)$ is the inverse variance, which is optimal for linear regressions, according to (II.65). In (7.23) the unknown noise spectrum $\Phi_v(\omega)$ is replaced by the model noise spectrum $\left|H(e^{i\omega}, \theta)\right|^2$. Consequently, the prediction-error methods can be seen as methods of fitting the ETFE to the model transfer function with a weighted norm, corresponding to the model signal-to-noise ratio at the frequency in question. For notational reasons, it is instructive to rewrite the sum (7.22) approximately as an integral:

$$V_N(\theta, Z^N) \approx \frac{1}{2\pi}\int_{-\pi}^{\pi}\frac{1}{2}\left|\hat{\hat{G}}_N(e^{i\omega}) - G(e^{i\omega}, \theta)\right|^2 Q_N(\omega, \theta)\, d\omega \tag{7.25}$$

The shift of integration interval from $(0, 2\pi)$ to $(-\pi, \pi)$ is possible since the integrand is periodic.

With this interpretation we have described the prediction-error estimate as an alternative way of smoothing the ETFE, showing a strong conceptual relationship to the spectral analysis methods of Section 6.4. See Problem 7G.2 for a direct tie.

When we specialize to the case of a time series [no input and $G(q, \theta) \equiv 0$], the criterion (7.25) takes the form

$$V_N(\theta, Z^N) = \frac{1}{2\pi} \int_{-\pi}^{\pi} \left| \frac{Y_N(\omega)}{H(e^{i\omega}, \theta)} \right|^2 d\omega \tag{7.26}$$

Such parametric estimators of spectra are known as "Whittle-type estimators," after Whittle (1951).

In Section 7.8 we shall return to frequency domain criteria. There, however, we take another viewpoint and assume that the observed data are in the frequency domain, being Fourier transforms of the input and output time domain signals.

Multivariable Systems (*)

For multioutput systems, the counterpart of the quadratic criterion is

$$\ell(\varepsilon) = \tfrac{1}{2}\varepsilon^T \Lambda^{-1} \varepsilon \tag{7.27}$$

for some symmetric, positive semidefinite $p \times p$ matrix Λ that weights together the relative importance of the components of ε.

One might discuss what is the best choice of norm Λ. We shall do that in some detail in Section 15.2. Here we only remark that, just as in (7.16), the parameter vector θ could be extended to include components of Λ, and the function ℓ will then be an appropriate function of θ.

As a variant of the criterion (7.11), where a scalar $\ell(\varepsilon)$ is formed for each t, we could first form the $p \times p$ matrix

$$Q_N(\theta, Z^N) = \frac{1}{N} \sum_{t=1}^{N} \varepsilon(t, \theta)\varepsilon^T(t, \theta) \tag{7.28}$$

and let the criterion be a scalar-valued function of this matrix:

$$V_N(\theta, Z^N) = h\left(Q_N(\theta, Z^N)\right) \tag{7.29}$$

The criterion (7.27) is then obtained by

$$h(Q) = \tfrac{1}{2}\text{tr}(Q\Lambda^{-1}) \tag{7.30}$$

7.3 LINEAR REGRESSIONS AND THE LEAST-SQUARES METHOD

Linear Regressions

We found in both Sections 4.2 and 5.2 that linear regression model structures are very useful in describing basic linear and nonlinear systems. The linear regression employs a predictor (5.67)

$$\hat{y}(t|\theta) = \varphi^T(t)\theta + \mu(t) \tag{7.31}$$

that is linear in θ. Here φ is the vector of regressors, the *regression vector*. Recall that for the ARX structure (4.7) we have

$$\varphi(t) = [-y(t-1)\ -y(t-2)\dots\ -y(t-n_a)\quad u(t-1)\dots u(t-n_b)]^T \tag{7.32}$$

In (7.31), $\mu(t)$ is a known data-dependent vector. For notational simplicity we shall take $\mu(t) = 0$ in the remainder of this section; it is quite straightforward to include it. See Problem 7D.1.

Linear regression forms a standard topic in statistics. The reader could consult Appendix II for a refresher of basic properties. The present section can, however, be read independently of Appendix II.

Least-squares Criterion

With (7.31) the prediction error becomes

$$\varepsilon(t,\theta) = y(t) - \varphi^T(t)\theta$$

and the criterion function resulting from (7.10) and (7.11), with $L(q) = 1$ and $\ell(\varepsilon) = \frac{1}{2}\varepsilon^2$, is

$$V_N(\theta, Z^N) = \frac{1}{N}\sum_{t=1}^{N} \tfrac{1}{2}\left[y(t) - \varphi^T(t)\theta\right]^2 \tag{7.33}$$

This is the *least-squares criterion* for the linear regression (7.31). The unique feature of this criterion, developed from the linear parametrization and the quadratic criterion, is that it is a quadratic function in θ. Therefore, it can be minimized analytically, which gives, provided the indicated inverse exists,

$$\hat{\theta}_N^{\text{LS}} = \arg\min V_N(\theta, Z^N) = \left[\frac{1}{N}\sum_{t=1}^{N}\varphi(t)\varphi^T(t)\right]^{-1} \frac{1}{N}\sum_{t=1}^{N}\varphi(t)y(t) \tag{7.34}$$

the *least-squares estimate (LSE)* (see Problem 7D.2).

Introduce the $d \times d$ matrix

$$R(N) = \frac{1}{N}\sum_{t=1}^{N}\varphi(t)\varphi^T(t) \tag{7.35}$$

and the d-dimensional column vector

$$f(N) = \frac{1}{N}\sum_{t=1}^{N}\varphi(t)y(t) \tag{7.36}$$

In the case (7.32), $\varphi(t)$ contains lagged input and output variables, and the entries of the quantities (7.35) and (7.36) will be of the form

$$[R(N)]_{ij} = \frac{1}{N}\sum_{t=1}^{N} y(t-i)y(t-j), \qquad 1 \le i, j \le n_a$$

and similar sums of $u(t-r) \cdot u(t-s)$ or $u(t-r) \cdot y(t-s)$ for the other entries of $R(N)$. That is, they will consist of estimates of the covariance functions of $\{y(t)\}$ and $\{u(t)\}$. The LSE can thus be computed using only such estimates and is therefore related to correlation analysis, as described in Section 6.1.

Properties of the LSE

The least-squares method is a special case of the prediction-error identification method (7.12). An analysis of its properties is therefore contained in the general treatment in Chapters 8 and 9. It is, however, useful to include a heuristic investigation of the LSE at this point.

Suppose that the observed data actually have been generated by

$$y(t) = \varphi^T(t)\theta_0 + v_0(t) \tag{7.37}$$

for some sequence $\{v_0(t)\}$. We may think of θ_0 as a "true value" of the parameter vector.

As in (1.14)–(1.15) we find that

$$\lim_{N\to\infty} \hat{\theta}_N^{\text{LS}} - \theta_0 = \lim_{N\to\infty} R^{-1}(N)\frac{1}{N}\sum_{t=1}^{N}\varphi(t)v_0(t) = (R^*)^{-1}f^*,$$

$$R^* = \overline{E}\varphi(t)\varphi^T(t), \quad f^* = \overline{E}\varphi(t)v_0(t) \tag{7.38}$$

provided v_0 and φ are quasi-stationary, so that Theorem 2.3 can be applied. For the LSE to be *consistent*, that is, for $\hat{\theta}_N^{\text{LS}}$ to converge to θ_0, we thus have to require:

i. R^* is non-singular. This will be secured by the input properties, as in (1.17)–(1.18), and discussed in much more detail in Chapter 13.

ii. $f^* = 0$. This will be the case if either:

(a) $\{v_0(t)\}$ is a sequence of independent random variables with zero mean values (white noise). Then $v_0(t)$ will not depend on what happened up to time $t-1$ and hence $E\varphi(t)v_0(t) = 0$.

(b) The input sequence $\{u(t)\}$ is independent of the zero mean sequence $\{v_0(t)\}$ and $n_a = 0$ in (7.32). Then $\varphi(t)$ contains only u-terms and hence $E\varphi(t)v_0(t) = 0$.

When $n_a > 0$ so that $\varphi(t)$ contains $y(k)$, $t - n_a \le k \le t-1$, and $v_0(t)$ is not white noise, then (usually) $E\varphi(t)v_0(t) \neq 0$. This follows since $\varphi(t)$ contains $y(t-1)$, while $y(t-1)$ contains the term $v_0(t-1)$ that is correlated with $v_0(t)$. Therefore, we may expect consistency only in cases (a) and (b).

Weighted Least Squares

Just as in (7.18) and (7.19), the different measurements could be assigned different weights in the least-squares criterion:

$$V_N(\theta, Z^N) = \frac{1}{N}\sum_{t=1}^{N}\alpha_t\left[y(t) - \varphi^T(t)\theta\right]^2 \tag{7.39}$$

or

$$V_N(\theta, Z^N) = \sum_{t=1}^{N}\beta(N,t)\left[y(t) - \varphi^T(t)\theta\right]^2 \tag{7.40}$$

The expression for the resulting estimate is quite analogous to (7.34):

$$\hat{\theta}_N^{\text{LS}} = \left[\sum_{t=1}^{N}\beta(N,t)\varphi(t)\varphi^T(t)\right]^{-1}\sum_{t=1}^{N}\beta(N,t)\varphi(t)y(t) \tag{7.41}$$

Multivariable Case (*)

If the output $y(t)$ is a p-vector and the norm (7.27) is used, the LS criterion takes the form

$$V_N(\theta, Z^N) = \frac{1}{N}\sum_{t=1}^{N}\frac{1}{2}\left[y(t) - \boldsymbol{\varphi}^T(t)\theta\right]^T\Lambda^{-1}\left[y(t) - \boldsymbol{\varphi}^T(t)\theta\right] \tag{7.42}$$

This gives the estimate

$$\hat{\theta}_N^{\text{LS}} = \left[\frac{1}{N}\sum_{t=1}^{N}\boldsymbol{\varphi}(t)\Lambda^{-1}\boldsymbol{\varphi}^T(t)\right]^{-1}\frac{1}{N}\sum_{t=1}^{N}\boldsymbol{\varphi}(t)\Lambda^{-1}y(t) \tag{7.43}$$

In case we use the particular parametrization (4.56) with $\boldsymbol{\theta}$ as an $r \times p$ matrix,

$$\hat{y}(t|\theta) = \boldsymbol{\theta}^T\varphi(t) \tag{7.44}$$

the LS criterion becomes

$$V_N(\boldsymbol{\theta}, Z^N) = \frac{1}{N}\sum_{t=1}^{N}\left\|y(t) - \boldsymbol{\theta}^T\varphi(t)\right\|^2 \tag{7.45}$$

with the estimate

$$\hat{\boldsymbol{\theta}}_N^{\text{LS}} = \left[\frac{1}{N}\sum_{t=1}^{N}\varphi(t)\varphi^T(t)\right]^{-1}\frac{1}{N}\sum_{t=1}^{N}\varphi(t)y^T(t) \tag{7.46}$$

(see problem 7D.2). The expression (7.46) brings out the advantages of the structure (7.44): To determine the $r \times p$ estimate $\hat{\theta}_N$, it is sufficient to invert an $r \times r$ matrix. In (7.43) θ is a $p \cdot r$ vector and the matrix inversion involves a $pr \times pr$ matrix.

Colored Equation-error Noise (*)

The LS method has many advantages, the most important one being that the global minimum of (7.33) can be found efficiently and unambiguously (no local minima other than global ones exist). Its main shortcoming relates to the asymptotic properties quoted previously: If, in a difference equation,

$$\begin{aligned} y(t) + a_1 y(t-1) + \cdots + a_{n_a} y(t-n_a) \\ = b_1 u(t-1) + \cdots + b_{n_b} u(t-n_b) + v(t) \end{aligned} \tag{7.47}$$

the equation error $v(t)$ is not white noise, then the LSE will not converge to the true values of a_i and b_i. To deal with this problem, we may incorporate further modeling of the equation error $v(t)$ as discussed in Section 4.2, let us say

$$v(t) = \kappa(q)e(t) \tag{7.48}$$

with e white and κ linear filter. Models employing (7.48) will typically take us out from the LS environment, except in two cases, which we now discuss.

Known noise properties: If in (7.47) and (7.48) a_i and b_i are unknown, but κ is a known filter (not too realistic a situation), we have

$$A(q)y(t) = B(q)u(t) + \kappa(q)e(t) \tag{7.49}$$

Filtering (7.49) through the filter $\kappa^{-1}(q)$ gives

$$A(q)y_F(t) = B(q)u_F(t) + e(t) \tag{7.50}$$

where

$$y_F(t) = \kappa^{-1}(q)y(t), \qquad u_F(t) = \kappa^{-1}(q)u(t) \tag{7.51}$$

Since e is white, the LS method can be applied to (7.50) without problems. Notice that this is equivalent to applying the filter $L(q) = \kappa^{-1}(q)$ in (7.10).

High-order models: Suppose that the noise v can be well described by $\kappa(q) = 1/D(q)$ in (7.48), where $D(q)$ is a polynomial of degree r. [That is, $v(t)$ is supposed to be an autoregressive (AR) process of order r.] This gives

$$A(q)y(t) = B(q)u(t) + \frac{1}{D(q)}e(t) \tag{7.52}$$

or

$$A(q)D(q)y(t) = B(q)D(q)u(t) + e(t) \tag{7.53}$$

Applying the LS method to (7.53) with orders $n_A = n_a + r$ and $n_B = n_b + r$ gives, since e is white, consistent estimates of AD and BD. Hence the transfer function from u to y,

$$\frac{B(q)D(q)}{A(q)D(q)} = \frac{B(q)}{A(q)}$$

is correctly estimated. This approach was called *repeated least squares* in Åström and Eykhoff (1971). See also Söderström (1975b)and Stoica (1976).

Estimating State Space Models Using Least Squares Techniques (Subspace Methods)

A linear system can always be represented in state space form as in (4.84):

$$\begin{aligned} x(t+1) &= Ax(t) + Bu(t) + w(t) \\ y(t) &= Cx(t) + Du(t) + \nu(t) \end{aligned} \tag{7.54}$$

with white noises w and ν. Alternatively we could just represent the input-output dynamics as in (4.80):

$$\begin{aligned} x(t+1) &= Ax(t) + Bu(t) \\ y(t) &= Cx(t) + Du(t) + v(t) \end{aligned} \tag{7.55}$$

where the noise at the output, v, very well could be colored. It should be noted that the input-output dynamics could be represented with a lower order model in (7.55) than in (7.54) since describing the noise character might require some extra states.

To estimate such a model, the matrices can be parameterized in ways that are described in Section 4.3 or Appendix 4A—either from physical grounds or as black boxes in canonical forms. Then these parameters can be estimated using the techniques dealt with in Section 7.4.

However, there are also other possibilities: We assume that we have no insight into the particular structure, and we would just estimate any matrices A, B, C, and D that give a good description of the input-output behavior of the system. Since there are an infinite number of such matrices that describe the same system (the similarity transforms), we will have to fix the coordinate basis of the state-space realization.

Let us for a moment assume that not only are u and y measured, but also the sequence of state vectors x. This would, by the way, fix the state-space realization coordinate basis. Now, with known u, y and x, the model (7.54) becomes a linear regression: the unknown parameters, all of the matrix entries in all the matrices, mix with measured signals in linear combinations. To see this clearly, let

$$Y(t) = \begin{bmatrix} x(t+1) \\ y(t) \end{bmatrix}, \quad \Theta = \begin{bmatrix} A & B \\ C & D \end{bmatrix}$$

$$\Phi(t) = \begin{bmatrix} x(t) \\ u(t) \end{bmatrix}, \quad E(t) = \begin{bmatrix} w(t) \\ \nu(t) \end{bmatrix}$$

Then, (7.54) can be rewritten as

$$Y(t) = \Theta\Phi(t) + E(t) \tag{7.56}$$

From this, all the matrix elements in Θ can be estimated by the simple least squares method (which in the case of Gaussian noise and known covariance matrix coincides with the maximum likelihood method), as described above in (7.44)–(7.46). The covariance matrix for $E(t)$ can also be estimated easily as the sample sum of the squared model residuals. That will give the covariance matrices as well as the cross covariance matrix for w and v. These matrices will, among other things, allow us to compute the Kalman filter for (7.54). Note that all of the above holds without changes for multivariable systems, i.e., when the output and input signals are vectors.

The problem is how to obtain the state vector sequence x. Some basic realization theory was reviewed in Appendix 4A, from which the essential results can be quoted as follows:

Let a system be given by the impulse response representation

$$y(t) = \sum_{j=0}^{\infty} [h_u(j)u(t-j) + h_e(j)e(t-j)] \tag{7.57}$$

where u is the input and e the innovations. Let the formal k-step ahead predictors be defined by just deleting the contributions to $y(t)$ from $e(j), u(j);\ j = t, \dots, t-k+1$:

$$\hat{y}(t|t-k) = \sum_{j=k}^{\infty} [h_u(j)u(t-j) + h_e(j)e(t-j)] \tag{7.58}$$

No attempt is thus made to predict the inputs $u(j);\ j = t, \dots, t-k+1$ from past data. Define

$$\hat{Y}_r(t) = \begin{bmatrix} \hat{y}(t|t-1) \\ \vdots \\ \hat{y}(t+r-1|t-1) \end{bmatrix} \tag{7.59a}$$

$$\hat{\mathbf{Y}} = \begin{bmatrix} \hat{Y}_r(1) & \dots & \hat{Y}_r(N) \end{bmatrix} \tag{7.59b}$$

Then the following is true as $N \to \infty$ (see Lemmas 4A.1 and 4A.2 and their proofs):

1. The system (7.57) has an nth order minimal state space description if and only if the rank $\hat{\mathbf{Y}}$ is equal to n for all $r \geq n$.

2. The state vector of any minimal realization in innovations form can be chosen as linear combinations of $\hat{Y}_r$ that form a row basis for $\hat{\mathbf{Y}}$, i.e.,

$$x(t) = L\hat{Y}_r(t) \tag{7.60}$$

where the $n \times pr$ matrix L is such that $L\hat{\mathbf{Y}}$ spans $\hat{\mathbf{Y}}$. (p is the dimension of the output vector $y(t)$.)

Note that the canonical state space representations described in Appendix 4A correspond to L matrices that just pick out certain rows of $\hat{Y}_n$. In general, we are not confined to such choices, but may pick L so that $x(t)$ becomes a well-conditioned basis.

It is clear that the facts above will allow us to find a suitable state vector from data. The only remaining problem is to estimate the k-step ahead predictors. The true predictor $\hat{y}(t+k-1|t-1)$ is given by (7.58). The innovation $e(j)$ can be written as a linear combination of past input-output data. The predictor can thus be expressed as a linear function of $u(i), y(i), i \le t-1$. For practical reasons the predictor is approximated so that it only depends on a fixed and finite amount of past data, like the s_1 past outputs and the s_2 past inputs. This means that it takes the form

$$\begin{aligned}\hat{y}(t+k-1|t-1) = \alpha_1 y(t-1) + \ldots + \alpha_{s_1} y(t-s_1) \\ + \beta_1 u(t-1) + \ldots + \beta_{s_2} u(t-s_2)\end{aligned} \tag{7.61}$$

This predictor can then efficiently be determined by another linear least squares projection directly on the input output data. That is, set up the model

$$y(t+k-1) = \theta_k^T \varphi_s(t) + \gamma_k^T U_\ell(t) + \varepsilon(t+k-1) \tag{7.62}$$

or, dealing with all r predictors simultaneously

$$Y_r(t) = \Theta\varphi_s(t) + \Gamma U_\ell(t) + E(t) \tag{7.63}$$

Here:

$$\varphi_s(t) = \left[y^T(t-1) \ldots y^T(t-s_1) \;\; u^T(t-1) \ldots u^T(t-s_2)\right]^T \tag{7.64a}$$

$$U_\ell(t) = \left[u^T(t) \ldots u^T(t+\ell-1)\right]^T \tag{7.64b}$$

$$Y_r(t) = \left[y^T(t) \ldots y^T(t+r-1)\right]^T \tag{7.64c}$$

$$\Theta = \left[\theta_1 \ldots \theta_r\right]^T, \quad \Gamma = \left[\gamma_1 \ldots \gamma_r\right]^T \tag{7.64d}$$

$$E(t) = \left[\varepsilon^T(t) \ldots \varepsilon^T(t+r-1)\right]^T \tag{7.64e}$$

Moreover, ℓ is the number, typically equal to r, of input values whose influence on $Y_r(t)$ is to be accounted for. Now, Θ and Γ in (7.63) can be estimated using least squares, giving $\hat{\Theta}_N$ and $\hat{\Gamma}_N$. The k-step ahead predictors are then given by

$$\hat{Y}_r(t) = \hat{\Theta}_N \varphi_s(t) \tag{7.65}$$

For large enough s, this will give a good approximation of the true predictors.

Remark 1: The reason for the term U_ℓ is as follows: The values of $u(t+1), \ldots, u(t+k)$ affect $y(t+k-1)$. If these values can be predicted from past measurements—which is the case if u is not white noise—then the predictions of $y(t+k-1)$ based on past data will account also for the influence of U_ℓ. If we estimate (7.61) directly, this influence will thus be included. However, as demanded by (7.58), the influence of U_ℓ should be ignored in the "formal" k-step ahead predictor we are seeking. This is the reason why this influence is explicitly estimated in (7.62) and then thrown away in the predictor (7.65).

Remark 2: If we seek a state-space realization like (7.55) that does not model the noise properties—*an output error model*—we would just ignore the terms $e(t-j)$ in (7.57)–(7.58). This implies that the predictor in (7.62) would be based on past inputs only, i.e. $s_1 = 0$ in (7.64).

The method thus consists of the following steps:

Basic Subspace Algorithm (7.66)

1. Choose s_1, s_2, r and ℓ and form $\hat{Y}_r(t)$ in (7.65) and $\mathbf{Y}$ as in (7.59).
2. Estimate the rank n of $\mathbf{Y}$ and determine L in (7.60) so that $x(t)$ corresponds to a well-conditioned basis for it.
3. Estimate A, B, C, D and the noise covariance matrices by applying the LS method to the linear regression (7.56).

What we have described now is the *subspace projection* approach to estimating the matrices of the state-space model (7.54), including the basis for the representation and the noise covariance matrices. There are a number of variants of this approach. See among several references, e.g. Van Overschee and DeMoor (1996), Larimore (1983), and Verhaegen (1994).

The approach gives very useful algorithms for model estimation, and is particularly well suited for multivariable systems. The algorithms also allow numerically very reliable implementations, and typically produce estimated models with good quality. If desired, the quality may be improved by using the model as an initial estimate for the prediction error method (7.12). Then the model first needs to be transformed to a suitable parameterization.

The algorithms contain a number of choices and options, like how to choose ℓ, s_i and r, and also how to carry out step number 3. There are also several "tricks" to do step 3 so as to achieve consistent estimates even for finite values of s_i. Accordingly, several variants of this method exist. In Section 10.5 we shall give more algorithmic details around this approach.

7.4 A STATISTICAL FRAMEWORK FOR PARAMETER ESTIMATION AND THE MAXIMUM LIKELIHOOD METHOD

So far we have not appealed to any statistical arguments for the estimation of θ. In fact, our framework of fitting models to data makes sense regardless of a stochastic setting of the data. It is, however, useful and instructive at this point to briefly describe basic aspects of statistical parameter estimation and relate them to our framework.

Estimators and the Principle of Maximum Likelihood

The area of statistical inference, as well as that of system identification and parameter estimation, deals with the problem of extracting information from observations that themselves could be unreliable. The observations are then described as realizations of stochastic variables. Suppose that the observations are represented by the random variable $y^N = (y(1), y(2), \ldots, y(N))$ that takes values in $\mathbf{R}^N$. The probability density function (PDF) of y^N is supposed to be

$$f(\theta; x_1, x_2, \ldots, x_N) = f_y(\theta; x^N) \tag{7.67}$$

That is,

$$P\left(y^N \in A\right) = \int_{x^N \in A} f_y(\theta; x^N)\, dx^N \tag{7.68}$$

In (7.67), θ is a d-dimensional parameter vector that describes properties of the observed variable. These are supposed to be unknown, and the purpose of the observation is in fact to estimate the vector θ using y^N. This is accomplished by an *estimator*,

$$\hat{\theta}(y^N) \tag{7.69}$$

which is a function from $\mathbf{R}^N$ to $\mathbf{R}^d$. If the observed value of y^N is y_*^N, then consequently the resulting estimate is $\hat{\theta}_* = \hat{\theta}(y_*^N)$.

Many such estimator functions are possible. A particular one that maximizes the probability of the observed event is the celebrated maximum likelihood estimator, introduced by Fisher (1912). It can be defined as follows: The joint probability density function for the random vector to be observed is given by (7.67). The probability that the realization (= observation) indeed should take the value y_*^N is thus proportional to

$$f_y(\theta; y_*^N)$$

This is a deterministic function of θ once the numerical value y_*^N is inserted. This function is called the *likelihood function*. It reflects the "likelihood" that the observed event should indeed take place. A reasonable estimator of θ could then be to select it so that the observed event becomes "as likely as possible." That is, we seek

$$\hat{\theta}_{\mathrm{ML}}(y_*^N) = \arg\max_{\theta} f_y(\theta; y_*^N) \tag{7.70}$$

where the maximization is performed for fixed y_*^N. This function is known as the *maximum likelihood estimator* (MLE).

An Example

Let $y(i)$, $i = 1, \ldots, N$, be independent random variables with normal distribution with (unknown) means θ_0 (independent of i) and (known) variances λ_i:

$$y(i) \in N(\theta_0, \lambda_i) \tag{7.71}$$

A common estimator of θ_0 is the sample mean:

$$\hat{\theta}_{\text{SM}}(y^N) = \frac{1}{N} \sum_{i=1}^{N} y(i) \tag{7.72}$$

To calculate the MLE, we start by determining the joint PDF (7.67) for the observations. Since the PDF for $y(i)$ is

$$\frac{1}{\sqrt{2\pi\lambda_i}} \exp\left[-\frac{(x_i - \theta)^2}{2\lambda_i}\right]$$

and the $y(i)$ are independent, we have

$$f_y(\theta; x^N) = \prod_{i=1}^{N} \frac{1}{\sqrt{2\pi\lambda_i}} \exp\left[-\frac{(x_i - \theta)^2}{2\lambda_i}\right] \tag{7.73}$$

The likelihood function is thus given by $f_y(\theta; y^N)$. Maximizing the likelihood function is the same as maximizing its logarithm. Thus

$$\begin{aligned} \hat{\theta}_{\text{ML}}(y^N) &= \arg\max_{\theta} \log f_y(\theta; y^N) \\ &= \arg\max_{\theta} \left\{ -\frac{N}{2} \log 2\pi - \sum_{i=1}^{N} \frac{1}{2} \log \lambda_i - \frac{1}{2} \sum_{i=1}^{N} \frac{(y(i) - \theta)^2}{\lambda_i} \right\} \end{aligned} \tag{7.74}$$

from which we find

$$\hat{\theta}_{\text{ML}}(y^N) = \frac{1}{\sum_{i=1}^{N} (1/\lambda_i)} \sum_{i=1}^{N} \frac{y(i)}{\lambda_i} \tag{7.75}$$

Relationship to the Maximum A Posteriori (MAP) Estimate

The *Bayesian approach* gives a related but conceptually different treatment of the parameter estimation problem. In the Bayesian approach the parameter itself is thought of as a random variable. Based on observations of other random variables that are correlated with the parameter, we may infer information about its value.

Suppose that the properties of the observations can be described in terms of a parameter vector θ. With a Bayesian view we thus consider θ to be a random vector with a certain prior distribution ("prior" means before the observations have been made). The observations y^N are obviously correlated with this θ. After the observations have been obtained, we then ask for the posterior PDF for θ. From this posterior PDF, different estimates of θ can be determined, for example, the value for which the PDF attains its maximum ("the most likely value"). This is known as the *maximum a posteriori (MAP) estimate.*

Suppose that the conditional PDF for y^N, given θ, is

$$f_y(\theta; x^N) = P(y^N = x^N|\theta)$$

and that the prior PDF for θ is

$$g_\theta(z) = P(\theta = z)$$

[Here $P(A|B) =$ the conditional probability of the event A given the event B. We also allowed somewhat informal notation.] Using Bayes's rule (I.10) and with some abuse of notation, we thus find the posterior PDF for θ, i.e., the conditional PDF for θ, given the observations:

$$P(\theta|y^N) = \frac{P(y^N|\theta) \cdot P(\theta)}{P(y^N)} \sim f_y(\theta; y^N) \cdot g_\theta(\theta) \tag{7.76}$$

The posterior PDF as a function of θ is thus proportional to the likelihood function multiplied by the prior PDF. Often the prior PDF has an insignificant influence. Then the MAP estimate

$$\hat{\theta}_{\text{MAP}}(y^N) = \arg\max_\theta \left\{ f_y(\theta; y^N) \cdot g_\theta(\theta) \right\} \tag{7.77}$$

is close to the MLE (7.70).

Cramér-Rao Inequality

The quality of an estimator can be assessed by its mean-square error matrix:

$$P = E\left[\hat{\theta}(y^N) - \theta_0\right]\left[\hat{\theta}(y^N) - \theta_0\right]^T \tag{7.78}$$

Here θ_0 denotes the "true value" of θ, and (7.78) is evaluated under the assumption that the PDF of y^N is $f_y(\theta_0; y^N)$.

We may be interested in selecting estimators that make P small. It is then interesting to note that there is a lower limit to the values of P that can be obtained with various unbiased estimators. This is the so called *Cramér-Rao inequality:*

Let $\hat{\theta}(y^N)$ be an estimator of θ such that $E\hat{\theta}(y^N) = \theta_0$, where E evaluates the mean, assuming that the PDF of y^N is $f_y(\theta_0; y^N)$ (to hold for all values of θ_0), and suppose that y^N may take values in a subset of $\mathbf{R}^N$, whose boundary does not depend on θ. Then

$$E\left[\hat{\theta}(y^N) - \theta_0\right]\left[\hat{\theta}(y^N) - \theta_0\right]^T \geq M^{-1} \tag{7.79}$$

where

$$\begin{aligned} M &= E\left[\frac{d}{d\theta}\log f_y(\theta; y^N)\right]\left[\frac{d}{d\theta}\log f_y(\theta; y^N)\right]^T\Bigg|_{\theta=\theta_0} \\ &= -E\frac{d^2}{d\theta^2}\log f_y(\theta; y^N)\Bigg|_{\theta=\theta_0} \end{aligned} \tag{7.80}$$

Since θ is a d-dimensional vector, $(d/d\theta)\log f_y(\theta; y^N)$ is a d-dimensional column vector and the Hessian $(d^2/d\theta^2)\log f_y(\theta; y^N)$ is a $d \times d$ matrix. This matrix M is known as the *Fisher information matrix.* Notice that the evaluation of M normally requires knowledge of θ_0, so the exact value of M may not be available to the user.

A proof of the Cramér-Rao inequality is given in Appendix 7A.

Asymptotic Properties of the MLE

It is often difficult to exactly calculate properties of an estimator, such as (7.78). Therefore, limiting properties as the sample size (in this case the number N) tends to infinity are calculated instead. Classical such results for the MLE in case of independent observations were obtained by Wald (1949)and Cramér (1946):

Suppose that the random variables $\{y(i)\}$ are independent and identically distributed, so that

$$f_y(\theta; x_1, \ldots, x_N) = \prod_{i=1}^{N} f_{y(i)}(\theta, x_i)$$

Suppose also that the distribution of y^N is given by $f_y(\theta_0; x^N)$ for some value θ_0. Then the random variable $\hat{\theta}_{\mathrm{ML}}(y^N)$ tends to θ_0 with probability 1 as N tends to infinity, and the random variable

$$\sqrt{N}\left[\hat{\theta}_{\mathrm{ML}}(y^N) - \theta_0\right]$$

converges in distribution to the normal distribution with zero mean and covariance matrix given by the Cramér-Rao lower bound [M^{-1} in (7.79) and (7.80)].

In Chapters 8 and 9 we will establish that these results also hold when the ML estimator is applied to dynamical systems. In this sense the MLE is thus the best possible estimator. Let it, however, also be said that the MLE sometimes has been criticized for less good small sample properties and that there are other ways to assess the quality of an estimator than (7.78).

Probabilistic Models of Dynamical Systems

Suppose that the models in the model structure we have chosen in Section 7.1 include both a predictor function and an assumed PDF for the associated prediction errors, as described in Section 5.7:

$$\begin{aligned} \mathcal{M}(\theta): \ \hat{y}(t|\theta) &= g(t, Z^{t-1}; \theta) \\ \varepsilon(t, \theta) &= y(t) - \hat{y}(t|\theta) \text{ are independent} \\ &\quad \text{and have the PDF } f_e(x, t; \theta) \end{aligned} \tag{7.81}$$

Recall that we term a model like (7.81) that includes a PDF for ε a (complete) *probabilistic model.*

Likelihood Function for Probabilistic Models of Dynamical Systems

We note that, according to the model (7.81), the output is generated by

$$y(t) = g(t, Z^{t-1}; \theta) + \varepsilon(t, \theta) \tag{7.82}$$

where $\varepsilon(t, \theta)$ has the PDF $f_e(x, t; \theta)$. The joint PDF for the observations y^N (given the deterministic sequence u^N) is then given by Lemma 5.1. By replacing the dummy variables x_i by the corresponding observations $y(i)$, we obtain the likelihood function:

$$\begin{aligned} \overline{f}_y(\theta; y^N) &= \prod_{t=1}^{N} f_e\big(y(t) - g(t, Z^{t-1}; \theta), t; \theta\big) \\ &= \prod_{t=1}^{N} f_e(\varepsilon(t, \theta), t; \theta) \end{aligned} \tag{7.83}$$

Maximizing this function is the same as maximizing

$$\frac{1}{N} \log \overline{f}_y(\theta; y^N) = \frac{1}{N} \sum_{t=1}^{N} \log f_e(\varepsilon(t, \theta), t; \theta) \tag{7.84}$$

If we define

$$\ell(\varepsilon, \theta, t) = -\log f_e(\varepsilon, t; \theta) \tag{7.85}$$

we may write

$$\hat{\theta}_{\text{ML}}(y^N) = \arg\min_{\theta} \frac{1}{N} \sum_{t=1}^{N} \ell\left(\varepsilon(t, \theta), \theta, t\right) \tag{7.86}$$

The maximum likelihood method can thus be seen as a special case of the prediction-error criterion (7.12).

It is worth stressing that (7.85) and (7.86) give the exact maximum likelihood method for the posed problem. It is sometimes pointed out that the exact likelihood function is quite complicated for time-series problems and that one has to resort to approximations of it (e.g., Kashyap and Rao, 1976; Akaike, 1973; Dzhaparidze and Yaglom, 1983). This is true in certain cases. The reason is that it may be difficult to put, say, an ARMA model in the predictor form (7.81) (it will typically require time-varying Kalman predictors). The problem is therefore related to finding the exact predictor and is not a problem with the ML method as such. When we employ time-invariant predictors, we implicitly assume all previous observations to be known [see (3.24)] and typically replace the corresponding initial values by zero or estimate them. Then it is appropriate to interpret the likelihood function as *conditional* w.r.t. these values and to call the method a *conditional ML method* (e.g., Kashyap and Rao, 1976).

Gaussian Special Case

When the prediction errors are assumed to be Gaussian with zero mean values and (t-independent) covariances λ, we have

$$\ell(\varepsilon, \theta, t) = -\log f_e(\varepsilon, t; \theta) = \text{const} + \frac{1}{2}\log\lambda + \frac{1}{2}\frac{\varepsilon^2}{\lambda} \tag{7.87}$$

If λ is known, then (7.87) is equivalent to the quadratic criterion (7.15). If λ is unknown, (7.87) is an example of a parameterized norm criterion (7.16). Depending on the underlying model structure, λ may or may not be parametrized independently of the predictor parameters. See Problem 7E.4 for an illustration of this. Compare also Problem 7E.7.

Fisher Information Matrix and the Cramér-Rao Bound for Dynamical Systems

Having established the log likelihood function in (7.84) for a model structure, we can compute the information matrix (7.80). For simplicity, we then assume that the PDF f_e is known (θ independent) and time invariant. Let $\ell_0(\varepsilon) = -\log f_e(\varepsilon)$. Hence

$$\frac{d}{d\theta}\log \overline{f}_y(\theta; y^N) = \sum_{t=1}^{N} \ell_0'(\varepsilon(t, \theta)) \cdot \psi(t, \theta)$$

where, as in (4.121),

$$\psi(t, \theta) = \frac{d}{d\theta}\hat{y}(t|\theta) = -\frac{d}{d\theta}\varepsilon(t, \theta), \qquad \text{[a } d\text{-dimensional column vector]}$$

Also, ℓ_0' is the derivative of $\ell_0(\varepsilon)$ w.r.t. ε. To find the Fisher information matrix, we now evaluate the expectation of

$$\frac{d}{d\theta}\log \overline{f}_y(\theta; y^N)\left[\frac{d}{d\theta}\log \overline{f}_y(\theta; y^N)\right]^T$$

at θ_0 under the assumption that the true PDF for y^N indeed is $\overline{f}_y(\theta_0; y^N)$. The latter statement means that $\varepsilon(t, \theta_0) = e_0(t)$ will be treated as a sequence of independent random variables with PDF's $f_e(x)$. Call this expectation M_N. Thus

$$M_N = E\sum_{t=1}^{N}\sum_{s=1}^{N} \ell_0'(e_0(t))\,\ell_0'(e_0(s))\,\psi(t,\theta_0)\psi^T(s,\theta_0)$$
$$= \sum_{t=1}^{N} E\left[\ell_0'(e_0(t))\right]^2 \cdot E\psi(t,\theta_0)\psi^T(t,\theta_0)$$

since $e_0(t)$ and $e_0(s)$ are independent for $s \neq t$. We also have $\ell_0'(x) = [\log f_e(x)]' = f_e'(x)/f_e(x)$, and

$$E\left[\ell_0'(e_0(t))\right]^2 = \int \frac{\left[f_e'(x)\right]^2}{f_e^2(x)} \cdot f_e(x)\,dx$$
$$= \int_{-\infty}^{\infty} \frac{\left[f_e'(x)\right]^2}{f_e(x)}\,dx \triangleq \frac{1}{\kappa_0} \tag{7.88}$$

If $e_0(t)$ is Gaussian with variance λ_0, it is easy to verify that $\kappa_0 = \lambda_0$. Hence

$$M_N = \frac{1}{\kappa_0} \cdot \sum_{t=1}^{N} E\psi(t,\theta_0)\psi^T(t,\theta_0) \tag{7.89}$$

Now the Cramér-Rao inequality tells us that for *any unbiased estimator* $\hat{\theta}_N$ of θ (i.e., estimators such that $E\hat{\theta}_N = \theta_0$ regardless of the true value θ_0) we must have

$$\operatorname{Cov}\hat{\theta}_N \geq M_N^{-1} \tag{7.90}$$

Notice that this bound applies for any N and for all parameter estimation methods. We thus have

$$\operatorname{Cov}\hat{\theta}_N \geq \kappa_0 \left[\sum_{t=1}^{N} E\psi(t,\theta_0)\psi^T(t,\theta_0)\right]^{-1}$$
$$\kappa_0 = \lambda_0 \text{ for Gaussian innovations} \tag{7.91}$$

Multivariable Gaussian Case (∗)

When the prediction errors are p-dimensional and jointly Gaussian with zero mean and covariance matrices Λ, we obtain from the multivariable Gaussian distribution

$$\ell(\varepsilon, t; \theta) = \text{const} + \tfrac{1}{2}\log\det\Lambda + \tfrac{1}{2}\varepsilon^T\Lambda^{-1}\varepsilon \tag{7.92}$$

Then the negative logarithm of the likelihood function takes the form

$$V_N(\theta, \Lambda, Z^N) = \text{const} + \frac{N}{2} \log \det \Lambda + \frac{1}{2} \sum_{t=1}^{N} \varepsilon^T(t, \theta) \Lambda^{-1} \varepsilon(t, \theta) \tag{7.93}$$

If the $p \times p$ covariance matrix Λ is fully unknown and not parametrized through θ, it is possible to minimize (7.93) analytically with respect to Λ for every fixed θ:

$$\arg\min_{\Lambda} V_N(\theta, \Lambda, Z^N) = \hat{\Lambda}_N(\theta) = \frac{1}{N} \sum_{t=1}^{N} \varepsilon(t, \theta) \varepsilon^T(t, \theta) \tag{7.94}$$

Then

$$\begin{aligned} \hat{\theta}_N &= \arg\min_{\theta} V_N(\theta, \hat{\Lambda}_N(\theta), Z^N) \\ &= \arg\min_{\theta} \left[\tfrac{1}{2} \log \det \hat{\Lambda}_N(\theta) + \tfrac{1}{2} p \right] \end{aligned} \tag{7.95}$$

(see problem 7D.3) where $p = \dim \varepsilon$. Hence we may in this particular case use the criterion

$$\hat{\theta}_N = \arg\min_{\theta} \det \left[\frac{1}{N} \sum_{t=1}^{N} \varepsilon(t, \theta) \varepsilon^T(t, \theta) \right] \tag{7.96}$$

With this we have actually been led to a criterion of the type (7.29) to (7.30) with $h(A) = \det A$.

Information and Entropy Measures (*)

In (5.69) and (5.70) we gave a general formulation of a model as an assumed PDF for the observations Z^t:

$$\overline{f}_m(t, Z^t) \tag{7.97}$$

Let $\overline{f}_0(t, Z^t)$ denote the true PDF for the observations. The agreement between two PDF's can be measured in terms of the *Kullback-Leibler information distance* (Kullback and Leibler, 1951):

$$I(\overline{f}_0; \overline{f}_m) = \int \overline{f}_0(t, x^t) \log \frac{\overline{f}_0(t, x^t)}{\overline{f}_m(t, x^t)} \, dx^t \tag{7.98}$$

Here we use x^t as an integration variable for Z^t. This distance is also the *negative entropy* of $\overline{f}_0$ with respect to $\overline{f}_m$:

$$S(\overline{f}_0; \overline{f}_m) = -I(\overline{f}_0; \overline{f}_m) \tag{7.99}$$

An attractive formulation of the identification problem is to *look for a model that maximizes the entropy with respect to the true system* or, alternatively, *minimizes the information distance to the true system.* This formulation has been pursued by Akaike in a number of interesting contributions Akaike (1972, 1974a, 1981).

With a parametrized set of models $\overline{f}_{\mathcal{M}(\theta)}(t, Z^t) = \overline{f}(\theta; t, Z^t)$, we would thus solve

$$\hat{\theta}_N = \arg\min_{\theta} I\left(\overline{f}_0(N, Z^N); \overline{f}(\theta; N, Z^N)\right) \tag{7.100}$$

The information measure can be written

$$\begin{aligned} I(\overline{f}_0; \overline{f}) &= -\int \log\left[\overline{f}(\theta; N, x^N)\right] \cdot \overline{f}_0(N, x^N)\, dx^N \\ &\quad + \int \log\left[\overline{f}_0(N, x^N)\right] \cdot \overline{f}_0(N, x^N)\, dx^N \\ &= -E_0 \log \overline{f}(\theta; N, Z^N) + \theta\text{-independent terms} \end{aligned}$$

where E_0 denotes expectation with respect to the true system.

The problem (7.100) is thus the same as

$$\hat{\theta}_N = \arg\min_{\theta} \left[-E_0 \log \overline{f}(\theta; N, Z^N)\right] \tag{7.101}$$

The problem here is of course that the expectation is not computable since the true PDF is unknown. A simple estimate of the expectation is to replace it by the observation

$$E_0 \log \overline{f}(\theta; N, Z^N) \approx \log \overline{f}(\theta; N, Z^N) \tag{7.102}$$

This gives the log likelihood function for the problem and (7.101) then equals the MLE. The ML approach to identification can consequently also be interpreted as a maximum entropy strategy or a minimum information distance method.

The distance between the resulting model and the true system thus is

$$I\left(\overline{f}_0(N, Z^N); \overline{f}(\hat{\theta}_N; N, Z^N)\right) \tag{7.103}$$

This is a random variable, since $\hat{\theta}_N$ depends on Z^N. As an ultimate criterion of fit, Akaike (1981)suggested the use of the average information distance, or average entropy

$$E_{\hat{\theta}_N} I\left(\overline{f}_0(N, Z^N); \overline{f}(\hat{\theta}_N; N, Z^N)\right) \tag{7.104}$$

This is to be minimized with respect to both the model set and $\hat{\theta}_N$. As an unbiased estimate of the quantity (7.104), he suggested

$$\log \overline{f}(\hat{\theta}_N; N, Z^N) - \dim \theta \tag{7.105}$$

Calculations supporting this estimate will be given in Section 16.4.

The expression (7.105) used in (7.101) gives, with (7.84) and (7.85),

$$\hat{\theta}_{\mathrm{AIC}}(Z^N) = \arg\min_{\theta} \left\{ \frac{1}{N} \sum_{t=1}^{N} \ell\left(\varepsilon(t,\theta), t, \theta\right) + \frac{\dim \theta}{N} \right\} \tag{7.106}$$

This is Akaike's information theoretic criterion (AIC). When applied to a given model structure, this estimate does not differ from the MLE in the same structure. The advantage with (7.106) is, however, that the minimization can be performed with respect to different model structures, thus allowing for a general identification theory. See Section 16.4 for a further discussion of this aspect.

An approach that is conceptually related to information measures is Rissanen's minimum description length (MDL) principle. This states that a model should be sought that allows the shortest possible code or description of the observed data. See Rissanen (1978, 1986). Within a given model structure, it gives estimates that coincide with the MLE. See also Section 16.4.

Regularization

Sometimes there is reason to consider the following modified version of the criterion (7.11):

$$W_N(\theta, Z^N) = \frac{1}{N} \sum_{t=1}^{N} \ell\left(\varepsilon_F(t,\theta)\right) + \delta|\theta - \theta^{\#}|^2 = V_N(\theta, Z^N) + \delta|\theta - \theta^{\#}|^2 \tag{7.107}$$

It differs from the basic criterion only by adding a cost on the squared distance between θ and $\theta^{\#}$. The latter is a fixed point in the parameter space, and is often taken as the origin, $\theta^{\#} = 0$. The reasons and interpretations for including such a term could be listed as follows:

- If θ contains many parameters, the problem of minimizing V_N may be ill-conditioned, in the sense that the Hessian V_N'' may be an ill-conditioned matrix. Adding the norm penalty will add δI to this matrix, to make it better conditioned. This is the reason why the technique is called *regularization*.
- If the model parameterization contains many parameters (like in the nonlinear black-box models of Section 5.4), it may not be possible to estimate several of them accurately. There are then advantages in pulling them towards a fixed point $\theta^{\#}$. The ones that have the smallest influence on V_N will be affected most by this pulling force. The advantages of this will be brought out more clearly in Section 16.4. We may think of δ as a knob by which we control the effective number of parameters that is used in the minimization. A large value of δ will lock more parameters to the vicinity of $\theta^{\#}$.
- Comparing with the MAP estimate (7.77) we see that this corresponds to minimizing $W_N(\theta, Z^N) = -(1/N)\log\left[\overline{f}_y(\theta, Z^N)\cdot g_\theta(\theta)\right]$ if we take

$$V_N(\theta, Z^N) = -\frac{1}{N}\log \overline{f}_y(\theta; Z^N) \tag{7.108a}$$

$$g_\theta(\theta) = (N\delta/\pi)^{d/2} e^{-N\delta|\theta-\theta^{\#}|^2}, \quad d = \dim\theta \tag{7.108b}$$

that is, we assign a prior probability to the parameters that they are Gaussian distributed with mean $\theta^{\#}$ and covariance matrix $\frac{1}{2N\delta}I$. This prior is clearly well in line with the second interpretation.

A Pragmatic Viewpoint

It is good and reassuring to know that general and sound basic principles, such as maximum likelihood, maximum entropy, and minimum information distance, lead to criteria of the kind (7.11). However, in the end we are faced with a sequence of figures that are to be compared with "guesses" produced by the model. It could then always be questioned whether a probabilistic framework and abstract principles are applicable, since we observe only a given sequence of data, and the framework relates to the thought experiment that the data collection can be repeated infinitely many times under "similar" conditions. It is thus an important feature that minimizing (7.11) makes sense, even without a probabilistic framework and without "alibis" provided by abstract principles.

7.5 CORRELATING PREDICTION ERRORS WITH PAST DATA

Ideally, the prediction error $\varepsilon(t, \theta)$ for a "good" model should be independent of past data Z^{t-1}. For one thing, this condition is inherent in a probabilistic model, such as (7.81). Another and more pragmatic way of seeing this condition is that if $\varepsilon(t, \theta)$ is correlated with Z^{t-1} then there was more information available in Z^{t-1} about $y(t)$ than picked up by $\hat{y}(t|\theta)$. The predictor is then not ideal. This leads to the characterization of a good model as one that produces prediction errors that are independent of past data.

A test if $\varepsilon(t, \theta)$ is independent of the whole (and increasing) data set Z^{t-1} would amount to testing whether all nonlinear transformations of $\varepsilon(t, \theta)$ are uncorrelated with all possible functions of Z^{t-1}. This is of course not feasible in practice.

Instead, we may select a certain finite-dimensional vector sequence $\{\zeta(t)\}$ derived from Z^{t-1} and demand a certain transformation of $\{\varepsilon(t, \theta)\}$ to be uncorrelated with this sequence. This would give

$$\frac{1}{N}\sum_{t=1}^{N}\zeta(t)\alpha(\varepsilon(t, \theta)) = 0 \tag{7.109}$$

and the θ-value that satisfies this equation would be the best estimate $\hat{\theta}_N$ based on the observed data. Here $\alpha(\varepsilon)$ is the chosen transformation of ε, and the typical choice would be $\alpha(\varepsilon) = \varepsilon$.

We may carry this idea into a somewhat higher degree of generality. In the first place, we could replace the prediction error with filtered versions as in (7.10). Second, we obviously have considerable freedom in choosing the sequence $\zeta(t)$. It is quite possible that what appears to be the best choice of $\zeta(t)$ may depend on properties of the system. In such a case we would let $\zeta(t)$ depend on θ, and we have the following method:

Choose a linear filter $L(q)$ and let

$$\varepsilon_F(t,\theta) = L(q)\varepsilon(t,\theta) \tag{7.110a}$$

Choose a sequence of correlation vectors

$$\zeta(t,\theta) = \zeta(t, Z^{t-1}, \theta) \tag{7.110b}$$

constructed from past data and, possibly, from θ. Choose a function $\alpha(\varepsilon)$. Then calculate

$$\hat{\theta}_N = \underset{\theta \in D_{\mathcal{M}}}{\text{sol}} \left[f_N(\theta, Z^N) = 0 \right] \tag{7.110c}$$

$$f_N(\theta, Z^N) = \frac{1}{N} \sum_{t=1}^{N} \zeta(t,\theta)\alpha(\varepsilon_F(t,\theta)) \tag{7.110d}$$

Here we used the notation

$$\text{sol}\,[f(x) = 0] = \text{ the solution(s) to the equation } f(x) = 0$$

Normally, the dimension of ζ would be chosen so that f_N is a d-dimensional vector (which means that ζ is $d \times p$ if the output is a p-vector). Then (7.110) has as many equations as unknowns. In some cases it may be useful to consider an augmented correlation sequence ζ of higher dimension than d so that (7.110) is an overdetermined set of equations, typically without any solution. Then the estimate is taken to be the value that minimizes some quadratic norm of f_N:

$$\hat{\theta}_N = \underset{\theta \in D_{\mathcal{M}}}{\arg\min} \left| f_N(\theta, Z^N) \right| \tag{7.111}$$

There are obviously formal links between these correlation approaches and the minimization approach of Section 7.2 (see, e.g., Problem 7D.6).

The procedure (7.110) is a conceptual method that takes different shapes, depending on which model structures it is applied to and on the particular choices of ζ. In the subsequent section we shall discuss the perhaps best known representatives of the family (7.110), the instrumental-variable methods. First, however, we shall discuss the pseudolinear regression models.

Pseudolinear Regressions

We found in Chapter 4 that a number of common prediction models could be written as

$$\hat{y}(t|\theta) = \varphi^T(t,\theta)\theta \tag{7.112}$$

[see (4.21) and (4.45)]. If the data vector $\varphi(t,\theta)$ does not depend on θ, this relationship would be a linear regression. From this the term pseudolinear regression for (7.112) is derived (Solo, 1978). For the model (7.112), the "pseudo-regression vector" $\varphi(t,\theta)$ contains relevant past data, partly reconstructed using the current model. It is thus reasonable to require from the model that the resulting prediction errors be uncorrelated with $\varphi(t, \theta)$. That is, we choose $\zeta(t, \theta) = \varphi(t, \theta)$ and $\alpha(\varepsilon) = \varepsilon$, in (7.110) and arrive at the estimate

$$\hat{\theta}_N^{\text{PLR}} = \text{sol}\left\{\frac{1}{N}\sum_{t=1}^{N}\varphi(t,\theta)\left[y(t) - \varphi^T(t,\theta)\theta\right] = 0\right\} \tag{7.113}$$

which we term the *PLR estimate.*

Models subject to (7.112) also lend themselves to a number of variants of (7.113), basically corresponding to replacing $\varphi(t, \theta)$ with vectors in which the "reconstructed" (θ-dependent) elements are determined in some other fashion. See Section 10.4.

7.6 INSTRUMENTAL-VARIABLE METHODS

Instrumental Variables

Consider again the linear regression model (7.31):

$$\hat{y}(t|\theta) = \varphi^T(t)\theta \tag{7.114}$$

Recall that this model contains several typical models of linear and nonlinear systems. The least-squares estimate of θ is given by (7.34) and can also be expressed as

$$\hat{\theta}_N^{\text{LS}} = \text{sol}\left\{\frac{1}{N}\sum_{t=1}^{N}\varphi(t)\left[y(t) - \varphi^T(t)\theta\right] = 0\right\} \tag{7.115}$$

An alternative interpretation of the LSE is consequently that it corresponds to (7.110) with $L(q) = 1$ and $\zeta(t, \theta) = \varphi(t)$.

Now suppose that the data actually can be described as in (7.37):

$$y(t) = \varphi^T(t)\theta_0 + v_0(t) \tag{7.116}$$

We then found in Section 7.3 that the LSE $\hat{\theta}_N$ will not tend to θ_0 in typical cases, the reason being correlation between $v_0(t)$ and $\varphi(t)$. Let us therefore try a general correlation vector $\zeta(t)$ in (7.115). Following general terminology in the system identification field, we call such an application of (7.110) to a linear regression an *instrumental-variable* method (IV). The elements of ζ are then called *instruments* or *instrumental variables*. This gives

$$\hat{\theta}_N^{\text{IV}} = \text{sol}\left\{\frac{1}{N}\sum_{t=1}^{N}\zeta(t)\left[y(t) - \varphi^T(t)\theta\right] = 0\right\} \tag{7.117}$$

or

$$\hat{\theta}_N^{\text{IV}} = \left[\frac{1}{N}\sum_{t=1}^{N}\zeta(t)\varphi^T(t)\right]^{-1}\frac{1}{N}\sum_{t=1}^{N}\zeta(t)y(t) \tag{7.118}$$

provided the indicated inverse exists. For $\hat{\theta}_N$ to tend to θ_0 for large N, we see from (7.117) that then $(1/N)\sum_{t=1}^{N}\zeta(t)v_0(t)$ should tend to zero. For the method (7.117) to be successfully applicable to the system (7.116), we would thus require the following properties of the instrumental variable $\zeta(t)$ (replacing sample means by expectation):

$$\overline{E}\zeta(t)\varphi^T(t) \text{ be nonsingular} \tag{7.119}$$

$$\overline{E}\zeta(t)v_0(t) = 0 \tag{7.120}$$

In words, we could say that the instruments must be correlated with the regression variables but uncorrelated with the noise. Let us now discuss possible choices of instruments that could be subject to (7.119) and (7.120).

Choices of Instruments

Suppose that (7.114) is an ARX model

$$\begin{aligned} y(t) + a_1 y(t-1) + \cdots + a_{n_a} y(t-n_a) \\ = b_1 u(t-1) + \cdots + b_{n_b} u(t-n_b) + v(t) \end{aligned} \tag{7.121}$$

Suppose also that the true description (7.116) corresponds to (7.121) with the coefficients indexed by "zero." A natural idea is to generate the instruments similarly to (7.121) so as to secure (7.119), but at the same time not let them be influenced by $\{v_0(t)\}$. This leads to

$$\begin{aligned} \zeta(t) = K(q)[-x(t-1) \quad -x(t-2)\ldots \\ -x(t-n_a) \quad u(t-1)\ldots u(t-n_b)]^T \end{aligned} \tag{7.122}$$

where K is a linear filter and $x(t)$ is generated from the input through a linear system

$$N(q)x(t) = M(q)u(t) \tag{7.123}$$

Here

$$\begin{aligned} N(q) &= 1 + n_1 q^{-1} + \cdots + n_{n_n} q^{-n_n} \\ M(q) &= m_0 + m_1 q^{-1} + \cdots + m_{n_m} q^{-n_m} \end{aligned} \tag{7.124}$$

Most instruments used in practice are generated in this way. Obviously, $\zeta(t)$ is obtained from past inputs by linear filtering and can be written, conceptually, as

$$\zeta(t) = \zeta(t, u^{t-1}) \tag{7.125}$$

If the input is generated in *open loop* so that it does not depend on the noise $v_0(t)$ in the system, then clearly (7.120) holds. Since both the φ-vector and the ζ-vector are generated from the same input sequence (φ contains in addition effects from v_0), it might be expected that (7.119) should hold "in general." We shall return to this question in Section 8.6.

A simple and appealing choice of instruments is to first apply the LS method to (7.121) and then use the LS-estimated model for N and M in (7.123). The instruments are then chosen as in (7.122) with $K(q) = 1$. Systems operating in closed loop and systems without inputs call for other ideas. See Problem 7G.3 for some suggestions.

As outlined in Problem 7D.5, the use of the instrumental vector (7.122) to (7.124) is equivalent to the vector

$$\zeta^*(t) = \frac{K(q)}{N(q)}[u(t-1) \quad u(t-2)\ldots u(t-n_a-n_b)]^T \tag{7.126}$$

The IV estimate $\hat{\theta}_N^{\mathrm{IV}}$ in (7.118) is thus the same for ζ^* as for ζ in (7.122) and does not, for example, depend on the filter M in (7.124).

Model-dependent Instruments (*)

The quality of the estimate $\hat{\theta}_N^{\mathrm{IV}}$ will depend on the choice of $\zeta(t)$. In Section 9.5 we shall derive general expressions for the asymptotic covariance of $\hat{\theta}_N^{\mathrm{IV}}$ and examine them further in Section 15.3. It then turns out that it may be desirable to choose the filter in (7.123) equal to those of the true system: $N(q) = A_0(q)$; $M(q) = B_0(q)$. These are clearly not known, but we may let the instruments depend on the parameters in the obvious way:

$$\begin{aligned} \zeta(t,\theta) &= K(q)\left[-x(t-1,\theta)\ldots -x(t-n_a,\theta) \quad u(t-1)\ldots u(t-n_b)\right]^T \\ A(q)x(t,\theta) &= B(q)u(t) \end{aligned} \tag{7.127}$$

In general, we could write the generation of $\zeta(t,\theta)$:

$$\zeta(t,\theta) = K_u(q,\theta)u(t) \tag{7.128}$$

where $K_u(q,\theta)$ is a d-dimensional column vector of linear filters.

Including a prefilter (7.110a) and a "shaping" function $\alpha(\cdot)$ for the prediction errors, the IV method could be summarized as follows:

$$\varepsilon_F(t,\theta) = L(q)\left[y(t) - \varphi^T(t)\theta\right] \tag{7.129a}$$

$$\hat{\theta}_N^{\mathrm{IV}} = \operatorname*{sol}_{\theta\in D_{\mathcal{M}}}\left[f_N(\theta, Z^N) = 0\right] \tag{7.129b}$$

where

$$f_N(\theta, Z^N) = \frac{1}{N}\sum_{t=1}^{N}\zeta(t,\theta)\alpha(\varepsilon_F(t,\theta)) \tag{7.129c}$$

$$\zeta(t,\theta) = K_u(q,\theta)u(t) \tag{7.129d}$$

Extended IV Methods (*)

So far in this section the dimension of ζ has been equal to dim θ. We may also work with augmented instrumental variable vectors with dimension dim $\zeta > d$. The resulting method, corresponding to (7.110) and (7.111), will be called an *extended IV method* and takes the form

$$\hat{\theta}_N^{\mathrm{EIV}} = \arg\min_{\theta} \left| \frac{1}{N} \sum_{t=1}^{N} \zeta(t,\theta)\alpha(\varepsilon_F(t,\theta)) \right|_Q^2 \tag{7.130}$$

The subscript Q denotes Q-norm:

$$|x|_Q^2 = x^T Q x \tag{7.131}$$

In case ζ does not depend on θ and $\alpha(\varepsilon) = \varepsilon$, (7.130) can be solved explicitly. See Problem 7D.7.

Frequency-domain Interpretation (*)

Quite analogously to (7.20) to (7.25) in the prediction error case, the criterion (7.129) can be expressed in the frequency domain using Parseval's relationship. We then assume that $\alpha(\varepsilon) = \varepsilon$, and that a linear generation of the instruments as in (7.128) is used. This gives

$$\begin{aligned} f_N(\theta, Z^N) \approx \frac{1}{2\pi} \int_{-\pi}^{\pi} \left[\hat{\hat{G}}_N(e^{i\omega}) - G(e^{i\omega},\theta) \right] |U_N(\omega)|^2 \\ \times A(e^{i\omega},\theta) L(e^{i\omega}) K_u(e^{i\omega},\theta)\, d\omega \end{aligned} \tag{7.132}$$

Here $A(q,\theta)$ is the A-polynomial that corresponds to θ in the model (7.121).

Multivariable Case (*)

Suppose now that the output is p-dimensional and the input m-dimensional. Then the instrument $\zeta(t)$ is a $d \times p$ matrix. A linear generation of $\zeta(t,\theta)$ could still be written as (7.128), with the interpretation that the ith column of $\zeta(t,\theta)$ is given by

$$\zeta^{(i)}(t,\theta) = K_u^{(i)}(q,\theta) u(t) \tag{7.133}$$

where $K_n^{(i)}(q,\theta)$ is a $d \times m$ matrix filter. [$K_u(q,\theta)$ in (7.128) is thus a tensor, a "three-index entity"]. With $\alpha(\varepsilon)$ being a function from $\mathbf{R}^p$ to $\mathbf{R}^p$ and $L(q)$ a $p \times p$ matrix filter, the IV method is still given by (7.129).

7.7 USING FREQUENCY DOMAIN DATA TO FIT LINEAR MODELS (*)

In actual practice, most data are of course collected as samples of the input and output time signals. There are occasions when it is natural and fruitful to consider the Fourier transforms of the inputs and the outputs to be the primary data. It could be, for example, that data are collected by a frequency analyzer, which provides the transforms to the user, rather than the original time domain data. It could also be that one subjects the measured data to Fourier transformation, before fitting them to models. In some applications, like microwave fields, impedances, etc., the raw measurements are naturally made directly in the frequency domain. This view has been less common in the traditional system identification literature, but has been of great importance in the Mechanical Engineering community, vibrational analysis, and so on. The possible advantages of this will be listed later in this section. The usefulness of such an approach has been made very clear in the work of Schoukens and Pintelon; see in particular the book Schoukens and Pintelon (1991)and the survey Pintelon et.al. (1994).

There is clearly a very close relationship between time domain methods and frequency domain methods for linear models. We saw in (7.25) that the prediction error method for time domain data can (approximately) be interpreted as a fit in the frequency domain. We shall in this section look at some aspects of working directly with data in the frequency domain.

Continuous Time Models

An important advantage to frequency domain data is that it is equally simple to build time continuous models as discrete time/sampled data ones. This means that we can work with models of the kind

$$y(t) = G(p, \theta)u(t) + H(p, \theta)e(t) \tag{7.134}$$

(where p denotes the differentiation operator) analogously to our basic discrete time model (7.3). See also (4.49) and the ensuing discussion. Note the considerable freedom in parameterizing (7.134): from black-box models in terms of numerator and denominator polynomials, or gain, time-delay, and time constant (see (4.50)), to physically parameterized ones like (4.64). In addition to these traditional time-domain parameterizations, one may also parameterize the transfer functions in a way that is more frequency domain oriented. A simple case (see Problem 7G.2) is to let

$$\begin{aligned} G(i\omega, \theta) &= \sum_{k=1}^{d} (g_k^R + i g_k^I) W_\gamma(k, \omega - \omega_k) \\ \theta &= \left[g_1^R, g_1^I, \dots, g_d^R, g_d^I\right] \end{aligned} \tag{7.135}$$

One should typically think of the functions $W_\gamma(k, \omega)$ as bandpass filters, with a width that may be scaled by γ. The parameter g_k would then describe the frequency response around the frequency value ω_k. If the width of the passband increases with frequency we obtain parameterizations linked to wavelet transforms. See, e.g., the insightful discussion by Ninness (1993).

Estimation from Frequency Domain Data

Suppose now that the original data are supposed to be

$$Z^N = \{Y(\omega_k), U(\omega_k), k = 1, \dots N\} \tag{7.136}$$

where $Y(\omega_k)$ and $U(\omega_k)$ either are the discrete Fourier transforms (2.37) of $y(t)$ and $u(t)$ or are considered as approximations of the Fourier transforms of the underlying continuous signals:

$$Y(\omega) \approx \int_0^\infty y(t)e^{-i\omega t}dt \tag{7.137}$$

Which interpretation is more suitable depends of course on the signal character, sampling interval, and so on.

How to estimate θ in (7.134) or its discrete time counterpart from (7.136)? In view of (7.25) it would be tempting to use

$$\begin{aligned} \hat{\theta}_N &= \arg\min_{\theta} V(\theta) \\ V(\theta) &= \sum_{k=1}^{N} \left|Y(\omega_k) - G(e^{i\omega_k T}, \theta)U(\omega_k)\right|^2 \cdot \frac{1}{\left|H(e^{i\omega_k T}, \theta)\right|^2} \end{aligned} \tag{7.138}$$

(replacing $e^{i\omega_k T}$ by $i\omega_k$ for the continuous-time model (7.134)). Here T is the sampling interval.

If H in fact does not depend on θ (the case of *fixed* or *known noise model*) experience shows that (7.138) works well. Otherwise the estimate $\hat{\theta}_N$ may not be consistent.

To find a better estimator we turn to the maximum likelihood (ML) method for advice. We give the expressions for the continuous time case; in the case of a discrete time model, just replace $i\omega_k$ by $e^{i\omega_k T}$. We will also be somewhat heuristic with the treatment of white noise.

If the data were generated by

$$y(t) = G(p, \theta)u(t) + H(p, \theta)e(t)$$

the Fourier transforms would be related by

$$Y(\omega) = G(i\omega, \theta)U(\omega) + H(i\omega, \theta)E(\omega) \tag{7.139}$$

To be true, (7.139) should in many cases contain an error term that accounts for finite time effects and the fact that the measured data $Y(\omega_k)$ often are not exact realizations of (7.137). For periodic signals, observed over an integer number of periods, (7.139) may however hold exactly for the input–output relation between u and y.

Now, if $e(t)$ is white noise, its Fourier transform (7.137) will have a complex Normal distribution (see (I.14)):

$$E(\omega) \in N_c(0, \lambda) \tag{7.140}$$

This means that the real and imaginary parts are each normally distributed, with zero means and variances $\lambda/2$. The real and imaginary parts are independent and, moreover, $E(\omega_1)$ and $E(\omega_2)$ are independent for $\omega_1 \neq \omega_2$. (For finite time there will remain some correlation for neighboring frequencies, which we will ignore here.)

This implies that

$$
\begin{aligned}
&Y(\omega_k) \in N_c(G(i\omega_k,\theta)U(\omega_k), \lambda\,|H(i\omega_k,\theta)|^2) \\
&Y(\omega_k) \text{ and } Y(\omega_\ell) \text{ independent for } \omega_k \neq \omega_\ell
\end{aligned}
\tag{7.141}
$$

according to the model, so that the negative logarithm of the likelihood function becomes

$$
\begin{aligned}
V_N(\theta) = {} & N\log\lambda + \sum_{k=1}^{N} 2\log|H(i\omega_k,\theta)| \\
& + \sum_{k=1}^{N} \frac{1}{\lambda}|Y(\omega_k) - G(i\omega_k,\theta)U(\omega_k)|^2 \cdot \frac{1}{|H(i\omega_k,\theta)|^2}
\end{aligned}
\tag{7.142}
$$

The Maximum Likelihood estimate is

$$
\hat{\theta}_N = \arg\min_{\theta} V_N(\theta) \tag{7.143}
$$

Remark: This is the ML criterion under the assumption (7.141). We noted above that the data might not be exactly subject to this condition, due to finite time effects when forming the Fourier transforms. It still makes sense to use the criterion (7.143), though.

If we perform analytical minimization of (7.143) w.r.t. λ, we obtain

$$
\hat{\theta}_N = \arg\min_{\theta}\left[N\cdot\log W_N(\theta) + 2\sum_{k=1}^{N}\log|H(i\omega_k,\theta)|\right] \tag{7.144}
$$

$$
W_N(\theta) = \frac{1}{N}\sum_{k=1}^{N}|Y(\omega_k) - G(i\omega_k,\theta)U(\omega_k)|^2 \cdot \frac{1}{|H(i\omega_k,\theta)|^2} \tag{7.145}
$$

$$
\hat{\lambda}_N = W_N(\hat{\theta}_N) \tag{7.146}
$$

Compared to (7.138) we thus have an extra term

$$
\sum_{k=1}^{N}\log|H(i\omega_k,\theta)|^2 \tag{7.147}
$$

If the noise model is given and fixed, H does not depend on θ, and the term (7.147) does not affect the estimate. This case of fixed noise models is very common in applications with frequency domain data (see Schoukens and Pintelon, 1991). One reason is that for a periodic input, we can obtain reasonable estimates of H in a preprocessing step. See Schoukens et.al. (1997)and (7.154) below.

We may also note that for any monic, stable, and inversely stable transfer function $H(q, \theta)$ we have

$$\int_{-\pi}^{\pi} \log \left| H(e^{i\omega}, \theta) \right|^2 d\omega \equiv 0 \tag{7.148}$$

The expression will also hold if the integral is replaced by summation over the frequencies $2\pi k/N$, $k = 1, \ldots, N$. This is the reason why (7.147) is missing from time domain criteria, like (7.25), which correspond to equally spaced frequencies ω_k.

Note that $W_N(\theta)$ in (7.144) can be rewritten as

$$W_N(\theta) = \frac{1}{N} \sum_{k=1}^{N} \left| \hat{\hat{G}}(i\omega_k) - G(i\omega_k, \theta) \right|^2 \cdot \frac{|U(\omega_k)|^2}{|H(i\omega_k, \theta)|^2} \tag{7.149}$$

in formal agreement with (7.25). Here $\hat{\hat{G}}$ is the empirical transfer function estimate, ETFE, defined in (6.24).

Some Variants of the Criterion

Weighted Nonlinear Least Squares Criterion. Given an estimate $\hat{\hat{G}}$ of the frequency function (the ETFE or anything else), it is natural to fit a parametric model to it by a (non-linear) least squares criterion

$$W_N^{\text{NLS}}(\theta) = \frac{1}{N} \sum_{k=1}^{N} \left| \hat{\hat{G}}(i\omega_k) - G(i\omega_k, \theta) \right|^2 W_k \tag{7.150}$$

with some weighting function W_k. We see that this corresponds to the ML criterion with $W_k = |U(\omega_k)|^2 / |H(i\omega_k, \theta)|^2$. In other words, (7.150) can be interpreted as the ML criterion with a fixed noise model

$$|H(i\omega_k)|^2 = \frac{|U(\omega_k)|^2}{W_k} \tag{7.151}$$

The numerical minimization of this criterion is typically carried out using a damped Gauss-Newton method, like for most of the other criteria discussed in this book. See Section 10.2.

A Linear Method. If we use an ARX-parameterization of the model (see (4.9))

$$G(p, \theta) = \frac{B(p)}{A(p)}, \qquad H(p, \theta) = \frac{1}{A(p)}$$

then the criterion (7.149) takes the form

$$W_N(\theta) = \frac{1}{N} \sum_{k=1}^{N} \left| A(i\omega_k)\hat{\hat{G}}(i\omega_k) - B(i\omega_k) \right|^2 |U(\omega_k)|^2 \tag{7.152}$$

This is a quadratic criterion in the coefficients of the polynomials A and B. It can therefore be minimized explicitly by the least squares solution.

A number of other variants can also be defined. We can, for example, define a frequency domain IV-method in analogy with (7.132). See Pintelon et.al. (1994)for a more complete survey.

Some Practical Features with Estimation from Frequency Domain Data

There are several distinct features with the direct frequency domain approach that could be quite useful. We shall list a few:

- **Prefiltering** is known as quite useful in the time-domain approach. See Section 14.4. For frequency domain data it becomes very simple: It just corresponds to assigning different weights to different frequencies in the weighted criterion (7.150). This, in turn, is the same as invoking a special noise model (7.151).

  Normally, it does not quite make sense to combine prefiltering with estimating a noise model, since a parameter-dependent weighting as in

$$W_N^{\mathrm{NLS}}(\theta) = \frac{1}{N}\sum_{k=1}^{N}\left|\hat{\hat{G}}(i\omega_k) - G(i\omega_k,\theta)\right|^2 W_k \frac{|U(\omega_k)|^2}{|H(i\omega_k,\theta)|^2}$$

  may undo any applied weighting from W.
- **Condensing Large Data Sets.** When dealing with systems with a fairly wide spread of time constants, large data sets have to be collected in the time domain. When converted to the frequency domain they can easily be condensed, so that, for example, logarithmically spaced frequencies are obtained. At higher frequencies one would thus decimate the data, which involves averaging over neighboring frequencies. Then the noise level (λ_k) is reduced accordingly.
- **Combining Experiments**. Nothing in the approach of (7.141)–(7.143) says that the frequency response data at different frequencies have to come from the same experiment, or even that the frequencies involved ($\omega_k, k = 1, \ldots, N$) all have to be different. It is thus very easy to combine data from different experiments.
- **Periodic Inputs**. The main drawback with the frequency domain approach is that the underlying frequency domain model (7.139) is strictly correct only for a periodic input and assuming all transients have died out. On the other hand, typical use of the time domain method assumes inputs and outputs prior to time $t = 0$ to be zero. Whichever assumption about past behavior is closer to the truth should thus affect the choice of approach. Note, though, that both the time-domain and the frequency-domain methods allow the possibility to estimate a finite number of parameters that pick up these transients, and thus give correct handling of these effects. See also Section 13.3.
- **Non-Parametric Noise Estimates from Periodic Inputs.** We have for the true system $y(t) = y_u(t) + v(t)$, where $y_u(t) = G_0(q)u(t)$. If $u(t)$ is periodic with period M, so will $y_u(t)$ be, after a transient. By averaging the output over K periods,

$$\overline{y}(t) = \frac{1}{K}\sum_{k=0}^{K-1} y(t + kM), \quad t = 1, \ldots, M \tag{7.153}$$

we can thus get a better estimate of $y_u(t)$, and also estimate the noise sequence as

$$\hat{v}(t) = y(t) - \overline{y}(t) \tag{7.154}$$

where the definition of $\overline{y}(t)$ has been extended to $1 \leq t \leq N$ by periodic continuation. Estimating the spectrum of $\hat{v}(t)$ with any (non-parametric) method gives a noise spectral model $|H(i\omega)|^2$ that can be used in (7.144).

- **Band-Limited Signals.** If the actual input signals are band-limited (like no power above the Nyquist frequency), the continuous time Fourier transform (7.137) can be well computed from sampled data. It is then possible to directly build continuous-time models without any extra work. Notice also that frequency contents above the Nyquist frequency can be eliminated from both input and output signals by anti-alias filtering (see Section 13.7) before sampling. Such filtering will not distort the input-output relationship, provided the input and output are subjected to exactly the same filters.
- **Continuous-Time Models.** The comment above shows that direct continuous-time system identification from "continuous-time data" can be dealt with in a rather straightforward fashion. Otherwise, continuous time data with continuous-time white noise descriptions are delicate mathematical objects.
- **Trade-off Noise/Frequency Resolution.** The approach also allows for a more direct and frequency dependent trade-off between frequency resolution and noise levels. That will be done as the original Fourier transform data are decimated to the selected range of frequencies ω_k, $k = 1, \ldots, N$.

7.8 SUMMARY

There are several ways to fit models in a given set to observed data. In this chapter we have pointed out two general procedures. Both deal with the sequence of prediction errors $\{\varepsilon(t, \theta)\}$ computed from the respective models using the observed data, and both could be said to aim at making this sequence "small."

The *prediction-error identification approach* (PEM) was defined by (7.10) to (7.12):

$$\hat{\theta}_N = \arg\min_{\theta \in D_{\mathcal{M}}} V_N(\theta, Z^N)$$

$$V_N(\theta, Z^N) = \frac{1}{N} \sum_{t=1}^{N} \ell\left(\varepsilon(t, \theta), \theta, t\right) \tag{7.155}$$

It contains well-known procedures, such as the least-squares (LS) method and the maximum-likelihood (ML) method and is at the same time closely related to Bayesian maximum a posteriori (MAP) estimation and Akaike's information criterion (AIC).

The *subspace approach to identifying state-space models* was defined by (7.66). It consists of three steps: (1) estimating the k-step ahead predictors using an LS-algorithm, and (2) selecting the state vector from these, and finally (3) estimating the state-space matrices using these states and the LS-method.

The *correlation approach* was defined by (7.110):

$$\varepsilon_F(t,\theta) = L(q)\varepsilon(t,\theta)$$
$$\hat{\theta}_N = \underset{\theta \in D_{\mathcal{M}}}{\text{sol}} \left[f_N(\theta, Z^N) = 0 \right]$$
$$f_N(\theta, Z^N) = \frac{1}{N}\sum_{t=1}^{N} \zeta(t,\theta)\alpha(\varepsilon_F(t,\theta)) \tag{7.156}$$

It contains the instrumental-variable (IV) technique, as well as several methods for rational transfer function models.

System identification has often been described as an area crowded with seemingly unrelated *ad hoc* methods and tricks. The list of names of available and suggested methods is no doubt a very long one. It is our purpose, however, with this chapter, as well as Chapters 8 to 11, to point out that the number of underlying basic ideas is really quite small, and that it indeed is quite possible to orient oneself in the area of system identification with these basic ideas as a starting point.

It might be added that for systems operating in closed loop some special identification techniques have been devised. We shall review these methods in Section 13.5, in connection with a discussion of the closed loop experiment situation. The bottom line is that a direct application of the prediction error methods of this chapter should be the prime choice, also for closed loop data.

7.9 BIBLIOGRAPHY

The parameter estimation methods described in this chapter all go back to basic methods in statistics. For general texts, we refer to Cramér (1946), Rao (1973), and Lindgren (1976).

Section 7.2: The term "prediction-error-methods" was perhaps first coined in Ljung (1974), but it had long been realized that the common methods of system identification had aimed at making the prediction error small. From an operational point of view, the criterion (7.155) can be viewed as a nonlinear regression method. See, for example, Jennrich (1969)and Hannan (1971b). Various norms have been discussed (see also Section 15.2). The ℓ_∞-norm, which is related to *unknown-but-bounded disturbances*, and *set membership identification*, is discussed in early contributions by Schweppe (1973)and Fogel and Huang (1982)(cf. Problem 7G.7). This approach has then been discussed in numerous papers, see the surveys in Milanese and Vicino (1991), Deller (1990), and Walter and Piet-Lahanier (1990). See also the bibliographies of Chapters 14 and 16 for contributions to this approach, that relate to identification for control and model validation.

The frequency-domain expressions for the prediction-error criteria go back to Whittle (1951), who dealt with the input-free case. Among many references, we may mention Hannan (1970), Chapter VI, for a detailed study (still with no input). Related formulas are also given by Solo (1978), Ljung and Glover (1981), and Wahlberg and Ljung (1986). An alternative approach based on interpolation of a given frequency function has become known as $\mathcal{H}_\infty$-identification. See e.g., Gu and Khargonekar (1992)and Helmicki, Jacobson, and Nett (1991).

Section 7.3: The statistical roots of the least-squares method are examined in Appendix II. The application to times series has its origin in the work of Yule (1927)and Walker (1931), with a first asymptotic analysis by Mann and Wald (1943). The application to dynamic systems with an input was made independently by several authors, with an early comprehensive description and analysis by Åström (1968), to some extent reprinted in Åström and Eykhoff (1971). A good account of different variants of the LS method is given in Hsia (1977). *The Subspace methods* really originate from classical realization theory as formulated in Ho and Kalman (1966)and Kung (1978). A basic treatment of the modern techniques is given in the book Van Overschee and DeMoor (1996). More references will be listed in Chapter 10 in connection with the algorithmic details of the approach.

Section 7.4: Whittle pioneered maximum likelihood methods for AR and ARMA models using frequency domain formulations, see e.g., Whittle (1951). The principle of maximum likelihood was then applied to dynamical systems by Åström and Bohlin (1965)(ARMAX model structures) and Box and Jenkins (1970)[model structure (4.31)]. Since then a long list of articles has dealt with this approach. Åström (1980)may be singled out for a survey.

Frequency-domain variants or approximations of the likelihood function have been extensively used by Whittle (1951), Hannan (1970), and others. The Bayesian MAP approach is comprehensively treated in Peterka (1981a), and Peterka (1981b). The calculations leading to (7.96) were first given by Eaton (1967)and Akaike (1973). Entropy and information theoretic criteria have been discussed extensively by Akaike and Rissanen. We may single out Akaike (1974a), Akaike (1981), Rissanen (1985), and Rissanen (1986)as recommended reading. A general reference on entropy and statistics is Kullback (1959). Regularization is really a general technique to solve ill-posed problems, Tikhonov and Arsenin (1977). Its applications to estimation is discussed in Wahba (1987), Wahba (1990), and Sjöberg, McKelvey, and Ljung (1993). Regularization is of particular importance for nonlinear black box models, where many parameters often are used. See, Sjöberg et.al. (1995)and Girosi, Jones, and Poggio (1995). The connection to Bayesian approaches in the latter context are pursued in Williams (1995), while Johansen (1996)describes how regularization can be used to incorporate prior knowledge in nonlinear models. See also (16.36) in Chapter 16.

Sections 7.5–7.6: The way to describe the "correlation approach" as presented here is novel, although the different methods are well known. The IV method was introduced into statistics and econometrics by Reiersøl (1941)and has been applied to many parameter estimation problems in statistics (see, e.g., Kendall and Stuart, 1961). Applications to dynamic systems in the control field have been pioneered by Wong and Polak (1967), Young (1965), and Mayne (1967). For applications to ARMA models see Stoica, Friedlander and Söderström (1986). A historical background is given by Young (1976). For comprehensive treatments, see Söderström and Stoica (1983)and Young (1984).

Section 7.8: An early reference for frequency-domain criteria of the type (7.152) is Levi (1959). A comprehensive treatment with many references is the book Schoukens and Pintelon (1991). See also the survey paper Pintelon et.al. (1994). There is a MATLAB Toolbox devoted to frequency domain identification methods, Kollàr (1994).

7.10 PROBLEMS

7G.1 *Input error and output error methods:* Consider a model structure

$$y(t) = G(q, \theta)u(t)$$

without a specified noise model. In the survey of Åström and Eykhoff (1971)identification methods that minimize "the output error"

$$\hat{\theta}_N = \arg\min \sum_{t=1}^{N} [y(t) - G(q, \theta)u(t)]^2$$

and the "input error"

$$\hat{\theta}_N = \arg\min \sum_{t=1}^{N} \left[u(t) - G^{-1}(q, \theta)y(t)\right]^2$$

are listed. Show that these methods are prediction error methods corresponding to particular choices of noise models $H(q, \theta)$.

7G.2 *Spectral analysis as a prediction error method:* Consider the model structure

$$G(e^{i\omega}, \theta) = \sum_{k=1}^{N} \left(g_k^R + i g_k^I\right) W_\gamma(\omega - \omega_k)$$

$$\theta = \left[g_1^R \quad g_1^I, \ldots, g_n^R \quad g_n^I\right]^T$$

and let $H(e^{i\omega}, \eta)$ be an arbitrary noise model parametrization. Let $\hat{\theta}_N$ be the prediction-error estimate obtained by minimization of (7.23) and (7.25):

$$\hat{\theta}_N = \arg\min_{\theta, \eta} \int_{-\pi}^{\pi} \frac{\left|\hat{\hat{G}}_N(e^{i\omega}) - G(e^{i\omega}, \theta)\right|^2 |U_N(\omega)|^2}{\left|H(e^{i\omega}, \eta)\right|^2} d\omega$$

(a) Consider the special case $H(e^{i\omega}, \eta) \equiv 1$ and

$$W_\gamma(\omega) = \begin{cases} 1, & |\omega| \le \dfrac{\pi}{2n} \\ 0, & |\omega| > \dfrac{\pi}{2n} \end{cases}$$

$$\omega_k = \frac{(k-1)\pi}{n}$$

Show that $\hat{G}(e^{i\omega_k}, \hat{\theta}_N)$ is then given by (6.46).

(b) Assume, in the general case, that

$$H(e^{i\omega}, \eta) \cdot W_\gamma(\omega - \omega_k) \approx H(e^{i\omega_k}, \eta) \cdot W_\gamma(\omega - \omega_k)$$

$$G(e^{i\omega}, \theta) \cdot W_\gamma(\omega - \omega_k) \approx G(e^{i\omega_k}, \theta) \cdot W_\gamma(\omega - \omega_k)$$

Show that (6.46) then holds approximately.

7G.3 *Instruments for closed-loop systems:* Consider a system

$$A_0(q)y(t) = B_0(q)u(t) + v_0(t)$$

under the output feedback

$$u(t) = F_1(q)r(t) - F_2(q)y(t)$$

(a) Let $x(t)$ and $\zeta(t)$ be given by

$$N(q)x(t) = M(q)r(t)$$

$$\zeta(t) = K(q)\big[-x(t-1)\ldots -x(t-n_a) \quad r(t-1)\ldots r(t-n_b)\big]^T$$

Show that (7.120) holds for these instruments, and verify that (7.119) holds for a simple first-order special case.

(b) Suppose that $v_0(t)$ is known to be an MA process of order s. Introduce the instruments

$$\zeta(t) = [-y(t-1-s)\ldots -y(t-n_a-s) \quad u(t-1-s)\ldots u(t-n_b-s)]^T$$

Show the same results as under part (a). See also Söderström, Stoica and Trulsson (1987).

7G.4 Suppose $Y_N = [y(1), \ldots, y(N)]^T$ is a Gaussian N-dimensional random vector with zero mean and covariance matrix $R_N(\theta)$. Let

$$R_N(\theta) = L_N(\theta)\Lambda_N(\theta)L_N^T(\theta)$$

where $L_N(\theta)$ is lower triangular with 1's along the diagonal and $\Lambda_N(\theta)$ a diagonal matrix with $\lambda_\theta(t)$ as the t, t element. Let

$$E_N(\theta) = L_N^{-1}(\theta)Y_N$$

$$E_N(\theta) = [\varepsilon(1,\theta), \ldots, \varepsilon(N,\theta)]^T$$

Show that, if θ is a parameter to be estimated, then the negative log likelihood function when Y_N is observed is

$$\frac{N}{2}\log 2\pi + \frac{1}{2}\log\det R_N(\theta) + \frac{1}{2}Y_N^T R_N^{-1}(\theta)Y_N$$

Show also that this can be rewritten as

$$\frac{N}{2}\log 2\pi + \frac{1}{2}\sum_{t=1}^{N}\log\lambda_\theta(t) + \frac{1}{2}\sum_{t=1}^{N}\frac{\varepsilon^2(t,\theta)}{\lambda_\theta(t)}$$

where $\varepsilon(t,\theta)$ are independent, normal random variables with variances $\lambda_\theta(t)$. How does this relate to our calculations (7.81) to (7.87)?

7G.5 Let the two random vectors X and Y be jointly Gaussian with

$$EX = m_X; \qquad EY = m_Y$$

$$E(X - m_X)(X - m_X)^T = P_X \qquad E(Y - m_Y)(Y - m_Y)^T = P_Y$$

$$E(X - m_X)(Y - m_Y)^T = P_{XY}$$

Show that the conditional distribution of X given Y is

$$(X|Y) \in N\left(m_X + P_{XY}P_Y^{-1}(Y - m_Y),\, P_X - P_{XY}P_Y^{-1}P_{XY}^T\right)$$

7G.6 Consider the model structure

$$\begin{aligned} X &= F(\theta)W \\ Y &= H(\theta)X + E \end{aligned} \tag{7.157}$$

where W and E are two independent, Gaussian random vectors with zero mean values and unit covariance matrices. Note that state-space models like (4.84), without input, can be written in this form by forming $X^T = [\,x^T(1) \quad x^T(2)\dots x^T(N)\,]$ and $Y^T = [\,y(1) \quad y(2)\dots y(N)\,]$. Let

$$R(\theta) = I + H(\theta)F(\theta)F^T(\theta)H^T(\theta)$$

Show the following:

(a) The negative log likelihood function for θ, (ignoring θ-independent terms) when Y is observed is

$$V(\theta) = -\log p(Y|\theta) = \tfrac{1}{2}Y^T R^{-1}(\theta)Y + \tfrac{1}{2}\log\det R(\theta)$$

Let

$$\hat{\theta}_{\text{ML}} = \arg\min_{\theta} V(\theta)$$

(cf. Problem 7G.4).

(b) Let the conditional expectation of X, given Y and θ be $\hat{X}^s(\theta)$. Show that

$$E(X|Y,\theta) = \hat{X}^s(\theta) = \left[F(\theta)F^T(\theta)H^T(\theta)\right]R^{-1}(\theta)Y \tag{7.158}$$

and that

$$-\log p(X|\theta, Y) = \tfrac{1}{2}\left(X - \hat{X}^s(\theta)\right)^T S^{-1}(\theta)\left(X - \hat{X}^s(\theta)\right) + \tfrac{1}{2}\log\det S(\theta)$$

$$S(\theta) = F(\theta)F^T(\theta) - F(\theta)F^T(\theta)H^T(\theta)R^{-1}(\theta)H(\theta)F(\theta)F^T(\theta) \tag{7.159}$$

(cf. Problem 7G.5) [$\hat{X}^s(\theta)$ gives the *smoothed* state estimate for the underlying state space model, see Anderson and Moore (1979)].

(c) Assume that the prior distribution of θ is flat. ($p(\theta) \approx$ independent of θ). Then show that the joint MAP estimate (7.77) of θ and X given Y,

$$(\hat{\theta}^s_{\text{MAP}}, \hat{X}^s_{\text{MAP}}) = \arg\max_{\theta, X} \; p(\theta, X|Y)$$

is given by

$$\arg\min_{X,\theta} \; [-\log p(Y, X|\theta)]$$

where

$$-\log p(Y, X|\theta) = \tfrac{1}{2}|Y - H(\theta)X|^2 + \tfrac{1}{2}\left|F^{-1}(\theta)X\right|^2 + \log\det F(\theta) \tag{7.160}$$

(d) Show that the value of X that minimizes (7.160) for fixed Y and θ is $\hat{X}^s(\theta)$, defined by (7.158). Hence

$$\hat{\theta}^s_{\text{MAP}} = \arg\min_{\theta} \left\{ \tfrac{1}{2}\left|Y - H(\theta)\hat{X}^s(\theta)\right|^2 + \tfrac{1}{2}\left|F^{-1}(\theta)\hat{X}^s(\theta)\right|^2 + \log\det F(\theta) \right\}$$

$$\hat{X}^s_{\text{MAP}} = \hat{X}^s(\hat{\theta}^s_{\text{MAP}})$$

(e) Establish that

$$-\log p(Y|\theta) = -\log p(Y, X|\theta) + \log p(X|\theta, Y) \tag{7.161}$$

(f) Establish that

$$-\log p(Y|\theta) = \left[\tfrac{1}{2}\left|Y - H(\theta)\hat{X}^s(\theta)\right|^2 + \tfrac{1}{2}\left|F^{-1}(\theta)\hat{X}^s(\theta)\right|^2 + \tfrac{1}{2}\log\det R(\theta)\right] \tag{7.162}$$

[*Hint:* Use the matrix identity (cf. (7.159))

$$S(\theta) = \left[F^{-T}(\theta)F^{-1}(\theta) + H^T(\theta)H(\theta)\right]^{-1}$$

and the determinant identity

$$\det(I_r + AB) = \det(I_s + BA)$$

for A and B being $r \times s$ and $s \times r$ matrices and I_r the $r \times r$ identity matrix.]

(g) Conclude that $\hat{\theta}_{\text{ML}} \neq \hat{\theta}^s_{\text{MAP}}$ in general.

Remark: The problem illustrates the relationships among various expressions for the likelihood function, the smoothing problem, and MAP-estimates. "Log likelihood functions" of the kind (7.160) have been discussed, e.g., in Sage and Melsa (1971)and Schweppe (1973), Section 14.3.2.

7G.7 Consider the linear regression structure

$$y(t) = \varphi^T(t)\theta + v(t)$$

Based on the theory of optimal algorithms for operator approximation, (Traub and Wozniakowski, 1980), Milanese and Tempo (1985), and Milanese, Tempo, and Vicino (1986)have suggested the following estimate:

For given δ, y_N and $\{\varphi(t)\}_1^N$, define the set

$$A_\delta = \left\{\theta \| y(t) - \varphi^T(t)\theta| < \delta \qquad \text{all } t = 1, \ldots, N\right\}$$

Assuming A_δ, to be bounded and non-empty define its "center" $\theta_c(A_\delta)$ as follows: The ith component is

$$[\theta_c(A_\delta)]^{(i)} = \tfrac{1}{2}\left[\sup_{\theta \in A_\delta} \theta^{(i)} + \inf_{\theta \in A_\delta} \theta^{(i)}\right]$$

(superscript (i) denoting i : th component). The estimate $\hat{\theta}_N^\delta$ is then taken as $\theta_c(A_\delta)$.

(a) Suppose that dim $\theta = 1$. Prove that $\hat{\theta}_N^\delta$ is independent of δ, as long as A_δ is nonempty and bounded.

(b) When dim $\theta > 1$, $\hat{\theta}_N^\delta$ may in general depend on δ. Suppose that as δ decreases to a value δ^*, A_δ reduces to a singleton

$$A_{\delta^*} = \{\theta^\dagger\}$$

Then clearly $\hat{\theta}_N^{\delta^*} = \theta^\dagger$. Show that

$$\hat{\theta}_N^{\delta^*} = \arg\min_\theta \max_t \left|y(t) - \varphi^T(t)\theta\right|$$

This "optimal estimate" thus corresponds to the prediction error estimate (7.12) with the ℓ_∞-norm

$$\ell_\infty\left(\varepsilon^N(\cdot, \theta)\right) = \max_t |\varepsilon(t, \theta)|$$

This in turn can be seen as the limit as $p \to \infty$ of the criterion functions

$$\ell(\varepsilon) = |\varepsilon|^p$$

in (7.11).

7E.1 *Estimating the AR Part of an ARMA model:* Consider the ARMA model

$$A(q)y(t) = C(q)e(t)$$

with orders n_a and n_c, respectively. A method to estimate the AR part has been given as follows. Let

$$\hat{R}_y^N(\tau) = \frac{1}{N}\sum_{t=\tau}^{N} y(t)y(t-\tau)$$

Then solve for $\hat{a}_i^N$ from

$$\hat{R}_y^N(\tau) + a_1 \hat{R}_y^N(\tau - 1) + \cdots + a_{na}\hat{R}_y^N(\tau - n_a) = 0$$

$$\tau = n_c + 1, n_c + 2, \ldots, n_c + n_a$$

Show that this (essentially) is an application of the IV method using specific instruments. Which ones? (See Cadzow, 1980, and Stoica, Söderström, and Friedlander, 1985.)

7E.2 *Sinusoids in noise:* Consider a sinusoid measured in white Gaussian noise:

$$y(t) = \alpha e^{i\omega t} + e(t)$$

For simplicity we use complex algebra. The constant α is thus complex-valued. The amplitude, phase and frequency are unknown: $\theta = (\alpha, \omega)$. The predictor thus is

$$\hat{y}(t|\theta) = \alpha e^{i\omega t}$$

If $e(t)$ has variance 1 (real and imaginary parts independent), the likelihood function gives the prediction-error criterion:

$$V_N(\theta, Z^N) = \frac{1}{2}\sum_{t=1}^{N} \left|y(t) - \hat{y}(t|\theta)\right|^2$$

Show that the MLE

$$\hat{\theta}_N = \begin{bmatrix} \hat{\alpha}_N \\ \hat{\omega}_N \end{bmatrix} = \arg\min_{\theta} V_N(\theta, Z^N)$$

obeys

$$\hat{\omega}_N = \arg\max_{\omega} |Y_N(\omega)|^2$$

where $Y_N(\omega)$ is the Fourier transform (2.37) of $y(t)$.

7E.3 *Error-in-variables models:* Econometric models often include disturbances both on inputs and outputs (compare our comment in Section 2.1 on Figure 2.2). Consider the model in Figure 7.1. The true inputs and outputs are thus s and x, while we measure u and y. In a first-order case, we have

$$x(t) + ax(t-1) = bs(t-1)$$

$$y(t) = x(t) + e(t)$$

$$u(t) = s(t) + w(t)$$

Suppose that w and e are independent white noises with unknown variances. Discuss how a, b, and these variances can be estimated using measurements of y and u.

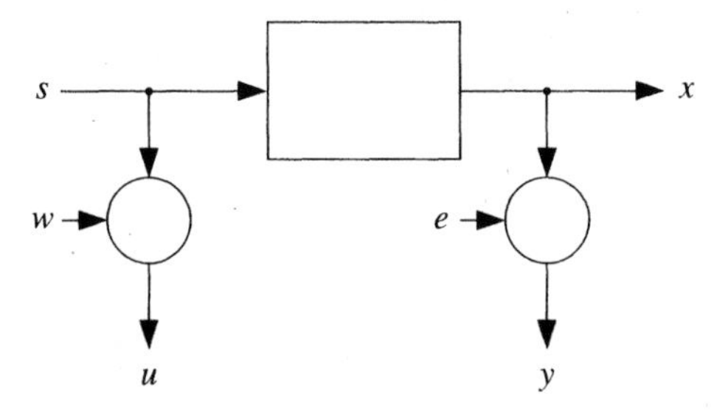

Figure 7.1 An error-in-variables model.

[*Remark:* With the assumption that the color of the noises are known, the problem is relatively simple. Without this assumption the problem is more difficult. See Kalman (1991), Anderson (1985), Söderström (1981), McKelvey (1996), and Stoica et.al. (1997)].

7E.4 Consider a probabilistic model, implicitly given in the state-space form

$$x(t+1) = ax(t) + w(t)$$
$$y(t) = x(t) + \nu(t) \tag{7.163a}$$

where $\{w(t)\}$ and $\{\nu(t)\}$ are assumed to be independent, white Gaussian noises, with variances

$$Ew^2(t) = r_1$$
$$E\nu^2(t) = 1 \quad \text{(assumed known)} \tag{7.163b}$$

Let the parameter vector be

$$\theta = \begin{bmatrix} a \\ r_1 \end{bmatrix} \tag{7.163c}$$

Assume initial conditions for $x(0)$ (mean and variance) such that the prediction $\hat{y}(t|\theta)$ becomes a stationary process for each θ (i.e., so that the steady-state Kalman filter can be used). Determine the log-likelihood function for this problem. Compare with the log-likelihood function for a directly parametrized innovations representation model (4.91).

7E.5 Consider the nonlinear model structure of Problem 5E.1. Discuss how the LS, ML, IV, and PLR methods can be applied to this structure. (Reference: Fnaiech and Ljung, 1986).

7E.6 Consider the model structure

$$y(t) = \varphi^T(t)\theta + v(t)$$

where the regression vector $\varphi(t)$ can only be measured with noise:

$$\eta(t) = \varphi(t) + w(t)$$

The noises $\{w(t)\}$ and $\{v(t)\}$ may be nonwhite and mutually correlated. Suppose a vector $\zeta(t)$ is known that is uncorrelated with $\{v(t)\}$ and $\{w(t)\}$ but correlated with $\varphi(t)$. Suggest how to estimate θ from $y(t)$, $\eta(t)$, and $\zeta(t)$, $t = 1, \ldots, N$.

7E.7 Suppose in (7.86) and (7.87) that λ does not depend on θ. Determine $\hat{\lambda}_N$.

7E.8 Consider the model structure

$$\hat{y}(t|\theta) = -ay(t-1) + bu(t-1)$$

and assume that the true system is given by

$$y(t) - 0.9y(t-1) = u(t-1) + e_0(t)$$

where $\{e_0(t)\}$ is white noise of unit variance. Determine the Cramér-Rao bound for the estimation of a and b. How does it depend on the properties of u?

7E.9 Suppose that $u(t)$ is periodic with period M, and that all transients have died out. We collect data over K periods: $[y(t), u(t)], t = 1, \ldots, KM$. We take the DFT of the signals and form (7.144) for a fixed noise model $H^*(i\omega)$. Show that this gives exactly the same results as if we just take the DFT over one period and use the averaged output (7.153). Is it essential that the noise model is fixed?

7T.1 Suppose that a true description of a certain system is given by

$$y(t) + a_1^0 y(t-1) + \ldots + a_{na}^0 y(t-n_a) = b_1^0 u(t-1) + \ldots + b_{n_b}^0 u(t-n_b) + v_0(t)$$

for a stationary process $\{v_0(t)\}$ independent of the input. Let $\varphi(t)$ be defined, as usual, by (7.32), and let $\tilde{\varphi}(t)$ be given by

$$\tilde{\varphi}(t) = [-y_0(t-1) \ldots -y_0(t-n_a) \quad u(t-1) \ldots u(t-n_b)]^T$$

where

$$y_0(t) + a_1^0 y_0(t-1) + \ldots + a_{na}^0 y_0(t-n_a) = b_1^0 u(t-1) + \ldots + b_{n_b}^0 u(t-n_b)$$

Prove that for any vector of instrumental variables of the general kind (7.122) we have

$$E\zeta(t)\varphi^T(t) = E\zeta(t)\tilde{\varphi}^T(t)$$

7D.1 Consider the ARX structure (4.7) where one parameter, say b_1, is known to have a certain value b_1^*. Show that the associated predictor can be written as

$$\hat{y}(t|\theta) = \theta^T\varphi(t) + \mu(t)$$

with proper definitions of θ, φ, and μ (φ and μ to be known variables at time t). Derive the LS estimate and the IV estimate for this model.

7D.2 Let A be a given, positive symmetric definite matrix and let B and C be given matrices. Establish that

$$\theta^T A\theta - \theta^T B - B^T\theta + C = [\theta - A^{-1}B]^T A[\theta - A^{-1}B] + C - B^T A^{-1}B$$
$$\geq C - B^T A^{-1} B$$

and use this result to prove all the expressions for the LSE in Section 7.3 [(7.34), (7.41), (7.43), and (7.46)]. The matrix inequality $D \geq B$ is to be interpreted as "$D - B$ is a positive semidefinite matrix."

Hint: For (7.46), rewrite (7.45) as

$$V_N(\theta, Z^N) = \operatorname{tr}\frac{1}{N}\sum_{t=1}^{N}\left[y(t) - \theta^T\varphi(t)\right]\left[y(t) - \theta^T\varphi(t)\right]^T$$

7D.3 Let Σ be an invertible square $p \times p$ matrix with elements σ_{ij}. Prove the differentiation formula

$$\frac{\partial}{\partial\alpha_{ij}}\det\Sigma = \det[\Sigma]\cdot\mu_{ij}$$

where μ_{ij} is the i, j element of Σ^{-1}. [*Hint:* Use $\det(I + \varepsilon A) = 1 + \varepsilon \operatorname{tr} A +$ higher-order terms in ε]. Use the result to prove (7.94) and (7.95).

7D.4 Show that the two instrumental variable vectors, of dimension d, $\zeta_1(t)$ and $\zeta_2(t)$, where $\rho_1(t) = T\rho_2(t)$ with T invertible, give the same estimate $\hat{\theta}_N^{\text{IV}}$ in (7.118).

7D.5 Show that if two variables x and u are associated as in (7.123) and (7.124) then we can write

$$\begin{bmatrix} -x(t-1) \\ \vdots \\ -x(t-n_n) \\ u(t-1) \\ \vdots \\ u(t-n_m) \end{bmatrix} = S(-M, N) \cdot \frac{1}{N(q)} \begin{bmatrix} u(t-1) \\ u(t-2) \\ \vdots \\ u(t-n_n-n_m) \end{bmatrix}$$

for an $(n_n + n_m) \times (n_n + n_m)$ matrix

$$S(-M, N) = \begin{bmatrix} -m_0 & -m_1 & \dots & -m_{n_m} & 0 & \dots & 0 \\ 0 & -m_0 & \dots & -m_{n_m-1} & -m_{n_m} & \dots & 0 \\ \vdots & \vdots & & \vdots & \vdots & & \vdots \\ 0 & 0 & \dots & -m_0 & -m_1 & \dots & -m_{n_m} \\ 1 & n_1 & \dots & n_{n_n} & 0 & \dots & 0 \\ 0 & 1 & \dots & n_{n_n-1} & n_{n_n} & \dots & 0 \\ \vdots & \vdots & & \vdots & \vdots & & \vdots \\ 0 & 0 & \dots & 1 & n_1 & \dots & n_{n_n} \end{bmatrix}$$

Such a matrix is called a *Sylvester matrix* (see, e.g., Kailath, 1980), and it will be nonsingular if and only if the polynomials in (7.124) have no common factor. Use this result to prove that the instruments (7.126) give the same IV estimate as the instruments (7.122). Reference: Söderström and Stoica (1983).

7D.6 Show that the prediction-error estimate obtained from (7.11) and (7.12) can also be seen as a correlation estimate (7.110) for a particular choice of L, ζ, and α.

7D.7 Give an explicit expression for the estimate $\hat{\theta}_N^{\text{EIV}}$ in (7.130) in the case ζ does not depend on θ, and $\alpha(\varepsilon) = \varepsilon$.

7D.8 Consider the symmetric matrix

$$H = \begin{bmatrix} A & B \\ B^T & C \end{bmatrix}$$

Show that if $H \geq 0$, then

$$A - BC^{-1}B^T \geq 0.$$

Hint: Consider xHx^T for

$$x = [x_1 \quad -x_1 BC^{-1}]$$

with x_1 arbitrary.

APPENDIX 7A: PROOF OF THE CRAMÉR-RAO INEQUALITY

The assumption $E\hat{\theta}(y^N) = \theta_0$ can be written

$$\theta_0 = \int_{\mathbf{R}^N} \hat{\theta}(x^N) f_y(\theta_0, x^N)\, dx^N \tag{7A.1}$$

By definition we also have

$$1 = \int_{\mathbf{R}^N} f_y(\theta_0, x^N)\, dx^N \tag{7A.2}$$

Differentiating these two expressions with respect to θ_0 gives

$$\begin{aligned} I &= \int_{\mathbf{R}^N} \hat{\theta}(x^N) \left[\frac{d}{d\theta_0} f_y(\theta_0, x^N)\right]^T dx^N \\ &= \int_{\mathbf{R}^N} \hat{\theta}(x^N) \left[\frac{d}{d\theta_0} \log f_y(\theta_0, x^N)\right]^T f_y(\theta_0, x^N)\, dx^N \\ &= E\hat{\theta}(y^N) \left[\frac{d}{d\theta_0} \log f_y(\theta_0, y^N)\right]^T \end{aligned} \tag{7A.3}$$

(I is the $d \times d$ unit matrix) and

$$\begin{aligned} 0 &= \int_{\mathbf{R}^N} \left[\frac{d}{d\theta_0} f_y(\theta_0, x^N)\right]^T dx^N = \int_{\mathbf{R}^N} \left[\frac{d}{d\theta_0} \log f_y(\theta_0, x^N)\right]^T f_y(\theta_0, x^N) dx^N \\ &= E\left[\frac{d}{d\theta_0} \log f_y(\theta_0, y^N)\right]^T \end{aligned} \tag{7A.4}$$

Expectation in these two expressions is hence w.r.t. y^N.

Now multiply (7A.4) by θ_0 and subtract it from (7A.3). This gives

$$E\left[\hat{\theta}(y^N) - \theta_0\right]\left[\frac{d}{d\theta_0} \log f_y(\theta_0, y^N)\right]^T = I \tag{7A.5}$$

Now denote

$$\alpha = \hat{\theta}(y^N) - \theta_0, \qquad \beta = \frac{d}{d\theta_0} \log f_y(\theta_0, y^N) \tag{7A.6}$$

(both d-dimensional column vectors) so that

$$E\alpha\beta^T = I \tag{7A.7}$$

Hence

$$E\begin{bmatrix} \alpha \\ \beta \end{bmatrix}\begin{bmatrix} \alpha \\ \beta \end{bmatrix}^T = \begin{bmatrix} E\alpha\alpha^T & I \\ I & E\beta\beta^T \end{bmatrix} \geq 0$$

where the positive semidefiniteness follows by construction. Hence Problem 7D.8 proves that

$$E\alpha\alpha^T \geq [E\beta\beta^T]^{-1}$$

which is (7.79). It only remains to prove the equality in (7.80). Differentiating the transpose of (7A.4) gives

$$0 = \int_{\mathbf{R}^N} \left[\frac{d^2}{d\theta_0^2} \log f_y(\theta_0, x^N) \right] f_y(\theta_0, x^N)\, dx^N$$

$$+ \int_{\mathbf{R}^N} \left[\frac{d}{d\theta_0} \log f_y(\theta_0, x^N) \right] \left[\frac{d}{d\theta_0} \log f_y(\theta_0, x^N) \right]^T f_y(\theta_0, x^N)\, dx^N$$

which gives (7.80).

8

CONVERGENCE AND CONSISTENCY

8.1 INTRODUCTION

In Chapter 7 we described a number of different methods to determine models from data. To use these methods in practice, we need insight into their properties: How well will the identified model describe the actual system? Are some identification methods better than others? How should the design variables associated with a certain method be chosen?

Such questions relate, from a formal point of view, to the mapping (7.7) from the data set Z^N to the parameter estimate $\hat{\theta}_N$:

$$Z^N \rightarrow \hat{\theta}_N \in D_{\mathcal{M}} \tag{8.1}$$

Questions about properties of this mapping can be answered basically in two ways:

1. Generate data Z^N with known characteristics. Apply the mapping (8.1) (corresponding to a particular identification method) and evaluate the properties of $\hat{\theta}_N$. This is known as *simulation* studies.
2. Assume certain properties of Z^N and try to calculate what the inherited properties of $\hat{\theta}_N$ are. This is known as *analysis*.

In this chapter we shall analyze the *convergence properties* of $\hat{\theta}_N$ as N tends to infinity. Since we will never encounter infinitely many data, such analysis has the character of a "thought experiment," and we must support it with some assumptions about a corresponding infinite data set Z^∞. There are some different possibilities for such assumptions (see Problem 8T.1). Here we shall adopt a stochastic framework for the observations, along the lines described in Chapter 2. We shall thus consider the data as realizations of a stochastic process with deterministic components. It might be worthwhile to contemplate what analysis under such assumptions actually amounts to. A probabilistic framework relates to the following questions: What would happen if I repeat the experiment? Should I then expect a very different result? Will the limit of $\hat{\theta}_N$ depend on the particular realization of the random variables? Even if the experiment is never repeated, it is clear that such questions are relevant for the confidence one should develop for the estimate, and this makes the analysis worthwhile. It is then another matter that the probabilistic framework that is set up to answer such questions may exist only in the mind of the analyzer and cannot be firmly tied to the real-world experiment.

It should also be remarked that a conventional stochastic description of disturbances is not without problems: For example, suppose we measure a distance with a crude measuring rod and describe the measurement error as a zero-mean random variable, which is independent of the error obtained when the experiment is repeated. This assumption implies, by the law of large numbers, that the distance can be determined with arbitrary accuracy, if only the measurements are repeated sufficiently many times. Clearly such a conclusion can be criticized from a practical point of view. Results from theoretical analysis must thus be interpreted with care when applied to a practical situation.

The question of how $\hat{\theta}_N$ behaves as N increases clearly relates to the question of how the corresponding criteria functions $V_N(\theta, Z^N)$ and $f_N(\theta, Z^N)$ behave. These are, with a stochastic framework, sums of random variables, and their convergence properties will be consequences of the law of large numbers. Our basic technical tool in this chapter will thus be Theorem 2B.1. In order not to conceal the basic ideas with too much technicalities, we shall only complete the proofs for linear, time-invariant models (such as those in Chapter 4) and quadratic criteria. The techniques and results, however, carry over also to more general cases.

The chapter is organized as follows. Assumptions about the infinite data set Z^∞ are given in Section 8.2. Convergence for prediction-error estimates is treated in Section 8.3. Consistency questions (i.e., whether the true system is retrieved in the limit) are discussed in Section 8.4. A frequency-domain characterization of the limit estimate is given in Section 8.5. In Section 8.6, the corresponding results are given for the correlation approach.

A Preview

In the chapter a general and natural result is derived: the estimate $\hat{\theta}_N$ obtained by the prediction-error method (7.155) will converge to the value that minimizes the average criterion $\bar{E}\ell(\varepsilon(t, \theta), \theta)$. Here $\bar{E}$ can heuristically be taken as averaging over time or ensembles (possible realizations) or both. The chapter deals both with the formal framework for establishing this "obvious" result and with characterizations of the limit value of $\hat{\theta}_N$. The reluctant reader of theory should concentrate on understanding the main result, Equation (8.29), and the frequency-domain characterization of the limit model in Section 8.5.

8.2 CONDITIONS ON THE DATA SET

The data set

$$Z^N = \{u(1), y(1), \ldots, u(N), y(N)\}$$

is the basic starting point. Analysis, we said, amounts to assuming certain properties about the data and computing the resulting properties of $\hat{\theta}_N$. Since the analysis of $\hat{\theta}_N$ will be carried out for $N \to \infty$, it is natural that the conditions on the data relate to the infinite set Z^∞. In this section we shall introduce such conditions, as well as some pertinent definitions.

A Technical Condition D1 (*)

We shall assume that the actual data are generated as depicted in Figure 8.1. The input u may be generated (partly) as output feedback or in open loop ($u = w$). The signal e_0 represents the disturbances that act on the process. [The subscript 0 distinguishes this "true" noise e_0 from the "dummy" noise e we have used in our model descriptions (7.3).] The prime objective with condition D1 is to describe the closed-loop system in Figure 8.1 as a stable system so that the dependence between far apart data decays. The most restrictive condition is the assumed linearity (8.2). It can be traded for more general conditions, at the price of more complicated analysis. See Ljung (1978a), condition S3. For our analysis, we shall use the following technical assumptions:

D1: The data set Z^∞ is such that for some filters $\left\{d_t^{(i)}(k)\right\}$

$$\begin{aligned} y(t) &= \sum_{k=1}^{\infty} d_t^{(1)}(k) r(t-k) + \sum_{k=0}^{\infty} d_t^{(2)}(k) e_0(t-k) \\ u(t) &= \sum_{k=0}^{\infty} d_t^{(3)}(k) r(t-k) + \sum_{k=0}^{\infty} d_t^{(4)}(k) e_0(t-k) \end{aligned} \tag{8.2}$$

where

1. $\{r(t)\}$ is a bounded, deterministic, external input sequence. (8.3)
2. $\{e_0(t)\}$ is a sequence of independent random variables with zero mean values and bounded moments of order $4 + \delta$ for some $\delta > 0$. (8.4)

Moreover,

3. The family of filters $\left\{d_t^{(i)}(k)\right\}_{k=1}^{\infty}$, $i = 1$–4; $t = 1, 2, \ldots$ is uniformly stable. (8.5)
4. The signals $\{y(t)\}$, $\{u(t)\}$ are jointly quasi-stationary. (8.6)

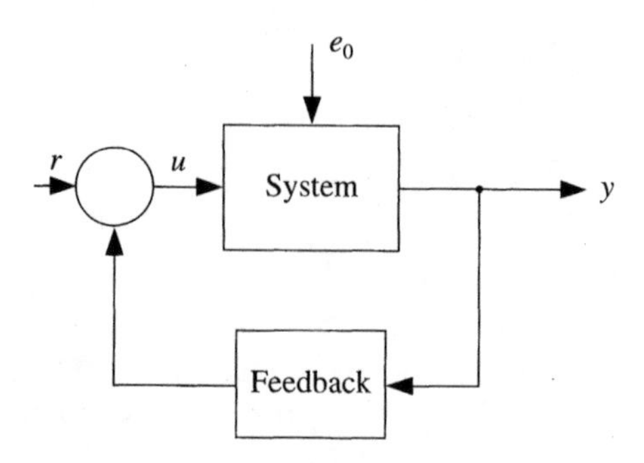

Figure 8.1 The data-generating configuration.

Recall the definitions of stability (2.29) and quasi-stationarity (2.58) to (2.62). (Problem 2T.4 showed that uniform stability holds, even if the closed-loop system goes through "unstable transients.")

Remark. When we say that $\{r(t)\}$ is "deterministic," we simply mean that we regard it as a given sequence that (in contrast to e_0) can be reproduced if the experiment is repeated. The stochastic operators and qualifiers, such as E, w.p.1, and AsN will thus average over the properties of $\{e_0(t)\}$ for the fixed sequence $\{r(t)\}$. Of course, this does not exclude that this particular sequence $\{r(t)\}$ actually is generated as a realization of a stochastic process, independent of the system disturbances. In that case it is sometimes convenient to let the expectation also average over the probabilistic properties of $\{r(t)\}$. We shall comment on how to do this below [Eq. (8.27)].

A True System S

We shall sometimes use a more specific assumption of a "true system":

S1: The data set Z^∞ is generated according to

$$S: \quad y(t) = G_0(q)u(t) + H_0(q)e_0(t) \tag{8.7}$$

where $\{e_0(t)\}$ is a sequence of independent random variables, with zero mean values, variances λ_0, and bounded moments of order $4+\delta$, some $\delta > 0$, and $H_0(q)$ is an inversely stable, monic filter.

We thus denote the true system by S. Given a model structure (4.4),

$$\mathcal{M}: \quad \{G(q,\theta), H(q,\theta) | \theta \in D_{\mathcal{M}}\} \tag{8.8}$$

it is natural to check whether the true system (8.7) belongs to the set defined by (8.8). We thus introduce

$$\begin{aligned} D_T(S,\mathcal{M}) = \{\theta \in D_{\mathcal{M}} |\ & G(e^{i\omega},\theta) = G_0(e^{i\omega}); \\ & H(e^{i\omega},\theta) = H_0(e^{i\omega}); -\pi \le \omega \le \pi\} \end{aligned} \tag{8.9}$$

This set is nonempty precisely when the model structure admits an exact description of the true system. We write this also as

$$\mathcal{S} \in \mathcal{M} \tag{8.10}$$

Although such an assumption is not particularly realistic in practical applications, it yields a quite useful insight into properties of the estimated models.

When S1 holds, a more explicit version of conditions D1 can be given:

Lemma 8.1. Suppose that S1 holds, and the input is chosen as

$$u(t) = -F(q)y(t) + r(t)$$

such that there is a delay in either G_0 or F and such that

$$[1 + G_0(q)F(q)]^{-1} G_0(q), \quad [1 + G_0(q)F(q)]^{-1} H_0(q),$$
$$F(q)[1 + G_0(q)F(q)]^{-1} G_0(q), \quad F(q)[1 + G_0(q)F(q)]^{-1} H_0(q)$$

are stable filters and that $\{w(t)\}$ is quasi-stationary. Then condition D1 holds.

Proof. We have, for the closed-loop system,

$$y(t) = [1 + G_0(q)F(q)]^{-1} G_0(q)r(t) + [1 + G_0(q)F(q)]^{-1} H_0(q)e_0(t) \tag{8.11}$$

and similarly for u. The stability condition means that the filters in (8.11) are stable. Thus (8.6) follows from Theorem 2.2. Moreover, (8.2), and (8.5), are immediate from (8.11) and the stability assumption. □

Information Content in the Data Set

The set Z^N is our source of information about the true system. This is to be fit to a model structure $\mathcal{M}$ of our choice. (The reader might at this point review Section 4.5, if necessary.) The structure $\mathcal{M}$ describes a set of models $\mathcal{M}^*$ within which the best one is sought for. Identifiability of model structures concerns the question whether different parameter vectors may describe the same model in the set $\mathcal{M}^*$. See Definitions 4.6 to 4.8. A related question is whether the data set Z^∞ allows us to distinguish between different models in the set. Recall that, according to Definition 4.1, a (linear time-invariant) model is given by a filter $W(q)$. We shall call a data set *informative* if it is capable of distinguishing between different models. We thus introduce the following concept:

Definition 8.1. A quasi-stationary data set Z^∞ is *informative enough with respect to the model set* $\mathcal{M}^*$ if, for any two models $W_1(q)$ and $W_2(q)$ in the set,

$$\bar{E}\left[(W_1(q) - W_2(q))\, z(t)\right]^2 = 0 \tag{8.12a}$$

implies that $W_1(e^{i\omega}) \equiv W_2(e^{i\omega})$ almost all ω.

We note that with

$$W_1(q) - W_2(q) = [\,\Delta W_u(q) \quad \Delta W_y(q)\,]$$

(8.12a) can be written

$$\overline{E}\left[\Delta W_u(q)u(t) + \Delta W_y(q)y(t)\right]^2 = 0 \tag{8.12b}$$

Note that the limit in (8.12) exists in view of (8.6) and Theorem 2.2. Recall also (4.112) and the definition of equality of models, (4.116).

Definition 8.2. A quasistationary data set Z^∞ is *informative* if it is informative enough with respect to the model set $\mathcal{L}^*$, consisting of all linear, time-invariant models.

The concept of informative data sets is very closely related to concepts of "persistently exciting" inputs, "general enough" inputs, and so on. We shall discuss the concept in detail in Chapter 13 in connection with experiment design. Here we give an immediate consequence of Definition 8.2.

Theorem 8.1. A quasi-stationary data set Z^∞ is informative if the spectrum matrix for $z(t) = [\,u(t) \quad y(t)\,]^T$ is strictly positive definite for almost all ω.

Proof. Consider (8.12) for arbitrary linear models W_1 and W_2. Let us denote $W_1(q) - W_2(q) = \tilde{W}(q)$. Then applying Theorem 2.2 to (8.12) gives

$$0 = \int_{-\pi}^{\pi} \tilde{W}(e^{i\omega})\Phi_z(\omega)\tilde{W}^T(e^{-i\omega})\,d\omega$$

where

$$\Phi_z(\omega) = \begin{bmatrix} \Phi_u(\omega) & \Phi_{uy}(\omega) \\ \Phi_{yu}(\omega) & \Phi_y(\omega) \end{bmatrix} \tag{8.13}$$

Since $\Phi_z(\omega)$ is positive definite, this implies that $\tilde{W}(e^{i\omega}) \equiv 0$ almost everywhere, which proves the theorem. □

Some Additional Concepts and Notations (∗)

In Definition 4.3 we defined a model structure as a differentiable mapping, such that the predictors and their gradients were stable for each $\theta \in D_{\mathcal{M}}$. To facilitate the analysis, we now strengthen this condition.

Definition 8.3. A model structure $\mathcal{M}$ is said to be *uniformly stable* if the family of filters $\{W(q,\theta),\ \Psi(q,\theta)$ and $(d/d\theta)\Psi(q,\theta);\ \theta \in D_{\mathcal{M}}\}$ is uniformly stable and if the set $D_{\mathcal{M}}$ is compact. [Recall the definition (2.29).]

Analogous to (4.109), we shall, when S1 holds, define

$$\chi_0(t) = \begin{bmatrix} u(t) \\ e_0(t) \end{bmatrix} \tag{8.14a}$$

and

$$T_0(q) = \begin{bmatrix} G_0(q) & H_0(q) \end{bmatrix} \tag{8.14b}$$

The system (8.7) can thus be written

$$y(t) = T_0(q)\chi_0(t)$$

The difference will be denoted

$$\tilde{T}(q, \theta) = T_0(q) - T(q, \theta) = [\, \tilde{G}(q, \theta) \quad \tilde{H}(q, \theta) \,] \tag{8.15}$$

8.3 PREDICTION-ERROR APPROACH

Basic Result

The prediction-error estimate is defined by (7.12)

$$\hat{\theta}_N = \underset{\theta \in D_{\mathcal{M}}}{\arg\min}\; V_N(\theta, Z^N) \tag{8.16}$$

To determine the limit to which $\hat{\theta}_N$ converges as N tends to infinity is obviously related to the limit properties of the function $V_N(\theta, Z^N)$. For a quadratic criterion and a linear, uniformly stable model structure $\mathcal{M}$, we have

$$V_N(\theta, Z^N) = \frac{1}{N} \sum_{t=1}^{N} \tfrac{1}{2}\varepsilon^2(t, \theta) \tag{8.17}$$

and, using (7.2),

$$\varepsilon(t, \theta) = \left[1 - W_y(q, \theta)\right] y(t) - W_u(q, \theta)u(t) \tag{8.18}$$

Under assumption D1 we can replace $y(t)$ and $u(t)$ in the preceding expression by (8.2), which gives

$$\varepsilon(t, \theta) = \sum_{k=1}^{\infty} d_t^{(5)}(k; \theta) r(t - k) + \sum_{k=0}^{\infty} d_t^{(6)}(k; \theta) e_0(t - k) \tag{8.19}$$

Now the filters in (8.18) are uniformly (in θ) stable since $\mathcal{M}$ is uniformly stable. Under assumption (8.5) the filters in (8.2) are uniformly (in t) stable. Hence the cascaded filters $\{d_t^{(i)}(k;\theta)\}$, $i = 5, 6$, in (8.19) are also uniformly (in both θ and t) stable (see Problem 8D.2). That is,

$$\left|d_t^{(i)}(k;\theta)\right| \le \beta_k, \forall t, \forall \theta \in D_{\mathcal{M}}, i = 5, 6 \qquad \sum_{1}^{\infty} \beta_k < \infty \tag{8.20}$$

Finally, under assumption (8.6), Theorem 2.2 implies that $\{\varepsilon(t,\theta)\}$ is quasi-stationary.

All conditions for Theorem 2B.1 are thus satisfied, and applying this theorem to (8.17) with (8.19) gives the following result.

Lemma 8.2. Consider a uniformly stable, linear model structure $\mathcal{M}$ (see Definitions 4.3 and 8.3). Assume that the data set Z^∞ is subject to D1. Then, with $V_N(\theta, Z^N)$ defined by (8.17),

$$\sup_{\theta \in D_{\mathcal{M}}} \left|V_N(\theta, Z^N) - \overline{V}(\theta)\right| \to 0, \qquad \text{w.p. 1 as } N \to \infty \tag{8.21}$$

where

$$\overline{V}(\theta) = \overline{E}\tfrac{1}{2}\varepsilon^2(t,\theta) \tag{8.22}$$

The criterion function $V_N(\theta, Z^N)$ thus converges uniformly in $\theta \in D_{\mathcal{M}}$ to the limit function $\overline{V}(\theta)$. This implies that the minimizing argument $\hat{\theta}_N$ of V_N also converges to the minimizing argument θ^* of $\overline{V}$ since $D_{\mathcal{M}}$ is compact. Notice that it is essential that the convergence is indeed uniform in θ for this to hold (see Problem 8D.1). It may happen that $\overline{V}(\theta)$ does not have a unique global minimum. In that case we define the set of minimizing values as

$$D_c = \arg\min_{\theta \in D_{\mathcal{M}}} \overline{V}(\theta) = \left\{\theta | \theta \in D_{\mathcal{M}}, \overline{V}(\theta) = \min_{\theta' \in D_{\mathcal{M}}} \overline{V}(\theta')\right\} \tag{8.23}$$

We can thus formulate this corollary to Lemma 8.2 as our main convergence result:

Theorem 8.2. Let $\hat{\theta}_N$ be defined by (8.16) and (8.17), where $\varepsilon(t,\theta)$ is determined from a uniformly stable linear model structure $\mathcal{M}$. Assume that the data set Z^∞ is subject to D1. Then

$$\hat{\theta}_N \to D_c, \qquad \text{w.p. 1 as } N \to \infty \tag{8.24}$$

where D_c is given by (8.22) and (8.23).

Remark. Convergence into a set as in (8.24) is to be interpreted as

$$\inf_{\overline{\theta} \in D_c} \left|\hat{\theta}_N - \overline{\theta}\right| \to 0, \qquad \text{as } N \to \infty \tag{8.25}$$

The function $\overline{V}(\theta)$ will in general depend both on the true system and the input properties. With a quadratic criterion and a linear model structure, it follows from Theorem 2.2 that it depends on the data only via the spectrum matrix $\Phi_z(\omega)$ in (8.13). [Explicit expressions will be given in (8.63) to (8.66).] This has the important consequence that it is only the second-order properties of the data that affect the convergence of the estimates.

Ensemble- and Time-averages

The signal sources for $\varepsilon(t,\theta)$ are r and e_0, as evidenced by (8.19). Recall that $r = u$ in case of open loop operation. The symbol $\overline{E}$ denotes as defined in (2.60) ensemble-averaging ("statistical expectation") over the stochastic process $\{e_0(t)\}$ and time-averaging over the deterministic signal $\{r(t)\}$. The function $\overline{V}(\theta)$ is thus "the average value" of $\varepsilon^2(t,\theta)$ in these two respects.

The reason for time-averaging over $\{r(t)\}$ is, as we have stated several times, that it might not always be suitable to describe this signal as a realization of a stochastic process. However, when indeed $\{r(t)\}$ is taken as a realization of a stationary stochastic process, (independent of e_0), Theorem 2.3 shows that, under weak conditions, time averages over $\{r(t)\}$ will, with probability 1, equal the ensemble averages:

$$\lim_{N\to\infty} \frac{1}{N} \sum_{t=1}^{N} r(t)r(t-\tau) = E_r r(t)r(t-\tau) \text{ w.p. } 1 \tag{8.26}$$

Here E_r denotes statistical expectation with respect to the r-process.

This means that e_0-ensemble- and r-time-averaging by $\overline{E}$ will, w.p. 1, be equivalent to taking total statistical expectation over both e_0 and r:

$$\text{"}\overline{E} = E_r E_{e_0} \qquad \text{w.p. 1"} \tag{8.27a}$$

i.e.,

$$\overline{V}(\theta) = \overline{E}\varepsilon^2(t,\theta) = E_r E_{e_0} \varepsilon^2(t,\theta) \qquad \text{w.p. } 1 \tag{8.27b}$$

For "hand calculation" it is often easier to apply this total expectation: See Examples 8.1 to 8.2.

[Conversely, one could also replace ensemble averages over e_0 by time averages to eliminate the probabilistic framework entirely: See Problem 8T.1.]

The General Case

With a little more technical effort, the results of Lemma 8.2 and Theorem 8.2 can also be established for general norms $\ell(\varepsilon,\theta)$ as in (7.16), in which case the limit is defined as

$$\overline{V}(\theta) = \overline{E}\ell\,(\varepsilon(t,\theta),\theta) \tag{8.28}$$

The result can also be extended to nonlinear, time-varying models and less restrictive assumptions on the data set than D1. See, for example, Ljung (1978a)for such results. In summary we thus have

$$\hat{\theta}_N \to \arg\min_{\theta \in D_{\mathcal{M}}} \overline{E}\ell\left(\varepsilon(t,\theta),\theta\right), \qquad \text{w.p. 1 as } N \to \infty \tag{8.29}$$

This convergence result is quite general and intuitively appealing. It states that *the estimate will converge to the best possible approximation of the system that is available in the model set.* The goodness of the approximation is then measured in terms of the criterion $\overline{V}(\theta)$ in (8.28). We shall dwell on what "best possible" actually means in more practical terms in the next two sections. First we give two examples.

Example 8.1 Bias in ARX Structures

Suppose that the system is given by

$$y(t) + a_0 y(t-1) = b_0 u(t-1) + e_0(t) + c_0 e_0(t-1) \tag{8.30}$$

where $\{u(t)\}$ and $\{e_0(t)\}$ are independent white noises with unit variances. Let the model structure be given by

$$\hat{y}(t|\theta) + ay(t-1) = bu(t-1), \qquad \theta = \begin{bmatrix} a \\ b \end{bmatrix} \tag{8.31}$$

The prediction-error variance is

$$\begin{aligned} \overline{V}(\theta) &= \overline{E}\left[y(t) + ay(t-1) - bu(t-1)\right]^2 \\ &= r_0(1 + a^2 - 2aa_0) + b^2 - 2bb_0 + 2ac_0 \end{aligned} \tag{8.32}$$

where

$$r_0 = Ey^2(t) = \frac{b_0^2 + c_0(c_0 - a_0) - a_0 c_0 + 1}{1 - a_0^2}$$

(see Problem 2E.7). It is easy to verify that the values of a and b that minimize (8.32) are $\theta^* = [\,a^* \quad b^*\,]^T$ given by

$$\begin{aligned} a^* &= a_0 - \frac{c_0}{r_0} \\ b^* &= b_0 \end{aligned} \tag{8.33}$$

These values give a prediction-error variance

$$\overline{V}(\theta^*) = 1 + c_0^2 - \frac{c_0^2}{r_0} \tag{8.34}$$

This variance is smaller than the "true values" $\theta_0 = [\,a_0 \quad b_0\,]^T$ inserted into (8.32) would give:

$$\overline{V}(\theta_0) = 1 + c_0^2 \tag{8.35}$$

When we apply the prediction error method to (8.30) and (8.31), the estimates $\hat{a}_N$ and $\hat{b}_N$ will converge, according to Theorem 8.2, to the values given by (8.33). The fact that $a^* \neq a_0$ is usually expressed as that the estimate is "biased." However, it is clear from (8.34) and (8.35) that the bias is beneficial for the prediction performance of the model (8.31). It gives a strictly better predictor for $\hat{a} = a^*$ than for $\hat{a} = a_0$. □

Example 8.1 stresses that the algorithm indeed gives us the best possible predictor, and it uses its parameters as vehicles for that. It is, however, important to keep in mind that what is the best approximate description of a system in general depends on the input used. We illustrate this by a simple example.

Example 8.2 Wrong Time Delay

Consider the system

$$y(t) = b_0 u(t-1) + e_0(t) \tag{8.36}$$

where

$$u(t) = d_0 u(t-1) + w(t) \tag{8.37}$$

and where $\{e_0(t)\}$ and $\{w(t)\}$ are independent white-noise sequences with unit variances. Let the model structure be given by

$$\hat{y}(t|\theta) = bu(t-2), \qquad \theta = b \tag{8.38}$$

The prediction-error variance associated with (8.38) is

$$\begin{aligned} E\,[y(t) - bu(t-2)]^2 &= E\,[b_0 u(t-1) - bu(t-2)]^2 + Ee_0^2(t) \\ &= E\,[(b_0 d_0 - b)u(t-2) + b_0 w(t-1)]^2 + 1 \\ &= \frac{(b_0 d_0 - b)^2}{1 - d_0^2} + b_0^2 + 1 \end{aligned}$$

Hence

$$\hat{b}_N \to b_0 d_0, \qquad \text{w.p. 1 as } N \to \infty$$

since this gives the smallest prediction-error variance. Now the predictor

$$\hat{y}(t|t-1) = b_0 d_0 u(t-2) \tag{8.39}$$

is a fairly reasonable one for the system (8.36) under the input (8.37). It yields the prediction-error variance $1 + b_0^2$, compared to the optimal value 1 for a correct model and the output variance

$$1 + \frac{b_0^2}{1 - d_0^2}$$

Notice, however, that the identified model is heavily dependent on the input that was used during the identification experiment. If (8.39) is applied to a white-noise input $\{u(t)\}$, the model (8.39) is useless: It yields the prediction-error variance $1 + b_0^2 + b_0^2 d_0^2$, which is larger than the output variance $1 + b_0^2$. □

8.4 CONSISTENCY AND IDENTIFIABILITY

Suppose now that assumption S1 holds so that we have a true system, denoted by $\mathcal{S}$. Let us discuss under what conditions it will be possible to recover this system using prediction-error identification.

Clearly, a first assumption must be that $\mathcal{S} \in \mathcal{M}$; that is, the set $D_T(\mathcal{S}, \mathcal{M})$ defined by (8.9) is nonempty.

$\mathcal{S} \in \mathcal{M}$: Quadratic Criteria

The basic consistency result is almost immediate.

Theorem 8.3. Suppose that the data set Z^∞ is subject to assumptions D1 and S1. Let $\mathcal{M}$ be a linear, uniformly stable model structure such that $\mathcal{S} \in \mathcal{M}$. Assume also that Z^∞ is informative enough with respect to $\mathcal{M}$. If the input contains output feedback then also assume that there is a delay either in the regulator or in both $G_0(q)$ and $G(q, \theta)$. Then

$$D_c = D_T(\mathcal{S}, \mathcal{M}) \tag{8.40}$$

where D_c is defined by (8.22) and (8.23) and $D_T(\mathcal{S}, \mathcal{M})$ by (8.9). If, in addition, the model structure is globally identifiable at $\theta_0 \in D_T(\mathcal{S}, \mathcal{M})$, then

$$D_c = \{\theta_0\} \tag{8.41}$$

Theorems 8.2 and 8.3 together consequently state that the estimated transfer functions obey

$$G(e^{i\omega}, \hat{\theta}_N) \to G_0(e^{i\omega});\ H(e^{i\omega}, \hat{\theta}_N) \to H_0(e^{i\omega}), \qquad \text{w.p. 1 as } N \to \infty \tag{8.42}$$

Proof of Theorem 8.3. Let $\theta_0 \in D_T$ and consider, for any $\theta \in D_{\mathcal{M}}$,

$$\overline{V}(\theta) - \overline{V}(\theta_0) = \overline{E}\,[\varepsilon(t,\theta) - \varepsilon(t,\theta_0)]\,\varepsilon(t,\theta_0) + \tfrac{1}{2}\overline{E}\,[\varepsilon(t,\theta) - \varepsilon(t,\theta_0)]^2 \tag{8.43}$$

Since $\theta_0 \in D_T$,

$$\varepsilon(t,\theta_0) = -H_0^{-1}(q)G_0(q)u(t) + H_0^{-1}(q)y(t) = e_0(t)$$

according to S1. Moreover, the difference

$$\varepsilon(t,\theta) - \varepsilon(t,\theta_0) = \hat{y}(t|\theta_0) - \hat{y}(t|\theta)$$

depends only on input-output data up to time $t-1$ and is therefore independent of $e_0(t)$ [cf. (8.2)]. (If there is no delay in the system/model, the term $u(t)$ will appear here. But then, according to the assumptions, there will be a delay in the regulator, so that $u(t)$ and $e_0(t)$ are independent.) The first term of (8.43) is therefore zero. The second term, which equals

$$\overline{E}\,[\hat{y}(t|\theta_0) - \hat{y}(t|\theta)]^2$$

is strictly positive if θ and θ_0 correspond to different models, since the data set is sufficiently informative; see (8.12). Hence (8.40) follows from (8.23). The result (8.41) follows since global identifiability of $\mathcal{M}$ at θ_0 implies that $D_T = \{\theta_0\}$ [see (4.135)].

$G_0 \in \mathcal{G}$: Quadratic Criteria

Often it is more important to have a good estimate of the transfer function G than of the noise filter H. We shall now study the situation where the set of model transfer functions

$$\mathcal{G} = \left\{G(e^{i\omega}, \theta) | \theta \in D_{\mathcal{M}}\right\}$$

is large enough to contain the true transfer function,

$$G_0 \in \mathcal{G} \tag{8.44}$$

but the true noise description H_0 cannot be exactly described within the model set. Hence $\mathcal{S} \notin \mathcal{M}$. We then have the following result:

Theorem 8.4. Suppose that the data set Z^∞ is subject to assumptions D1 and S1. Let $\mathcal{M}$ be a linear uniformly stable model structure, such that G and H are independently parametrized:

$$\theta = \begin{bmatrix} \rho \\ \eta \end{bmatrix} \quad G(q, \theta) = G(q, \rho), \quad H(q, \theta) = H(q, \eta) \tag{8.45}$$

and such that the set

$$D_G(\mathcal{S}, \mathcal{M}) = \left\{\rho | G(e^{i\omega}, \rho) = G_0(e^{i\omega}) \,\forall \omega\right\} \tag{8.46}$$

is nonempty. Assume that Z^∞ is informative enough with respect to $\mathcal{M}$ and that the system operates in open loop; that is,

$$\{u(t)\} \text{ and } \{e_0(t)\} \text{ are independent} \tag{8.47}$$

Let

$$\hat{\theta}_N = \begin{bmatrix} \hat{\rho}_N \\ \hat{\eta}_N \end{bmatrix}$$

be obtained by the prediction-error method (8.16) and (8.17). Then

$$\hat{\rho}_N \to D_G(\mathcal{S}, \mathcal{M}) \qquad \text{w.p. 1 as } N \to \infty \tag{8.48}$$

The result (8.48) can be written more suggestively as

$$G(e^{i\omega}, \hat{\theta}_N) \to G_0(e^{i\omega}), \qquad \text{w.p. 1 as } N \to \infty \tag{8.49}$$

Proof. Consider the function $\overline{V}(\theta)$ given by (8.22). We have from S1

$$\begin{aligned}\varepsilon(t,\theta) &= H^{-1}(q,\eta)\left[y(t) - G(q,\rho)u(t)\right] \\ &= H^{-1}(q,\eta)\left[(G_0(q) - G(q,\rho))\,u(t) + H_0(q)e_0(t)\right] \\ &= u_F(t,\eta,\rho) + e_F(t,\eta)\end{aligned}$$

with obvious notation. Since u an e_0 are independent, we have that

$$\overline{V}(\theta) = \overline{V}(\rho,\eta) = \tfrac{1}{2}\left[\overline{E}u_F^2(t,\rho,\eta) + Ee_F^2(t,\eta)\right]$$

The first term is zero precisely when $\rho \in D_G(S,\mathcal{M})$, and the second term is independent of ρ. Hence

$$\arg\min_{\rho} \overline{V}(\rho,\eta) = D_G(S,\mathcal{M})$$

irrespective of H, which, together with Theorem 8.2, concludes the proof. □

We may add that both assumptions (8.45) and (8.47) are essential for the result to hold. See Example 8.1 and Problem 8E.3.

The case of independent parametrization (8.45) covers the output error model (4.25) along with variants with fixed noise models

$$y(t) = G(q,\theta)u(t) + H_*(q)e(t) \tag{8.50}$$

[which alternatively can be regarded as the output error model used with a prefilter $L(q) = 1/H_*(q)$; see (7.13) and (7.14)]. It also covers the Box-Jenkins model structure (4.31). These model structures consequently have the important advantage that the transfer function G can be consistently estimated, even when the noise model set is too simple to admit a completely correct description of the system.

Example 8.3 First Order Output Error Model

Consider the system (8.30) of Example 8.1, and let the model structure be a first-order output error model:

$$\hat{y}(t|\theta) = \frac{bq^{-1}}{1 + aq^{-1}}u(t)$$

In this case it follows from Theorem 8.4 that the estimates $\hat{a}_N$ and $\hat{b}_N$ will converge to the true values a_0 and b_0. □

$S \in \mathcal{M}$: General Norm $\ell(\varepsilon)$ (*)

With a general, θ-independent norm $\ell(\varepsilon)$, the estimate converges into the set D_c:

$$\hat{\theta}_N \to D_c = \arg\min_{\theta \in D_{\mathcal{M}}} \overline{E}\ell\left(\varepsilon(t,\theta)\right) \tag{8.51}$$

according to (8.29). In general, the set D_c will depend on ℓ. However, when $S \in \mathcal{M}$ it is desirable that $D_c = D_T(S, \mathcal{M})$ for all reasonable choices of ℓ. Clearly, some conditions must be imposed on ℓ, and Problem 8D.3 shows that it is not sufficient to require $\ell(\varepsilon)$ to be increasing with $|\varepsilon|$. We have to require $\ell(\varepsilon)$ to be convex in order to prove a result that holds for all distributions of the innovation $e_0(t)$. We thus have the following extension of Theorem 8.3.

Theorem 8.5. Let $\ell(x)$ be a twice differentiable function such that

$$E\ell'(e_0(t)) = 0 \tag{8.52}$$

$$\ell''(x) \geq \delta > 0, \qquad \forall x \tag{8.53}$$

Here $e_0(t)$ are the innovations in assumption S1. Then, under the assumptions of Theorem 8.3,

$$D_c = D_T(S, \mathcal{M})$$

with D_c defined by (8.28) and (8.23).

Proof. Let $\theta_0 \in D_T$ and denote as usual

$$\varepsilon(t, \theta_0) = e_0(t)$$

Then for any $\theta \notin D_T$

$$\varepsilon(t, \theta) = e_0(t) + \tilde{y}(t, \theta)$$

where $\overline{E}\,[\tilde{y}(t, \theta)]^2 > 0$ since the data set is sufficiently informative. Hence, by Taylor's expansion,

$$\ell\,(\varepsilon(t, \theta)) = \ell\,(e_0(t)) + \tilde{y}(t, \theta)\ell'(e_0(t)) + \tfrac{1}{2}\,[\tilde{y}(t, \theta)]^2\,\ell''(\xi(t))$$

where $\xi(t)$ is a value between $e_0(t)$ and $\varepsilon(t, \theta)$. Since $e_0(t)$ and $\tilde{y}(t, \theta)$ are independent, this expression gives

$$\begin{aligned} \overline{E}\ell\,(\varepsilon(t, \theta)) &= \overline{E}\ell\,(e_0(t)) + 0 + \tfrac{1}{2}\overline{E}\left\{[\tilde{y}(t, \theta)]^2\,\ell''\,(\xi(t))\right\} \\ &\geq \overline{E}\ell\,(e_0(t)) + \frac{\delta}{2} \cdot \overline{E}\,[\tilde{y}(t, \theta)]^2 > \overline{E}\ell\,(e_0(t)) \end{aligned}$$

using (8.52) and (8.53) and $\overline{E}\,[\tilde{y}(t, \theta)]^2 > 0$, respectively. This concludes the proof. □

Clearly, an analogous extension of Theorem 8.4 can also be given.

In the maximum likelihood method the norm ℓ is chosen as the negative logarithm of the PDF of the innovations; (7.85):

$$\ell(x) = -\log f_e(x) \tag{8.54}$$

It can be shown that (8.52) automatically holds for this norm, and that Theorem 8.5 holds without condition (8.53). See Problem 8G.3.

$S \in \mathcal{M}$: General Norm $\ell(\varepsilon, \alpha)$ (*)

We consider now the case where the norm is parametrized by an α that is independent of the parametrization of the predictor as in (7.17). We thus have that the limit values of θ and α are given by

$$(\theta^*, \alpha^*) = \arg\min_{\theta,\alpha} \overline{V}(\theta, \alpha) = \arg\min_{\theta,\alpha} \overline{E}\ell\left(\varepsilon(t, \theta), \alpha\right) \tag{8.55}$$

If $S \in \mathcal{M}$ and the conditions of Theorem 8.5 are satisfied for all α, then it is clear that $\theta^* \in D_T(S, \mathcal{M})$, regardless of α. This means that

$$\alpha^* = \arg\min_{\alpha} \overline{E}\ell\left(e_0(t), \alpha\right) \tag{8.56}$$

We shall study what (8.56) tells us about the limit value α^*. We first have the following result.

Lemma 8.3. Consider a norm (7.17), normalized so that

$$\int_{-\infty}^{\infty} e^{-\ell(x,\alpha)}\, dx = 1 \ \forall \alpha \tag{8.57}$$

Let the PDF of $e_0(t)$ be $f_e(x)$, and assume that for some α_0

$$\ell(x, \alpha_0) = -\log f_e(x) \tag{8.58}$$

Then $\alpha^* = \alpha_0$ in (8.56).

Proof. Let

$$f_\alpha(x) = e^{-\ell(x,\alpha)}$$

Hence

$$\overline{E}\ell\left(e_0(t), \alpha\right) - \overline{E}\ell\left(e_0(t), \alpha_0\right) = -\overline{E}\log\frac{f_\alpha(e_0(t))}{f_e(e_0(t))}$$

$$\geq -\log E\frac{f_\alpha(e_0(t))}{f_e(e_0(t))} = -\log\int\left[\frac{f_\alpha(x)}{f_e(x)}\right] f_e(x)dx = -\log\int f_\alpha(x)dx = 0$$

The inequality is Jensen's inequality (see Chung, 1974) since $-\log x$ is a convex function, and equality holds if and only if $f_\alpha(x) = \text{const} \cdot f_e(x)$. This proves the lemma. □

Heuristically, we could thus say that

the minimization with respect to α in (8.55) tries to make the norm $\ell(\varepsilon, \alpha)$ look like the negative logarithm of the PDF of the true innovations. (8.59)

Example 8.4 Estimating the Innovations Variance

Let $\ell(\varepsilon, \alpha)$ be given by (7.87):

$$\ell(\varepsilon, \alpha) = \frac{1}{2}\left[\frac{\varepsilon^2}{\alpha} + \log \alpha\right]$$

We find that

$$\overline{E}\left[\ell\left(e_0(t), \alpha\right)\right] = \frac{1}{2}\left[\frac{Ee_0^2(t)}{\alpha} + \log \alpha\right] = \frac{1}{2}\left[\frac{\lambda_0}{\alpha} + \log \alpha\right]$$

which is minimized by $\alpha = \lambda_0$. The estimate $\hat{\alpha}_N$ will thus converge to the innovation variance as N tends to infinity. See also Problem 7E.7. □

When the parametrizations of the predictor and of the norm $\ell(\varepsilon, \theta)$ have common parameters, the conclusion is that

$$\theta^* = \arg\min_{\theta \in D_{\mathcal{M}}} \overline{E}\ell\left(\varepsilon(t, \theta), \theta\right)$$

will give a compromise between making the prediction errors $\{\varepsilon(t, \theta)\}$ equal to the true innovations $\{e_0(t)\}$ and (8.59), that is, making the norm look like $-\log f_e(x)$. In case these two objectives cannot be reached simultaneously, consistency may be lost even if $D_T(\mathcal{S}, \mathcal{M})$ is nonempty. See Problem 8E.2.

Multivariable Case (*)

The convergence and consistency results for multivariable systems are entirely analogous to the scalar case. The result (8.29) holds without notational changes for the multivariable case. The counterparts of Theorems 8.3 and 8.4 with quadratic criteria

$$\overline{E}\ell\left(\varepsilon(t, \theta)\right) = \tfrac{1}{2}\overline{E}\varepsilon^T(t, \theta)\Lambda^{-1}\varepsilon(t, \theta) \tag{8.60}$$

hold as stated, with only obvious notational changes in the proofs. For Theorem 8.5, the condition (8.53) takes the form that the $p \times p$ matrix $\ell''(\varepsilon)$ should be positive definite.

8.5 LINEAR TIME-INVARIANT MODELS: A FREQUENCY-DOMAIN DESCRIPTION OF THE LIMIT MODEL

Theorem 8.2 describes the limiting estimate θ^*, $\theta^* \in D_c$, as the one that minimizes the prediction error variance among all models in the structure $\mathcal{M}$. In case $\mathcal{S} \in \mathcal{M}$, this means that $\theta^* = \theta_0$ is a true description of the system (see Theorem 8.3), but otherwise the model will differ from the true system. In this section we shall develop some expressions that characterize this misfit between the limiting model and the true system for the case of linear time-invariant models. See also Problem 8G.4.

A Note on Data from Arbitrary Systems (*)

Even though the approximation expressions below take their starting point in a "true linear system" according to assumption S1, it is of interest to note that they are equally applicable to data from arbitrary, nonlinear, time-varying systems. Suppose that the input and output signals are jointly quasi-stationary, so that their spectra are defined. Then we can determine the optimal predictor Wiener filter for predicting $y(t)$ from past data:

$$\hat{y}(t|t-1) = W_u(q)u(t) + W_y(q)y(t) \tag{8.61}$$

where W_y and W_u are computed from the cross spectra Φ_u, Φ_{yu} and Φ_y in a well defined way. (See Wiener (1949)and also Problem 8G.5). The error $e_0(t) = y(t) - \hat{y}(t|t-1)$ will by construction be uncorrelated with past inputs and outputs. This also means that $e_0(k)$, $k < t$ will be uncorrelated with $e_0(t)$, since it is constructed from past input-output data. If we introduce $H_0(q) = (1 - W_y(q))^{-1}$ and $G_0(q) = H_0(q)W_u(q)$ (cf (4.114)), we can rewrite (8.61) as

$$y(t) = G_0(q)u(t) + H_0(q)e_0(t) \tag{8.62}$$

where $e_0(t)$ is an uncorrelated sequence, and uncorrelated also with past input-output data. In general, independence will not hold, but in the calculations to follow we will only utilize the second-order properties of the data. All this means that (8.62) *is a correct description of the observed data, if only their second-order properties are considered: "the best linear time-invariant approximation of the true system."* (It may still be useless for control and decision-making, though, since this approximation depends on the actual data spectra.)

An Expression for $\overline{V}(\theta)$

By the fundamental expression (2.65), we may write

$$\overline{V}(\theta) = \overline{E}\tfrac{1}{2}\varepsilon^2(t,\theta) = \frac{1}{4\pi}\int_{-\pi}^{\pi} \Phi_\varepsilon(\omega,\theta)d\omega \tag{8.63}$$

where $\Phi_\varepsilon(\omega,\theta)$ is the spectrum of the prediction errors $\{\varepsilon(t,\theta)\}$. Under assumptions S1 we have

$$y(t) = G_0(q)u(t) + H_0(q)e_0(t) \tag{8.64}$$

where the noise source e_0 has variance λ_0. Then, for a linear model structure, we obtain the prediction errors

$$\begin{aligned}
\varepsilon(t,\theta) &= H^{-1}(q,\theta)\left[y(t) - G(q,\theta)u(t)\right] = H_\theta^{-1}\left[(G_0 - G_\theta)u(t) + H_0 e_0(t)\right] \\
&= H_\theta^{-1}\left[(G_0 - G_\theta)u(t) + (H_0 - H_\theta)e_0(t)\right] + e_0(t) \\
&= H_\theta^{-1}\left[\begin{matrix}(G_0 - G_\theta) & (H_0 - H_\theta)\end{matrix}\right]\begin{bmatrix} u(t) \\ e_0(t)\end{bmatrix} + e_0(t)
\end{aligned} \tag{8.65}$$

Here, we introduced the shorter notation $H_\theta = H(q, \theta)$ and similarly for G_θ. Assume now, as in Theorem 8.3, that the system may be under feedback control, but is such that there is a delay either in the system and the model (i.e. G_0 and G_θ both contain a delay), or in the regulator (so that $u(t)$ may depend only on $y(t-1)$ and earlier values). Note also that, since both H_0 and H_θ are monic, the term $(H_0 - H_\theta)e_0(t)$ is independent of $e_0(t)$. All this means that the innovation $e_0(t)$ will be uncorrelated with the first terms in (8.65). Then, according to (2.95), the spectrum of ε can be written

$$\Phi_\varepsilon(\omega, \theta) = \frac{1}{|H_\theta|^2}\left[(G_0 - G_\theta) \quad (H_0 - H_\theta)\right] \times \begin{bmatrix} \Phi_u & \Phi_{ue} \\ \Phi_{eu} & \lambda_0 \end{bmatrix} \begin{bmatrix} (\overline{G}_0 - \overline{G}_\theta) \\ (\overline{H}_0 - \overline{H}_\theta) \end{bmatrix} + \lambda_0 \tag{8.66}$$

Overbar here means complex conjugation. $\Phi_{eu} = \overline{\Phi}_{ue}$ is the cross spectrum between e_0 and u, which of course is zero if the system operates in open loop. Note that the data spectrum can be factorized as

$$\begin{bmatrix} \Phi_u & \Phi_{ue} \\ \Phi_{eu} & \lambda_0 \end{bmatrix} = \begin{bmatrix} I & 0 \\ \frac{\Phi_{eu}}{\Phi_u} & I \end{bmatrix} \begin{bmatrix} \Phi_u & 0 \\ 0 & \lambda_0 - \frac{|\Phi_{eu}|^2}{\Phi_u} \end{bmatrix} \begin{bmatrix} I & \frac{\Phi_{ue}}{\Phi_u} \\ 0 & I \end{bmatrix} \tag{8.67}$$

Let us introduce

$$B(e^{i\omega}, \theta) = \frac{\left(H_0(e^{i\omega}) - H(e^{i\omega}, \theta)\right)\Phi_{ue}(\omega)}{\Phi_u(\omega)} \tag{8.68}$$

Then, using (8.67), (8.66) can be rewritten

$$\Phi_\varepsilon(\omega, \theta) = \frac{|G_0 + B_\theta - G_\theta|^2 \Phi_u}{|H_\theta|^2} + \frac{|H_0 - H_\theta|^2 \left(\lambda_0 - \frac{|\Phi_{ue}|^2}{\Phi_u}\right)}{|H_\theta|^2} + \lambda_0 \tag{8.69}$$

We now have a characterization of

$$D_c = \arg\min_\theta \overline{V}(\theta) \tag{8.70}$$

in the frequency domain. We see that if there is a parameter value θ_0 such that $G_{\theta_0} = G_0$ and $H_{\theta_0} = H_0$, then this value will minimize the integral of (8.69), since the first two terms then vanish. This we knew before; it is a restatement of Theorem 8.3. We have however also obtained a more explicit picture of how the model is approximating the true system in case it cannot be exactly described. We see that G_θ will be pulled towards $G_0 + B_\theta$ and H_θ to H_0 with the indicated weightings. To be more specific, let us investigate some special cases.

Open Loop Case

If the system operates in open loop, u and e will be independent, so $\Phi_{ue} \equiv 0$, and hence $B = 0$. If the noise model is fixed to $H(q, \theta) = H_*(q)$, (8.70) and (8.69) specialize to

$$D_c = \arg\min_{\theta} \int_{-\pi}^{\pi} \left|G_0(e^{i\omega}) - G(e^{i\omega}, \theta)\right|^2 Q_*(\omega) d\omega \tag{8.71a}$$

$$Q_*(\omega) = \frac{\Phi_u(\omega)}{\left|H_*(e^{i\omega})\right|^2} \tag{8.71b}$$

where we disposed of the θ-independent terms. Let $\theta^* \in D_c$. In this case the limiting model $G(e^{i\omega}, \theta^*)$ is a clear-cut best mean-square approximation of $G_0(e^{i\omega})$, with a frequency weighting Q_*. This depends on the noise model H_* and the input spectrum Φ_u, and can be interpreted as the model signal-to-noise ratio.

Consider now an independently parameterized noise model with $\theta = [\rho \quad \eta]$ as in (8.45), (4.128). We can define the spectrum of the output error

$$(G_0(q) - G(q, \rho))u(t) + H_0(q)e_0(t)$$

for each value ρ as

$$\begin{aligned} \Phi_{\text{ER}}(\omega, \rho) &= \left|G_0(e^{i\omega}) - G(e^{i\omega}, \rho)\right|^2 \Phi_u(\omega) + \lambda_0 \left|H_0(e^{i\omega})\right|^2 \\ &= \beta_\rho \left|R(e^{i\omega}, \rho)\right|^2 \end{aligned} \tag{8.72}$$

where the last equation is a definition of the spectral factor R. This is a monic function; see the spectral factorization result in Section 2.3. Then the integral of (8.69) takes the form

$$\begin{aligned} &\int_{-\pi}^{\pi} \frac{|G_0 - G_\rho|^2 \Phi_u}{|H_\eta|^2} + \lambda_0 \left(\left|\frac{H_0}{H_\eta} - 1\right|^2 + 1 \right) d\omega \\ &= \int_{-\pi}^{\pi} \frac{\Phi_{\text{ER}}(\omega, \rho)}{\left|H(e^{i\omega}, \eta)\right|^2} d\omega \\ &= \int_{-\pi}^{\pi} \beta_\rho \frac{\left|R(e^{i\omega}, \rho)\right|^2}{\left|H(e^{i\omega}, \eta)\right|^2} d\omega = \int_{-\pi}^{\pi} \beta_\rho \left(\left|\frac{R(e^{i\omega}, \rho)}{H(e^{i\omega}, \eta)} - 1\right|^2 + 1 \right) d\omega \\ &= \int_{-\pi}^{\pi} \left|\frac{1}{H(e^{i\omega}, \eta)} - \frac{1}{R(e^{i\omega}, \rho)}\right|^2 \Phi_{\text{ER}}(\omega, \rho) d\omega + 2\pi\beta_\rho \end{aligned}$$

where we twice used the identity $\int |R|^2 d\omega = \int \left(|R - 1|^2 + 1\right) d\omega$ for a monic function R.

Now, we can characterize the parameters that minimize the criterion function. Let $\theta^* \in D_c$. Then we can write $\theta^* = [\,\rho^* \quad \eta^*\,]$, with

$$\rho^* = \arg\min_{\rho} \int_{-\pi}^{\pi} \left|G_0(e^{i\omega}) - G(e^{i\omega}, \rho)\right|^2 Q_*(\omega, \eta^*) d\omega \tag{8.73a}$$

$$Q_*(\omega, \eta) = \frac{\Phi_u(\omega)}{\left|H(e^{i\omega}, \eta)\right|^2} \tag{8.73b}$$

$$\eta^* = \arg\min_{\eta} \int_{-\pi}^{\pi} \left|\frac{1}{H(e^{i\omega}, \eta)} - \frac{1}{R(e^{i\omega}, \rho^*)}\right|^2 \Phi_{\text{ER}}(\omega, \rho^*) d\omega \tag{8.73c}$$

Here Φ_{ER} and R_ρ are defined by (8.72). In this case we see that $G(e^{i\omega}, \rho^*)$ is fitted to $G_0(e^{i\omega})$ in the $Q(\omega, \eta^*)$ norm. This norm is not known *a priori*, but will be known after the minimization, once η^* is computed. We also see that the noise model $H(e^{i\omega}, \eta)$ is fitted to describe the spectrum of the output error.

In the general case, where G and H share parameters, no clear-cut formal characterization of the fit of G can be given. It is useful, though, and intuitively appealing to see the resulting estimate θ^* as a compromise between fitting G_θ to G_0 in the frequency norm $\Phi_u/|H_{\theta^*}|^2$ and fitting the model spectrum $|H_\theta|^2$ to the output error spectrum $\Phi_{\text{ER}}(\omega, \theta^*)$.

Closed Loop Case

Let us assume that the regulator is linear, so that u will be a linear function of the reference signal r and the noise source e_0 as in Assumption D1. We can then split up the input spectrum Φ_u into that part that originates from r and that part that comes from e_0:

$$\Phi_u(\omega) = \Phi_u^r(\omega) + \Phi_u^e(\omega) \tag{8.74}$$

If the linear filters that define the input are time-invariant, $u(t) = K_1(q)r(t) + K_2(q)e_0(t)$, we find that $\Phi_{ue}(\omega) = \lambda_0 K_2(e^{i\omega})$ and hence

$$|\Phi_{ue}(\omega)|^2 = \lambda_0 \Phi_u^e(\omega) \tag{8.75}$$

This means that we can characterize the size of the "bias-pull" term B in (8.68) more explicitly as

$$\left|B(e^{i\omega}, \theta)\right|^2 = \frac{\lambda_0}{\Phi_u(\omega)} \cdot \frac{\Phi_u^e(\omega)}{\Phi_u(\omega)} \cdot \left|H_0(e^{i\omega}) - H(e^{i\omega}, \theta)\right|^2 \tag{8.76}$$

and the resulting parameter θ^* will be the value that minimizes

$$\begin{aligned} &\int_{-\pi}^{\pi} |G_0(e^{i\omega}) + B(e^{i\omega}, \theta) - G(e^{i\omega}, \theta)|^2 \frac{\Phi_u(\omega)}{\left|H(e^{i\omega}, \theta)\right|^2} d\omega \\ &+ \lambda_0 \int_{-\pi}^{\pi} \frac{\left|H_0(e^{i\omega}) - H(e^{i\omega}, \theta)\right|^2}{\left|H(e^{i\omega}, \theta)\right|^2} \frac{\Phi_u^r(\omega)}{\Phi_u(\omega)} d\omega \end{aligned} \tag{8.77}$$

We shall investigate this expression further in Section 13.4.

Example 8.5 Approximation in the Frequency Domain

Consider the system

$$y(t) = G_0(q)u(t)$$

with

$$G_0(q) = \frac{0.001q^{-2}(10 + 7.4q^{-1} + 0.924q^{-2} + 0.1764q^{-3})}{1 - 2.14q^{-1} + 1.553q^{-2} - 0.4387q^{-3} + 0.042q^{-4}} \tag{8.78}$$

No disturbances act on the system. The input is a PRBS (see Section 13.3) with basic period one sample, which gives $\Phi_u(\omega) \approx 1$ all ω.

This system was identified with the prediction-error method using a quadratic criterion and prefilter $L(q) \equiv 1$ in the output error model structure

$$\hat{y}(t|\theta) = \frac{b_1 q^{-1} + b_2 q^{-2}}{1 + f_1 q^{-1} + f_2 q^{-2}} u(t) \tag{8.79}$$

Bode plots of the true system and of the resulting model are given in Figure 8.2a. We see that the model gives a good description of the low-frequency properties but is bad at high frequencies. According to (8.71), the limiting model is characterized by

$$\theta^* = \arg\min_{\theta} \int_{-\pi}^{\pi} \left| G_0(e^{i\omega}) - G(e^{i\omega}, \theta) \right|^2 d\omega \tag{8.80}$$

since $H^*(q) = 1$ and $\Phi_u(\omega) \equiv 1$. Since the amplitude of the true system falls off by a factor of 10^{-2} to 10^{-3} for $\omega > 1$, it is clear that errors at higher frequencies contribute only marginally to the criterion (8.80); hence the good low-frequency fit.

Consider now instead an ARX structure

$$y(t) = \frac{b_1 q^{-1} + b_2 q^{-2}}{1 + a_1 q^{-1} + a_2 q^{-2}} u(t) + \frac{1}{1 + a_1 q^{-1} + a_2 q^{-2}} e(t) \tag{8.81}$$

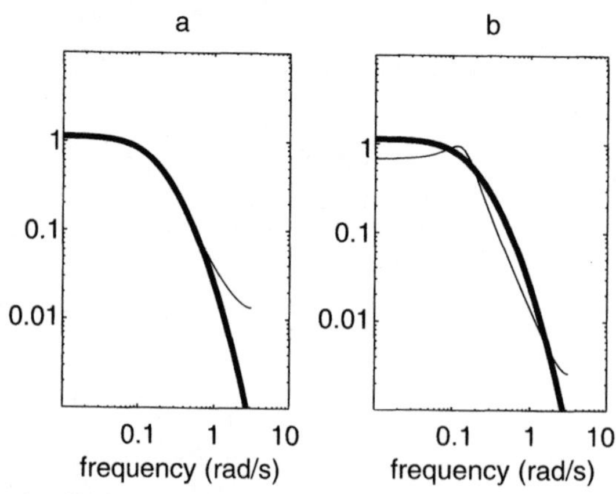

Figure 8.2 Amplitude Bode plots of true system (thick lines) and model (thin lines). (a) OE model estimated in (8.79). (b) ARX model estimated in (8.81).

corresponding to the linear regression predictor

$$\hat{y}(t|\theta) = -a_1 y(t-1) - a_2 y(t-2) + b_1 u(t-1) + b_2 u(t-2)$$

When applied to the same data, this structure gives the model description in Figure 8.2b, with a much worse low-frequency fit. According to our discussion in this section, this limit model is a compromise between fitting $1/|1 + a_1 e^{i\omega} + a_2 e^{2i\omega}|^2$ to the error spectrum and minimizing (a_1^* and a_2^* correspond to the limit estimate θ^*).

$$\int_{-\pi}^{\pi} \left|G_0(e^{i\omega}) - G(e^{i\omega}, \theta)\right|^2 \cdot \left|1 + a_1^* e^{i\omega} + a_2^* e^{2i\omega}\right|^2 d\omega \tag{8.82}$$

The function $\left|A_*(e^{i\omega})\right|^2 = \left|1 + a_1^* e^{i\omega} + a_2^* e^{2i\omega}\right|^2$ is plotted in Figure 8.3. It assumes values at high frequencies that are 10^4 times those at low frequencies. Hence, compared to (8.80), the criterion (8.82) penalizes high-frequency misfit much more. This explains the different properties of the limit models obtained in the model structures (8.79) and (8.81), respectively. □

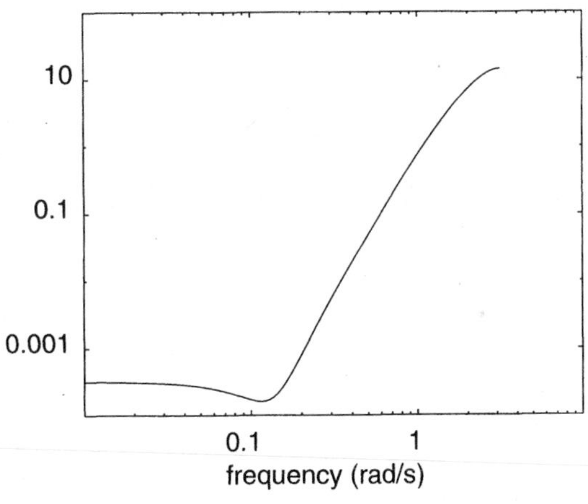

Figure 8.3 The weighting function $\left|A_*(e^{i\omega})\right|^2$ in (8.82).

8.6 THE CORRELATION APPROACH

In Section 7.5 we defined the correlation approach to identification, with the special cases of PLR and IV methods. The convergence analysis for these methods is quite analogous to the prediction-error approach as given in the previous few sections.

Basic Convergence Result

Consider the function

$$f_N(\theta, Z^N) = \frac{1}{N} \sum_{t=1}^{N} \zeta(t, \theta) \varepsilon_F(t, \theta) \tag{8.83}$$

where ε_F is given by

$$\varepsilon_F(t, \theta) = L(q)\varepsilon(t, \theta) \tag{8.84}$$

and the correlation vector $\zeta(t, \theta)$ is obtained by linear filtering of past data:

$$\zeta(t, \theta) = K_y(q, \theta)y(t) + K_u(q, \theta)u(t) \tag{8.85}$$

(both filters contain a delay). Determining the estimate $\hat{\theta}_N$ by solving $f_N(\theta, Z^N) = 0$ gives the correlation approach (7.110). We have here specialized the general instruments (7.110b) to a linear case (8.85).

The convergence analysis of (8.83) is entirely analogous to the prediction-error case. Thus we have from Theorem 2B.1:

Lemma 8.4. Let the data set Z^∞ be subject to D1, and let the prediction errors be computed using a uniformly stable linear model structure. Assume that the family of filters

$$\left\{K_y(q, \theta), K_u(q, \theta); \theta \in D_{\mathcal{M}}\right\}$$

is uniformly stable. Then

$$\sup_{\theta \in D_{\mathcal{M}}} \left| f_N(\theta, Z^N) - \overline{f}(\theta) \right| \to 0, \qquad \text{w.p. 1 as } N \to \infty \tag{8.86}$$

where

$$\overline{f}(\theta) = \overline{E}\zeta(t, \theta)\varepsilon_F(t, \theta) \tag{8.87}$$

For the estimate $\hat{\theta}_N$, we thus have the following result.

Theorem 8.6. Let $\hat{\theta}_N$ be defined by

$$\hat{\theta}_N = \underset{\theta \in D_{\mathcal{M}}}{\text{sol}} \left[f_N(\theta, Z^N) = 0 \right]$$

Then, subject to the assumptions of Lemma 8.4,

$$\hat{\theta}_N \to D_{cf}, \qquad \text{w.p. 1 as } N \to \infty \tag{8.88}$$

where

$$D_{cf} = \left\{ \theta | \theta \in D_{\mathcal{M}}, \overline{f}(\theta) = 0 \right\} \tag{8.89}$$

The theorem is here given for the special choice $\alpha(\varepsilon_F) = \varepsilon_F$ in (7.110). The extension to general $\alpha(\cdot)$ is straightforward.

This convergence result is quite general and also quite natural. The limiting estimate $\theta^* \in D_{cf}$ will be characterized by the property that the filtered prediction errors $\{\varepsilon_F(t, \theta^*)\}$ indeed are uncorrelated with the instruments $\{\zeta(t, \theta^*)\}$. This was also our guideline when selecting $\hat{\theta}_N$. We shall now characterize D_{cf} in more practical terms for some special cases.

$S \in \mathcal{M}$: Consistency (*)

An assumption $S \in \mathcal{M}$ would in this case be that there exists a value $\theta_0 \in D_{\mathcal{M}}$ such that $\{\varepsilon(t, \theta_0) = e_0(t)\}$ is a white noise. With $L(q) = 1$, we thus find that

$\overline{f}(\theta_0) = 0$, since $e_0(t)$ is independent of past data and in particular of $\zeta(t, \theta_0)$. Hence, as expected.

$$\theta_0 \in D_{cf} \tag{8.90}$$

Whether this set contains more elements when the data are informative and the model structure globally identifiable at θ_0 is not so easy to analyze in the general case.

$G_0 \in \mathcal{G}$: Instrumental-variable Methods

Consider the IV method with instruments

$$\zeta(t, \theta) = K_u(q, \theta)u(t) \tag{8.91}$$

The underlying model is

$$A(q)y(t) = B(q)u(t) + v(t)$$

for which the predictor

$$\hat{y}(t|\theta) = \varphi^T(t)\theta$$

is determined as in (4.11) and (4.12).

Under assumption S1, the true system is given by (8.7). If there exists a θ_0 corresponding to $(A_0(q), B_0(q))$ such that

$$G_0(q) = \frac{B_0(q)}{A_0(q)}, \qquad [G_0 \in \mathcal{G}; \text{ cf. } (8.44)]$$

we can consequently write (8.7) as

$$y(t) = \frac{B_0(q)}{A_0(q)}u(t) + H_0(q)e_0(t) \tag{8.92}$$

or

$$y(t) = \varphi^T(t)\theta_0 + w_0(t) \tag{8.93}$$

where

$$w_0(t) = A_0(q)H_0(q)e_0(t) \tag{8.94}$$

Suppose now that the system operates in open loop so that $\{w_0(t)\}$ and $\{u(t)\}$ are independent. Then

$$\begin{aligned} \overline{f}(\theta) &= \overline{E}\zeta(t, \theta)L(q)\left[\varphi^T(t)(\theta_0 - \theta) + w_0(t)\right] \\ &= \left[\overline{E}\zeta(t, \theta)\varphi_F^T(t)\right](\theta_0 - \theta) \end{aligned} \tag{8.95}$$

where

$$\varphi_F^T(t) = L(q)\varphi^T(t) \tag{8.96}$$

The second equality in (8.95) follows since $\zeta(t, \theta)$ is entirely constructed from past $u(t)$, while $L(q)w_0(t)$ is independent of $\{u(t)\}$. Under the stated assumptions, we thus have that $\theta_0 \in D_{cf}$, and whether this set contains more θ-values depends on whether the matrix $\overline{E}\zeta(t, \theta)\varphi_F^T(t)$ is singular.

Suppose now that the instruments ζ do not depend on θ and are generated according to (7.122) to (7.124). The matrix

$$R = \overline{E}\zeta(t)\varphi_F^T(t) \tag{8.97}$$

is then a constant matrix that depends only on the filters $L(q)$, $K(q)$, $N(q)$, and $M(q)$, on the true system, and on the properties of $\{u(t)\}$. A thorough discussion of the nonsingularity if R is given in Söderström and Stoica (1983). We first note the following facts. Let n_a^0, n_b^0 be the orders of the true description (8.92), and let n_a, n_b be the corresponding model orders. Let the orders of the instrument filters (7.122) to (7.124) be n_n and n_m. Then

1. If $\min(n_a - n_a^0, n_b - n_b^0) > 0$, then R is singular. (8.98a)

2. If $\min(n_a - n_n, n_b - n_m) > 0$, then R is singular. (8.98b)

To see this, let

$$z_0(t) = \frac{B_0(q)}{A_0(q)}u(t) \tag{8.99}$$

$$\tilde{\varphi}(t) = \left[-z_0(t-1)\ldots -z_0(t-n_a) \quad u(t-1)\ldots u(t-n_b)\right]^T$$

Let $\tilde{\varphi}_F(t) = L(q)\tilde{\varphi}(t)$. If $n_a > n_a^0$ and $n_b > n_b^0$, then (8.99) implies that there exists an $(n_a + n_b)$-dimensional vector S such that $\tilde{\varphi}^T(t)S \equiv 0$. Then also $\tilde{\varphi}_F^T(t)S \equiv 0$. Now, since $\{w_0(t)\}$ and $\{u(t)\}$ are independent, we have

$$R = \overline{E}\zeta(t)\varphi_F^T(t) = \overline{E}\zeta(t)\tilde{\varphi}_F^T(t) \tag{8.100}$$

which shows that $RS = 0$ and R is singular. Similarly, (8.98b) implies the existence of a vector S such that $S^T\zeta(t) \equiv 0$.

When neither of (8.98) hold, the matrix R is "generically" nonsingular. To show this, the reasoning goes as follows: For a given true system and a given input, denote the coefficients of the filters L, K, N, and M by ρ. The matrix R is thus a function of $\rho : R(\rho)$. Now consider the scalar-valued function $\det R(\rho)$. This is an analytic function of ρ (see Finigan and Rowe, 1974). If such a function is zero for ρ in a set of positive Lebesque measure, then it must be identically zero. If we can find a value ρ^* such that $\det R(\rho^*) \neq 0$, we thus can conclude that $\det R(\rho) \neq 0$ for almost all ρ (in the set where $\det R$ is analytic). Such ρ^* can be found if the input spectrum $\Phi_u(\omega) > 0$ for all ω and the orders of the filters N and M are chosen at least as large as the corresponding model orders n_a and n_b (see Problem 8T.2). We thus have the following result.

> Suppose that the system is given by (8.92), that $\Phi_u(\omega) > 0$, and that u and e_0 are independent. Let the instruments $\zeta(t)$ be given by (7.122) to (7.124). Assume that neither of the conditions in (8.98) holds. Then R in (8.97) is nonsingular for almost all such choices of N, M, L, and K. (8.101)

Frequency-domain Characterization of D_{cf} for the IV Method (∗)

The prediction errors, under assumption S1, can be written

$$\begin{aligned}\varepsilon(t,\theta) &= A(q)y(t) - B(q)u(t)\\ &= A(q)\left[G_0(q)u(t) + H_0(q)e_0(t) - \frac{B(q)}{A(q)}u(t)\right]\end{aligned}$$

using (8.92). With the instruments given by (8.91), we thus have, analogous to (8.63) to (8.71),

$$\begin{aligned}\overline{f}(\theta) &= \overline{E}\zeta(t,\theta)\varepsilon_F(t,\theta)\\ &= \frac{1}{2\pi}\int_{-\pi}^{\pi}\left[G_0(e^{i\omega}) - G(e^{i\omega},\theta)\right]\\ &\quad\times \Phi_u(\omega)A(e^{i\omega})L(e^{i\omega})K_u(e^{-i\omega},\theta)\,d\omega\end{aligned} \tag{8.102}$$

with

$$G(e^{i\omega},\theta) = \frac{B(e^{i\omega})}{A(e^{i\omega})}$$

Here $K_u(e^{-i\omega},\theta)$ is a d-dimensional column vector.

The limiting estimates $\theta_{IV}^* \in D_{cf}$ are thus characterized by the fact that certain scalar products with the error $G_0(e^{i\omega}) - G(e^{i\omega},\theta_{IV}^*)$ over the frequency domain are zero.

8.7 SUMMARY

In this chapter we have answered the question of what would happen with the estimates if more and more observed data become available. The answer is natural: We have for the prediction-error approach (7.155) that

$$\hat{\theta}_N \to \arg\min_{\theta\in D_{\mathcal{M}}} \overline{E}\ell\left(\varepsilon(t,\theta),\theta\right), \qquad \text{w.p. 1 as } N\to\infty \tag{8.103}$$

and for the correlation approach (7.156) that

$$\hat{\theta}_N \to \operatorname*{sol}_{\theta\in D_{\mathcal{M}}}\left[\overline{E}\zeta(t,\theta)\alpha(\varepsilon_F(t,\theta)) = 0\right], \qquad \text{w.p. 1 as } N\to\infty \tag{8.104}$$

These results were proved in Theorems 8.2 and 8.6, respectively.

The limiting models are thus the same as those that we could have computed as the best system approximations, in case the probabilistic properties of the system had been known to begin with.

In case a true description of the system is available in the model set, the model will converge to this description under certain natural conditions on ℓ, provided the data set is informative enough. This was shown in Theorems 8.3 to 8.5.

When no exact description can be obtained, the model will be fitted to the system in a way that for linear time-invariant models can be characterized in the frequency domain as follows [see (8.71) to (8.77)]:

The limiting transfer function estimate $G^*(e^{i\omega})$ is partly or entirely determined as the closest function to the true transfer function, measured in a quadratic norm over the frequency range, with a weighting function $\Phi_u(\omega)/\left|H(e^{i\omega},\theta^*)\right|^2$, while the resulting noise model $\left|H(e^{i\omega},\theta^*)\right|^2$ resembles the output error spectrum $\Phi_{ER}(\omega)$ as much as possible. (8.105)

We have not specifically addressed the convergence and consistency of the *subspace method* (7.66) for estimating state-space models. That will be done in connection with the algorithmic details in Section 10.6. However, a heuristic convergence analysis follows from the results of this chapter: As s and N increase to infinity (s much slower than N) the k-step ahead predictors in $\mathbf{Y}$ will converge to the true ones. This means that the procedure will extract a correct state vector, which in turn means that the second LS-step will estimate the system matrices A, B, C, D and K consistently.

8.8 BIBLIOGRAPHY

The convergence and consistency analysis of parameter estimates has, of course, a long history in the statistical literature. The analysis given here has some roots in consistency analysis of the maximum likelihood method, starting with Wald (1949)for the case of independent samples and Åström and Bohlin (1965)for applications to dynamical systems. Analysis of the maximum likelihood method applied to time series is, essentially, equivalent. This problem is described in Walker (1964), Hannan (1970), Anderson (1971), Rissanen and Caines (1979), Hannan and Deistler (1988), and many other contributions. Another study that is relevant for our investigation is Jennrich's (1969)analysis of nonlinear least squares. Asymptotic analysis, which does not utilize a stochastic framework, is given in Niederlinski (1984)and Ljung (1985d).

Analysis in the prediction-error case, where the distribution of the disturbances is not necessarily Gaussian, has been given in Hannan (1973), Ljung (1974, 1976c), and Caines (1976a). The situation, studied in the present chapter, where a true description is not necessarily available in the model set, was, perhaps, first analyzed in Ljung (1976a, 1978a). Similar results are also given in Caines (1978), Anderson, Moore, and Hawkes (1978), and Kabaila and Goodwin (1980).

The frequency-domain characterization of the limiting model is further discussed in Wahlberg and Ljung (1986). Analysis of the frequency-domain based methods of Section 7.7 is carried out, e.g., in McKelvey and Ljung (1997).

Convergence and consistency of IV estimates are treated in a very comprehensive manner in Söderström and Stoica (1983). A variant of the "generic" consistency result for IV estimates (8.101) was first given by Finigan and Rowe (1974). The first consistency analysis of the IV method when applied to dynamic systems was apparently given by Wong and Polak (1967).

It could be noted that, when the prediction-error method is applied to linear regression so that the least-squares method results, the analysis simplifies considerably. The first analysis of this situation was by Mann and Wald (1943). Other contributions on the consistency of the LS method include Ljung (1976b), Anderson and Taylor (1979), Lai and Wei (1982), and Rootzén and Sternby (1984).

8.9 PROBLEMS

8G.1 *Local minima:* Consider the matrix $\Gamma_2(\theta)$ defined in Problem 4G.4 and recall that the condition

$$\Gamma_2(\theta) > 0 \tag{8.106a}$$

implies local identifiability of the model structure at θ. Show that this condition together with the condition

$$\Phi\chi_0(\omega) > 0, \qquad \forall\omega \tag{8.106b}$$

[$\Phi\chi_0(\omega)$ is the spectrum of $\chi_0(t)$ defined in (8.14)] implies that

$$\overline{E}\psi(t,\theta)\psi^T(t,\theta) > 0$$

where $\psi(t,\theta)$ as usual is $(d/d\theta)\hat{y}(t|\theta)$. Show also that if $\varepsilon(t,\theta_0) = e_0(t)$ is white noise and if

$$\overline{E}\ell'(e_0(t)) = 0, \qquad \overline{E}\ell''(e_0(t)) > 0,$$

then the conditions (8.106) (at $\theta = \theta_0$) imply that $\overline{V}(\theta) = \overline{E}\ell(\varepsilon(t,\theta))$ has a strict local minimum at $\theta = \theta_0$.

8G.2 Suppose that the transfer-function model set $\{G(q,\theta)\}$ consists of nth-order linear transfer functions:

$$G(q,\theta) = \frac{b_1 q^{-1} + \cdots + b_n q^{-n}}{1 + a_1 q^{-1} + \cdots + a_n q^{-n}}$$

and suppose that the input consists of n sinusoids of different frequencies, ω_i, $i = 1,\ldots,n$, $0 < \omega_i < \pi$.

(a) Suppose that the noise model is independently parametrized. Show that the limit model $G^*(e^{i\omega})$ fits exactly to the true system $G_0(e^{i\omega})$ at the frequencies in question. It is consequently the same result as if we applied frequency analysis with these input frequencies.

(b) Suppose that the noise model has parameters in common with $G(q,\theta)$, but that the system is noise-free: $\Phi_v(\omega) \equiv 0$. Show that the result under part (a) still holds.

8G.3 *Consistency of the maximum likelihood method:* Suppose that the conditions of Theorem 8.5, except (8.52) and (8.53), hold. Let

$$\ell(x) = -\log f_e(x)$$

where $f_e(x)$ is the PDF of $e_0(t)$. Show that (8.52) holds and that the theorem holds even if (8.53) is not satisfied.

8G.4 Suppose that the system is controlled by output feedback: $u(t) = r(t) - F(q)y(t)$. Show that $\overline{V}(\theta)$ in (8.63) then also can be written

$$V(\theta) = \int \frac{|G(e^{i\omega},\theta) - G_0(e^{i\omega})|^2}{|H(e^{i\omega},\theta)|^2}\Phi_u^r(\omega)d\omega$$

$$+ \int \left|\frac{1 + G(e^{i\omega},\theta)F(e^{i\omega})}{1 + G_0(e^{i\omega})F(e^{i\omega})}\right|^2 \frac{1}{|H(e^{i\omega},\theta)|^2}\Phi_v(\omega)d\omega$$

where Φ_u^r is defined by (8.74). Suppose that we have parameterized G, so that there is a θ_0 for which $G(q, \theta_0) = G_0(q)$, and that no noise model is of interest. Use the above expression for $\overline{V}(\theta)$ to suggest a parameterization of $H(q, \theta)$, such that $\hat{\theta}_N$ will converge to θ_0 even though no correct noise model is obtained.

8G.5 Let $\{u, y\}$ be an arbitrary, quasistationary input-output data set, and consider the output error linear model with prediction errors $\varepsilon(t, \theta) = y(t) - G(q, \theta)u(t)$. Assume that G starts with a delay.

(a) Show that

$$\overline{E}\varepsilon^2(t, \theta) = \frac{1}{2\pi}\int_{-\pi}^{\pi}\left|\frac{\Phi_{yu}(\omega)}{\Phi_u(\omega)} - G(e^{i\omega}, \theta)\right|^2 \Phi_u(\omega)d\omega + \theta\text{—independent terms}$$

(b) Let Φ_u be factorized as $\Phi_u(\omega) = L(e^{i\omega})L(e^{-i\omega})$ for L such that both L and L^{-1} are stable, and causal. (That L is casual means that

$$L(e^{i\omega}) = \sum_{k=1}^{\infty} \ell_k e^{-ik\omega}$$

Now define,

$$R(e^{i\omega}) = \frac{\Phi_{yu}(\omega)}{L(e^{-i\omega})} = \sum_{k=-\infty}^{\infty} r_k e^{-i\omega}$$

$$= \sum_{k=-\infty}^{0} r_k e^{-i\omega} + \sum_{k=1}^{\infty} r_k e^{-i\omega} \triangleq R_a(e^{i\omega}) + R_c(e^{i\omega})$$

where we have split R into an anti-causal and a causal part. Define $G_0(e^{i\omega}) = R_c(e^{i\omega})L(e^{i\omega})$, and show that

$$\overline{E}\varepsilon^2(t, \theta) = \frac{1}{2\pi}\int_{-\pi}^{\pi} |G_0(e^{i\omega}) - G(e^{i\omega}, \theta)|^2 \Phi_u(\omega)d\omega + \theta\text{—independent terms}$$

(c) Show that if a fixed noise model H_* is used, the corresponding expression becomes

$$G_0(e^{i\omega}) = \left[\frac{\Phi_{yu}(\omega)}{L(e^{-i\omega})H_*(e^{i\omega})}\right]_{\text{causal}} \cdot \frac{H_*(e^{i\omega})}{L(e^{i\omega})}$$

where $[\cdot]_{\text{causal}}$ means the causal part. Show also that $e(t) \triangleq H_*^{-1}(q)(y(t) - G_0(q)u(t))$ will be uncorrelated with past inputs, for the G_0 thus defined. (Hint: Derive the cross spectrum between u and e and show that is an anti-causal function.)

8G.6 Consider (8.75) and assume that both $G_0(e^{i\omega})$ and $G(e^{i\omega}, \theta)$ are causal. Show that the "bias-pull"-function $B(e^{i\omega}, \theta)$ can be replaced by

$$G_B(e^{i\omega}, \theta) = \left[\frac{(H_0(e^{i\omega}) - H(e^{i\omega}, \theta))\Phi_{ue}(\omega)}{L(e^{-i\omega})H(e^{i\omega}, \theta)}\right]_{\text{causal}} \cdot \frac{H(e^{i\omega}, \theta)}{L(e^{i\omega})}$$

with notation as in the previous problem.

8E.1 Apply Theorem 8.2 to the LS criterion (7.33) and verify the heuristically derived result (7.38).

8E.2 Consider Problem 7E.4. Here the criterion function $\ell(\varepsilon(t,\theta),\theta)$ is not parametrized independently from the parametrization of the predictor. Suppose that the true system is given by

$$x(t+1) = a_0 x(t) + w_0(t)$$

$$y(t) = x(t) + v_0(t)$$

where $\{w_0(t)\}$ and $\{v_0(t)\}$ are independent, white Gaussian noises with variances $Ew_0^2(t) = 0.1$ and $Ev_0^2(t) = 10$, respectively.

(a) Show that there exists a value θ^* in the parametrization (7.163) such that

$$\varepsilon(t,\theta^*) = e_0(t) = \text{the true innovations}$$

(b) Show that the maximum likelihood estimate $\hat{\theta}_N^{\text{ML}}$ does *not* converge to θ^* as N tends to infinity.

(c) Explain the paradox.

8E.3 Consider the output error model structure

$$\hat{y}(t|\theta) = \frac{b}{1+fq^{-1}}u(t-1), \qquad \theta = \begin{bmatrix} b \\ f \end{bmatrix}$$

Suppose the true system is as in Example 8.1:

$$y(t) + a_0 y(t-1) = b_0 u(t-1) + e_0(t) + c_0 e_0(t-1)$$

and that the input is generated as

$$u(t) = -k_0 y(t) + r(t)$$

where $\{r(t)\}$ is white noise and

$$|a_0 + k_0 b_0| < 1$$

Give an expression that characterizes the limit of $\hat{\theta}_N$. Will it be equal to $[\,a_0 \quad b_0\,]^T$? On what point are the assumptions of Theorem 8.4 violated?

8E.4 Consider the model structure

$$y(t) + ay(t-1) = bu(t-1) + e(t), \qquad \theta = \begin{bmatrix} a \\ b \end{bmatrix}$$

Suppose that the true system and input are given as in Problem 8E.3. Let $\hat{\theta}_N$ be the IV estimate of θ with instruments

$$\zeta(t) = \begin{bmatrix} u(t-1) \\ u(t-2) \end{bmatrix}$$

Does $\hat{\theta}_N$ converge to $\theta_0 = \begin{bmatrix} a_0 \\ b_0 \end{bmatrix}$?

8E.5 Consider the expression for Φ_ε in (8.69). Suppose that the feedback can be described as

$$u(t) = K(q)e(t) + r(t)$$

Show that the error spectrum can then be written as

$$\lambda_0 + \frac{\left\{ \left|\tilde{G}(e^{i\omega},\theta)\right|^2 \Phi_r(\omega) + \left|\tilde{G}(e^{i\omega},\theta)K(e^{i\omega}) + \tilde{H}(e^{i\omega},\theta)\right|^2 \lambda_0 \right\}}{\left|H(e^{i\omega},\theta)\right|^2}$$

8T.1 Consider a quasi-stationary sequence $\{u(t)\}$ and a uniformly stable family of filters $\{G(q,\theta), \theta \in D_{\mathcal{M}}\}$. Let

$$z(t,\theta) = G(q,\theta)u(t)$$

Use the corollary to Theorem 2.2 (Appendix 2A) to note: For each $\theta \in D_{\mathcal{M}}$, $\{z(t,\theta)\}$ is a quasi-stationary sequence and

$$\sup_{\theta \in D_{\mathcal{M}}} \left| \frac{1}{N} \sum_{t=1}^{N} z^2(t,\theta) - R_\theta(0) \right| \to 0, \qquad \text{as } N \to \infty$$

Here

$$R_\theta(0) = \overline{E}z^2(t,\theta) = \frac{1}{2\pi} \int_{-\pi}^{\pi} \left|G(e^{i\omega},\theta)\right|^2 \Phi_u(\omega)\, d\omega$$

Use this result to give a probabilistic-free counterpart of Theorem 8.2: For any quasi-stationary deterministic sequences $\{y(t)\}$ and $\{u(t)\}$,

$$\hat{\theta}_N \to \arg\min_{\theta \in D_{\mathcal{M}}} \overline{E}\varepsilon^2(t,\theta)$$

where

$$\overline{E}\varepsilon^2(t,\theta) = \lim_{N\to\infty} \frac{1}{N} \sum_{t=1}^{N} \varepsilon^2(t,\theta)$$

See Ljung (1985d)for a related discussion.

8T.2 Consider the situation of (8.101). Let the true system be given by (8.92) and suppose that the filter choices are as follows:

$$L(q) = K(q) = 1$$

$$N(q) = A_0(q)$$

$$M(q) = B_0(q)$$

Suppose $\Phi_u(\omega) > 0$ for all ω. Show that $R(\rho)$ is positive definite (and hence nonsingular) for this choice of filters.

8T.3 Let the system be given by

$$\mathcal{S}: \quad y(t) = G_0(q)u(t) + H_0(q)e_0(t)$$

and an underlying model by

$$\mathcal{M}: \quad y(t) = G_0(q,\theta)u(t) + H_0(q,\theta)e(t) \tag{8.107}$$

Suppose that

$$D_T(S, \mathcal{M}) = \{\theta_0\}$$

and that the data are informative. Now let

$$\hat{y}(t|t-k;\theta)$$

be the k-step-ahead predictor computed from (8.107), and let $\hat{\theta}_N$ be determined by

$$\hat{\theta}_N = \arg\min \frac{1}{N}\sum_{t=1}^{N}[y(t) - \hat{y}(t|t-k;\theta)]^2$$

Show that $\hat{\theta}_N \to \theta_0$ as $N \to \infty$, provided there is no output feedback in the generation of the input. What happens if there is feedback? What happens if there is feedback, but there is a k-step time delay between input and output? (*Hint:* Note that the k-step-ahead predictor is a special case of the general linear model so that Theorem 8.2 is applicable. Try to copy the technique of Theorem 8.3.)

8D.1 Show that if

$$\sup_{-1\le x\le 1} \left|f_N(x) - \overline{f}(x)\right| \to 0, \qquad N \to \infty$$

and

$$x_N = \arg\min_{-1\le x\le 1} f_N(x)$$

then

$$x_N \to \arg\min \overline{f}(x)$$

8D.2 Show that (8.20) follows from (8.18), (8.2), and (8.5).

8D.3 To show that it is not sufficient to require $\ell(x)$ to be increasing with $|x|$ in Theorem 8.5, consider the following counterexample:

$$\ell(x) = \begin{cases} 0, & x = 0 \\ 2|x|, & 0 \le |x| \le \frac{1}{2} \\ 1, & |x| \ge \frac{1}{2} \end{cases}$$

$$\varepsilon(t,\theta_0) = e_0(t) = \begin{cases} +1, & \text{with probability } 1/2 \\ -1, & \text{with probability } 1/2 \end{cases}$$

Suppose there exists a value $\tilde{\theta}$ such that $\hat{y}(t|\tilde{\theta}) - \hat{y}(t|\theta_0)$ has the following distribution:

$$\begin{cases} +1, & \text{with probability } 1/2 \\ -1, & \text{with probability } 1/2 \end{cases}$$

Check that

(a) Condition (8.52) of Theorem 8.5 is satisfied, but not (8.53).

(b) $E\varepsilon^2(t,\tilde{\theta}) > E\varepsilon^2(t,\theta_0)$.

(c) $E\ell\left(\varepsilon(t,\tilde{\theta})\right) < E\ell\left(\varepsilon(t,\theta_0)\right)$ so that $\hat{\theta}_N \not\to \theta_0$.

9

ASYMPTOTIC DISTRIBUTION OF PARAMETER ESTIMATES

9.1 INTRODUCTION

Once we have understood the convergence properties of the estimate $\hat{\theta}_N$, the question arises as to how fast the estimate approaches this limit. If θ^* denotes the limit, we are thus interested in the variable $\hat{\theta}_N - \theta^*$. We shall impose the same stochastic framework on the estimation problem as in the previous chapter. This means that $\hat{\theta}_N - \theta^*$ will be a random variable, and its "size" can be characterized by its covariance matrix or, more completely, by its probability distribution. It would be a difficult task to compute the distribution in the general case for any N, and we will have to be content with asymptotic expressions for large N. It will turn out that $(\hat{\theta}_N - \theta^*)$ typically decays as $1/\sqrt{N}$, so the distribution of the random variable

$$\sqrt{N}(\hat{\theta}_N - \theta^*)$$

will then converge to a well-defined distribution, which will turn out to be Gaussian under weak assumptions.

In this chapter we shall derive expressions for these asymptotic distributions. The chapter is structured analogously to Chapter 8. Thus, in Section 9.2 we give the basic result for prediction-error methods. Section 9.3 gives explicit expressions for the asymptotic covariance matrices in cases where the limit θ^* gives a correct description of the true transfer function. Frequency-domain expressions for the variances, as well as the variance of the resulting transfer function, are derived in Section 9.4. The correlation approach to parameter estimation is treated in Section 9.5, and Section 9.6 deals with the practical use of the results derived in the chapter. The reader might find it useful to first review the analysis in Appendix II, Section II.2, which could serve as a simplified preview of the techniques and results of the present chapter.

9.2 THE PREDICTION-ERROR APPROACH: BASIC THEOREM

Heuristic Analysis

Consider, as before,

$$\hat{\theta}_N = \arg\min_{\theta \in D_{\mathcal{M}}} V_N(\theta, Z^N) \tag{9.1}$$

$$V_N(\theta, Z^N) = \frac{1}{N}\sum_{t=1}^{N} \tfrac{1}{2}\varepsilon^2(t, \theta) \tag{9.2}$$

Then, with prime denoting differentiation with respect to θ,

$$V_N'(\hat{\theta}_N, Z^N) = 0$$

Suppose that the set D_c, in (8.23) consists of only one point, θ^*. Expanding the preceding expression into Taylor series around θ^* gives

$$0 = V_N'(\theta^*, Z^N) + V_N''(\xi_N, Z^N)(\hat{\theta}_N - \theta^*) \tag{9.3}$$

where ξ_N is a value "between" $\hat{\theta}_N$ and θ^*. We know that $\hat{\theta}_N \to \theta^*$ w.p. 1. By arguments analogous to Lemma 8.2, it should be possible to show that $V_N''(\theta, Z_N)$ converges uniformly in θ to $\overline{V}''(\theta)$. Then

$$V_N''(\xi_N, Z^N) \to \overline{V}''(\theta^*), \qquad \text{as } N \to \infty \text{ w.p. 1} \tag{9.4}$$

Provided this matrix ($d \times d$) is invertible, (9.3) suggests that for large N the difference is given by

$$(\hat{\theta}_N - \theta^*) = -[\overline{V}''(\theta^*)]^{-1} V_N'(\theta^*, Z^N) \tag{9.5}$$

The second factor is given by

$$-V_N'(\theta^*, Z^N) = \frac{1}{N}\sum_{t=1}^{N} \psi(t, \theta^*)\varepsilon(t, \theta^*) \tag{9.6}$$

where, as usual,

$$\psi(t, \theta^*) = -\frac{d}{d\theta}\varepsilon(t, \theta)|_{\theta=\theta^*} = \frac{d}{d\theta}\hat{y}(t|\theta)|_{\theta=\theta^*}$$

is a d-dimensional column vector. By definition

$$\overline{V}'(\theta^*) = -\overline{E}\psi(t, \theta^*)\varepsilon(t, \theta^*) = 0$$

Apart from the difference

$$D_N = E\left[\frac{1}{N}\sum_{t=1}^{N}[\psi(t,\theta^*)\varepsilon(t,\theta^*) - \overline{E}\psi(t,\theta^*)\varepsilon(t,\theta^*)]\right] \tag{9.7}$$

which we assume to tend to zero "quickly," the expression (9.6) is thus a sum of random variables $\psi(t,\theta^*)\varepsilon(t,\theta^*)$ with zero mean values. Had they been independent, it would have been a direct consequence of the central limit theorem that

$$\frac{1}{\sqrt{N}}\sum_{t=1}^{N}\psi(t,\theta^*)\varepsilon(t,\theta^*) \in AsN(0, Q) \tag{9.8}$$

with

$$Q = \lim_{N\to\infty} N \cdot E\left\{[V_N'(\theta^*, Z^N)][V_N'(\theta^*, Z^N)]^T\right\} \tag{9.9}$$

Here (9.8) means that the random variable on the left converges in distribution to the normal distribution with zero mean and covariance matrix Q [see (I.17)]. The terms of V_N' are not independent, but with assumptions D1 and (8.5) the dependence between distant terms will decrease. It seems reasonable that (9.8) will still hold.

If (9.8) holds, we have directly from (9.5) that

$$\sqrt{N}(\hat{\theta}_N - \theta^*) \in AsN(0, P_\theta) \tag{9.10}$$

$$P_\theta = [\overline{V}''(\theta^*)]^{-1}Q[\overline{V}''(\theta^*)]^{-1} \tag{9.11}$$

Asymptotic Normality

The preceding heuristic analysis can be rigorously justified as shown in Appendix 9A. The result is summarized as follows.

Theorem 9.1. Consider the estimate $\hat{\theta}_N$ determined by (9.1) and (9.2). Assume that the model structure is linear and uniformly stable (see Definitions 4.3 and 8.3) and that the data set Z^∞ is subject to D1 (see Section 8.2). Assume also that for a unique value θ^* interior to $D_{\mathcal{M}}$ we have

$$\hat{\theta}_N \to \theta^*, \qquad \text{w.p. 1 as } N \to \infty$$

$$\overline{V}''(\theta^*) > 0$$

and that

$$\sqrt{N}D_N \to 0, \qquad \text{as } N \to \infty$$

with D_N defined by (9.7). Then

$$\sqrt{N}(\hat{\theta}_N - \theta^*) \in AsN(0, P_\theta)$$

where P_θ is given by (9.11) and (9.9).

With more technical effort the same result can also be shown under relaxed assumptions on the data set and the model structure (see, e.g., Ljung and Caines, 1979). Notice in particular that the result holds, without changes, for general norms $\ell(\varepsilon, \theta, t)$, provided ℓ is sufficiently smooth in ε, and θ. The calculations of the derivatives V_N' and $\overline{V}''$ become more laborous, though. For

$$V_N(\theta, Z^N) = \frac{1}{N}\sum_{t=1}^{N} \ell(\varepsilon(t,\theta), \theta, t) \tag{9.12}$$

we have

$$V_N'(\theta, Z^N) = \frac{1}{N}\sum_{t=1}^{N} \left[-\psi(t,\theta)\ell_\varepsilon'(\varepsilon(t,\theta),\theta,t) + \ell_\theta'(\varepsilon(t,\theta),\theta,t)\right] \tag{9.13}$$

where ℓ_ε' is $(\partial/\partial\varepsilon)\ell(\varepsilon,\theta,t)$ and ℓ_θ' is the $d \times 1$ vector $(\partial/\partial\theta)\ell(\varepsilon,\theta,t)$. Similarly,

$$\begin{aligned} \overline{V}''(\theta) &= \overline{E}\psi(t,\theta)\ell_{\varepsilon\varepsilon}''(\varepsilon(t,\theta),\theta,t)\psi^T(t,\theta) \\ &- \overline{E}\frac{\partial}{\partial\theta}\psi(t,\theta)\ell_\varepsilon'(\varepsilon(t,\theta),\theta,t) - \overline{E}\psi(t,\theta)\ell_{\varepsilon\theta}''(\varepsilon(t,\theta),\theta,t) \\ &- \overline{E}\ell_{\theta\varepsilon}''(\varepsilon(t,\theta),\theta,t)\psi^T(t,\theta) + \overline{E}\ell_{\theta\theta}''(\varepsilon(t,\theta),\theta,t) \end{aligned} \tag{9.14}$$

with obvious notation for the second-order derivatives. We shall allow ourselves to use the asymptotic normality result in this more general form, whenever required.

The matrix P_θ in (9.11) is thus the covariance matrix of the asymptotic distribution. When our main interest is in the covariance, we shall write

$$\text{Cov}\,\hat{\theta}_N \sim \frac{1}{N}P_\theta \tag{9.15}$$

In Appendix 9B a formal verification of (9.15) is given.

The expression for the asymptotic covariance matrix P_θ via (9.9), (9.11), (9.13), and (9.14) is rather complicated in general. In the next section we shall evaluate it in some special cases.

9.3 EXPRESSIONS FOR THE ASYMPTOTIC VARIANCE

In this section we shall consider the asymptotic covariance matrix for prediction-error estimates in case, essentially, a true description of the system is available in the model set.

$S \in \mathcal{M}$: Quadratic Criterion

Assume that the conditions of Theorem 8.3 hold. Then $D_c = \{\theta_0\}$ and $\varepsilon(t, \theta_0) = e_0(t)$ is a sequence of independent random variables with zero mean values and variances λ_0. From (9.6) and (9.9) we then have

$$\begin{aligned} Q &= \lim_{N\to\infty} \frac{1}{N} \sum_{t=1}^{N} \sum_{s=1}^{N} E\psi(t, \theta_0) e_0(t) e_0(s) \psi^T(s, \theta_0) \\ &= \lim_{N\to\infty} \frac{1}{N} \sum_{t=1}^{N} \lambda_0 E\psi(t, \theta_0)\psi^T(t, \theta_0) = \lambda_o \bar{E}\psi(t, \theta_0)\psi^T(t, \theta_0) \end{aligned} \tag{9.16}$$

since $\{e_0(t)\}$ is white (see also Problem 9D.1). Similarly,

$$\bar{V}''(\theta_0) = \bar{E}\psi(t, \theta_0)\psi^T(t, \theta_0) - \bar{E}\psi'(t, \theta_0)e_0(t) = \bar{E}\psi(t, \theta_0)\psi^T(t, \theta_0)$$

The last equality follows since $\psi'(t, \theta_0)$ is formed from Z^{t-1} only. Hence, from (9.11), we obtain

$$P_\theta = \lambda_0 \left[\bar{E}\psi(t, \theta_0)\psi^T(t, \theta_0)\right]^{-1} \tag{9.17}$$

This result has a natural interpretation. Since ψ is the gradient of $\hat{y}$, we see that the asymptotic accuracy of a certain parameter is related to how sensitive the prediction $\hat{y}(t|\theta)$ is with respect to this parameter. Clearly, the more a parameter affects the prediction, the easier it will be to determine its value. A very important and useful aspect of expressions for the asymptotic covariance matrix like (9.17) is that it can be estimated from data. Having processed N data points and determined $\hat{\theta}_N$, we may use

$$\hat{P}_N = \hat{\lambda}_N \left[\frac{1}{N} \sum_{t=1}^{N} \psi(t, \hat{\theta}_N)\psi^T(t, \hat{\theta}_N)\right]^{-1} \tag{9.18}$$

$$\hat{\lambda}_N = \frac{1}{N} \sum_{t=1}^{N} \varepsilon^2(t, \hat{\theta}_N) \tag{9.19}$$

as an estimate of P_θ. See Problem 9G.3 for the case in where S "almost" belongs to $\mathcal{M}$.

Note also that according to Problem 9G.5, the estimate $\hat{\lambda}_N$ will be asymptotically uncorrelated with $\hat{\theta}_N$.

Example 9.1 Covariance of LS Estimates

Consider the system

$$S: \quad y(t) + a_0 y(t-1) = u(t-1) + e_0(t) \tag{9.20}$$

The input $\{u(t)\}$ is white noise with variance μ, independent of $\{e_0(t)\}$ which is white noise with variance λ_0. We suppose that the coefficient for $u(t-1)$ is known and the system is identified in the model structure

$$\mathcal{M}: \quad y(t) + ay(t-1) = u(t-1) + e(t), \qquad \theta = a$$

or

$$\hat{y}(t|\theta) = -ay(t-1) + u(t-1) \tag{9.21}$$

using a quadratic prediction-error criterion. We thus obtain, in this case, the LS estimate (7.34). We have

$$\psi(t,\theta) = \frac{d}{d\theta}\hat{y}(t|\theta) = -y(t-1) \tag{9.22}$$

Hence

$$P = \frac{\lambda_o}{\overline{E}y^2(t-1)} \tag{9.23}$$

To compute the covariances, square (9.20) and take expectation. This gives

$$R_y(0) + a_0^2 R_y(0) + 2a_0 R_y(1) = \mu + \lambda_0$$

with the standard notation (2.61). (We used (8.27) for the evaluation of $\overline{E}$.) Multiplying (9.20) by $y(t-1)$ and taking expectation gives

$$R_y(1) + a_0 R_y(0) = R_{yu}(0) = 0$$

where the last equality follows, since $u(t)$ does not affect $y(t)$ (due to the time delay). Hence

$$\overline{E}y^2(t-1) = R_y(0) = \frac{\mu + \lambda_0}{1 - a_0^2} \tag{9.24}$$

and

$$\operatorname{Cov}\hat{a}_N \sim \frac{\lambda_0}{N}\frac{1-a_0^2}{\mu+\lambda_0} \tag{9.25}$$

□

Example 9.2 Covariance of an MA(1) Parameter

Consider the system

$$y(t) = e_0(t) + c_0 e_0(t-1)$$

where $\{e_0(t)\}$ is white noise with variance λ_0. The MA(1) model structure

$$y(t) = e(t) + ce(t-1)$$

is used, giving the predictor (4.18):

$$\hat{y}(t|c) + c\hat{y}(t-1|c) = cy(t-1) \tag{9.26}$$

Differentiation w.r.t. c gives

$$\psi(t,c) + c\psi(t-1,c) = -\hat{y}(t-1,c) + y(t-1)$$

At $c = c_0$, we have

$$\psi(t,c_0) = \frac{1}{1 + c_0 q^{-1}} e_0(t-1) \tag{9.27}$$

If $\hat{c}_N$ is the PEM estimate of c, we have, according to (9.17),

$$\operatorname{Cov} \hat{c}_N \sim \frac{\lambda_0}{N} \cdot \frac{1}{E\psi^2(t,c_0)} = \frac{1 - c_0^2}{N} \tag{9.28}$$

□

$S \in \mathcal{M}$: θ-independent General Norm $\ell(\varepsilon)$

Consider now the general criterion (9.12), with $\ell(\varepsilon, \theta, t) = \ell(\varepsilon)$ (no explicit θ- and t-dependence). Assume that $E\ell'_\varepsilon(e_0(t)) = 0$. Then, under the assumption that $S \in \mathcal{M}$, we find after straightforward calculations from (9.13) and (9.14) that

$$P_\theta = \kappa(\ell)[\overline{E}\psi(t,\theta_0)\psi^T(t,\theta_0)]^{-1} \tag{9.29}$$

$$\kappa(\ell) = \frac{\overline{E}\,[\ell'(e_0(t))]^2}{\left[\overline{E}\ell''(e_0(t))\right]^2} \tag{9.30}$$

Here ℓ' and ℓ'' are the first and second derivatives of ℓ with respect to its argument.

Clearly, for $\ell(\varepsilon) = \frac{1}{2}\varepsilon^2$, $\ell'(e) = e$, and $\ell''(e) = 1$, so $\kappa(\ell) = Ee_0^2(t) = \lambda_0$ for quadratic ℓ. This confirms (9.17) as a special case of (9.29) and (9.30). It is interesting to note that the choice of ℓ in the criterion only acts as *scaling* of the covariance matrix.

Asymptotic Cramér-Rao Bound

The Cramér-Rao bound (7.91) applies to all unbiased estimators and all N. It can be rewritten

$$\text{Cov}\left(\sqrt{N}(\hat{\theta}_N - \theta_0)\right) \geq \kappa_0 \left[\frac{1}{N}\sum_{t=1}^{N} E\psi(t,\theta_0)\psi^T(t,\theta_0)\right]^{-1} \tag{9.31}$$

With $f_e(\cdot)$ denoting the PDF of $e_0(t)$, it can be verified that

$$\kappa(-\log f_e) = \kappa_0$$

where κ_0 is defined by (7.88) and the function $\kappa(\ell)$ by (9.30). We thus find that *the asymptotic covariance matrix P_θ in (9.29) equals the limit (as $N \to \infty$) of the Cramér-Rao bound* if $\ell(\cdot)$ is chosen as $-\log f_e(\cdot)$. In this sense the prediction-error estimate (= the maximum likelihood estimate for this choice of ℓ) has the best possible asymptotic properties one can hope for.

$G_0 \in \mathcal{G}$**: Quadratic Criterion,** G **and** H **Independently Parametrized (*)**

We now turn to the case of Theorem 8.4, where G and H are independently parametrized and G_0 can be exactly described in the model set, but not necessarily so for H_0. Assume that all the conditions of Theorem 8.4 hold, and assume in addition that

$$D_G(S, \mathcal{M}) = \{\rho_0\} \tag{9.32}$$

and

$$D_c = \{\theta^*\}, \qquad \theta^* = \begin{bmatrix}\rho_0 \\ \eta^*\end{bmatrix} \tag{9.33}$$

Let

$$F(q) = \frac{H_0(q)}{H(q,\eta^*)} = \sum_{i=0}^{\infty} f_i q^{-i} \tag{9.34}$$

where $H(q,\eta^*)$ is the limiting noise model and $H_0(q)$ the true one according to (8.7). We then have

$$\varepsilon(t,\theta^*) = H^{-1}(q,\eta^*)[y(t) - G(q,\rho_0)u(t)] = F(q)e_0(t) \tag{9.35}$$

and

$$\psi(t,\theta^*) = -\frac{d}{d\theta}\varepsilon(t,\theta)|_{\theta=\theta^*} = \begin{bmatrix} H^{-1}(q,\eta^*)\, G'_\rho(q,\rho_0)u(t) \\ H^{-1}(q,\eta^*)H'_\eta(q,\eta^*)F(q)e_0(t)\end{bmatrix} \tag{9.36}$$

Moreover, with $\ell(\varepsilon) = \frac{1}{2}\varepsilon^2$

$$\overline{V}''(\theta^*) = \overline{E}\psi(t,\theta^*)\psi^T(t,\theta^*) - \overline{E}[(\psi'(t,\theta^*)F(q)e_0(t)] \tag{9.37}$$

Since $\{u(t)\}$ and $\{e_0(t)\}$ are independent, the off-diagonal blocks of the matrix (9.37) will be zero (the blocks corresponding to the decomposition of θ into ρ and η).

Similarly, Q in (9.9) will be block diagonal and hence so will P_θ. We now concentrate on the interesting block corresponding to ρ. Denote

$$\psi_\rho(t) = H^{-1}(q, \eta^*)G'_\rho(q, \rho_0)u(t) \tag{9.38}$$

which is a deterministic (e_0-independent) quantity. Here G'_ρ is the gradient of G w.r.t. ρ. Then

$$\begin{aligned} Q_\rho &= \lim_{N\to\infty} \frac{1}{N} \sum_{t=1}^{N} \sum_{s=1}^{N} \psi_\rho(t)\psi_\rho^T(s) E F(q)e_0(t) \cdot F(q)e_0(s) \\ &= \lim_{N\to\infty} \frac{1}{N} \sum_{t=1}^{N} \sum_{s=1}^{N} \sum_{i=0}^{\infty} \sum_{j=0}^{\infty} \psi_\rho(t) f_i E e_0(t-i)e_0(s-j) f_j \psi_\rho^T(s) \\ &= [t - i = s - j = \tau] = \lim_{N\to\infty} \frac{1}{N} \lambda_0 \sum_{\tau=1}^{N} \sum_{i=0}^{N-\tau} \sum_{j=0}^{N-\tau} f_i \psi_\rho(\tau + i) f_j \psi_\rho^T(\tau + j) \end{aligned}$$

Define

$$\tilde{\psi}(\tau) = \sum_{i=0}^{\infty} f_i \psi_\rho(\tau + i) \tag{9.39}$$

Then

$$Q_\rho = \lambda_0 \overline{E} \tilde{\psi}(t) \tilde{\psi}^T(t) \tag{9.40}$$

Similarly, for the upper left block of (9.37),

$$[\overline{V}''(\theta^*)]_\rho = \overline{E} \psi_\rho(t) \psi_\rho^T(t) \tag{9.41}$$

We can summarize the result as follows:

> Under the assumptions of Theorem 8.4 and (9.33)
>
> $$\sqrt{N}(\hat{\rho}_N - \rho_0) \in AsN(0, P_\rho)$$
>
> where
>
> $$P_\rho = \lambda_0 [\overline{E} \psi_\rho(t) \psi_\rho^T(t)]^{-1} [\overline{E} \tilde{\psi}(t) \tilde{\psi}^T(t)] [\overline{E} \psi_\rho(t) \psi_\rho^T(t)]^{-1} \tag{9.42}$$
>
> The estimates $\hat{\rho}_N$ and $\hat{\eta}_N$ are asymptotically uncorrelated.

This result can be extended to more general norms $\ell(\varepsilon)$ analogously to the case $S \in \mathcal{M}$.

One might ask what (fixed or estimated) noise model $H(q, \eta^*)$ minimizes P_ρ. Not surprisingly, the result is $H(q, \eta^*) = H_0(q)$ so that $F(q) = 1$ in (9.34). See Problem 9G.6.

Example 9.3 Covariance of Output Error Estimate

Consider again the system (9.20), now together with the model structure

$$\mathcal{M}: \qquad y(t) = \frac{q^{-1}}{1 + fq^{-1}} u(t) + e(t), \qquad \theta = f \tag{9.43}$$

or

$$\hat{y}(t|\theta) = \frac{q^{-1}}{1 + fq^{-1}} u(t)$$

This is an output error model. Clearly, $\hat{f} = a_0$ will give a correct description of the transfer function, while the noise properties of (9.20) cannot be correctly described within (9.43), since

$$H_0(q) = \frac{1}{1 + a_0 q^{-1}} \quad \text{and} \quad H(q, \theta) \equiv 1$$

We are thus in a situation, to which the result (9.42) applies. We find that

$$\psi_\rho(t, \theta) = \frac{d}{df}\hat{y}(t|\theta) = \frac{-q^{-2}}{(1 + fq^{-1})^2} u(t) = \frac{-1}{(1 + a_0 q^{-1})^2} u(t-2) \tag{9.44}$$

when evaluated at $\theta^* = a_0$. For $F(q)$ in (9.34), we have

$$F(q) = \frac{1/(1 + a_0 q^{-1})}{1} = \sum_{i=0}^{\infty} (-a_0)^i q^{-i} \tag{9.45}$$

Hence

$$\tilde{\psi}(t) = \sum_{i=0}^{\infty} (-a_0)^i \psi_\rho(t+1) = \left[\sum_{i=0}^{\infty} (-a_0)^i q^i\right] \psi_\rho(t) = \frac{1}{1 + a_0 q} \psi_\rho(t)$$

The spectra are

$$\Phi_{\psi_\rho}(\omega) = \frac{\mu}{|1 + a_0 e^{i\omega}|^4}, \qquad \Phi_{\tilde{\psi}}(\omega) = \frac{\mu}{|1 + a_0 e^{i\omega}|^6}$$

from which the variances can be computed as in Appendix 2C. This gives

$$\overline{E}\psi_\rho^2(t) = \mu \frac{1 + a_0^2}{(1 - a_0^2)^3}, \qquad \overline{E}\tilde{\psi}^2(t) = \mu \frac{(1 + a_0^2)(1 - a_0^2) + 2a_0^2(2 - a_0^2)}{(1 - a_0^2)^5}$$

and thus the variance is, according to (9.42),

$$\operatorname{Cov} \hat{f}_N \sim \frac{1}{N} \cdot P_\rho = \frac{1}{N} \frac{\lambda_0}{\mu} (1 - a_0^2) \left[\frac{1 - a_0^2}{1 + a_0^2} + 2a_0^2 \frac{2 - a_0^2}{(1 + a_0^2)^2}\right] \tag{9.46}$$

□

Multivariable Case (*)

Consider the case where $\varepsilon(t)$ is a p-dimensional vector, and

$$V_N(\theta, Z^N) = \frac{1}{N}\sum_{t=1}^{N} \ell(\varepsilon(t,\theta))$$

The formulas (9.9) to (9.14) are still applicable. Assuming $\theta^* = \theta_0$ is such that $\varepsilon(t,\theta_0) = e_0(t)$ is a sequence of independent, zero mean vectors with $E\ell'_\varepsilon(e_0(t)) = 0$ then straightforward calculations give

$$P_\theta = \left[\overline{E}\psi(t,\theta_0)\Xi\psi^T(t,\theta_0)\right]^{-1} \times \left[\overline{E}\psi(t,\theta_0)\Sigma\psi^T(t,\theta_0)\right]\left[\overline{E}\psi(t,\theta_0)\Xi\psi^T(t,\theta_0)\right]^{-1} \tag{9.47}$$

Here $\psi(t,\theta_0)$ is the $d \times p$ matrix $(d/d\theta)\hat{y}(t|\theta)$ and Ξ and Σ are $p \times p$ matrices:

$$\Xi = \overline{E}\ell''(e_0(t)) \tag{9.48a}$$

$$\Sigma = \overline{E}\ell'(e_0(t))[\ell'(e_o(t))]^T \tag{9.48b}$$

For the quadratic criterion (7.27),

$$\ell(\varepsilon) = \tfrac{1}{2}\varepsilon^T\Lambda^{-1}\varepsilon$$

we find that

$$\Xi = \Lambda^{-1}$$

and

$$\Sigma = \Lambda^{-1}\Lambda_0\Lambda^{-1}$$

where $\Lambda_0 = \overline{E}e_0(t)e_0^T(t)$.

If the choice of weighting matrix Λ in the criterion is $\Lambda = \Lambda_0$ we thus obtain from (9.47)

$$P_\theta = [\overline{E}\psi(t,\theta_0)\Lambda_0^{-1}\psi^T(t,\theta_0)]^{-1} \tag{9.49}$$

See also Problem 9E.4 and the discussion in Section 15.2.

9.4 FREQUENCY-DOMAIN EXPRESSIONS FOR THE ASYMPTOTIC VARIANCE

Variance When $S \in \mathcal{M}$

We found in the previous section that the asymptotic variance is given by (9.17) when $S \in \mathcal{M}$. This can be expressed in terms of frequency functions if Parseval's relation (or Theorem 2.2) is applied. First we need a convenient expression for $\psi(t,\theta)$.

With $\mathbf{T}'$ being the gradient of $T = [G\ H]$ w.r.t. θ, we have from (4.121) and (4.124)

$$\psi(t,\theta) = \frac{1}{H(q,\theta)}\mathbf{T}'(q,\theta)\begin{bmatrix} 1 & 0 \\ -H^{-1}(q,\theta)G(q,\theta) & H^{-1}(q,\theta) \end{bmatrix}\begin{bmatrix} u(t) \\ y(t) \end{bmatrix} \tag{9.50}$$

At $\theta = \theta_0$, using $\varepsilon(t,\theta_0) = e_0(t)$, we thus obtain

$$\psi(t,\theta_0) = H^{-1}(q,\theta_0)\mathbf{T}'(q,\theta_0)\chi_0(t) \tag{9.51}$$

where $\chi_0(t)$ is defined by (8.14). Let $\Phi_{\chi_0}(\omega)$ be the spectrum of χ_0 (a 2×2 matrix):

$$\Phi_{\chi_0}(\omega) = \begin{bmatrix} \Phi_u(\omega) & \Phi_{ue}(\omega) \\ \Phi_{eu}(\omega) & \lambda_0 \end{bmatrix} \tag{9.52}$$

Here $\Phi_{ue}(\omega)$ is the cross-spectrum between the input and the innovations (which is identically zero when the system operates in open loop).

From Theorem 2.2 and Parseval's relation we thus have

$$\begin{aligned} &\overline{E}\psi(t,\theta_0)\psi^T(t,\theta_0) \\ &\quad = \frac{1}{2\pi}\int_{-\pi}^{\pi} \left|H(e^{i\omega},\theta_0)\right|^{-2}\mathbf{T}'(e^{i\omega},\theta_0)\Phi_{\chi_0}(\omega)\mathbf{T}'^T(e^{-i\omega},\theta_0)d\omega \end{aligned} \tag{9.53}$$

Now, using

$$\Phi_v(\omega) = \lambda_0\left|H_0(e^{i\omega})\right|^2$$

for the noise spectrum, we find that

$$\operatorname{Cov}\hat{\theta}_N \sim \frac{1}{N}\left[\frac{1}{2\pi}\int_{-\pi}^{\pi}\frac{1}{\Phi_v(\omega)}\mathbf{T}'(e^{i\omega},\theta_0)\Phi_{\chi_0}(\omega)\mathbf{T}'^T(e^{-i\omega},\theta_0)d\omega\right]^{-1} \tag{9.54}$$

under the assumption that $S \in \mathcal{M}$ and that a quadratic criterion is used. Recall the definition of $\mathbf{T}'$ in (4.125).

Variance when $G_0 \in \mathcal{G}$ (*)

The expression (9.42) can be given in the frequency domain by quite analogous calculations. We obtain

$$\text{Cov}\,\rho_N \sim \frac{1}{N} R^{-1} Q R^{-1}$$

$$R = \frac{1}{2\pi} \int_{-\pi}^{\pi} \frac{1}{|H(e^{i\omega}, \theta^*)|^2} G'_\rho(e^{i\omega}, \rho_0) \Phi_u(\omega) G'_\rho(e^{-i\omega}, \rho_0)^T d\omega$$

$$Q = \frac{1}{2\pi} \int_{-\pi}^{\pi} \frac{\Phi_v(\omega)}{|H(e^{i\omega}, \theta^*)|^4} G'_\rho(e^{i\omega}, \rho_0) \Phi_u(\omega) G'_\rho(e^{-i\omega}, \rho_0)^T d\omega \tag{9.55}$$

Variance of the Transfer Functions

So far, we have discussed the covariance matrix of the parameters θ. For linear, black-box models these parameters are only means to describe various input-output properties, like the frequency function, the impulse response, the poles and zeros, etc. It is of interest to translate the covariance matrix of the parameters to the covariance of these properties.

First, let us consider in general how the covariance properties transform. Let the d-dimensional vector $\hat{\theta}$ be an estimate with mean θ_0 and covariance matrix P. We are interested in the p-dimensional random variable $f(\hat{\theta})$. Asymptotically, as $\hat{\theta}$ becomes sufficiently close to θ_0, we have with good accuracy

$$f(\hat{\theta}) \approx f(\theta_0) + f'(\theta_0)(\hat{\theta} - \theta_0)$$

where f' is the $p \times d$ derivative of f with respect to θ. This means that we asymptotically have

$$\begin{aligned}
\text{Cov} f(\hat{\theta}) &= E\left(f(\hat{\theta}) - Ef(\hat{\theta})\right)\left(f(\hat{\theta}) - Ef(\hat{\theta})\right)^T \\
&\approx E\left(f(\hat{\theta}) - f(\theta_0)\right)\left(f(\hat{\theta}) - f(\theta_0)\right)^T \\
&\approx f'(\theta_0) P \left(f'(\theta_0)\right)^T
\end{aligned} \tag{9.56}$$

This expression is also known as *Gauss' approximation formula.*

Let us now focus on the frequency functions:

$$\hat{G}_N(e^{i\omega}) = G(e^{i\omega}, \hat{\theta}_N), \quad \hat{H}_N(e^{i\omega}) = H(e^{i\omega}, \hat{\theta}_N)$$

These are complex-valued functions, and in accordance with (I.13) we define

$$\text{Cov}\hat{G}(e^{i\omega}) = E\left|\hat{G}(e^{i\omega}) - E\hat{G}(e^{i\omega})\right|^2 \tag{9.57}$$

and

$$\operatorname{Cov}\begin{bmatrix}\hat{G}(e^{i\omega})\\ \hat{H}(e^{i\omega})\end{bmatrix} = E\left(\begin{bmatrix}\hat{G}(e^{i\omega})\\ \hat{H}(e^{i\omega})\end{bmatrix} - E\begin{bmatrix}\hat{G}(e^{i\omega})\\ \hat{H}(e^{i\omega})\end{bmatrix}\right) \times \left(\begin{bmatrix}\hat{G}(e^{-i\omega})\\ \hat{H}(e^{-i\omega})\end{bmatrix} - E\begin{bmatrix}\hat{G}(e^{-i\omega})\\ \hat{H}(e^{-i\omega})\end{bmatrix}\right)^T \tag{9.58}$$

If the asymptotic covariance matrix of $\hat{\theta}_N$ is $\frac{1}{N}P_\theta$, then Gauss' approximation formula directly gives

$$\operatorname{Cov}\begin{bmatrix}\hat{G}_N(e^{i\omega})\\ \hat{H}_N(e^{i\omega})\end{bmatrix} \approx \frac{1}{N}\left(\mathbf{T}'(e^{i\omega},\theta_0)\right)^T P_\theta \mathbf{T}'(e^{-i\omega},\theta_0) \tag{9.59}$$

where $\mathbf{T}'$ is defined by (4.125). By plugging in the expression for P_θ from (9.54), a nicely symmetric expression for the variance of the frequency functions is obtained.

We can be more specific for certain common choice of model structures. Consider an ARX-model

$$G(q,\theta) = \frac{B(q)}{A(q)}, \quad H(q,\theta) = \frac{1}{A(q)}$$

with the same order of the A- and B-polynomials; $A(q) = 1 + a_1 q^{-1} + \ldots + a_n q^{-n}$, $B(q) = b_1 q^{-1} + \ldots + b_n q^{-n}$. Let $\theta_k = [\, a_k \quad b_k \,]$. Then the derivative with respect to θ_k, i.e. the rows $2(k-1)$ and $2k-1$ of $\mathbf{T}'(q,\theta)$, will be

$$\begin{bmatrix}\frac{d}{d\theta_k}\frac{B(q)}{A(q)} & \frac{d}{d\theta_k}\frac{1}{A(q)}\end{bmatrix} = \begin{bmatrix}-\frac{B(q)}{A^2(q)}q^{-k} & -\frac{1}{A^2(q)}q^{-k}\\ \frac{1}{A(q)}q^{-k} & 0\end{bmatrix} = q^{-k}M(q,\theta)$$

where the last step is a definition of $M(q,\theta)$. This means that

$$\mathbf{T}'(e^{i\omega},\theta) = W_n(e^{-i\omega})M(e^{i\omega},\theta) \tag{9.60a}$$

$$M(e^{i\omega},\theta) = \begin{bmatrix}-\frac{B(e^{i\omega})}{A^2(e^{i\omega})} & -\frac{1}{A^2(e^{i\omega})}\\ \frac{1}{A(e^{i\omega})} & 0\end{bmatrix} \tag{9.60b}$$

$$W_n(e^{-i\omega}) = \begin{bmatrix}e^{-i\omega}I & e^{-i2\omega}I & \ldots & e^{-in\omega}I\end{bmatrix}^T \tag{9.60c}$$

Here I is the 2×2 identity matrix. This shows how we can easily build $\mathbf{T}'$ for arbitrary orders n. It is also straightforward to show that all the black-box models in the family (4.33) obey the same structure, with different definitions of M. This matrix will have dimension $s \times 2$, where s is the number of different polynomials used by the structure, i.e., 3 for ARMAX, 2 for OE and 4 for BJ. Moreover the identity matrix will then be of dimension $s \times s$.

Asymptotic Black-Box Theory

The expressions (9.60), (9.59), and (9.54) together constitute an explicit way of computing the frequency function covariances. It is not easy, though, to get an intuitive feeling for the result. It turns out, however, that as the model order increases, $n \to \infty$, the expressions simplify drastically. The simplification is based on the following result: Let W_n be defined by (9.60c) with the identity matrix of size s, and let $L(\omega)$ be a $s \times s$ invertible matrix function, with a well defined inverse Fourier transform. Then

$$\lim_{n\to\infty} \frac{1}{n} W_n^T(e^{-i\omega}) \left[\frac{1}{2\pi} \int_{-\pi}^{\pi} W_n(e^{i\xi}) L(\xi) W_n^T(e^{-i\xi}) d\xi \right]^{-1} W_n(e^{i\omega_1})$$
$$= [L(\omega)]^{-1} \delta_{\omega,\omega_1} \tag{9.61}$$

where $\delta_{s,t}$ is 1 if the subscripts coincide and 0 otherwise. This result is proved as Lemma 4.3 in Yuan and Ljung (1985). See also Lemma 4.2 in Ljung and Yuan (1985).

We can now combine (9.59) with (9.60) and (9.54):

$$N \cdot \text{Cov} \begin{bmatrix} \hat{G}_N(e^{i\omega}) \\ \hat{H}_N(e^{i\omega}) \end{bmatrix} = M^T(e^{i\omega}, \theta_0) W_n^T(e^{-i\omega})$$
$$\times \left[\frac{1}{2\pi} \int_{-\pi}^{\pi} W_n(e^{i\xi}) \{ M(e^{-i\xi}, \theta_0) \frac{\Phi_{\chi_0}(-\xi)}{\Phi_v(\xi)} M^T(e^{i\xi}, \theta_0) \} W_n^T(e^{-i\xi}) d\xi \right]^{-1}$$
$$\times W_n(e^{i\omega}) M(e^{-i\omega}, \theta_0)$$
$$\approx n \cdot M^T(e^{i\omega}, \theta_0) \left[M(e^{-i\omega}, \theta_0) \frac{\Phi_{\chi_0}(-\omega)}{\Phi_v(\omega)} M^T(e^{i\omega}, \theta_0) \right]^{-1} M(e^{-i\omega}, \theta_0)$$
$$= n \Phi_v(\omega) \Phi_{\chi_0}^{-1}(-\omega)$$

where the approximate equality follows from (9.61). We complex-conjugated (9.54) first, which is allowed since the parameter covariance matrix is real. In summary we have

$$\text{Cov} \begin{bmatrix} \hat{G}_N(e^{i\omega}) \\ \hat{H}_N(e^{i\omega}) \end{bmatrix} \approx \frac{n}{N} \Phi_v(\omega) \begin{bmatrix} \Phi_u(\omega) & \Phi_{ue}(-\omega) \\ \Phi_{ue}(\omega) & \lambda_0 \end{bmatrix}^{-1} \tag{9.62}$$

where n is the model order and N is the number of data, and where the approximation improves as the model order increases. Note that (9.61) also implies that estimates at different frequencies are uncorrelated. We showed the result here for the ARX-model, but the only requirement on the model structure is the shift property (9.60). This holds for many black-box parameterizations, including the family (4.33). The result (9.62) is thus fundamental. However, note that it is asymptotic in the model order n, and corresponds to the case where the model structure is so flexible that the estimates at different frequencies become decoupled.

In open loop operation, where $\Phi_{ue} \equiv 0$, we find that the estimates $\hat{G}_N$ and $\hat{H}_N$ are asymptotically uncorrelated, *even when they are parameterized with common parameters.* Then

$$\text{Cov}\hat{G}_N(e^{i\omega}) \approx \frac{n}{N}\frac{\Phi_v(\omega)}{\Phi_u(\omega)} \tag{9.63a}$$

$$\text{Cov}\hat{H}_N(e^{i\omega}) \approx \frac{n}{N}|H_0(e^{i\omega})|^2 \tag{9.63b}$$

In Part III we shall find (9.62) to be of considerable use for many issues in experiment design. It is therefore of interest to check how well (9.62) describes the exact expression for moderate values of n.

Example 9.4 Comparing (9.62) with Exact Expressions for Finite n

Consider a second-order system

$$y(t) + a_1^0 y(t-1) + a_2^0 y(t-2) = b_1^0 u(t-1) + b_2^0 u(t-2) + e_0(t) \tag{9.64}$$

We assume that this system is identified using the least-squares method in a model structure given by

$$y(t) + a_1 y(t-1) + \cdots + a_n y(t-n) = b_1 u(t-1) + \cdots + b_n u(t-n) + e(t) \tag{9.65}$$

The input $\{u(t)\}$ is taken as white noise with variance 1, and the disturbance $e_0(t)$ to the system (9.64) is also supposed to be white noise with unit variance.

Let $\hat{G}_N(e^{i\omega}, n)$ be the estimate of the transfer function obtained with model order n. Since the true system (9.64) is of second order, we can use (9.54), (9.59), and (9.60) to evaluate $P_n(\omega)$ in

$$\text{Cov}\,\hat{G}_N(e^{i\omega}, n) \sim \frac{1}{N}P_n(\omega)$$

for $n \geq 2$. This gives an explicit, but complicated expression for $P_n(\omega)$, which is exact in n (but asymptotic in N). According to (9.63a) we have, asymptotically in n,

$$\text{Cov}\,\hat{G}_N(e^{i\omega}, n) \sim \frac{n}{N}P_L(\omega) = \frac{n}{N}\frac{\Phi_v(\omega)}{\Phi_u(\omega)} = \frac{n/N}{|1 + a_1^0 e^{i\omega} + a_2^0 e^{2i\omega}|^2} \tag{9.66}$$

We shall thus compare how well $n \cdot P_L(\omega)$ approximates P_n. We do this by plotting $P_L(\omega)$ and $(1/n)P_n(\omega)$ as functions of ω. First consider the "Åström system,"

$$a_1^0 = -1.5, \qquad a_2^0 = 0.7, \qquad b_1^0 = 1, \qquad b_2^0 = 0.5 \tag{9.67}$$

In Figure 9.1(a) we plot $\log(1/n)P_n(\omega)$ for $n = 2, 10$, and $\log P_L(\omega)$ against $\log \omega$. The figure illustrates well the convergence to the limit as n increases. The convergence could be said to be rather slow. Of greater importance, however, is that, even for small n, the limit expression gives a reasonable feel for how the exact variance changes with frequency. To test this feature, we evaluate the expression for a number of second-order systems with different characteristics. The results are shown in Figures 9.1(b) to 9.1(d).

According to the asymptotic expression (9.62), the variance should not depend on the number of estimated parameters, but only on the model order. This is a slightly surprising result. To test whether this result also has relevance for low-order models, we identified the system (9.64) and (9.67) also in the ARMAX-structure,

$$y(t) + a_1 y(t-1) + a_2 y(t-2)$$
$$= b_1 u(t-1) + b_2 u(t-2) + e(t) + c_1 e(t-1) + c_2 e(t-2) \quad (9.68)$$

which employs 50% more parameters than (9.65) with the same model order. In Figure 9.1(e) we show the variance of the transfer function when estimated in the model structures (9.65) and (9.68), respectively, with $n = 2$. The agreement is striking.

Finally, we identified the system (9.64) and (9.67) in the ARMAX model structure (9.68), with a low-frequency input with spectrum

$$\Phi_u(\omega) = \frac{0.25}{1.25 - \cos\omega} \quad (9.69)$$

The results are shown in Figure 9.1(f). Comparing with Figure 9.1(a), we see that the agreement between the asymptotic and exact expressions is now worse, especially at low frequencies. □

We may conclude from the example that the asymptotic variance expressions give a good feel for character of the true variance, but must be used with care for quantitative calculations.

9.5 THE CORRELATION APPROACH

Basic Theorem

The correlation estimate $\hat{\theta}_N$ is defined by (7.110). We shall confine ourselves to the case studied in Theorem 8.6, that is, $\alpha(\varepsilon) = \varepsilon$ and linearly generated instruments.

We thus have

$$\hat{\theta}_N = \text{sol}[f_N(\theta, Z^N) = 0] \quad (9.70a)$$

$$f_N(\theta, Z^N) = \frac{1}{N}\sum_{t=1}^{N} \zeta(t,\theta)\varepsilon_F(t,\theta) \quad (9.70b)$$

$$\varepsilon_F(t,\theta) = L(q)\varepsilon(t,\theta) \quad (9.70c)$$

$$\zeta(t,\theta) = K_y(q,\theta)y(t) + K_u(q,\theta)u(t) \quad (9.70d)$$

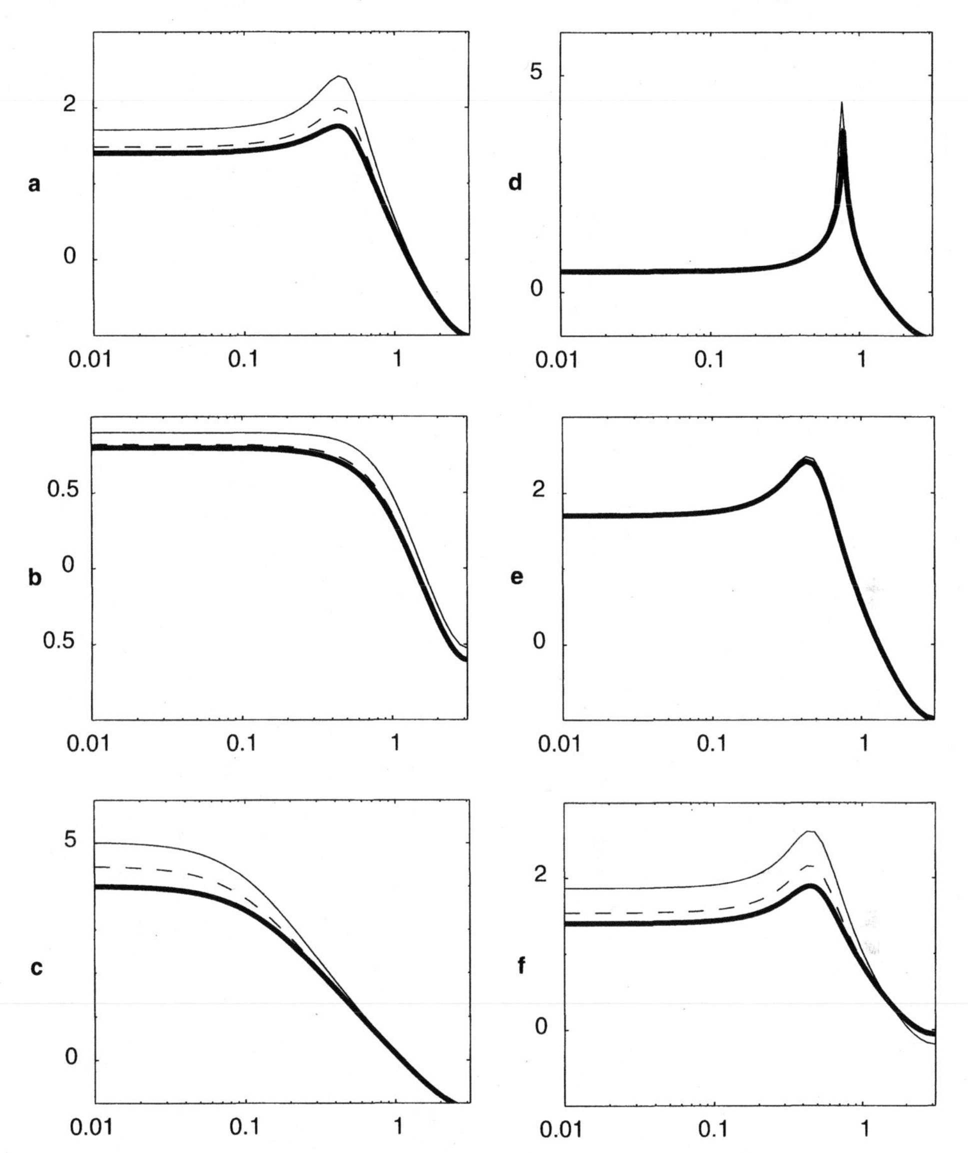

Figure 9.1 Plots of $\log(1/n)P_n(\omega)$ and of $\log\ P_L(\omega)$ versus ω for different systems, model structures, and inputs. Thick line = the asymptotic expression P_L. Thin lines: Normalized true variance for $n = 2$ (solid) and $n = 10$ (dashed).
(a) System: (9.64), (9.67); model: (9.65); input: white noise.
(b) System: (9.64), $a_1^0 = -0.8$, $a_1^0 = 0.2$, $b_1^0 = 1$, $b_2^0 = -0.9$; model: (9.65); input: white noise.
(c) System: (9.64), $a_1^0 = -1.8$, $a_2^0 = 0.81$, $b_1^0 = 1$, $b_2^0 = 0$; model: (9.65); input: white noise.
(d) System: (9.64), $a_1^0 = -1.4$, $a_2^0 = 0.98$, $b_1^0 = 1$, $b_2^0 = 0.5$; model: (9.65); input: white noise.
(e) System: (9.64), (9.67); Thick line: model (9.65). Thin line: model (9.68). $n = 2$. Input: white noise.
(f) System: (9.64), (9.67); model: (9.68); input: (9.69).

By Taylor's expansion, we obtain

$$0 = f_N(\hat{\theta}_N, Z^N) = f_N(\theta^*, Z^N) + (\hat{\theta}_N - \theta^*) f_N'(\xi_N, Z^N) \tag{9.71}$$

This is entirely analogous to (9.3) with the difference that $\psi(t, \theta^*)$ in $V_N'(\theta^*, Z^N)$ is replaced by $\zeta(t, \theta^*)$ in $f_N(\theta^*, Z^N)$. The analysis of (9.71) is therefore essentially identical to the case already studied in Section 9.2, and the result can be formulated as follows.

Theorem 9.2. Consider $\hat{\theta}_N$ determined by (9.70). Assume that $\varepsilon(t, \theta)$ is computed from a linear, uniformly stable model structure, and that

$$\left\{ K_y(q, \theta), K_u(q, \theta), \frac{d}{d\theta} K_y(q, \theta), \frac{d}{d\theta} K_u(q, \theta); \theta \in D_{\mathcal{M}} \right\}$$

is a uniformly stable family of filters. Assume that the data set Z^∞ is subject to D1 (see Section 8.2). Assume also that $\hat{\theta}_N \to \theta^*$ w.p. 1 as $N \to \infty$ that $\overline{f}'(\theta^*)$ is nonsingular [$\overline{f}$ defined by (8.87)], and that

$$\sqrt{N} E f_N(\theta^*, Z^N) \to 0, \qquad \text{as } N \to \infty$$

Then

$$\sqrt{N}(\hat{\theta}_N - \theta^*) \in AsN(0, P_\theta) \tag{9.72}$$

with

$$P_\theta = [\overline{f}'(\theta^*)]^{-1} Q [\overline{f}'(\theta^*)]^{-T} \tag{9.73}$$

$$Q = \lim_{N \to \infty} N \cdot E f_N(\theta^*, Z^N) f_N^T(\theta^*, Z^N) \tag{9.74}$$

We find from (9.70) that

$$\overline{f}'(\theta) = \overline{E} \zeta_\theta'(t, \theta) \varepsilon_F(t, \theta) - \overline{E} \zeta(t, \theta) [L(q) \psi(t, \theta)]^T \tag{9.75}$$

where ψ as before denotes the negative gradient of ε, with respect to θ.

Variance Expression when $S \in \mathcal{M}$, $L(q) \equiv 1$ (*)

Under the assumption $S \in \mathcal{M}$, there exists a value θ_0 such that $\varepsilon(t, \theta_0) = e_0(t)$ is a sequence of independent random variables with zero mean values and variances λ_0. We thus obtain, for $L(q) = 1$,

$$\overline{f}'(\theta_0) = -\overline{E} \zeta(t, \theta_0) \psi^T(t, \theta_0) + \overline{E} \zeta'(t, \theta_0) e_0(t) = -\overline{E} \zeta(t, \theta_0) \psi^T(t, \theta_0) \tag{9.76}$$

and

$$Q = \lim_{N \to \infty} E \frac{1}{N} \sum_{t=1}^{N} \sum_{s=1}^{N} \zeta(t, \theta_0) e_0(t) e_0^T(s) \zeta^T(s, \theta_0)$$

$$= \lambda_0 \overline{E} \zeta(t, \theta_0) \zeta^T(t, \theta_0) \tag{9.77}$$

so

$$P_\theta = \lambda_0 \left[\overline{E}\zeta(t,\theta_0)\psi^T(t,\theta_0) \right]^{-1} \times \left[\overline{E}\zeta(t,\theta_0)\zeta^T(t,\theta_0) \right] \left[\overline{E}\psi(t,\theta_0)\zeta^T(t,\theta_0) \right]^{-1} \tag{9.78}$$

See also Problem 9G.4.

Example 9.5 Covariance of Pseudolinear Regression Estimate

Consider again the system and the model of Example 9.2, but suppose that the c-estimate is determined by the PLR method (7.113); that is,

$$\hat{c}_N^{\text{PLR}} = \text{sol}\left\{ \frac{1}{N}\sum_{t=1}^{N} \varepsilon(t-1,c)\varepsilon(t,c) = 0 \right\}$$

Here $\zeta(t,\theta) = \varphi(t,\theta) = \varepsilon(t-1,c) = y(t-1) - \hat{y}(t-1|c), \theta = c$. At $c = c_0$ we have [see (9.27)]

$$\zeta(t,\theta_0) = e_0(t-1), \qquad \psi(t,\theta_0) = \frac{1}{1 + c_0 q^{-1}} e_0(t-1)$$

Hence

$$\overline{E}\zeta(t,\theta_0)\zeta^T(t,\theta_0) = \lambda_0$$

$$\overline{E}\zeta(t,\theta_0)\psi^T(t,\theta_0) = \lambda_0$$

so, according to (9.78),

$$\text{Cov}\,\hat{c}_N^{\text{PLR}} \sim \frac{1}{N}$$

Compare with (9.28). Note that $|c_0| < 1$ always. □

Instrumental-variable Methods: $G_0 \in \mathcal{G}$

It is of interest to specialize Theorem 9.2 to the instrumental-variable case of Section 7.6. We then have the model

$$\hat{y}(t|\theta) = \varphi^T(t)\theta \tag{9.79}$$

and the procedure (7.129). Suppose now that the true system is given as in (8.92) to (8.94) by

$$y(t) = \varphi^T(t)\theta_0 + H_0(q)A_0(q)e_0(t) \tag{9.80}$$

where $\{e_0(t)\}$ is white noise with variance λ_0, independent of $\{u(t)\}$. Then

$$\varepsilon_F(t,\theta_0) = L(q)H_0(q)A_0(q)e_0(t)$$

is independent of $\{u(t)\}$ and hence of $\zeta(t,\theta_0)$ if the system operates in open loop. Thus θ_0 is a solution to

$$\overline{E}\zeta(t,\theta)\varepsilon_F(t,\theta) = 0$$

as we also found in (8.95). To get an asymptotic distribution, we shall assume it is the only solution in $D_{\mathcal{M}}$:

$$\overline{E}\zeta(t,\theta)\varepsilon_F(t,\theta) = 0 \quad \text{and} \quad \theta \in D_{\mathcal{M}} \Rightarrow \theta = \theta_0 \tag{9.81}$$

Introduce also the monic filter

$$F(q) = L(q)H_0(q)A_0(q) = \sum_{i=0}^{\infty} f_i q^{-i} \tag{9.82}$$

Inserting these expressions into (9.72) to (9.74) gives, by entirely analogous calculation as in (9.42), the following result:

For the IV estimate $\hat{\theta}_N$ we have, under assumptions (9.80) and (9.81), that

$$P_\theta = \lambda_0[\overline{E}\zeta(t,\theta_0)\varphi_F^T(t)]^{-1}[\overline{E}\zeta_F(t,\theta_0)\zeta_F^T(t,\theta_0)][\overline{E}\zeta(t,\theta_0)\varphi_F^T(t)]^{-T} \tag{9.83}$$

with

$$\varphi_F^T(t) = L(q)\varphi^T(t)$$

$$\zeta_F(t,\theta) = \sum_{i=0}^{\infty} f_i\zeta(t+i,\theta)$$

Example 9.6 Covariance of an IV Estimate

Consider again the system (9.20) of Example 9.1. Let the model be the linear regression (9.21) and let a be estimated by the IV method using

$$\zeta(t) = -\frac{q^{-1}}{1 + a_* q^{-1}}u(t-1)$$

as the instrument and $L(q) = 1$. To evaluate (9.83), we find by comparing (9.80) and (9.20) that

$$F(q) = H_0(q)A_0(q) = 1.$$

Hence

$$\varphi_F(t) = -y(t-1) = -\frac{q^{-1}}{1 + a_0 q^{-1}}u(t-1) - \frac{1}{1 + a_0 q^{-1}}e_0(t-1)$$

$$\zeta_F(t) = \zeta(t)$$

and

$$
\begin{aligned}
\overline{E}\zeta_F(t)\varphi_F(t) &= \frac{1}{2\pi}\int_{-\pi}^{\pi} \frac{1}{1 + a_0 e^{i\omega}} \cdot \frac{1}{1 + a_* e^{-i\omega}} \Phi_u(\omega) d\omega \\
&= \frac{\mu}{2\pi}\int_{-\pi}^{\pi} \frac{1}{1 + a_0 e^{i\omega}} \frac{1}{1 + a_* e^{-i\omega}} d\omega = \frac{\mu}{1 - a_0 a_*} \\
\overline{E}\zeta_F^2(t) &= \frac{1}{2\pi}\int_{-\pi}^{\pi} \frac{1}{|1 + a_* e^{i\omega}|^2} \Phi_u(\omega) d\omega \\
&= \frac{\mu}{2\pi}\int_{-\pi}^{\pi} \frac{1}{|1 + a_* e^{i\omega}|^2} d\omega = \frac{\mu}{1 - a_*^2}
\end{aligned}
$$

Hence

$$
\operatorname{Cov} \hat{a}_N \sim \frac{1}{N}\frac{\lambda_0}{\mu}\frac{(1 - a_0 a_*)^2}{(1 - a_*^2)} \tag{9.84}
$$

□

Frequency-domain Expressions for the Variance (*)

Frequency-domain expressions for the covariance matrices in (9.78) and (9.83) can be developed along the same lines as for (9.54) and (9.55). For example, for (9.83) we find that

$$
\begin{aligned}
\operatorname{Cov} \hat{\theta}_N &\sim \frac{1}{N} R^{-1} Q R^{-T} \\
R &= \frac{1}{2\pi}\int_{-\pi}^{\pi} K_u(e^{i\omega}) \cdot K_0^T(e^{-i\omega}) \cdot L(e^{-i\omega}) \Phi_u(\omega) d\omega \\
Q &= \frac{\lambda}{2\pi}\int_{-\pi}^{\pi} K_u(e^{i\omega}) \cdot K_u^T(e^{-i\omega}) \cdot |F(e^{i\omega})|^2 \Phi_u(\omega) d\omega
\end{aligned} \tag{9.85}
$$

Here K_u and F are given by (7.128) and (9.82), respectively, while

$$
K_0(q) = \begin{bmatrix} \frac{B_0(q)}{A_0(q)} q^{-1} \\ \vdots \\ \frac{B_0(q)}{A_0(q)} q^{-n_a} \\ q^{-1} \\ \vdots \\ q^{-n_b} \end{bmatrix}
$$

Note that

$$\varphi(t) = K_0(q)u(t) + e_0\text{-dependent terms}$$

An "asymptotic black-box analysis" as in (9.62) can be developed also for IV and PLR estimates. The result is that, for open-loop operation, (9.63) still holds. See Ljung (1985c).

9.6 USE AND RELEVANCE OF ASYMPTOTIC VARIANCE EXPRESSIONS

In this chapter we have developed a number of expressions for the asymptotic covariance of the estimated quantities. Now, what are these good for? There are two basic uses for such results. One is to use the covariance matrix P_θ, (or some derived quantity) as a quality measure for resulting estimates. Our expressions can then be used for analytical or numerical studies of how various design variables affect the identification result. We shall make extensive use of this approach in Part III when discussing the user's choices.

The other application of the asymptotic covariance results is to compute confidence intervals to assess the reliability of particular estimates obtained from an observed data set. We shall comment on this application shortly.

The derived expressions are asymptotic in N, the number of observed data. Our theory has not told us how large N has to be for the results to be applicable. Clearly, this is an important question in order to evaluate the relevance of the covariance expression.

Confidence Intervals

If the vector $\hat{\theta}_N$ obeys

$$\sqrt{N}(\hat{\theta}_N - \theta_0) \in AsN(0, P_\theta) \tag{9.86}$$

then the kth component obeys

$$\sqrt{N}(\hat{\theta}_N^{(k)} - \theta_0^{(k)}) \in AsN(0, P_\theta^{(kk)}) \tag{9.87}$$

$P_\theta^{(kk)}$ being the k, k diagonal element of P_θ. This means that the probability distribution of the random variable $\sqrt{N}(\hat{\theta}_N^{(k)} - \theta_0^{(k)})$ converges to the Gaussian distribution, and we can use the limit to evaluate, for example, the probability

$$P(|\hat{\theta}_N^{(k)} - \theta_0^{(k)}| > \alpha) \approx \frac{\sqrt{N}}{\sqrt{2\pi P_\theta^{(kk)}}} \int_{|x|>\alpha} e^{-x^2 \cdot N/2P_\theta^{(kk)}} dx \tag{9.88}$$

In this way we may form *confidence intervals* for the true parameter; that is, we can tell, with a certain degree of confidence, that the true value $\theta_0^{(k)}$ is to be found in a certain interval around the estimate (provided the underlying assumptions are satisfied, that is).

The vector expression (9.86) tells us more. If a random vector η has a Gaussian distribution

$$\eta \in N(0, P)$$

then the scalar

$$z = \eta^T P^{-1} \eta$$

has a χ^2 distribution with dim $\eta = d$ degrees of freedom,

$$z \in \chi^2(d).$$

From (9.86) we thus draw the conclusion that

$$\eta_N = N \cdot (\hat{\theta}_N - \theta_0)^T P_\theta^{-1} (\hat{\theta}_N - \theta_0) \in As\chi^2(d) \tag{9.89}$$

That is, η_N converges in distribution to the $\chi^2(d)$ distribution as N tends to infinity. Using χ^2 tables, we can thus derive confidence intervals for η_N, which correspond to *confidence ellipsoids* for $\hat{\theta}_N$ (see also Section II.2).

Estimating the Covariance Matrix

The expressions (9.86) to (9.89) require the knowledge of P_θ to be used in practice. Typical expressions for P_θ, *assuming* $\mathcal{S} \in \mathcal{M}$, such as (9.17), (9.42), (9.78), and (9.83) involve both θ_0, λ_0 and the symbol $\overline{E}$, and as such are unknown to the user. Natural estimates $\hat{P}_N$ of P_θ are, however, straightforward, replacing θ_0 by $\hat{\theta}_N$ and $\overline{E}$ by the sample sum, as in (9.18) and (9.19). Under weak conditions, Theorem 2.3 will then imply that $\hat{P}_N$ converges w.p.1 to P_θ, so it is still reasonable to apply the asymptotic results with $\hat{P}_N$ replacing P_θ. (If $\hat{\theta}_N$ has an exact Gaussian distribution for finite N, more sophisticated calculations, involving t- and F-distribution, can be carried out to find more accurate nonasymptotic, confidence intervals. See Section II.2.)

It could be noted that it is more cumbersome to find estimates of P_θ in the more general case when $\mathcal{S} \notin \mathcal{M}$. Replacing θ^* by $\hat{\theta}_N$ in (9.9), gives, trivially, $Q = 0$, which is a useless estimate. Hjalmarsson and Ljung (1992)have shown how to circumvent this problem, by devising a consistent estimator of Q also in this case. For the output error case, note that the expressions (9.42) and (9.83) contain the noise filter $H_0(q)$, which typically would not be known to the user.

Relevance of the Asymptotic Expressions for Finite N

From the asymptotic expressions we know the behavior of $\sqrt{N}(\hat{\theta}_N - \theta_0)$ for "large" N. This information is, as we saw, crucial for the confidence we develop in the model. The question remains, how large N has to be for the asymptotic expressions to be reliable. No general answer can be given to this question. Monte-Carlo studies have been performed, in which systems have been simulated many times with different noise realizations. The statistics over the estimates so obtained have been compared with the estimated covariance matrices, (9.18)–(9.19). Such studies show that for typical system identification applications, the asymptotic variance expressions are reliable within 10% or so for $N \gtrsim 300$. Such Monte-Carlo tests depend however also on the properties of the random generators used. Despite the slight vagueness of these statements, the conclusion is that the estimated asymptotic variance expression gives a reliable picture of the actual variability and uncertainty of the estimates.

An alternative picture of the estimate uncertainty, that does not rely upon analytical, asymptotic expressions is obtained using *bootstrap techniques.* The bootstrap principle starts from the Monte-Carlo idea just described: If we know how our data have been generated, we may regenerate them many times and compute the statistics over the resulting estimates, to determine their reliability. Now, in practice, we do not know the mechanism behind the data generation. Instead, the only information is the data set Z^N itself. The bootstrap idea is now to generate new data sets that resemble Z^N in the sense that they are drawn from the empirical distribution that the observed data Z^N defines. Monte-Carlo studies over such collection of data sets will now yield confidence intervals and covariance properties. See e.g. Efron and Tibshirani (1993)for a thorough treatment and Zoubir and Boashash (1998) and Politis (1998)for tutorials focused on signal processing applications.

9.7 SUMMARY

In this chapter we have shown that the parameter estimates obtained by the prediction error or the correlation approach are asymptotically normal distributed as the number of observed data tends to infinity. This was established in Theorems 9.1 and 9.2. Several explicit expressions for the asymptotic covariance matrix have been developed under varying assumptions. The archetypal result is (9.17):

$$\operatorname{Cov}\hat{\theta}_N = \frac{\lambda_0}{N}[\bar{E}\psi(t,\theta_0)\psi^T(t,\theta_0)]^{-1} \tag{9.90}$$

which is valid for a quadratic prediction-error criterion estimate under the assumption that $S \in \mathcal{M}$. It tells us that the covariance matrix is given by the inverse of the covariance matrix of the predictor gradients ψ, normalized by the innovations variance λ_0 divided by the number of observed data N. This expression is also the limit of the Cramér-Rao lower bound, in case the innovations are normally distributed. This means that the estimate $\hat{\theta}_N$ then is *asymptotically efficient.*

The corresponding expression for the correlation approach is (9.78):

$$\begin{aligned}\operatorname{Cov}\hat{\theta}_N \sim \frac{\lambda_0}{N}&[\bar{E}\zeta(t,\theta_0)\psi^T(t,\theta_0)]^{-1} \\ &\times [\bar{E}\zeta(t,\theta_0)\zeta^T(t,\theta_0)][\bar{E}\zeta(t,\theta_0)\psi^T(t,\theta_0)]^{-T}\end{aligned} \tag{9.91}$$

We have also developed results for the asymptotic distribution of the resulting transfer-function estimates in black-box parametrizations. They tell us that

$$\operatorname{Cov}\begin{bmatrix}\hat{G}(e^{i\omega},\hat{\theta}_N)\\ \hat{H}(e^{i\omega},\hat{\theta}_N)\end{bmatrix} \sim \frac{n}{N}\cdot\Phi_v(\omega)\begin{bmatrix}\Phi_u(\omega) & \Phi_{ue}(-\omega)\\ \Phi_{ue}(\omega) & \lambda_0\end{bmatrix}^{-1} \tag{9.92}$$

asymptotically both in model order n and number of data N.

The distribution of the random vector $\hat{\theta}_N$ stresses that the result of the parameter identification phase is not a model; rather it is a set of models. Guided by the data, we have narrowed the original model set to, it is hoped, a much smaller set of possible descriptions around $\mathcal{M}(\hat{\theta}_N)$. Expressions (9.90) and (9.91) form a "quality tag" with which the model $\mathcal{M}(\hat{\theta}_N)$ is delivered to the user.

9.8 BIBLIOGRAPHY

The asymptotic distribution of parameter estimates is, as we have seen, basically an application of a suitable central limit theorem (CLT). The probability literature offers an abundant supply of CLTs under varying conditions (see, e.g., Chung, 1974; Brown, 1971; Ibragimov and Linnik, 1975; Withers, 1981; and Dvoretzky, 1972). A key assumption in these results is that the dependence between samples far apart should be decaying at at least a certain rate. *Mixing conditions* are used to describe this. In our treatment, the stability assumption (8.5) played this role.

The asymptotic normality of ML estimates in the case of independent observations is discussed in, for example, Kendall and Stuart (1961). This result was extended by Åström and Bohlin (1965)to ARMAX models.

In the statistical literature, asymptotic normality for parameter estimates in ARMA models of time series has been treated in a number of articles. Let us in particular mention the work of Hannan (1970, 1973, 1979), Dunsmuir and Hannan (1976), Hannan and Deistler (1988), and Anderson (1975).

The situation where $S \notin \mathcal{M}$ was studied in Kabaila and Goodwin (1980)and in Ljung and Caines (1979). Kabaila (1983)has studied the OE-result (9.42).

Frequency-domain expressions for the covariance matrices, such as (9.54), have been used in, for example, Hannan (1970)and Kabaila and Goodwin (1980). The black-box expression (9.62) has been derived in Ljung (1985b). Results where the order n is a function of the number of data N and increases to infinity are given for FIR models in Ljung and Yuan (1985)and for the ARX model (4.7) in Ljung and Wahlberg (1992). Related results for AR modeling of spectra were obtained by Berk (1974).

Extensions of the results (9.62) to the multivariable case are given in Zhu (1989), and to recursively identified models in Gunnarsson and Ljung (1989).

The asymptotic normality for IV estimates was shown by Caines (1976b)and a comprehensive treatment is contained in Söderström and Stoica (1983). For PLR methods, results are given by Stoica et.al. (1984).

Discussions of the use of confidence intervals are standard in many textbooks (see, e.g., Draper and Smith, 1981).

9.9 PROBLEMS

9G.1 Consider the asymptotic expression (9.62) and let $\Phi_{ue}(\omega) \equiv 0$. Note that the real and imaginary parts are uncorrelated. (This follows as in (6.62) since the estimates are uncorrelated at different frequencies.) Let

$$\hat{A}_N(\omega) = |\hat{G}_N(e^{i\omega})|, \qquad \hat{\varphi}_N(\omega) = \arg\hat{G}_N(e^{i\omega}) \quad \text{(radians)}$$

Show that

$$\operatorname{Cov} \hat{A}_N(\omega) \sim \frac{1}{2}\frac{n}{N}\frac{\Phi_v(\omega)}{\Phi_u(\omega)}$$

$$\operatorname{Cov} \hat{\varphi}_N(\omega) \sim \frac{1}{2}\frac{n}{N}\frac{\Phi_v(\omega)}{\Phi_u(\omega)|\hat{G}_0(e^{i\omega})|^2}$$

Show also that

$$E|\hat{G}_N(e^{i\omega})|^2 - |E\hat{G}_N(e^{i\omega})|^2 \sim \frac{n}{N}\frac{\Phi_v(\omega)}{\Phi_u(\omega)}$$

9G.2 From a model $H(e^{i\omega}, \hat{\theta}_N)$ and an estimate $\hat{\lambda}_N$ of the innovation variance, we can form an estimate of the noise spectrum

$$\hat{\Phi}_v^N(\omega) = \hat{\lambda}_N |H(e^{i\omega}, \hat{\theta}_N)|^2$$

Use the asymptotic expression (9.62) to show that

$$\operatorname{Var} \hat{\Phi}_v^N(\omega) \sim \frac{2n}{N}\Phi_v^2(\omega), \qquad \omega \neq 0, 2\pi$$

in case $\Phi_{ue}(\omega) \equiv 0$. (Note that the variance of $\hat{\lambda}_N$ does not increase with the model order n; cf. Problem 9E.1.) Compare with (6.75).

9G.3 Suppose that $S \notin \mathcal{M}$ but that S "is close to" $\mathcal{M}$ in the sense

$$\varepsilon(t, \theta^*) = e_0(t) + \rho(t)$$

where $\{e_0(t)\}$ is white noise with variance λ_0 and $E\rho^2(t) = \sigma^2$. Show that the matrix in Theorem 9.1 then can be written

$$P_\theta = \lambda_0[\overline{E}\psi(t, \theta^*)\psi^T(t, \theta^*)]^{-1} + R(\sigma)$$

where

$$||R(\sigma)|| \leq C \cdot \sigma$$

What quantities does the constant C depend on?

9G.4 Consider the correlation estimate $\hat{\theta}_N$ defined by (7.110) for a general, differentiable function $\alpha(\cdot)$. Use the expressions (9.73) and (9.74) to show that, in case $S \in \mathcal{M}$, the asymptotic covariance is given by (9.78), with λ_0 replaced by

$$\frac{E[\alpha(e_0(t))]^2}{[E\alpha'(e_0(t))]^2}$$

9G.5 Show that the estimate $\hat{\lambda}_N$ in (9.19) is asymptotically uncorrelated with $\hat{\theta}_N$ in case $S \in \mathcal{M}$.

Hint: Use the parameterization of (7.87). Note that the minimizing λ is (9.19). Note also that (9.15) is applicable to arbitrary criteria (9.12).

9G.6 Consider the result (9.42). Show that

$$P_\rho \geq \lambda_0[\overline{E}\psi_\rho(t)\psi_\rho^T(t)]^{-1}$$

and that equality is obtained for $F(q) = 1$, i.e., $H(q, \eta^*) = H_0(q)$.

Hint: Use $\sum_{t=1}^N \psi_\rho(t)\psi_\rho^T(t) = \Psi^T\Psi$, for a suitably defined matrix Ψ, and similarly for $\tilde{\Psi}$. Note that $\tilde{\Psi} = \mathbf{F}\Psi$ for a suitably defined matrix $\mathbf{F}$. Then apply Lemma II.2.

9G.7 The result (9.61) can be extended as follows. Let

$$R_n(A) = \frac{1}{2\pi}\int_{-\pi}^{\pi} W_n(e^{i\xi})A(\xi)W_n^T(e^{-i\xi})d\xi$$

with W_n defined by (9.60), and similarly for $R_n(B)$. Then

$$\lim_{n\to\infty}\frac{1}{n}W_n^T(e^{-i\omega})R_n(A)R_n(B)W_n(e^{i\omega}) = A(\omega)B(\omega) \tag{9.93}$$

(see Ljung and Yuan, 1985). Use this result together with (9.55) to prove that (9.63a) still holds if an output error model is applied, despite the fact that a noise model is not estimated. (This shows that asymptotically, as the model order tends to infinity, estimating a correct noise model gives no gain in the accuracy of G. This may be counterintuitive, and is not true for finite n. The reason is, however, that the asymptotic results refer to the case where the estimates at different frequencies are decoupled. Then there is nothing to gain by using information at different frequencies, weighted together according to the true noise model, to form the estimate of G.)

9E.1 Consider the case with a θ-parametrized criterion function

$$\ell(\varepsilon, \theta)$$

Assume that $S \in \mathcal{M}$, and derive an expression for the asymptotic covariance matrix using (9.9) to (9.14). Apply this expression to

$$\begin{bmatrix}\hat{\overline{\theta}}_N \\ \hat{\lambda}_N\end{bmatrix} = \arg\min_{\overline{\theta},\lambda} \frac{1}{N}\sum_{t=1}^{N}\frac{1}{2}\frac{\varepsilon^2(t,\overline{\theta})}{\lambda} + \frac{1}{2}\log\lambda$$

$$\theta = \begin{bmatrix}\overline{\theta} \\ \lambda\end{bmatrix}$$

Assume that $\varepsilon(t,\overline{\theta}_0) = e_0(t)$, where $e_0(t)$ is white noise with $Ee_0^2(t) = \lambda_0$, $Ee_0^3(t) = 0$, and $Ee_0^4(t) = \mu_0$. Show that

$$\operatorname{Cov}\hat{\lambda}_N \sim \frac{1}{N}(\mu_0 - \lambda_0^2) \tag{9.94}$$

[We know from Problem 7E.7 that the minimization gives

$$\hat{\lambda}_N = \frac{1}{N}\sum_{t=1}^{N}\varepsilon^2(t,\hat{\theta}_N)$$

so (9.94) gives the variance of the estimate in (9.19). Compare also (II.73).]

9E.2 Consider a signal given by

$$y(t) - a_0 y(t-1) = e_0(t)$$

where $\{e_0(t)\}$ is white noise with variance λ_0. The k-step-ahead predictor model structure

$$\hat{y}(t|t-k,a) = a^k y(t-k)$$

is used ($\theta = a$). Determine the asymptotic variance of $\hat{a}_N$. Which k minimizes this variance? (Try $k = 1$, $k = 2$ if the general case seems too difficult.)

9E.3 Consider Example 9.1, and assume that the following ARMAX model structure is used:

$$\mathcal{M}: \quad y(t) + ay(t-1) = u(t-1) + e(t) + ce(t-1)$$

$$\theta = [a \quad c]^T$$

Let θ be estimated by a quadratic prediction-error method. Show that

$$\operatorname{Cov} \hat{a}_N \sim \frac{\lambda_0}{N} \frac{1 - a_0^2}{\mu + \lambda_0 a_0^2}$$

What is the variance if instead the PLR method (7.113) is used?

9E.4 Apply the multivariable expression (9.47) to the quadratic criterion (7.27) $\ell(\varepsilon) = \frac{1}{2}\varepsilon^T \Lambda^{-1} \varepsilon$, assuming $\Lambda_0 = \overline{E} e_0(t) e_0^T(t)$. Let $P_\theta(\Lambda)$ denote the resulting covariance matrix, and show that

$$P_\theta(\Lambda) = [\overline{E}\psi\Lambda^{-1}\psi^T]^{-1}[\overline{E}\psi\Lambda^{-1}\Lambda_0\Lambda^{-1}\psi^T][\overline{E}\psi\Lambda^{-1}\psi^T]^{-1}$$

Use Problem 7D.8 to show that

$$P_\theta(\Lambda) \geq P_\theta(\Lambda_0), \qquad \text{for all symmetric positive definite } \Lambda$$

9E.5 Let

$$V_N(\theta, Z^N) = \det \frac{1}{N} \sum_{t=1}^{N} \varepsilon(t, \theta)\varepsilon^T(t, \theta)$$

and define $\hat{\theta}_N = \arg\min V_N(\theta, Z^N)$. Apply the formulas (9.9) and (9.11) to determine the asymptotic covariance of $\hat{\theta}_N$, assuming $S \in \mathcal{M}$ and $E e_0(t) e_0^T(t) = \Lambda_0$. Compare with Problem 9E.4.

9T.1 Suppose that we are using time-varying norms $\ell(\varepsilon, t)$ in the criterion function [as in (7.18); assume no θ-dependence, though]. Show that if $S \in \mathcal{M}$ and $E\ell'_\varepsilon(e_0(t), t) = 0$ for all t, then (9.29) and (9.30) still hold with

$$\kappa(\ell) = \frac{\overline{E}[\ell'_\varepsilon(e_0(t), t)]^2}{[\overline{E}\ell''_{\varepsilon\varepsilon}(e_0(t), t)]^2}$$

$$= \frac{\lim_{N\to\infty} \frac{1}{N} \sum_{t=1}^{N} \int [\ell'_\varepsilon(x, t)]^2 f_e(x, t) dx}{\left[\lim_{N\to\infty} \frac{1}{N} \sum_{t=1}^{N} \int \ell''_{\varepsilon\varepsilon}(x, t) f_e(x, t) dx\right]^2}$$

where $f_e(x, t)$ is the possibly time varying PDF of $e_0(t)$.

9D.1 Let $\{e(t)\}$ be a sequence of independent random variables with zero means, and let $z(t)$ depend on $e(s)$, $s \leq t-1$. Thus $z(t)$ is independent of $e(s)$, $s \geq t$. Show that

$$Ez(t)z(s)e(t)e(s) = Ez(t)z(s) \cdot Ee(t)e(s)$$

9D.2 Suppose that, in Theorem 9.1, the condition $\sqrt{N} D_N \to 0$ does not hold. Define a nonrandom parameter value θ_N^* suitably, and show that the result of Theorem 9.1 then holds when applied to $\sqrt{N}(\hat{\theta}_N - \theta_N^*)$.

APPENDIX 9A: PROOF OF THEOREM 9.1

In this appendix we shall employ a technique for proving asymptotic normality that is quite useful when dealing with signals related to identification experiments. The idea is to split the sum (9.8) into one part that satisfies a certain independence condition (M-dependence) among its terms and one part that is small. When studying these parts, the following two lemmas are instrumental.

Lemma 9.A1. (Orey, 1958; Rosén, 1967). Consider the sum of doubly indexed random variables $x_N(k)$:

$$Z_N = \sum_{t=1}^{N} x_N(t) \tag{9A.1}$$

where

$$E x_N(t) = 0 \tag{9A.2}$$

Suppose that

$$\{x_N(1), \cdots, x_N(s)\} \quad \text{and} \quad \{x_N(t), x_N(t+1), \cdots, x_N(n)\} \tag{9A.3}$$

are independent if $t - s > M$, where M is an integer, that

$$\limsup_{N\to\infty} \sum_{k=1}^{N} E|x_N(k)|^2 < \infty \tag{9A.4}$$

and that

$$\lim_{N\to\infty} \sum_{k=1}^{N} E|x_N(k)|^{2+\delta} = 0 \qquad \text{some } \delta > 0 \tag{9A.5}$$

Let

$$Q = \lim_{N\to\infty} E Z_N Z_N^T \tag{9A.6}$$

Then

$$Z_N \in AsN(0, Q) \tag{9A.7}$$

Proof. See Orey (1958)or Rosén (1967). Here we note that (9A.5) (Lyapunov's condition) implies Lindeberg's condition, which is used in the quoted references. A sequence $\{x_N(k)\}$ subject to (9A.3) is said to be M-dependent. □

Lemma 9A.2 (Diananda, 1953; Anderson, 1959). Let

$$S_N = Z_M(N) + X_M(N), \qquad M, N = 1, 2 \cdots$$

such that

$$E X_M^2(N) \le C_M, \qquad \lim_{M\to\infty} C_M = 0 \tag{9A.8}$$

$$P\{Z_M(N) \le z\} = F_{M,N}(z) \tag{9A.9}$$

$$\lim_{N\to\infty} F_{M,N}(z) = F_M(z) \tag{9A.10}$$

$$\lim_{M\to\infty} F_M(z) = F(z) \tag{9A.11}$$

Then

$$\lim_{N\to\infty} P\{S_N \le z\} = F(z) \tag{9A.12}$$

Proof. See Diananda (1953)or Anderson (1959). □

We now turn to the proof of Theorem 9.1. Write for short $\psi(t) = \psi(t, \theta^*)$, $\varepsilon(t) = \varepsilon(t, \theta^*)$ and let

$$S_N = \frac{1}{\sqrt{N}} \sum_{t=1}^{N} (\psi(t)\varepsilon(t) - E\psi(t)\varepsilon(t)) \tag{9A.13}$$

Then, according to (9.6) and (9.7),

$$-V_N'(\theta^*, Z^N) = \frac{1}{\sqrt{N}} \cdot S_N + D_N \tag{9A.14}$$

From (8.19),

$$\varepsilon(t) = \sum_{k=1}^{\infty} d_t^{(5)}(k) r(t-k) + \sum_{k=0}^{\infty} d_t^{(6)}(k) e_0(t-k)$$

Similarly,

$$\psi(t) = \sum_{k=1}^{\infty} d_t^{(7)}(k) r(t-k) + \sum_{k=0}^{\infty} d_t^{(8)}(k) e_0(t-k)$$

Here

$$|d_t^{(i)}(k)| \le \beta_k, \qquad \text{all } t, i, \quad \text{and} \quad \sum_{1}^{\infty} \beta_k < \infty \tag{9A.15}$$

according to the assumptions of uniform stability. Now let

$$\varepsilon^M(t) = \sum_{k=1}^{\infty} d_t^{(5)}(k) r(t-k) + \sum_{k=0}^{M} d_t^{(6)}(k) e_0(t-k) \tag{9A.16}$$

$$\tilde{\varepsilon}^M(t) = \varepsilon(t) - \varepsilon^M(t) = \sum_{k=M+1}^{\infty} d_t^{(6)}(k) e_0(t-k) \tag{9A.17}$$

Define ψ^M and $\tilde{\psi}^M(t)$ analogously. Now

$$S_N = Z_M(N) + X_M(N) \tag{9A.18}$$

where

$$Z_M(N) = \sum_{t=1}^{N} \left(\frac{1}{\sqrt{N}} \psi^M(t)\varepsilon^M(t) - E\frac{1}{\sqrt{N}} \psi^M(t)\varepsilon^M(t) \right) \tag{9A.19}$$

$$\begin{aligned} X_M(N) = \frac{1}{\sqrt{N}} \sum_{t=1}^{N} \Big\{ & \left(\tilde{\psi}^M(t)\varepsilon(t) + \psi^M(t)\tilde{\varepsilon}^M(t) \right) \\ & - E\left(\tilde{\psi}^M(t)\varepsilon(t) + \psi^M(t)\tilde{\varepsilon}^M(t) \right) \Big\} \end{aligned} \tag{9A.20}$$

We shall apply Lemma 9A.2 to (9A.18). First we show that $Z_M(N)$ is asymptotically normal according to Lemma 9A.1. The terms of $Z_M(N)$ are clearly zero mean and M-dependent by construction. Thus conditions (9A.1) to (9A.3) of Lemma 9A.1 are satisfied. For (9A.4) and (9A.5), we find that

$$\begin{aligned} & E\left| \frac{1}{\sqrt{N}} (\psi^M(t)\varepsilon^M(t) - E\psi^M(t)\varepsilon^M(t)) \right|^{2+\delta} \\ & \leq \frac{1}{N} \cdot N^{-\delta/2} \left(E|\psi^M(t)|^{4+2\delta} \cdot E|\varepsilon^M(t)|^{4+2\delta} \right)^{1/2} \\ & \leq \frac{1}{N} \cdot N^{-\delta/2} \cdot C \end{aligned} \tag{9A.21}$$

where the last inequality follows from the fact that ψ^M and ε^M are finite sums of random variables with bounded $(4 + \delta)$ moments. The expression (9A.21) proves both (9A.4) and (9A.5). With

$$Q_M = \lim_{N \to \infty} E Z_M(N) Z_M^T(N) \tag{9A.22}$$

it thus follows from Lemma 9A.1 that

$$Z_M(N) \in \underset{N \to \infty}{AsN}(0, Q_M) \tag{9A.23}$$

Consider now the term X_M in (9A.20). Let

$$\sqrt{N} X_M^{(1)}(N) = \sum_{t=1}^{M} \left\{ \tilde{\psi}^M(t)\varepsilon(t) - E\tilde{\psi}^M(t)\varepsilon(t) \right\}$$

From the corollary to Lemma 2B.1, we have

$$E|\sqrt{N}X_M^{(1)}(N)|^2 \leq C \cdot C_{\psi M}^2 \cdot C_\varepsilon^2 \cdot N$$

where

$$C_{\tilde{\psi}M} = \sum_{k=M+1}^{\infty} \sup_t |d_t^{(8)}(k)| \leq \sum_{k=M+1}^{\infty} \beta_k$$

and

$$C_\varepsilon = \sum_{k=1}^{\infty} \sup_t |d_t^{(6)}(k)| \leq \sum_{1}^{\infty} \beta_k \leq C$$

A similar bound applies to the other terms in $X_M(N)$. Hence

$$E|X_M(N)|^2 \leq C \cdot \left[\sum_{k=M+1}^{\infty} \beta_k\right]^2$$

which tends to zero as M tends to infinity, since the sum over β_k is convergent. Lemma 9A.2 now tells us that the asymptotic distribution of S_N is given by (9A.23) in the limit $M \to \infty$. Hence

$$\begin{aligned} S_N &\in AsN(0, Q) \\ Q &= \lim_{M\to\infty} Q_M \end{aligned} \tag{9A.24}$$

Since $\sqrt{N}D_N \to 0$, the asymptotic distribution of $\sqrt{N}V_N'(\theta^*, Z^N)$ coincides with that of S_N; that is,

$$\sqrt{N}V_N'(\theta^*, Z^N) \in AsN(0, Q) \tag{9A.25}$$

Also, $\sqrt{N}D_N \to 0$ together with

$$\lim_{M\to\infty} \lim_{N\to\infty} E|X_M(N)|^2 = 0$$

implies that

$$\begin{aligned} Q &= \lim_{M\to\infty} \lim_{N\to\infty} EZ_M(N)Z_M^T(N) \\ &= \lim_{N\to\infty} N \cdot EV_N'(\theta^*, Z^N) \cdot [V_N'(\theta^*, Z^N)]^T \end{aligned} \tag{9A.26}$$

We have now completed the proof of (9.8) and (9.9). It remains only to verify (9.4). We have

$$V_N''(\theta, Z^N) = \frac{1}{N}\sum_{t=1}^{N}[\psi(t,\theta)\psi^T(t,\theta) - \psi'(t,\theta)\varepsilon(t,\theta)] \tag{9A.27}$$

Since the model set and the system are uniformly stable, we find, as in (8.19) and (8.20), that ε, ψ, and ψ' are obtained by uniformly stable filtering of white noise. Hence, as in Lemma 8.2, Theorem 2B.1 implies that

$$\sup_{\theta \in D_{\mathcal{M}}} |V_N''(\theta, Z^N) - \overline{V}''(\theta)| \to 0, \qquad \text{w.p. 1 as } N \to \infty \tag{9A.28}$$

Consequently, since $\hat{\theta}_N \to \theta^*$,

$$V_N''(\xi_N, Z^N) \to \overline{V}''(\theta^*), \qquad \text{w.p. 1 as } N \to \infty \tag{9A.29}$$

if ξ_N belongs to a neighborhood of θ^* with radius $|\hat{\theta}_N - \theta^*|$. [*Technical note*: The application of the Taylor expansion in (9.3) actually may give "different ξ_N" in different rows of this vector expression. The result (9A.29) is, however, not affected by this.]

Now

$$\sqrt{N}(\hat{\theta}_N - \theta^*) = [V_N''(\xi_N, Z^N)]^{-1}\sqrt{N}V_N'(\theta^*, Z^N)$$

and (9A.29) together with (9A.25) complete the proof of Theorem 9.1.

Remark. If $\hat{\theta}_N$ is on the boundary of $D_{\mathcal{M}}$, (9.3) does not apply. However, it follows from (9B.11) that this event has a probability that decays sufficiently fast not to affect the asymptotic distribution.

APPENDIX 9B: THE ASYMPTOTIC PARAMETER VARIANCE

The asymptotic distribution result of Theorem 9.1 does not necessarily imply that

$$\text{Cov}(\sqrt{N}\hat{\theta}_N) = N \cdot E(\hat{\theta}_N - E\hat{\theta}_N)(\hat{\theta}_N - E\hat{\theta}_N)^T \to P_\theta \qquad \text{as } N \to \infty \tag{9B.1}$$

with P_θ given by (9.11) and (9.9), or that the left side of (9B.1) even exists. In this appendix we shall deal with this question.

We introduce the following notation:

$$V_N(\theta, Z^N), \qquad \hat{\theta}_N \text{ and } \theta^* \text{ as in (9.1) to (9.3)}$$

$$\overline{V}_N(\theta) = EV_N(\theta, Z^N)$$

$$\theta_N^* = \arg\min_{\theta \in D_{\mathcal{M}}} \overline{V}_N(\theta)$$

$$\overline{\theta}_N = E\hat{\theta}_N$$

We then have, as in (9.3),

$$\hat{\theta}_N - \theta_N^* = -[V_N''(\xi_N, Z^N)]^{-1}V_N'(\theta_N^*, Z_N) \tag{9B.2}$$

The assumption of Theorem 9.1 that $\overline{V}''(\theta^*) > 0$ together with the continuity of $\overline{V}''(\theta)$ implies that there exists a $\delta > 0$ such that

$$\overline{V}''(\theta) > \delta I, \qquad \forall\theta, \qquad |\theta - \theta^*| < \delta \tag{9B.3}$$

Similarly, the assumption that θ^* is the unique minimizing point implies that there exist $\delta > 0$, $\delta^* > 0$ such that

$$|\theta_1 - \theta^*| < \frac{\delta}{2} \Rightarrow \overline{V}(\theta_1) < \overline{V}(\theta) - \delta^* \qquad \forall\theta, \qquad |\theta - \theta^*| > \delta \tag{9B.4}$$

(The deltas in (9B.3) and (9B.4) are as such unrelated; for convenience they are taken to be the same, without loss of generality.) Introduce the following subsets of the event space:

$$\Omega_1^N = \left\{\omega \,\middle|\, |\hat{\theta}_N - \theta^*| < \frac{\delta}{2}\right\} \tag{9B.5}$$

$$\Omega_2^N = \left\{\omega \,\middle|\, V_N''(\xi_N, Z^N) \geq \frac{\delta}{2} \cdot I \text{ all } \xi_N, \quad |\xi_N - \theta^*| < \frac{\delta}{2}\right\} \tag{9B.6}$$

Let $\overline{\Omega}_i^N$ be the complement events. Clearly, in view of (9B.4),

$$\overline{\Omega}_1^N \subset \Omega_3^N = \left\{\omega \,\middle|\, \sup_\theta \left|V_N(\theta, Z^N) - \overline{V}(\theta)\right| > \frac{\delta^*}{2}\right\} \tag{9B.7}$$

$$\overline{\Omega}_2^N \subset \Omega_4^N = \left\{\omega \,\middle|\, \sup_\theta \left\|V_N''(\theta, Z^N) - \overline{V}''(\theta)\right\| > \frac{\delta}{2}\right\} \tag{9B.8}$$

(ω here is the elementary event variable, of which the random variables Z^N and $\hat{\theta}_N$ are functions.) Let $P(\Omega_1)$ denote the probability of the event Ω_1. Then $P(\Omega_1^N) \to 1$ as $N \to \infty$ since $\hat{\theta}_N \to \theta^*$, w.p.1, and $P(\Omega_2^N) \to 1$ as $N \to \infty$ since (9B.3) holds and $V_N''(\theta, Z^N)$ converges uniformly to $\overline{V}''(\theta)$, w.p.1 [see (9A.28)]. Let us compute bounds for the probabilities. First note the following strengthening of Lemma 2B.2: Under the assumption of Theorem 2B.1, strengthened so that the eighth moments of $\{e(t)\}$ are bounded, we have

$$E(R_r^N)^4 \leq C(N - r)^2 \tag{9B.9}$$

See Ljung (1984), Corollary 2, with $\gamma(n) = 1/n$ for a proof. Applying Chebyshev's inequality to the fourth moments then gives

$$P(\Omega_3^N) \leq \frac{16}{(\delta^*)^4} \cdot \frac{C}{N^2} \tag{9B.10a}$$

and

$$P(\Omega_4^N) \leq \frac{16}{(\delta)^4} \cdot \frac{C}{N^2} \tag{9B.10b}$$

Let

$$\Omega^N = \Omega_1^N \cap \Omega_2^N$$

Then, for the complement event, we have from (9B.10)

$$P(\overline{\Omega}^N) \leq \frac{C}{N^2} \tag{9B.11}$$

Now consider for (9B.2)

$$E|\sqrt{N}(\hat{\theta}_N - \theta_N^*)|^4 \leq E\left\{\left\|\left[V_N''(\xi_N, Z^N)\right]^{-1}\right\|^4 \cdot |\sqrt{N}V_N'(\theta_N^*, Z_N)|^4\right\}$$

The right side is an integration over ω. Splitting the integration into the subsets Ω^N and $\overline{\Omega}^N$, we find that on Ω^N

$$V_N''(\xi_N, Z^N) \geq \frac{\delta}{2}I$$

(this is indeed the rationale behind defining the set Ω^N). Hence, with symbolic notation,

$$\begin{aligned} E|\sqrt{N}(\hat{\theta}_N - \theta_N^*)|^4 &\leq \int_{\Omega^N} \frac{16}{\delta^4} \cdot |\sqrt{N}V_N'(\theta_N^*, Z^N)|^4 d\omega \\ &\quad + \int_{\overline{\Omega}^N} N^2|\hat{\theta}_N - \theta_N^*|^4 d\omega \\ &\leq C \cdot E|\sqrt{N}V_N'(\theta_N^*, Z^N)|^4 + N^2 \cdot C \cdot P(\overline{\Omega}_N) \leq C \end{aligned}$$

The second inequality follows since $\hat{\theta}_N$ and θ_N^* belong to a bounded set $D_{\mathcal{M}}$. The last inequality follows from (9B.11) and from Lemma 2B.2 applied in its strengthened version (9B.9) to the sum of zero mean variables $N \cdot V_N'(\theta_N^*, Z^N)$.

Theorem 4.5.2 of Chung (1974)now implies that

$$N \cdot E(\hat{\theta}_N - \theta_N^*)(\hat{\theta}_N - \theta_N^*)^T \to P_\theta, \qquad \text{as } N \to \infty \tag{9B.12}$$

with P_θ being the variance of the asymptotic distribution.

Finally, rewrite (9B.2) as

$$V_N'(\theta_N^*, Z^N) = -\overline{V}''(\theta_N^*)(\hat{\theta}_N - \theta_N^*) - [V_N''(\xi_N, Z^N) - \overline{V}''(\theta_N^*)](\hat{\theta}_N - \theta_N^*)$$

Taking expectation gives

$$0 = \overline{V}''(\theta_N^*)(\overline{\theta}_N - \theta_N^*) + E\left\{[V_N''(\xi_N, Z^N) - \overline{V}''(\theta_N^*)] \cdot (\hat{\theta}_N - \theta_N^*)\right\}$$

or, using Schwartz's inequality,

$$\begin{aligned} |\overline{\theta}_N - \theta_N^*| \leq \left\|[\overline{V}''(\theta_N^*)]^{-1}\right\| &\cdot [E|V_N''(\xi_N, Z^N) - \overline{V}''(\xi_N)|^2 \\ &+ E|\overline{V}''(\xi_N) - \overline{V}''(\theta_N^*)|^2]^{1/2} \cdot [E|\hat{\theta}_N - \theta_N^*|^2]^{1/2} \end{aligned}$$

For the middle factor, we apply Lemma 2B.2 to the first term, showing it decays like C/N, and we use (assuming $\overline{V}''$ to be differentiable)

$$|\overline{V}''(\xi_N) - \overline{V}''(\theta_N^*)| \leq C|\xi_N - \theta_N^*| \leq C|\hat{\theta}_N - \theta_N^*|$$

for the second term, showing that it decays like $E|\hat{\theta}_N - \theta_N^*|^2$ [i.e., C/N according to (9B.12)]. Collecting this gives

$$E|\hat{\theta}_N - \theta_N^*|^2 \leq \frac{C}{N} \tag{9B.13}$$

Clearly, (9B.12) and (9B.13) imply (9B.1). We can thus sum up the discussion in this appendix as follows:

> *Consider the estimate $\hat{\theta}_N$ under the conditions of Theorem 9.1, strengthened so that the eighth moments of $\{e_0(t)\}$ in (8.4) are bounded, and that $\overline{V}(\theta)$ is three times continuously differentiable. Dispense with the assumption that $\sqrt{N}D_N \to 0$ as $N \to \infty$. Then (9B.1) and (9B.13) hold.*

Remark. The assumption in Theorem 9.1 that $\sqrt{N}D_N$ tends to zero implies that $\theta_N^* \to \theta^*$ sufficiently fast so as to allow θ^* to be used in the asymptotic expressions instead of θ_N^*.

10

COMPUTING THE ESTIMATE

In Chapter 7 we gave three basic procedures for parameter estimation:

1. The prediction-error approach in which a certain function $V_N(\theta, Z^N)$ is minimized with respect to θ.
2. The correlation approach, in which a certain equation $f_N(\theta, Z^N) = 0$ is solved for θ.
3. The subspace approach to estimating state space models.

In this chapter we shall discuss how these problems are best solved numerically.

At time N, when the data set Z^N is known, the functions V_N and f_N are just ordinary functions of a finite-dimensional real parameter vector θ. Solving the problems therefore amounts to standard questions of nonlinear programming and numerical analysis. Nevertheless, it is worthwhile to consider the problems in our parameter estimation setting, since this adds a certain amount of structure to the functions involved.

The subspace methods (7.66) can be implemented in several different ways, and contain many options. In Section 10.6 we give the details of this, together with an independent derivation of the techniques.

10.1 LINEAR REGRESSIONS AND LEAST SQUARES

The Normal Equations

For linear regressions, the prediction is given as

$$\hat{y}(t|\theta) = \varphi^T(t)\theta \tag{10.1}$$

The prediction-error approach with a quadratic norm applied to (10.1) gives the least-squares method described in Section 7.3. The minimizing element $\hat{\theta}_N^{\mathrm{LS}}$ can then be written as in (7.34):

$$\hat{\theta}_N^{\mathrm{LS}} = R^{-1}(N) f(N) \tag{10.2}$$

with

$$R(N) = \frac{1}{N} \sum_{t=1}^{N} \varphi(t)\varphi^T(t) \tag{10.3}$$

$$f(N) = \frac{1}{N} \sum_{t=1}^{N} \varphi(t) y(t) \tag{10.4}$$

An alternative is to view $\hat{\theta}_N^{\mathrm{LS}}$ as the solution of

$$R(N)\hat{\theta}_N^{\mathrm{LS}} = f(N) \tag{10.5}$$

These equations are known as the *normal equations.* Note that the basic equation (7.118) for the IV method is quite analogous to (10.5), and most of what is said in this section about the LS method also applies to the IV method with obvious adaptation.

The coefficient matrix $R(N)$ in (10.5) may be ill conditioned, in particular if its dimension is high. There exist methods to find $\hat{\theta}_N$ that are much better numerically behaved, which do not have the normal equations as a starting point. This has been discussed in an extensive literature on numerical studies of linear least-squares problems. Lawson and Hanson (1974)can be mentioned as a basic reference for the problems under consideration. The underlying idea in these methods is that the matrix $R(N)$ should not be formed, since it contains products of the original data. Instead, a matrix R is constructed with the property

$$R R^T = R(N)$$

Therefore, this class of methods is commonly known as "square-root algorithms" in the engineering literature. The term is not quite adequate, since no square roots are ever taken. It would be more appropriate to use the term "quadratic methods" when solving (10.5).

Solving for the LS Estimate by QR Factorization

There are some different approaches to the construction of R, such as Householder transformations, Householder (1964), the Gram-Schmidt procedure, Björck (1967), and Cholesky decomposition. We shall here describe an efficient way using QR-factorizations. See, e.g., Golub and van Loan (1996)for a thorough description of the method. The QR-factorization of an $n \times d$ matrix A is defined as

$$A = QR, \quad QQ^T = I, \qquad R \text{ upper triangular} \tag{10.6}$$

Here Q is $n \times n$ and R is $n \times d$.

To apply this to the LS parameter estimation case, we rewrite the general multivariable case (4.59) in matrix terms, by introducing

$$\begin{aligned} \mathbf{Y}^T &= \begin{bmatrix} y^T(1) & \dots & y^T(N) \end{bmatrix}, \quad \mathbf{Y} \text{ is } Np \times 1 \\ \Phi^T &= \begin{bmatrix} \varphi(1) & \dots & \varphi(N) \end{bmatrix}, \quad \Phi \text{ is } Np \times d \end{aligned} \tag{10.7}$$

(Here $p = \dim y$). Then the LS criterion can be written (cf. (II.13))

$$V_N(\theta, Z^N) = |\mathbf{Y} - \Phi\theta|^2 = \sum_{t=1}^{N} \left| y(t) - \varphi^T(t)\theta \right|^2 \tag{10.8}$$

The norm is obviously not affected by any orthonormal transformation applied to the vector $\mathbf{Y} - \Phi\theta$. Therefore if Q $(pN \times pN)$ is orthonormal, that is $QQ^T = I$, then

$$V_N(\theta, Z^N) = |Q(\mathbf{Y} - \Phi\theta)|^2$$

Now, introduce the QR-factorization

$$[\Phi \quad \mathbf{Y}] = QR, \quad R = \begin{bmatrix} R_0 \\ \dots \\ 0 \end{bmatrix} \tag{10.9}$$

Here R_0 is an upper triangular $(d + 1) \times (d + 1)$ matrix, which we decompose as

$$R_0 = \begin{bmatrix} R_1 & R_2 \\ 0 & R_3 \end{bmatrix}, \quad R_1 \text{ is } d \times d, \;\; R_2 \text{ is } d \times 1, \;\; R_3 \text{ is scalar} \tag{10.10}$$

This means that

$$V_N(\theta, Z^N) = \left|Q^T(\mathbf{Y} - \Phi\theta)\right|^2 = \left\| \begin{bmatrix} R_2 \\ R_3 \end{bmatrix} - \begin{bmatrix} R_1\theta \\ 0 \end{bmatrix} \right\|^2 = |R_2 - R_1\theta|^2 + |R_3|^2$$

which clearly is minimized for

$$R_1\hat{\theta}_N = R_2, \quad \text{giving} \quad V_N(\hat{\theta}_N, Z^N) = |R_3|^2 \tag{10.11}$$

There are three important advantages with this way of solving for the LS estimate:

1. With $R(N)$ as in (10.3), $R(N) = \Phi^T\Phi = R_1^T R_1$ so the conditioning number (the ratio between the largest and smallest singular value) of R_1 is the square root of that of $R(N)$. Therefore (10.11) is much better conditioned than its counterpart (10.5).
2. R_1 is a triangular matrix, so the equation is easy to solve.
3. If the QR-factorization is performed for a regressor size d^*, then the solutions and loss functions for all models with fewer parameters—obtained by setting trailing parameters in θ to zero—are easily obtained from R_0.

Note that the big matrix Q is never required to find $\hat{\theta}_N$ and the loss function. All information is contained in the "small" matrix R_0. In MATLAB considerable computational saving is obtained by taking **R=triu(qr(A))** to compute (10.6) if Q is of no interest, and A has many more rows than columns.

Initial Conditions: "Windowed" Data

A typical structure of the regression vector $\varphi(t)$ is that it consists of shifted data (possibly after some trivial reordering):

$$\varphi(t) = \begin{bmatrix} z(t-1) \\ \vdots \\ z(t-n) \end{bmatrix} \tag{10.12}$$

Here $z(t-1)$ is an r-dimensional vector. For example, the ARX model (4.11) with $n_a = n_b = n$ gives (10.12) with

$$z(t) = \begin{bmatrix} -y(t) \\ u(t) \end{bmatrix}$$

while an AR-model for a p-dimensional process $\{y(t)\}$ [cf. (4.57), $n_b = 0$] obeys (10.12) with $z(t) = -y(t)$.

With the structure (10.12), the matrix $R(N)$ in (10.3) will be an $n \times n$ block matrix, whose ij block is the $r \times r$ matrix

$$R_{ij}(N) = \frac{1}{N} \sum_{t=1}^{N} z(t-i) z^T(t-j) \tag{10.13}$$

If we have knowledge only of $z(t)$ for $1 \le t \le N$, the question arises of how to deal with the unknown initial conditions for $t \le 0$ in (10.13). Two approaches can be taken:

1. Start the summation in (10.3) and (10.4) at $t = n+1$ rather than at $t = 1$. Then all sums (10.13) will involve only known data. [After a suitable redefinition of N and the time origin, we can of course stick to our usual expressions, assuming $z(t)$ to be known for $t \ge -n$.]

2. Replace the unknown initial values by zeros ("prewindowing"). For symmetry, the trailing values $z(t), t = N+1, \ldots, N+n$, could also be replaced by zeros ("postwindowing") and the summation in (10.3) is extended to $N+n$. In this case (10.5) are also known as the Yule-Walker equations. Often additional data windows ("tapering"; cf. Problem 6G.5) are applied to both ends of the data record to soften the effects of the appended zeros.

In speech processing these approaches are known as the *covariance method* and *autocorrelation method*, respectively (Makhoul and Wolf, 1972). No doubt, from a logical point of view, approach 1 seems the most natural. Approach 2, however, gives the special feature that the blocks (10.13) will depend only on the difference between the indexes:

$$R_{ij}(N) = R_\tau(N), \qquad \tau = i - j$$

$$R_\tau(N) = \frac{1}{N}\sum_{t=\tau}^{N} z(t-\tau)z^T(t) \qquad \tau \geq 0, \text{ analogously for } \tau < 0 \tag{10.14}$$

This makes $R(N)$ a *block Toeplitz matrix*, which gives distinct advantages when solving (10.5), as we shall demonstrate shortly.

Clearly, when $N \gg n$, the difference between the two approaches becomes insignificant.

Levinson Algorithm (*)

The shift structure (10.12) gives a specific structure for the matrix $R(N)$. There is an extensive literature on *fast algorithms* that utilize such structures. The simplest, but generic, example of these methods is the *Levinson algorithm* (Levinson, 1947), which we shall now derive.

Consider the case of an AR model of a signal

$$\hat{y}^n(t|\theta) = -a_1^n y(t-1) - \cdots - a_n^n y(t-n) \tag{10.15}$$

(the upper index n indicates that we are fitting an nth-order model). This corresponds to a linear regression with $\varphi(t)$ subject to (10.12) with $z(t) = -y(t)$. If we apply the autocorrelation method, we should solve (10.5); that is,

$$\begin{bmatrix} R_0 & R_1 \ldots & R_{n-1} \\ R_1 & R_0 \ldots & R_{n-2} \\ \vdots & \vdots \quad \vdots & \\ R_{n-1} & R_{n-2} \ldots R_0 & \end{bmatrix} \begin{bmatrix} a_1^n \\ a_2^n \\ \vdots \\ a_n^n \end{bmatrix} = \begin{bmatrix} -R_1 \\ -R_2 \\ \vdots \\ -R_n \end{bmatrix} \tag{10.16}$$

for a_i^n. Here

$$R_\tau = \hat{R}_y^N(\tau) = \frac{1}{N}\sum_{t=\tau}^{N} y(t-\tau)y(t), \qquad \tau \geq 0 \tag{10.17}$$

and we have dropped the argument N. Equation (10.16) can be rewritten as

$$\begin{bmatrix} R_0 & R_1 \dots & R_n \\ R_1 & R_0 \dots & R_{n-1} \\ R_2 & R_1 \dots & R_{n-2} \\ \vdots & \vdots & \vdots \\ R_n & R_{n-1} \dots & R_0 \end{bmatrix} \begin{bmatrix} 1 \\ a_1^n \\ a_2^n \\ \vdots \\ a_n^n \end{bmatrix} = \begin{bmatrix} V_n \\ 0 \\ 0 \\ \vdots \\ 0 \end{bmatrix} \tag{10.18}$$

Here the n last rows are identical to (10.16), while the first row is a definition of V_n.

Suppose now that we have solved (10.18) for a_i^n and seek a solution for a higher-order model (10.15) with order $n+1$. The estimates a_i^{n+1} will then be defined analogously to (10.18). To find these, we first note that

$$\begin{bmatrix} R_0 & R_1 \dots & R_n & R_{n+1} \\ R_1 & R_0 \dots & R_{n-1} & R_n \\ \vdots & \vdots & \vdots & \vdots \\ R_n & R_{n-1} \dots & R_0 & R_1 \\ R_{n+1} & R_n \dots & R_1 & R_0 \end{bmatrix} \begin{bmatrix} 1 \\ a_1^n \\ \vdots \\ a_n^n \\ 0 \end{bmatrix} = \begin{bmatrix} V_n \\ 0 \\ \vdots \\ 0 \\ \alpha_n \end{bmatrix} \tag{10.19}$$

Here the first $n+1$ rows are identical to (10.18), while the last row is a definition of α_n. The definition of $\hat{a}_i^{n+1}$ looks quite like (10.19), the only difference being that all but the first row of the right side should be zero. We thus seek to remove α_n. A moment's reflection on (10.19) shows that it can also be written as

$$\begin{bmatrix} R_0 & R_1 \dots & R_n & R_{n+1} \\ R_1 & R_0 \dots & R_{n-1} & R_n \\ \vdots & \vdots & \vdots & \vdots \\ R_n & R_{n-1} \dots & R_0 & R_1 \\ R_{n+1} & R_n \dots & R_1 & R_0 \end{bmatrix} \begin{bmatrix} 0 \\ a_n^n \\ \vdots \\ a_1^n \\ 1 \end{bmatrix} = \begin{bmatrix} \alpha_n \\ 0 \\ \vdots \\ 0 \\ V_n \end{bmatrix} \tag{10.20}$$

since the coefficient matrix is a symmetric Toeplitz matrix. We can also view the last $n+1$ rows of (10.20) as the normal equations for the regression

$$\check{y}(t-n-1|\theta) = -a_1^n y(t-n) - a_2^n y(t-n+1) - \dots - a_n^n y(t-1) \tag{10.21}$$

This is a reversed time model for the signal $y(t)$. Since the second order properties of a scalar stationary signal are symmetric with respect to the direction of time, the coefficients in (10.21) coincide with those of (10.15). This is a signal theoretic reason for the equality between (10.19) and (10.20). See also the Remark at the end of this subsection.

Now multiply (10.20) by $\rho_n = -\alpha_n / V_n$ and add it to (10.19). This gives

$$\begin{bmatrix} R_0 & R_1 \dots & R_n & R_{n+1} \\ R_1 & R_0 \dots & R_{n-1} & R_n \\ \vdots & \vdots & \vdots & \vdots \\ R_n & R_{n-1} \dots & R_0 & R_1 \\ R_{n+1} & R_n \dots & R_1 & R_0 \end{bmatrix} \begin{bmatrix} 1 \\ a_1^n + \rho_n a_n^n \\ \vdots \\ a_n^n + \rho_n a_1^1 \\ \rho_n \end{bmatrix} = \begin{bmatrix} V_n + \rho_n \cdot \alpha_n \\ 0 \\ \vdots \\ 0 \\ 0 \end{bmatrix} \tag{10.22}$$

This is the defining relationship for $\hat{a}_i^{n+1}$. Hence

$$\begin{aligned} \hat{a}_k^{n+1} &= \hat{a}_k^n + \hat{\rho}_n \hat{a}_{n-k+1}^n, \quad k = 1, \dots, n \\ \hat{a}_{n+1}^{n+1} &= \hat{\rho}_n \\ V_{n+1} &= V_n + \hat{\rho}_n \alpha_n \\ \hat{\rho}_n &= \frac{-\alpha_n}{V_n} \\ \alpha_n &= R_{n+1} + \sum_{k=1}^{n} \hat{a}_k^n R_{n+1-k} \end{aligned} \tag{10.23}$$

(The hat here indicates the actual estimate, based on N data, as opposed to the general model parameters a_k^n.) This expression allows us to easily compute $\hat{a}_k^{n+1}$ from $\hat{a}_k^n$. With the initial conditions

$$\begin{aligned} V_1 &= R_0 - \frac{R_1^2}{R_0} \\ \hat{a}_1^1 &= \frac{-R_1}{R_0} \end{aligned} \tag{10.24}$$

we have a scheme for computing estimates of arbitrary orders. We note that going from $\hat{a}_k^n$ to $\hat{a}_k^{n+1}$ in (10.23) requires $4n + 2$ additions and multiplications and one division. The computation of $\hat{a}_k^n$ thus requires proportional to $2n^2$ operations, which is of an order of magnitude (in n) less than the general procedures (10.8) to (10.11). Hence the term "fast algorithms."

The Levinson algorithm (10.23) has been widely applied and extended to the case of vector-valued z, as well as to "the covariance method." See, for example, Whittle (1963), Wiggins and Robinson (1965), and Morf et.al. (1977).

Remark. The most important change when dealing with vector-valued z is that the corresponding reversed-time model (10.21)

$$\check{z}(t-n-1|\theta) = -b_1^n z(t-n) - \cdots - b_n^n z(t-1) \tag{10.25}$$

gives b_i estimates that differ from a_i. The scheme (10.23) must then be complemented with an analogous scheme for updating the b_i. See Problem 10G.1.

Lattice Filters (∗)

Consider the predictors (10.15) for orders n and $n+1$ evaluated at

$$\theta = \hat{\theta} = \hat{\theta}_N$$

$$\begin{aligned} \hat{y}_n(t|\hat{\theta}^n) &= -\hat{a}_1^n y(t-1) - \cdots - \hat{a}_n^n y(t-n) \\ \hat{y}_{n+1}(t|\hat{\theta}^{n+1}) &= -\hat{a}_1^{n+1} y(t-1) - \cdots - \hat{a}_n^{n+1} y(t-n) \\ &\quad - \hat{a}_{n+1}^{n+1} y(t-n-1) \end{aligned} \tag{10.26}$$

Subtracting these expressions from each other gives, using (10.23),

$$\hat{y}_{n+1}(t|\hat{\theta}^{n+1}) = \hat{y}_n(t|\hat{\theta}^n) - \hat{\rho}_n \hat{r}_n(t-1) \tag{10.27}$$

where

$$\hat{r}_n(t-1) = y(t-n-1) + \hat{a}_1^n y(t-n) + \cdots + \hat{a}_n^n y(t-1) \tag{10.28}$$

We recognize in (10.28) the error in the reversed-time predictor (10.21). Let us, with the definition (10.28), consider

$$\hat{r}_{n+1}(t) = y(t-n-1) + \hat{a}_1^{n+1} y(t-n) + \cdots + \hat{a}_n^{n+1} y(t-1) + \hat{a}_{n+1}^{n+1} y(t)$$

Subtracting (10.28) from this gives, again using (10.23),

$$\hat{r}_{n+1}(t) - \hat{r}_n(t-1) = \hat{\rho}_n \left[y(t) + \hat{a}_1^n y(t-1) + \cdots + \hat{a}_n^n y(t-n) \right] \tag{10.29}$$

With the prediction error

$$\hat{e}_n(t) = y(t) - \hat{y}_n(t|\hat{\theta}^n)$$

the expressions (10.27) to (10.29) can be summarized as

$$\hat{e}_{n+1}(t) = \hat{e}_n(t) + \hat{\rho}_n \hat{r}_n(t-1) \tag{10.30a}$$

$$\hat{r}_{n+1}(t) = \hat{r}_n(t-1) + \hat{\rho}_n \hat{e}_n(t) \tag{10.30b}$$

$$\hat{e}_0(t) = \hat{r}_0(t) = y(t) \tag{10.30c}$$

This simple representation of the prediction errors [and the predictions $\hat{y}_n(t|\hat{\theta}^n) = y(t)-\hat{e}_n(t)$] can be graphically represented as in Figure 10.1. Because of the structure of this representation, (10.30) is usually called a *lattice filter* (sometimes a *ladder filter*).

An important feature of the representation (10.30) is that the variables $\hat{e}_n$ and $\hat{r}_n$ obey the following orthogonality relationships:

$$\frac{1}{N}\sum_{t=1}^{N}\hat{e}_n(t)\hat{e}_{n-k}(t-k) = \begin{cases} V_n, & \text{if } k = 0 \\ 0, & \text{if } k \neq 0 \end{cases}$$

$$\frac{1}{N}\sum_{t=1}^{N}\hat{r}_n(t)\hat{r}_k(t) = \begin{cases} V_n, & \text{if } n = k \\ 0, & \text{if } n \neq k \end{cases} \tag{10.31}$$

$$\frac{1}{N}\sum_{t=1}^{N}\hat{e}_n(t)\hat{r}_k(t-1) = \begin{cases} \alpha_n, & \text{if } n = k \\ 0, & \text{if } n > k \end{cases}$$

(see Problem 10D.1). Hence the *reflection coefficients* $\hat{\rho}_n$ can easily be computed as

$$\hat{\rho}_n = \frac{-\sum_{t=1}^{N}\hat{e}_n(t)\hat{r}_n(t-1)}{\sum_{t=1}^{N}\hat{e}_n^2(t)} \tag{10.32}$$

The scheme (10.30) together with (10.32) also forms an efficient way of estimating the reflection coefficients $\hat{\rho}_n$, as well as the predictions, as an alternative to the Levinson algorithm. An important aspect is that the scheme produces all lower-order predictors as a by-product. Lattice filters have been used extensively in signal-processing applications. See, for example, Makhoul (1977), Griffiths (1977), and Lee, Morf, and Friedlander (1981). See also Section 11.7 for recursive versions.

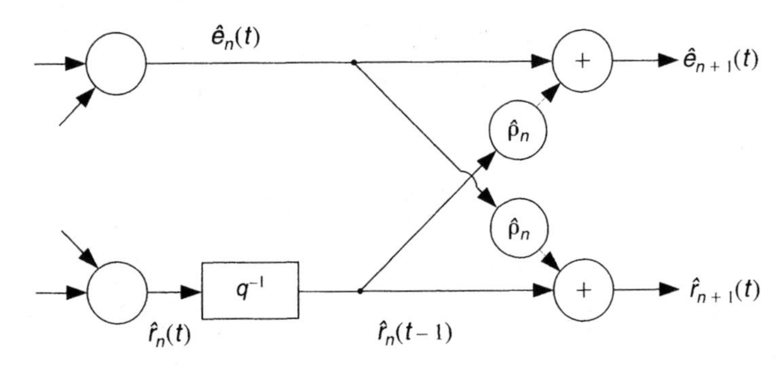

Figure 10.1 A lattice filter representation.

10.2 NUMERICAL SOLUTION BY ITERATIVE SEARCH METHODS

In general, the function

$$V_N(\theta, Z^N) = \frac{1}{N}\sum_{t=1}^{N} \ell\left(\varepsilon(t,\theta),\theta\right) \tag{10.33}$$

cannot be minimized by analytical methods. Neither can the equation

$$0 = f_N(\theta, Z^N) = \frac{1}{N}\sum_{t=1}^{N} \zeta(t,\theta)\alpha(\varepsilon(t,\theta)) \tag{10.34}$$

be solved by direct means in general. The solution then has to be found by iterative, numerical techniques. There is an extensive literature on such numerical problems. See, for example, Luenberger (1973), Bertsekas (1982), or Dennis and Schnabel (1983)for general treatments.

Numerical Minimization

Methods for numerical minimization of a function $V(\theta)$ update the estimate of the minimizing point iteratively. This is usually done according to

$$\hat{\theta}^{(i+1)} = \hat{\theta}^{(i)} + \alpha f^{(i)} \tag{10.35}$$

where $f^{(i)}$ is a search direction based on information about $V(\theta)$ acquired at previous iterations, and α is a positive constant determined so that an appropriate decrease in the value of $V(\theta)$ is obtained. Depending on the information supplied by the user to determine $f^{(i)}$, numerical minimization methods can be divided into three groups:

1. Methods using function values only.
2. Methods using values of the function V as well as of its gradient.
3. Methods using values of the function, of its gradient, and of its Hessian (the second derivative matrix).

The typical member of group 3 corresponds to *Newton algorithms*, where the correction in (10.35) is chosen in the "Newton" direction:

$$f^{(i)} = -\left[V''(\hat{\theta}^{(i)})\right]^{-1} V'(\hat{\theta}^{(i)}) \tag{10.36}$$

The most important subclass of group 2 consists of *quasi-Newton methods*, which somehow form an estimate of the Hessian and then use (10.36). Algorithms of group 1 either form gradient estimates by difference approximations and proceed as quasi-Newton methods or have other specific search patterns. See Powell (1964).

Many standard programs implementing these ideas are available. The easiest way for the identification user could be to supply such a program with necessary information and leave the search for the minimum to the program. In any case, it will be necessary to compute the function values of (10.33) for any required value of θ. The major burden for this lies in the calculation of the sequence of prediction errors $\varepsilon(t, \theta)$, $t = 1, \ldots, N$. This itself could be a simple or complicated task. Compare, for example, the model structures of Section 4.2 with the one in Example 4.1.

The gradient of (10.33) is

$$V_N'(\theta, Z^N) = -\frac{1}{N}\sum_{t=1}^{N}\left\{\psi(t,\theta)\ell_\varepsilon'(\varepsilon(t,\theta),\theta) - \ell_\theta'(\varepsilon(t,\theta),\theta)\right\} \tag{10.37}$$

Here, as usual, $\psi(t, \theta)$ is the $d \times p$ gradient matrix of $\hat{y}(t|\theta)$ ($p = \dim y$) with respect to θ. The major computational burden in (10.37) lies in the calculation of the sequence $\psi(t, \theta)$, $t = 1, 2, \ldots, N$. We discuss in Section 10.3 how this gradient is computed for some common model structures. However, for some models, direct calculation of ψ could be forbidding, and then one has to resort to the minimization methods of the group 1 or to form estimates of ψ by difference approximations.

Some Explicit Search Schemes

Consider the special case of scalar output and quadratic criterion

$$V_N(\theta, Z^N) = \frac{1}{N}\sum_{t=1}^{N}\tfrac{1}{2}\varepsilon^2(t,\theta) \tag{10.38}$$

This problem is known as "the nonlinear least-squares problem" in numerical analysis. An excellent and authoritative account of this problem is given in Chapter 10 of Dennis and Schnabel (1983). The criterion (10.38) has the gradient

$$V_N'(\theta, Z^N) = -\frac{1}{N}\sum_{t=1}^{N}\psi(t,\theta)\varepsilon(t,\theta) \tag{10.39}$$

A general family of search routines is then given by

$$\hat{\theta}_N^{(i+1)} = \hat{\theta}_N^{(i)} - \mu_N^{(i)}[R_N^{(i)}]^{-1}V_N'(\hat{\theta}_N^{(i)}, Z^N) \tag{10.40}$$

where $\hat{\theta}_N^{(i)}$ denotes the ith iterate. $R_N^{(i)}$ is a $d \times d$ matrix that modifies the search direction (it will be discussed later), and the step size $\mu_N^{(i)}$ is chosen so that

$$V_N(\hat{\theta}_N^{(i+1)}, Z^N) < V_N(\hat{\theta}_N^{(i)}, Z^N) \tag{10.41}$$

We should also keep in mind that the minimization problem normally is a constrained one: $\theta \in D_{\mathcal{M}}$. Often, however, $\partial D_{\mathcal{M}}$, the boundary of $D_{\mathcal{M}}$, corresponds to the stability boundary of the predictor (cf. Definition 4.1) so that $V_N(\theta, Z^N)$ increases rapidly as θ approaches $\partial D_{\mathcal{M}}$. Then the constraint can easily be obeyed by proper selection of μ.

The simplest choice of $R_N^{(i)}$ is to take it as the identity matrix,

$$R_N^{(i)} = I \tag{10.42}$$

which makes (10.40) the *gradient* or steepest-descent method. This method is fairly inefficient close to the minimum. Newton methods typically perform much better there. For (10.38), the Hessian is

$$V_N''(\theta, Z^N) = \frac{1}{N}\sum_{t=1}^{N} \psi(t,\theta)\psi^T(t,\theta) - \frac{1}{N}\sum_{t=1}^{N} \psi'(t,\theta)\varepsilon(t,\theta) \tag{10.43}$$

where $\psi'(t,\theta)$ is the $d \times d$ Hessian of $\varepsilon(t,\theta)$.

Choosing

$$R_N^{(i)} = V_N''(\hat{\theta}_N^{(i)}, Z^N) \tag{10.44}$$

makes (10.40) a Newton method. It may, however, be quite costly to compute all the terms of ψ'. Suppose now that there is a value θ_0 such that the prediction errors $\varepsilon(t,\theta_0) = e_0(t)$ are independent. Then this value yields the global minimum of $EV_N(\theta, Z^N)$. Close to θ_0 the second sum of (10.43) will then be close to zero since $E\psi'(t,\theta_0)e_0(t) = 0$. We thus have

$$V_N''(\theta, Z^N) \approx \frac{1}{N}\sum_{1}^{N} \psi(t,\theta)\psi^T(t,\theta) \triangleq H_N(\theta) \tag{10.45}$$

If we apply a Newton method to the minimization problem, we need a good estimate of the Hessian only in the vicinity of the minimum. The reason for this is that Newton methods are designed to give one-step convergence for quadratic functions. When the function values between the current iterate and the minimum cannot be approximated very well by a quadratic function, the effect of the Hessian in (10.36) is not so important. Moreover, by omitting the last sum in (10.43) the estimate of the Hessian is always assured to be positive semidefinite. This makes the numerical procedure a descent algorithm and guarantees convergence to a stationary point. The conclusion is consequently that

$$R_N^{(i)} = H_N(\hat{\theta}_N^{(i)}) \tag{10.46}$$

is a quite suitable choice for our problem. This is also known as the *Gauss-Newton method.* In the statistical literature the technique is called "the method of scoring," (Rao, 1973). In the control literature the terms modified Newton-Raphson and quasi-linearization have also been used. Dennis and Schnabel (1983)reserve the term Gauss-Newton for (10.40) and (10.46) with the particular choice $\mu_N^{(i)} = 1$, and suggest the term *damped Gauss-Newton* when an adjusted step size μ is applied.

Even though the expression (10.45) is assured to be positive semidefinite, it may be singular or close to singular. This is the case, for example, if the model is over-parametrized or the data not informative enough. Then some numerical problems arise in (10.40). Various ways to overcome this problem exist and are known as *regularization* techniques. One common way is the *Levenberg-Marquardt* procedure (Levenberg, 1944; Marquardt, 1963). Then an approximation

$$R_N^{(i)}(\lambda) = \frac{1}{N}\sum_{t=1}^{N} \psi(t, \hat{\theta}_N^{(i)})\psi^T(t, \hat{\theta}_N^{(i)}) + \lambda I \tag{10.47}$$

is used for the Hessian. Here λ is a positive scalar that is used to control the convergence in the iterative scheme rather than the step size parameter. We thus have $\mu_N^{(i)} = 1$ in (10.40). With $\lambda = 0$ we have the Gauss-Newton case. Increasing λ means that the step size is decreased and the search direction is turned towards the gradient. Several schemes for manipulating λ based on the test (10.41) have been suggested, see, e.g., Scales (1985).

If the minimum does not give independent prediction errors, the second sum of (10.43) then need not be negligible close to the minimum, and (10.45) need not be a good approximation of the Hessian. A typical method then is to make use of the known first sum of the Hessian and estimate the second sum with some secant technique (see Dennis, Gay, and Welsch, 1981).

Correlation Equation

Solving the equation (10.34) is quite analogous to the minimization of (10.33) (see, e.g., Dennis and Schnabel, 1983). Standard numerical procedures are the *substitution method* (corresponding to (10.40) and (10.42)):

$$\hat{\theta}_N^{(i)} = \hat{\theta}_N^{(i-1)} - \mu_N^{(i)} f_N(\hat{\theta}_N^{(i-1)}, Z^N) \tag{10.48}$$

and the *Newton-Raphson method* [corresponding to (10.40) and (10.44)]:

$$\hat{\theta}_N^{(i)} = \hat{\theta}_N^{(i-1)} - \mu_N^{(i)} \left[f_N'(\hat{\theta}_N^{(i-1)}, Z^N) \right]^{-1} f_N(\hat{\theta}_N^{(i-1)}, Z^N) \tag{10.49}$$

10.3 COMPUTING GRADIENTS

To use the formulas in the previous section, we need expressions for $\psi(t, \theta)$, the gradient of the prediction. The amount of work required to compute $\psi(t, \theta)$ is highly dependent on the model structure, and sometimes one may have to resort to numerical differentiation. In this section we shall provide expressions for some common model structures.

Example 10.1 The ARMAX Model Structure

Consider the ARMAX model (4.14). The predictor is given by (4.18):

$$C(q)\hat{y}(t|\theta) = B(q)u(t) + [C(q) - A(q)]\, y(t) \tag{10.50}$$

Differentiating this expression with respect to a_k gives

$$C(q)\frac{\partial}{\partial a_k}\hat{y}(t|\theta) = -q^{-k} y(t) \tag{10.51}$$

Similarly,

$$C(q)\frac{\partial}{\partial b_k}\hat{y}(t|\theta) = -q^{-k}u(t)$$

and

$$q^{-k}\hat{y}(t|\theta) + C(q)\frac{\partial}{\partial c_k}\hat{y}(t|\theta) = -q^{-k}y(t)$$

With the vector $\varphi(t, \theta)$ defined by (4.20), these expressions can be rewritten conveniently as

$$C(q)\psi(t, \theta) = \varphi(t, \theta) \tag{10.52}$$

The gradient is thus obtained by filtering the "regression vector" $\varphi(t, \theta)$ through the filter $1/C(q)$. This filter is stable for all θ for which the predictor (10.50) is stable. □

SISO Black-box Model (*)

Most formulas for SISO black-box models will be contained in a treatment of the general model (4.33). The predictor for this model is given by (4.35):

$$\hat{y}(t|\theta) = \frac{D(q)B(q)}{C(q)F(q)}u(t) + \left[1 - \frac{D(q)A(q)}{C(q)}\right]y(t) \tag{10.53}$$

from which we find, as in Example 10.1, that

$$\frac{\partial}{\partial a_k}\hat{y}(t|\theta) = -\frac{D(q)}{C(q)}y(t-k) \tag{10.54a}$$

$$\frac{\partial}{\partial b_k}\hat{y}(t|\theta) = -\frac{D(q)}{C(q)F(q)}u(t-k) \tag{10.54b}$$

$$\begin{aligned}\frac{\partial}{\partial c_k}\hat{y}(t|\theta) &= -\frac{D(q)B(q)}{C(q)C(q)F(q)}u(t-k) + \frac{D(q)A(q)}{C(q)C(q)}y(t-k) \\ &= \frac{1}{C(q)}\varepsilon(t-k, \theta)\end{aligned} \tag{10.54c}$$

$$\begin{aligned}\frac{\partial}{\partial d_k}\hat{y}(t|\theta) &= \frac{B(q)}{C(q)F(q)}u(t-k) - \frac{A(q)}{C(q)}y(t-k) \\ &= -\frac{1}{C(q)}v(t-k, \theta)\end{aligned} \tag{10.54d}$$

$$\begin{aligned}\frac{\partial}{\partial f_k}\hat{y}(t|\theta) &= -\frac{D(q)B(q)}{C(q)F(q)F(q)}u(t-k) \\ &= \frac{D(q)}{C(q)F(q)}w(t-k, \theta)\end{aligned} \tag{10.54e}$$

where we used ε, v, and w as defined by (4.37) to (4.39). The gradient $\psi(t, \theta)$ is thus also in this case obtained by filtering the regression vector $\varphi(t, \theta)$ [defined in (4.40)] through linear filters, although different parts of φ are associated with different filters in the general case. It is clear that the filters involved here are all stable for $\theta \in D_{\mathcal{M}}$, defined in Lemma 4.1, which are also the θ for which the predictors are stable.

In the special case of an output error model, $A(q) = C(q) = D(q) = 1$, [see also (4.25)], we obtain from (10.54b, e)

$$F(q)\psi(t, \theta) = \varphi(t, \theta) \tag{10.55}$$

General Finite-dimensional Linear Time-invariant Models (*)

A linear time-invariant finite-dimensional model can always be represented as

$$\begin{aligned} \varphi(t+1, \theta) &= \mathcal{F}(\theta)\varphi(t, \theta) + \mathcal{G}(\theta)\begin{bmatrix} y(t) \\ u(t) \end{bmatrix} \\ \hat{y}(t|\theta) &= \mathcal{H}(\theta)\varphi(t, \theta) \end{aligned} \tag{10.56}$$

with proper choices of the matrices $\mathcal{F}$, $\mathcal{G}$, and $\mathcal{H}$ and with dim $\varphi = n$. This is true for the general SISO model (4.35) for which $\varphi(t, \theta)$ can be chosen as (4.40), as well as for the general state-space model (4.86) for which

$$\mathcal{F}(\theta) = A(\theta) - K(\theta)C(\theta) \tag{10.57a}$$

$$\mathcal{G}(\theta) = \begin{bmatrix} K(\theta) & B(\theta) \end{bmatrix} \tag{10.57b}$$

$$\mathcal{H}(\theta) = C(\theta), \quad \hat{x}(t, \theta) = \varphi(t, \theta) \tag{10.57c}$$

Stability of the predictor (10.56) requires θ to belong to

$$D_{\mathcal{M}} = \{\theta | \mathcal{F}(\theta) \text{ has all eigenvalues inside the unit circle}\} \tag{10.58}$$

The equation (10.56) can now be differentiated with respect to θ. Introducing

$$\xi(t, \theta) = [\varphi^T(t, \theta) \quad \tfrac{\partial}{\partial\theta_1}\varphi^T(t, \theta) \quad \dots \quad \tfrac{\partial}{\partial\theta_d}\varphi^T(t, \theta)\,]^T \tag{10.59}$$

we may, for some matrices $\mathcal{A}(\theta)$, $\mathcal{B}(\theta)$, and $\mathcal{C}(\theta)$, write

$$\begin{aligned} \xi(t+1, \theta) &= \mathcal{A}(\theta)\xi(t, \theta) + \mathcal{B}(\theta)\begin{bmatrix} y(t) \\ u(t) \end{bmatrix} \\ \begin{bmatrix} \hat{y}(t|\theta) \\ \psi(t, \theta) \end{bmatrix} &= \mathcal{C}(\theta)\xi(t, \theta) \end{aligned} \tag{10.60}$$

It can readily be verified that the $(d+1)n \times (d+1)n$ matrix $\mathcal{A}(\theta)$ will contain the matrix $\mathcal{F}(\theta)$ in each of its $d+1$ block-diagonal entries and has all zeros above the block diagonal. Hence the stability properties of $\mathcal{A}(\theta)$ coincide with those of $\mathcal{F}(\theta)$.

We will find (10.60) useful for developing general algorithms. Clearly, however, these equations, as given, will not be suited for practical calculations of the gradient, since the filter is of order $d(n+1)$. It has been shown, though, by Gupta and Mehra (1974)that the dimension of the controllability subspace of (10.60) does not exceed $4n$ regardless of the parametrization and regardless of the value of θ. By proper transformations, the necessary calculations involved in (10.60) can thus be substantially reduced.

Finally, we note that when the methods of Section 10.2 are used the expressions of the present section have to be used between each iteration to compute $\psi(t, \hat{\theta}_N^{(i)})$. This means that we have to run the data from $t = 1$ to N through filters like (10.54) or (10.60) for each iteration in (10.40).

Nonlinear Black-Box Models and Back-Propagation

In connection with neural networks, the celebrated *Back-Propagation (BP) algorithm* is used to compute the gradients of the predictor. Back-propagation has been described in several contexts, see e.g., Werbos (1974), Rumelhart, Hinton, and Cybenko (1986). Sometimes in the neural network literature the entire search algorithm is called Back-Propagation. It is, however, more consistent to keep this notation just for the algorithm used to calculate the gradient.

Consider the general nonlinear black-box structure (5.43):

$$g(\varphi, \theta) = \sum_{k=1}^{n} \alpha_k \kappa(\beta_k(\varphi - \gamma_k)) \tag{10.61}$$

To find the gradient $\psi(t, \theta) = (d/d\theta) g(\varphi, \theta)$ for this one hidden layer network is quite simple. We just need to compute

$$\frac{d}{d\alpha} \alpha\kappa(\beta\varphi - \gamma) = \kappa(\beta\varphi - \gamma)$$

$$\frac{d}{d\gamma} \alpha\kappa(\beta\varphi - \gamma) = -\alpha\kappa'(\beta\varphi - \gamma)$$

$$\frac{d}{d\beta} \alpha\kappa(\beta\varphi - \gamma) = \alpha\kappa'(\beta\varphi - \gamma)\varphi$$

The BP algorithm in this case means that the factor $\alpha\kappa'(\beta\varphi - \gamma)$ from the derivative with respect to γ is re-used in the calculation of the derivative with respect to β.

The Back-Propagation algorithm is however very general and not limited to one-hidden-layer sigmoid neural network models. Instead, it applies to all network models and it can be described as the chain rule for differentiation applied to the expression (5.47) with a smart re-use of intermediate results which are needed at several places in the algorithm. For ridge construction models (5.42) where β_i is a parameter vector, the only complicated thing with the algorithm is actually to keep track of all indexes. When β_i is a parameter matrix, like in the radial approach (5.40), then the calculation becomes somewhat more complicated, but the basic procedure remains the same. See Saarinen, Bramley, and Cybenko (1993) for an illuminating description of these general aspects.

When shifting to multi-layer network models, the possibilities of re-using intermediate result increase and so does the importance of the BP algorithm.

For recurrent models the calculation of the gradient becomes more complicated. The gradient $\psi(t, \theta)$ at one time instant does not only depend on the regressor $\varphi(t, \theta)$ but also on the gradient at the previous time instant $\psi(t-1, \theta)$. See Nerrand et.al. (1993)for a discussion on this topic. The additional problem to calculate the gradient does, however, not change anything essential in the minimization algorithm. In the neural network literature this is often referred to as *Back-propagation through time*.

Recursive Techniques for Off-line Problems

An idea to save work in the minimization procedure could be to combine the methods of Sections 10.2 and 10.3 so as to modify the estimate $\hat{\theta}_N^{(i)}$ in (10.40) at the same time as the prediction error gradients are computed in (10.60) [i.e., to "link the index (i) to N"]. We shall develop such *recursive* algorithms in Chapter 11 for on-line applications. These are, however, quite useful also for off-line problems as alternatives to (10.40). Then typically the data record is run through the recursive algorithm a couple of times, and it can be shown that such a procedure will have the same convergence properties as (10.40). See Ljung and Söderström (1983), Section 7.2, and Solbrand, Ahlén, and Ljung (1985)for further details.

10.4 TWO-STAGE AND MULTISTAGE METHODS

The techniques described in Sections 10.2 and 10.3 should be regarded as the basic numerical methods for parameter estimation. They have the advantages of guaranteed convergence (to a local minimum), efficiency, and applicability to general model structures. Nevertheless, the literature is abundant with alternative techniques, mostly related to special cases of the general linear model structure (4.33):

$$A(q)y(t) = \frac{B(q)}{F(q)}u(t) + \frac{C(q)}{D(q)}e(t) \tag{10.62}$$

(or multivariable counterparts), and to the general nonlinear black-box structure (10.61). A basic idea is to rephrase the problem as a linear regression problem or a sequence of such problems, so that the efficient methods of Section 10.1 can be applied. For (10.61) it may involve fixing the parameters that enter nonlinearly (i.e. β and γ) and estimate the α:s as a linear regression.

The algorithms typically involve two or several LS stages (or IV stages) applied to different substructures, and we therefore call them *two-stage* or *multi-stage* methods.

In this section we shall give a short description of the building blocks of such procedures. Mixing techniques (IV, LS, PEM, PLR) and models (FIR, ARX, ARMAX, etc.) into procedures involving several stages leads to a myriad of "identification methods." There will be no need to list all these. They can, however, be understood and analyzed by our techniques, applied to the different stages (see Problems 10G.2

and 10E.1). Our interest in this topic is twofold: it helps us to understand the identification literature, and the techniques may be useful for providing initial estimates for the basic schemes of Section 10.2.

The subspace method (7.66) can also be regarded as a two-stage method, being built up from two LS-steps. Due to the rather complex nature of this algorithm, we will treat it separately in Section 10.6. For the rich possibilities offered for nonlinear black-box models, we refer to Sjöberg et.al. (1995).

Bootstrap Methods

Consider the correlation formulation (7.110): Solve

$$f_N(\theta, Z^N) = \frac{1}{N}\sum_{t=1}^{N} \zeta(t,\theta)\varepsilon(t,\theta) = 0 \tag{10.63a}$$

in the special case where the prediction error can be written

$$\varepsilon(t,\theta) = y(t) - \varphi^T(t,\theta)\theta \tag{10.63b}$$

This formulation contains a number of common situations:

- IV methods with $\zeta(t,\theta)$ as in (7.127) and $\varphi(t,\theta) = \varphi(t)$ as in (7.114).
- PLR methods with $\zeta(t,\theta) = \varphi(t,\theta)$ as in (7.113).
- Minimizing the quadratic criterion (10.38) for models that can be written as (7.112), taking $\zeta(t,\theta) = \psi(t,\theta)$.

With a nominal iterate $\hat{\theta}_N^{(i-1)}$ at hand, it is then natural to determine the next one by solving

$$\frac{1}{N}\sum_{t=1}^{N} \zeta(t, \hat{\theta}_N^{(i-1)}) \left[y(t) - \varphi^T(t, \hat{\theta}_N^{(i-1)})\theta\right] = 0$$

for θ. This is a linear problem and can be solved as

$$\hat{\theta}_N^{(i)} = \left[\frac{1}{N}\sum_{t=1}^{N} \zeta(t, \hat{\theta}_N^{(i-1)})\varphi^T(t, \hat{\theta}_N^{(i-1)})\right]^{-1} \left[\frac{1}{N}\sum_{t=1}^{N} \zeta(t, \hat{\theta}_N^{(i-1)})y(t)\right] \tag{10.64}$$

Solving (10.64) is essentially a least-squares problem (10.2) with proper definitions of $R(N)$ and $f(N)$. The techniques described in Section 10.1 thus apply also to (10.64).

The algorithm (10.64) is known as a bootstrap method, since it alternates between computing θ and forming new vectors φ and ζ. It should be noted that it does not necessarily converge to a solution of (10.63). A convergence analysis is given by Stoica and Söderström (1981b), and Stoica et.al. (1985).

Bilinear Parametrizations

For some model structures, the predictor is bilinear in the parameters. That is, the parameter vector θ can be split up into two parts,

$$\theta = \begin{bmatrix} \rho \\ \eta \end{bmatrix}$$

such that

$$\hat{y}(t|\theta) = \hat{y}(t|\rho, \eta) \tag{10.65}$$

is linear in ρ for fixed η and linear in η for fixed ρ. A typical such situation is the ARARX structure (4.22):

$$\hat{y}(t|\theta) = B(q)D(q)u(t) + [1 - A(q)D(q)]\, y(t) \tag{10.66}$$

Clearly, by associating ρ with the A- and B-parameters and η with the D-parameters the preceding bilinear situation is at hand.

With this situation, a natural way of minimizing

$$V_N(\theta, Z^N) = V_N(\rho, \eta, Z^N) = \frac{1}{N}\sum_{t=1}^{N}\left(y(t) - \hat{y}(t|\rho, \eta)\right)^2 \tag{10.67}$$

would be to treat it as a sequence of least-squares problems. Let

$$\hat{\rho}_N^{(i)} = \arg\min_{\rho} V_N(\rho, \hat{\eta}_N^{(i-1)}, Z^N) \tag{10.68a}$$

$$\hat{\eta}_N^{(i)} = \arg\min_{\eta} V_N(\hat{\rho}_N^{(i)}, \eta, Z^N) \tag{10.68b}$$

Each of these problems is a pure least-squares problem and can be solved efficiently. Although this procedure bears some resemblance to the bootstrap methods, it is indeed a minimization method that will lead to a local minimum (cf. Problems 10T.3 and 10E.9).

Separable Least Squares

A more general situation than the bilinear case is when one set of parameters enter linearly and another set nonlinearly in the predictor:

$$\hat{y}(t|\theta, \eta) = \theta^T \varphi(t, \eta) \tag{10.69}$$

The identification criterion then becomes

$$V_N(\theta, \eta, Z^N) = \sum_{t=1}^{N}\left|y(t) - \theta^T\varphi(t, \eta)\right|^2 = |\mathbf{Y} - \Phi(\eta)\theta|^2 \tag{10.70}$$

where we introduced matrix notation, analogously to (10.7). For given η this criterion is an LS criterion and minimized w.r.t. θ by

$$\hat{\theta} = \left[\Phi^T(\eta)\Phi(\eta)\right]^{-1}\Phi^T(\eta)\mathbf{Y} \tag{10.71}$$

We can thus insert this into (10.70) and define the problem as

$$\min_{\eta} |\mathbf{Y} - P(\eta)\mathbf{Y}|^2 = |(I - P(\eta))\,\mathbf{Y}|^2,$$

$$P(\eta) = \Phi(\eta)\left[\Phi^T(\eta)\Phi(\eta)\right]^{-1}\Phi^T(\eta) \tag{10.72}$$

The estimate θ is then obtained by inserting the minimizing η into (10.71). Note that the matrix P is a projection matrix: $P^2 = P$. The method is called *separable least squares* since the LS-part has been separated out, and the problem reduced to a minimization problem of lower dimension. See Golub and Pereyra (1973)for a thorough treatment of this approach. It is known to give numerically well-conditioned calculations, but does not necessary give faster convergence than applying a damped Gauss-Newton method to (10.70) without utilizing the particular structure.

High-Order AR(X) Models

Suppose the true system is given as

$$y(t) = G_0(q)u(t) + H_0(q)e_0(t)$$

and an ARX structure

$$A^M(q)y(t) = B^M(q)u(t) + e(t)$$

of order M is used. Then it can be shown (e.g., Hannan and Kavalieris, 1984, and Ljung and Wahlberg, 1992) that as the number of data N tends to infinity, as well as the model M (N "faster than" M), the model $\hat{A}_N^M$, $\hat{B}_N^M$ will converge to the true system in the following sense:

$$\frac{\hat{B}_N^M(e^{i\omega})}{\hat{A}_N^M(e^{i\omega})} \to G_0(e^{i\omega}), \qquad \text{uniformly in } \omega \text{ as } N \gg M \to \infty$$

$$\frac{1}{\hat{A}_N^M(e^{i\omega})} \to H_0(e^{i\omega}), \qquad \text{uniformly in } \omega \text{ as } N \gg M \to \infty$$

This means that a high-order ARX model is capable of approximating any linear system arbitrarily well. It is of course desirable to reduce this high-order model to more tractable versions within the structure (10.62), and for that purpose a number of different possibilities are at hand:

1. Find $G = B/A$ as a rational structure by eliminating common factors in $\hat{A}_N^M$ and $\hat{B}_N^M$ (Söderström, 1975b).

2. Apply model reduction techniques based on balanced realizations to $\hat{B}_N^M/\hat{A}_N^M$ (Wahlberg, 1986, Zhu and Backx, 1993).

3. Let $z(t)$ be the output of the model $\hat{B}_N^M/\hat{A}_N^M$ driven by the actual input u and apply an ARX model to the input-output pair (z, u) (Pandaya, 1974; Hsia, 1977, Chapter 7).

4. Let $\hat{e}_M(t)$ be the residuals associated with the model $\hat{A}_N^M$, $\hat{B}_N^M$. Use a model structure

$$A(q)y(t) = B(q)u(t) + [C(q) - 1]\hat{e}_M(t) + e(t) \tag{10.73}$$

to estimate A, B, and C. Since $\hat{e}_M(t)$ is a known sequence, this structure is an ARX structure with two inputs, and the estimates are thus determined by the LS method (Mayne and Firoozan, 1982).

5. The subspace method (7.66) (which is described in detail in Section 10.6) should rightly be included in this family too: The k-step ahead predictors computed from (7.62) are the high order ARX-models, while (7.60) corresponds to the model reduction step, and (7.56) is the second LS-stage.

Separating Dynamics and Noise Models

In the general linear model (10.62), we can always determine the dynamic part from u to y using the IV method. Splitting the denominator estimate thus obtained into one factor $\hat{A}(q)$ that is supposed to be in common with the noise description and one factor $\hat{F}(q)$ that is particular to the dynamics (typically one would postulate one of A and F to be unity), we can then determine

$$\hat{v}(t) = \hat{A}(q)y(t) - \frac{\hat{B}(q)}{\hat{F}(q)}u(t) \tag{10.74}$$

as an estimate of the equation noise [cf. (4.38)]. This noise can then be regarded as a measured signal, and an ARMA model

$$\hat{v}(t) = \frac{C(q)}{D(q)}e(t) \tag{10.75}$$

can be constructed as a separate step. Young has developed this technique in a number of papers (see, e.g., Young and Jakeman, 1979).

Determining ARMA Models

The parameters of the ARMA model (10.75) can of course be estimated using the prediction-error approach. Two alternatives that avoid iterative search procedures are as follows:

1. Apply a high-order AR model to $\hat{v}(t)$ in (10.75) to form estimates of the innovations $\hat{e}$. Then form the ARX model

$$D(q)\hat{v}(t) = [C(q) - 1]\hat{e}(t) + e(t) \tag{10.76}$$

with $\hat{v}(t)$ as output and $\hat{e}(t)$ as input, and estimate D and C with the LS method [cf. (10.73)].

2. Estimate the AR parameters $D(q)$ using the IV method as explained in Problem 7E.1. Then model $\hat{w}(t) = D(q)\hat{v}(t)$ as an MA model. See Durbin (1959)and Walker (1961)for related tεchniques, and Broersen (1997)for a discussion of order selection.

10.5 LOCAL SOLUTIONS AND INITIAL VALUES

Local Minima

The general numerical schemes for minimization and equation solution that we discussed in Section 10.2 typically have the property that, with suitably chosen step length μ, they will converge to a solution of the posed problem. This means that (10.48) and (10.49) will converge to a point θ_N^* such that

$$f_N(\theta_N^*, Z^N) = 0 \tag{10.77}$$

while (10.40), with positive definite R, converges to a *local minimum* of $V_N(\theta, Z^N)$.

For the minimization problem, it is the *global minimum* that interests us. The theoretical results of Chapters 8 and 9 dealt with properties of the globally minimizing estimate $\hat{\theta}_N$. Similarly, the equation (10.77) may have several solutions. It is obviously an inherent feature of the iterative search routines of Section 10.2 that only convergence to a *local solution* of the problem can be guaranteed. To find the global solution, there is usually no other way than to start the iterative minimization routine at different feasible initial values and compare the results. An important possibility is to use some preliminary estimation procedure to produce a good initial value for the minimization. See the following discussion.

When validating the model as we shall discuss in Sections 16.5 and 16.6, the model is judged according to its performance, though. Therefore, local minima do not necessarily create problems in practice. If a model passes the validation tests, it should be an acceptable model, even if it does not give the global minimum of the criterion function.

The problem of "false" local solutions has two aspects. Let us concentrate on the problem of local minima. It may be that the limit of the criterion function as N tends to infinity, $\overline{V}(\theta)$, has such local minima. Then also $V_N(\theta, Z^N)$ will have such minima for large N, according to Lemma 8.2. The existence of local minima of $\overline{V}(\theta)$ can be analyzed, but only few results are available as yet. Some of them will be given later. The other aspect is that, even if $\overline{V}(\theta)$ has only one local minimum (= the global one), the function $V_N(\theta, Z^N)$ may have other local minima due to the randomness in data. This is a much harder problem to treat analytically. The one exception is the linear regression least-squares method, where by construction the criterion function has no nonglobal local minima regardless of the properties of the data.

Results for SISO Black-box Models

The only analytical results available on local solutions are for black-box models *under the assumption that the system can be described within the model set:* $\mathcal{S} \in \mathcal{M}$. We list these results here with references for proofs. They all concern the general SISO model set (10.62), and refer to

$$\overline{V}(\theta) = \overline{E}\tfrac{1}{2}\varepsilon^2(t, \theta)$$

For simplicity we call a nonglobal, local minimum a "false minimum."

- For ARMA models ($B \equiv 0, D \equiv F \equiv 1$) all stationary points of $\overline{V}(\theta)$ are global minima (Åström and Söderström, 1974).
- For ARARX models ($C \equiv F \equiv 1$) there are no false local minima if the signal-to-noise ratio is large enough. If it is very small, false local minima do exist (Söderström, 1974).
- If $A \equiv 1$, there are no false local minima if $n_f = 1$ (Söderström, 1975c).
- If $A \equiv C \equiv D \equiv 1$, there are no false local minima if the input is white noise. For other inputs, however, false local minima can exist (Söderström, 1975c).

For the ARMAX model ($F \equiv D \equiv 1$), it is not known whether false local minima exist. For the pseudolinear regression approach (7.113), it can, however, be shown that

$$\overline{E}\varphi(t,\theta)\varepsilon(t,\theta) = 0 \Rightarrow \theta = \theta_0 \tag{10.78}$$

in the case of an ARMAX model, with θ_0 denoting the true parameters (Ljung, Söderström, and Gustavsson, 1975).

The practical experience with different model structures is that the global minimum is usually found without too much problem for ARMAX models. See, for example, Bohlin (1971)for a discussion of these points. For output error structures, on the other hand, convergence to false local minima is not uncommon.

Initial Parameter Values

Due to the possible occurrence of undesired local minima in the criterion function, it is worthwhile to spend some effort on producing good initial values for the iterative search procedures. Also, since the Newton-type methods described in Section 10.2 have good local convergence rates, but not necessarily fast convergence far from the minimum, these efforts usually pay off in fewer iterations and shorter total computing time.

For a *physically parametrized model structure*, it is most natural to use our physical insight to provide reasonable initial values. Also, it allows us to monitor and interact with the iterative search scheme.

For a *linear black-box model structure* several possibilities exist. It is our experience that the following is a good start-up procedure for the general model structure (10.62):

1. Apply the IV method to estimate the dynamic transfer function B/AF. Most often one of A and F is unity. For a system that has operated in open loop, first a LS estimate of an ARX model can be determined, to be used in the generation of instruments as in (7.123). (10.79a)
2. Determine an estimate of the equation noise as in (10.74). (10.79b)

3. Determine C and/or D in (10.75) by (10.76) after a first high-order AR step to find $\hat{e}$ (which is unnecessary if $C = 1$). The order of the AR model can be chosen as the sum of all model orders in (10.62) so as to balance the computational effort. (10.79c)

In case $\mathcal{S} \in \mathcal{M}$ this will bring the initial parameter estimate arbitrarily close to the true values as N increases. From there, the methods of Section 10.2 will efficiently bring us to the global minimum of the criterion. In this case we thus have a procedure that is globally convergent to the global minimum for large enough N.

For a *nonlinear black-box structure* (10.61) several techniques exist. A simple one is to "seed" a large number of fixed values of the non-linear parameters β_k and γ_k, and estimate the corresponding α:s by linear least squares. The estimates α_k that are most significant (relative to their estimated standard deviations) are then selected and the corresponding γ_k and β_k are used as initial values for the ensuing Gauss-Newton iterative search.

Initial Filter Conditions

The filters (10.53)–(10.54) as well as (10.56) require initial values $\varphi(0, \theta)$ to be initialized. In case the filters have finite impulse response, which happens only for the ARX special case, we can wait to initialize the filters until enough past data are known. This is what we called approach 1 following (10.13) in Section 10.1. In the general case we need a strategy to deal with the unknown initial conditions. This has not been extensively discussed in the literature, but we can point to the following approaches:

1. Take $\varphi(0, \theta) = 0$.
2. Select $\varphi(0, \theta)$ so that the first $\hat{y}(t|\theta)$, $t = 1, \ldots, \dim\varphi$ match $y(t)$ exactly.
3. Introduce $\varphi(0, \theta) = \eta$ as a parameter, and estimate it along with θ.
4. Estimate or "backforecast" $\varphi(0, \theta)$ from the data by running suitable filters backwards in time, Knudsen (1994).

For a model where the predictor filter transient is short compared to the data record, it does not matter so much which approach is taken. However, with slowly decaying transients, the two first methods may have a very negative influence on the model quality. This is particularily pronounced for OE models, where no noise model will pick up the residuals from bad transient behaviour.

10.6 SUBSPACE METHODS FOR ESTIMATING STATE SPACE MODELS

Let us now consider how to estimate the system matrices A, B, C, and D in a state space model

$$\begin{aligned} x(t+1) &= Ax(t) + Bu(t) + w(t) \\ y(t) &= Cx(t) + Du(t) + v(t) \end{aligned} \tag{10.80}$$

We assume that the output $y(t)$ is a p-dimensional column vector, while the input $u(t)$ is an m-dimensional column vector. The order of the system, i.e., the dimension of $x(t)$, is n. We also assume that this state-space representation is a minimal realization. It is well known that the same input-output relationship can also be described by

$$\begin{aligned} \tilde{x}(t+1) &= T^{-1}AT\tilde{x}(t) + T^{-1}Bu(t) + \tilde{w}(t) \\ y(t) &= CT\tilde{x}(t) + Du(t) + v(t) \end{aligned} \tag{10.81}$$

for any invertible matrix T. This corresponds to the change of basis $\tilde{x}(t) = T^{-1}x(t)$ in the state space.

In Section 7.3, algorithm (7.66), we presented an archetypical algorithm for this. We shall here describe a family of related algorithms which all address this problem. The discussion will be quite technical and a reader who primarily is interested in the result could go directly to (10.125). In summary the algorithms are based on the following observations:

- If $\hat{A}$ and $\hat{C}$ are known, it is an easy linear least squares problem to estimate B and D from

$$y(t) = \hat{C}(qI - \hat{A})^{-1}Bu(t) + Du(t) + v(t) \tag{10.82}$$

  using the predictor

$$\hat{y}(t|B, D) = \hat{C}(qI - \hat{A})^{-1}Bu(t) + Du(t) \tag{10.83}$$

  (The initial state $x(0)$ can also be estimated; see (10.86) below.)

- If the (extended) observability matrix for the system

$$O_r = \begin{bmatrix} C \\ CA \\ \vdots \\ CA^{r-1} \end{bmatrix} \tag{10.84}$$

  is known, then it is easy to determine C and A. Use the first block row of O_r and the shift property, respectively. This is really the key step.

- The extended observability matrix can be consistently estimated from input-output data by direct least-squares like (projection) steps.

- Once the observability matrix has been estimated, the states $x(t)$ can be constructed and the statistical properties of the noise contributions $w(t)$ and $v(t)$ can be established.

We shall now deal with each of these steps in somewhat more detail.

Estimating B and D

For given and fixed $\hat{A}$ and $\hat{C}$ the model structure (10.83):

$$\hat{y}(t|B, D) = \hat{C}(qI - \hat{A})^{-1}Bu(t) + Du(t) \tag{10.85a}$$

is clearly linear in B and D. The predictor is also formed entirely from past inputs, so it is an *output error* model structure. If the system operates in open loop, we can thus consistently estimate B and D according to Theorem 8.4, even if the noise sequence

$$v(t) = C(qI - A)^{-1}w(t) + \nu(t)$$

in (10.82) is non-white.

Let us write the predictor (10.83) in the standard linear regression form

$$\hat{y}(t) = \varphi(t)\theta = \varphi(t)\begin{bmatrix} \text{Vec}(B) \\ \text{Vec}(D) \end{bmatrix} \tag{10.85b}$$

with a $p \times (mn + mp)$ matrix $\varphi(t)$. Here "Vec" is the operation that builds a vector from a matrix, by stacking its columns on top of each other. Let $r = (k-1)n + j$. To find the r:th ($r \leq mn$) column of $\varphi(t)$, which corresponds to the r:th element of θ, i.e., the element $B_{j,k}$, we differentiate (10.85b) w.r.t. this element and obtain

$$\varphi_r(t) = \hat{C}(qI - \hat{A})^{-1}E_j u_k(t)$$

where E_j is the column vector with the j:th element equal to 1 and the others equal to 0. The rows for $r > nm$ are handled in a similar way.

If desired, also the initial state $x_0 = x(0)$ can be estimated in an analogous way, since the predictor with initial values taken into account is

$$\hat{y}(t|B, D, x_0) = \hat{C}(qI - \hat{A})^{-1}x_0\delta(t) + \hat{C}(qI - \hat{A})^{-1}Bu(t) + Du(t) \tag{10.86}$$

which is linear also in x_0. Here $\delta(t)$ is the unit pulse at time 0. Moreover, the estimates can be improved by estimating the color of v in (10.82) and prefiltering the data accordingly.

Remark: If $\hat{A}$ and $\hat{C}$ are the correct values, the least squares estimates of B and D will also converge to their true values, according to Theorem 8.4. If consistent estimates $\hat{A}_N$ and $\hat{C}_N$ are used instead, convergence of $\hat{B}_N$ and $\hat{D}_N$ to their true values still holds. This follows by fairly straightforward calculations. See Vandersteen, Van hamme, and Pintelon (1996) for a general treatment of such issues.

Finding A and C from the Extended Observability Matrix

Suppose that a $pr \times n^*$ dimensional matrix G is given, that is related to the extended observability matrix of the system, (10.84). We have to determine A and C from G, and we shall here consider cases of increased complexity.

Known System Order. Suppose first we know that

$$G = \mathcal{O}_r$$

so that $n^* = n$. To find C is then immediate:

$$\hat{C} = O_r(1 : p, 1 : n) \tag{10.87}$$

Here we used MATLAB notation, meaning that $M(s : \ell, j : k)$ is the matrix obtained by extracting the rows $s, s+1, \ldots, \ell$ and the columns $j, j+1, \ldots, k$ from the matrix M.

Similarly we can find $\hat{A}$ from the equation

$$O_r(p+1 : pr, 1 : n) = O_r(1 : p(r-1), 1 : n)\hat{A} \tag{10.88}$$

which is easily seen from the definition (10.84). Under the assumption of observability, O_{r-1} has rank n, so $\hat{A}$ can be determined uniquely. Normally (10.88) is an overdetermined set of equations (n^2 unknowns in A and $npr - np$ equations; recall that $r \geq n+1$). This is of no consequence if O_r is exactly of the form (10.84), since any full rank subset of equations will give the same $\hat{A}$.

Role of State-Space Basis. The extended observability matrix depends on the choice of basis in the state-space representation. For the representation (10.81) it is easy to verify that the observability matrix would be

$$\tilde{O}_r = O_r T \tag{10.89}$$

Applying (10.87) and (10.88) to $\tilde{O}_r$ would thus give the system matrices associated with (10.81). Consequently, *multiplying the extended observability matrix from the right by any invertible matrix before applying (10.87) and (10.88) will not change the system estimate—just the basis of representation.*

Unknown System Order. Suppose now that the true order of the system is unknown, and that n^*—the number of columns of G—is just an upper bound for the order. This means that we have

$$G = O_r \tilde{T} \tag{10.90}$$

for some unknown, but full rank, $n \times n^*$ matrix $\tilde{T}$, where also n is unknown to us. The rank of G is n. A straightforward way to deal with this would be to determine this rank, delete the last $n^* - n$ columns of G and then proceed as above. A more general and numerically sound way of reducing the column space is to use singular value decomposition (SVD):

$$G = USV^T = U \begin{bmatrix} \sigma_1 & 0 & 0 & \ldots & 0 \\ 0 & \sigma_2 & 0 & \ldots & 0 \\ 0 & 0 & \sigma_3 & \ldots & 0 \\ \vdots & \vdots & \vdots & \ddots & \vdots \\ 0 & 0 & 0 & \ldots & \sigma_{n^*} \\ 0 & 0 & 0 & \ldots & 0 \\ \vdots & \vdots & \vdots & \ddots & \vdots \\ 0 & 0 & 0 & \ldots & 0 \end{bmatrix} V^T \tag{10.91}$$

Here U and V are orthonormal matrices ($U^T U = I,\ V^T V = I$) of dimensions $pr \times pr$ and $n^* \times n^*$, respectively. S is a $pr \times n^*$ matrix with the singular values of G along the diagonal and zeros elsewhere. If G has rank n, only the first n singular values σ_k will be non-zero. This means that we can rewrite

$$G = USV^T = U_1 S_1 V_1^T \tag{10.92}$$

where U_1 is a $pr \times n$ matrix containing the first n columns of U, while S_1 is the $n \times n$ upper left part of S, and V_1 consists of the first n columns of V. (We still have $V_1^T V_1 = I$.) From (10.90) we find $\mathcal{O}_r \tilde{T} = U_1 S_1 V_1^T$. Multiplying this by V_1 from the right gives

$$\mathcal{O}_r \tilde{T} V_1 = \mathcal{O}_r T = U_1 S_1 \tag{10.93}$$

for some invertible matrix $T = \tilde{T} V_1$. We are now in the situation (10.89) that we know the observability matrix up to an invertible matrix T—or equivalently, we know the observability matrix in some state-space basis. Consequently we can use $\hat{\mathcal{O}}_r = U_1 S_1$ or $\hat{\mathcal{O}}_r = U_1$ or any matrix that can be written as

$$\hat{\mathcal{O}}_r = U_1 R \tag{10.94}$$

for some invertible R in (10.87) and (10.88) to determine the $p \times n$ matrix $\hat{C}$ and the $n \times n$ matrix $\hat{A}$.

Using a Noisy Estimate of the Extended Observability Matrix. Let us now assume that the given $pr \times n^*$ matrix G is a noisy estimate of the true observability matrix

$$G = \mathcal{O}_r \tilde{T} + E_N \tag{10.95}$$

where E_N is small and tends to zero as $N \to \infty$. The rank of $\mathcal{O}_r$ is not known, while the "noise matrix" E_N is likely to be of full rank. It is reasonable to proceed as above and perform an SVD on G:

$$G = USV^T \tag{10.96}$$

Due to the noise, S will typically have all singular values $\sigma_k;\ k = 1, \ldots, \min(n^*, pr)$ non-zero. The first n will be supported by $\mathcal{O}_r$, while the remaining ones will stem from E_N. If the noise is small, one should expect that the latter are significantly smaller than the former. Therefore determine $\hat{n}$ as the number of singular values that are significantly larger than 0. Then keep those and replace the others in S by zeros, and proceed as in (10.92) to determine U_1 and S_1. Then use $\hat{\mathcal{O}}_r$ in (10.94) to determine $\hat{A}$ and $\hat{C}$ as before. However, in this noisy case, $\hat{\mathcal{O}}_r$ will not be exactly subject to the shift structure (10.88), so this system of equations should be solved in a least-squares sense.

The *consistency* of this process as $N \to \infty$ and $E_N \to 0$ is rather easy to establish by a continuity argument: As E_N tends to zero, the corresponding estimates of A and C will tend to the values that are obtained from (10.90) with $E_N = 0$.

It is more difficult to analyze how the variance of E_N will influence the variance of $\hat{A}$ and $\hat{C}$. Some results about this are given in Viberg et.al. (1993), based on work by T.W. Anderson.

Using Weighting Matrices in the SVD. For more flexibility we could pre- and post-multiply G as $\hat{G} = W_1 G W_2$ before performing the SVD

$$\hat{G} = W_1 G W_2 = USV^T \approx U_1 S_1 V_1^T \tag{10.97}$$

and then instead of (10.94) use

$$\hat{O}_r = W_1^{-1} U_1 R \tag{10.98}$$

to determine $\hat{C}$ and $\hat{A}$ in (10.87) and (10.88). Here R is an arbitrary matrix, that will determine the coordinate basis for the state representation. The post-multiplication by W_2 just corresponds to a change of basis in the state-space and the pre-multiplication by W_1 is eliminated in (10.98), so in the noiseless case $E = 0$, these weightings are without consequence. However, when noise is present, they have an important influence on the space spanned by U_1, and hence on the quality of the estimates $\hat{C}$ and $\hat{A}$. We may remark that post-multiplying W_2 by an orthonormal matrix does not effect the U_1-matrix in the decomposition. See Problem 10E.10. We shall return to these questions below.

Estimating the Extended Observability Matrix

The Basic Expression. From (10.80) we find that

$$\begin{aligned}
y(t+k) &= Cx(t+k) + Du(t+k) + v(t+k) \\
&= CAx(t+k-1) + CBu(t+k-1) + Cw(t+k-1) \\
&\quad + Du(t+k) + v(t+k) \\
&= \dots \\
&= CA^k x(t) + CA^{k-1}Bu(t) + CA^{k-2}Bu(t+1) + \dots \\
&\quad + CBu(t+k-1) + Du(t+k) \\
&\quad + CA^{k-1}w(t) + CA^{k-2}w(t+1) + \dots \\
&\quad + Cw(t+k-1) + v(t+k)
\end{aligned} \tag{10.99}$$

Now, form the vectors

$$Y_r(t) = \begin{bmatrix} y(t) \\ y(t+1) \\ \vdots \\ y(t+r-1) \end{bmatrix}, \qquad U_r(t) = \begin{bmatrix} u(t) \\ u(t+1) \\ \vdots \\ u(t+r-1) \end{bmatrix} \tag{10.100}$$

and collect (10.99) as

$$Y_r(t) = O_r x(t) + S_r U_r(t) + V(t) \tag{10.101}$$

with

$$S_r = \begin{bmatrix} D & 0 & \cdots & 0 & 0 \\ CB & D & \cdots & 0 & 0 \\ \vdots & \vdots & \ddots & \vdots & \vdots \\ CA^{r-2}B & CA^{r-3}B & \cdots & CB & D \end{bmatrix}$$

and the k:th block component of $V(t)$

$$\begin{aligned} V_k(t) &= CA^{k-2}w(t) + CA^{k-3}w(t+1) + \dots \\ &\quad + Cw(t+k-2) + v(t+k-1) \end{aligned} \tag{10.102}$$

We shall use (10.101) to estimate O_r, or rather a matrix O_rT for some (unknown) T. The idea is to correlate both sides of (10.101) with quantities that eliminate the term with $U_r(t)$ and make the noise influence from V disappear asymptotically. For this, we have the measurements $y(t), u(t), t = 1, \dots, N+r-1$ available. It will be easier to describe the correlation operations as matrix multiplications, and we therefore introduce

$$\begin{aligned} \mathbf{Y} &= \begin{bmatrix} Y_r(1) & Y_r(2) & \dots & Y_r(N) \end{bmatrix} \\ \mathbf{X} &= \begin{bmatrix} x(1) & x(2) & \dots & x(N) \end{bmatrix} \\ \mathbf{U} &= \begin{bmatrix} U_r(1) & U_r(2) & \dots & U_r(N) \end{bmatrix} \\ \mathbf{V} &= \begin{bmatrix} V(1) & V(2) & \dots & V(N) \end{bmatrix} \end{aligned} \tag{10.103}$$

These quantities depend on r and N, but this dependence is suppressed. We can now rewrite (10.101) as the basic expression

$$\mathbf{Y} = O_r\mathbf{X} + S_r\mathbf{U} + \mathbf{V} \tag{10.104}$$

Remark. Define as in (7.59) the vector $\hat{\mathbf{Y}}$ of true k-step ahead predictors. Then it follows from (10.104) that

$$\hat{\mathbf{Y}} = O_r\hat{\mathbf{X}} \tag{10.105}$$

where $\hat{\mathbf{X}}$ is made up from the predicted (Kalman-filter) states $\hat{x}(t|t-1)$, i.e. the best estimate of $x(t)$ based on past input-output data.

Removing the U-term. Form the $N \times N$ matrix

$$\Pi^{\perp}_{\mathbf{U}^{\mathrm{T}}} = \mathbf{I} - \mathbf{U}^{\mathrm{T}}(\mathbf{U}\mathbf{U}^{\mathrm{T}})^{-1}\mathbf{U} \tag{10.106}$$

(if $\mathbf{U}\mathbf{U}^{\mathrm{T}}$ is singular, use pseudoinverse instead). This matrix performs projection, orthogonal to the matrix $\mathbf{U}$, i.e.,

$$\mathbf{U}\Pi^{\perp}_{\mathbf{U}^{\mathrm{T}}} = \mathbf{U} - \mathbf{U}\mathbf{U}^{\mathrm{T}}(\mathbf{U}\mathbf{U}^{\mathrm{T}})^{-1}\mathbf{U} = \mathbf{0}$$

Multiplying (10.104) from the right by $\Pi^{\perp}_{\mathbf{U}^{\mathrm{T}}}$ will thus eliminate the term with $\mathbf{U}$:

$$\mathbf{Y}\Pi^{\perp}_{\mathbf{U}^{\mathrm{T}}} = O_r\mathbf{X}\Pi^{\perp}_{\mathbf{U}^{\mathrm{T}}} + \mathbf{V}\Pi^{\perp}_{\mathbf{U}^{\mathrm{T}}} \tag{10.107}$$

Removing the Noise Term. The next problem is to eliminate the last term. Since this term is made up of noise contributions, the idea is to correlate it away with a suitable matrix. Define the $s \times N$ matrix ($s \geq n$)

$$\Phi = [\, \varphi_s(1) \quad \varphi_s(2) \quad \ldots \quad \varphi_s(N)\,] \tag{10.108}$$

where $\varphi_s(t)$ is a yet undefined vector. Multiply (10.107) from the right by Φ^T and normalize by N:

$$G = \frac{1}{N}\mathbf{Y}\Pi^{\perp}_{\mathbf{U}^T}\Phi^T = O_r\frac{1}{N}\mathbf{X}\Pi^{\perp}_{\mathbf{U}^T}\Phi^T + \frac{1}{N}\mathbf{V}\Pi^{\perp}_{\mathbf{U}^T}\Phi^T \stackrel{\text{def}}{=} O_r\tilde{T}_N + V_N \tag{10.109}$$

Here $\tilde{T}_N$ is an $n \times s$ matrix. Suppose now that we can find $\varphi_s(t)$ so that

$$\lim_{N\to\infty} V_N = \lim_{N\to\infty}\frac{1}{N}\mathbf{V}\Pi^{\perp}_{\mathbf{U}^T}\Phi^T = 0 \tag{10.110a}$$

$$\lim_{N\to\infty} \tilde{T}_N = \lim_{N\to\infty}\frac{1}{N}\Pi^{\perp}_{\mathbf{U}^T}\Phi^T = \tilde{T} \quad \text{has full rank } n \tag{10.110b}$$

Then (10.109) would read

$$\begin{aligned} G &= \frac{1}{N}\mathbf{Y}\Pi^{\perp}_{\mathbf{U}^T}\Phi^T = O_r\tilde{T} + E_N \\ E_N &= O_r(\tilde{T}_N - \tilde{T}) + V_N \to 0 \text{ as } N \to \infty \end{aligned} \tag{10.111}$$

The $pr \times s$ matrix G can thus be seen as a noisy estimate (10.95) and we can subject it to the treatment (10.96)–(10.98) to obtain estimates of A and C.

Finding Good Instruments. The only remaining question is how to achieve (10.110). Notice that these requirements are just like (7.119) and (7.120) for the *instrumental variable method*. Using the expression (10.106) for $\Pi^{\perp}_{\mathbf{U}^T}$ and writing out the matrix multiplications as sums gives

$$\begin{aligned} \frac{1}{N}\mathbf{V}\Pi^{\perp}_{\mathbf{U}^T}\Phi^T &= \frac{1}{N}\sum_{t=1}^{N} V(t)\varphi_s^T(t) - \frac{1}{N}\sum_{t=1}^{N} V(t)U_r^T(t) \\ &\times \left[\frac{1}{N}\sum_{t=1}^{N} U_r(t)U_r^T(t)\right]^{-1} \frac{1}{N}\sum_{t=1}^{N} U_r(t)\varphi_s^T(t) \end{aligned} \tag{10.112}$$

Under mild conditions, the law of large numbers states that the sample sums converge to their respective expected values

$$\lim_{N\to\infty}\frac{1}{N}\mathbf{V}\Pi^{\perp}_{\mathbf{U}^T}\Phi^T = \overline{E}V(t)\varphi_s^T(t) - \overline{E}V(t)U_r^T(t)R_u^{-1}\overline{E}U_r(t)\varphi_s^T(t) \tag{10.113}$$

where

$$R_u = \overline{E}U_r(t)U_r^T(t)$$

is the $r \times r$ covariance matrix of the input. Now, assume that the input u is generated in open loop, so that it is independent of the noise terms in V (see (10.102)). Then $EV(t)U_r^T(t) = 0$. Assume also that R_u is invertible. (This means that the input is *persistently exciting* of order r—see Section 13.2.) Then the second term of (10.113) will be zero. (If the pseudo-inverse is used in (10.106), this is still true, even if R_u is not invertible.) For the first term to be zero, we must require $V(t)$ and $\varphi_s(t)$ to be uncorrelated. Since $V(t)$ according to (10.102) is made up of white noise terms from time t and onwards, any choice $\varphi_s(t)$ built up from data prior to time t will satisfy (10.110a). A typical choice would be

$$\varphi_s(t) = \begin{bmatrix} y(t-1) \\ \vdots \\ y(t-s_1) \\ u(t-1) \\ \vdots \\ u(t-s_2) \end{bmatrix} \tag{10.114}$$

Now, turning to (10.110b) we find by a similar argument that

$$\tilde{T} = \overline{E}x(t)\varphi_s^T(t) - \overline{E}x(t)U_r^T(t)R_u^{-1}\overline{E}U_r(t)\varphi_s^T(t) \tag{10.115}$$

A formal proof that $\tilde{T}$ has full rank is not immediate and will involve properties of the input. See Problem 10G.6 and Van Overschee and DeMoor (1996).

Summing up, forming $G = \frac{1}{N}\mathbf{Y}\Pi_{\mathbf{U}^T}^{\perp}\Phi^T$ with Φ defined by (10.114) and (10.108) gives the properties (10.111), which allows us to consistently determine A and C, via (10.97), (10.98), and (10.88), (10.87). We shall sum up all the steps in the algorithm later.

Finding the States and Estimating the Noise Statistics

In (10.104) we constructed a direct relationship between future outputs and the states. Let us now shift the perspective somewhat and return to the prediction approach of Section 7.3. This will give an expression which is closely related to (10.104) and shows the links between states and predictors.

Let us estimate the k-step ahead predictors. Recall (7.63).

$$Y_r(t) = \Theta\varphi_s(t) + \Gamma U_r(t) + E(t) \tag{10.116}$$

Collecting all t as in (10.104) gives

$$\mathbf{Y} = \Theta\Phi + \Gamma\mathbf{U} + \mathbf{E} \tag{10.117}$$

The Least Squares estimate of the parameters is

$$[\hat{\Theta} \quad \hat{\Gamma}] = [\mathbf{Y}\Phi^T \quad \mathbf{Y}\mathbf{U}^T] \begin{bmatrix} \Phi\Phi^T & \Phi\mathbf{U}^T \\ \mathbf{U}\Phi^T & \mathbf{U}\mathbf{U}^T \end{bmatrix}^{-1} \tag{10.118}$$

Using the expression for inverting block matrices gives (see Problem 10E.11)

$$\hat{\Theta} = \mathbf{Y}\Pi^{\perp}_{\mathbf{U}^T}\Phi^T\left(\Phi\Pi^{\perp}_{\mathbf{U}^T}\Phi^T\right)^{-1} \tag{10.119}$$

This means that the matrix of predicted outputs can be written

$$\begin{aligned}\hat{\mathbf{Y}} &= \begin{bmatrix}\hat{Y}_r(1) & \dots & \hat{Y}_r(N)\end{bmatrix} \\ &= \mathbf{Y}\Pi^{\perp}_{\mathbf{U}^T}\Phi^T(\Phi\Pi^{\perp}_{\mathbf{U}^T}\Phi^T)^{-1}\Phi \end{aligned}\tag{10.120}$$

which we recognize as $\hat{G}$ in (10.97), with G as in (10.109) and the weightings $W_1 = I$ and $W_2 = (\Phi\Pi^{\perp}_{\mathbf{U}^T}\Phi^T)^{-1}\Phi$. Performing the SVD and deleting small singular values thus gives

$$\hat{\mathbf{Y}} \approx U_1 S_1 V_1^T \tag{10.121}$$

Here V_1^T is an $n \times N$ matrix. We know from (10.98) that U_1 is related to the observability matrix by some invertible matrix R as $U_1 = O_r R^{-1}$. Introduce $\hat{\mathbf{X}} = R^{-1}S_1V_1$. Then (10.121) can be written as

$$\hat{\mathbf{Y}} \approx O_r R^{-1} S_1 V_1^T = O_r\hat{\mathbf{X}} \tag{10.122}$$

Comparing with (10.105) shows that $\hat{\mathbf{X}}$ must be the matrix of the correct state estimates if $\hat{\mathbf{Y}}$ is the matrix of true predicted outputs. The true predicted outputs are, however, normally obtained only when the orders s_i in (10.114) tend to infinity. In such a case we have then found the true state estimates—in the state-space basis in question—from the SVD. Alternatively we can write

$$\begin{aligned}\hat{\mathbf{X}} &= L\hat{\mathbf{Y}} = \begin{bmatrix}\hat{x}(1) & \dots & \hat{x}(N)\end{bmatrix} \\ L &= R^{-1}U_1^T\end{aligned}\tag{10.123}$$

since $U_1^T U_1 = I$ from the SVD-properties. With the states given, we can estimate the process and measurement noises as

$$\begin{aligned} w(t) &= \hat{x}(t+1) - \hat{A}\hat{x}(t) - \hat{B}u(t) \\ v(t) &= y(t) - \hat{C}\hat{x}(t) - \hat{D}u(t)\end{aligned}\tag{10.124}$$

and estimate their covariance matrices in a straightforward fashion. Here $\hat{A}$, $\hat{B}$, $\hat{C}$, $\hat{D}$ are the estimates of the system matrices obtained as described above. Alternatively, these could be directly estimated by the least squares procedure (7.66), once the states are known.

Putting It All Together

We have now described all the basic steps of the algorithm, although in somewhat reverse order. The complete subspace algorithm can be summarized as follows:

The Family of Subspace Algorithms (10.125)

1. From the input-output data, form

$$G = \frac{1}{N}\mathbf{Y}\Pi^{\perp}_{\mathbf{U}^T}\Phi^T \tag{10.126}$$

with the involved matrices defined by (10.103), (10.100), (10.106), (10.114), and (10.108).

2. Select weighting matrices W_1 ($rp \times rp$ and invertible) and W_2 ($(ps_1 + ms_2) \times \alpha$) and perform SVD

$$\hat{G} = W_1 G W_2 = USV^T \approx U_1 S_1 V_1^T \tag{10.127}$$

where the last approximation is obtained by keeping the n most significant values of the singular values in S and setting the remaining ones to zero. (U_1 is now $rp \times n$. S_1 is $n \times n$ and V_1^T is $n \times \alpha$.) As remarked above, post-multiplying W_2 by any $\alpha \times k$ orthonormal matrix (with $k \geq rp$) will not change U_1. (See Problem 10E.10.)

3. Select a full rank matrix R and define the $rp \times n$ matrix $\hat{O}_r = W_1^{-1} U_1 R$. Solve

$$\hat{C} = \hat{O}_r(1 : p, 1 : n) \tag{10.128a}$$

$$\hat{O}_r(p+1 : pr, 1 : n) = O_r(1 : p(r-1), 1 : n)\hat{A} \tag{10.128b}$$

for $\hat{C}$ and $\hat{A}$. The latter equation should be solved in a least squares sense.

4. Estimate $\hat{B}$, $\hat{D}$ and $\hat{x}_0$ from the linear regression problem:

$$\arg\min_{B,D,x_0} \frac{1}{N}\sum_{t=1}^{N} \left\| y(t) - \hat{C}(qI - \hat{A})^{-1}Bu(t) - Du(t) - \hat{C}(qI - \hat{A})^{-1}x_0\delta(t) \right\|^2 \tag{10.129}$$

5. If a noise model is sought, form $\hat{\mathbf{X}}$ as in (10.123) and estimate the noise contributions as in (10.124).

Numerical Implementation. It should be mentioned that the most efficient numerical implementation of the above steps is to apply QR-factorization of the data matrix $[\mathbf{U}^T \;\; \Phi^T \;\; \mathbf{Y}^T]^T = LQ$. ($L$ is here a lower triangular, $pr + mr + s$ square matrix, while Q is an orthonormal $(pr + mr + s) \times N$ matrix.) Then the crucial SVD-factor U_1 in (10.127) can be found entirely from the "L-part" of this factorization, i.e., the "small" matrix. See Problem 10G.5 as well as Van Overschee and DeMoor (1996) and Viberg, Wahlberg, and Ottersten (1997).

The family of methods contains a number of design variables. Different algorithms described in the literature correspond to different choices of these variables. It is at present not fully understood how to choose them optimally. The choices are:

- The choice of correlation vector $\varphi_s(t)$ in (10.114). The requirement is that (10.110) holds. Notice that this choice also determines the ARX-model that is used to approximate the system when finding the k-step ahead predictors in (7.62). Most algorithms choose $\varphi_s(t)$ to consist of past inputs and outputs as in (10.114) with $s_1 = s_2$. The scalar s is then the only design variable.

If $s_1 = 0$ only past inputs are used, and an *output-error* variant of the algorithm is obtained. Then the noise is ignored when forming the predictors, which leads to models like (7.55) that do not attempt to describe the noise properties, but therefore also may describe the input-output properties with fewer states. The OE-MOESP algorithm of Verhaegen (1994)uses this choice of regressors.

- The scalar r, which is the maximal prediction horizon used. Many algorithms use $r = s$, but there is no particular reason for such a choice.
- The weighting matrices W_1 and W_2. This is the perhaps most important choice. Existing algorithms employ the following choices:
  - MOESP, Verhaegen (1994): $W_1 = I$, $W_2 = (\frac{1}{N}\Phi\Pi^{\perp}_{\mathbf{U}^T}\Phi^T)^{-1}\Phi\Pi^{\perp}_{\mathbf{U}^T}$
  - N4SID, Van Overschee and DeMoor (1994): $W_1 = I$, $W_2 = (\frac{1}{N}\Phi\Pi^{\perp}_{\mathbf{U}^T}\Phi^T)^{-1}\Phi$ (see also (10.120).)
  - IVM, Viberg (1995): $W_1 = (\frac{1}{N}\mathbf{Y}\Pi^{\perp}_{\mathbf{U}^T}\mathbf{Y})^{-1/2}$, $W_2 = (\frac{1}{N}\Phi\Phi^T)^{-1/2}$
  - CVA, Larimore (1990): $W_1 = (\frac{1}{N}\mathbf{Y}\Pi^{\perp}_{\mathbf{U}^T}\mathbf{Y})^{-1/2}$, $W_2 = (\frac{1}{N}\Phi\Pi^{\perp}_{\mathbf{U}^T}\Phi^T)^{-1/2}$

  (Note that W_2 can be expressed in several ways, since post-multiplying with any orthonormal matrix does not change the resulting estimates.) The effects of the weightings are discussed in several papers. See the bibliography.
- The matrix R in step 3. Typical choices are $R = I$, $R = S_1$ or $R = S_1^{1/2}$.

10.7 SUMMARY

Determining a parameter estimate from data has two aspects. First, one has to decide how to characterize the sought estimate: as the solution of a certain equation or as the minimizing argument of some function. Second, one has to devise a numerical method that calculates that estimate. It is important to keep these issues separate. The combination of several different approaches to characterize the desired estimate with many techniques to actually compute it has lead to a wide, and sometimes confusing, variety of identification methods. Our aim in this chapter, as well as in Chapter 7, has been to point to underlying basic ideas.

For linear regression problems (LS and IV methods), we have recommended QR factorization-type methods (10.11) and also pointed to the possibilities of using Levinson and/or lattice methods [(10.23) and (10.30), (10.32) respectively] for special structures.

For general PEM, we have recommended the damped Gauss-Newton iterative method (10.40), (10.41), and (10.46) as the basic choice, complemented with (10.79) to find initial values for linear black-box models.

For subspace methods the essential parts of the numerical calculations consist of a QR-factorization step and an SVD. This allows very robust numerical methods for the calculation of the estimates.

10.8 BIBLIOGRAPHY

The computation of estimates is of course a topic that is covered in many articles and books on system identification. The basic techniques are also the subject of many studies in numerical analysis.

For the linear least-squares problem of Section 10.1, an excellent overview is given in Lawson and Hanson (1974). An account of the Levinson algorithm and its ramifications is given, for example, in Kailath (1974). An early application of the Levinson algorithm for estimation problems was given by Durbin (1960). The algorithm with estimated $R(\tau)$ is therefore sometimes referred to as the Levinson-Durbin algorithm. The multivariable Levinson algorithm was given in Whittle (1963)and Wiggins and Robinson (1965). A Levinson algorithm for the "covariance method" [for which $R(N)$ is not a Toepliz matrix, but deviates "slightly" from this structure] was derived by Morf et.al. (1977). Levinson algorithm has been widely applied in geophysics (e.g., Robinson, 1967; Burg, 1967) and speech processing (e.g., Markel and Gray, 1976), while it has been less used in control applications. Its numerical properties are investigated in Cybenko (1980).

Lattice filters are used extensively in Markel and Gray (1976), Honig and Messerschmitt (1984), and Rabiner and Schafer (1978), in addition to the references mentioned in the text. The numerical stability of the calculation of reflection coefficients in (10.30), and (10.32) has been analyzed by Cybenko (1984). Considerable attention has been paid to the recursive updating of the reflection coefficients, which we shall return to in Chapter 11.

For the methods of Section 10.2, Dennis and Schnabel (1983)serves as a basic references. It contains many additional references and also pseudocode for typical algorithms. Variants of the Newton methods for system identification applications have been discussed in, for example, Åström and Bohlin (1965), Gupta and Mehra (1974), Kashyap and Nasburg (1974). The gradients $\psi = (\partial/\partial\theta)\hat{y}$, $(\partial/\partial\theta)\hat{x}$, and so on, are known as sensitivity functions or sensitivity derivatives. These have been studied also in connection with sensitivity analysis of control design. Simple methods for calculating these gradients in state-space models have been discussed in, for example, Denery (1971)and Neuman and Sood (1972), as well as in Gupta and Mehra (1974) and Hill (1985). The use of Lagrangian multipliers to reduce the computational burden is described in Kashyap (1970)and van Zee and Bosgra (1982). Another possibility is to apply Parseval's relationship to V_N' and H_N in (10.39) and (10.45) and evaluate these in the frequency-domain in terms of the Fourier transforms of the signals. See Hannan (1969)and Akaike (1973). The expressions follow easily from (7.25) and (9.53). A special technique for maximizing the likelihood functions, the EM algorithm, has been developed by Dempster, Laird, and Rubin (1977). See Problem 10G.3.

Further results on the uniqueness of solutions are given in Chapter 12 of Söderström and Stoica (1989).

Analysis of bootstrap methods is carried out in Stoica et.al. (1985). Two- and multistage methods have been discussed in many variants. In addition to those described in Section 10.4, there are, for example, the well-known methods of Durbin (1959)and Walker (1961). Both start with a high-order AR model. The coefficients of these models (the corresponding covariance functions in the Walker case) lead to a system of equations for solving for MA parameters. See also Anderson (1971), Section 5.7.2. Other techniques to build models by reduction of high order ARX-models are described in Wahlberg (1989)and Zhu and Backx (1993).

The Subspace methods really originate from classical realization theory as formulated in Ho and Kalman (1966)and Kung (1978). These algorithms pointed to the essential relationships (10.88) and (10.87). The extended observability matrix can also be found by factorizing the Hankel matrix of impulse responses, and several identification methods based on this have been devised, like King, Desai, and Skelton (1988), Liu and Skelton (1992). Larimore developed his algorithms in Larimore (1983), Larimore (1990)inspired by Akaike's work on canonical correlation, Akaike (1974b)and Akaike (1976). Related algorithms were developed by Aoki in Aoki (1987)for the time-series case.

The family of methods developed with the related approaches by Moonen et.al. (1989), Verhaegen (1991), lead to the basic presentations Verhaegen (1994)and Van Overschee and DeMoor (1994). The relationships between the approaches have been pointed out by Viberg (1995)and Van Overschee and DeMoor (1996), which can be recommended as general overviews. Statistical analysis is presented in Peternell, Scherrer, and Deistler (1996). Special techniques to handle closed loop data are described in Chou and Verhaegen (1997). The presentation in Section 10.6 is largely based on Viberg, Wahlberg, and Ottersten (1997), with a similar perspective described in Jansson and Wahlberg (1996).

Subspace methods to fit frequency-domain data are treated in McKelvey, Akcay, and Ljung (1996).

10.9 PROBLEMS

10G.1 Let $z(t)$ be a p-dimensional signal, and let $\hat{a}_i^n$ and $\hat{b}_i^n$ ($p \times p$ matrices) be the least-squares estimates of the linear regression models

$$\begin{aligned} \hat{z}^n(t|\theta) &= -a_1^n z(t-1) - \cdots - a_n^n z(t-n) \\ \check{z}^n(t-n-1|\theta) &= -b_1^n z(t-n) - \cdots - b_n^n z(t-1) \end{aligned} \tag{10.130}$$

based on data $z(t)$, $1 \le t \le N$. Show, by arguments analogous to (10.15) to (10.24), that these estimates can be computed as

$$\hat{a}_k^{n+1} = \hat{a}_k^n + \rho_n^a \hat{b}_{n-k+1}^n, \quad \hat{a}_{n+1}^{n+1} = \rho_n^a$$

$$\hat{b}_k^{n+1} = \hat{b}_k^n + \rho_n^b \hat{a}_{n-k+1}^n, \quad \hat{b}_{n+1}^{n+1} = \rho_n^b$$

$$\alpha_n = R_{n+1} + \sum_{k=1}^{n} \hat{a}_k^n R_{n+1-k}, \quad \beta_n = R_{-n-1} + \sum_{k=1}^{n} \hat{b}_k^n R_{-n-1+k}, \ (= \alpha_n^T)$$

$$V_{n+1}^a = V_n^a - \alpha_n [V_n^b]^{-1} \beta_n, \quad V_{n+1}^b = V_n^b - \beta_n [V_n^a]^{-1} \alpha_n$$

$$\rho_n^a = -\alpha_n [V_n^b]^{-1}, \quad \rho_n^b = -\beta_n [V_n^a]^{-1}$$

Here

$$R_k = \frac{1}{N} \sum_{t=-n}^{N+n} z(t+k)z^T(t)$$

with $z(t) = 0$ outside the interval $1 \le t \le N$. (This is the multivariable Levinson algorithm, as derived by Whittle, 1963.)

10G.2 *Steiglitz-McBride method:* Steiglitz and McBride (1965) have suggested the following approach to identify a linear system subject to white measurement errors: Consider the OE model (4.25)

$$y(t) = \frac{B(q)}{F(q)}u(t) + e(t)$$

Step 1. Apply the LS method to the ARX model

$$F(q)y(t) = B(q)u(t) + e(t)$$

This gives $\hat{B}_N(q)$ and $\hat{F}_N(q)$.

Step 2. Filter the data through the prefilter

$$y_F(t) = \frac{1}{\hat{F}_N(q)}y(t), \qquad u_F(t) = \frac{1}{\hat{F}_N(q)}u(t)$$

Step 3. Apply the LS method to the ARX model

$$F(q)y_F(t) = B(q)u_F(t) + e(t)$$

Repeat from step 2 with the new $\hat{F}$ estimate. Stop when $\hat{F}_N$ and $\hat{B}_N$ have converged.

(a) This method can be interpreted as a way of solving for a correlation estimate as in (7.110) with a θ-dependent prefilter. What is the correlation vector $\zeta(t, \theta)$ and what is the prefilter $L(q, \theta)$?

(b) By what numerical technique (according to the classification of this chapter) is the estimate computed?

(c) Suppose the numerical scheme converges to a unique solution $\hat{B}_N$, $\hat{F}_N$ of the correlation equation. Use Theorem 8.6 to discuss whether these estimates will be consistent if the true system is described by

$$y(t) = \frac{B_0(q)}{F_0(q)}u(t) + v_0(t)$$

where $\{v_0(t)\}$ is white or colored, respectively, noise. In case $\{v_0(t)\}$ is white, what does Theorem 9.2 say about the asymptotic variance of the estimate? (See, also, Stoica and Söderström, 1981a, for the analysis.)

10G.3 *The EM algorithm:* Consider Problem 7G.6 and the expression (7.161) for the negative log likelihood function:

$$V(\theta) = -\log p(Y|\theta) = -\log p(Y, X|\theta) + \log p(X|\theta, Y)$$

This expression holds for all X, and can thus be integrated over any measure for X, $f(X)$, without affecting the X-independent left side:

$$V(\theta) = -\int_{X\in\mathbf{R}^n} \log p(Y, X|\theta) f(X)\, dX + \int_{X\in\mathbf{R}^n} \log p(X|\theta, Y) f(X)\, dX$$

Let now in particular $f(X)$ be the conditional PDF for X, given Y and assuming $\theta = \alpha$:

$$f(X) = p(X|Y, \alpha)$$

Then

$$V(\theta) = H_1(Y, \theta, \alpha) + H_2(Y, \theta, \alpha)$$

$$H_1(Y, \theta, \alpha) = -\int_{X\in\mathbf{R}^n} \log p(Y, X|\theta) p(X|Y, \alpha)\, dX = E\left(-\log p(Y, X|\theta)|Y, \alpha\right)$$

$$H_2(Y, \theta, \alpha) = \int \log p(X|Y, \theta) \cdot p(X|Y, \alpha)\, dX$$

The EM-algorithm (Dempster, Laird, and Rubin, 1977) for minimizing $V(\theta)$ consists of the following steps:

1. Fix α_k and determine the conditional mean of $-\log p(Y, X|\theta)$ with respect to X, given Y under the assumption that the true value of θ is α_k. This gives $H_1(Y, \theta, \alpha_k)$. (Note that the θ in $p(Y, X|\theta)$ is left as a free variable.)
2. Minimize

$$H_1(Y, \theta, \alpha_k)$$

with respect to θ giving $\hat{\theta}_k$.

3. Set $\alpha_{k+1} = \hat{\theta}_k$ and repeat from 1.

(a) Now, show that

$$H_2(Y, \hat{\theta}_k, \alpha_k) \leq H_2(Y, \alpha_k, \alpha_k)$$

and that hence

$$V(\hat{\theta}_k) \leq V(\alpha_k)$$

The algorithm thus produces decreasing values of the negative log likelihood function.

(b) Write out the EM-algorithm applied to the case of Problem 7G.6.

[Step 1 is the Estimation and step 2 the Minimization step in the EM-algorithm. The algorithm is useful when the likelihood function given Y is complicated and the addition of some auxiliary measurements X would have given a much simpler likelihood function. We thus expand the problem with these fake measurements and average over them using their conditional density given the actual observations and that the system is described by the current θ-estimate. Note that $H_2(Y, \theta, \alpha)$ is never formed in the algorithm.]

10G.4 Let $\mathbf{Y}$ and $\mathbf{U}$ be defined by (10.103). Consider the following LQ-factorization

$$\begin{bmatrix} \mathbf{U} \\ \mathbf{Y} \end{bmatrix} = \begin{bmatrix} L_{11} & 0 \\ L_{21} & L_{22} \end{bmatrix} \begin{bmatrix} Q_1^T \\ Q_2^T \end{bmatrix}$$

where

$$\begin{bmatrix} Q_1^T \\ Q_2^T \end{bmatrix} [\, Q_1 \quad Q_2 \,] = \begin{bmatrix} I & 0 \\ 0 & I \end{bmatrix}$$

Show that

$$\mathbf{Y}\Pi^{\perp}_{\mathbf{U^T}} = \mathbf{Y}(\mathbf{I} - \mathbf{U^T}(\mathbf{U}\mathbf{U^T})^{-1}\mathbf{U}) = L_{22}Q_2^T$$

Hint: Just plug in the factorized expressions for $\mathbf{U}$ and $\mathbf{Y}$.

10G.5 Let $\mathbf{Y}$, $\mathbf{U}$ and Φ be defined by (10.103) and (10.108). Consider the LQ-factorization

$$\begin{bmatrix} \mathbf{U} \\ \Phi \\ \mathbf{Y} \end{bmatrix} = \begin{bmatrix} L_{11} & 0 & 0 \\ L_{21} & L_{22} & 0 \\ L_{31} & L_{32} & L_{33} \end{bmatrix} \begin{bmatrix} Q_1^T \\ Q_2^T \\ Q_3^T \end{bmatrix}, \quad Q^T Q = I.$$

Show that

(a) $\mathbf{Y}/_{\mathbf{U}}\Phi = \mathbf{Y}\Pi^{\perp}_{\mathbf{U^T}}\Phi^T(\Phi\Pi^{\perp}_{\mathbf{U^T}}\Phi^T)^{-1}\Phi = L_{32}L_{22}^{-1}[\, L_{21} \quad L_{22} \,]\begin{bmatrix} Q_1^T \\ Q_2^T \end{bmatrix}$

(b) $\mathbf{Y}/_{\mathbf{U}}\Phi\Pi^{\perp}_{U^T} = L_{32}Q_2^T$

Here the big 0-part of L and the corresponding rows of Q of the original LQ-factorization have been thrown away. You may assume that indicated inverses exist.

Hint: Compare with the previous exercise.

(The notation $\mathbf{Y}/_{\mathbf{U}}\Phi$ for (10.120) is due to Van Overschee and DeMoor (1994)and is to be read: "The oblique projection of $\mathbf{Y}$ onto the space Φ, along the row space of $\mathbf{U}$".)

Note that (a) corresponds to the N4SID choice of $\hat{G}$ according to (10.120). Moreover (b) corresponds to the matrix $\hat{G}$ used in MOESP. Finally, note that according to Problem 10E.10, you can always post-multiply the matrices with an orthonormal matrix, without affecting the factor U_1 in the SVD (10.127). Hence we can work entirely with the "L-parts" of the factorizations above, and throw away the (big) Q-parts, when performing the calculations in (10.125).

10G.6 Show that (10.115) has full rank n provided

1. $\mathrm{E}\varphi_s(t)\varphi_s^T(t)$ is positive definite.
2. $\mathrm{E}\begin{bmatrix} x(t) \\ U_r(t) \end{bmatrix} [\, x^T(t) \quad U_r^T(t) \,]$ is positive definite. This means that the r future inputs should not be linearly dependent on the current state.
3. s_1 and s_2 are sufficiently large so that $\hat{x}(t) \approx L_x\varphi_s(t)$ for some L_x. (See (10.123)).

Hint: Use that $\mathrm{E}x(t)\varphi_s^T(t) = \mathrm{E}\hat{x}(t)\varphi_s^T(t)$, where $\hat{x}(t)$ is that part of the state that can be reconstructed from past input-outputs. Similarly $\mathrm{E}x(t)U_r^T(t) = \mathrm{E}\hat{x}(t)U_r^T(t)$.

10E.1 In Hsia (1977), Section 6.7, the following identification procedure is described. Let

$$\varphi_1^T(t) = \left[-y(t-1)\dots -y(t-n_a) \quad u(t-1)\dots u(t-n_b)\right]$$

$$\rho^T = \left[a_1 \dots a_{na}\, b_1 \dots b_{nb}\right]$$

The model is then written as

$$y(t) = \varphi_1^T(t)\rho + \varepsilon(t)$$

The estimate $\hat{\rho}_N$ is computed as a "bias correction"

$$\hat{\rho}_N = \hat{\rho}_N^{\text{LS}} - \hat{\rho}_N^{\text{BIAS}} \tag{10.131}$$

where $\hat{\rho}_N^{\text{LS}}$ and $\hat{\rho}_N^{\text{BIAS}}$ are computed iteratively as follows:
Step 1. Let

$$R_N^{(1)} = \frac{1}{N}\sum_{t=1}^{N} \varphi_1(t)\varphi_1^T(t)$$

and

$$\hat{\rho}_N^{\text{LS}} = \left[R_N^{(1)}\right]^{-1} \frac{1}{N}\sum_{t=1}^{N} \varphi_1(t)y(t)$$

$$\hat{\rho}_N^{\text{BIAS}} = 0$$

Step 2. Let

$$\hat{\rho}_N = \hat{\rho}_N^{\text{LS}} - \hat{\rho}_N^{\text{BIAS}}$$

Step 3. Let

$$\varepsilon(t) = y(t) - \varphi_1^T(t)\hat{\rho}_N$$

and define

$$\varphi_2^T(t) = [-\varepsilon(t), \dots, -\varepsilon(t-n_d)]$$

Let

$$R_N^{(2)} = \frac{1}{N}\sum_{t=1}^{N} \varphi_2(t)\varphi_2^T(t)$$

$$R_N^{(12)} = \frac{1}{N}\sum_{t=1}^{N} \varphi_1(t)\varphi_2^T(t)$$

Compute

$$D_N = R_N^{(2)} - \left[R_N^{(12)}\right]^T \left[R_N^{(1)}\right]^{-1} R_N^{(12)}$$

$$\hat{\rho}_N^{\text{BIAS}} = \left[R_N^{(1)}\right]^{-1} R_N^{(12)} D_N^{-1} \left[\frac{1}{N}\sum_{t=1}^{N} \varphi_2(t)y(t) - \left[R_N^{(12)}\right]^T \left[R_N^{(1)}\right]^{-1} \frac{1}{N}\sum_{t=1}^{N} \varphi_1(t)y(t)\right]$$

and repeat from step 2 until convergence.

Now show that this procedure is the bootstrap algorithm (10.64) for the PLR method (7.113) using the ARARX model (4.22).

10E.2 Consider the model structure of Problem 5E.1. Give an expression for how to compute the gradient

$$\psi(t,\theta) = \frac{d}{d\theta}\hat{y}(t|\theta)$$

10E.3 Apply the Gauss-Newton method (10.40) and (10.46) to the linear regression problem (10.1) with quadratic criterion. Take $\mu = 1$.

10E.4 Introduce the approximation

$$\varepsilon(t,\theta) \approx \varepsilon(t,\hat{\theta}^{(i-1)}) + \psi^T(t,\hat{\theta}^{(i-1)})(\theta - \hat{\theta}^{(i-1)})$$

Use this approximation in

$$V_N(\theta, Z^N) = \frac{1}{2}\sum_{t=1}^{N}\varepsilon^2(t,\theta) \tag{10.132}$$

and solve for the minimizing θ. Show that this gives the (undamped) Gauss-Newton method (10.40) and (10.46).

10E.5 Consider Problem 10E.4. Minimize (10.132) subject to the constraint

$$\left\|\theta - \hat{\theta}^{(i-1)}\right\| \le C_\delta$$

Discuss the relationship to the Levenberg-Marquardt method (10.47).

10E.6 Let V_n be defined by (10.23) and (10.24). Show that

$$V_n = \frac{1}{N}\sum_{t=1}^{N}\left(y(t) - \hat{y}_n(t|\theta)\right)^2 = \frac{1}{N}\sum_{t=1}^{N}e_n^2(t)$$

with $\hat{y}^n$ given by (10.15).

10E.7 Consider the ARX model

$$y(t) + a_1 y(t-1) + \cdots + a_n y(t-n) = b_1 u(t-1) + \cdots + b_n u(t-n) + e(t)$$

Introduce

$$z(t) = \begin{bmatrix} -y(t) \\ u(t) \end{bmatrix}$$

and show how the estimates of a_i and b_i can be computed using the multivariable Levinson algorithm of Problem 10G.1.

10E.8 Show that, in the lattice filter, we have

$$\hat{y}_n(t|\hat{\theta}^n) = -\hat{\rho}_1\hat{r}_1(t-1) - \hat{\rho}_2\hat{r}_2(t-1) - \cdots - \hat{\rho}_n\hat{r}_n(t-1)$$

Compute the covariance matrix of $\hat{\rho}_i$, $\quad i = 1, \ldots, n$.

10E.9 Apply the method (10.68) to the ARARX model (10.66). Spell out the steps explicitly and show that they consist of a sequence of LS problems mixed with simple filtering operations. [This is the generalized least squares (GLS) method, developed by Clarke (1967). See also Söderström (1974).]

10E.10 Let the $p \times N$ matrix G be given, with $p < N$. Let its SVD be

$$G = USV^T$$

(U is $p \times p$ with $U^T U = I$, and V is $N \times N$ with $V^T V = I$, and S is a $p \times N$ matrix with the singular values along the diagonal, and zeros elsewhere.) Suppose $p \leq r < N$ and W is a $N \times r$ matrix such that $W^T W = I$. Let

$$GW = U_1 S_1 V_1^T$$

be the SVD of GW. Show that $U = U_1$. (Hint: Note that, with MATLAB notation, `S*V'=S(:,1:r)*V(:,1:r)'`. Then use that `W'*V(:,1:r)` will be orthogonal.)

10E.11 The block matrix inversion lemma says:

$$\begin{bmatrix} A & D \\ C & B \end{bmatrix}^{-1} = \begin{bmatrix} \Delta^{-1} & -\Delta^{-1} D B^{-1} \\ -B^{-1} C \Delta^{-1} & B^{-1} + B^{-1} C \Delta^{-1} D B^{-1} \end{bmatrix}$$

where

$$\Delta = A - DB^{-1}C$$

Apply this to the matrix in (10.118); show that Δ becomes $\Phi \Pi_{\mathbf{U^T}}^{\perp} \Phi^T$ and that (10.119) holds.

10T.1 *Householder transformations:* A Householder transformation is a matrix

$$Q = I - 2ww^T$$

where w is a column vector with norm 1. Show the following

(a) Q is symmetric and orthogonal.

(b) Let x be an arbitrary vector. Then there exists a Q as in part (a) such that

$$Qx = |x| \begin{bmatrix} 1 \\ 0 \\ \vdots \\ 0 \end{bmatrix}$$

(c) Let A be an $n \times m$ matrix, $n \geq m$. Then there exists an orthogonal matrix

$$\overline{Q} = Q_m Q_{m-1} \cdots Q_1$$

being a product of Householder transformations such that

$$\overline{Q} A = \begin{bmatrix} R \\ 0 \end{bmatrix}$$

where R is a square ($m \times m$) upper triangular matrix (see Lawson and Hanson, 1974).

10T.2 Consider the system

$$A_0(q)y(t) = B_0(q)u(t) + C_0(q)e_0(t)$$

and an ARMAX model structure

$$\theta = \left[a_1 \dots a_{n_a}\, b_1 \dots b_{n_b}\, c_1 \dots c_{n_c}\right]^T$$

$$A(q)y(t) = B(q)u(t) + C(q)e(t)$$

with polynomial orders larger or equal to those of the true system. Let

$$\tilde{D}_{\mathcal{M}} = \left\{\theta \left| \operatorname{Re}\frac{C(e^{i\omega})}{C_0(e^{i\omega})} > 0 \,\forall\omega \right.\right\}$$

Show that the prediction-error criterion

$$\overline{V}(\theta) = \overline{E}\varepsilon^2(t,\theta)$$

has no false local minimum in $\theta \in \tilde{D}_{\mathcal{M}}$; that is,

$$\overline{V}'(\theta) = 0 \text{ and } \theta \in \tilde{D}_{\mathcal{M}} \Rightarrow \frac{B(q)}{A(q)} = \frac{B_0(q)}{A_0(q)}, \qquad \frac{C(q)}{A(q)} = \frac{C_0(q)}{A_0(q)}$$

10T.3 Consider the method (10.68) to minimize (10.67) for a bilinear parametrization. Write (10.68) as an update step (10.40) with a block-diagonal $R_N^{(i)}$ matrix. It is thus indeed a descent method that will converge to a local minimum.

10D.1 Verify the relationships (10.31). *Hint:* By definition,

$$\frac{1}{N}\sum_{t=1}^{N} \hat{e}_n(t)y(t-k) = 0 \qquad 1 \le k \le n$$

[the residuals are uncorrelated with the regressors; see Figure II.1].

10D.2 Use Problem 10G.1 to derive a lattice filter for a multivariable signal $z(t)$.

11

RECURSIVE ESTIMATION METHODS

11.1 INTRODUCTION

In many cases it is necessary, or useful, to have a model of the system available on-line while the system is in operation. The model should then be based on observations up to the current time. The need for such an on-line model construction typically arises since a model is required in order to take some decision about the system. This could be

- Which input should be applied at the next sampling instant?
- How should the parameters of a matched filter be tuned?
- What are the best predictions of the next few outputs?
- Has a failure occurred and, if so, of what type?

Methods coping with such problems using an on-line adjusted model of some sort are usually called *adaptive* (see Figure 11.1). We thus talk about adaptive control, adaptive filtering, adaptive signal processing, and adaptive prediction.

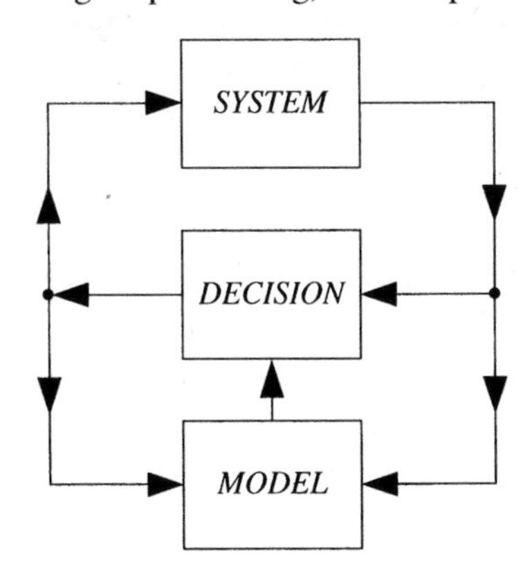

Figure 11.1 Adaptive methods.

The on-line computation of the model must also be done in such a way that the processing of the measurements from one sample can, with certainty, be completed during one sampling interval. Otherwise the model building cannot keep up with the information flow.

Identification techniques that comply with this requirement will here be called *recursive identification methods*, since the measured input-output data are processed recursively (sequentially) as they become available. Other commonly used terms for such techniques are on-line or real-time identification, adaptive parameter estimation, or sequential parameter estimation. Apart from the use of recursive methods in adaptive schemes, they are of importance also for the following two reasons:

1. Typically, as we shall see, they will carry their own estimate of the parameter variance. This means that data can be collected from the system and processed until a sufficient degree of model accuracy has been reached.
2. The algorithms of this chapter will also turn out to be quite competitive alternatives for parameter estimation in off-line situations. See Section 10.3.

In this chapter we shall discuss how recursive identification algorithms can be constructed, what their properties are, and how to deal with some practical issues. We start by formally describing the requirement of finite-time computability.

Algorithm Format

We defined a general identification method as a mapping from the data set Z^t to the parameter space in (7.7):

$$\hat{\theta}_t = F(t, Z^t) \tag{11.1}$$

where the function F may be implicitly defined (e.g., as the minimizing argument of some function). Such a general expression (11.1) cannot be used in a recursive algorithm, since the evaluation of F may involve an unforeseen amount of calculations, which perhaps may not be terminated at the next sampling instant. Instead, a recursive algorithm must comply with the following format:

$$\begin{aligned} X(t) &= H\,(t, X(t-1), y(t), u(t)) \\ \hat{\theta}_t &= h\,(X(t)) \end{aligned} \tag{11.2}$$

Here $X(t)$ is a vector of fixed dimension that represents some "information state." The functions H and h are explicit expressions that can be evaluated with a fixed and a priori known amount of calculations. In that way it can be secured that $\hat{\theta}_t$, can be evaluated during a sampling interval.

Since the information content in the latest pair of measurements, $y(t)$, $u(t)$, normally is small compared to the information already accumulated from previous measurements, the algorithm (11.2) typically takes a more specific form:

$$\begin{aligned} \hat{\theta}_t &= \hat{\theta}_{t-1} + \gamma_t Q_\theta\left(X(t), y(t), u(t)\right) \\ X(t) &= X(t-1) + \mu_t Q_X\left(X(t-1), y(t), u(t)\right) \end{aligned} \tag{11.3}$$

where γ and μ are small numbers reflecting the relative information value in the latest measurement.

11.2 THE RECURSIVE LEAST-SQUARES ALGORITHM

In this section we shall consider the least-squares method as a simple but archetypal case. The obtained algorithms and insights will then also serve as a preview of the following sections.

Weighted LS Criterion

In Section 7.3 we computed the estimate that minimizes the *weighted least-squares* criterion:

$$\hat{\theta}_t = \arg\min_\theta \sum_{k=1}^{t} \beta(t,k)\left[y(k) - \varphi^T(k)\theta\right]^2 \tag{11.4}$$

This is given by (7.41):

$$\hat{\theta}_t = \overline{R}^{-1}(t) f(t) \tag{11.5a}$$

$$\overline{R}(t) = \sum_{k=1}^{t} \beta(t,k)\varphi(k)\varphi^T(k) \tag{11.5b}$$

$$f(t) = \sum_{k=1}^{t} \beta(t,k)\varphi(k)y(k) \tag{11.5c}$$

To compute (11.5) as given, we would at time t form the indicated matrix and vector from Z^t and then solve (11.5a). If we had computed $\hat{\theta}_{t-1}$ previously, this would not be of any immediate help. However, it is clear from the expressions that $\hat{\theta}_t$ and $\hat{\theta}_{t-1}$ are closely related. Let us therefore try to utilize this relationship.

Recursive Algorithm

Suppose that the weighting sequence has the following property:

$$\begin{aligned} \beta(t,k) &= \lambda(t)\beta(t-1,k), \qquad 0 \le k \le t-1 \\ \beta(t,t) &= 1 \end{aligned} \tag{11.6}$$

This means that we may write

$$\beta(t, k) = \prod_{k+1}^{t} \lambda(j) \tag{11.7}$$

We shall later discuss the significance of this assumption. We note, though, that it implies that

$$\overline{R}(t) = \lambda(t)\overline{R}(t-1) + \varphi(t)\varphi^T(t) \tag{11.8a}$$

$$f(t) = \lambda(t)f(t-1) + \varphi(t)y(t) \tag{11.8b}$$

Now

$$\begin{aligned}
\hat{\theta}_t &= \overline{R}^{-1}(t)f(t) = \overline{R}^{-1}(t)\left[\lambda(t)f(t-1) + \varphi(t)y(t)\right] \\
&= \overline{R}^{-1}(t)\left[\lambda(t)\overline{R}(t-1)\hat{\theta}_{t-1} + \varphi(t)y(t)\right] \\
&= \overline{R}^{-1}(t)\left\{\left[\overline{R}(t) - \varphi(t)\varphi^T(t)\right]\hat{\theta}_{t-1} + \varphi(t)y(t)\right\} \\
&= \hat{\theta}_{t-1} + \overline{R}^{-1}(t)\varphi(t)\left[y(t) - \varphi^T(t)\hat{\theta}_{t-1}\right]
\end{aligned}$$

We thus have

$$\hat{\theta}_t = \hat{\theta}_{t-1} + \overline{R}^{-1}(t)\varphi(t)\left[y(t) - \varphi^T(t)\hat{\theta}_{t-1}\right] \tag{11.9a}$$

$$\overline{R}(t) = \lambda(t)\overline{R}(t-1) + \varphi(t)\varphi^T(t) \tag{11.9b}$$

which is a recursive algorithm, complying with the requirement (11.2): At time $t-1$ we store only the finite-dimensional information vector $X(t-1) = [\hat{\theta}_{t-1}, \overline{R}(t-1)]$. Since $\overline{R}$ is symmetric, the dimension of X is $d + d(d+1)/2$. At time t this vector is updated using (11.9), which is done with a given, fixed amount of operations.

Version with Efficient Matrix Inversion

To avoid inverting $\overline{R}(t)$ at each step, it is convenient to introduce

$$P(t) = \overline{R}^{-1}(t)$$

and apply the matrix inversion lemma

$$[A + BCD]^{-1} = A^{-1} - A^{-1}B[DA^{-1}B + C^{-1}]^{-1}DA^{-1} \tag{11.10}$$

to (11.9b). Taking $A = \lambda(t)\overline{R}(t-1)$, $B = D^T = \varphi(t)$, and $C = 1$ gives

$$P(t) = \frac{1}{\lambda(t)}\left[P(t-1) - \frac{P(t-1)\varphi(t)\varphi^T(t)P(t-1)}{\lambda(t) + \varphi^T(t)P(t-1)\varphi(t)}\right] \tag{11.11}$$

Moreover, we have

$$\overline{R}^{-1}(t)\varphi(t) = \frac{1}{\lambda(t)}P(t-1)\varphi(t) - \frac{1}{\lambda(t)}\frac{P(t-1)\varphi(t)\varphi^T(t)P(t-1)\varphi(t)}{\lambda(t)+\varphi^T(t)P(t-1)\varphi(t)}$$
$$= \frac{P(t-1)\varphi(t)}{\lambda(t)+\varphi^T(t)P(t-1)\varphi(t)}$$

We can thus summarize this version of the algorithm as

$$\hat{\theta}(t) = \hat{\theta}(t-1) + L(t)\left[y(t) - \varphi^T(t)\hat{\theta}(t-1)\right] \tag{11.12a}$$

$$L(t) = \frac{P(t-1)\varphi(t)}{\lambda(t)+\varphi^T(t)P(t-1)\varphi(t)} \tag{11.12b}$$

$$P(t) = \frac{1}{\lambda(t)}\left[P(t-1) - \frac{P(t-1)\varphi(t)\varphi^T(t)P(t-1)}{\lambda(t)+\varphi^T(t)P(t-1)\varphi(t)}\right] \tag{11.12c}$$

Here we switched to the notation $\hat{\theta}(t)$ rather than $\hat{\theta}_t$ to account for certain differences due to initial conditions (see the following).

Normalized Gain Version

The "size" of the matrix $\overline{R}(t)$ in (11.5b) and (11.9) will depend on the $\lambda(t)$. To clearly bring out the amount of modification inflicted on $\hat{\theta}_{t-1}$ in (11.9a), it is instructive to normalize $\overline{R}(t)$ so that

$$R(t) = \gamma(t)\overline{R}(t), \qquad \gamma(t) = \left[\sum_{k=1}^{t}\beta(t,k)\right]^{-1} \tag{11.13}$$

Notice that

$$\frac{1}{\gamma(t)} = \frac{\lambda(t)}{\gamma(t-1)} + 1 \tag{11.14}$$

according to (11.6), and that $R(t)$ now is a weighted arithmetic mean of $\varphi(k)\varphi^T(k)$. From (11.9b) and (11.14),

$$\begin{aligned} R(t) &= \gamma(t)\left[\lambda(t)\frac{1}{\gamma(t-1)}R(t-1) + \varphi(t)\varphi^T(t)\right] \\ &= R(t-1) + \gamma(t)\left[\varphi(t)\varphi^T(t) - R(t-1)\right] \end{aligned} \tag{11.15}$$

Then (11.9) can be rewritten as

$$\begin{aligned}
\varepsilon(t) &= y(t) - \varphi^T(t)\hat{\theta}(t-1) \\
\hat{\theta}(t) &= \hat{\theta}(t-1) + \gamma(t)R^{-1}(t)\varphi(t)\varepsilon(t) \\
R(t) &= R(t-1) + \gamma(t)\left[\varphi(t)\varphi^T(t) - R(t-1)\right]
\end{aligned} \tag{11.16}$$

Notice that $\varepsilon(t)$ is the prediction error according to the current model. Since $R(t)$ is a normalized matrix, the variable $\gamma(t)$ can be viewed as an updating *step size* or *gain* in the algorithm (11.16). Compare also with (11.3).

Initial Conditions

To use the recursive algorithms, initial values for their start-up are required. The correct initial conditions in (11.9) at time $t = 0$ would be $\overline{R}(0) = 0$, $\hat{\theta}_0$ arbitrary, according to the definition of $\overline{R}$. These cannot, however, be used. A possibility could then be to initialize only at a time instant t_0 when $\overline{R}(t_0)$ has become invertible (typically $t_0 \geq d$) and then use

$$P^{-1}(t_0) = \overline{R}(t_0) = \sum_{k=1}^{t_0} \beta(t_0, k)\varphi(k)\varphi^T(k) \tag{11.17}$$

$$\hat{\theta}_{t_0} = P(t_0)\sum_{k=1}^{t_0} \beta(t_0, k)\varphi(k)y(k) \tag{11.18}$$

A simpler alternative is, however, to use $P(0) = P_0$ and $\hat{\theta}(0) = \theta_I$ in (11.12). This gives

$$\begin{aligned}
\hat{\theta}(t) &= \left[\beta(t,0)P_0^{-1} + \sum_{k=1}^{t} \beta(t,k)\varphi(k)\varphi^T(k)\right]^{-1} \\
&\quad \times \left[\beta(t,0)P_0^{-1}\theta_I + \sum_{k=1}^{t} \beta(t,k)\varphi(k)y(k)\right]
\end{aligned} \tag{11.19}$$

where $\beta(t, 0)$ is defined by (11.7). Clearly, if P_0 is large or t is large, then the difference between (11.19) and (11.5) is insignificant.

Multivariable Case (*)

Consider now the weighted multivariable case [cf. (7.42) and (7.43)]

$$\hat{\theta}_t = \arg\min_{\theta} \frac{1}{2}\sum_{k=1}^{t} \beta(t,k)\left[y(k) - \varphi^T(k)\theta\right]^T \Lambda_k^{-1}\left[y(k) - \varphi^T(k)\theta\right] \tag{11.20}$$

where $\beta(t, k)$ is subject to (11.6). Entirely analogous calculations as before give the multivariable counterpart of (11.12),

$$\hat{\theta}(t) = \hat{\theta}(t-1) + L(t)\left[y(t) - \varphi^T(t)\hat{\theta}(t-1)\right]$$
$$L(t) = P(t-1)\varphi(t)\left[\lambda(t)\Lambda_t + \varphi^T(t)P(t-1)\varphi(t)\right]^{-1} \tag{11.21}$$
$$P(t) = \frac{P(t-1) - P(t-1)\varphi(t)\left[\lambda(t)\Lambda_t + \varphi^T(t)P(t-1)\varphi(t)\right]^{-1}\varphi^T(t)P(t-1)}{\lambda(t)}$$

and of (11.16):

$$\varepsilon(t) = y(t) - \varphi^T(t)\hat{\theta}(t-1)$$
$$\hat{\theta}(t) = \hat{\theta}(t-1) + \gamma(t)R^{-1}(t)\varphi(t)\Lambda_t^{-1}\varepsilon(t) \tag{11.22}$$
$$R(t) = R(t-1) + \gamma(t)\left[\varphi(t)\Lambda_t^{-1}\varphi^T(t) - R(t-1)\right]$$

Notice that these expressions are useful also for a scalar system when a weighted norm with

$$\beta(t,k) = \alpha_k \prod_{k+1}^{t} \lambda(j) \tag{11.23}$$

is used in (11.4). The scalar α_k then corresponds to Λ_k^{-1}.

Asymptotic Properties of the Estimate

Since $\hat{\theta}(t)$ computed using recursive least squares (RLS) differs from the off-line counterpart at most by the initial effects, as shown in (11.19), the asymptotic properties will coincide with those discussed in Chapters 8 and 9.

Kalman Filter Interpretation

The Kalman filter for estimating the state of the system

$$\begin{aligned} x(t+1) &= F(t)x(t) + w(t) \\ y(t) &= H(t)x(t) + \nu(t) \end{aligned} \tag{11.24}$$

is given by (4.94) and (4.95). The linear regression model

$$\hat{y}(t|\theta) = \varphi^T(t)\theta$$

underlying our calculations can be cast into the form (11.24) by

$$\begin{aligned} \theta(t+1) &= \theta(t), \qquad (= \theta) \\ y(t) &= \varphi^T(t)\theta(t) + v(t) \end{aligned} \tag{11.25}$$

Applying the Kalman filter to (11.25) with $F(t) = I$, $H(t) = \varphi^T(t)$, $R_1(t) = 0$ $\left[= Ew(t)w^T(t)\right]$, and $Ev(t)v^T(t) = R_2(t)$ now gives exactly (11.21), with $\lambda(t) \equiv 1$ and $\Lambda_t = R_2(t)$.

This gives important information, as well as some practical hints:

1. If the noise $v(t)$ in (11.25) is white and Gaussian, then the Kalman filter theory tells us that *the posterior distribution of $\theta(t)$, given Z^{t-1}, is Gaussian with mean value $\hat{\theta}(t)$ and covariance matrix $P(t)$*, given by (11.21), with $\lambda(t) = 1$ and $\Lambda_t = R_2(t)$.

2. Moreover, the initial conditions can be interpreted so that $\hat{\theta}(0)$ is the mean and $P(0)$ is the covariance matrix of the prior distribution. In plain words, this means that $\hat{\theta}(0)$ is what we guess the parameter vector to be before we have seen the data, and $P(0)$ reflects our confidence in this guess.

3. In addition, the natural choice of the norm Λ_t in the multivariable case is to let it equal the equation error noise covariance matrix. If, in the scalar case, $\alpha_t^{-1} = Ev^2(t)$ is time varying, we should use $\beta(k, k) = \alpha_k$ in the weighted criterion (11.4) [cf. (11.23)].

Coping with Time-varying Systems

An important reason for using adaptive methods and recursive identification in practice is that the properties of the system may be time varying and that we want the identification algorithm to track the variations. This is handled in a natural way in the weighted criterion (11.4) by assigning less weight to older measurements that are no longer representative for the system. This means, in terms of (11.6), that we choose $\lambda(j) < 1$. In particular, if $\lambda(j) \equiv \lambda$, then

$$\beta(t, k) = \lambda^{t-k} \tag{11.26}$$

and old measurements in the criterion are exponentially discounted. In that case, λ is often called the *forgetting factor*. The corresponding $\gamma(t)$ will then, according to (11.14), be

$$\gamma(t) \equiv \gamma = 1 - \lambda \tag{11.27}$$

These choices have the natural effect on the algorithm (11.12) or (11.16) that the step size or gain will not decrease to zero. This issue is discussed in more detail in Section 11.6.

Another and more formal alternative to deal with time-varying parameters is to postulate that the true parameter vector in (11.25) is not constant, but varies like a random walk:

$$\theta(t + 1) = \theta(t) + w(t), \qquad Ew(t)w^T(t) = R_1(t) \tag{11.28}$$

with w white Gaussian and $Ev^2(t) = R_2(t)$. The Kalman filter then still gives the conditional expectation and covariance of $\hat{\theta}(t)$ as

$$\hat{\theta}(t) = \hat{\theta}(t-1) + L(t)\left[y(t) - \varphi^T(t)\hat{\theta}(t-1)\right] \tag{11.29a}$$

$$L(t) = \frac{P(t-1)\varphi(t)}{R_2(t) + \varphi^T(t)P(t-1)\varphi(t)} \tag{11.29b}$$

$$P(t) = P(t-1) - \frac{P(t-1)\varphi(t)\varphi^T(t)P(t-1)}{R_2(t) + \varphi^T(t)P(t-1)\varphi(t)} + R_1(t) \tag{11.29c}$$

We see in this formulation that it is the additive $R_1(t)$ term in (11.29c) that prevents the gain $L(t)$ from tending to zero.

11.3 THE RECURSIVE IV METHOD

The IV estimate for fixed (not model dependent) instruments is given by (7.118). Including weights as in (11.5) gives

$$\hat{\theta}_t^{\mathrm{IV}} = \overline{R}^{-1}(t)f(t) \tag{11.30}$$

with

$$\overline{R}(t) = \sum_{k=1}^{t} \beta(t,k)\zeta(k)\varphi^T(k)$$

$$f(t) = \sum_{k=1}^{t} \beta(t,k)\zeta(k)y(k) \tag{11.31}$$

This is closely related to the formulation (11.5), and the recursive computation of $\hat{\theta}_t^{\mathrm{IV}}$ is quite analogous to that of $\hat{\theta}_t^{\mathrm{LS}}$. The counterpart of (11.12) is

$$\hat{\theta}(t) = \hat{\theta}(t-1) + L(t)\left[y(t) - \varphi^T(t)\hat{\theta}(t-1)\right] \tag{11.32a}$$

$$L(t) = \frac{P(t-1)\zeta(t)}{\lambda(t) + \varphi^T(t)P(t-1)\zeta(t)} \tag{11.32b}$$

$$P(t) = \frac{1}{\lambda(t)}\left[P(t-1) - \frac{P(t-1)\zeta(t)\varphi^T(t)P(t-1)}{\lambda(t) + \varphi^T(t)P(t-1)\zeta(t)}\right] \tag{11.32c}$$

Asymptotic Properties

Apart from possible initial-value effects, $\hat{\theta}(t)$ computed as in (11.32) coincides with its off-line counterpart (11.30). Hence its asymptotic properties are given by the analysis in Chapters 8 and 9.

11.4 RECURSIVE PREDICTION-ERROR METHODS

Analogous to the weighted LS case, let us consider a weighted quadratic prediction-error criterion

$$V_t(\theta, Z^t) = \gamma(t)\frac{1}{2}\sum_{k=1}^{t}\beta(t,k)\varepsilon^2(k,\theta) \tag{11.33}$$

with β and γ given by (11.6) and (11.13). Note that

$$\sum_{k=1}^{t}\gamma(t)\beta(t,k) = 1$$

and that the gradient w.r.t θ obeys

$$\begin{aligned} V_t'(\theta, Z^t) &= -\gamma(t)\sum_{k=1}^{t}\beta(t,k)\psi(k,\theta)\varepsilon(k,\theta) \\ &= \gamma(t)\left[\lambda(t)\frac{1}{\gamma(t-1)}V_{t-1}'(\theta, Z^{t-1}) - \psi(t,\theta)\varepsilon(t,\theta)\right] \qquad (11.34) \\ &= V_{t-1}'(\theta, Z^{t-1}) + \gamma(t)\left[-\psi(t,\theta)\varepsilon(t,\theta) - V_{t-1}'(\theta, Z^{t-1})\right] \end{aligned}$$

just as in (11.15).

For the prediction-error approach, we developed the general search algorithm (10.40):

$$\hat{\theta}_t^{(i)} = \hat{\theta}_t^{(i-1)} - \mu_t^{(i)}\left[R_t^{(i)}\right]^{-1}V_t'(\hat{\theta}_t^{(i-1)}, Z^t) \tag{11.35}$$

Here the subscript t denotes that the estimate is based on t data (i.e., Z^t). The superscript (i) denotes the ith iteration of the minimization procedure.

Suppose now that, for each iteration i, we also collect one more data point. This would give an algorithm

$$\hat{\theta}_t^{(t)} = \hat{\theta}_{t-1}^{(t-1)} - \mu_t^{(t)}\left[R_t^{(t)}\right]^{-1}V_t'(\hat{\theta}_{t-1}^{(t-1)}, Z^t) \tag{11.36}$$

For easier notation, we introduce

$$\hat{\theta}(t) = \hat{\theta}_t^{(t)}, \qquad R(t) = R_t^{(t)} \tag{11.37}$$

We now make the induction assumption that $\hat{\theta}(t-1)$ actually minimized $V_{t-1}(\theta, Z^{t-1})$ so that

$$V_{t-1}'(\hat{\theta}(t-1), Z^{t-1}) = 0 \tag{11.38}$$

(this will of course be an approximation). Then, we have, from (11.34),

$$V_t'(\hat{\theta}(t-1), Z^t) = -\gamma(t)\psi(t, \hat{\theta}(t-1))\varepsilon(t, \hat{\theta}(t-1)) \tag{11.39}$$

With this approximation [and taking $\mu(t) = 1$], we thus arrive at the algorithm

$$\hat{\theta}(t) = \hat{\theta}(t-1) + \gamma(t)R^{-1}(t)\psi(t, \hat{\theta}(t-1))\varepsilon(t, \hat{\theta}(t-1)) \tag{11.40}$$

We shall discuss the choice of $R(t)$ shortly, but our main concern now is with the variables $\psi(t, \hat{\theta}(t-1))$ and $\varepsilon(t, \hat{\theta}(t-1))$. These are derived from the prediction $\hat{y}(t|\hat{\theta}(t-1))$. In general , the computation of $\hat{y}(t|\theta)$ for any given value of θ requires the knowledge of all the data Z^{t-1}. For finite-dimensional linear models, this means that $\hat{y}(t|\theta)$ is obtained as the output of a linear filter whose coefficients depend on θ. See (10.56) and (10.60) for a conceptual expression. This means that $\psi(t, \hat{\theta}(t-1))$ and $\hat{y}(t|\hat{\theta}(t-1))$ cannot be computed " recursively" (i.e., with fixed-size memory). Instead we have to use some approximation of these variables. The following approach is natural:

In the time recursions defining $\psi(t, \theta)$ and $\hat{y}(t|\theta)$ from Z^t for any given θ, replace, at time k, the parameter θ by the currently available estimate $\hat{\theta}(k)$. Denote the resulting approximation of $\psi(t, \hat{\theta}(t-1))$ and $\hat{y}(t|\hat{\theta}(t-1))$ by $\psi(t)$ and $\hat{y}(t)$. *(11.41)*

For a finite-dimensional, linear, and time-invariant model (10.60), the approximation (11.41) takes the form

$$\xi(t+1) = \mathcal{A}(\hat{\theta}(t))\xi(t) + \mathcal{B}(\hat{\theta}(t))z(t) \tag{11.42a}$$

$$\begin{bmatrix} \hat{y}(t) \\ \psi(t) \end{bmatrix} = C(\hat{\theta}(t-1))\xi(t) \tag{11.42b}$$

For the Gauss-Newton choice (10.45) and (10.46) of $R(N)$, the rule (11.41) suggests the following approximation:

$$R(t) = \gamma(t) \sum_{k=1}^{t} \beta(t, k)\psi(k)\psi^T(k) \tag{11.43}$$

Using (11.41) and (11.43) in (11.40) now gives the recursive scheme

$$\varepsilon(t) = y(t) - \hat{y}(t) \tag{11.44a}$$

$$\hat{\theta}(t) = \hat{\theta}(t-1) + \gamma(t)R^{-1}(t)\psi(t)\varepsilon(t) \tag{11.44b}$$

$$R(t) = R(t-1) + \gamma(t)\left[\psi(t)\psi^T(t) - R(t-1)\right] \tag{11.44c}$$

The resulting scheme (11.44) together with (11.42) is a *recursive Gauss-Newton prediction-error algorithm.*

Family of Recursive Prediction-error Methods

Depending on the underlying model structure, as well as on the choice of $R(t)$, the scheme (11.44b) corresponds to specific algorithms in a wide family of methods, which we shall call *recursive prediction-error methods* (RPEM). For example, the linear regression

$$\hat{y}(t|\theta) = \varphi^T(t)\theta$$

gives $\psi(t,\theta) = \psi(t) = \varphi(t)$, and (11.44) is indeed the recursive least-squares method (11.16). A gradient variant $[R(t) = I]$ applied to the same structure gives

$$\hat{\theta}(t) = \hat{\theta}(t-1) + \gamma(t)\varphi(t)\varepsilon(t) \tag{11.45}$$

where the gain $\gamma(t)$ could be a given sequence or normalized as

$$\gamma(t) = \frac{\gamma'(t)}{|\varphi(t)|^2} \tag{11.46}$$

This scheme has been widely used, in particular for various adaptive signal-processing problems, under the name LMS (least mean squares) by Widrow and co-workers. See Widrow and Stearns (1985). For ARMAX models, we have the following example:

Example 11.1 Recursive Maximum Likelihood

Consider the ARMAX model (4.15). Introduce $\varphi(t,\theta)$ as in (4.20). Then

$$\hat{y}(t|\theta) = \varphi^T(t,\theta)\theta; \qquad \varepsilon(t,\theta) = y(t) - \hat{y}(t|\theta)$$

$$\psi(t,\theta) + c_1\psi(t-1,\theta) + \cdots + c_{n_c}\psi(t-n_c,\theta) = \varphi(t,\theta)$$

[see (10.52)]. The rule (11.41) then gives the following approximations:

$$\begin{aligned} \overline{\varepsilon}(t) &= y(t) - \varphi^T(t)\hat{\theta}(t) \\ \varphi(t) &= \left[-y(t-1)\ldots-y(t-n_a) \quad u(t-1)\ldots u(t-n_b) \quad \overline{\varepsilon}(t-1)\ldots\overline{\varepsilon}(t-n_c)\right]^T \end{aligned} \tag{11.47}$$

$$\begin{aligned} \hat{y}(t) &= \varphi^T(t)\hat{\theta}(t-1); \qquad \varepsilon(t) = y(t) - \hat{y}(t) \\ \psi(t) &+ \hat{c}_1(t-1)\psi(t-1) + \cdots + \hat{c}_{n_c}(t-1)\psi(t-n_c) = \varphi(t) \end{aligned} \tag{11.48}$$

and the algorithm becomes

$$\hat{\theta}(t) = \hat{\theta}(t-1) + \gamma(t)R^{-1}(t)\psi(t)\varepsilon(t) \tag{11.49}$$

This scheme is known as recursive maximum likelihood (RML).

Notice the difference between (the "prediction error") $\varepsilon(t)$ and (the "residual") $\bar{\varepsilon}(t)$. The latter enters the vector $\varphi(t+1)$ and is not required until after $\hat{\theta}(t)$ is computed. Hence it is natural to distinguish between the quantities as indicated (cf. Section 5.11, Ljung and Söderström, 1983). □

Similarly, the Gauss-Newton RPEM applied to the ARARX structure (4.22) gives another recursive ML method derived by Gertler and Bànyàsz (1974), while the same algorithm with an enforced block diagonal structure of $R(t)$ is called recursive generalized least squares (RGLS) and was introduced by Hasting-James and Sage (1969). See also Table 11.1 in Section 11.5.

Applied to state-space models, the RPEM is closely related to the well-known extended Kalman filter (EKF), as pointed out in Ljung (1979a). The algorithm (11.44b) thus contains a rich collection of specific, "named," methods as special cases. One of its main advantages is also its general applicability. The only requirement on the model structure is the computability of the gradient ψ.

Projection into $D_{\mathcal{M}}$

The model structure is well defined only for $\theta \in D_{\mathcal{M}}$, giving stable predictors [corresponding to the set of θ, for which the matrix $\mathcal{A}$ in (11.42) is stable]. In off-line minimization of the criterion function, this must be kept in mind as a constraint. The same is true for the recursive minimization (11.44). The simplest way of handling this problem is to project the estimates into $D_{\mathcal{M}}$, for example, by

$$\hat{\theta}'(t) = \hat{\theta}(t-1) + \gamma(t)R^{-1}(t)\psi(t)\varepsilon(t)$$
$$\hat{\theta}(t) = \begin{cases} \hat{\theta}'(t) & \text{if } \hat{\theta}'(t) \in D_{\mathcal{M}} \\ \hat{\theta}(t-1) & \text{if } \hat{\theta}'(t) \notin D_{\mathcal{M}} \end{cases} \tag{11.50}$$

The extra computational burden involved in (11.50) is the stability test of whether $\hat{\theta}'(t) \in D_{\mathcal{M}}$. It turns out that for successful operation of (11.44) a test of the kind (11.50) is necessary. However, experience also shows that the projection typically takes place at only a few samples in the beginning of the data record. The information loss by ignoring certain samples, as in (11.50), is therefore moderate.

Asymptotic Properties

The recursive prediction-error method (11.44) is designed to make updates of $\hat{\theta}$ in a direction that "on the average" is a modified negative gradient of

$$\bar{V}(\theta) = \tfrac{1}{2}\bar{E}\varepsilon^2(t,\theta)$$

i.e.

$$\frac{d}{d\theta}\bar{V}(\theta) = -\bar{E}\psi(t,\theta)\varepsilon(t,\theta)$$

It is thus reasonable to expect that $\hat{\theta}(t)$ would converge to a local minimum of $\bar{V}(\theta)$. This is in fact the case (Ljung, 1981) under certain regularity conditions. Moreover, for a Gauss-Newton RPEM, with $\gamma(t) = 1/t$, it can be shown that $\hat{\theta}(t)$ has an asymptotically normal distribution, which coincides with that of the corresponding off-line estimate [see (9.17)]. We thus have

- If $R(t) \geq \delta I$, $\delta > 0$, and $\gamma(t) \to 0$ as $t \to \infty$, then, w.p. 1, $\hat{\theta}(t)$ converges to a local minimum of $\overline{V}(\theta) = \frac{1}{2}\overline{E}\varepsilon^2(t, \theta)$. [Measures to ensure $R(t) \geq \delta I$ are called regularization and are discussed in Section 11.7.]
- Suppose that $S \in \mathcal{M}$ [see (8.10)] and that $\hat{\theta}(t)$ converges to the true parameter θ_0. Suppose that the Gauss-Newton RPEM (11.44b and c) is used with $\gamma(t) = 1/t$. Then

$$\sqrt{t}\left(\hat{\theta}(t) - \theta_0\right) \in AsN(0, P_\theta)$$
$$P_\theta = \lambda_0 \left[\overline{E}\psi(t, \theta_0)\psi^T(t, \theta_0)\right]^{-1} \tag{11.51}$$

See Appendix 11A for techniques of proof and more insights and results.

General Norms, Multivariable Case (*)

Starting with a general criterion

$$V_t(\theta, Z^t) = \gamma(t) \sum_{k=1}^{t} \beta(t, k) \ell\left(\varepsilon(k, \theta), k\right)$$

where $\dim \varepsilon = \dim y = p$, leads to a Gauss-Newton RPEM

$$\begin{aligned} \hat{\theta}(t) &= \hat{\theta}(t-1) + \gamma(t) R^{-1}(t) \psi(t) \ell'_\varepsilon(\varepsilon(t), t) \\ R(t) &= R(t-1) + \gamma(t)\left[\psi(t) \ell''_{\varepsilon\varepsilon}(\varepsilon(t), t)\, \psi^T(t) - R(t-1)\right] \end{aligned} \tag{11.52}$$

Here ℓ'_ε is a $p \times 1$ column vector, $\psi(t)$ is a $d \times p$ matrix, and $\ell''_{\varepsilon\varepsilon}$ is a $p \times p$ matrix. The case with explicit θ-dependence in ℓ is analogous.

11.5 RECURSIVE PSEUDOLINEAR REGRESSIONS

Consider the pseudolinear representation of the prediction (7.112):

$$\hat{y}(t|\theta) = \varphi^T(t, \theta)\theta \tag{11.53}$$

and recall that this model structure contains, among other models, the general linear SISO model (4.33). A bootstrap method for estimating θ in (11.53) was given by (10.64):

$$\hat{\theta}_t^{(i)} = \hat{\theta}_t^{(i-1)} + \left[R_t^{(i-1)}\right]^{-1} f_t(\hat{\theta}_t^{(i-1)}, Z^t) \tag{11.54a}$$

$$R_t^{(i-1)} = \gamma(t) \sum_{k=1}^{t} \beta(t, k) \varphi(k, \hat{\theta}_t^{(i-1)}) \varphi^T(k, \hat{\theta}_t^{(i-1)}) \tag{11.54b}$$

$$f_t(\theta, Z^t) = \gamma(t) \sum_{k=1}^{t} \beta(t, k) \varphi(k, \theta) \varepsilon(k, \theta) \tag{11.55}$$

Here we replaced the equally weighted sums in (10.64) by generally weighted ones, analogous to (11.33).

With the same approach as for the recursive prediction-error method (making one new iteration at the same time a new measurement is brought in, and assuming the previous estimate was a solution $[f_{t-1}(\hat{\theta}(t-1), Z^{t-1}) = 0]$), we obtain from (11.54)

$$\hat{\theta}(t) = \hat{\theta}(t-1) + \gamma(t)R^{-1}(t)\varphi(t, \hat{\theta}(t-1))\varepsilon(t, \hat{\theta}(t-1)) \tag{11.56a}$$

$$\begin{aligned} R(t) = {} & R(t-1) \\ & + \gamma(t)\left[\varphi(t, \hat{\theta}(t-1))\varphi^T(t, \hat{\theta}(t-1)) - R(t-1)\right] \end{aligned} \tag{11.56b}$$

This algorithm suffers from the same problem as (11.40): The computations of $\varphi(t, \hat{\theta}(t-1))$ and $\varepsilon(t, \hat{\theta}(t-1))$ cannot usually be performed recursively. This problem can, however, be solved in the same way as for RPEMs, see (11.41). We thus form as an approximation of $\varphi(t, \hat{\theta}(t-1))$ a vector $\varphi(t)$ in which all θ-dependent entries are replaced by recursively computed quantities, analogous to (11.47). We then have the *recursive pseudolinear regression* (RPLR):

$$\begin{aligned} \hat{y}(t) &= \varphi^T(t)\hat{\theta}(t-1) \\ \varepsilon(t) &= y(t) - \hat{y}(t) \\ \hat{\theta}(t) &= \hat{\theta}(t-1) + \gamma(t)R^{-1}(t)\varphi(t)\varepsilon(t) \\ R(t) &= R(t-1) + \gamma(t)\left[\varphi(t)\varphi^T(t) - R(t-1)\right] \end{aligned} \tag{11.57}$$

This algorithm looks exactly like the RLS algorithm (11.16). The same software can thus be used for RPLR as for RLS. The operational difference lies in the fact that $\varphi(t)$ in (11.57) contains entries that are constructed from data using past models. This also affects the convergence properties of the scheme (see the following).

Notice that the only difference between RPLR compared to a RPEM for the model structure (11.53) is that ψ in (11.44b and c) has been replaced by φ. For the general SISO structure (4.33), the relationship between ψ and φ is given by (10.54).

Family of RPLRs

The RPLR scheme (11.57) represents a family of well-known algorithms when applied to different special cases of (11.53). The ARMAX case is perhaps the best known of these. With $\varphi(t)$ defined by (11.47), the algorithm (11.57) constitutes a scheme for estimating the parameters of an ARMAX model. This scheme is known as *extended least squares* (ELS). Other special cases are displayed in Table 11.1.

TABLE 11.1 Classification of Some Recursive Identification Schemes *

Model Structure	RPEM	RPLR
ARX	RLS	RLS
ARMAX	RML (Söderström, 1973)	ELS (Young, 1968; Panuska, 1968)
ARARX	RGLS (Hasting-James and Sage, 1969); Gertler and Bànyàsz, 1974)	Bethoux, 1976
ARARMAX	—	EMM (Talmon and van den Boom, 1973)
OE	White, 1975	Landau, 1976
BJ	Young and Jakeman, 1979	—

* Compare with Table 4.1.

Asymptotic Properties

Convergence results for the RPLR scheme (11.57) have been given only for some of the special cases in Table 11.1. Since it differs from RPEM in that ψ is replaced by φ, one might guess that the convergence properties will depend on the relationship between these two vectors.

For ARMAX structures, we have (11.48)

$$\psi(t) = \frac{1}{C(q)}\varphi(t)$$

In fact, it turns out that a sufficient condition for the ELS estimate to converge to the true parameter values ($S \in \mathcal{M}$) is that

$$\text{Re}\frac{1}{C_0(e^{i\omega})} \geq \frac{1}{2}, \qquad \forall\omega \tag{11.58}$$

where $C_0(q)$ is the C-polynomial of the true system description. The condition (11.58) is often expressed as "the filter $(1/C_0(q)) - \frac{1}{2}$ is positive real" and can be seen as a condition that $C_0(q)$ is close to unity (see Problem 11E.4). When RPLR is applied to the OE structure (4.25) (Landau's scheme), the corresponding condition for convergence is

$$\text{Re}\frac{1}{F_0(e^{i\omega})} - \frac{1}{2} \geq 0$$

See Appendix 11A for references and further insights and results.

11.6 THE CHOICE OF UPDATING STEP

The recursive identification algorithms (11.44) and (11.57) are largely given by their off-line counterparts. The calculation of the prediction is derived from the corresponding model structure and the selection of $\varphi(t)$ or $\psi(t)$ has its roots in the choice between the prediction-error or correlation approaches. What remains is the quantity $\gamma(t)R^{-1}(t)$ that modifies the update direction and determines the length of the update step. In this section we shall discuss some aspects of how to determine $\gamma(t)$ and $R(t)$. [For notational convenience, we give the expressions for the RPEM (11.44b). RPLR is analogous with $\varphi(t)$ replacing $\psi(t)$.]

Update Direction

There are two basic choices of update directions:

1. The "Gauss-Newton" direction, corresponding to $R(t)$ being an approximation of the Hessian of the underlying identification criterion:

$$R(t) = R(t-1) + \gamma(t)\left[\psi(t)\psi^T(t) - R(t-1)\right] \tag{11.59}$$

2. The "gradient" direction, corresponding to $R(t)$ being a scaled version of the identity matrix:

$$R(t) = |\psi(t)|^2 \cdot I \tag{11.60}$$

or

$$R(t) = R(t-1) + \gamma(t)\left[|\psi(t)|^2 \cdot I - R(t-1)\right] \tag{11.61}$$

The choice between the two directions can be characterized as a trade-off between convergence rate and algorithm complexity. Clearly, the Gauss-Newton direction requires more computations. Updating $\gamma(t)R^{-1}(t)$ as in (11.12c) (see also Section 11.7) requires proportional to d^2 operations, which will typically constitute the dominating part of the computational burden in (11.44). The gradient direction can be implemented with proportional to d operations per update.

On the other hand, the convergence rate can often be drastically faster with the Gauss-Newton direction. For the constant-parameter case, analysis shows that this update direction will yield estimates whose asymptotic distribution has a variance equal to the Cramér-Rao lower bound [see (11.51)]. This is not true for other update directions. Notice, though, that this theoretical result holds for the time-invariant-system case only. When the true system parameters are drifting, it will typically be better to use another update direction adapted to the parameter drift as in (11.67) (see Benveniste and Ruget, 1982). An interesting possibility to speed up convergence by averaging is described in Polyak and Juditsky (1992).

Update Step: Adaptation Gain

An important aspect of recursive algorithms is, as we noted in Section 11.2, their ability to cope with time-varying systems. There are two different ways of achieving this:

1. Selecting an appropriate forgetting profile $\beta(t, k)$ in the criterion (11.33) or selecting a suitable gain $\gamma(t)$ in (11.44) or (11.57). These two approaches are equivalent in view of the relationships (11.7), (11.13), or (11.14), which may be summarized as

$$\beta(t, k) = \prod_{j=k+1}^{t} \lambda(j) = \frac{\gamma(k)}{\gamma(t)} \prod_{j=k+1}^{t} (1 - \gamma(j)) \tag{11.62a}$$

$$\lambda(t) = \frac{\gamma(t-1)}{\gamma(t)} (1 - \gamma(t)) \tag{11.62b}$$

$$\gamma(t) = \frac{1}{1 + \dfrac{\lambda(t)}{\gamma(t-1)}} \tag{11.62c}$$

2. Introducing an assumed covariance matrix $R_1(t)$ for the parameter changes per sample as in (11.29c). This will increase the matrix $P(t)$ and hence the gain vector $L(t)$.

In either case, the choice of update step or "gain" in the algorithm is a trade-off between *tracking ability* and *noise sensitivity*. A high gain means that the algorithm is alert in tracking parameter changes but at the same time sensitive to disturbances in the data, since these are erroneously interpreted as signs of parameter changes. This trade-off can be discussed more precisely in terms of the quantities $\lambda(t)$, $\gamma(t)$, and $R_1(t)$.

Choice of Forgetting Factors $\lambda(t)$

The choice of forgetting profile $\beta(t, k)$ is conceptually simple: Select it so that the criterion essentially contains those measurements that are relevant for the current properties of the system. For a system that changes gradually and in a "stationary manner," the most common choice is to take a constant forgetting factor:

$$\beta(t, k) = \lambda^{t-k}; \text{ i.e., } \lambda(t) \equiv \lambda \tag{11.63}$$

The constant λ is always chosen slightly less than 1 so that

$$\beta(t, k) = e^{(t-k)\log \lambda} \approx e^{-(t-k)(1-\lambda)} \tag{11.64}$$

This means that measurements that are older than $T_0 = 1/(1 - \lambda)$ samples are included in the criterion with a weight that is $e^{-1} \approx 36\%$ of that of the most recent measurement. We could call

$$T_0 = \frac{1}{1 - \lambda} \tag{11.65a}$$

the *memory time constant* of the criterion. If the system remains approximately constant over T_0 samples, a suitable choice of λ can then be made from (11.65a). Since the sampling interval typically reflects the natural time constants of the system dynamics, we could thus select λ so that $1/(1-\lambda)$ reflects the ratio between the time constants of variations in the dynamics and those of the dynamics itself. Typical choices of λ are in the range between 0.98 and 0.995.

We can also consider the response to a sudden change in the true system. If the change occurred k samples ago, the ratio of relevant-to-obsolete entries in the criterion is $1-\lambda^k$. The response to a step change in the system is thus like that of a first-order system with time constant (11.65a). For a constant system belonging to the model set, it follows from Problem 11A.6 that the deviation of the estimate from the true value behaves like

$$E(\hat{\theta}(t)-\theta_0)(\hat{\theta}(t)-\theta_0)^T \sim \frac{1-\lambda}{2}\lambda_0\left[\overline{E}\psi(t,\theta_0)\psi^T(t,\theta_0)\right]^{-1} \quad (11.65b)$$

Here λ_0 is the true innovations variance. The two expressions (11.65a and b) describe in formal terms the trade-off in λ between tracking alertness and noise sensitivity.

For a system that undergoes abrupt and sudden changes, rather than steady and slow ones, an adaptive choice of λ could be conceived. When an abrupt system change has been detected, it is suitable to decrease $\lambda(t)$ to a small value for one sample, thereby "cutting off" past measurements from the criterion, and then to increase it to a value close to 1 again. Such adaptive choices of λ are discussed, for example, in Fortesque, Kershenbaum, and Ydstie (1981)and Hägglund (1984).

Choice of Gain $\gamma(t)$

The choice of gain can be translated from the corresponding choice of forgetting factor using (11.62). A constant forgetting factor λ gives, after a transient, a constant gain

$$\gamma = 1-\lambda$$

Similarly, a sudden decrease at time t_0 in $\lambda(t)$ to a small value and then back to 1 corresponds to a sudden increase in $\gamma(t)$ to a value close to 1 [see (11.62)], and then $\gamma(t) \approx 1/(t-t_0)$.

It is, however, also instructive to discuss the choice of gain in direct terms. Intuitively, *the gain should reflect the relative information contents in the current observations.* An observation with important information (compared to what is already known) deserves a high gain, and vice versa. This is a useful principle that can be applied to a variety of situations: For a constant system the relative importance of a single observation decays like $1/t$. After a substantial change in system dynamics, the relative information in the observation increases. A measurement with a large noise component has low information contents, and so on. See also Problem 11E.3.

Including a Model of Parameter Changes

Analogously to the Kalman filter version (11.29), we could introduce an assumption that the true parameters vary according to

$$\theta(t) = \theta(t-1) + w(t) \quad (11.66a)$$

$$Ew(t)w^T(t) = R_1(t) \quad (11.66b)$$

If we assume that the innovations variance is $R_2(t)$ we get the following version of the general algorithm (11.44):

$$\begin{aligned}
\hat{\theta}(t) &= \hat{\theta}(t-1) + L(t)\varepsilon(t) \\
\varepsilon(t) &= y(t) - \hat{y}(t) \\
L(t) &= \frac{P(t-1)\psi(t)}{R_2(t) + \psi^T(t)P(t-1)\psi(t)} \\
P(t) &= P(t-1) - \frac{P(t-1)\psi(t)\psi^T(t)P(t-1)}{R_2(t) + \psi^T(t)P(t-1)\psi(t)} + R_1(t)
\end{aligned} \tag{11.67}$$

In the case of a linear regression model, this algorithm does give the *optimal* trade-off between tracking ability and noise sensitivity, in terms of a minimal *a posteriori* parameter error covariance matrix. (This follows from the original derivation of the Kalman filter, Kalman and Bucy, 1961, as pointed out in Bohlin, 1970, and Åström and Wittenmark, 1971). However, for other models the algorithm (11.67) is somewhat *ad hoc*. See Problem 11T.2 for a heuristic derivation of it. Nevertheless, it is a very useful alternative, in particular if we have some insight into how the parameters might vary (e.g., if certain parameters vary more rapidly than others). A fringe benefit of the algorithm is that $P(t)$ is an estimate of the variance of the parameter error, also taking into account the variation of the true system. For a linear regression with normal disturbances and normal drift, $P(t)$ is exactly the covariance matrix of the posterior distribution of $\theta(t)$, the mean being $\hat{\theta}(t)$. See (11.29).

The case where the parameters are subject to variations that themselves are of a nonstationary nature [i.e., $R_1(t)$ in (11.66) varies substantially with t] can be dealt with in a parallel algorithm structure, as described in Andersson (1985).

Constant Systems

For a time-invariant system, the natural forgetting profile is $\lambda(t) \equiv 1$ or $\gamma(t) = 1/t$. However, it turns out that for many recursive algorithms (not including RLS) the transient convergence rate is significantly improved with a forgetting factor that increases from, say, 0.95 to 1 over the first 500 data or so:

$$\lambda(t) = 1 - (0.05) \cdot (0.98)^t \tag{11.68}$$

The reason apparently is that early information is somewhat misused and should therefore carry a lower weight in the criterion compared to later measurements, whose information contents are processed in a better way (the filters corresponding to (11.42) are then more accurate).

Asymptotic Behavior in the Time-varying Case

A heuristic analysis of the asymptotic behavior of (11.67) can be carried out as follows.

Suppose that the true parameters vary according to (11.66), and that we consider just the linear regression case:

$$y(t) = \psi^T(t)\theta(t-1) + e(t), \quad Ee^2(t) = \lambda_0$$

Let us also study a simplified algorithm

$$\begin{aligned}\hat{\theta}(t) &= \hat{\theta}(t-1) + P\psi(t)\varepsilon(t) \\ &= \hat{\theta}(t-1) + P\psi(t)\left(y(t) - \psi^T(t)\hat{\theta}(t-1)\right)\end{aligned} \tag{11.69}$$

for some constant "small" matrix P. For the LMS-case (11.45) with constant gain, this corresponds to $P = \gamma$, while the forgetting factor case, (11.19) or (11.59), with a constant forgetting factor $\lambda = 1 - \gamma$ corresponds to

$$P(t) = \left(\sum_{k=1}^{t} \lambda^{t-k}\psi(k)\psi^T(k)\right)^{-1} \approx (1-\lambda)S^{-1} = P$$

$$S = \overline{E}\psi(t)\psi^T(t)$$

where the approximation holds for λ close to 1 and large t, so that we can average over the many non-vanishing terms. Equation (11.69) gives

$$\tilde{\theta}(t) = \theta(t) - \hat{\theta}(t)$$

$$\tilde{\theta}(t) = \left(I - P\psi(t)\psi^T(t)\right)\tilde{\theta}(t-1) - P\psi(t)e(t) + w(t)$$

If we square both sides and take expectation, disregarding any correlation between $\tilde{\theta}$ and ψ and assuming stationarity, we obtain

$$\Pi = \overline{E}\tilde{\theta}(t)\tilde{\theta}^T(t)$$

$$\Pi = \Pi - PS\Pi - \Pi SP + PS\Pi SP + PSP\lambda_0 + R_1, \quad \text{or}$$

$$PS\Pi + \Pi SP = PSP\lambda_0 + R_1$$

where in the last step we ignored the term $PS\Pi SP$ (since P and Π are "small"). If we insert the RLS-choice $P = (1 - \lambda)S^{-1}$ we obtain from this the parameter error variance

$$\Pi = \frac{1}{2}\left[(1 - \lambda)\lambda_0 S^{-1} + \frac{1}{1 - \lambda} R_1\right] = \frac{1}{2}\left[\gamma\lambda_0 S^{-1} + \frac{1}{\gamma} R_1\right] \tag{11.70}$$

which is a generalization of (11.65b) to the time-varying case. Similarly, the LMS-choice gives that the parameter error is the solution Π to

$$S\Pi + \Pi S = \gamma\lambda_0 S + \frac{1}{\gamma} R_1 \tag{11.71}$$

These expressions clearly show how the choice of gain $\gamma = 1 - \lambda$ is a trade-off between the tracking ability R_1/γ, which favors large γ, and the noise sensitivity $\gamma\lambda_0$, which favors small γ.

In Guo and Ljung (1995)it is formally verified that Π indeed is the actual covariance matrix of the parameter errors, up to terms that tend to zero in a well-defined way as γ tends to zero. In that paper the general algorithm (11.67) is treated in the linear regression case. See also Ljung and Gunnarsson (1990), which describes the extension to non-linear parameterizations.

11.7 IMPLEMENTATION

The basic, general Gauss-Newton algorithm was given in the form (11.44) or (11.57). It is clearly not suited for direct implementation as it stands, since a $d \times d$ matrix $R(t)$ would have to be inverted at each time step. In this section we shall discuss some aspects on how to best implement recursive algorithms. A more thorough discussion is given in Chapter 6 of Ljung and Söderström (1983).

Using the Matrix Inversion Lemma

By applying the matrix inversion lemma (11.10) to (11.44), we obtain, analogously to (11.11), the algorithm (for the case of vector outputs)

$$\hat{\theta}(t) = \hat{\theta}(t - 1) + L(t)\varepsilon(t) \tag{11.72a}$$

$$\varepsilon(t) = y(t) - \hat{y}(t) \tag{11.72b}$$

$$L(t) = P(t - 1)\eta(t)\left[\lambda(t)\Lambda_t + \eta^T(t)P(t - 1)\eta(t)\right]^{-1} \tag{11.72c}$$

$$P(t) = \frac{P(t-1) - P(t - 1)\eta(t)\left[\lambda(t)\Lambda_t + \eta^T(t)P(t-1)\eta(t)\right]^{-1}\eta^T(t)P(t-1)}{\lambda(t)} \tag{11.72d}$$

Here $\eta(t)$ (a $d \times p$ matrix) represents either φ or ψ depending on the approach. In this form the dimension of the matrix to be inverted is only $p \times p$, so the saving in computations compared to (11.44) is substantial. Unfortunately, the P-recursion (11.72d) (which in fact is a Riccati equation) is not numerically sound: the equation is sensitive to round-off errors that can accumulate and make $P(t)$ indefinite, since $P(t)$ is essentially computed by successive subtractions.

Using Factorization

As we discussed in Section 10.1, it is useful to represent the data matrices in factorized form [see (10.11)] so as to work with better-conditioned matrices. For recursive identification, this means that we represent $P(t)$ as a product of matrices and rearrange (11.72d) so as to update these matrices instead of P itself. Useful representations are

$$P(t) = Q(t)Q^T(t) \tag{11.73}$$

which, for triangular Q, is the Cholesky decomposition and the UD-factorization:

$$P(t) = U(t)D(t)U^T(t) \tag{11.74}$$

with $U(t)$ as an upper triangular matrix with all diagonal elements equal to 1 and $D(t)$ as a diagonal matrix. Potter (1963)has given an algorithm for updating $Q(t)$ in (11.73) and Bierman (1977)has developed numerically sound algorithms for U and D in (11.74). Here we shall give some details of a related algorithm, which is directly based on Householder transformation (see Problem 10T.1). It was given by Morf and Kailath (1975).

Step 1. At time $t-1$, let $Q(t-1)$ be the lower triangular square root of $P(t-1)$ as in (11.73). Let $\mu(t)$ be a square root of $\lambda(t)\Lambda_t$. Form the $(p+d) \times (p+d)$ matrix

$$\mathcal{L}(t-1) = \begin{bmatrix} \mu(t) & 0 \\ Q^T(t-1)\eta(t) & Q^T(t-1) \end{bmatrix} \tag{11.75}$$

Step 2. Apply an orthogonal $(p+d) \times (p+d)$ transformation

$$\mathrm{T} \quad (\mathrm{T}^T\mathrm{T} = I)$$

to $\mathcal{L}(t-1)$ so that $\mathrm{T}\mathcal{L}(t-1)$ becomes an upper triangular matrix. T can, for example, be found by QR-factorization. Let $\Pi(t)$, $\tilde{L}(t)$, and $\overline{Q}(t)$ be the $p \times p$, $d \times p$, and $d \times d$ matrices defined by

$$\mathrm{T}\mathcal{L}(t-1) = \begin{bmatrix} \Pi^T(t) & \tilde{L}^T(t) \\ 0 & \overline{Q}^T(t) \end{bmatrix} \tag{11.76}$$

(Clearly, Π and $\overline{Q}$ are lower triangular.)

Step 3. Now with $L(t)$ and $P(t)$ as in (11.72c and d), we have

$$L(t) = \tilde{L}(t)\Pi^{-1}(t)$$

$$P(t) = \frac{\overline{Q}(t)\overline{Q}^T(t)}{\lambda(t)}$$

$$\Pi(t)\Pi^T(t) = \lambda(t)\Lambda_t + \eta^T(t)P(t-1)\eta(t) \tag{11.77}$$

Hence

$$Q(t) = \frac{\overline{Q}(t)}{\sqrt{\lambda(t)}} \tag{11.78}$$

Verification. Multiplying (11.76) by its transpose gives

$$\begin{bmatrix} \Pi(t) & 0 \\ \tilde{L}(t) & \overline{Q}(t) \end{bmatrix} \begin{bmatrix} \Pi^T(t) & \tilde{L}^T(t) \\ 0 & \overline{Q}^T(t) \end{bmatrix}$$

$$= \begin{bmatrix} \Pi(t)\Pi^T(t) & \Pi(t)\tilde{L}^T(t) \\ \tilde{L}(t)\Pi^T(t) & \overline{Q}(t)\overline{Q}^T(t) + \tilde{L}(t)\tilde{L}^T(t) \end{bmatrix}$$

$$= \mathcal{L}^T(t-1)\mathrm{T}^T\mathrm{T}\mathcal{L}(t-1) = \mathcal{L}^T(t-1)\mathcal{L}(t-1)$$

$$= \begin{bmatrix} \mu^T(t)\mu(t) + \eta^T(t)Q(t-1)Q^T(t-1)\eta(t) & \eta^T(t)Q(t-1)Q^T(t-1) \\ Q(t-1)Q^T(t-1)\eta(t) & Q(t-1)Q^T(t-1) \end{bmatrix}$$

Using the facts that $Q(t-1)Q^T(t-1) = P(t-1)$ and $\mu^T(t)\mu(t) = \lambda(t)\Lambda_t$, it is now immediate to verify the equalities in (11.77) by a comparison with (11.72c and d).

There are several advantages with this particular way of performing (11.72c and d). First, the only essential computation to perform is the triangularization step (or QR-factorization) (11.76), for which several good numerical procedures exist. This step both gives the new Q and the gain L after simple additional calculations. Note that $\Pi(t)$ is a triangular $p \times p$ matrix, so it is a simple matter to invert it. Second, in the update (11.76) we only deal with square roots of P. Hence the conditioning number of the matrix $\mathcal{L}(t-1)$ is much better than that of P. Third, with the triangular square root $Q(t)$ it is easy to introduce *regularization*, that is, measures to ensure that the eigenvalues of P stay bounded, at the same time as P remains positive definite.

Lattice Algorithms (∗)

In (10.30) and (10.32), we gave a lattice filter scheme for computing predictions that can be applied to certain model structures (see also Problems 10E.8, 10D.2). This scheme is not recursive, since the variables $\hat{\rho}_n$, $\hat{e}_n(t)$, and $\hat{r}_n(t)$ rightly should carry also an index N, reflecting their dependence on the whole data record via (10.32). A tempting approach would be to approximate $\hat{e}_n^N(t)$ and $\hat{r}_n^N(t-1)$ in (10.32) by the past, already computed residuals $\hat{e}_n^t(t)$ and $\hat{r}_n^{t-1}(t-1)$. Then $\hat{\rho}_n$ could be recursively computed, and we obtain a scheme

$$
\begin{aligned}
\hat{e}_{n+1}(t) &= \hat{e}_n(t) + \hat{\rho}_n(t-1)\hat{r}_n(t-1) \\
\hat{r}_{n+1}(t) &= \hat{r}_n(t-1) + \hat{\rho}_n(t-1)\hat{e}_n(t) \\
R_n^{er}(t) &= R_n^{er}(t-1) + \hat{e}_n(t)\hat{r}_n(t-1) \\
R_n^{e}(t) &= R_n^{e}(t-1) + \hat{e}_n^2(t) \\
\hat{\rho}_n(t) &= -\frac{R_n^{er}(t)}{R_n^{e}(t)} \\
\hat{e}_0(t) &= \hat{r}_0(t) = y(t)
\end{aligned}
\tag{11.79}
$$

This algorithm was developed by Griffiths (1977)and Makhoul (1977) and has been called a *gradient lattice algorithm*. It involves approximation, as we pointed out, and consequently does not exactly implement the off-line estimate. It is interesting to note, though, that by slightly modifying the R^{er} and R^e updates an exact version is obtained. This was proved by Lee, Morf, and Friedlander (1981). We here give the resulting algorithm for the case of a multivariate signal $z(t)$ in (10.12) (which includes applications to dynamic systems; see Problem 10E.7). We also include an arbitrary constant forgetting factor λ.

1. Initialize at $t = 0$: Let

$$R_n^e(0) := \delta I,\ R_n^r(-1) := \delta \mathrm{I},\ R_n^{re}(0) := 0,\ r_n(0) := 0, \qquad n = 0, \dots, M-1$$

2. At time $t-1$, store

$$R_n^e(t-1),\ R_n^{re}(t-1),\ R_n^r(t-2),\ r_n(t-1), \qquad n = 0, \dots, M-1$$

3. At time $t-1$, compute for $n = 0, \dots, M-1$:

$$
\begin{aligned}
\hat{\rho}_n^a(t-1) &:= -R_n^{re}(t-1)^T\left[R_n^r(t-2)\right]^{-1} \\
\hat{\rho}_n^b(t-1) &:= -R_n^{re}(t-1)\left[R_n^e(t-1)\right]^{-1}
\end{aligned}
\tag{11.80}
$$

4. For $n = 0, \ldots, M-1$, update

$$\hat{e}_0(t) = \hat{r}_0(t) = z(t)$$

$$\hat{e}_{n+1}(t) = \hat{e}_n(t) + \hat{\rho}_n^a(t-1)\hat{r}_n(t-1)$$

$$\hat{r}_{n+1}(t) = \hat{r}_n(t-1) + \hat{\rho}_n^b(t-1)\hat{e}_n(t)$$

5. Update for $n = 0, \ldots, M-1$:

$$R_n^r(t-1) = \lambda R_n^r(t-2) + [1 - \beta_n(t)]\, r_n(t-1) r_n^T(t-1)$$

$$R_n^e(t) = \lambda R_n^e(t-1) + [1 - \beta_n(t)]\, e_n(t) e_n^T(t)$$

$$R_n^{re}(t) = \lambda R_n^{re}(t-1) + [1 - \beta_n(t)]\, r_n(t-1) e_n^T(t)$$

$$\beta_{n+1}(t) = \beta_n(t) + [1 - \beta_n(t)]^2 r_n^T(t-1) \left[R_n^r(t-1)\right]^{-1} r_n(t-1)$$

$$\beta_0(t) = 0$$

6. Go to step 2.

The prediction of $z(t)$ based on $z(t-1), \ldots, z(t-n)$ is thus given by

$$\hat{z}_n(t) = z(t) - \hat{e}_n(t)$$

It can be computed before $z(t)$ is received by rearranging (11.80b) as

$$\hat{z}_n(t) = \hat{z}_{n-1}(t) - \hat{\rho}_n^a(t-1)\hat{r}_{n-1}(t-1), \quad \hat{z}_0(t) = 0$$

11.8 SUMMARY

Recursive identification algorithms are instrumental for most adaptation schemes. A recursive algorithm can be derived from an off-line counterpart using the philosophy of performing one iteration in the numerical search at the same time as a new observation is included in the criterion. In some special cases (e.g., the recursive least squares and the recursive instrumental variable cases) this leads to algorithms that exactly calculate the off-line estimates in a recursive fashion. In general, though, the recursive constraint means that the data record is not maximally utilized.

With the described philosophy, three basic and common classes of recursive methods can be distinguished:

1. Recursive prediction-error methods (RPEM), (11.44)
2. Recursive pseudolinear regressions (RPLR), (11.57)
3. Recursive instrumental-variable methods (RIV), (11.32)

The asymptotic properties obtained by a RPEM for a constant system, using a Gauss-Newton update direction, coincide with those of the corresponding off-line prediction-error method. This is a very important result, which means that the discussion and results of Chapters 8 and 9, as well as their consequences for user's choices in Part III, apply equally well to RPEMs. Convergence of PLRs is tied to positive realness of certain transfer functions associated with the true system.

The identification problem includes a variety of design variables to be chosen by the user. Many of these are common to both off-line and recursive methods. Recursive algorithms include, in addition, two important quantities that may have a considerable effect on the quality of the estimates: the update direction and the gain. Apparently, the Gauss-Newton direction is a very good choice of update direction for constant systems, even though it may include more calculations than other choices. As a guiding principle for the choice of gain, it can be said that it should reflect the relative information contents in the current measurement.

The principles of recursive identification as outlined in this chapter can be applied to "stochastic" and "deterministic" systems equally well. The distinction, which is partly semantic, shows up on two occasions: (1) when forming the prediction model one can apply some probabilistic machinery or simply "guess," and (2) when selecting the adaptation gain, the "relative information contents in the current measurement" can be interpreted as a formal signal-to-noise ratio or in a more intuitive manner.

11.9 BIBLIOGRAPHY

A thorough treatment of recursive identification following the framework of the present chapter is given in Ljung and Söderström (1983). The monograph by Young (1984)gives a comprehensive study of recursive estimation techniques with a focus on instrumental variable methods, and Solo and Kong (1995)contains a comprehensive study of the asymptotic behavior. A survey of algorithms for adaptation and tracking is given in Ljung and Gunnarsson (1990). Widrow and Stearns (1985)focuses on adaptive signal-processing applications. Adaptive control is treated, for example, in Åström and Wittenmark (1989), Landau (1979), and Goodwin and Sin (1984).

Section 11.2: The derivation of the RLS algorithm goes back to Gauss (1809), as discussed in Appendix 2 of Young (1984). A more recent "early" reference is Plackett (1950). The relationship between RLS and the Kalman filter has been discussed by Ho (1963), and Bohlin (1970).

Section 11.3: RIV has been extensively discussed in Young (1968)and Young (1984).

Section 11.4: The general RPEM as presented here was derived in Ljung (1981). The RML (11.49) was derived by Söderström (1973)and Åström (1972). A similar algorithm was independently suggested by Fuhrt (1973). Moore and Weiss (1979)suggested a general RPEM, based on a slightly different framework. A comprehensive bibliography of the ramifications of RPEM is given in Ljung and Söderström (1983). The asymptotic distribution of the RML algorithm has been discussed also by Hannan (1980a)and Solo (1981).

Section 11.5: Apparently, the first RPLR algorithm was the ELS scheme developed by Panuska (1968)and Young (1968). The term PLR is adopted from Solo (1978). Variants of RPLR have been referred to in the text. The convergence analysis of such schemes has been developed by Ljung (1977a), Moore and Ledwich (1980), and Solo (1979). The classification of Table 11.1 goes back to Ljung (1979b).

Section 11.6: This section is based on Chapter 5 in Ljung and Söderström (1983). Gain adaptation and step-size selection have also been discussed in Bohlin (1976), Benveniste and Ruget (1982), and Kulhavy and Kraus (1996). Relations to detection are treated in, e.g., Basseville and Nikiforov (1993)and Gustafsson (1996).

Section 11.7: Chapter 6 of Ljung and Söderström (1983)gives more details of the implementation. The numerical properties of the schemes have been studied in Ljung and Ljung (1985), Graupe, Jain, and Salahi (1980), Mueller (1981), and Samson and Reddy (1983). Algorithms for fast calculation of the vector $L(t)$ in (11.72) have been derived in Ljung, Morf, and Falconer (1978)and discussed in Carayannis, Manolakis, and Kalouptsidis (1983), Cioffi and Kailath (1984), and Lin (1984). The literature on lattice algorithms is extensive. See, for example, Lee, Morf, and Friedlander (1981), Friedlander (1982), Samson (1982), and the monograph by Honig and Messerschmitt (1984).

Appendix 11A: In addition to the references in the text, the ODE approach to the convergence analysis of recursive algorithms has also been discussed in Ljung (1978b), Ljung (1984), Ljung, Pflug, and Walk (1992), Kushner and Clark (1978), and Benveniste, Métivier, and Priouret (1990). The link between simple recursive algorithms and an associated ODE was established by Khasminski (1966)for non-decreasing gain algorithms. Related techniques to study the asymptotic distribution are described in Kushner and Huang (1979)and Benveniste and Ruget (1982).

For RPLRs, a martingale-based convergence technique, Moore and Ledwich (1980), Solo (1979), has been quite successful. See Goodwin and Sin (1984) for numerous applications of this technique to adaptive control algorithms.

11.10 PROBLEMS

11E.1 Apply RPEM to a first-order ARMA model

$$y(t) + ay(t-1) = e(t) + ce(t-1)$$

Derive an explicit expression for the difference

$$\hat{y}(t) - \hat{y}(t|\hat{\theta}(t-1))$$

Discuss when this difference will be small.

11E.2 Suppose that the algorithm (11.12) is used with $\lambda(t) \equiv 1$. Suppose also that $S \in \mathcal{M}$. Show that the variance of $\hat{\theta}(t)$ (neglecting initial-value effects) is then given by $\lambda_0 P(t)$, where λ_0 is the innovations variance.

11E.3 Formalize the notion that the gain $\gamma(t)$ should "reflect the relative information contents in the current observation," as follows: Decompose, for the RLS method, the prediction error $\varepsilon(t) = y(t) - \hat{\theta}^T(t-1)\varphi(t) = e_0(t) + \tilde{\theta}^T(t-1)\varphi(t)$, where $\tilde{\theta}(t)$ is the parameter error $\theta_0 - \hat{\theta}(t)$. Interpret the scalar $\varphi^T(t)L(t)$ in the algorithm (11.29) as the signal to signal + noise ratio for the "measured" quantity $\varepsilon(t)$.

11E.4 Prove that

$$\text{Re}\frac{1}{C_0(e^{i\omega})} > \frac{1}{2} \Leftrightarrow \left|1 - C_0(e^{i\omega})\right| < 1$$

11T.1 Convert the Newton-Raphson algorithm (10.49) for the correlation approach into a recursive algorithm, and compare with the RPLR algorithm (11.57).

11T.2 Consider a general model structure and assume that the true parameter vector is varying according to (11.66). Suppose that it is known that $\theta(t)$ is close to a known value $\theta_*(t)$ (like, e.g., the previous estimate). With the true innovation $e_0(t)$, we thus have

$$y(t) = \hat{y}(t|\theta(t)) + e_0(t)$$

Now approximate

$$\hat{y}(t|\theta(t)) \approx \hat{y}(t|\theta_*(t)) + \psi^T(t, \theta_*(t))[\theta(t) - \theta_*(t)]$$

Introduce the variable

$$y_*(t) = y(t) - \hat{y}(t|\theta_*(t)) + \psi^T(t, \theta_*(t))\theta_*(t)$$

whose value is known at time t, and write the model as

$$\theta(t) = \theta(t-1) + w(t)$$
$$y_*(t) = \theta(t)\psi^T(t, \theta_*(t)) + e_0(t)$$

With $\psi(t, \theta_*(t)) = \psi(t)$ a known vector, this is now of the linear regression type (11.25) and (11.28). Apply the optimal algorithm (11.29) to this approximate description and show that (11.67) is obtained.

11D.1 Let $\beta(t, k)$ be defined by (11.6) and let $\gamma(t)$ be defined by (11.13). Show that

$$\beta(t, k) = \frac{\gamma(k)}{\gamma(t)} \prod_{j=k+1}^{t} [1 - \gamma(j)]$$

11D.2 Verify the expression (11.19).

APPENDIX 11A: TECHNIQUES FOR ASYMPTOTIC ANALYSIS OF RECURSIVE ALGORITHMS

Methods to prove convergence and analytically assess the quality of recursively computed estimates tend to be technical. A comprehensive treatment is given in Chapter 4 of Ljung and Söderström (1983), and we provide in this appendix some insights and guidelines.

Most of the existing results deal with the convergence properties of $\hat{\theta}(t)$ as t tends to infinity and the gain $\gamma(t)$ tends to zero. There are also some results on the asymptotic distribution of $\hat{\theta}(t)$ under the same assumption. Results for the tracking case, when the true parameter changes and the gain $\gamma(t)$ does not tend to zero, have been studied by Benveniste and Ruget (1982), Kushner and Huang (1981), Guo and Ljung (1995), Solo and Kong (1995), and others.

An Algorithm Format

The recursive algorithms can all conceptually be given in the form

$$\hat{\theta}(t) = \hat{\theta}(t-1) + \gamma(t)R^{-1}(t)\eta(t)\varepsilon(t) \tag{11A.1a}$$

$$\varepsilon(t) = y(t) - \hat{y}(t) \tag{11A.1b}$$

$$\xi(t+1) = \mathcal{A}(\hat{\theta}(t))\xi(t) + \mathcal{B}(\hat{\theta}(t))\begin{bmatrix} y(t) \\ u(t) \end{bmatrix} \tag{11A.1c}$$

$$\begin{bmatrix} \hat{y}(t+1) \\ \eta(t+1) \end{bmatrix} = C(\hat{\theta}(t))\xi(t+1) \tag{11A.1d}$$

The choice of R could be any positive definite matrix. Most common is, however, the Gauss-Newton choice

$$R(t) = R(t-1) + \gamma(t)\left[\eta(t)\eta^T(t) - R(t-1)\right] \tag{11A.2}$$

Here $\eta(t)$ corresponds to $\psi(t)$, $\varphi(t)$, or $\zeta(t)$, depending on the particular algorithm used.

In some special cases the recursive algorithm (11A.1) can be solved to yield an explicit expression for $\hat{\theta}(t)$. This happens for the RLS algorithm (11.16) [see (11.19)] and for the RIV algorithm (11.32) with given instruments [see (11.30)]. In these cases an analysis can be carried out based on the explicit expressions. This analysis will then coincide with the off-line analysis of Chapters 8 and 9.

In most cases, however, no explicit expression for $\hat{\theta}(t)$ can be obtained. Indeed, $\hat{\theta}(t)$ will be a fairly complicated function of the data set Z^t, partly as a result of the time-varying, estimate-dependent filtering in (11A.1cd). See (11.41). This means that it is a difficult problem to determine the asymptotic properties of the estimate $\hat{\theta}(t)$. In this appendix we shall give some insights into the behavior of (11A.1), and we shall also state some basic results on convergence and the asymptotic distribution of $\hat{\theta}(t)$. For a formal analysis, we refer to Ljung and Söderström (1983).

Error Models

Regarding $\gamma(t)$ and $R^{-1}(t)$ as modifications of the basic update information $\eta(t)\varepsilon(t)$, it follows that the relationship between the crucial quantities

$$\theta - \eta - \varepsilon \tag{11A.3}$$

will be a key to the convergence properties of (11A.1). Such a relationship is often called an *error model* (in particular in connection with adaptive control applications).

We can write, symbolically,

$$\Delta\hat{\theta}(t) \sim \eta(t)\varepsilon(t) \tag{11A.4}$$

for (11A.1a). The right side is a random variable. If $\gamma(t)$ is small so that a noticeable change in $\hat{\theta}$ is the result of many steps like (11A.4), then this change is likely to occur in the direction of the expected value:

$$\Delta\hat{\theta} \sim E\eta\varepsilon \tag{11A.5}$$

All asymptotic analysis of (11A.1) is based on the concept (11A.5), but there are several ways of formalizing the analysis. Here we shall briefly describe an approach based on an associated ordinary differential equation (ODE).

An Associated ODE

Let $\hat{y}(t|\theta)$ and $\eta(t, \theta)$ be defined by

$$\begin{aligned} \xi(t+1, \theta) &= \mathcal{A}(\theta)\xi(t, \theta) + \mathcal{B}(\theta)\begin{bmatrix} y(t) \\ u(t) \end{bmatrix} \\ \begin{bmatrix} \hat{y}(t|\theta) \\ \eta(t, \theta) \end{bmatrix} &= C(\theta)\xi(t, \theta) \end{aligned} \tag{11A.6}$$

Define

$$f(\theta) = \overline{E}\eta(t, \theta)\varepsilon(t, \theta) \tag{11A.7}$$

as the average update direction, associated with the parameter value θ. Then (11A.1) is associated with the following ODE:

$$\frac{d}{d\tau}\theta_D(\tau) = R^{-1} f(\theta_D(\tau)) \tag{11A.8}$$

If $R(t)$ is determined as in (11A.2), then define

$$G(\theta) = \overline{E}\eta(t, \theta)\eta^T(t, \theta) \tag{11A.9}$$

and the associated ODE becomes

$$\begin{aligned} \frac{d}{d\tau}\theta_D(\tau) &= R_D^{-1}(\tau) f(\theta_D(\tau)) \\ \frac{d}{d\tau}R_D(\tau) &= G(\theta_D(\tau)) - R_D(\tau) \end{aligned} \tag{11A.10}$$

We shall shortly describe in what sense (11A.10) is "associated" with (11A.1) and (11A.2), but first let us discuss heuristically how it is obtained.

Suppose that $R(t)$ in (11A.1a) is fixed to be R, that $\hat{\theta}(t_0) = \overline{\theta}$, and that $t > t_0$ is chosen so that $\hat{\theta}(t)$ and $\hat{\theta}(t_0)$ are close. Then

$$\hat{\theta}(t) - \hat{\theta}(t_0) = R^{-1}\sum_{k=t_0}^{t}\gamma(k)\eta(k)\varepsilon(k) \approx R^{-1}\sum_{k=t_0}^{t}\gamma(k)\eta(k,\overline{\theta})\varepsilon(k,\overline{\theta})$$

$$= R^{-1}\overline{E}\eta(k,\overline{\theta})\varepsilon(k,\overline{\theta})\sum_{k=t_0}^{t}\gamma(k)$$

$$+ R^{-1}\sum_{k=t_0}^{t}\gamma(k)\left[\eta(k,\overline{\theta})\varepsilon(k,\overline{\theta}) - \overline{E}\eta(k,\overline{\theta})\varepsilon(k,\overline{\theta})\right] \tag{11A.11}$$

$$\approx \Delta\tau R^{-1}\overline{E}\eta(k,\overline{\theta})\varepsilon(k,\overline{\theta})$$

where

$$\Delta\tau \triangleq \sum_{k=t_0}^{t}\gamma(k) \tag{11A.12}$$

Here the first approximation comes from $\hat{\theta}(k)$ being close to $\overline{\theta}$ for $t_0 \leq k \leq t$ and thus replacing $\hat{\theta}(k)$ by $\overline{\theta}$ in (11A.1c d), yielding (11A.6). The second approximation amounts to neglecting the second term, being a sum of zero-mean random variables, compared to the mean value. This is an application of a law of large numbers. Now, changing the time scale from t to τ so that

$$\theta_D(\tau_t) \leftrightarrow \hat{\theta}(t), \qquad \tau_t = \sum_{k=1}^{t}\gamma(k) \tag{11A.13}$$

we can write (11A.11) as

$$\theta_D(\tau_t) = \theta_D(\tau_{t_0}) + (\tau_t - \tau_{t_0})R^{-1}f\left(\theta_D(\tau_{t_0})\right) \tag{11A.14}$$

This expression is Euler's method for solving the ODE (11A.8), and hence the link between (11A.1) and (11A.8) is heuristically established. Including (11A.2) leads to (11A.10) in a similar way.

The heuristic discussion suggests that the trajectories of (11A.10) describe the behavior of (11A.1) and (11A.2) if γ is small enough. With a certain amount of technical work, the following links can be formally established (under certain regularity conditions):

- If $\gamma(t) \to 0$ as $t \to \infty$ and all trajectories of (11A.10) converge to a set $D_c \subset D_{\mathcal{M}}$, then the estimates $\hat{\theta}(t)$ converge to D_c with probability 1 as $t \to \infty$ provided they are constrained to $D_{\mathcal{M}}$. (11A.15)
- If $\gamma(t) \to 0$ as $t \to \infty$ and $\hat{\theta}(t) \to \theta^*$ with positive probability, then θ^* must be a stable stationary point of (11A.10). (11A.16)

- If $\hat{\theta}(t_0) = \overline{\theta}$ and $\theta_D(\tau)$ denotes the solution to (11A.10) with $\theta_D(\tau_{t_0}) = \overline{\theta}$, then

$$P\left[\sup_{t_0 \le t \le T} \left|\hat{\theta}(t) - \theta_D(\tau_t)\right| > \varepsilon\right] \le K(\varepsilon, p) \sum_{k=t_0}^{T} \gamma^p(k) \tag{11A.17}$$

where τ_t, is defined by (11A.13).

For proofs of these statements, see Ljung (1977b).

Convergence of RPEM

For the RPEM family, we have

$$\eta = \boxed{\psi \approx -\frac{d}{d\theta}\varepsilon} \tag{11A.18}$$

This defining relationship can indeed be seen as an error model (11A.3) that holds regardless of the properties of the data.

We also have

$$f(\theta) = \overline{E}\psi(t,\theta)\varepsilon(t,\theta) = -\frac{d}{d\theta}\frac{1}{2}\overline{E}\varepsilon^2(t,\theta) \tag{11A.19}$$

Let

$$\overline{V}(\theta) = \tfrac{1}{2}\overline{E}\varepsilon^2(t,\theta) \tag{11A.20}$$

Then along trajectories of (11A.8) (recall that $R > 0$)

$$\begin{aligned} \frac{d}{d\tau}\overline{V}(\theta_D(\tau)) &= \frac{d}{d\theta}\overline{V}(\theta)\bigg|_{\theta=\theta_D(\tau)} R^{-1} f(\theta_D(\tau)) \\ &= -f^T(\theta_D(\tau))\, R^{-1} f(\theta_D(\tau)) \end{aligned} \tag{11A.21}$$

which shows that $\overline{V}$ is decreasing outside the set

$$D_c = \{\theta | f(\theta) = 0\} \tag{11A.22}$$

$\overline{V}(\theta)$ is thus a Lyapunov function for (11A.8), showing that its trajectories (among those that stay in $D_{\mathcal{M}}$) will converge to D_c as $\tau \to \infty$. According to (11A.15), this implies that

$$\hat{\theta}(t) \to D_c, \qquad \text{w.p. 1 as } t \to \infty \tag{11A.23}$$

(or that $\hat{\theta}(t)$ tends to the boundary of $D_{\mathcal{M}}$, which cannot be excluded unless the particular way $\hat{\theta}(t)$ is projected into $D_{\mathcal{M}}$ [see (11.50)] is specified). In view of (11A.16), points that are not local minima of $\overline{V}(\theta)$ can be excluded from D_c. We thus find that the RPEM estimate will converge w.p. 1 to a local minimum of $\overline{V}(\theta)$, which is exactly the same result as for the off-line counterpart (10.40). We summarize this as follows:

Result 11.1. Consider the algorithm (11A.1) with $\eta(t) = \psi(t)$ and with a positive definite $R(t)$. Assume that $\gamma(t) \to 0$ as $t \to \infty$ and that $\hat{\theta}(t)$ is constrained to a subset of $D_{\mathcal{M}}$. Then $\hat{\theta}(t)$ converges w.p. 1 to a local minimum of $\overline{V}(\theta)$ (or to the boundary of $D_{\mathcal{M}}$) as $t \to \infty$.

Asymptotic Distribution of RPE Estimates

Consider the Gauss-Newton RPEM (11.44), and assume that there exists a θ_0 such that $\varepsilon(t, \theta_0) = e_0(t)$ is white noise. Let

$$\overline{R}(t) = \frac{1}{\gamma(t)} R(t) = \sum_{k=1}^{t} \beta(t,k)\psi(k)\psi^T(k) \tag{11A.24}$$

[cf. (11.5b) and (11.43)]. Then, analogous to (11.9b),

$$\overline{R}(t) = \lambda(t)\overline{R}(t-1) + \psi(t)\psi^T(t) \tag{11A.25}$$

Introduce $\tilde{\theta}(t) = \hat{\theta}(t) - \theta_0$ and rewrite (11.44b) as

$$\begin{aligned}\overline{R}(t)\tilde{\theta}(t) &= \overline{R}(t)\tilde{\theta}(t-1) + \psi(t)\varepsilon(t) \\ &= \lambda(t)\overline{R}(t-1)\tilde{\theta}(t-1) + \psi(t)\psi^T(t)\tilde{\theta}(t-1) + \psi(t)\varepsilon(t)\end{aligned}$$

This expression can be summed from $t = 0$ to t, giving

$$\begin{aligned}\overline{R}(t)\tilde{\theta}(t) = {} & \beta(t,0)\overline{R}(0)\tilde{\theta}(0) \\ & + \sum_{k=1}^{t} \beta(t,k)\psi(k)\left[\psi^T(k)\tilde{\theta}(k-1) + \varepsilon(k)\right]\end{aligned} \tag{11A.26}$$

Consider the sum

$$S_t = \sum_{k=1}^{t} \beta(t,k)\psi(k)\left[\psi^T(k)\tilde{\theta}(k-1) + \varepsilon(k) - \varepsilon(k,\theta_0)\right]$$

By definition,

$$\varepsilon(k) - \varepsilon(k,\theta_0) \approx \varepsilon(k,\hat{\theta}(k)) - \varepsilon(k,\theta_0) \approx -\psi^T(k)\left[\hat{\theta}(k) - \theta_0\right] \tag{11A.27}$$

where the first approximation follows since $\varepsilon(k)$ is filtered using estimates approximately equal to $\hat{\theta}(k)$ [see (11A.1c d)] and the second approximation is the mean-value theorem. This means that the sum S_t is likely to be negligible compared to other terms in (11A.26). Also, $\beta(t,0)\overline{R}(0)\tilde{\theta}(0)$ should be negligible. Hence

$$\tilde{\theta}(t) \approx \overline{R}^{-1}(t)\sum_{k=1}^{t} \beta(t,k)\psi(k)e_0(k) \tag{11A.28}$$

Assuming that $\hat{\theta}(k)$ is close to θ_0 "asymptotically most of the time," we can replace $\psi(k)$ by $\psi(k, \theta_0)$ in (11A.24) and (11A.28) without too much error. This gives

$$\left(\hat{\theta}(t) - \theta_0\right) \approx \overline{R}_0^{-1}(t) \sum_{k=1}^{t} \beta(t,k)\psi(k,\theta_0)e_0(k)$$
$$= -\left[V_t''(\theta_0, Z^t)\right]^{-1} V_t'(\theta_0, Z^t) \tag{11A.29}$$

where

$$V_t(\theta, Z^t) = \frac{1}{2}\sum_{k=1}^{t} \beta(t,k)\varepsilon^2(k,\theta) \tag{11A.30}$$

From Section 9.2 we know that (11A.29) is exactly the same asymptotic expression as we would get for the off-line estimate

$$\hat{\theta}_t = \arg\min_{\theta} V_t(\theta, Z^t) \tag{11A.31}$$

This discussion has been heuristic, neglecting terms without formal justification. However, with some technical labor, the approximations can be verified formally. This is done in Theorem 4.5 in Ljung and Söderström (1983). The result thus is that

The asymptotic distribution of estimates obtained by a Gauss-Newton RPEM is the same as for the corresponding off-line estimate. *(11A.32)*

In particular, for $\gamma(t) = 1/t$, which implies that $\beta(t,k) = 1/t\ \forall k$, and we obtain from Theorem 9.1 and (9.17) the following result:

Result 11.2. Consider the Gauss-Newton RPEM (11.44) with $\gamma(t) = 1/t$. Assume that there exists a θ_0 such that $\varepsilon(t,\theta_0) = e_0(t)$ is white noise with variance λ_0. Then, if $\hat{\theta}(t) \to \theta_0$,

$$\sqrt{t}\left(\hat{\theta}(t) - \theta_0\right) \in AsN\left(0, \lambda_0\left[\overline{E}\psi(t,\theta_0)\psi^T(t,\theta_0)\right]^{-1}\right) \tag{11A.33}$$

Convergence of RPLRs

First we postulate a certain error model (11A.3):

$$\begin{aligned} \varepsilon(t,\theta) &= \tilde{\varphi}^T(t,\theta)(\theta_0 - \theta) + e_0(t) \\ \tilde{\varphi}(t,\theta) &= H_0(q)\varphi(t,\theta) \\ & e_0(t) \text{ independent of } \varphi(s,\theta) \quad s \le t \end{aligned} \tag{11A.34}$$

Let us investigate what the convergence properties of (11A.1) and (11A.2), with $\eta(t) = \varphi(t)$, might be under the assumption (11A.34). The right side of the associated ODE (11A.10) becomes

$$\begin{aligned} f(\theta) &= \overline{E}\varphi(t,\theta)\tilde{\varphi}^T(t,\theta)(\theta_0 - \theta) + \overline{E}\varphi(t,\theta)e_0(t) \\ &= \tilde{G}(\theta)(\theta_0 - \theta) \end{aligned} \tag{11A.35}$$

$$\tilde{G}(\theta) = E\varphi(t,\theta)\tilde{\varphi}^T(t,\theta) \tag{11A.36}$$

since $\varphi(t,\theta)$ and $e_0(t)$ are independent. With

$$G(\theta) = E\varphi(t,\theta)\varphi^T(t,\theta) \tag{11A.37}$$

the ODE thus is

$$\begin{aligned} \frac{d}{d\tau}\theta_D(\tau) &= R_D^{-1}(\tau)\tilde{G}(\theta_D(\tau))\,[\theta_0 - \theta_D(\tau)] \\ \frac{d}{d\tau}R_D(\tau) &= G\,(\theta_D(\tau)) - R_D(\tau) \end{aligned} \tag{11A.38}$$

Trying the Lyapunov function

$$V(\theta, R) = (\theta - \theta_0)^T R(\theta - \theta_0) \tag{11A.39}$$

for (11A.38) gives

$$\begin{aligned} &\frac{d}{d\tau}V\,(\theta_D(\tau), R_D(\tau)) \\ &= -(\theta - \theta_0)^T\left[\tilde{G}(\theta) + \tilde{G}^T(\theta) - G(\theta) + R(\tau)\right](\theta - \theta_0) \end{aligned} \tag{11A.40}$$

Suppose now that the matrix

$$\overline{G}(\theta) = \tilde{G}(\theta) + \tilde{G}^T(\theta) - G(\theta) \tag{11A.41}$$

is positive semidefinite for all θ. Then (11A.40) shows that all trajectories of (11A.38) end up in

$$D_c = \left\{\theta | \overline{G}(\theta)(\theta - \theta_0) = 0\right\} \tag{11A.42}$$

$\overline{G}(\theta)$ being positive semidefinite is thus a sufficient condition for convergence of $\hat{\theta}(t)$ into D_c. In Problem 11A.5, the reader is asked to prove that

$$\mathrm{Re}H_0(e^{i\omega}) > \frac{1}{2}\ \forall\omega \Rightarrow \begin{cases} \overline{G}(\theta) \geq 0 \\ \overline{G}(\theta)(\theta - \theta_0) = 0 \Rightarrow \varepsilon(t,\theta) = e_0(t) \end{cases} \tag{11A.43}$$

This condition as H_0 is usually expressed as "$H_0(q) - \frac{1}{2}$ is strictly positive real." We can summarize the result as follows:

Result 11.3. Consider the RPLR (11A.1) and (11A.2), with $\eta(t) = \varphi(t)$ and $\gamma(t) \to 0$ as $t \to \infty$. Assume that the relationship (11A.34) holds between $\varepsilon(t, \theta)$, θ, and $\varphi(t, \theta)$ and that the transfer function $H_0(q) - \frac{1}{2}$ is strictly positive real. Then

$$\hat{\theta}(t) \to D_c = \{\theta | \varepsilon(t, \theta) = e_0(t)\}$$

with probability 1 as $t \to \infty$.

Notice that we had to assume that the true system belongs to the model set [θ_0 appears in (11A.34)]. This was not the case for Result 11.1.

For the ARMAX model, it can readily be shown that (11A.34) holds with

$$H_0(q) = \frac{1}{C_0(q)}$$

where $C_0(q)$ is the polynomial associated with the noise $C_0(q)e_0(t)$ for the true system. See Ljung (1977a). The condition on positive realness will thus be

$$\text{Re}\frac{1}{C_0(e^{i\omega})} - \frac{1}{2} > 0 \; \forall \omega \tag{11A.44}$$

Since the condition (11A.44) relates to the true system, it cannot be guaranteed a priori. With some prior knowledge about the properties of the noise, (11A.44) can, however, be somewhat relaxed (see Problem 11A.1).

When the RPLR algorithm is applied to the output error model (4.25), the analysis is quite analogous. The condition for convergence is that

$$\text{Re}\frac{1}{F_0(e^{i\omega})} - \frac{1}{2} > 0 \; \forall \omega \tag{11A.45}$$

where $F_0(q)$ is the true denominator polynomial of the system (Problem 11A.2).

Local Convergence of RPLR

The ODE (11A.38) can be linearized around the desired convergence point θ_0. It can be verified (Problem 11A.3) that the stability properties of the linearized equation are entirely determined by the matrix $-G^{-1}(\theta_0)\tilde{G}(\theta_0)$. If this matrix has eigenvalues in the right half-plane, then, according to (11A.16), $\hat{\theta}(t)$ cannot converge to θ_0. In some special cases, the eigenvalues of this matrix can be explicitly calculated, and cases can be constructed for which the RPLR estimate $\hat{\theta}(t)$ cannot converge to its true value. See Ljung and Söderström (1983), Example 4.11 and Stoica, Holst, and Söderström (1982).

Conclusions

In this appendix we have given three basic results relating to the asymptotic properties of recursive identification algorithms. We should point out that the results have not been phrased with full rigor and with all technical conditions. See, for example, Ljung and Söderström (1983)for such an account.

11A PROBLEMS

11A.1 Consider an RPLR for an ARMAX model (the ELS algorithm). Suppose that the vector $\varphi(t)$ in (11.57) is replaced by $\varphi^*(t) = L(q)\varphi(t)$. Show that the convergence condition (11.58) is changed to

$$\text{Re}\frac{1}{L(e^{i\omega})C_0(e^{i\omega})} - \frac{1}{2} > 0 \; \forall\omega$$

11A.2 Carry out the convergence analysis for the RPLR method applied to the output error model (4.25) and show that the condition is given by (11A.45).

11A.3 Consider the ODE (11A.38) and linearize it around $\theta = \theta_0$. Show that the linearized equation is asymptotically stable if and only if the eigenvalues of the matrix

$$-G^{-1}(\theta_0)\tilde{G}(\theta_0)$$

are in the left half-plane. [G and $\tilde{G}$ are defined by (11A.36) and (11A.37).]

11A.4 Linearize the ODE associated with an RPEM around a true parameter value θ_0, and show that the linearized equation is always stable.

11A.5 Verify (11A.43). (Reference: Ljung, 1977a).

11A.6 Consider the recursive Gauss-Newton prediction error algorithm (11.44) with a constant small gain $\gamma(t) \equiv \gamma_0$. Assume that the true system is constant and corresponds to the parameter vector θ_0. Let $Ee_0^2 = E\varepsilon^2(t, \theta_0) = \lambda_0$. Use expression (11A.29) to show that

$$E\left(\hat{\theta}(t) - \theta_0\right)\left(\hat{\theta}(t) - \theta_0\right)^T \approx \frac{\gamma_0}{2} \cdot \lambda_0 \left[\overline{E}\psi(t, \theta_0)\psi^T(t, \theta_0)\right]^{-1}$$

[*Hints:* Recall (11.62), use the approximation

$$\overline{R}_0^{-1}(t) \approx \left[E\overline{R}_0(t)\right]^{-1}, \text{ and assume } \psi(t, \theta_0) \text{ to be stationary.}]$$

III user's choices

12

OPTIONS AND OBJECTIVES

The goal of the identification procedure is, in loose terms, to *obtain a good and reliable model with a reasonable amount of work*. To this end a number of different techniques have been developed, as we saw in Part II. They also contain a number of design variables to be chosen by the user. In this part we shall discuss how to make these choices so as to achieve our goals. In the present chapter we start by pinpointing the options that are available to the user and by formalizing the objectives of the identification exercise. In the latter context we concentrate on linear time-invariant structures.

12.1 OPTIONS

When confronted with a process with unknown dynamical properties, the user has to take a number of decisions as we discussed in Section 1.4: An experiment has to be designed, a model structure must be chosen, a criterion of fit must be selected, and a procedure for validating the obtained model has to be devised. As Figure 1.10 indicates, these choices may also have to be revised a number of times during the identification procedure.

The options include how to perform the identification experiment, what model structures to choose, what identification algorithms to apply, and how to validate the obtained model. We shall collectively refer to all these options or design variables as

$$\mathcal{D} = \{\text{all design variables}\}$$

Each of the many choices will have an influence on the quality of the resulting model. Using the asymptotic theory of Chapters 8 and 9, it is possible to evaluate the effects, and give advice about suitable choices of $\mathcal{D}$. Indeed, this is the objective of the theory. This sets the scenario for the chapters to follow.

12.2 OBJECTIVES

What do we mean by "a good and reliable model" and what is "a reasonable amount of work"? Both issues have a certain amount of subjective flavor, and it is not possible and desirable to give a fully formalized discussion on this topic. We shall, however, in this section study the quality of a model as related to its intended use. Our discussion will be confined to the case of linear single-input, single-output systems and models.

The True System and the Model

To assess the quality of a certain model, it is unavoidable that some assumptions, implicit or explicit, about the true properties of the process have to be invoked. For the present purposes we shall suppose that the true system is subject to assumption S1 of Chapter 8; that is,

$$y(t) = G_0(q)u(t) + H_0(q)e_0(t) \tag{12.1}$$

where $\{e_0(t)\}$ is white noise with variance λ_0.

Clearly, the realism of such an assumption can be questioned. However, let us reiterate the point we made in Section 8.1: Analysis pertains to assuming certain properties of the true data-generation mechanism and subsequently calculating the resulting properties of the models. Such calculations turn out to be useful and suggestive, even when the underlying assumption may not be verifiable.

For simpler notation, we shall use, as before,

$$T_0(q) = [\, G_0(q) \quad H_0(q)\,] \tag{12.2}$$

Suppose that we have decided on all the design variables $\mathcal{D}$, and as a result obtained the model

$$\hat{T}(q, \mathcal{D}) = [\, \hat{G}(q, \mathcal{D}) \quad \hat{H}(q, \mathcal{D})\,] \tag{12.3}$$

Recall that $\mathcal{D}$ contains, among other things, N, the number of data, and the model orders.

Scalar Design Criterion

It is desirable that the model $\hat{T}(q, \mathcal{D})$ be close to $T_0(q)$. The difference

$$\tilde{T}(e^{i\omega}, \mathcal{D}) \triangleq \hat{T}(e^{i\omega}, \mathcal{D}) - T_0(e^{i\omega}) \tag{12.4}$$

should, in other words, be small. Let us develop a formal measure of the size of $\tilde{T}$. Depending on the intended use of the model, a good fit in some frequency ranges may be more important than in others. To capture this fact, we introduce a frequency weighted scalar criterion

$$J_1(\tilde{T}(\cdot, \mathcal{D})) = \int_{-\pi}^{\pi} \tilde{T}(e^{i\omega}, \mathcal{D})C(\omega)\tilde{T}^T(e^{-i\omega}, \mathcal{D})\, d\omega \tag{12.5}$$

where the 2×2 matrix function

$$C(\omega) = \begin{bmatrix} C_{11}(\omega) & C_{12}(\omega) \\ C_{21}(\omega) & C_{22}(\omega) \end{bmatrix} \tag{12.6}$$

describes the relative importance of a good fit at different frequencies, as well as the relative importance of the fit in G and H, respectively. We shall generally assume that $C(\omega)$ is Hermitian; that is,

$$C_{21}(\omega) = \overline{C_{12}(\omega)} \, [= C_{12}(-\omega)]$$

(The last equality follows when the dependence on ω is via $e^{i\omega}$.) We shall shortly give examples of how such weighting functions can be determined.

The scalar $J_1(\tilde{T}(\cdot, \mathcal{D}))$ is a random variable due to the randomness of $\hat{T}$. To obtain a realization independent quality measure, it is natural to take the expectation of J_1, and form the criterion

$$\begin{aligned} \overline{J}(\mathcal{D}) &= \int_{-\pi}^{\pi} E\tilde{T}(e^{i\omega}, \mathcal{D})C(\omega)\tilde{T}^T(e^{-i\omega}, \mathcal{D}) \, d\omega \\ &= \int_{-\pi}^{\pi} \operatorname{tr}[\Pi(\omega, \mathcal{D})C(\omega)] \, d\omega \end{aligned} \tag{12.7}$$

where the 2×2 matrix Π is given by

$$\Pi(\omega, \mathcal{D}) = E\tilde{T}^T(e^{-i\omega}, \mathcal{D})\tilde{T}(e^{i\omega}, \mathcal{D}) \tag{12.8}$$

The problem of choosing design variables can now be stated as

$$\min_{\mathcal{D} \in \Delta} \overline{J}(\mathcal{D}) \tag{12.9}$$

where Δ denotes the constraints associated with our desire to do at most "a reasonable amount of work." These will typically include a maximum number of samples, signal power constraints, not too complex numerical procedures, and so on. The constraints Δ could also include that certain design variables simply are not available to the user in the particular application in question.

The problem (12.9) will be discussed in Chapters 13 to 16. First, we give some examples of model applications that lead to different functions $C(\omega)$ in (12.7).

Model Applications

In Chapter 3 we listed some typical uses of linear models. They all give rise to different weighting functions $C(\omega)$ in the objective criterion (12.7).

Example 12.1 Simulation

Suppose that the transfer function G is used to simulate the input-output part of the system with input $u^*(t)$ as in (3.2). The model $\hat{G}(q, \mathcal{D})$ then produces the output

$$y_{\mathcal{D}}(t) = \hat{G}(q, \mathcal{D})u^*(t)$$

while the true system would give the correct output

$$y_0(t) = G_0(q)u^*(t)$$

The error signal

$$\tilde{y}_{\mathcal{D}}(t) = y_{\mathcal{D}}(t) - y_0(t) = \left[\hat{G}(q, \mathcal{D}) - G_0(q)\right]u^*(t)$$

has the spectrum

$$\Phi_{\tilde{y}}(\omega, \mathcal{D}) = \left|\hat{G}(e^{i\omega}, \mathcal{D}) - G_0(e^{i\omega})\right|^2 \Phi_u^*(\omega) \tag{12.10}$$

where $\Phi_u^*(\omega)$ is the spectrum of $\{u^*(t)\}$. This, again, is a random function, and its expectation w.r.t $\hat{G}$,

$$\Psi_{\tilde{y}}(\omega, \mathcal{D}) = E\left|\hat{G}(e^{i\omega}, \mathcal{D}) - G_0(e^{i\omega})\right|^2 \Phi_u^*(\omega) \tag{12.11}$$

is a measure of the *average performance degradation* due to errors in the model $\hat{G}$. Note that, with (12.8) and

$$C(\omega) = \begin{bmatrix} \Phi_u^*(\omega) & 0 \\ 0 & 0 \end{bmatrix} \tag{12.12}$$

we can rewrite (12.11) as

$$\Psi_{\tilde{y}}(\omega, \mathcal{D}) = \text{tr}\Pi(\omega, \mathcal{D})C(\omega) \tag{12.13}$$

Finally, the average variance $\overline{E}\tilde{y}^2(t)$ (averaged over $\{u^*(t)\}$, as well as over $\hat{G}$) will be

$$2\pi\overline{E}\tilde{y}^2(t) = \overline{J}(\mathcal{D}) = \int_{-\pi}^{\pi} \Psi_{\tilde{y}}(\omega, \mathcal{D})\, d\omega \tag{12.14}$$

which is a special case of (12.7). This illustrates how the quadratic design criterion (12.7) may have an explicit physical interpretation. □

Example 12.2 Prediction

The one-step-ahead prediction is given by (3.20):

$$\hat{y}(t|t-1) = H^{-1}(q)G(q)u^*(t) + \left[1 - H^{-1}(q)\right]y^*(t) \tag{12.15}$$

when applied to input-output data $\{u^*(t), y^*(t)\}$ generated by the system. The discrepancy between the prediction $\hat{y}_{\mathcal{D}}(t|t-1)$ obtained by the model $\hat{T}(q, \mathcal{D})$ and the true prediction $\hat{y}_0(t|t-1)$ thus is, suppressing arguments,

$$\begin{aligned} \tilde{y}_{\mathcal{D}}(t|t-1) &= \left[\hat{H}^{-1}\hat{G} - H_0^{-1}G_0\right]u^* + \left[H_0^{-1} - \hat{H}^{-1}\right]y^* \\ &= H_0^{-1}(y^* - G_0u^*) - \hat{H}^{-1}(y^* - \hat{G}u^*) \end{aligned} \tag{12.16}$$

The input-output data obey

$$y^*(t) = G_0(q)u^*(t) + H_0(q)e_0(t) \tag{12.17}$$

which gives

$$\begin{aligned} \tilde{y}_{\mathcal{D}}(t|t-1) &= \hat{H}^{-1}\tilde{G}u^* + (1 - \hat{H}^{-1}H_0)e_0 \\ &= \hat{H}^{-1}\left[\tilde{G}u^* + \tilde{H}e_0\right] = \hat{H}^{-1}(q, \mathcal{D})\tilde{T}(q, \mathcal{D})\begin{bmatrix} u^*(t) \\ e_0(t) \end{bmatrix} \end{aligned} \tag{12.18}$$

This signal has the spectrum

$$\Phi_{\tilde{y}}(\omega, \mathcal{D}) = \frac{1}{\left|\hat{H}(e^{i\omega}, \mathcal{D})\right|^2}\tilde{T}(e^{i\omega}, \mathcal{D})\begin{bmatrix} \Phi_u^*(\omega) & \Phi_{ue}^*(\omega) \\ \Phi_{ue}^*(-\omega,) & \lambda_0 \end{bmatrix}\tilde{T}^T(e^{-i\omega}, \mathcal{D})$$

where $\Phi_u^*(\omega)$ is the spectrum of $\{u^*(t)\}$ and $\Phi_{ue}^*(\omega)$ the cross spectrum between $\{u^*(t)\}$ and $\{e_0(t)\}$. Due to the appearance of $\hat{H}$ in the denominator, this expression is not quadratic in the model error. However, assuming the error to be small so that higher-order terms of $\tilde{T}$ are neglected, we can replace $\hat{H}$ by H_0. This gives the (approximate) average spectrum of the error signal

$$\Psi_{\tilde{y}}(\omega, \mathcal{D}) = \operatorname{tr}\Pi(\omega, \mathcal{D})C(\omega) \tag{12.19}$$

with

$$C(\omega) = \frac{1}{\left|H_0(e^{i\omega})\right|^2}\begin{bmatrix} \Phi_u^*(\omega) & \Phi_{ue}^*(\omega) \\ \Phi_{ue}^*(-\omega) & \lambda_0 \end{bmatrix} \tag{12.20}$$

The average variance of the error signal $E\tilde{y}_{\mathcal{D}}^2(t|t-1)$ is thus approximately given by the criterion (12.7) with (12.20) for small errors. □

In this way different intended model uses lead to the criterion (12.7) and (12.8) with different weighting functions $C(\omega)$. See Problem 12E.1 for a general interpretation of (12.7) and Problem 12E.3 for its connection to the accuracy of the underlying parameter estimate.

12.3 BIAS AND VARIANCE

In this section we shall discuss the frequency-domain objective criterion (12.7) and (12.8) a bit more closely for parametric identification methods. As in (4.111), let

$$T(q,\theta) = [\,G(q,\theta) \quad H(q,\theta)\,] \tag{12.21}$$

and let $\hat{\theta}_N(\mathcal{D})$ be the estimate resulting from one of the methods described in Chapter 7. Here we choose to show the N-dependence explicitly. Thus the transfer function estimate (12.3) is

$$\hat{T}_N(e^{i\omega}, \mathcal{D}) = T(e^{i\omega}, \hat{\theta}_N(\mathcal{D})) \tag{12.22}$$

Let us develop an expression for the mean-square error (MSE) $\Pi_N(\omega, \mathcal{D})$ in (12.8). According to Chapter 8, $\hat{\theta}_N(\mathcal{D})$ converges w.p. 1 to a value $\theta^*(\mathcal{D})$. With $\mathbf{T}'(e^{i\omega}, \theta)$ as the $d \times 2$ derivative of T w.r.t θ, defined in (4.125), the mean-value theorem gives

$$\begin{aligned} T(e^{i\omega}, \hat{\theta}_N(\mathcal{D})) \approx\ & T\left(e^{i\omega}, \theta^*(\mathcal{D})\right) \\ & + \left[\hat{\theta}_N(\mathcal{D}) - \theta^*(\mathcal{D})\right]^T \mathbf{T}'\left(e^{i\omega}, \theta^*(\mathcal{D})\right) \end{aligned} \tag{12.23}$$

Introduce the notation

$$B(e^{i\omega}, \mathcal{D}) = T\left(e^{i\omega}, \theta^*(\mathcal{D})\right) - T_0(e^{i\omega})$$

for the model discrepancy in the limit $N = \infty$. Then for (12.8) we obtain

$$\Pi_N(\omega, \mathcal{D}) \approx B^T(e^{-i\omega}, \mathcal{D}) B(e^{i\omega}, \mathcal{D}) + \frac{1}{N} P(\omega, \mathcal{D}) \tag{12.24}$$

where

$$P(\omega, \mathcal{D}) = \mathbf{T}'^T(e^{-i\omega}, \theta^*(\mathcal{D})) \left[N \cdot \operatorname{Cov} \hat{\theta}_N(\mathcal{D})\right] \mathbf{T}'(e^{i\omega}, \theta^*(\mathcal{D})) \tag{12.25}$$

In (12.24) we have neglected the term

$$2\operatorname{Re} B^T(e^{i\omega}, \mathcal{D}) \left[E\hat{\theta}_N(\mathcal{D}) - \theta^*(\mathcal{D})\right] \mathbf{T}'(e^{i\omega}, \theta^*(\mathcal{D}))$$

which, in view of (9B.13) will be dominated by either of the terms of (12.24) for large N, under the assumptions of Theorem 9.1. Notice that this theorem also gives us an expression (9.15) for determining $P(\omega, \mathcal{D})$ explicitly. With (12.24), the design criterion (12.7) and (12.8) takes the asymptotic form

$$\begin{aligned}
\overline{J}(\mathcal{D}) &\approx J(\mathcal{D}) \triangleq J_P(\mathcal{D}) + J_B(\mathcal{D}) \\
J_P(\mathcal{D}) &= \frac{1}{N}\int_{-\pi}^{\pi} \operatorname{tr}\left[P(\omega, \mathcal{D})C(\omega)\right] d\omega \\
J_B(\mathcal{D}) &= \int_{-\pi}^{\pi} B(e^{i\omega}, \mathcal{D})C(\omega)B^T(e^{-i\omega}, \mathcal{D})\, d\omega
\end{aligned} \tag{12.26}$$

Notice that, using (12.25),

$$\begin{aligned}
J_P(\mathcal{D}) &= \frac{1}{N} \operatorname{tr} P_\theta(\mathcal{D})C_\theta \\
C_\theta &= \int_{-\pi}^{\pi} \mathbf{T}'(e^{i\omega}, \theta^*(\mathcal{D}))C(\omega)\mathbf{T}'^T(e^{-i\omega}, \theta^*(\mathcal{D}))\, d\omega \\
P_\theta(\mathcal{D}) &\sim N \cdot \operatorname{Cov} \hat{\theta}_N
\end{aligned} \tag{12.27}$$

The expression for $J(\mathcal{D})$ emphasizes the basic feature of a "variance contribution" J_P and a "bias contribution" J_B to the objective criterion. These two components are typically affected by the design variables in somewhat different ways. The bias term is mostly affected by the model set (a large, flexible, and/or well-adapted model set gives small bias) and is typically unaffected by data record length, signal powers, and so on. The variance term, on the other hand, typically decreases with increasing amounts of data and input signal power, while it increases with the number of estimated parameters.

We shall discuss various aspects of the subproblem

$$\min_{\mathcal{D}\in\Delta} J_P(\mathcal{D}) \tag{12.28}$$

in more detail in Chapters 13 and 15, while the subproblem

$$\min_{\mathcal{D}\in\Delta} J_B(\mathcal{D}) \tag{12.29}$$

is dealt with in Section 14.5, where we shall also comment on how to combine the results on the subproblems to solve

$$\min_{\mathcal{D}\in\Delta} J(\mathcal{D}) \tag{12.30}$$

Asymptotic Expression for $J_P(\mathcal{D})$

With the expression (9.92) we have, asymptotically in the model order n and the data record length N

$$\frac{1}{N}P(\omega, \mathcal{D}) \approx \frac{n}{N}\Phi_v(\omega)\begin{bmatrix} \Phi_u(\omega) & \Phi_{ue}(-\omega) \\ \Phi_{ue}(\omega) & \lambda_0 \end{bmatrix}^{-1} \tag{12.31}$$

Here n, N, the input spectrum Φ_u, and the cross spectrum Φ_{ue} are all variables contained in $\mathcal{D}$. On the other hand, these design variables are the *only ones* that affect this asymptotic expression for the covariance matrix. There are many important design variables (such as prefilters, noise models, and prediction horizons) that *do not affect* $P(\omega, \mathcal{D})$ in this asymptotic form.

Inserting (12.31) into (12.26) gives the explicit expression

$$J_P(\mathcal{D}) \approx \frac{n}{N}\int_{-\pi}^{\pi} \frac{\{\lambda_0 C_{11}(\omega) - 2\text{Re}\,[C_{12}(\omega)\Phi_{ue}(-\omega)] + C_{22}(\omega)\Phi_u(\omega)\}\,\Phi_v(\omega)}{\lambda_0\Phi_u(\omega) - |\Phi_{ue}(\omega)|^2} \tag{12.32}$$

using (12.6).

We shall discuss this expression for the variance contribution in more detail in Section 13.6.

12.4 SUMMARY

The many identification methods potentially available for a particular application can be described as a list of choices and options ("design variables" $\mathcal{D}$).

An ideal route to determining these design variables would be to pose a criterion for what a "good model" (for the application in question) is and to list the constraints that are imposed on the design by limited time and cost, as well as the availability of the system. In that case the "best" identification result can be secured. We have sketched such a route in this chapter [see (12.26) to (12.31)], and we shall pursue the choice of design variables in more detail in the following chapters. In practical application, a less formal attitude will of course be taken, but our formalization is useful to bring out the character of the considerations that have to be made.

12.5 BIBLIOGRAPHY

The particular way of describing the available design variables and the formalization of a mean-square identification objective are further described in Ljung (1985b), Ljung (1986), and Yuan and Ljung (1985).

12.6 PROBLEMS

12E.1 Let $s(t)$ be a signal derived from a model application. It could, for example, be the output of a system when a minimum variance regulator, computed using the model, is applied. Conceptually, we may write

$$s(t) = f\,(T(q))\,w(t)$$

to denote that the transfer function T, as well as some additional signals $w(t)$ (reference signals and/or noises), are used to determine $s(t)$. Assume that the difference $\hat{T}(q, \mathcal{D}) - T_0(q)$ is small so that effects higher than second order can be neglected. Then derive an expression for the expected spectrum of the "performance degradation signal"

$$\Delta s(t) = \left[f(\hat{T}(q, \mathcal{D})) - f(T_0(q)) \right] w(t)$$

(see Ljung, 1985b).

12E.2 Assume that the obtained model is going to be used for prediction on input-output data with the same second-order properties as those used during the identification experiment (see Example 12.2). Assume also that the asymptotic variance expression (12.31) is approximately applicable and that the bias contribution can be neglected. Give an expression for the expected spectrum of the error between the ideal prediction and the one obtained using the model.

12E.3 Suppose that $\mathcal{S} \in \mathcal{M}$ so that, for some θ_0,

$$T_0(q) = T(q, \theta_0)$$

Then derive an expression for the criterion (12.7) and (12.8) in terms of the mean-square parameter error

$$\Pi_\theta = E(\hat{\theta}_N - \theta_0)(\hat{\theta}_N - \theta_0)^T$$

13

EXPERIMENT DESIGN

The design of an identification experiment includes several choices, such as which signals to measure and when to measure them and which signals to manipulate and how to manipulate them. It also includes some more practical aspects, such as how to condition the signals before sampling them (choice of presampling filters).

In a sense, the design variables associated with the identification experiment are more crucial than many of the other variables described in Section 12.1. While several different design variables associated with models and methods can be tried out at the computer, the experimental data can be changed only by a new experiment, which could be a costly and time-consuming procedure. Therefore, it is worthwhile to design the experiment thoughtfully so as to generate data that are sufficiently informative.

In this chapter we shall discuss the different choices that concern experiment design. Some basic principles are discussed in Section 13.1, while the concept of informative experiments is treated in Section 13.2. Open loop input design is studied in Section 13.3. Identifiability issues for closed loop data are discussed in Section 13.4 while a review of methods for identification of systems operating in a closed loop is given in Section 13.5. Experiment design based on the asymptotic expression (12.32) is treated in Section 13.6. The choice of sampling interval and sampling filters is discussed in Section 13.7.

13.1 SOME GENERAL CONSIDERATIONS

Design Variables

When confronted with a physical system whose dynamics is to be identified, there are a number of questions to be answered. First, the system definition may not be given: Which signals are to be considered as outputs and which are to be considered as inputs? This is the question of where in the process the sensors should be placed (outputs) and which signals should be manipulated (inputs) so as to "excite" the system during the experiment. It should also be stressed that there may be signals associated with the process that rightly are to be considered as inputs (in the sense that they affect the system), even though it is not possible, feasible, or allowed to manipulate them. If they are measurable, it is then still highly desirable to include them among the measured input signals and treat them as such when building models, even though from an operational point of view they should rather be considered as (measurable) disturbances. See Figure 1.1.

When it has been decided upon *where* and *what* to measure, the next question is *when* to measure. Most often the signals are sampled using a constant sampling interval T, and then this quantity has to be chosen.

The choice of input signals has a very substantial influence on the observed data. The input signals determine the operating point of the system and which parts and modes of the system are excited during the experiment. The user's freedom in choosing the input characteristics may vary considerably with the application. In process industry, it may not be allowed at all to manipulate a system in continuous production mode. For other systems, such as economic and ecological ones, it is simply not possible to affect the system for the purpose of an identification experiment. In laboratory applications and during development phases of new equipment, on the other hand, the choice of inputs is perhaps not restricted other than by power limitations.

Two different aspects are associated with the choice of input. One concerns the second-order properties of u, such as its spectrum $\Phi_u(\omega)$ and the cross spectrum $\Phi_{ue}(\omega)$ between input and driving noise (realized by output feedback). The other concerns the "shape" of the signal. We can work with inputs being sums of sinusoids, or filtered white noise, or pseudorandom signals, or binary signals (assuming only two values), and so on.

As a final choice for the identification experiment, let us list N, the number of input-output measurements to be collected.

Basic Guiding Principles

Several of these listed choices will be dealt with in more detail in the ensuing sections. We shall here, however, point to some guiding principles.

Let us denote all the design variables associated with the experiment by $\mathcal{X}$. ($\mathcal{X}$ is thus a subset of $\mathcal{D}$, defined in Section 12.1.) The asymptotic properties of the resulting estimate can then be described by

$$\theta^*(\mathcal{X}) \tag{13.1}$$

the limit to which $\hat{\theta}_N$ converges, and by

$$P_\theta(\mathcal{X}) \tag{13.2}$$

the asymptotic covariance matrix of the parameter estimate (see Chapters 8 and 9). These expressions can then be translated to other quantities of interest, such as the resulting transfer-function estimate (see Chapter 12).

Stretching the formal results obtained in Chapters 8 and 9 to more suggestive formulations, we could say that, for PEMs

The model $\mathcal{M}(\theta^*(\mathcal{X}))$ is the best approximation of the system under the chosen $\mathcal{X}$. (Note that what is the "best approximation" of a system normally depends on the applied input, see Example 8.2). (13.3)

$\mathcal{M}(\theta^*(\mathcal{X})) = \mathcal{S}$ if $\mathcal{M}$ is large enough to contain $\mathcal{S}$ and $\mathcal{X}$ is such that no other model is equivalent to the system under $\mathcal{X}$. (13.4)

$$P_\theta(\mathcal{X}) \sim \lambda_0 \left[E\left(\frac{d}{d\theta}\hat{y}(t|\theta)\right)\left(\frac{d}{d\theta}\hat{y}(t|\theta)\right)^T \right]^{-1} \tag{13.5}$$

See Theorems 8.2, 8.3, and 9.1 and (9.17), respectively.

Bias

The formulation (13.3) suggests that when the bias may be significant it is wise to let the experiment resemble the situation under which the model is to be used. This may of course be difficult to accomplish, since often the objective with identification is to find out suitable operating conditions. If the true system is suspected to be nonlinear and a linear model is sought, then the result (13.3) gives the reasonable advice that the experiment should be carried out around the nominal operating point for the plant. For a linear system, the issue of bias distribution and how it depends on the input will be further discussed in Section 14.4.

Informative Experiments

The issue (13.4) relates to the concept of informative data sets defined in Definitions 8.1 and 8.2. Clearly, a primary goal is to design experiments that lead to data by which we can discriminate between different models in intended model sets. This problem will be further discussed in Section 13.2. Notice that when $\mathcal{S} \in \mathcal{M}$ then the issues (13.3) and (13.4) leave the choice of $\mathcal{X}$ open within the set of sufficiently informative ones.

Minimizing Variances

Once $\mathcal{X}$ is chosen so that the limiting model $\theta^*(\mathcal{X})$ is acceptable, but only then, it becomes interesting to further select $\mathcal{X}$ so that the covariance matrix $P_\theta(\mathcal{X})$ is minimized. Formally, the problem of optimal input design could be stated as

$$\min_{\mathcal{X} \in X} \alpha\left(P_\theta(\mathcal{X})\right) \tag{13.6}$$

where $\alpha(P)$ is a scalar measure of how large the matrix P is and X is a set of admissible designs subject also to the constraints that (13.3) and (13.4) might impose. We gave an explicit example of $\alpha(P)$ in (12.27), of the kind

$$\alpha(P) = \operatorname{tr} CP$$

The expression (13.5) gives a suggestive hint for the choice of $\mathcal{X}$:

A small variance in a certain component of θ results if the predictor is sensitive to that component. Hence choose the outputs $y(t)$ and the inputs $u(t)$ so that the predicted output becomes sensitive with respect to parameters that are important for the application in question. (13.7)

This advice could be used as a general mental picture for experiment design. It applies to the sensor location problem, the selection of input variables as well as their characteristics, and to other design issues. The mathematical formalization of this observation is conceptually straightforward but may be technically involved. In Sections 13.3 and 13.6 we shall illustrate the formalization for the open-loop input design problem.

Validation Power

Another leading principle in experiment design is that the input—the "probing signal"—should be *rich*. It should excite the system and force it to show its properties, even the ones that may be unknown. This desired property may be in conflict with the bias and variance aspects, though. For example, if we seek a variance-optimal input to identify a second order system, the solution may be a signal consisting of two sinusoids. Such a signal will never reveal if the system would be of higher order; we cannot invalidate a second order model. Similarly, an optimal input for a linear system may be one that assumes only two values, and shifts between those is a certain fashion. With such an input we can never find out if there is a static nonlinearity at the input side, since this would simply shift the two levels. All this illustrates that the input should have validation (and invalidation) power to test possible properties of interest in the system.

13.2 INFORMATIVE EXPERIMENTS

In Section 8.2 we introduced the concept of data sets Z^∞ that are "informative enough" with respect to a model set $\mathcal{M}^*$, meaning that the data allow discrimination between any two different models in the set. We shall transfer this terminology to identification experiments by calling an *experiment* "informative enough" if it generates a data set that is informative enough.

Clearly, it is a very basic requirement on the design that the experiment should be informative enough with respect to all model sets that are likely to be used. In Theorem 8.1 we gave a general result on informative experiments. We shall develop more detailed and specific characterizations of such experiments in this section.

Open-Loop Experiments

Consider a yet unspecified model set of single-input, single-output linear models:

$$\mathcal{M}^* = \{G(q, \theta), H(q, \theta) | \theta \in D_{\mathcal{M}}\} \tag{13.8}$$

Suppose that θ_1 and θ_2 correspond to different models in $\mathcal{M}^*$, let $\varepsilon_i(t) = \varepsilon(t, \theta_i)$, $G_i(q) = G(q, \theta_i)$; $\Delta G(q) = G_2(q) - G_1(q)$, and H_i analogously. Then

$$\Delta\varepsilon(t) = \varepsilon_1(t) - \varepsilon_2(t) = \frac{1}{H_1(q)}[\Delta G(q)u(t) + \Delta H(q)\varepsilon_2(t)]$$

Now

$$\varepsilon_2(t) = \frac{1}{H_2(q)}[(G_0(q) - G_2(q))\,u(t) + H_0(q)e_0(t)]$$

where G_0, H_0 is the true description (8.7) of the system, which need not belong to (13.8). Suppose that the experiment is carried out in *open loop* so that $\{u(t)\}$ and $\{e_0(t)\}$ are independent. Then, using (2.65) and Theorem 2.2,

$$\overline{E}\,[\Delta\varepsilon(t)]^2 = \frac{1}{2\pi}\int_{-\pi}^{\pi} \frac{1}{\left|H_1(e^{i\omega})\right|^2}\left[\left|\Delta G(e^{i\omega}) + \frac{G_0(e^{i\omega}) - G_2(e^{i\omega})}{H_2(e^{i\omega})}\cdot \Delta H(e^{i\omega})\right|^2 \Phi_u(\omega) + \left|\Delta H(e^{i\omega})\right|^2 \cdot \left|\frac{H_0(e^{i\omega})}{H_2(e^{i\omega})}\right|^2 \lambda_0\right] d\omega \tag{13.9}$$

where $\lambda_0 = Ee_0^2(t)$. According to our standard assumptions on invertibility of the noise model, $\left|H_0(e^{i\omega})\right|^2 > 0,\ \forall\omega$. Suppose now that the data are not informative with respect to $\mathcal{M}^*$ so that

$$\overline{E}\,[\Delta\varepsilon(t)]^2 = 0 \tag{13.10}$$

even though $\Delta G(e^{i\omega})$ and $\Delta H(e^{i\omega})$ are not both identically zero. Equation (13.10) implies that both the terms within square brackets in (13.9) are identically zero, so

$$\Delta H(e^{i\omega}) \equiv 0$$

which means that the first term takes the form

$$\left|\Delta G(e^{i\omega})\right|^2 \Phi_u(\omega) \equiv 0 \tag{13.11}$$

This is the crucial condition on the open-loop input spectrum $\Phi_u(\omega)$, which we shall develop further. If (13.11) implies that $\Delta G(e^{i\omega}) \equiv 0$, then it follows that (13.10) implies that the two models are equal, and hence that the data are sufficiently informative with respect to $\mathcal{M}^*$.

Persistence of Excitation

Inspired by (13.11), we introduce the following concept:

Definition 13.1. A quasi-stationary signal $\{u(t)\}$, with spectrum $\Phi_u(\omega)$, is said to be **persistently exciting of order n** if, for all filters of the form

$$M_n(q) = m_1 q^{-1} + \ldots + m_n q^{-n} \tag{13.12}$$

the relation

$$\left|M_n(e^{i\omega})\right|^2 \Phi_u(\omega) \equiv 0 \quad \text{implies that} \quad M_n(e^{i\omega}) \equiv 0 \tag{13.13}$$

The concept can be given more explicit interpretations. Clearly, the function $M_n(z)M_n(z^{-1})$ can have at most $n-1$ different zeros on the unit circle (since one zero is always at the origin) taking symmetry into account. Hence $u(t)$ is persistently exciting of order n, if $\Phi_u(\omega)$ *is different from zero on at least n points in the interval* $-\pi < \omega \leq \pi$. This is a direct consequence of the definition. Consider, for example, with u consisting of n different sinusoids:

$$u(t) = \sum_{k=1}^{n} \mu_k \cos(\omega_k t), \quad \omega_k \neq \omega_j, \; k \neq j, \quad \omega_k \neq 0, \; \omega_k \neq \pi \tag{13.14}$$

According to Example 2.4, each sinusoid gives rise to a spectral line at ω_k and $-\omega_k$. This signal is thus persistently exciting of order $2n$. If one of the frequencies equals 0, the order drops to $2n-1$, since this only gives one spectral line. Similarly, if one of the frequencies equals π (the Nyquist frequency), the order drops by (another) 1.

We also notice that, according to Theorem 2.2, $\left|M_n(e^{i\omega})\right|^2 \Phi_u(\omega)$ is the spectrum of the signal $v(t) = M_n(q)u(t)$. Hence a signal that is persistently exciting of order n *cannot be filtered to zero* by an $(n-1)th$-order moving-average filter (13.12).

Another characterization can be given in terms of the covariance function $R_u(\tau)$:

Lemma 13.1. Let $u(t)$ be a quasi-stationary signal, and let the $n \times n$ matrix $\overline{R}_n$ be defined by

$$\overline{R}_n = \begin{bmatrix} R_u(0) & R_u(1)\ldots & R_u(n-1) \\ R_u(1) & R_u(0)\ldots & R_u(n-2) \\ \vdots & \vdots & \vdots \\ R_u(n-1) & R_u(n-2)\ldots & R_u(0) \end{bmatrix} \tag{13.15}$$

Then $u(t)$ is persistently exciting of order n if and only if $\overline{R}_n$ is nonsingular.

Proof. Let $m = [m_1 \;\; m_2 \;\; \ldots \;\; m_n]^T$. Then $\overline{R}_n$ is nonsingular if and only if

$$m^T \overline{R}_n m = 0 \Rightarrow m = 0 \tag{13.16}$$

It is easy to verify that

$$m^T \overline{R}_n m = \overline{E}\,[M_n(q)u(t)]^2$$

with M_n defined by (13.12). Hence

$$m^T \overline{R}_n m = \frac{1}{2\pi} \int_{-\pi}^{\pi} \left|M_n(e^{i\omega})\right|^2 \Phi_u(\omega) d\omega$$

according to (2.65) and Theorem 2.2, so (13.16) can be rephrased as

$$\left|M_n(e^{i\omega})\right|^2 \Phi_u(\omega) \equiv 0 \Rightarrow M_n(e^{i\omega}) \equiv 0$$

which is the definition of persistence of excitation. □

It is useful to consider also a strengthened version of this concept.

Definition 13.2. A quasi-stationary signal $\{u(t)\}$ with spectrum $\Phi_u(\omega)$ is said to be **persistently exciting** if

$$\Phi_u(\omega) > 0, \qquad \text{for almost all } \omega \tag{13.17}$$

"Almost all" means that the spectrum may be zero on a set of measure zero (like a countable number of points). Note that a persistently exciting signal cannot loose this property by standard linear filtering, since a filter that is an analytic function can have at most a finite number of zeros on the unit circle.

Informative Open-Loop Experiments

With these concepts, it is now easy to characterize sufficiently informative open-loop experiments. We have the following results.

Theorem 13.1. Consider a set $\mathcal{M}^*$ of SISO models given by (13.8) such that the transfer functions $G(z, \theta)$ are rational functions:

$$G(q, \theta) = \frac{B(q, \theta)}{F(q, \theta)} = \frac{q^{-n_k}(b_1 + b_2 q^{-1} + \ldots + b_{n_b} q^{-n_b+1})}{1 + f_1 q^{-1} + \ldots + f_{n_f} q^{-n_f}}$$

Then an open-loop experiment with an input that is persistently exciting of order $n_b + n_f$ is sufficiently informative with respect to $\mathcal{M}^*$.

Proof. For two different models, we have

$$\Delta G(q) = \frac{B_1(q)F_2(q) - B_2(q)F_1(q)}{F_1(q)F_2(q)}$$

Hence (13.11) implies that

$$\left|B_1(e^{i\omega})F_2(e^{i\omega}) - B_2(e^{i\omega})F_1(e^{i\omega})\right|^2 \Phi_u(\omega) \equiv 0$$

Since this numerator is a polynomial of degree at most $n_b + n_f - 1$ (we can always shift $n_k - 1$ steps, so as to conform with (13.12)), it follows from Definition 13.1 that it is identically zero, and hence $\Delta G(q) \equiv 0$. The theorem now follows from the discussion following (13.11). □

Corollary. An open-loop experiment is *informative* if the input is persistently exciting.

Note that the necessary order of persistent excitation equals the number of parameters to be estimated in this case. If the numerator and denominator of the model have the same number of parameters n, then the input should be persistently exciting of order $2n$. This means that $\Phi_u(\omega)$ should be nonzero at $2n$ points, which is achieved for the input (13.14). It is thus sufficient to use n sinusoids to identify an nth order system, a result that ties in nicely with the frequency analysis described in Section 6.2.

Theorem 13.1 covers, for example, the general model set (4.33) with its several special cases. It should also be clear that by analogous techniques other structures can be treated, including multivariable ones. See Problem 13E.2.

Remark. Persistence of excitation of an m-dimensional signal is defined analogously to Definition 13.1: Let $M_n(q)$ be defined by (13.12) with m_i as $1 \times m$ row-matrices. Then $\{u(t)\}$ is said to be persistently exciting of order n if

$$M_n(e^{i\omega})\Phi_u(\omega)M_n^T(e^{-i\omega}) \equiv 0 \quad \text{implies that } M_n(e^{i\omega}) \equiv 0 \tag{13.18}$$

Lemma 13.1 has an immediate multivariable counterpart.

13.3 INPUT DESIGN FOR OPEN LOOP EXPERIMENTS

The requirement from the previous section that the data should be informative means for open loop operation that the input should be persistently exciting (p.e.) of a certain order; i.e., that it contains sufficiently many distinct frequencies. This leaves a substantial amount of freedom for the actual choice, and we shall in this section discuss good and typical choices of input signals.

For the identification of linear systems, there are three basic facts that govern the choices:

1. The asymptotic properties of the estimate (bias and variance) depend only on the *input spectrum*—not the actual waveform of the input.
2. The input must have limited amplitude: $\underline{u} \leq u(t) \leq \overline{u}$.
3. Periodic inputs may have certain advantages.

The first fact follows from (8.71) and (9.54). The second one is obvious from practical considerations. The advantages and disadvantages of periodic inputs will be discussed later in this section.

The Crest Factor

The covariance matrix is typically inversely proportional to the input power. We would thus like to have as much input power as possible. In practice, the actual input limitation concerns amplitude constraints $\underline{u}$ and $\overline{u}$. The desired property of the waveform therefore is defined in terms of the *crest factor* C_r, which for a zero mean signal is defined as

$$C_r^2 = \frac{\max_t u^2(t)}{\lim_{N\to\infty} \frac{1}{N}\sum_{t=1}^{N} u^2(t)} \tag{13.19}$$

(More sophisticated definitions only use the signal power in a certain frequency band of interest in the denominator.) A good signal waveform is consequently one that has a small crest factor. The theoretic lower bound of C_r clearly is 1, which is achieved for *binary, symmetric signals*: $u(t) = \pm\overline{u}$.

This gives a theoretical advantage for binary signals, and indeed, several of the signals that we will discuss in this section will be binary. However, the following *caution* should be mentioned: A binary input will not allow validation against non-linearities. For example, if the true system has a static non-linearity at the input (as in the Hammerstein model of Figure 5.1) and a binary input is used, the input is still binary after the non-linearity, and just corresponds to a scaling. There is consequently no way to detect that such a non-linearity is present from a binary input.

The Frequency Contents of the Input

Consider, as before, the general SISO model structure

$$y(t) = G(q,\theta)u(t) + H(q,\theta)e(t) \tag{13.20}$$

The asymptotic covariance matrix that results when a prediction error method is applied to (13.20) was computed in (9.29) and (9.30):

$$P_\theta(\mathcal{X}) = \kappa(\ell) \cdot \left[\overline{E}\psi(t,\theta_0)\psi^T(t,\theta_0)\right]^{-1} \tag{13.21a}$$

$$\kappa(\ell) = \frac{E\left[\ell'(e_0(t))\right]^2}{\left[E\ell''(e_0(t))\right]^2} \tag{13.21b}$$

From (7.89) we can define the average *information matrix per sample*, $\overline{M}$, as

$$\overline{M}(\mathcal{X}) = \lim_{N\to\infty} \frac{1}{N} M_N = \frac{1}{\kappa_0}\overline{E}\psi(t,\theta_0)\psi^T(t,\theta_0) \tag{13.22}$$

where

$$\frac{1}{\kappa_0} = \int_{-\infty}^{\infty} \frac{\left[f_e'(x)\right]^2}{f_e(x)} dx \tag{13.23}$$

and $f_e(x)$ is the PDF of the true innovations. The important consequence of these expressions is that the choice of norm $\ell(\varepsilon)$ and the distribution of the innovations *act only as input-independent scaling* of the covariance matrix and the information matrix. Optimization measures based on $\overline{M}$ in (13.22) thus cover the Cramér-Rao lower bound for $P_\theta(\mathcal{X})$ as well as all asymptotic expressions for $P_\theta(\mathcal{X})$ obtained with prediction error methods, up to an $\mathcal{X}$-independent scaling, that is immaterial for the experiment design.

In (9.54) we gave an expression for the covariance matrix. This can be rewritten in terms of $\overline{M}(\mathcal{X})$ as

$$\overline{M}(\mathcal{X}) = \frac{\lambda_0}{2\pi\kappa_0}\int_{-\pi}^{\pi}\Big\{G_\theta'(e^{i\omega},\theta_0)\left[G_\theta'(e^{-i\omega},\theta_0)\right]^T\Phi_u(\omega)$$

$$+ H_\theta'(e^{i\omega},\theta_0)\left[H_\theta'(e^{-i\omega},\theta_0)\right]^T\lambda_0\Big\}/\Phi_v(\omega)d\omega \tag{13.24}$$

provided u and e_0 are independent. Here G'_θ and H'_θ are the $d \times 1$ gradients of G and H. Introduce

$$\tilde{M}(\omega) = \frac{\lambda_0 G'_\theta(e^{i\omega}, \theta_0) \left[G'_\theta(e^{-i\omega}, \theta_0)\right]^T}{2\pi\kappa_0 \Phi_v(\omega)} \tag{13.25a}$$

$$M_e = \frac{\lambda_0^2}{2\pi\kappa_0} \int_{-\pi}^{\pi} \frac{H'_\theta(e^{i\omega}, \theta_0) \left[H'_\theta(e^{-i\omega}, \theta_0)\right]^T}{\Phi_v(\omega)} d\omega \tag{13.25b}$$

Then we have

$$\overline{M}(\mathcal{X}) = \overline{M}(\Phi_u) = \int_{-\pi}^{\pi} \tilde{M}(\omega)\Phi_u(\omega) d\omega + M_e \tag{13.26}$$

This expression gives a good impression of how the input spectrum affects the information matrix in the open loop case. It ties nicely with the intuitive advice (13.7). To achieve a large information matrix, we should spend the input power at frequencies where $\tilde{M}(\omega)$ is large, that is, where the Bode plot is sensitive to parameter variations (G'_θ large). Put more leisurely, if a parameter is of special interest, then vary it and check where the Bode plot moves, and put the input power there. In many cases this may give sufficient guidance for good input design. In Section 13.6 we shall use a more formal approach for high order models, to obtain similar results.

Notice also that (13.26) manifests the first fact listed in this section: The information matrix/covariance matrix of the parameters depends only on the input spectrum—not the particular waveforms.

Optimal Frequency Contents: It is quite clear that formal optimal design problems can be formulated from (13.26): Pick a scalar measure of the size of the matrix $\overline{M}^{-1}(\Phi_u)$, like its (weighted) trace, its determinant or its matrix norm and minimize this with respect to the input spectrum Φ_u. There is quite an extensive literature on this, e.g. Goodwin and Payne (1977) and Zarrop (1979). One important issue in this context is how to consider a restricted, finitely parameterized set of inputs that still covers the set of achievable information matrices. A typical result is that it is sufficient to consider inputs that are a finite sum of sinusoids, as in (13.14). The number of necessary sinusoids depends on the system/model order. See, e.g., Stoica and Söderström (1982a). An efficient algorithm for selecting the frequencies on the DFT-grid is given in Section 4.3.4 of Schoukens and Pintelon (1991).

It should be noted that the optimal input design will depend on the (unknown) system, so this optimality approach is worthwhile, only when good prior knowledge is available about the system. The optimum number of sinusoids may also correspond to the minimum one for identifying a system of this order. To allow for validation against higher order models, it is thus wise to use an input with more frequencies. In practice, it is suitable to decide upon an important and interesting frequency band to identify the system in question, and then select a signal with a more or less flat spectrum over this band. We will discuss such designs in the remainder of this section.

Common Input Signals

The basic issue for input signal design is now clear: For linear system identification, *achieve a desired input spectrum for a signal with as small crest factor as possible.* Unfortunately these properties are somewhat in conflict: If it is easy to manipulate a signal's spectrum, it tends to have a high crest factor and *vice versa*. We shall now describe typical choices of waveforms, and how to achieve desired spectra.

A general comment is that it is always advisable to generate the signal and study its properties, off-line, before using it as an input in an identification experiment.

Filtered Gaussian White Noise. A simple choice is to let the signal be generated as white Gaussian noise, filtered through a linear filter. With this we can achieve virtually any signal spectrum (that does not have too narrow pass bands) by proper choice of filters. Since the signal is generated off-line, non-causal filters can be applied and transient effects can be eliminated, which gives even better spectral behavior. See any book on filter design, like Parks and Burrus (1987). The Gaussian signal is theoretically unbounded, so it has to be saturated ("clipped") at a certain amplitude. Picking that, e.g., to be at 3 standard deviations gives a crest factor of 3, and at the same time, only an average of 1% of the time points are affected. This should lead to quite minor distortions of the spectrum.

Random Binary Signal. A random binary signal is a random process which assumes only two values. It can be generated in a number of different ways. The *telegraph signal* is generated as a random process which at any given sample has a certain probability to change from the current level to the other one. Apparently, the most common way is to simply generate white, zero mean Gaussian noise, filter it by an appropriately chosen linear filter, and then just take the sign of the filtered signal. It can then be adjusted to any desired binary levels. The crest factor is thus the ideal 1. The problem is that taking the sign of the filtered Gaussian signal will change its spectrum. We therefore do not have full control of shaping the spectrum. In the off-line situation we can however always check the spectrum of the signal before using it as input to the process to see if it is acceptable.

Example 13.1 Band-limited Gaussian and Binary Signals

Suppose we seek an input with power concentrated to the band $1 \leq \omega \leq 2$ (rad/s). Let e be generated as white Gaussian noise. Filter this signal through a 5th order Butterworth filter with the indicated pass band. This gives the signal in Figure 13.1. Its spectrum is shown in Figure 13.2. Taking the sign of this signal gives the random binary signal of Figure 13.1. Its spectrum is shown in Figure 13.2. The distortion of the desired spectrum is clear. □

Pseudo-Random Binary Signal, PRBS. A *Pseudo-Random Binary Signal* is a periodic, deterministic signal with white-noise-like properties. It is generated by the difference equation

$$u(t) = \operatorname{rem}(A(q)u(t), 2) = \operatorname{rem}(a_1 u(t-1) + \ldots + a_n u(t-n), 2) \quad (13.27)$$

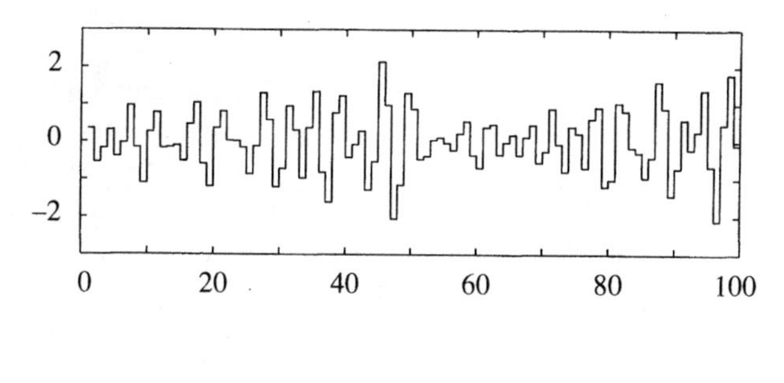

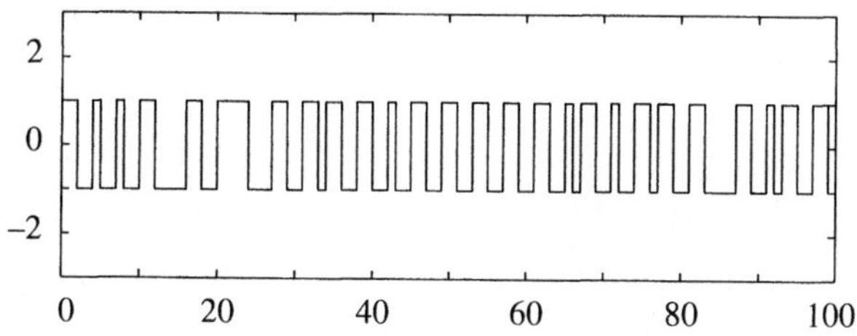

Figure 13.1 Upper plot: Random Gaussian noise, filtered through a pass-band of $1 \le \omega \le 2$. Lower plot: Random binary noise obtained by taking the sign of the signal in the upper plot. Both signals are plotted as piecewise constant signals.

Here $\text{rem}(x, 2)$ is the remainder as x is divided by 2, i.e., the calculations in (13.27) should be carried out modulo 2. $u(t)$ thus only assumes the values 0 and 1. After u is generated, we can of course change that to any two levels. The vector of past inputs $[\,u(t-1) \quad \dots \quad u(t-n)\,]$ can only assume 2^n different values. The sequence u must thus be periodic with a period of at most 2^n. In fact, since n consecutive zeros would make further u's identically zero, we can eliminate that state, and the maximum period length is $M = 2^n - 1$. Now the actual period of the signal will

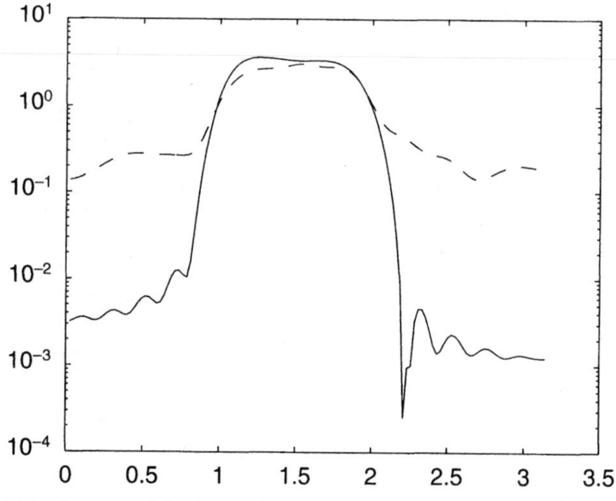

Figure 13.2 Spectra of the signals in Figure 13.1. Solid line: The spectrum of the Gaussian signal. Dashed line: The spectrum of the binary signal.

TABLE 13.1 A-polynomials that generate maximum length ($=M$) PRBS for different orders n. $a_k = 1$ for the indicated k and 0 otherwise. Several other choices may exist for each n.

Order n	$M = 2^n - 1$	a_k non-zero for k
2	3	1,2
3	7	2,3
4	15	1,4
5	31	2,5
6	63	1,6
7	127	3,7
8	255	1,2,7,8
9	511	4,9
10	1023	7,10
11	2047	9,11

depend on the choice of $A(q)$, but it can be shown that for each n there exists choices of $A(q)$ that give this maximum length. Such choices are shown in Table 13.1 and the corresponding inputs are called *Maximum length PRBS*. See Davies (1970). The interest in maximum length PRBS follows from the following property:

Any maximum length PRBS shifting between $\pm\bar{u}$ *has the first and second order properties*

$$\left|\frac{1}{M}\sum_{t=1}^{M} u(t)\right| = \frac{\bar{u}}{M}$$

$$R_u(k) = \frac{1}{M}\sum_{t=1}^{M} u(t)u(t+k) = \begin{cases} \bar{u}^2 & k = 0, \pm M, \pm 2M \ldots \\ -\frac{\bar{u}^2}{M} & \text{else} \end{cases} \tag{13.28}$$

Here $M = 2^n - 1$ is the (maximum length) period, and the summation is performed with periodic continuation of the signal. Note that the signal does not have exactly zero mean. Its covariance function thus differs from the second moment function (13.28). To compute the spectrum, we proceed as in Example 2.3. In the notation of that example, we find that

$$\Phi_u^p(\omega) = \sum_{k=0}^{M-1} R_u(k)e^{-ik\omega} = \bar{u}^2\left[1 - \frac{1}{M}\sum_{k=1}^{M-1} e^{-ik\omega}\right]$$

$$= \bar{u}^2\left[1 - \frac{1}{M}\frac{e^{-i\omega} - e^{-iM\omega}}{1 - e^{-i\omega}}\right]$$

$$= \begin{cases} \bar{u}^2\frac{1}{M} & \text{for } \omega = 0 \\ \bar{u}^2(1 + \frac{1}{M}) & \text{for } \omega = 2\pi k/M,\ k = 1, \ldots, M-1 \end{cases}$$

The expression (2.67) now gives the spectrum

$$\Phi_u(\omega) = \frac{2\pi\overline{u}^2}{M} \sum_{k=1}^{M-1} \delta(\omega - 2\pi k/M), \quad 0 \le \omega < 2\pi \tag{13.29}$$

where we ignored terms proportional to $1/M^2$. In the region $-\pi \le \omega < \pi$ there will be $M - 1$ frequency peaks ($\omega = 0$ excluded). This shows that maximum length PRBS behaves like "periodic white noise," and is persistently exciting of order $M - 1$. Figure 13.3 shows one period of a PRBS and its spectrum.

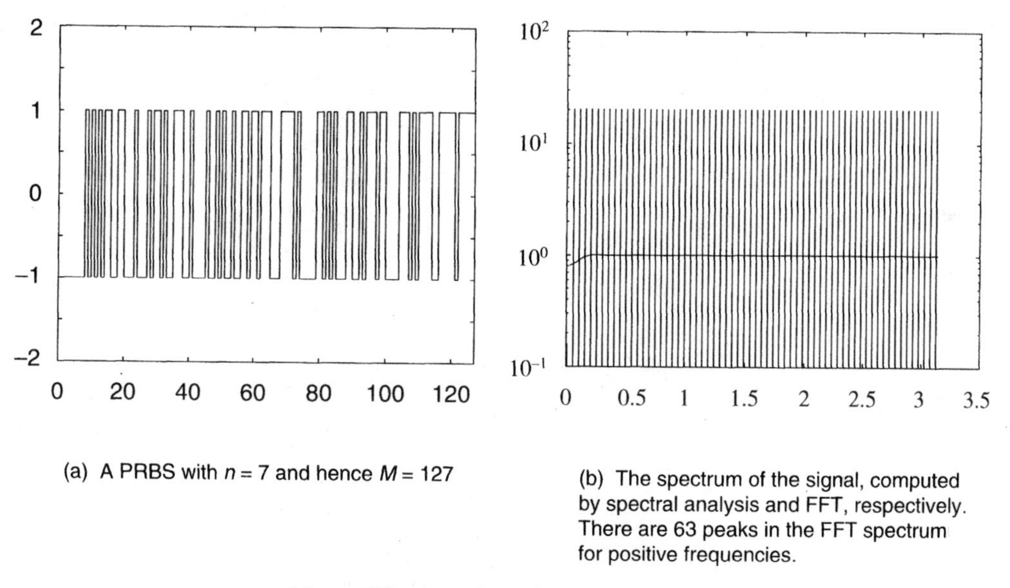

(a) A PRBS with $n = 7$ and hence $M = 127$

(b) The spectrum of the signal, computed by spectral analysis and FFT, respectively. There are 63 peaks in the FFT spectrum for positive frequencies.

Figure 13.3 A maximum length PRBS signal.

Notice that it is essential to perform these calculations over whole periods. Generating just a part of a period of a PRBS will not give a signal with properties (13.28).

Like white random binary noise, PRBS has an optimal crest factor. The advantages and disadvantages of PBRS compared to binary random noise can be summarized as follows:

- If the PRBS contains whole periods, its covariance matrix will have a very special pattern according to (13.28). It can be analytically inverted, which will facilitate certain computations.
- As a deterministic signal, PRBS has its second order properties secured when evaluated over whole periods. For random signals, one must rely upon the law of large numbers to have good second order properties for finite samples.

- There is essentially only one PRBS for each choice of $A(q)$. Different initial values when generating (13.27) only correspond to shifting the sequence. It is therefore not straightforward to generate mutually uncorrelated sequences with PRBS. A simple way would be to excite one input at a time, or variants thereof. See Problem 13E.7.
- For PRBS one should work with an integer number of periods to enjoy its good properties, which limits the choice of experiment length.

Low-Pass Filtering by Increasing the Clock Period. It is easy to generate binary signals with white noise second order properties, either as PRBS or by a random generator. To give the signal a more low-frequency character, we could filter it through a low pass filter. This would make the signal non-binary, though, with a worse crest factor. An alternative is to sample faster, i.e., from the given PRBS, create a new signal u by taking P samples over each sampling period of the original signal e. The new signal will thus always stay constant over at least P samples. We have thus increased the sampling frequency to be P times faster than the frequency at which the PRBS is generated. This is usually expressed as having a *clock period* of P. It can be shown (see Example 5.10 in Söderström and Stoica, 1989) that the new signal u has the same covariance function as

$$\tilde{u}(t) = \frac{1}{P}\left(e(t) + \ldots + e(t - P + 1)\right)$$

obtained by simple moving average low pass filtering of e.

A typical advice is to let the clock frequency in the PRBS be about 2.5 times the bandwidth to be covered by the signal, Schoukens, Guillaume, and Pintelon (1993). Another advice is to sample about 10 times faster than the bandwidth to be modeled (see Section 13.7). Together this shows that a good choice is to take the clock period $P = 4$. A spectrum for a PRBS with $P = 4$ is shown in Figure 13.4. We see that the frequencies up to about 1/5 of the Nyquist frequency are well covered.

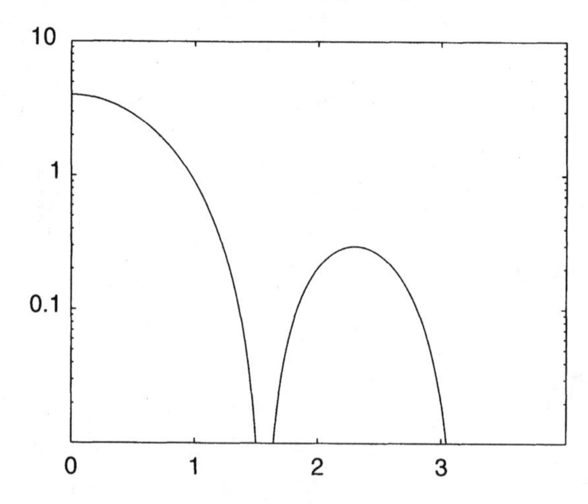

Figure 13.4 The spectrum of a PRBS signal with $P = 4$.

Multi-Sines. A natural choice of input is to form it as a sum of sinusoids:

$$u(t) = \sum_{k=1}^{d} a_k \cos(\omega_k t + \phi_k) \tag{13.30}$$

Apart from transient effects this gives a spectrum according to Example 2.4:

$$\Phi_u(\omega) = 2\pi \sum_{k=1}^{d} \frac{a_k^2}{4} [\delta(\omega - \omega_k) + \delta(\omega + \omega_k)] \tag{13.31}$$

With d, a_k and ω_k we can thus place the signal power very precisely to desired frequencies. In addition, we mentioned in Section 13.2 that all possible information matrices can be obtained within the family (13.30) for large enough d. The only problem with this input is the crest factor. The power of the signal is $\sum a_k^2/2$. If all sinusoids are in phase, the squared amplitude will be $(\sum a_k)^2$. The crest factor can thus be up to $\sqrt{2d}$ (if all a_k are equal). The way to control the crest factor is to choose the phases ϕ_k so that the cosines are "as much out of phase" as possible. A simple solution is the so-called *Schroeder phase* choice, Schroeder (1970), which means that the phases are spread as follows when the amplitudes a_k are equal:

$$\begin{aligned} &\phi_1 \text{ arbitrary} \\ &\phi_k = \phi_1 - \frac{k(k-1)}{d}\pi; \quad 2 \le k \le d. \end{aligned} \tag{13.32}$$

Chirp Signals or Swept Sinusoids. A chirp signal is a sinusoid with a frequency that changes continuously over a certain band $\Omega : \omega_1 \le \omega \le \omega_2$ over a certain time period $0 \le t \le M$:

$$u(t) = A \cos\left(\omega_1 t + (\omega_2 - \omega_1)t^2/(2M)\right) \tag{13.33}$$

The "instantaneous frequency" ω_i in this signal is obtained by differentiating the argument w.r.t. time t:

$$\omega_i = \omega_1 + \frac{t}{M}(\omega_2 - \omega_1)$$

and we see that it increases from ω_1 to ω_2. This signal has the same crest factor as a pure sinusoid, i.e., $\sqrt{2}$, and it gives good control over the excited frequency band. Due to the sliding frequency, there will however also be power contributions outside the band Ω.

Example 13.2 Sinusoids and Swept Sinuoids

In Figure 13.5 we show the signal which is obtained from (13.30) with 10 frequencies of equal amplitude (= 1), equally spread over the frequency band $1 \le \omega \le 2$ rad/sec. The three cases correspond to $\phi_k \equiv 0$, the Schroeder choice (13.32), and randomized phases, respectively. They all have the same spectrum, shown in Figure 13.5d. Figure 13.6a shows the chirp signal (13.33) over the chosen band, while its spectrum is shown in Figure 13.6b. □

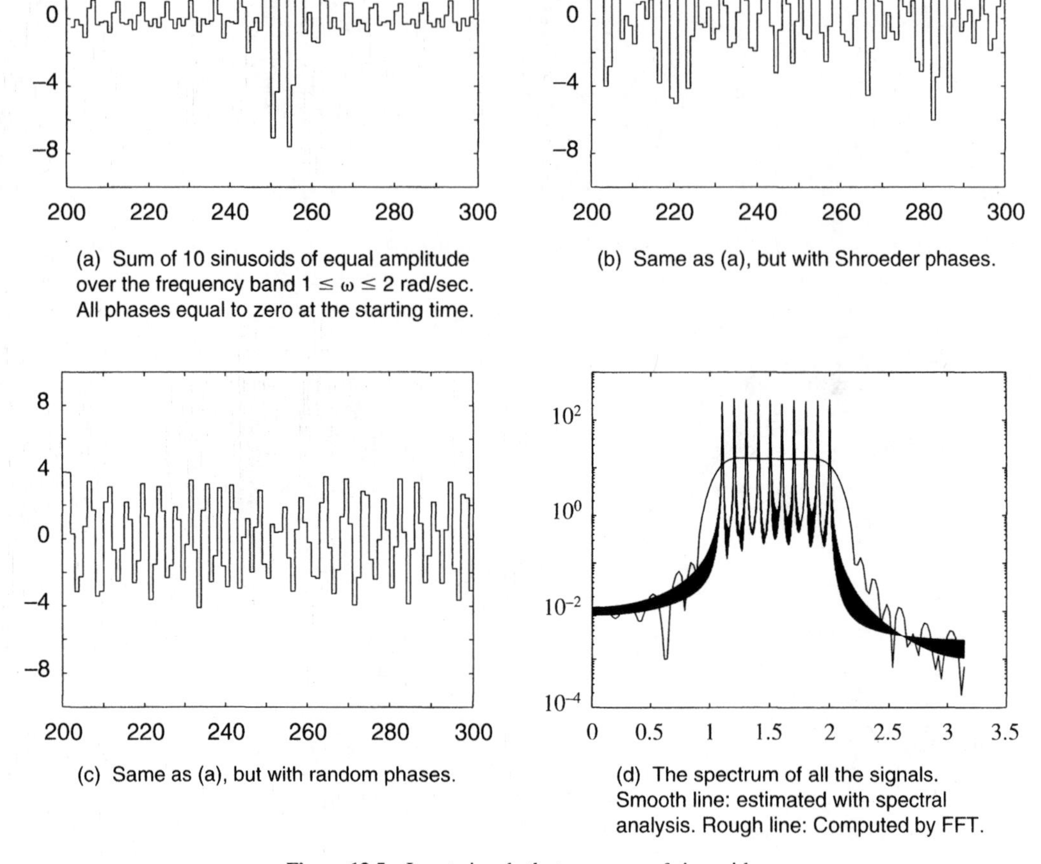

Figure 13.5 Input signals that are sums of sinusoids.

Periodic Inputs

Some of the signals above are inherently periodic, like the PRBS, or the sum of sinusoids. All of them can in any case be made periodic by simple repetition. To retain the nice frequency properties they have been designed for, the following facts must be taken into account when creating periodic signals:

- The PRBS signal must be generated over one full period, $M = 2^n - 1$, and then be repeated. This follows from the discussion of its second order properties.
- To create a multi-sine of period M, the frequencies ω_k in (13.30) must be chosen from the DFT-grid $\omega_\ell = 2\pi\ell/M$, $\ell = 0, 1, \ldots, M-1$.

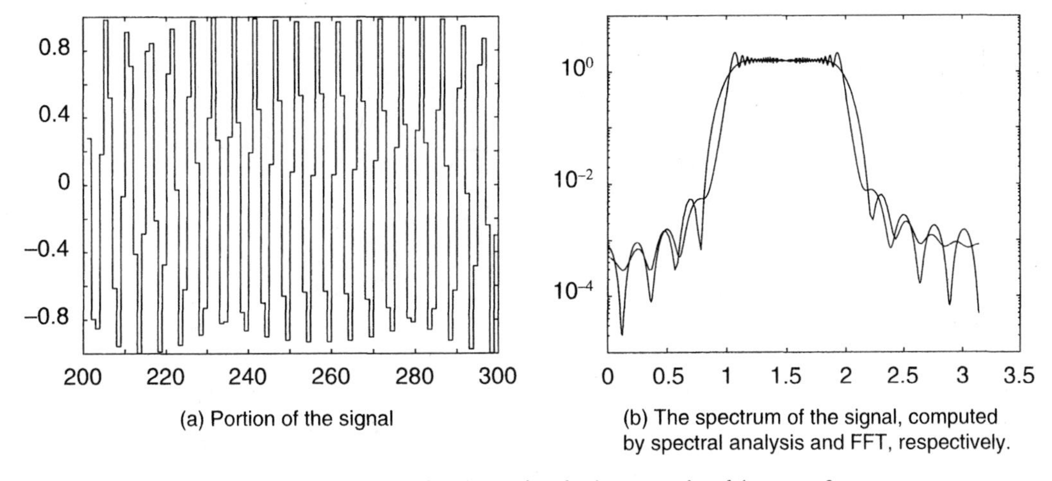

(a) Portion of the signal

(b) The spectrum of the signal, computed by spectral analysis and FFT, respectively.

Figure 13.6 A chirp signal covering the frequency band $1 \leq \omega \leq 2$.

- To make the chirp signal (13.33) nicely periodic with period M, ω_1 and ω_2 must be chosen as $2\pi k_i/M$ for some integers k_1 and k_2. The signal generated by (13.33) can then be repeated an arbitrary number of times.
- To display the spectrum of a periodic signal of length N with an even number of periods without any leakage, it should be computed for the (DFT) frequencies $\omega_k = 2\pi k/N$. ("Leakage" means that the Fourier transform is distorted by boundary effects. Note that the "ringing" in Figures 13.5d and 13.6b is due to leakage.)

What are the advantages and disadvantages with periodic inputs?

- A signal with period M can have at most M distinct frequencies in its spectrum. It is thus persistently exciting of, at most, order M. In this sense, non-periodic inputs inject more excitation into the system over a given time span.
- When a periodic input has been applied, say K periods each of length M ($N = KM$), it is usually advisable to average the output over the periods, and work only with one period of input-output data in the model building session. This gives less data to handle. The signal to noise ratio is improved by a factor of K by this operation, at the same time as the data record is reduced by the same factor. No difference in asymptotic properties should thus result from this (unless the noise model and the dynamics model share parameters). However, several methods have a performance threshold for finite samples and poor signal-to-noise ratios, so in practice there might also be an accuracy benefit from averaging the measurement over the periods.
- A periodic input allows both formal and informal estimates of the noise level in the system. After transient effects have disappeared, the differences in the output response over the different periods must be attributed to the noise sources. This could be quite helpful in the model validation process for the

important distinction between model errors and noise (cf. Chapter 16). More formally, let the output be

$$y(t) = y_u(t) + v(t)$$

where $y_u(t)$ is the noise-free part that originates from the input. It is thus periodic and a natural estimate is

$$\hat{y}_u(t) = \frac{1}{K} \sum_{k=0}^{K-1} y(t + kM); \; 1 \leq t \leq M \tag{13.34}$$

and periodically continued for larger t. This gives the noise estimates $\hat{v}(t) = y(t) - \hat{y}_u(t)$ from which both noise levels and noise colors can be estimated. The noise variance λ_v, e.g., is estimated as

$$\hat{\lambda}_v = \frac{1}{(K-1)M} \sum_{t=1}^{KM} \hat{v}^2(t)$$

- When the models are estimated in terms of Fourier transformed data (see Section 7.7), periodic signals give no leakage when forming the Fourier transforms.

Intersample Behavior of the Input (∗)

So far we have just considered *the discrete-time properties* of the input $u(t)$, $t = 1, 2, \ldots$, assuming a unit sampling interval. What will be applied to the actual process is of course a continuous-time signal $u(t)$ defined for all real t in a certain interval. To stress this point, we shall in this subsection use the notation u_k, $k = 1, 2, \ldots$ for the input sequence and use $u(t)$ to denote the continuous time input. The spectrum Φ_u^d of the sequence u_k, $k = 1, 2, \ldots$ is defined by (2.63):

$$\Phi_u^d(\omega) = \sum_{\ell=-\infty}^{\infty} R_\ell e^{-i\ell\omega} \tag{13.35a}$$

$$R_\ell = \lim_{N\to\infty} \frac{1}{N} \sum_{k=1}^{N} u_{k-\ell} u_k \tag{13.35b}$$

This is the spectrum we have designed in this chapter, and this is also the spectrum that determines the model quality as in (13.26). It is only relevant to consider this spectrum over $|\omega| \leq \pi$, since $\Phi_u^d(\omega)$ by definition will be periodic with period 2π.

The spectrum of the continuous time signal is defined analogously as

$$\Phi_u^c(\omega) = \int_{-\infty}^{\infty} R_u^c(\tau) e^{-i\tau\omega} d\tau \tag{13.36a}$$

$$R_u^c(\tau) = \lim_{N\to\infty} \frac{1}{N} \int_0^N u(t-\tau)u(t)dt \tag{13.36b}$$

If we choose a sampling interval T, it is natural to construct a continuous time signal $u(t)$, such that $u(kT) = u_k$, $k = 1, 2, \ldots$. This can be done in several different ways, depending on how we select the *intersample behavior*. The simplest case is to let the input be constant between the sampling instants

$$u(t) = u_k \quad \text{if} \quad kT \le t < (k+1)T \tag{13.37}$$

where T is the *sampling interval*. Such an input is called *zero order hold*, ZOH. If the continuous-time input is defined by linear interpolation between the sampling instants, we have the *first order hold*, FOH, case:

$$u(t) = \frac{(t - kT)u_{k+1} + (kT + T - t)u_k}{T} \quad \text{if} \quad kT \le t < (k+1)T \tag{13.38}$$

The third common choice is to let the continuous time signal be *band-limited*. This means that the continuous time signal has no power above the Nyquist frequency, and that its spectrum coincides with the spectrum of the discrete time signal, up to this frequency:

$$\Phi_u^c(\omega) = \Phi_u^d(\omega T) \quad \text{for} \quad |\omega| \le \pi/T, \qquad \Phi_u^c(\omega) = 0 \quad \text{else} \tag{13.39}$$

We may think of this signal as obtained as a sum of sinusoids (up to frequency πT) $u(t)$ adjusted so that $u(kT) = u_k$, $k = 1, 2, \ldots$. This is trigonometric interpolation.

Note the following aspects:

- For first and zero order hold, the discrete signal spectrum $T\Phi_u^d(\omega T)$ does not coincide with the continuous one $\Phi_u^c(\omega)$ even for $|\omega| \le \pi/T$. The reason is, loosely speaking, that the abrupt changes in the ZOH signal create new frequencies in the continuous signal.
- In all cases where the continuous time input can be constructed exactly from its values at the sampling points, it will be possible to form an exact discrete time model for how $u(kT)$, $k = 1, 2, \ldots$ affect $y(kT)$, $k = 1, 2, \ldots$ (apart from the noise contributions, of course). The actual discrete time model will however depend on the intersample behavior, i.e., a model which has a perfect fit for ZOH input will not describe the system under a FOH input.
- The formulas for translating a discrete time model to continuous time will depend on the intersample behavior of the input for which the model was fitted. For a ZOH input we should use the inverse of (4.67)–(4.71).

  Note that the methods of this book also allow a direct fit of a continuous time model to discrete time data. It is just a matter of parameterization as in (4.65). However, the parameterization must be done using a sampling formula that is consistent with the true intersample behavior of the input.
- When we build a discrete time model from $u(kT) = u_k$ to $y(kT)$, it is the discrete signal spectrum Φ_u^d that determines the model quality (uncertainty) as in (13.24). This follows from the analysis in Chapters 8 and 9 where the intersample behavior does not enter. However, as noted above, the translation of the discrete time model to continuous time should depend on the input intersample behavior. Therefore, the quality of the resulting continuous-time model will depend both on the input's discrete signal spectrum and its intersample properties.

Which aspects should then guide the choice of input intersample behavior?

- *Affinity to the model use.* If the model is to be used for ZOH inputs, as is typical in all computer controlled applications, it is natural to use a ZOH input experiment to generate the identification data. Then the resulting model can be used directly in discrete time.
- *Ease and accuracy of input generation.* The practical experiment equipment will of course also decide what is the natural choice of input character. Note that if the model is to be constructed as, or transformed to, continuous time, it is necessary that the assumed intersample behavior coincides with the actual one. Whatever choice is easier to implement accurately is then to be preferred.

Finally, we may note that if the sampling rate is fast compared to the bandwidth of the system (as will be suggested in Section 13.7), the difference between various intersample behaviors may be insignificant.

13.4 IDENTIFICATION IN CLOSED LOOP: IDENTIFIABILITY

It is sometimes necessary to perform the identification experiment under output feedback, i.e., in closed loop. The reason may be that the plant is unstable, or that it has to be controlled for production, economic, or safety reasons, or that it contains inherent feedback mechanisms.

In this section we shall study problems and possibilities with identification data from closed loop operation. In many cases we will not need to know the feedback mechanism, but for some of the analytic treatment we shall work with the following linear output feedback setup: The true system is

$$y(t) = G_0(q)u(t) + v(t) = G_0(q)u(t) + H_0(q)e(t) \tag{13.40a}$$

Here $\{e(t)\}$ is white noise with variance λ_0. We shall also use the notation $v(t) = H_0(q)e(t)$. The regulator is as in Figure 13.7:

$$u(t) = r(t) - F_y(q)y(t) \tag{13.40b}$$

Here $\{r(t)\}$ is a reference signal (filtered version of a setpoint, or any other external signal) that is independent of the noise $\{e(t)\}$. The model is

$$y(t) = G(q,\theta)u(t) + H(q,\theta)e(t) \tag{13.40c}$$

We also assume that the closed loop is well defined in the sense that

$$\text{Either } F_y(q) \text{ or both } G(q,\theta) \text{ and } G_0(q) \text{ contain a delay} \tag{13.40d}$$

$$\text{The closed loop system is stable} \tag{13.40e}$$

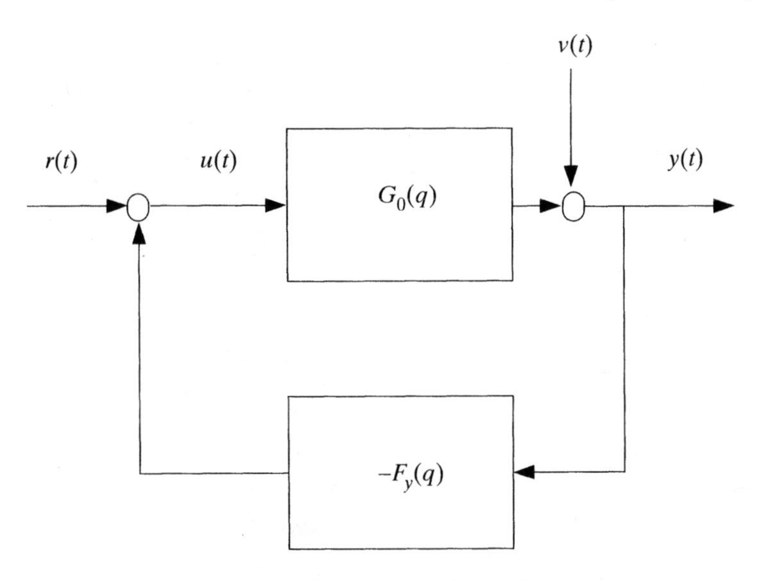

Figure 13.7 Block diagram of a typical feedback system.

The closed loop equations become

$$y(t) = G_0(q)S_0(q)r(t) + S_0(q)v(t) \tag{13.41a}$$

$$u(t) = S_0(q)r(t) - F_y(q)S_0(q)v(t) \tag{13.41b}$$

where $S_0(q)$ is the sensitivity function

$$S_0(q) = \frac{1}{1 + F_y(q)G_0(q)} \tag{13.42}$$

In the sequel we shall omit arguments ω, q, $e^{i\omega}$, and t whenever there is no risk of confusion.

The input spectrum is

$$\Phi_u = |S_0|^2\Phi_r + |F_y|^2|S_0|^2\Phi_v \tag{13.43}$$

Here Φ_r and Φ_v are the spectra of the reference signal and the noise, respectively. We shall use the notation

$$\Phi_u^r = |S_0|^2\Phi_r, \quad \Phi_u^e = |F_y|^2|S_0|^2\Phi_v \tag{13.44}$$

to show the two components of the input spectrum, originating from the reference signal and the noise respectively. See also (8.74).

Some Basic Good News

Most of the analytical development in Chapters 8 and 9 was done under general conditions that include closed loop data. The basic convergence theorem, Theorem 8.2, applies also to closed loop data, and Theorem 8.3 tells us that a prediction error method will consistently estimate the system if

- The data is informative (Definition 8.1)
- The model set contains the true system

regardless if the data $\{u, y\}$ have been collected under feedback. Since the variance results of Chapter 9 (Theorem 9.1, equations (9.17) and (9.31)) apply also to closed loop data, we know that under the above assumptions the straightforward prediction error estimate will have optimal accuracy.

We therefore need only look into what constitutes informative experiments under closed loop, and what the approximation aspects are when the model set does not contain the true system.

Some Fallacies with Closed Loop Identification

There are some fallacies associated with closed loop data:

- The closed loop experiment may be non-informative even if the input in itself is persistently exciting. The reason then is that the regulator is too simple. See Example 13.3 below.
- Spectral analysis applied in a straightforward fashion, as described in Chapter 6, will give erroneous results. According to Problem 6G.1 the estimate of G will converge to

$$G_*(e^{i\omega}) = \frac{G_0(e^{i\omega})\Phi_r(\omega) - F_y(e^{-i\omega})\Phi_v(\omega)}{\Phi_r(\omega) + \left|F_y(e^{i\omega})\right|^2 \Phi_v(\omega)}$$

- Correlation analysis, as described by (6.7)–(6.11) will give a biased estimate of the impulse response, since the assumption $\bar{E}u(t)v(t-\tau) = 0$ is violated.
- For open loop data, output error models (see (4.25) and (4.117)) will give consistent estimates of G, even if the additive noise is not white. This follows from Theorem 8.4. This is not true for closed loop data. See (13.53) below.
- The subspace method (7.66) will typically not give consistent estimates when applied to closed loop data.

Example 13.3 Proportional Feedback

Consider the first-order model structure

$$y(t) + ay(t-1) = bu(t-1) + e(t) \tag{13.45}$$

and suppose that the system is controlled by a proportional regulator during the experiment:

$$u(t) = -fy(t) \tag{13.46}$$

Inserting the feedback law into the model gives

$$y(t) + (a + bf)y(t - 1) = e(t) \tag{13.47}$$

which is the model of the closed-loop system. From this we conclude that *all models* $(\hat{a}, \hat{b})$ subject to

$$\begin{aligned} \hat{a} &= a + \gamma f \\ \hat{b} &= b - \gamma \end{aligned} \tag{13.48}$$

with γ an arbitrary scalar, give the same input-output description of the system as the model (a, b) under the feedback (13.46). There is consequently no way to distinguish between these models. Notice in particular that it is of no help to know the regulator parameter f. The experimental condition (13.46) is consequently not informative enough with respect to the model structure (13.45). It is true, though, that the input signal $u(t)$ is persistently exciting since it consists of filtered white noise. Persistence of excitation is thus not a sufficient condition on the input in closed-loop experiments.

If the model structure (13.45) is restricted by, for example, constraining b to be 1

$$y(t) + ay(t - 1) = u(t - 1) + e(t)$$

then it is clear that the data generated by (13.46) are sufficiently informative to distinguish between values of the a-parameter. □

Informative, Closed Loop Experiments

It could consequently be problematic to obtain relevant information from closed-loop experiments. Conditions for informative data sets must also involve the feedback mechanisms. To get a feeling for the problem, consider (8.12) in Definition 8.1. If (8.12) holds for two different models, then

$$\overline{E}\left[\Delta W_y(q)y(t) + \Delta W_u(q)u(t)\right]^2 = 0 \tag{13.49}$$

for some filters ΔW_y and ΔW_u that are not both zero and that are of about the same complexity as the models in the model set. For the regulator (13.40b) this would mean that

$$0 = \overline{E}\left|[\,\Delta W_y \quad \Delta W_u\,]\begin{bmatrix} y \\ u \end{bmatrix}\right|^2 = \overline{E}\left|[\,\Delta W_y \quad \Delta W_u\,]\begin{bmatrix} G_0 & S_0 \\ 1 & -F_y S_0 \end{bmatrix}\begin{bmatrix} S_0 r \\ v \end{bmatrix}\right|^2$$

using (13.41). Let

$$\tilde{W} = [\,\tilde{W}_y \quad \tilde{W}_u\,] = [\,\Delta W_y \quad \Delta W_u\,]\begin{bmatrix} G_0 & S_0 \\ 1 & -F_y S_0 \end{bmatrix}$$

The determinant of the last matrix is $-G_0 F_y S_0 - S_0 = -1$, so it is always invertible, which means that $\tilde{W} = 0$ will imply that both ΔW_y and ΔW_u are zero. Recall that r and v are uncorrelated by assumption, so

$$0 = \overline{E}\left|\tilde{W}\begin{bmatrix} S_0 r \\ v \end{bmatrix}\right|^2 = \overline{E}\left|\tilde{W}_y S_0 r\right|^2 + \overline{E}\left|\tilde{W}_u v\right|^2$$

The filtered noise v is persistently exciting, and if $S_0 r$ is that too, then the conclusion is that $\tilde{W} = 0$. Under that assumption we have thus shown that (13.49) implies $\Delta W_y = 0$ and $\Delta W_u = 0$. Note also that $S_0 r$ will be persistently exciting if r is, since the analytical function S_0 can be zero at at most finitely many points. We summarize this result as a theorem.

Theorem 13.2. The closed loop experiment (13.40) is informative if and only if r is persistently exciting.

As a matter of fact (13.49) contains more information. This equation essentially implies that there is a linear, time-invariant, and noise-free relationship between y and u:

$$\Delta W_u(q)u(t) \approx -\Delta W_y(q)y(t) \tag{13.50}$$

Therefore, only if there is a feedback like (13.50) during the experiment is the data set not informative enough. Thus not only external signals like r in the regulator will assure an informative experiment, but also nonlinear or time-varying or complex (high-order) regulators should, in general, yield experiments that are informative enough. This is the most general statement that can be made about informative experiments.

A general result for the multivariable case with time-varying regulators can be formulated as follows:

Let the input signal be given by output feedback plus an extra signal:

$$u(t) = -F_i(q)y(t) + K_i(q)r(t), \qquad i = 1, 2, \dots, r \tag{13.51}$$

Here F_i and K_i are linear filters that are changed during the experiment between r (different) ones. The changes are made so that each regulator is used a nonzero proportion of the total time, and so seldom that any high-frequency contributions to the signal spectra that arise from the shifts can be neglected. The dimension of the filters F_i are $m \times p$ ($m = \dim u$, $p = \dim y$), and of K_i are $m \times s$ ($s = \dim r$).

Assume that the signal r is persistently exciting. Then the experiment is informative if and only if

$$\det \begin{bmatrix} \sum_{i=1}^{r} \left[K_i(e^{i\omega})K_i^T(e^{-i\omega}) + F_i(e^{i\omega})F_i^T(e^{-i\omega})\right] & -\sum_{i=1}^{r} F_i(e^{i\omega}) \\ -\sum_{i=1}^{r} F_i^T(e^{-i\omega}) & r \cdot I \end{bmatrix} > 0 \tag{13.52}$$

for almost all frequencies. See Söderström, Ljung, and Gustavsson (1976)for a proof. An interesting consequence of this result is that, even if no extra input is allowed, informative experiments result if we shift between different linear regulators. By checking (13.52) in the SISO case, we find that it is sufficient to use two regulators

$$u(t) = -F_1(q)y(t) \quad \text{and} \quad u(t) = -F_2(q)y(t)$$

subject to

$$\left[F_1(e^{i\omega}) - F_2(e^{i\omega})\right] \neq 0, \qquad \forall\omega$$

which is a very mild condition. This could be a useful way to achieving informative experiments when the requirement of good control is stringent.

Bias Distribution

We shall now characterize in what sense the model approximates the true system, when it cannot be exactly described within the model class. The discussion will be based on the frequency domain expressions for the limiting criterion function, developed in Section 8.5, see (8.76)–(8.77).

Let us focus on the case with a fixed noise model $H(q, \theta) = H_*(q)$. This case can be extended to the case of independently parameterized G and H, analogously to (8.73). Recall that any prefiltering of the data or prediction errors is equivalent to changing the noise model. The expressions below therefore contain the case of arbitrary prefiltering. For a fixed noise model, only the first term of (8.77) matters in the minimization, and we find that the limiting model is obtained as

$$G_* = \arg\min_G \int_{-\pi}^{\pi} \left|G_0(e^{i\omega}) + B(e^{i\omega}) - G(e^{i\omega}, \theta)\right|^2 \times \frac{\Phi_u(\omega)}{\left|H_*(e^{i\omega})\right|^2} d\omega \tag{13.53a}$$

$$\left|B(e^{i\omega})\right|^2 = \frac{\lambda_0}{\Phi_u(\omega)} \cdot \frac{\Phi_u^e(\omega)}{\Phi_u(\omega)} \cdot \left|H_0(e^{i\omega}) - H_*(e^{i\omega})\right|^2 \tag{13.53b}$$

This is identical to the open loop expression (8.71), except for the bias term B. See also Problem 8G.6. Within the chosen model class, the model G will approximate the biased transfer function $G_0 + B$ as well as possible, according to the weighted frequency domain function above. The weighting function $\Phi_u/|H_*|^2$ is the same as in the open loop case. The major difference is thus that an erroneous noise model (or unsuitable prefiltering) may cause the model to approximate a biased transfer function.

Let us comment on the bias function B. From (13.53b) we see that the bias-inclination will be small in frequency ranges where either (or all) of the following holds

- The noise model is good ($H_0 - H_*$ is small)
- The feedback contribution to the input spectrum (Φ_u^e/Φ_u) is small
- The signal to noise ratio is good (λ_0/Φ_u is small)

In particular, it follows that if a reasonably flexible, independently parameterized noise model is used, then the bias-inclination of the G-estimate can be negligible.

Variance and Information Contents in Closed Loop Data

Let us now consider the asymptotic variance of the estimated transfer function $\hat{G}_N$ using the asymptotic black-box theory of Section 9.4. Note that the basic result (9.62) applies also to the closed loop case.

From the general expression (9.62) we can directly solve for the upper left element:

$$\operatorname{Cov} \hat{G}_N = \frac{n}{N} \Phi_v(\omega) \cdot \frac{\lambda_0}{\lambda_0 \Phi_u(\omega) - |\Phi_{ue}(\omega)|^2} \tag{13.54}$$

From (8.75) we easily find that the denominator is equal to Φ_u^r so

$$\operatorname{Cov} \hat{G}_N = \frac{n}{N} \frac{\Phi_v(\omega)}{|S_0|^2 \Phi_r(\omega)} = \frac{n}{N} \frac{\Phi_v(\omega)}{\Phi_u^r(\omega)} \tag{13.55}$$

The denominator of (13.55) is the spectrum of that part of the input that originates from the reference signal r. The open loop expression has the total input spectrum here.

The expression (13.55) is derived from the covariance matrix obtained for the maximum likelihood method for Gaussian noise, see Chapter 9. This means that it is also the asymptotic Cramér-Rao lower limit (for Gaussian noise; see (7.79)), so it tells us precisely "the value of information" of closed loop experiments. It is the noise-to-signal ratio (where "signal" is what derives from the injected reference) that determines how well the open loop transfer function can be estimated. From this perspective, the part of the input that originates from the feedback has no information value when estimating G.

The expression (13.55) also clearly points to *a fundamental property in closed loop identification*: The purpose of feedback is to make the sensitivity function S_0 small, especially at frequencies with disturbances and poor system knowledge. Feedback will thus worsen the measured data's information about the system at these frequencies.

However, this is not the whole truth. Feedback will also allow us to inject more input in certain frequency ranges, without increasing the output power. We shall see in Section 13.6 that for experiment design that involve output variance constraints it is always optimal to use closed loop experiments.

Finally, let us stress that the basic result (13.55) is asymptotic when the orders of both G and H tend to infinity, as well as N.

13.5 APPROACHES TO CLOSED LOOP IDENTIFICATION

As noted in Section 13.4, a directly applied prediction error method—applied as if any feedback did not exist—will work well and give optimal accuracy if the true system can be described within the chosen model structure (both regarding the noise model and the dynamics model). Nevertheless, due to the pitfalls in closed loop identification, several alternative methods have been suggested. Although these methods, as such, are not experiment design issues, it is natural to discuss them in the context of the current chapter.

One may distinguish between methods that

1. Assume no knowledge about the nature of the feedback mechanism, and do not use r even if known.

2. Assume the signal r and the regulator to be known (and typically of the linear form (13.40b)).
3. Assume the regulator to be unknown. Use the measured r to infer information about it, and use the estimate of the regulator to recover the system.

If the regulator indeed has the form (13.40b), there is no major difference between (1), (2), and (3): This noise-free relationship can be exactly determined based on a fairly short data record, and then also r carries no further information about the system, if u is measured. The problem in industrial practice is rather that no regulator has this simple, linear form: Various delimiters, anti-windup functions and other non-linearities will have the input deviate from (13.40b), even if the regulator parameters (e.g. PID-coefficients) are known. This strongly disfavors the second approach. The methods correspondingly fall into the following main groups:

1. *The Direct Approach*: Apply the basic prediction error method (7.12) in a straightforward manner: use the output y of the process and the input u in the same way as for open loop operation, ignoring any possible feedback, and not using the reference signal r.
2. *The Indirect Approach*: Identify the closed loop system from reference input r to output y, and retrieve from that the open loop system, making use of the known regulator.
3. *The Joint Input-Output Approach*: Consider y and u as outputs of a system driven by r (if measured) and noise. Recover knowledge of the system and the regulator from this joint model.

We shall in this section treat each of these approaches.

Direct Identification

The Direct Identification approach should be seen as the natural approach to closed loop data analysis. The main reasons for this are:

- It works regardless of the complexity of the regulator, and requires no knowledge about the character of the feedback.
- No special algorithms and software are required.
- Consistency and optimal accuracy are obtained if the model structure contains the true system (including the noise properties).
- Unstable systems can be handled without problems, as long as the closed loop system is stable and the *predictor* is stable. This means that any unstable poles of G must be shared by H, like in ARX, ARMAX and state-space models (see (4.23) and (4.91)).

The only drawback with the direct approach is that we will need good noise models. In open loop operation we can use output error models (and other models with fixed or independently parameterized noise models) to obtain consistent estimates (but not of optimal accuracy) of G even when the noise model H is not sufficiently flexible. See Theorem 8.4.

This shows up when a simple model is sought that should approximate the system dynamics in a pre-specified frequency norm. In open loop we can do so with the output error method and a fixed prefilter/noise model that matches the specifications. See (8.71) and Examples 8.5 and 14.2. For closed loop data, a prefilter/noise model that deviates considerably from the true noise characteristics will introduce bias, according to (13.53).

All this means that we cannot handle model approximation issues with full control in the feedback case. However, consistency and optimal accuracy is guaranteed, just as in the open loop case, if the true system is contained in the model structure.

A natural solution to this would be to first build a higher order model $\hat{\hat{G}}$ using the direct approach, with small bias, and then reduce this model to lower order with the proper frequency weighting. While many model reduction schemes now exist, based on balanced realizations and the like, an identification-based way to achieve it is as follows: First simulate the model with an input u of suitable spectrum, thus generating noise free output $\hat{y} = \hat{\hat{G}}u$. Then subject the input-output data $\hat{y}$, u to an output error model of desired complexity. This gives the model

$$G^* = \arg\min_{\hat{G}} = \int \left|\hat{\hat{G}}(e^{i\omega}) - \hat{G}(e^{i\omega})\right|^2 \Phi_u(\omega)d\omega \tag{13.56}$$

Note, though, that reduction of unstable models may contain difficulties.

Indirect Identification

The closed loop system under (13.40b) is

$$\begin{aligned} y(t) &= G_{cl}(q)r(t) + v_{cl}(t) \\ &= \frac{G_0(q)}{1 + F_y(q)G_0(q)} r(t) + \frac{1}{1 + F_y(q)G_0(q)} v(t) \end{aligned} \tag{13.57}$$

The indirect approach means that G_{cl} is estimated from measured y and r, giving $\hat{G}_{cl}$, and then the open loop transfer function estimate $\hat{G}$ is retrieved from the equation

$$\hat{G}_{cl} = \frac{\hat{G}}{1 + \hat{G}F_y} \tag{13.58}$$

An advantage with the indirect approach is that any identification method can be applied to (13.57) to estimate $\hat{G}_{cl}$, since this is an open loop problem. Therefore methods like spectral analysis, instrumental variables, and subspace methods, that may have problems with closed loop data, also can be applied.

The major disadvantage with indirect identification is that any error in F_y (including deviations from a linear regulator, due to, e.g., input saturations or anti-windup measures) will be transported directly to $\hat{G}$.

For methods, like the prediction error method, that allow arbitrary parameterizations $G_{cl}(q,\theta)$ it is natural to let the parameters θ relate to properties of the open loop system G, so that

$$G_{cl}(q,\theta) = \frac{G(q,\theta)}{1 + F_y(q)G(q,\theta)} \tag{13.59}$$

That will make the task to retrieve the open loop system from the closed loop one more immediate.

We shall now assume that G_{cl} is estimated using a prediction error method with a fixed noise model/prefilter H_*:

$$y(t) = G_{cl}(q,\theta)r(t) + H_*(q)e(t) \tag{13.60}$$

The parameterization can be arbitrary, and we shall comment on it below. It is quite important to realize that *as long as the parameterization describes the same set of G, the resulting transfer function $\hat{G}(q,\hat{\theta}_N)$ will be the same, regardless of the parameterizations.* The choice of parameterization may thus be important for numerical and algebraic issues, but it does not affect the statistical properties of the estimated transfer function.

Let us now discuss bias and variance aspects of $\hat{G}$ estimated from (13.60) and (13.59). We start with the variance. According to the open loop result (9.63) (which holds also if the noise is not modelled; see Problem 9G.7), the asymptotic variance of $\hat{G}_{cl,N}$ will be

$$\text{Cov}\,\hat{G}_{cl,N} = \frac{n}{N}\frac{\Phi_{v,cl}(\omega)}{\Phi_r(\omega)} = \frac{n}{N}\frac{|S_0|^2\Phi_v}{\Phi_r} \tag{13.61}$$

regardless of the noise model H_*. Here $\Phi_{v,cl}$ is the spectrum of the additive noise v_{cl} in the closed loop system (13.57), which equals the open loop additive noise, filtered through the true sensitivity function. To transform this result to the variance of the open loop transfer function, we use Gauss' approximation formula (see (9.56)):

$$\text{Cov}\,\hat{G} = \frac{dG}{dG_{cl}}\text{Cov}\,\hat{G}_{cl}\left(\frac{dG}{dG_{cl}}\right)^* \tag{13.62}$$

It is easy to verify that

$$\frac{dG}{dG_{cl}} = \frac{1}{|S_0|^2}, \quad \text{so} \quad \text{Cov}\,\hat{G}_N = \frac{n}{N}\frac{\Phi_v}{|S_0|^2\Phi_r} = \frac{n}{N}\frac{\Phi_v}{\Phi_u^r}$$

which—not surprisingly—equals what the direct approach gives, (13.55).

For the bias, we know from (8.71) that the limiting estimate θ^* is given by (we write G_θ as short for $G(e^{i\omega},\theta)$)

$$\begin{aligned}\theta^* &= \arg\min_\theta \int_{-\pi}^{\pi}\left|\frac{G_0}{1+F_yG_0} - \frac{G_\theta}{1+F_yG_\theta}\right|^2\frac{\Phi_r}{|H_*|^2}d\omega \\ &= \arg\min_\theta \int_{-\pi}^{\pi}\left|\frac{G_0 - G_\theta}{1+F_yG_\theta}\right|^2\frac{|S_0|^2\Phi_r}{|H_*|^2}d\omega\end{aligned}$$

Now, this is no clear cut minimization of the distance $G_0 - G_\theta$. The estimate θ^* will be a compromise between making G_θ close to G_0 and making $1/(1 + F_y G_\theta)$ (the model sensitivity function) small. There will thus be a "bias-pull" towards transfer functions that give a small sensitivity for the given regulator, but unlike (13.53) it is not easy to quantify this bias component. However, if the true system can be represented within the model set, this will always be the minimizing model, so there is no bias in this case.

Parameterizations. The above results are independent of how the closed loop system is parameterized. A nice and interesting parameterization for this indirect identification of closed loop systems has been suggested by Hansen, Franklin, and Kosut (1989)and Schrama (1991). It is based on so-called dual Youla-Kucera parameterization. See Problem 13G.3.

Indirect Identification with Nonlinear, Known Regulator. The indirect technique can also be applied when the regulator is non-linear, but with considerably more work: One will have to compute the model output $\hat{y}(t|\theta) = f(\theta, \mathcal{R}, r^t)$ as a function of the open loop dynamic parameters θ, the known regulator $\mathcal{R}$, and the past reference signal values r^t, and then form an output error criterion.

Joint Input-Output Identification

Assume that there possibly is a non-measured signal w in the regulator in addition to r:

$$u(t) = r(t) + w(t) - F_y(q)y(t) \tag{13.63}$$

We assume that w is independent of r and v. The closed loop from v, w and r can be written similarly to (13.41) as:

$$y = G_0 S_0 r + S_0 v + G_0 S_0 w = G_{cl} r + \nu_1 \tag{13.64a}$$

$$u = S_0 r - F_y S_0 v + S_0 w = G_{ru} r + \nu_2 \tag{13.64b}$$

Identification methods that use models of how both y and u are generated are termed *joint input-output* techniques. This leaves a number of variants open, which fall into the following groups:

1. Allow correlation between ν_1 and ν_2, and work with a model:

$$\begin{bmatrix} y \\ u \end{bmatrix} = \mathcal{G} r + \mathcal{H} \nu \tag{13.65}$$

2. Disregard the correlation in the noise sources and treat (13.64a) and (13.64b) as separate models.

The first approach works also when there is no measurable reference signal r. It can be shown that this approach in essence is equivalent to the direct approach of estimating G in (13.40a) and F_y in (13.63). See Problem 13G.4.

The second approach in turn has some variants, but they all have in common that the system dynamics is estimated as

$$\hat{G} = \frac{\hat{G}_{cl}}{\hat{G}_{ru}} \tag{13.66}$$

where $\hat{G}_{cl}$ and $\hat{G}_{ru}$ are estimated from the two open loop systems (13.64). The case when spectral analysis estimates are used was described and analyzed by Akaike (1967, 1968). Various parametric approaches have also been suggested.

From (13.64) we see that

$$G_{cl} = G_0 G_{ru} \tag{13.67}$$

so cancellations should take place when (13.66) is formed. This will however not happen for the estimated models, due to the model uncertainties, so $\hat{G}$ will then be of unnecessarily high order. It is thus natural to enforce (13.67) in the model parameterization:

$$G_{cl}(q, \Theta) = G(q, \theta)S(q, \eta), \quad G_{ru}(q, \Theta) = S(q, \eta), \quad \Theta = \begin{bmatrix} \theta \\ \eta \end{bmatrix} \tag{13.68}$$

If we assume ν_1 and ν_2 to be independent white noises with variances 1 and $1/\alpha$, we obtain the following identification criterion for (13.64), (13.68):

$$V(\Theta) = \sum |y(t) - G(q, \theta)S(q, \eta)r(t)|^2 + \sum \alpha\, |u(t) - S(q, \eta)(t)|^2 \tag{13.69}$$

The question still remains how to parameterize G and S. Some possibilities, including the so-called coprime factor method, are described in Van den Hof et.al. (1995b).

The Two-Stage Method. We shall turn to the case where $\alpha \to \infty$ in (13.69). If α is very large, the second sum will dominate the criterion when η is determined to give $\hat{S}$. Since θ only enters the first sum, and S is given from the second term, this procedure is the same as first estimating $\hat{S}$ in (13.64b) and then using

$$\hat{u}(t) = S(q, \hat{\eta})r(t) \tag{13.70}$$

in

$$y(t) = G(q, \theta)\hat{u}(t) + \nu_1(t) \tag{13.71}$$

to estimate G. This is the two-stage method suggested by Van den Hof and Schrama (1993). A variant with a non-causal S is described in Forssell and Ljung (1998d). Let us analyze the properties of the latter variant: Suppose $S(\eta)$ is parameterized as a non-causal FIR filter:

$$S(q, \eta) = \sum_{k=-M}^{M} s_k q^{-k}$$

and take M so large that any correlation between $u(t)$ and $r(s)$, $|s - t| > M$ can be ignored. The model

$$u(t) = S(q, \eta)r(t) + \nu_2(t)$$

can then be estimated using the least squares method giving

$$\hat{u}(t) = S(q, \hat{\eta})r(t)$$

such that the sequence $\tilde{u} = u - \hat{u}$ is uncorrelated with the sequence r. Notice that *this holds irrespectively of the true relationship between r and u, which very well may* *be non-linear*! (In this latter case $\tilde{u}$ and r will be dependent, but still uncorrelated.) Suppose that the true system is given by (13.40a). Then, inserting $\hat{u}$ gives

$$y = G_0\hat{u} + v + G_0\tilde{u}$$

where the "noise" sequence $v + G_0\tilde{u}$ is uncorrelated with the "input" sequence $\hat{u}$. Suppose now that we estimate G from (13.71) with a fixed noise model H_*. From (8.71), which is valid if the noise and input are uncorrelated, we then have the model converging to $G^* = G(\theta^*)$ where

$$\theta^* = \arg\min_{\theta} \int \left|G_0(e^{i\omega}) - \hat{G}(e^{i\omega}, \theta)\right|^2 \Phi_{\hat{u}}(\omega) / \left|H_*(e^{i\omega})\right|^2 d\omega \tag{13.72}$$

Here the spectrum $\Phi_{\hat{u}}$ is fixed and known to us. So, if $\hat{u}$ is persistently exciting, we can consistently estimate G_0 by using a large enough model structure G_θ. In any case we can achieve an approximation to G_0 in a known frequency norm $\Phi_{\hat{u}}/|H_*|^2$ that we can affect by a proper choice of noise model (prefilter). We have thus gained something over the direct identification method, which gives a possible bias as in (13.53). The price is the increased variance caused by the extra "noise" $G_0\tilde{u}$. Note again, that this result is valid *even if the regulator is non-linear.*

Summarizing Remarks

We may summarize the basic issues on closed loop identification as follows:

- The basic problem with closed loop data is that it typically has less information about the open loop system—an important purpose of feedback is to make the closed loop system less sensitive to changes in the open loop system.
- Prediction error methods, applied in a direct fashion, with a noise model that can describe the true noise properties still give consistent estimates and optimal accuracy. No knowledge of the feedback is required. This should be regarded as a prime choice of methods.
- Several methods that give consistent estimates for open loop data may fail when applied in a direct way to closed loop identification. This includes spectral and correlation analysis, the instrumental variable method, the subspace methods, and output error methods with incorrect noise model.
- If the regulator mechanism is correctly known, indirect identification can be applied. Its basic advantage is that the dynamics model G can be correctly estimated without estimating any noise model, even when G_0 is unstable. However, any error in the assumed regulator will directly cause a corresponding error in the estimate of G. Since most regulators contain non-linearities, this means that indirect identification has fallacies.
- The joint input-output approach in the two-stage variant offers the advantage that model approximation in a known and user-chosen frequency weighting norm can be achieved (see (13.72)) at the price of higher variance.

13.6 OPTIMAL EXPERIMENT DESIGN FOR HIGH-ORDER BLACK-BOX MODELS

In this section we shall use the asymptotic variance expression (9.62) to design optimal experiments. In Chapter 12 we derived a general design criterion based on this expression, (12.32):

$$J_P(\mathcal{X}) = \int_{-\pi}^{\pi} \Psi(\omega, \mathcal{X}) d\omega \tag{13.73a}$$

$$\Psi(\omega, \mathcal{X}) = \frac{\lambda_0 C_{11}(\omega) - 2\text{Re}\left[C_{12}(\omega)\Phi_{ue}(-\omega)\right] + C_{22}(\omega)\Phi_u(\omega)}{\lambda_0 \Phi_u(\omega) - |\Phi_{ue}(\omega)|^2} \cdot \Phi_v(\omega) \tag{13.73b}$$

Here, we dispensed with the scaling n/N, which is immaterial for the choice of experimental condition $\mathcal{X}$. Recall that the variance expression is asymptotic in the model order.

For design variables, we can work with different equivalent setups. An immediate choice is to design the input spectrum and the cross spectrum:

$$\mathcal{X} = \{\Phi_u, \ \Phi_{ue}\} \tag{13.74}$$

A more explicit way is to work directly with the regulator and the reference signal spectrum:

$$u(t) = -F_y(q)y(t) + r(t)$$

and regard $\mathcal{X} = \{F_y, \ \Phi_r\}$ as the design variables. In this case we have

$$\lambda_0 \Phi_u(\omega) - |\Phi_{ue}(\omega)|^2 = \lambda_0 \Phi_u^r(\omega)$$

where Φ_u^r is that part of the input spectrum that originates from the reference signal (see (13.44)).

Criterion Involving $\hat{G}$ Only.

We shall first consider the case where $C_{12} = C_{22} = 0$, that is the design criterion involves only the dynamics part G:

$$J_p(\mathcal{X}) = \int_{-\pi}^{\pi} \text{Cov}\, \hat{G}_N(e^{i\omega}) C_{11}(\omega) d\omega \sim \int_{-\pi}^{\pi} \frac{\Phi_v(\omega)}{\Phi_u^r(\omega)} C_{11}(\omega) d\omega \tag{13.75}$$

This is no doubt the most common special case, where the quality of the noise model is of less importance.

We shall minimize this design criterion, subject to constrained variance of the input and the output in the general form

$$\alpha E u^2 + \beta E y^2 \leq 1 \tag{13.76}$$

Here, we assume that α and β are chosen so that it is at all possible to achieve this constraint with the given disturbances. The solution is given by the following theorem.

Theorem 13.3. Consider the problem to minimize J_p in (13.75) with respect to the design variables $\mathcal{X} = \{F_y, \ \Phi_r\}$ under the constraint (13.76). This is achieved by selecting the regulator $u(t) = -F_y(q)y(t)$ that solves the standard LQG problem

$$F_y^{\text{opt}} = \arg\min_{F_y}[\alpha E u^2 + \beta E y^2], \quad y = G_0 u + H_0 e \tag{13.77}$$

The reference signal spectrum shall be chosen as

$$\Phi_r^{\text{opt}}(\omega) = \mu\sqrt{\Phi_v(\omega)C_{11}(\omega)}\frac{\left|1 + G_0(e^{i\omega})F_y^{\text{opt}}(e^{i\omega})\right|^2}{\sqrt{\alpha + \beta\left|G_0(e^{i\omega})\right|^2}} \tag{13.78}$$

where μ is a constant, adjusted so that equality is met in (13.76).

Proof. We first establish the following straighforward result. For two positive functions $A(t)$ and $X(t)$ we have that

$$\begin{aligned} &\text{The integral} \quad \int \frac{A(t)}{X(t)}dt \quad \text{with constraint} \quad \int X(t) \leq K \\ &\text{is minimized by} \quad X(t) = \mu\sqrt{A(t)}, \quad \mu = \frac{K}{\int \sqrt{A(t)}dt} \end{aligned} \tag{13.79}$$

To prove this we have by Schwarz's inequality

$$\begin{aligned} \left[\int \sqrt{A(t)}dt\right]^2 &= \left[\int \sqrt{\frac{A(t)}{X(t)}}\sqrt{X(t)}dt\right]^2 \\ &\leq \int \frac{A(t)}{X(t)}dt \cdot \int X(t)dt \leq K \cdot \int \frac{A(t)}{X(t)}dt \end{aligned}$$

or

$$\int \frac{A(t)}{X(t)}dt \geq \frac{\left[\int \sqrt{A(t)}dt\right]^2}{K}$$

which gives (13.79).

For the closed loop we have (see (13.42)–(13.44) and (8.75))

$$\begin{aligned} S_0(q) &= \frac{1}{1 + F_y(q)G_0(q)}, \quad \Phi_u(\omega) = |S_0|^2\Phi_r(\omega) + |F_y|^2|S_0|^2\Phi_v(\omega) \\ \Phi_y(\omega) &= |G_0|^2|S_0|^2\Phi_r(\omega) + |S_0|^2\Phi_v(\omega) \\ \Phi_u^r &= |S_0|^2\Phi_r, \quad \Phi_u^e = |F_y|^2|S_0|^2\Phi_v, \quad |\Phi_{ue}(\omega)|^2 = \lambda_0\Phi_u^e(\omega) \end{aligned} \tag{13.80}$$

This means that it is convenient to use

$$\mathcal{X} = \{\Phi_u^r, \ F_y\} \tag{13.81}$$

as design variables. With this, we can rewrite the constraint (13.76) as

$$\begin{aligned}1 &\geq \int_{-\pi}^{\pi} \alpha\Phi_u(\omega) + \beta\Phi_y(\omega)d\omega \\ &= \int_{-\pi}^{\pi} \left[(\alpha + \beta|G_0|^2)\Phi_u^r + \frac{\alpha|F_y|^2 + \beta}{|1 + G_0F_y|^2}\Phi_v \right] d\omega\end{aligned}$$

and the criterion is

$$\min_{F_y, \Phi_u^r} \int_{-\pi}^{\pi} \frac{\Phi_v}{\Phi_u^r} C_{11} d\omega$$

The criterion does not depend explicitly on F_y, and is a question of making Φ_u^r as large as allowed. From the constraint we see that this means that F_y should be chosen by solving the LQ problem

$$\min_{F_y} \int_{-\pi}^{\pi} \frac{\alpha|F_y|^2 + \beta}{|1 + G_0F_y|^2}\Phi_v d\omega = \min_{F_y} E(\alpha u^2 + \beta y^2)$$

$$\text{for} \quad y(t) = G_0(q)u(t) + v(t), \quad u(t) = -F_y(q)y(t)$$

Define the constant γ as

$$\gamma = 1 - \int_{-\pi}^{\pi} \frac{\alpha|F_y^{\text{opt}}|^2 + \beta}{|1 + G_0F_y^{\text{opt}}|^2}\Phi_v d\omega$$

which is assumed to be positive.

The minimization problem now reads

$$\min_{\Phi_u^r} \int_{-\pi}^{\pi} \frac{\Phi_v C_{11}}{\Phi_u^r} d\omega, \quad \text{with} \int_{-\pi}^{\pi} (\alpha + \beta|G_0|^2)\Phi_u^r d\omega \leq \gamma$$

In (13.79) we take $X = (\alpha + \beta|G_0|^2)\Phi_u^r)$ and $A = \Phi_v C_{11}(\alpha + \beta|G_0|^2)$ which gives

$$\Phi_u^{r,\text{opt}} = \mu\sqrt{\frac{\Phi_v C_{11}}{\alpha + \beta|G_0|^2}}$$

which, via (13.80), concludes the proof. □

The theorem tells us a number of useful things:

- The optimal experiment design depends on the (unknown) true system and noise characteristics. This is the normal situation for optimality results, and in practice it has to be handled by using the best prior information available about the system.

- If we have a pure input power constraint, $Eu^2 \leq K$ and the system is stable, then it is always *optimal to use open loop operation*, $F_y = 0$. The input spectrum is then proportional to $\sqrt{\Phi_v C_{11}}$, which shows that the input power should be spent in frequency bands where *there is substantial noise* (Φ_v large) and/or where *a good model is particularly important* (C_{11} large).
- If the constraint involves any limitation on the output variance ($\beta > 0$), then it is always *optimal to use closed loop experiments*. In these cases the regulator does not depend on the criterion function C_{11}, but only the constraint and the system.

Input Variance Constraint Only

Consider now the situation were also the noise model quality enters the criterion ($C_{22}(\omega) > 0$), but there is no cross term ($C_{12} = C_{21} = 0$). Suppose also that the constraint incolves the input variance only. That is, we have

$$\min_{\mathcal{X}} \int_{-\pi}^{\pi} \frac{\lambda_0 C_{11}(\omega) + C_{22}(\omega)\Phi_u(\omega)}{\lambda_0 \Phi_u(\omega) - |\Phi_{ue}(\omega)|^2} \cdot \Phi_v(\omega) d\omega \tag{13.82a}$$

$$\mathcal{X} = \{\Phi_u, \Phi_{ue}\} \tag{13.82b}$$

$$Eu^2 \leq K \tag{13.82c}$$

Since the design variable Φ_{ue} does not enter the constraint, we realize immediately that the optimum choice of this variable is $\Phi_{ue} = 0$, since this minimizes the integrand in the criterion pointwise. This means that open loop operation is optimal. This, in turn, means that the variance of $\hat{H}$ is not affected by the design. In other words, the integrand reads

$$\frac{\Phi_v C_{11}}{\Phi_u} + \frac{\Phi_v C_{22}}{\lambda_0}$$

so this case is solved by Theorem 13.3: The optimal design is an open loop experiment, with an input with spectrum

$$\Phi_u = \mu\sqrt{\Phi_v C_{11}} \tag{13.83}$$

where μ is adjusted so that the input power constraint is met.

Other cases of the general problem (13.73) are treated in the Problem Section, and in the references mentioned in the bibliography.

13.7 CHOICE OF SAMPLING INTERVAL AND PRESAMPLING FILTERS

The procedure of sampling the data that are produced by the system is inherent in computer-based data-acquisition systems. It is unavoidable that sampling as such leads to information losses, and it is important to select the sampling instances so that these losses are insignificant. In this section we shall assume that the sampling is carried out with equidistant sampling instants, and we shall discuss the choice of the sampling interval T.

Aliasing

The information loss incurred by sampling is best described in the frequency domain. Suppose that a signal $s(t)$ is sampled with the *sampling interval* T:

$$s_k = s(kT), \qquad k = 1, 2, \ldots$$

Denote by $\omega_s = 2\pi/T$ the *sampling frequency*, and then $\omega_N = \omega_s/2$ is the *Nyquist frequency*. Now, it is well known that a sinusoid with frequency higher than ω_N cannot, when sampled, be distinguished from one in the interval $[-\omega_N, \omega_N]$:

With $|\omega| > \omega_N$ there exists a $\overline{\omega}$; $-\omega_N \leq \overline{\omega} \leq \omega_N$ so that

$$\begin{aligned} \cos \omega kT &= \cos \overline{\omega} kT \\ \sin \omega kT &= \sin \overline{\omega} kT \end{aligned} \qquad k = 0, 1, \ldots \tag{13.84}$$

This follows from simple manipulations with trigonometric formulas. Consequently, the part of the signal spectrum that corresponds to frequencies higher than ω_N will be interpreted as contributions from lower frequencies. This is the *alias* phenomenon; the frequencies appear under assumed names. It also means that the spectrum of the sampled signal will be a superposition of different parts of the original spectrum:

$$\Phi_s^{(T)}(\omega) = \sum_{r=-\infty}^{\infty} \Phi_s^c(\omega + r\omega_s) \tag{13.85}$$

Here Φ_s^c is the spectrum of the continuous-time spectrum, defined by (13.36) and $\Phi_s^{(T)}(\omega)$ is the spectrum of the sampled signal:

$$\begin{aligned} R_T(\ell T) &= \overline{E} s_k s_{k+\ell} = \lim_{N\to\infty} \frac{1}{N} \sum_{k=1}^{N} E s(kT) s(kT + \ell T) \\ \Phi_s^{(T)}(\omega) &= T \sum_{\ell=-\infty}^{\infty} R_T(\ell T) e^{-i\omega\ell T} \end{aligned} \tag{13.86}$$

The effect of (13.85) is often called *folding*: the original spectrum is "folded" (and added) to give the sampled spectrum.

Antialiasing Presampling Filters

The information about frequencies higher than the Nyquist one is thus lost by sampling. It is then important not to make bad worse by letting the folding effect distort the interesting part of the spectrum below the Nyquist frequency. This is achieved by a presampling filter $\kappa(p)$:

$$s_F(t) = \kappa(p) s(t) \tag{13.87}$$

(p is here the differentiation operator). Analogous to the formulas of Theorem 2.2, the spectrum of the filtered signal $s_F(t)$ will be

$$\Phi^c_{s_F}(\omega) = |\kappa(i\omega)|^2 \, \Phi^c_s(\omega) \tag{13.88}$$

Ideally, $\kappa(i\omega)$ should have a characteristic so that

$$\begin{aligned} |\kappa(i\omega)| &= 1, \qquad |\omega| \leq \omega_N \\ |\kappa(i\omega)| &= 0, \qquad |\omega| > \omega_N \end{aligned} \tag{13.89}$$

This can be realized only approximately. In the ideal case (13.89), we would have

$$\Phi^c_{s_F}(\omega) = \begin{cases} \Phi^c_s(\omega), & \omega \leq |\omega_N| \\ 0, & \omega > |\omega_N| \end{cases}$$

which means that the sampled signal

$$s^F_k = s_F(kT)$$

will have a spectrum, according to (13.85),

$$\Phi^{(T)}_{s_F}(\omega) = \Phi^c_s(\omega), \qquad -\omega_N \leq \omega \leq \omega_N \tag{13.90}$$

With the filter (13.87) and (13.89) we thus achieve a sampled spectrum with no alias effects. Therefore, this filter is also called an *antialiasing filter*. According to what we said, such a filter should always be applied before sampling if we suspect that the signal has nonnegligible energy above the Nyquist frequency.

Noise-reduction Effect of Antialiasing Filters

A typical situation is that the signal consists of a useful part and a disturbance part, and that the spectrum of the disturbances is more broadband than that of the signal. Then the sampling interval is usually chosen so that most of the spectrum of the useful part is below ω_N. The antiasiasing filter then essentially cuts away the high-frequency noise contributions. Suppose we have

$$s(t) = m(t) + v(t)$$

where $m(t)$ is the useful signal and $v(t)$ is the noise. Let $\Phi^v_c(\omega)$ be the spectrum of $v(t)$. The sampled, prefiltered signal then is

$$s^F_k = m^F_k + v^F_k, \qquad s^F_k = s_F(kT)$$

where the variance of the noise is

$$E(v^F_k)^2 = \int_{-\omega_N}^{\omega_N} \Phi^{(T)}_{v_F}(\omega)\, d\omega = \sum_{r=-\infty}^{\infty} \int_{-\omega_N}^{\omega_N} \Phi^c_{v_F}(\omega + r\omega_s)\, d\omega$$

From this expression we see how the noise effects from higher frequencies are folded into the region $[-\omega_N, \omega_N]$ and are thus contributing to the noise power. By eliminating the high-frequency noise by an antialiasing filter (13.89), the variance of v_k^F is thus reduced by the term

$$\sum_{r \neq 0} \int_{-\omega_N}^{\omega_N} \Phi_v^c(\omega + r\omega_s)d\omega = \int_{|\omega|>\omega_N} \Phi_v^c(\omega)d\omega$$

compared to no presampling filter at all. This is a significant noise reduction if the noise spectrum has considerable energy above the Nyquist frequency.

Antialiasing Filters during Data Acquisition

Let us comment on the role of the antialiasing filters in system identification applications. Suppose first that the system is *not* under sampled-data control so that the continuous-time input is not piecewise constant. This may be the case when we collect data from a process in normal operation. If the input then is band limited and has no energy above the frequency ω_B, this means that all useful information in the output also lies below ω_B, provided the process is linear. We could then apply an antialiasing filter with cutoff frequency ω_B and sample with $T = \pi/\omega_B$ with no loss of information. If the input is not bandlimited, the antialiasing filter will destroy useful information at the same time as the noise is reduced. If T is chosen so that the Nyquist frequency (= the cutoff frequency for the filter) is above the bandwidth of the system, the loss of useful information is insignificant. Notice that in this case the antialiasing presampling filter should be applied also to the input signal.

Consider now the case that the input is piecewise constant over the sampling interval. Then, clearly, the sampled input equals the piecewise constant values, and no presampling filtering should be applied to this sequence. The stepwise changes in the process input do, though, contain high frequencies that could travel through the process to the output. An antialiasing filter applied to the process output could thus distort useful information. There are three ways to handle this problem:

1. Sample fast enough that the process is well damped above the Nyquist frequency. Then the high-frequency components in the output that originate from the input are insignificant.
2. Consider the antialiasing output filter as part of the process and model the system from input to filtered output (this might increase the necessary model orders, though).
3. Since the antialiasing filter is known, include it as a known part of the model, and let the predicted output pass through the filter before being used in the identification criterion [this approach is illustrated in (13.95) and (13.96)].

Solution 1 is the most natural; it is conceptually depicted in Figure 13.8.

Remark: For control purposes it might be a good idea to apply a low-pass filter to the piecewise constant sampled-data input sequence. This will also be helpful for solution 1.

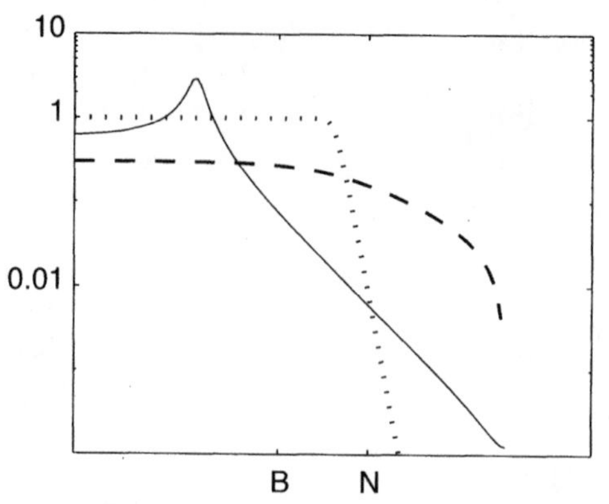

Figure 13.8 Sampling depicted in the frequency domain. Solid line: frequency characteristics of the process; dashed line: noise spectrum; dotted line: Frequency characteristics of the antialiasing filter. N: the Nyquist frequency. B: Band-width.

Some General Aspects on the Choice of T and N

If the total experiment time $0 \leq t \leq T_N$ is limited, but the acquisition of data within this time is costless, it is clearly advantageous from an information theoretic point of view to sample as fast as possible. Slower sampling leads to data sets that are subsets of the maximal one, and hence is less informative. The cost effectiveness of the new information will, however, typically decrease as we sample faster and faster (cf. Figure 13.9). In this idealized case, where adding new data points is costless, there are only two aspects that may prevent us from sampling as fast as technically possible: One is that building sampled models with very small sampling interval compared to the natural time constants is a numerically sensitive procedure (all poles cluster around the point 1). See Problem 13G.1. The other is that the model fit may be concentrated to the high-frequency band (see the following discussion on bias). The latter problem should be dealt with by prefiltering the data so as to redistribute bias, as explained in Section 14.4. The former problem should probably be handled by fitting continuous-time models directly to the fast sampled data with models of the type (2.23).

Another idealized situation is that the experiment time $0 \leq t \leq T_N$ as such is costless, and all cost is associated with acquiring and handling the data. We could then settle for collecting, say, N data and select T (then $T_N = N \cdot T$) so that the data set becomes as informative as possible. A T that is much larger than the interesting time constants of the system would then yield data with little information about the dynamics. A small T, on the other hand, would not allow for much noise reduction, and the data might be less informative for that reason. A good choice of T should thus be a trade-off between noise reduction and relevance for the dynamics.

If the model should be used for control purposes, certain other aspects will enter. The sampling interval for which we build the model should be the same as for the control application (unless we want to recalculate it from one sampling interval to another). A fast sampled model will often be nonminimum phase (Åström and Wittenmark, 1984), and a system with dead time may be modeled with delay of many sampling periods. Such effects may cause problems for the control design and will therefore influence the choice of T.

For the choice of N, it is useful to have the asymptotic result (9.92) in mind. It describes how small the model order to sample size ratio has to be in order to achieve a certain accuracy for a given design and given noise spectrum.

Bias Considerations

We know from Chapter 8 that the fit between the model transfer function and the true one is related to a quadratic norm [see (8.71)]:

$$\theta^* = \arg\min_{\theta} \int_{-\pi/T}^{\pi/T} \left| G_0(e^{i\omega T}) - G(e^{i\omega T}, \theta) \right|^2 Q(\omega, \theta)\, d\omega \tag{13.91}$$

Here $Q(\omega, \theta)$ is the filtered input spectrum divided by the noise spectrum:

$$Q(\omega, \theta) = \frac{\Phi_u(\omega)}{\left| H(e^{i\omega T}, \theta) \right|^2}$$

We also marked where the T-dependence enters. As T tends to zero, the frequency range over which the fit is made in (13.91) increases. Normally, though, the natural dynamics of the system and the model are such that $G_0(e^{i\omega T}) - G(e^{i\omega T}, \theta)$ is well damped for high frequencies so that the contribution from higher values of ω in (13.91) will be insignificant even if the input is broadband. An important exception is the case where the noise model is coupled to the dynamics, as in the ARX structure (4.9), where $H(e^{i\omega T}) = 1/A(e^{i\omega T})$. Then the product

$$\frac{\left| G_0(e^{i\omega T}) - G(e^{i\omega T}, \theta) \right|^2}{\left| H(e^{i\omega T}, \theta) \right|^2}$$

does not tend to zero as ω increases and the fit in (13.91) is pushed into very high frequency bands as T decreases. This may lead to quite curious results, as illustrated in Wahlberg and Ljung (1986). In such cases, very fast sampling is thus undesirable, even apart from the numerical difficulties that may arise. The effects can be counteracted by proper prefiltering, as described in Section 14.4. In any case, it is important to keep in mind the influence of T on the bias distribution.

Variance Considerations

The variance of an estimated parameter based on a given number of data will depend on the average information per sample. This, as mentioned previously, is a trade-off between the noise reduction that slow sampling may offer and the poor information about the dynamics that slowly sampled data contain. To pinpoint this trade-off, let us consider a simple example.

Example 13.4 Optimal Sampling

Consider a continuous-time system

$$y(t) = \frac{1}{1 + p\tau} u(t) + v(t)$$

or

$$\begin{aligned} \tau\dot{x}(t) + x(t) &= u(t) \\ y(t) &= x(t) + v(t) \end{aligned} \tag{13.92}$$

where $v(t)$ is a very broadband noise (= "almost white noise") with variance λ/T_0, where $1/T_0$ is its bandwidth [i.e., T_0 is the smallest sampling interval under which the sampled version of $v(t)$ is truly white]. As a simple presampling filter, we employ an integrator

$$y_k = \overline{y}(kT) = \frac{1}{T}\int_{t=(k-1)T}^{kT} y(t)dt = \overline{x}(kT) + v_T(kT) \tag{13.93}$$

Here $\overline{x}(kT)$ is the mean of the useful signal $x(t)$ over the sampling interval and $\{v_T(kT)\}$ is a sequence of independent random variables with variance λ/T (if $T > T_0$). We use an output error model set

$$x(kT + T) = e^{-aT}x(kT) + (1 - e^{-aT})u(kT) \tag{13.94a}$$

$$\hat{y}(kT + T|kT, a) = x(kT + T) = \frac{q(1 - e^{-aT})}{q - e^{-aT}} u(kT) \tag{13.94b}$$

Here, the model parameter a corresponds to $1/\tau$ with τ as in (13.92). We let the input signal be a sinusoid (piecewise constant) of frequency ω_0:

$$u(kT) = \alpha \cdot \cos(\omega_0 kT)$$

When calculating the predictor (13.94), we ignored the presampling filter, which may be reasonable when T is small enough. To allow for a fair treatment also of larger values of T, we could take the presampling filter (13.93) into account and let the prediction be

$$\frac{d}{dt}x(t, a) = ax(t, a) + au(t) \tag{13.95a}$$

$$\begin{aligned} \hat{y}_T(kT + T|kT, a) &= \frac{1}{T}\int_{t=kT}^{(k+1)T} x(t, a)dt \\ &= \frac{\beta q - e^{-aT}\left[1 - (1/aT)(1 - e^{-aT})\right]}{q - e^{-aT}} u(kT) \end{aligned} \tag{13.95b}$$

$$\beta = \left(1 - \frac{1}{aT}(1 - e^{-aT}) + \frac{1}{aT}(1 - e^{-aT})^2\right) \tag{13.96}$$

(Notice that the second term in the numerator is $\approx aT/2$ for small T.) The asymptotic variance of $\hat{a}_N$ is now given by Theorem 9.1 as

$$E(\hat{a}_N - 1/\tau)^2 \sim \frac{1}{N} \frac{\text{Var}\, \overline{v}_T(t)}{\overline{E}\,(\psi_T(t))^2} \tag{13.97}$$

where

$$\text{Var}\, \overline{v}_T(t) = \frac{\lambda}{T} \quad \text{and} \quad \psi_T(t) = \frac{d}{da}\hat{y}_T(kT + T|kT, a)|_{a=(1/\tau)}$$

For the simplified expression (13.94), we have

$$\psi_T(t) = \frac{Te^{-T/\tau}(q-1)}{(q - e^{-T/\tau})^2}\alpha \cos \omega_0 t$$

and

$$\overline{E}\,(\psi_T(t))^2 = \alpha^2 \frac{T^2 e^{-2T/\tau}(2 - 2\cos \omega_0 T)}{\left[1 - 2e^{-T/\tau}\cos \omega_0 T + e^{-2T/\tau}\right]^2} \tag{13.98}$$

We thus have

$$\text{Var}\, \hat{a}_N \sim \frac{\lambda}{NT \cdot \overline{E}\,(\psi_T(t))^2} \tag{13.99}$$

This expression tends to infinity as

$$\frac{1}{T^3}e^{2T/\tau}$$

as T increases to infinity; this is the effect of poor information about τ with slow sampling. Also, some calculations reveal that it tends to infinity as $1/T$ when T tends to zero; this is the effect of poor noise rejection at fast sampling. We have thus formalized the earlier mentioned trade-off.

The exact predictor (13.96) gives similar, but more complicated expressions. In Figure 13.9 the expression (13.99) and the exact counterpart are plotted as funtions of T for $\omega_0 = 1/\tau$. The figure reveals two things:

1. The optimal choice of sampling interval lies around the time constant of the system.
2. It is far worse to use a too large T than a too small one: $T = 10\tau$ gives a variance more than 10^5 times the optimal one, while $T = 0.1\tau$ gives a variance that is less than 10 times the optimal one.

□

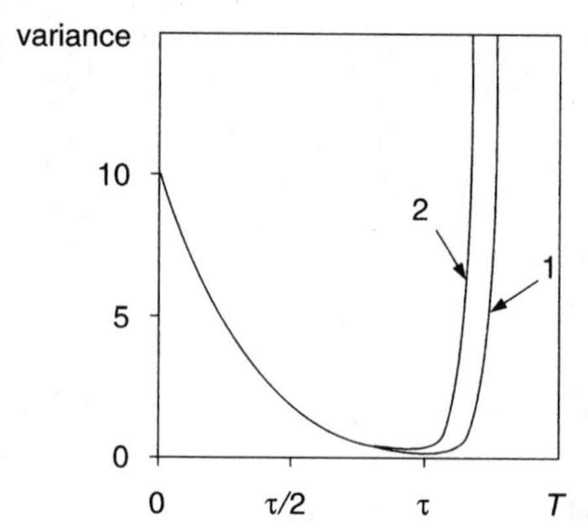

Figure 13.9 Variance of $\hat{a}_N$ plotted as a function of the sampling interval $T(\omega_0 = 1/\tau)$ (1.) Expression (13.99). (2.) Using (13.96).

Conclusions

Let us summarize the discussion on sampling rates as follows.

- Very fast sampling leads to numerical problems, model fits in high-frequency bands, and poor returns for extra work.
- As the sampling interval increases over the natural time constants of the system, the variance increases drastically.
- Optimal choices of T for a fixed number of samples will lie in the range of the time constants of the system. These are, however, not exactly known, and overestimating them may lead to very bad results.

All these aspects point to the advice that a sampling frequency that is about ten times the bandwidth of the system should be a good choice in most cases. Note that this discussion concerns the sampling rate chosen for the model building. With "cheap" data acquisition we can always sample as fast as possible during the experiment and leave the actual choice of T for later by digitally prefiltering and decimating the original data record.

13.8 SUMMARY

Careful experiment design, yielding data with good information is the basis of a successful identification application. In this chapter we have discussed how to design good experiments. The leading principles are as follows:

- Let the experimental condition resemble the situation for which the model is going to be used.
- Identifiability is secured by persistently exciting input and not allowing too simple feedback mechanisms (Theorem 13.2).

- To minimize the parameter variance with respect to the experiment design variables, the expression

$$P_\theta(\mathcal{X}) \sim \lambda_0 \left[E \left(\frac{d}{d\theta} \hat{y}(t|\theta) \right) \left(\frac{d}{d\theta} \hat{y}(t|\theta) \right)^T \right]^{-1}$$

shows that interesting parameters must have a clear effect on the output predictions.

- Periodic inputs have certain advantages, in particular, for single input systems. Sum of sinusoids and PRBS signals may then be good choices. To make use of the advantages, an integer number of periods should be applied.
- For systems operating in closed loop, the basic choice of method is to apply a prediction error method in a direct fashion, using a flexible noise model.
- To minimize a quadratic frequency-domain fit of the estimated transfer function $\hat{G}_N(e^{i\omega})$ in a (high order) linear model set, use, in open loop,

$$\Phi_u(\omega) \sim \sqrt{\Phi_v(\omega) \cdot C_{11}(\omega)}$$

where C_{11} is the weighting function in the criterion of fit (Theorem 13.3).

- A suitable choice of sampling frequency lies in the range of ten times the guessed bandwidth of the system. In practice, it is useful to first record a step response from the system, and then select the sampling interval so that it gives 4–6 samples during the rise time.

13.9 BIBLIOGRAPHY

Design of statistical experiments is a subject that has received considerable attention in the statistical literature. Fedorov (1972)can be mentioned as a basic reference. Applications to identification of dynamic systems have been treated comprehensively in Goodwin and Payne (1977), Goodwin (1987), Mehra (1981), and Zarrop (1979).

Section 13.2: The concept of persistence of excitation was introduced in Åström and Bohlin (1965)as a condition for consistency. Multivariable extensions have been discussed e.g., in Stoica (1981). Results on how the persistence property changes by time-invariant and time-varying filtering are given in Ljung (1971), Moore (1983), and Bai and Sastry (1985).

Section 13.3: Godfrey (1993)gives a comprehensive overview of input signal design. General treatments are also given in Mehra (1974), Wahba (1980), and in Chapter 4 of Schoukens and Pintelon (1991). Qureshi, Ng, and Goodwin (1980)discuss design for distributed parameter systems. The question of how to minimize the crest factor for sums of sinusoids has been treated in many papers. See, e.g., Van der Ouderaa, Schoukens, and Renneboog (1998). Issues around periodic, piecewise constant, and band-limited input signals are discussed in the illuminating paper Schoukens, Pintelon, and Van hamme (1994).

Section 13.4: A survey of identifiability in closed-loop experiments is given in Gustavsson, Ljung, and Söderström (1977)and Gustavsson, Ljung, and Söderström (1981). The result (13.52) was shown in Söderström, Ljung, and Gustavsson (1976). Extensions to sufficiently informative experiments for finite-order ARMAX models are given in Söderström, Gustavsson, and Ljung (1975). See also, for example, Anderson and Gevers (1982), Caines and Chan (1975), and Sin and Goodwin (1980).

Section 13.5: A general reference for this section is Gustavsson, Ljung, and Söderström (1977)and Chapter 10 of Söderström and Stoica (1989). A large number of papers appeared in the 1990's, spurred by the interest in identification for control, typically dealing with various parametrizations of the indirect approach. See, e.g., Van den Hof and Schrama (1993), Schrama (1991), and the survey Forssell and Ljung (1998a). The latter also describes the use of a non-causal FIR filter in (13.70). The application of direct output error methods to identify unstable systems is discussed in Forssell and Ljung (1998c). For the subspace methods of Section 10.6, several algorithms have been specifically devised to handle closed loop data: The algorithm in Chou and Verhaegen (1997)can be described as a direct approach, while Van Overschee and DeMoor (1997)and Verhaegen (1993)can be described as indirect and joint input/output, respectively.

Section 13.6: The section is based on Forssell and Ljung (1998b), Ljung (1985b), and Ljung (1986). Open-loop experiments are treated in Yuan and Ljung (1985)and closed-loop ones in Gevers and Ljung (1986).

Section 13.7: For alias effects and anti-alias filtering see, e.g., Oppenheim and Willsky (1983)or Åström and Wittenmark (1984). A study of the choice of sampling interval similar to our Example 13.4 is given in Åström (1969). A frequency-domain-based study is given in Payne, Goodwin, and Zarrop (1975). The bias considerations are further discussed in Wahlberg and Ljung (1986).

13.10 PROBLEMS

13G.1 *Effects of round-off errors:* Suppose that the data are measured with a precision of 10 bits (which is quite reasonable for typical A/D converters) and suppose that the bandwidth of the system (the highest frequency of interest) is ω_B. Follow the advice of Section 13.7 and choose the sampling frequency $\omega_s = 10\omega_B$. What is the lowest frequency that can be adequately modeled with the finite precision data? (*Hint:* A natural mode with time constants behaves, approximately, as

$$y(t+T) = \left(1 - \frac{T}{\tau}\right) y(t) + \frac{T}{\tau} u(t)$$

if $T \ll \tau$, where T is the sampling interval $T = 2\pi/\omega_s$.) *Answer:* Lowest frequency $\gtrsim 0.01\omega_B$. Notice the implications on how wide (i.e., narrow) are the frequency ranges that can be adequately modeled! Reference: Goodwin (1985).

13G.2 *Singular criterion matrix:* Suppose that the criterion matrix $C(\omega)$ in (12.6) and (13.73) is singular, and rewrite it

$$C(\omega) = S(\omega) \begin{bmatrix} |M(e^{i\omega})|^2 & M(e^{i\omega}) \\ M(e^{-i\omega}) & 1 \end{bmatrix}$$

Show that, in (13.73b),

$$\Psi(\omega, X) = \frac{S(\omega) \cdot \Phi_v(\omega)}{\lambda_0} \left[1 + \frac{\left|\lambda_0 M(e^{i\omega}) - \Phi_{ue}(\omega)\right|^2}{\lambda_0 \Phi_u(\omega) - |\Phi_{ue}(\omega)|^2} \right]$$

Conclude from this that the criterion

$$\min_X \int_{-\pi}^{\pi} \Psi(\omega, X)\, d\omega \tag{13.100}$$

subject to *any* constraint is minimized by the closed-loop design

$$u(t) = \frac{M(q)}{G_0(q)M(q) + H_0(q)} y(t) + r(t) \tag{13.101}$$

for *any* extra input $r(t)$ [including $r(t) \equiv 0$], provided it is an admissible $\{u(t)\}$. That is, (13.101) gives *the global minimum* of (13.100) w.r.t X in (13.73) (reference: Ljung, 1986).

13G.3 Youla-Parameterization is a way of parameterizing all stabilizing regulators for a given systems—or equivalently parameterizing all systems that are stabilized by a given regulator F_y. (See, e.g., Vidyasagar, 1985.) In the SISO case it works as follows. Let $F_y = X/Y$ (X, Y stable, coprime) and let $G_{nom} = N/D$ (N, D stable, coprime) be any system that is stabilized by F_y. Then, as R ranges over all stable transfer functions, the set

$$\mathcal{G} = \left\{ G : G(q, \theta) = \frac{N(q) + Y(q)R(q,\theta)}{D(q) - X(q)R(q,\theta)} \right\}$$

describes all systems that are stabilized by F_y. This idea can now be used for identification (see, e.g., Hansen, Franklin, and Kosut, 1989, Van den Hof and Schrama, 1991): Let u and y be input-output data from a plant controlled by the regulator F_y. Define X, Y, N, D as above and let

$$z(t) = y(t) - L(q)N(q)Y(q)r(t)$$

$$x(t) = L(q)Y^2(q)r(t)$$

where $L = 1/(YD+NX)$, which is stable and inversely stable (since G_{nom} is stablized by F_y). Then estimate $R(q, \theta)$ from the open loop identification problem

$$z(t) = R(q, \theta)x(t) + H(q, \theta)e(t) \tag{13.102}$$

Use the resulting estimate $\hat{R}$ in $\mathcal{G}$ to find the corresponding estimate $\hat{G}$ of the transfer function from u to y. Show that this method is the same as applying indirect identification for the model parameterization $\mathcal{G}$. The main advantage of this method is that the obtained estimate $\hat{G}$ is guaranteed to be stabilized by F_y.

13G.4 Consider the joint input-output model (13.65) with the parameterization

$$\mathcal{G}(q, \theta) = \frac{1}{1 + G(q,\theta)F_y(q,\theta)} \begin{bmatrix} G(q,\theta) \\ 1 \end{bmatrix}$$

$$\mathcal{H}(q, \theta) = \frac{1}{1 + G(q,\theta)F_y(q,\theta)} \begin{bmatrix} H(q,\theta) & G(q,\theta) \\ -F(q,\theta)H(q,\theta) & 1 \end{bmatrix}$$

This is the structure we obtain if we think in terms of the regulator (13.63) and take $\nu = [e \;\; w]^T$. Assume that the covariance matrix of ν is $\begin{bmatrix} \lambda_1 & 0 \\ 0 & \lambda_2 \end{bmatrix}$. Apply the multivariable prediction error/maximum likelihood method to this parameterization and show that the criterion becomes

$$V_N(Z^N, \theta) = \frac{1}{\lambda_1} \sum_{t=1}^{N} [H^{-1}(q, \theta)(y(t) - G(q, \theta)u(t))]^2$$

$$+ \frac{1}{\lambda_2} \sum_{t=1}^{N} [u(t) - r(t) + F_y(q, \theta)y(t)]^2$$

If G and F_y are independently parameterized, this is the same as estimating G by the direct method and separately determining F_y from (13.63).

13E.1 Suppose that the signal $u(t)$ is persistently exciting of order n. Give a condition on the stable filter $L(q)$ that guarantees that $u_F(t) = L(q)u(t)$ is also persistently exciting of order n.

13E.2 Consider the ARX structure

$$A(q)y(t) = B(q)u(t) + e(t)$$

where the degree of B is n_b. Show that an open-loop input that is persistently exciting of order n_b is sufficiently informative with respect to this set, regardless of the order of A, provided the process noise is persistently exciting.

13E.3 Consider a model structure

$$y(t) = G(q, \rho)u(t) + H(q, \eta)e(t)$$

with independent parametrization of G and H. Show that it is not possible to affect the accuracy of $\hat{\eta}$ in an open-loop experiment.

13E.4 Consider the FIR model structure

$$y(t) = b_1 u(t-1) + \ldots + b_m u(t-m) + e(t)$$

Determine the input spectrum that minimizes det $P_\theta(\mathcal{X})$ subject to the constraint $\bar{E}u^2(t) \leq 1$.

13E.5 Consider the model structure

$$y(t) + ay(t-1) = bu(t-1) + e(t)$$

and determine the open-loop input spectrum that minimizes det $P_\theta(\mathcal{X})$ subject to $Eu^2(t) \leq 1$. What is the optimal input in case b is fixed to the value 1? Assume that the true system is given by

$$y(t) - 0.5y(t-1) = u(t-1) + e_0(t)$$

where $e_0(t)$ is white noise.

13E.6 Suppose we are interested in the dynamics from propeller velocity to speed of a ship. We may measure both these signals but may only affect the torque of the propeller axis from the engine. This axis is also affected by forces from the water resistance that depend on the ship's speed in a complex manner. Discuss the identifiability of the loop from propeller velocity to ship speed based on such experiments.

13E.7 As mentioned in the text, it is not straightforward to use PRBS for multi-input systems. One idea for a two-input system is as follows: Let $s^M = s(t), t = 1, \dots, M$ be a maximum length (M) PRBS signal. Make an identification experiment over a multiple of $2M$ samples by letting $u_1 = u_2 = s^M$ for the first M samples, and $u_1 = -u_2 = s^M$ for the next M samples. Show that this gives the same input covariance matrix as exciting one input at a time with $\sqrt{2}s^M$, i.e., $u_1 = \sqrt{2}s^M$, $u_2 = 0$ for the first M samples and then $u_1 = 0$, $u_2 = \sqrt{2}s^M$ for the next M samples.

13T.1 Minimize $J_P(\mathcal{X})$ in (13.73) with respect to Φ_u with $\Phi_{ue}(\omega) \equiv 0$, subject to the constraint $\bar{E}y^2(t) \leq \alpha$. Assume that the true system is given by

$$y(t) = G_0(q)u(t) + H_0(q)e_0(t)$$

13T.2 *Experiment design for minimum variance control:* Suppose that the intended model application is minimum variance control. Use Problem 13G.2 to compare the performance degradation $J(\mathcal{X})$ obtained with an optimal *open-loop* input to that obtained for the overall optimal input (reference: Gevers and Ljung, 1986).

13D.1 Consider the time-varying system

$$y(t) = G_t(q)u(t), \qquad \text{all } t$$

such that

$$G_t(q) = G_1(q), \qquad kN \leq t < kN + \alpha N$$

$$G_t(q) = G_2(q), \qquad kN + \alpha N \leq t < (k+1)N$$

Let the spectrum of u be $\Phi_u(\omega)$. Show that, as N increases, the spectrum of y becomes

$$\Phi_y(\omega) = \alpha G_1(e^{i\omega})\Phi_u(\omega)G_1^T(e^{-i\omega}) + (1-\alpha)G_2(e^{i\omega})\Phi_u(\omega)G_2^T(e^{-i\omega})$$

14

PREPROCESSING DATA

When the data have been collected from the identification experiment, they are not likely to be in shape for immediate use in identification algorithms. There are several possible deficiencies in the data that should be attended to:

1. High-frequency disturbances in the data record, above the frequencies of interest to the system dynamics
2. Occasional bursts and outliers, missing data, non-continuous data records
3. Drift and offset, low-frequency disturbances, possibly of periodic character

It must be stressed that in off-line applications, one should always first plot the data in order to inspect them for these deficiencies. In this section we shall discuss how to preprocess the data so as to avoid problems in the identification procedures later.

14.1 DRIFTS AND DETRENDING

Low-frequency disturbances, offsets, trends, drift, and periodic (seasonal) variations are not uncommon in data. They typically stem from external sources that we may or may not prefer to include in the modeling. There are basically two different approaches to dealing with such problems:

1. Removing the disturbances by explicit pretreatment of the data
2. Letting the noise model take care of the disturbances

The first approach involves removing trends and offsets by direct subtraction, while the second relies on noise models with poles on or close to the unit circle, like the ARIMA models (I for integration) much used in the Box and Jenkins (1970)approach.

Signal Offsets

We shall illustrate the two approaches applied to the offset problem. The standard linear models that we use, like

$$A(q)y(t) = B(q)u(t) + v(t) \tag{14.1}$$

describe the relationship between u and y. That covers the dynamic properties, i.e., how does a change in u cause changes in y, as well as the static properties, i.e., the static relationship between a constant $u(t) \equiv \overline{u}$ and the resulting steady-state value of $y(t)$, say $\overline{y}$:

$$A(1)\overline{y} = B(1)\overline{u} \tag{14.2}$$

In practice, the raw input-output measurements, say $u^m(t)$, $y^m(t)$, are collected and recorded in physical units, the levels of which may be quite arbitrary. The equation (14.1) that describes the dynamic properties may therefore have very little to do with the equation (14.2) that relates the levels of the signals. In other words, (14.2) is quite an unnecessary constraint for (14.1). There are at least six ways to deal with this problem:

1. **Let $y(t)$ and $u(t)$ be deviations from a physical equilibrium:** The most natural approach is to determine the level $\overline{y}$ that corresponds to a constant $u^m(t) \equiv \overline{u}$ close to the desired operating point. Then define

$$y(t) = y^m(t) - \overline{y} \tag{14.3a}$$

$$u(t) = u^m(t) - \overline{u} \tag{14.3b}$$

as the deviations from this equilibrium. These translated variables will automatically satisfy (14.2), making both members equal to zero, and (14.2) will thus not influence the fit in (14.1). This approach emphasizes the physical interpretation of (14.1) as a linearization around the equilibrium.

2. **Subtract sample means:** A sound approach is to define

$$\overline{y} = \frac{1}{N}\sum_{t=1}^{N} y^m(t), \qquad \overline{u} = \frac{1}{N}\sum_{t=1}^{N} u^m(t) \tag{14.4}$$

and then use (14.3). If an input $u^m(t)$ that varies around $\overline{u}$ leads to an output that varies around $\overline{y}$, then $(\overline{u}, \overline{y})$ is likely to be close to an equilibrium point of the system. Approach 2 is thus closely related to the first approach.

3. **Estimate the offset explicitly:** One could also model the system using variables in the original physical units and add a constant that takes care of the offsets:

$$A(q)y^m(t) = B(q)u^m(t) + \alpha + v(t) \tag{14.5}$$

Comparing with (14.1) to (14.3), we see that α corresponds to $A(1)\overline{y} - B(1)\overline{u}$. The value α is then included in the parameter vector θ and estimated from data. It turns out that this approach in fact is a slight variant of the second approach. See Problem 14E.1.

4. **Using a noise model with integration (= differencing the data):** In (14.5) the constant α could be viewed as a constant disturbance, which is modeled as

$$\frac{\alpha}{1 - q^{-1}}\delta(t) \tag{14.6}$$

where $\delta(t)$ is the unit pulse at time zero. The model then reads

$$y^m(t) = \frac{B(q)}{A(q)}u^m(t) + \frac{1}{(1 - q^{-1})A(q)}w(t) \tag{14.7}$$

where $w(t)$ is the combined noise source $\alpha\delta(t) + v(t) - v(t-1)$. The offset α can thus be described by changing the noise model from $1/A(q)$ to $1/\left[(1 - q^{-1})A(q)\right]$. According to what we noted in (7.14) *this is equivalent to prefiltering* the data through the filter $L(q) = 1 - q^{-1}$, that is, differencing the data:

$$\begin{aligned} y_F^m(t) &= L(q)y^m(t) = y^m(t) - y^m(t-1) \\ u_F^m(t) &= L(q)u^m(t) = u^m(t) - u^m(t-1) \end{aligned} \tag{14.8}$$

5. **Extending the noise model:** Notice that the model (14.7) becomes a special case of (14.1) if the orders of the A and B polynomials in (14.1) are increased by 1. Then a common factor $1 - q^{-1}$ can be included in $A(q)$ and $B(q)$. This means that a higher-order model, when applied to the raw data y^m, u^m, will converge to a model like (14.7).

6. **High pass filtering:** Differencing data is a rather drastic filter for removing a static component. Any high-pass filter that has gain (close to) zero at frequency zero will have the same effect. See Section 14.4 for further discussion of prefiltering.

Evaluation of the Approaches

In an off-line application with offsets, the approach to recommend would be the first one or, if a steady-state experiment is not feasible, the second one. Estimating the offset explicitly (approach 3) is an unnecessarily complicated way to subtract the sample mean. Differencing the data as in (14.8) corresponds to a prefilter (inverse noise model) that has a very high gain at high frequencies. According to (8.71), this will push the model fit into a high-frequency region, which is unsuitable for many applications. Approach 5 has the additional drawback that more parameters have to be estimated. Approach 6 may be a quite useful alternative, especially if the offset is slowly time-varying.

It is especially important to remove offsets (trends and drifts) when output error models are employed. The discrepancy in levels will then be dominating the criterion of fit, and the dynamic properties become secondary. For methods that use flexible noise models (such as the least-squares method), the problem is less pronounced, since the effects of approach 5 will automatically de-emphasize the importance of signal levels.

Drift, Trends, Seasonal Variations

Methods to cope with other slow disturbances in the data are quite analogous to the approaches we discussed previously. Drifts and trends can be seen as time-varying equilibria. Straight lines or curve segments can be fitted to the data in the same manner as the constant offset levels in (14.4), and deviations from these time-varying means are considered. For seasonal variations, several techniques of this character have been developed for economic time series. Periodic signals are adjusted to data, and then subtracted. See, e.g., Box and Jenkins (1970).

Another approach would be to difference the data, analogously to (14.8) or, equivalently, to use ARIMA model structures, which include an integrator in the noise model. Alternatively, the noise model could be given extra flexibility to find the integrator or a complex pair of poles on the unit circle to account for periodic variations. In Goodwin et.al. (1986)a comprehensive discussion of the latter problem is given. With some knowledge of the frequencies of these slow variations, a better alternative may be to high-pass filter the data. This has the same effect of removing off-sets and slow drifts, but does not push the model fit into the high frequency range as differencing does. See Section 14.4 for further comments on this.

14.2 OUTLIERS AND MISSING DATA

In practice, the data acquisition equipment is not perfect. It may be that single values or portions of the input-output data are missing, due to malfunctions in the sensors or communication links. It may also be that certain measured values are in obvious error due to measurement failures. Such bad values are often called *outliers*, and may have a substantial negative effect on the estimate. Bad values are often much easier to detect in a residual plot (see Section 16.6).

Example 14.1 Outliers

Consider simulated data from the system

$$\begin{aligned} y(t) - 2.85y(t-1) + 2.717y(t-2) - 0.865y(t-3) \\ = u(t-1) + u(t-2) + u(t-3) + e(t) + 0.7e(t-1) + 0.2e(t-3) \end{aligned} \tag{14.9}$$

The values $y(313), \ldots, y(320)$ were then artificially changed to zero. The resulting output plot is shown in Figure 14.1. Although visual inspection shows a possible glitch around these values, no serious errors seem to be at hand. An ARMAX model with

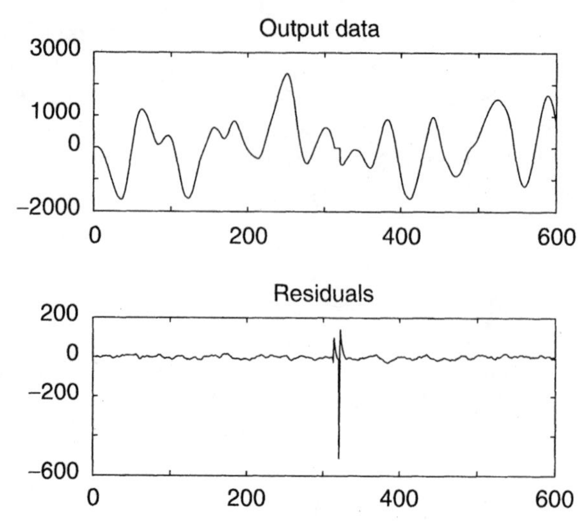

Figure 14.1 Data with outliers. Upper plot: the output signal. Lower plot: The residuals from model $\hat{\theta}_1$.

the correct orders is estimated giving the estimate $\hat{\theta}_1$. The residuals (prediction errors) from this model are also shown in Figure 14.1. Now, there is no doubt about the data problems around $t = 318$. Another model was then estimated based only on the first 300 data points. This is denoted by $\hat{\theta}_2$. Finally, the whole data record was used to estimate an ARMAX model applying a robust norm in the criterion according to (15.9)–(15.10). That gave the estimate $\hat{\theta}_3$. The estimates of the A- and B-polynomials are summarized as follows, where θ_0 denotes the true values:

$$\begin{array}{lrrrrrr} \theta_0 & -2.8500 & 2.7170 & -0.8650 & 1.0000 & 1.0000 & 1.0000 \\ \hat{\theta}_1 & -2.8523 & 2.7200 & -0.8668 & -0.1669 & 2.6418 & 0.4159 \\ \hat{\theta}_2 & -2.8504 & 2.7165 & -0.8652 & 0.9726 & 1.0496 & 1.0221 \\ \hat{\theta}_3 & -2.8557 & 2.7267 & -0.8701 & 1.0250 & 1.0185 & 0.8842 \end{array} \tag{14.10}$$

We see that the few outliers have made the estimate of the B-polynomial in $\hat{\theta}_1$ quite bad. It should be added that the estimates "are aware" of this: The estimated standard deviations of the 3 B-parameters in $\hat{\theta}_1$ are given as 0.9997, 1.7157, and 1.1421, while those of $\hat{\theta}_2$ have standard deviations 0.0602, 0.0750, and 0.0611, respectively. □

To deal with outliers and missing data, there are a few possibilities. One is to cut out segments of the data sequence so that portions with bad data are avoided. The segments can then be merged using the techniques of Section 14.3. For a data set with many inputs and outputs it might be difficult—in certain applications—to find data segments that are "clean" in all variables. It is then better to treat outliers, both in inputs and outputs, as missing data and view them as unknown parameters.

Dealing With Missing Data

Assume, for the moment, that we have a model $\mathcal{M}(\theta)$ that describes the relationship between the input-output data. In the basic linear predictor form (4.6) it is given by

$$\hat{y}(t|\theta) = \sum_{k=1}^{t} g(t-k,\theta)u(k) + \sum_{k=1}^{t} h(t-k,\theta)y(k) \tag{14.11}$$

Suppose now that some of the input-output data are missing. One must distinguish between missing inputs and missing outputs, since they should be handled differently. Let us first consider *missing inputs.*

Missing Input Data. If the input is a determistic sequence, it is natural to consider missing inputs as unknown parameters. Since the above expression is linear in the data, it is clear that for a given model $\mathcal{M}(\theta)$ the missing data can be estimated using a linear regression, least squares procedure. If we denote the missing data with the vector η we have

$$\hat{y}(t|\theta,\eta) = \sum_{k\in K_u} g(t-k,\theta)u(k) + \varphi^T(t,\theta)\eta + \sum_{k=1}^{t} h(t-k,\theta)y(k) \tag{14.12}$$

where $k \in K_u$ is the set of non-missing inputs $u(k)$, and $\varphi(t,\theta)$ is made up from $g(t-k_i,\theta)$, $k_i \notin K_u$ in an obvious way. The parameters θ and η can then be estimated by a prediction error criterion in the usual way. Note that for fixed θ, (14.12) is a linear regression for η, so missing input data can easily be estimated for any given model. It may then be natural (but not necessarily numerically efficient) to iterate between estimating the missing data, using the current model, i.e., estimating η for fixed θ, and estimating the model θ using the currently reconstructed missing data. To start up the iterations, the first model can be built using linearly interpolated values for the missing data.

Missing Output Data. It is not natural to regard missing output data as unknown parameters, since they are treated as random variables in the prediction framework. The correct prediction error criterion will be to minimize the error between $y(t)$ and $\hat{y}(t|\theta, Y_{K_y})$, where the prediction is based on those past $y(k)$ that actually have been observed ($k \in K_y$). To compute this prediction correctly we can use the time-varying Kalman filter (4.94)–(4.95) and deal with the missing data as irregular sampling. To be more specific, suppose that the underlying, discrete time model is given in innovations form as (4.91):

$$x(t+1,\theta) = A(\theta)x(t,\theta) + B(\theta)u(t) + K(\theta)e(t) \tag{14.13}$$

$$y(t) = C(\theta)x(t,\theta) + e(t) \tag{14.14}$$

The cross-covariance between process noise and measurement noise (see (4.85) will be $R_{12}(\theta) = K(\theta)R_2$. Now, if some or all components of $y(t)$ are missing at a certain time t, this is treated as *time-varying* $C_t(\theta)$ and $R_{t,12}(\theta)$, where only those rows of

C and R_{12} are extracted, that correspond to measured outputs. If all outputs are missing at time t, C_t and $R_{t,12}$ will be the empty matrices. The time varying Kalman filter (4.94)–(4.95) with $C_t(\theta)$ and $R_{t,12}(\theta)$ inserted into (4.95) will now produce the correct predictors

$$\hat{y}(t|\theta) = C(\theta)\hat{x}(t,\theta) = \hat{y}(t|\theta, Y_{K_y})$$

Working with the time varying predictor of course leads to much more computations, and approximate alternatives sometimes may be preferrable. One approximation would be to replace any missing $y(k)$ in (14.11) by $\hat{y}(k|\theta)$, i.e., the predictor based on data up to time $k-1$. (The correct replacement would be to use the smoothed estimate of $y(k)$ using measured data up to time $t-1$.) Another approximation would be to treat also missing outputs as unknown parameters. This corresponds to replacing missing $y(k)$ in (14.11) by their smoothed estimates, using the whole data record. A third possibility is to carry out the minimization of the prediction error criterion by the EM-method, see Problem 10G.3. The missing data then correspond to the auxiliary measurements X. See Isaksson (1993).

14.3 SELECTING SEGMENTS OF DATA AND MERGING EXPERIMENTS

Selecting Data Segments

When data from an identification experiment or, in particular, from normal operating records are plotted, it often happens that there are portions of bad data or non-relevant information. The reason could be that there are long stretches of missing data which will be difficult or computationally costly to reconstruct. There could be portions with disturbances that are considered to be non-representative, or that take the process into operating points that are of less interest. In particular for normal operating records, there could also be long periods of "no information;" nothing seems to happen that carries any information about the process dynamics. In these cases it is natural to select segments of the original data set which are considered to contain relevant information about dynamics of interest. The procedure of how to select such segments will basically be subjective and will have to rely mostly upon intuition and process insights.

Merging Data Sets

It is also a very common situation in practice that a number of separate experiments have been performed. The reason could be that the plant is not available for long, continuous experiments, or that only one input at a time is allowed to be manipulated in separate experiments. A further reason is, as described above, that bad data have forced us to split up the data record into several separate segments. How shall such separate records be treated? We cannot simply concatenate the data segments, because the connection points would cause transients that may destroy the estimate.

Suppose we build a model for each of the data segments, all with the same structure. Let the parameter estimate for segment i be denoted by $\hat{\theta}^{(i)}$, and let its estimated covariance matrix be $P^{(i)}$. Assume also that the segments are so well separated that the different estimates can be regarded as independent. It is then well known from basic statistics that the optimal way to combine these estimates (giving a resulting estimate of smallest variance) is to weigh them according to their inverse covariance matrices.

$$\hat{\theta} = P \sum_{i=1}^{n} [P^{(i)}]^{-1} \hat{\theta}^{(i)}, \quad P = \left[\sum_{i=1}^{n} [P^{(i)}]^{-1} \right]^{-1} \tag{14.15}$$

P will then also be the covariance matrix of the resulting estimate $\hat{\theta}$. This can be proved in different ways. See Problem 14E.2 for a matrix-based proof. Compare also Lemma II.2 in Appendix II and Problem 6E.3.

One might ask if this estimate could not be obtained directly from the data segments. To see this, let us seek guidance from the linear regression case. Assume we are treating the model

$$y(t) = \varphi^T(t)\theta \tag{14.16}$$

The estimate for any segment will be

$$\hat{\theta}^{(i)} = \left[\sum_{t \in T^i} \varphi(t)\varphi^T(t) \right]^{-1} \sum_{t \in T^i} \varphi(t)y(t), \quad P^{(i)} = \hat{\lambda}^{(i)} \left[\sum_{t \in T^i} \varphi(t)\varphi^T(t) \right]^{-1}$$

$$\hat{\lambda}^{(i)} = \frac{1}{|T^i|} \sum_{t \in T^i} \left(y(t) - \varphi^T(t)\hat{\theta}^{(i)} \right)^2$$

Here T^i is the index set of the ith segment, excluding those t for which $\varphi(t)$ is not fully known. This means that the first $\max(n_a, n_b)$ samples are excluded from each segment for an ARX-model (4.7). It was called the covariance method or the non-windowed case in the discussion following (10.13). By $|T^i|$ we mean the number of time indices in T^i.

If we apply (14.15) to these estimates, and assume that $\hat{\lambda}^{(i)}$ is independent of i, it is easy to see that the resulting estimate is the same as (14.16) would give if the summation was carried out over the union $\cup_i T^i$ of segments. This is of course most natural. Note, however, that for a dynamic model, this is not the same as first concatenating the data segments and then applying an ARX-model. Cutting away the first observations in each segment eliminates the transient problems at the merging points.

For the general case, with predictor models that have infinite impulse responses, this suggests that a criterion should be formed as

$$V(\theta) = \sum_{t \in T^1} \left(y(t) - \hat{y}(t|\theta) \right)^2 + \ldots + \sum_{t \in T^n} \left(y(t) - \hat{y}(t|\theta) \right)^2 \tag{14.17}$$

where the filters that compute $\hat{y}(t|\theta)$ for each of the segments should be reinitialized with zero initial conditions (or associated with a separate set of initial conditions to be estimated). The actual minimization algorithm is however entirely analogous to the one described in Sections 10.2 and 10.3.

What are the advantages of (14.17) compared to (14.15)? In the first place, it is more efficient to use just one minimization run rather than n separate ones. Second, if each of the experiments (segments) are poorly exciting but provide good excitation taken together, then minimizing (14.17) will be better conditioned than minimizing each of the sub-sums. Such a situation arises, e.g., when safety or production reasons require separate experiments, where one input at a time is manipulated.

Averaging over Periodic Data

A different way of "merging" data sets is at hand when an experiment with a periodic input has been conducted. We noted in Section 13.3 that it is then advantageous to average the output signal over the periods, so that the condensed set consists of just one period of input-output data. See (13.34). This allows shorter data records and independent noise estimates.

14.4 PREFILTERING

Prefiltering the input and the output data through the same filter will not change the input-output relation for a linear system:

$$y(t) = G_0(q)u(t) + H_0(q)e(t) \Rightarrow L(q)y(t) = G_0(q)L(q)u(t) + L(q)H_0(q)e(t)$$

(In the multivariable case all signals must be subjected to the same filter, so that $L(q)$ is a multiple of the identity matrix.) The filtering however changes the noise characteristics, so the estimated model will still be affected by the prefiltering. In this section we shall discuss the role and use of this feature.

From an estimation point of view, filtering the prediction errors before making the fit, as in (7.10), is an important option:

$$\begin{aligned} \varepsilon_F(t, \theta) &= L(q)\varepsilon(t, \theta) = \frac{L(q)}{H(q, \theta)}(y(t) - G(q, \theta)u(t)) \\ &= \frac{1}{H(q, \theta)}(L(q)y(t) - G(q, \theta)L(q)u(t)) \end{aligned} \tag{14.18}$$

From these expressions we see a few things:

- Filtering prediction errors is the same as filtering the observed input-output data. In the multivariable case the same filter must then be applied to all signals.
- A prefilter $L(q)$ is equivalent to a noise model $H(q) = 1/L(q)$. We can thus interchangeably talk about prefilters and noise models.

The noise model/prefilter has three functions:

1. We know from Section 8.5 that it will affect the bias distribution of the resulting model. We shall shortly review these results.
2. The discussion in Section 9.3 shows that if the transfer function estimate $\hat{G}$ is unbiased, then the best accuracy is obtained for a prefilter that corresponds to the true noise characteristics: $L(q) = 1/H_0(q)$. (See discussion following (9.42).)
3. The function of the prefilter may also be to remove disturbances of high or low frequencies that we do not want to include in the modeling.

The second function is the classical statistical one: In order to attain the Cramér-Rao bound, we need a correct noise model. Since this is typically unknown, it is natural to estimate that too, by including parameters in the noise model/prefilter. For purposes 1 and 3 there is no reason to let the prefilter contain adjustable parameters: On the contrary, a parameterized noise model will pull $H(q, \theta)/L(q)$ to resemble the error spectrum (see (8.73)), and this may undo what $L(q)$ was intended to achieve for these purposes. The three functions of the prefilter may consequently be conflicting.

While the second purpose of the prefilter really is a noise modeling issue, the two others correspond to pure data preprocessing. We shall now review the use of prefiltering for each of these two purposes.

Affecting the Bias Distribution

In general, it is not possible to describe the true system exactly within the chosen model set, so that the model will be biased. Prefiltering the data may have a substantial influence on the distribution of this bias. As we found in Section 8.5, the limiting model can be interpreted as a compromise between minimizing

$$
\begin{aligned}
\theta^*(\mathcal{D}) &\approx \arg\min_{\theta \in D_{\mathcal{M}}} \int_{-\pi}^{\pi} \left| G_0(e^{i\omega}) - G(e^{i\omega}, \theta) \right|^2 Q(\omega, \theta^*) d\omega \\
Q(\omega, \theta) &= \frac{\left| L(e^{i\omega}) \right|^2 \Phi_u(\omega)}{|H(e^{i\omega}, \theta)|^2}
\end{aligned}
\tag{14.19}
$$

on the one hand and fitting $\left| H(e^{i\omega}, \theta)/L(e^{i\omega}) \right|^2$ to the error spectrum $\Phi_{ER}(\omega, \theta^*)$ on the other. See (8.73). This means that $Q(\omega, \theta^*)$ will be taken as the weighting function that determines the bias distribution of G. This weighting function can in turn be affected by properly selecting the

- Input spectrum $\Phi_u(\omega)$
- Noise model set $H(q, \theta)$ (14.20)
- Prefilter $L(q)$

Notice that it is only the ratio $\Phi_u|L|^2/|H|^2$ that determines the bias distribution; the values of the individual functions Φ_u, H, and L are immaterial. Note also that the interpretation of the role of the prefilter is clear cut only if the noise model H does not depend on θ. In the general case (14.19) is somewhat heuristic, but it still is a quite useful tool to understand and manipulate the bias distribution. We illustrate that in the following example.

Example 14.2 Affecting the Bias Distribution

Consider the system (8.78) of Example 8.5. The resulting model in the OE-structure (8.79) gave the Bode plot of Figure 8.2a. This corresponds to $Q(\omega) \equiv 1$ (θ-independent) in (14.19). Since $|G(e^{i\omega})|$ decays very rapidly for high frequencies (has a rapid roll-off), this means that high frequencies play very little role in the Bode plot fit.

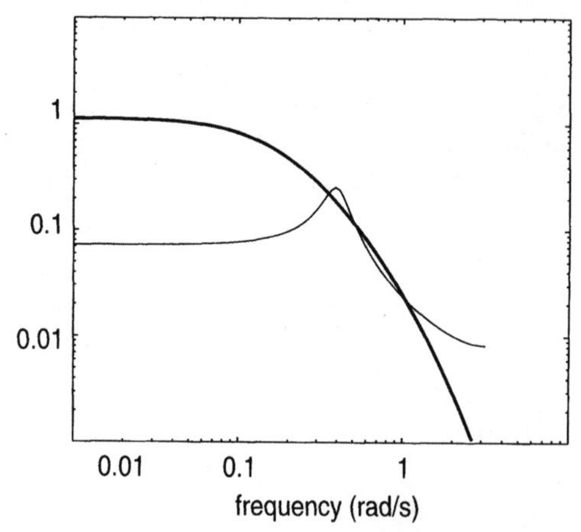

Figure 14.2 Amplitude Bode plot of the true system and model identified in the OE-structure (8.79), with an HP prefilter $L_1(q)$ (cf. Figure 8.2). Thick line: true system; thin line: model.

To enhance the high-frequency fit, we filter the prediction errors through a fifth-order high-pass (HP) Butterworth filter $L_1(q)$ with a cut-off frequency of 0.5 rad/sec. This changes $Q(\omega)$ to this HP-filter. The Bode plot of the resulting estimate is given in Figure 14.2. The fit has now moved into high frequencies, but clearly the second-order model has problems describing the fourth-order roll-off.

Consider now the estimate obtained by the least-squares method in the ARX-model structure (8.81). This was depicted in Figure 8.2b. If we want a better low-frequency fit, it seems reasonable to counteract the HP weighting function $Q(\omega, \theta^*)$ in Figure 8.3 by low pass (LP) filtering of the prediction errors. We thus construct $L_2(q)$ as a fifth-order LP Butterworth filter with cut-off frequency 0.5 rad/sec. The ARX model is then estimated for the input-output data filtered through $L_2(q)$. Equivalently, we could say that the prediction-error method is used for the model structure

$$y(t) = \frac{b_1 q^{-1} + b_2 q^{-2}}{1 + a_1 q^{-1} + a_2 q^{-2}} u(t) + \frac{1}{L_2(q)(1 + a_1 q^{-1} + a_2 q^{-2})} e(t) \tag{14.21}$$

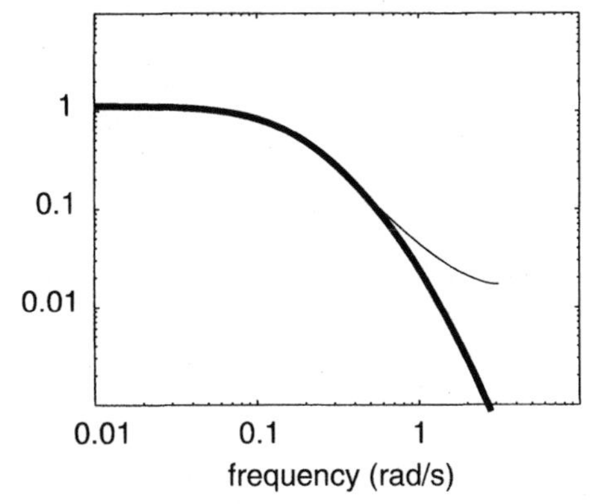

Figure 14.3 Bode plots of the true system and model identified in the ARX-structure (8.81) with an LP prefilter $L_2(q)$ (cf. Figure 8.2b). Legend as in Figure 14.2.

The resulting estimate is shown in Figure 14.3, and the corresponding weighting function $Q(\omega, \theta^*)$ in Figure 14.4. The resulting models in Figures 8.2a and 14.3 are quite similar. One should then realize that the ARX estimate for filtered data of Figure 14.3 is much easier to obtain than the output error estimate of Figure 8.2a, which requires iterative search. □

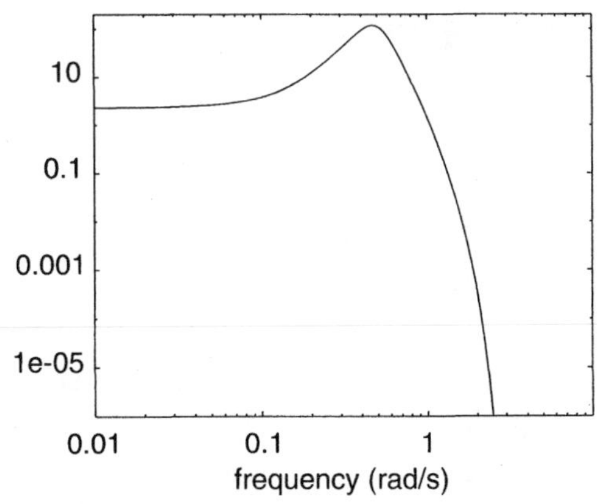

Figure 14.4 The weighting function $Q(\omega, \theta^*) = \left|L_2(e^{i\omega})\right|^2 \cdot \left|1 + a_1^* e^{-i\omega} + a_2^* e^{-2\omega}\right|^2$ corresponding to the estimate in Figure 14.3.

Dealing with Disturbances

The third purpose of prefiltering, as listed in the beginning of this section, is to remove disturbances in the data that we do not want to include in the modeling. This actually goes hand in hand with the noise modeling aspect of prefiltering: Removing, say, a seasonal variation of a certain frequency by a band-stop filter, can also be interpreted as fixing a noise model with very high gain in this frequency band, which is a way of expressing the presence of the seasonal variation.

High-Frequency Disturbances. High frequency disturbances in the data, above the frequencies of interest for the system dynamics, indicate that the choices of sampling interval and presampling filters were not thoughtful enough. This can however be remedied by low pass filtering of the data. Also, if it turns out that the sampling interval was unnecessarily short, one may always *resample* the data by picking every sth sample from the original record. Then, however, a digital antialias filter must be applied before the resampling, in the same manner as discussed in Section 13.7.

Low-Frequency Disturbances. Low frequency disturbances in terms of offset, drift, and slow seasonal variations were discussed in Section 14.1. A very suitable method to deal with such problems is to apply high pass filtering. This must be considered as a clearly better alternative to data differencing.

14.5 FORMAL DESIGN OF PREFILTERING AND INPUT PROPERTIES

To secure good models, we have a number of design variables, as discussed in Chapter 12. In addition to designing the experiment according to the advice of Chapter 13, we also have the prefilter as an important design variable. We shall in this section return to the formal design problem (12.7)–(12.9) and consider the joint design of prefilter and input signal properties.

Optimizing the Bias Distribution

Let us first turn to the formal design problem (12.29) for the bias error, in the special case

$$C(\omega) = \begin{bmatrix} C_{11}(\omega) & 0 \\ 0 & 0 \end{bmatrix} \tag{14.22}$$

With (14.22), (12.29) can be rewritten

$$\mathcal{D}_{\text{opt}} = \arg\min_{\mathcal{D}\in\Delta} \int_{-\pi}^{\pi} \left|G\left(e^{i\omega}, \theta^*(\mathcal{D})\right) - G_0(e^{i\omega})\right|^2 C_{11}(\omega)d\omega \tag{14.23}$$

Let us also specialize to open-loop operation: $\Phi_{ue}(\omega) \equiv 0$ and the noise model to be fixed to 1: $H(q,\theta) \equiv 1$. (Since we allow prefiltering, this includes the case of any given, fixed noise model.) We have the input spectrum and prefilter

$$\mathcal{D} = \{\Phi_u(\cdot), L(q)\} \tag{14.24}$$

as design variables. In this case (14.19) specializes to

$$\theta^*(\mathcal{D}) = \arg\min_{\theta\in D_{\mathcal{M}}} \int_{-\pi}^{\pi} \left|G_0(e^{i\omega}) - G(e^{i\omega},\theta)\right|^2 \Phi_u(\omega) \left|L(e^{i\omega})\right|^2 d\omega \tag{14.25}$$

We are thus faced with the minimization problem (14.23) with the function $\theta^*(\mathcal{D})$ defined by (14.24) and (14.25). This problem has an explicit solution.

Theorem 14.1. Consider the optimization problem (14.23) with $\theta^*(\mathcal{D})$ defined by (14.24) and (14.25). The minimum is obtained for any $\mathcal{D}$ such that

$$\Phi_u(\omega) \left|L(e^{i\omega})\right|^2 = \alpha \cdot C_{11}(\omega) \tag{14.26}$$

provided this is an admissible design for some positive scalar α.

Before turning to the proof, we note that the theorem says that the signal-to-noise model ratio should be chosen proportional to the criterion weighting function, and this can be accomplished either by input design or by prefilter/noise model selection. Some extensions to this theorem are outlined in Problems 14G.1 to 14G.3. Note in particular that for (14.22), open-loop operation indeed is optimal. That is, if $\Phi_{ue}(\omega)$ is included among the design variables (14.24), then the optimal solution is $\Phi_{ue}(\omega) \equiv 0$, together with (14.26).

Proof. First we establish the following lemma.

Lemma 14.1. Let $V(x, y)$ be a scalar-valued function of two variables such that each may take values in some general Hilbert space. For a fixed y, let

$$x^*(y) = \arg\min_x V(x, y) \tag{14.27}$$

and for a fixed z, let

$$y^*(z) = \arg\min_y V\left(x^*(y), z\right) \tag{14.28}$$

assuming that these minimizing values are unique and well defined. Then

$$y^*(z) = z \tag{14.29}$$

Proof of Lemma 14.1: By the definition (14.27),

$$V\left(x^*(z), z\right) \le V(x, z), \qquad \forall x, \forall z$$

Hence

$$V\left(x^*(z), z\right) \le V\left(x^*(y), z\right), \qquad \forall y \tag{14.30}$$

From (14.28), by definition,

$$V\left(x^*(y^*(z)), z\right) \le V\left(x^*(y), z\right) \qquad \forall y \tag{14.31}$$

Now, (14.30) and (14.31) imply

$$x^*\left(y^*(z)\right) = x^*(z)$$

which implies (14.29) since the mapping x^* is assumed to be injective (follows from the assumed uniqueness of the minimum in (14.28)). End proof of Lemma 14.1. □

Remark. The assumption of uniqueness in (14.28) can be relaxed by considering instead the *set*

$$y^*(z) = \left\{ y | V\left(x^*(y), z\right) = \min_y V\left(x^*(y), z\right) \right\}$$

The statement corresponding to (14.29) then is $z \in y^*(z)$ which is sufficient for our purposes.

To return to the problem (14.23), introduce the function

$$W(\theta, \kappa(\cdot)) = \int_{-\pi}^{\pi} \left|G(e^{i\omega}, \theta) - G_0(e^{i\omega})\right|^2 \kappa(\omega) d\omega \tag{14.32}$$

and define

$$\overline{\theta}\,(\kappa(\cdot)) = \arg\min_{\theta} W\,(\theta, \kappa(\cdot)) \tag{14.33}$$

We thus have, from (14.25),

$$\theta^*(\mathcal{D}) = \overline{\theta}\left(\Phi_u(\omega)\left|L(e^{i\omega})\right|^2\right) \tag{14.34}$$

Since the limiting estimate θ^* depends on the functions in $\mathcal{D}$ only via the product

$$\kappa(\omega, \mathcal{D}) = \Phi_u(\omega)\left|L(e^{i\omega})\right|^2$$

the optimal design problem (14.23) is in fact a search for the best κ:

$$\kappa_{\text{opt}} = \arg\min_{\mathcal{D}} W\left(\theta^*(\mathcal{D}), C_{11}(\cdot)\right) = \arg\min_{\kappa} W\left(\overline{\theta}(\kappa(\cdot)), C_{11}(\cdot)\right) \tag{14.35}$$

Applying Lemma 14.1 to (14.35), we find that

$$\kappa_{\text{opt}}(\omega) = \alpha \cdot C_{11}(\omega) \tag{14.36}$$

where α is any positive scalar such that the design is admissible ($\mathcal{D} \in \Delta$). This follows since scaling κ does not affect the problem (14.35). This concludes the proof of Theorem 14.1. □

Optimizing the Mean Square Error

Theorem 14.1 solves the problem of minimizing the bias contribution to the criterion (14.23). In (13.83) we minimized the variance contribution to the same criterion. It is easy to see that both criteria can be minimized simultaneously, which means that we can solve the total error problem (12.30). This gives the following result.

Theorem 14.2. Consider the problem of minimizing the mean square error

$$\arg\min_{\mathcal{D}\in\Delta} \int_{-\pi}^{\pi} E\left|G(e^{i\omega}, \hat{\theta}_N(\mathcal{D})) - G_0(e^{i\omega})\right|^2 C_{11}(\omega) d\omega \tag{14.37}$$

with respect to the design variables prefilter, input spectrum, and cross spectrum between input and noise (i.e. feedback mechanism)

$$\mathcal{D} = \{L(q), \Phi_u(\cdot), \Phi_{ue}(\cdot)\} \tag{14.38}$$

under the constraints

$$\int_{-\pi}^{\pi} \Phi_u(\omega) d\omega \leq \alpha, \quad H(q, \theta) = 1 \tag{14.39}$$

The solution is

$$\mathcal{D}_{\text{opt}} = \begin{cases} \Phi_u(\omega) & = \mu_1 \sqrt{C_{11}(\omega)\Phi_v(\omega)} \\ \Phi_{ue}(\omega) & \equiv 0 \\ \left|L(e^{i\omega})\right|^2 & = \mu_2 \sqrt{\frac{C_{11}(\omega)}{\Phi_v(\omega)}} \end{cases} \tag{14.40}$$

Here μ_1 is adjusted so that the input power constraint is met, while μ_2 is a constant, such that the filter $L(q)$ is monic.

This result could be obtained easily since the two components of the mean square error in (12.26) could be minimized simultaneously with respect to $\mathcal{D}$, and no compromise had to be made. For other design variables, typically the model order, bias and variance are conflicting and an optimal trade-off has to be met.

Example 14.3 Pole Placement

Suppose we intend to use the model to design a regulator that achieves a certain closed loop system $R(q)$. Suppose the true system is G_0 and we use the model $\hat{G}$. A regulator that gives the desired closed loop with the model is

$$u(t) = F_r(q)r(t) - F_y(q)y(t)$$

where F_y, F_r are subject to

$$\hat{G}(q)F_r(q) = R(q)\left(1 + \hat{G}(q)F_y(q)\right)$$

Here r is the reference signal, and we want to achieve $y = Rr$. The difference between the desired and actual closed loop is

$$\begin{aligned} \tilde{G}_{\text{cl}} &= \frac{G_0 F_r}{1 + G_0 F_y} - R = \frac{G_0 F_r}{1 + G_0 F_y} - \frac{\hat{G} F_r}{1 + \hat{G} F_y} \\ &= \frac{(G_0 - \hat{G})F_r}{(1 + G_0 F_y)(1 + \hat{G} F_y)} = \frac{(G_0 - \hat{G})R}{\hat{G}(1 + G_0 F_y)} \approx \frac{(G_0 - \hat{G})R}{G_0(1 + G_0 F_y)} \end{aligned}$$

where the last step holds if $\hat{G}$ is close to G_0. The size (variance) of the error $\tilde{y} = \tilde{G}_{\text{cl}} r(t)$ is

$$\int_{-\pi}^{\pi} |G_0 - \hat{G}|^2 \cdot C_{11} d\omega, \quad C_{11} = \left|\frac{R}{G_0(1 + G_0 F_y)}\right|^2 \Phi_r$$

Consequently the experiment that gives the best model for this regulator design under the constraints (14.39) is

$$\Phi_u^{\text{opt}}(\omega) = \mu \frac{\left|R(e^{i\omega})\right| \sqrt{\Phi_r(\omega)} \left|H_0(e^{i\omega})\right|}{\left|G_0(e^{i\omega})\right| \left|1 + G_0(e^{i\omega})F_y(e^{i\omega})\right|}$$

$$L^{\text{opt}}(e^{i\omega}) = \frac{\left|R(e^{i\omega})\right| \sqrt{\Phi_r(\omega)}}{\left|G_0(e^{i\omega})\right| \left|H_0(e^{i\omega})\right| \left|1 + G_0(e^{i\omega})F_y(e^{i\omega})\right|}$$

We see that the characteristics of the true system are required in order to compute the optimal design. Even though these may not be known in detail, the expressions are still useful. They tell us to spend the input power where

1. A gain increase is desired: $\left|\frac{R(e^{i\omega})}{G_0(e^{i\omega})}\right| \gg 1$
2. The reference signal is going to have energy: $\Phi_r(\omega)$ large
3. The disturbances are significant: $\left|H_0(e^{i\omega})\right|$ large
4. The sensitivity due to feedback is poor: $\left|1 + G_0(e^{i\omega})F_y(e^{i\omega})\right|$ small

These suggestions are as such quite natural, but their formalization is useful. □

Identification for Control

Control design is one of the most important uses of identified models. Considerable effort has therefore been spent on designing experiments and methods that give models well suited for control design. Feedback control is both forgiving and demanding in the sense that we can have good control even with a mediocre model, as long as it is reliable in certain frequency ranges. Loosely speaking, the model has to be reliable around the cross-over frequency ($\approx$ the bandwidth of the closed loop system), and it may be bad where the closed loop sensitivity function is small. The required accuracy of the model therefore depends on the (unknown and to-be-designed) sensitivity function. We saw the basic features of the experiment design in Example 14.3, and even though the optimal choices depend on unknown facts, it may be sufficient for a successful application. The interplay between the model and the criterion C_{11}, which depends on the regulator, which in turn depends on the model, has led to a substantial literature on "identification for control." This frequently involves iterative approaches, where a sequence of experiments are performed, interleaved with evaluations of the preliminary regulator designs. For an overview, see Gevers (1993).

14.6 SUMMARY

Preprocessing of data is an important prerequisite for the estimation phase. It may involve "repair" of the data in terms of replacing missing or obviously wrong data as well as merging disjunct data sets. It typically also involves data polishing by removing undesired disturbance features in the data. This is accomplished primarily by low-pass or high-pass prefiltering and/or subtracting offsets and trends from the data. Notice that if the data are prefiltered, the noise model is affected. If a very specific effect is desired with the prefiltering, it may therefore be wise not to let the noise model be flexible.

Prefiltering also affects the distribution of bias over the frequency range, together with other design variables. We showed in this chapter how, in some cases, formal design criteria could be optimized with respect to the design variables. More important, however, is that insights into the mechanisms that govern the bias distribution allow thoughtful design that secures a good fit in important frequency ranges, even when the optimality results are not directly applicable.

14.7 BIBLIOGRAPHY

Various approaches to deal with the offset problem are illustrated in Åström (1980). Modeling using marginally unstable noise models to cope with slow disturbances is studied, e.g., in Box and Jenkins (1970)and Goodwin et.al. (1986). Practical aspects of the data handling are given in several treatments, like Isermann (1980).

A general treatment of missing data is given in Little and Rubin (1987). Applications to time-series and dynamical systems are treated in, e.g., Isaksson (1993), Ansley and Kohn (1983), Kohn and Ansley (1986), and Goodwin and Feuer (1998). The outlier problem is specifically addressed in Abraham and Chuang (1993), and will commented upon again in Chapter 15.

More aspects on bias shaping by prefiltering are given in Wahlberg and Ljung (1986), while a comprehensive discussion of the issues of Section 14.5 is given in Ljung (1986a). The use of prefilters for control-relevant identification has been discussed also by Rivera, Pollard, and Garcia (1992).

Identification for control has been discussed extensively in the 1990's. See, e.g., Hjalmarsson, Gevers, and De Bruyne (1996), Lee et.al. (1995), Schrama (1992), Van den Hof and Schrama (1995), and Zang, Bitmead, and Gevers (1995). The topic has also spurred ideas of different approaches to identification and frequency function interpolation based on $\mathcal{H}_\infty$ and other techniques. See the special issue Kosut, Goodwin, and Polis (1992)for several such ideas.

14.8 PROBLEMS

14G.1 Consider the design problem (14.23) with $\Phi_{ue}(\omega) \equiv 0$ and a given noise model set

$$\mathcal{H} = \{H(q, \eta)\}$$

that is parametrized independently from the transfer function model set $\mathcal{G}$. Show that the solution to (14.23) then is

$$\mathcal{D}_{\text{opt}} : \qquad \frac{\Phi_u(\omega)}{\left|H(e^{i\omega}, \eta^*)\right|^2} = \alpha \cdot C_{11}(\omega)$$

where η^* denotes the resulting noise model parameter.

Hint: Establish first the following corollary to Lemma 14.1: Let $x^*(y)$ be defined as in the lemma. Let $f(y)$ be an arbitrary function, and let

$$y^*(z) = \arg\min_y V\left(x^*\left(f(y)\right), z\right)$$

Then $f\left(y^*(z)\right) = z$ (reference: Yuan and Ljung, 1985).

14G.2 Consider the design problem (12.29) with a general matrix $C(\omega)$. Assume that the noise model is fixed to H_* and chosen a priori (thus it does not belong to $\mathcal{D}$). The design variables consist of Φ_u and Φ_{ue}. Show that the solution to (12.29) is to select Φ_u and Φ_{ue} so that

$$\frac{1}{\left|H_*(e^{i\omega})\right|^2}\begin{bmatrix} \Phi_u(\omega) & \Phi_{ue}(\omega) \\ \Phi_{ue}(-\omega) & * \end{bmatrix} = \alpha \begin{bmatrix} C_{11}(\omega) & C_{12}(\omega) \\ C_{21}(\omega) & * \end{bmatrix}$$

Here $*$ means that the element in question does not affect the optimal design. α is an arbitrary positive scalar (reference: Ljung, 1986a).

14G.3 Consider the design problem (14.23) subject to the design variables

$$\mathcal{D} = \{\Phi_u(\cdot), \Phi_{ue}(\cdot), H_*(q)\}$$

Show that the optimal design is

$$\Phi_{ue}^{\text{opt}}(\omega) \equiv 0, \qquad \frac{\Phi_u^{\text{opt}}(\omega)}{\left|H_*^{\text{opt}}(e^{i\omega})\right|^2} = \alpha \cdot C_{11}(\omega)$$

Hint: Use Problem 14G.2 (reference: Ljung, 1986a).

14E.1 Consider the model (14.5) and introduce

$$\varphi^m(t) = \left[-y^m(t-1)\dots -y^m(t-n_a) \quad u^m(t-1)\dots u^m(t-n_b) \quad 1\right]^T$$

$$\theta = \left[a_1 \dots a_{n_a}\, b_1 \dots b_{n_b}\, \alpha\right]^T$$

Derive an expression for the LS estimate $\hat{\theta}_N$. Show that with the approximations

$$\frac{1}{N}\sum_{t=1}^{N} y(t-k) \approx \overline{y}, \qquad \text{for all } 1 \le k \le n_a$$

$$\frac{1}{N}\sum_{t=1}^{N} u(t-k) \approx \overline{u}, \qquad \text{for all } 1 \le k \le n_b$$

the estimates of a_i and b_i are *identical* to those obtained by subtracting sample means as in (14.4) and (14.3) and then applying the LS method to (14.1).

14E.2 Prove the following matrix identities:

$$(I-Z)P_1(I-Z)^T + ZP_2Z^T = P_1 - P_1R^{-1}P_1 + (Z - P_1R^{-1})R(Z - P_1R^{-1})^T$$

$$P_1R^{-1} = (P_1^{-1} + P_2^{-1})^{-1}P_2^{-1}, \quad \text{where } R = (P_1 + P_2)$$

Use this to prove that if θ_i are unbiased estimates of θ_0 (i.e. $\mathrm{E}\theta_i = \theta_0$) with variances P_i, then the variance of

$$\theta = \alpha_1\theta_1 + \alpha_2\theta_2, \quad \mathrm{E}\theta = \theta_0$$

is minimized for $\alpha_i = (P_1^{-1} + P_2^{-1})^{-1}P_i^{-1}$.

15

CHOICE OF IDENTIFICATION CRITERION

The selection of an identification method is one of the important decisions to be taken in system identification. In Chapter 7, we have structured the abundant supply of candidates. Different aspects and results on various methods have been mentioned in earlier chapters. It is the purpose of the present chapter to sum up these and also discuss how different options affect the properties of the resulting estimates.

A general discussion is given in Section 15.1. The particular problem of choosing the norm $\ell(\cdot)$ for the prediction-error method (7.155) or the shaping function $\alpha(\cdot)$ in (7.156) is studied in Section 15.2. Section 15.3 deals with optimal instruments for the IV method and their approximate implementation. Section 15.4 summarizes some basic advice to the user.

15.1 GENERAL ASPECTS

We have described three basic approaches to identification in this book, each associated with some design variables:

1. The prediction-error approach (7.12)
   - $\ell(\cdot)$: norm
   - $\mathcal{H}$: noise model set, including prefilter $L(q)$
2. The correlation approach (7.110)
   - $\alpha(\cdot)$: shaping function
   - $L(q)$: prefilter
   - $\zeta(t, \theta)$: correlation vector
3. The subspace approach to estimating state-space models (7.66), Section 10.6
   - $\varphi_s(t)$: correlation vector, corresponding to the regressors for which the k-step ahead predictors are determined
   - r: Maximum prediction horizon
   - W_1, W_2: The weighting matrices in (10.127)
   - R: The "post-multiplication matrix" in (10.128).

The choice between the approaches and the choices of design variables within the approaches are guided by a number of issues:

Applicability

The prediction error approach has the advantage that it is applicable to all model structures, linear and nonlinear, tailor-made and black-box-parameterized. It is also valid for systems operating in open as well as closed loop. The minimization code is essentially the same, only the computation of the predictor and its gradient is specific for the model structure.

The correlation approach is also, in principle, generally applicable, but it is most naturally used only for the linear black-box family (4.33). Most of its use is really as an IV-method for the ARX-model. Closed-loop applications require special care in the choice of instruments.

The subspace method is specifically designed for black-box linear systems in state-space form. Also for this method, it is necessary to use special solutions for closed-loop operation.

Bias Considerations

Will the method give unbiased estimates in case $\mathcal{S} \in \mathcal{M}$? All the listed methods for all generic choices of design variables will give consistency in the case of open loop operation. The prediction error approach has the added advantage of guaranteeing consistency for systems operating in closed loop as well, in case the model structure (including noise model) contains the true system.

If $\mathcal{S} \notin \mathcal{M}$, can the bias distribution be clearly explained and affected? Here prediction-error methods have a clear advantage. Expressions (8.71) and related ones describe quite clearly in what sense the model approximates the true system in the linear case. For fixed noise models and open loop systems, it is easy to control the frequency emphasis by prefiltering. The approximation aspects of the correlation methods can be written down, but are less transparent. The exact nature of the approximation properties of subspace methods and how these are affected by the design variables are not yet fully understood.

Variance and Robustness Considerations

Optimizing (=minimizing) the variance is a simpler problem than optimizing bias. The reason is that we have explicit expressions, from Chapter 9, for how the variance is affected by the design variables in the prediction-error and correlation cases. For the subspace method, it is currently not fully known how the design variables affect the variance. Some partial results were quoted in Section 10.6.

We know from Chapter 9 that the theoretical Cramér-Rao lower bound is asymptotically achievable by the maximum likelihood method, so in a sense we know the answers beforehand to all variance optimization questions:

- Best choice of $\ell(x) = -\log f_e(x)$ ($f_e(\cdot)$ being the PDF of the true innovations)
- Best choice of noise model/prefilter = true noise description (possibly estimated)
- Best choice of IV method = make it equal to the preceeding prediction error method

Still, as we shall see, the MLE may not be the best approach in all cases. The reasons may be the bias effects discussed previously, or that the estimates are sensitive to prior knowledge that may be imprecise. We shall, in Section 15.2, look into robustness issues for the choice of norm ℓ in the prediction error approach. In Section 15.3 we shall study variance-optimal instruments for the IV method, and see if optimal accuracy can be obtained even without iterative search.

Ease of Computation

The subspace methods have the important advantage that the algorithms do not contain iterative search. They can also be implemented using numerically robust algorithms. The IV-method has a similar advantage that it can estimate the dynamics of a linear system (but not the noise properties) without iterative search. The prediction error methods, except in the linear regression case, must rely upon iterative search methods, and may be trapped in false solutions that correspond to local minima.

15.2 CHOICE OF NORM: ROBUSTNESS

From (9.29) and (9.30) we know how the choice of $\ell(\cdot)$ in the prediction-error approach affects the asymptotic variance, provided $S \in \mathcal{M}$. *The covariance matrix (9.29) is scaled by the scalar*

$$\kappa(\ell) = \frac{\overline{E}[\ell'(e_0(t))]^2}{[\overline{E}\ell''(e_0(t))]^2} \tag{15.1}$$

According to Problem 9G.4, the shaping function $\alpha(x)$ in the correlation approach scales the covariance matrix (9.78) by the same scalar (15.1) with $\ell'(x) = \alpha(x)$. Hence the choices of ℓ and α can be discussed simultaneously. The scalar κ depends only on the function $\ell(x)$ and on the distribution of the true innovations $e_0(t)$. Let the PDF of these be denoted by $f_e(x)$. Then

$$\kappa(\ell) = \kappa(\ell, f_e) = \frac{\int (\ell'(x))^2 f_e(x)dx}{(\int \ell''(x) f_e(x)dx)^2} \tag{15.2}$$

Here prime and double prime denote differentiation with respect to the argument x.

Optimal Norm

If we focus on variance aspects. our objective is to select ℓ so that $\kappa(\ell, f_e)$ is minimized. Notice that this problem, as such, is quite independent of the underlying system identification problem, the particular model structure used, and so on. We have the following result.

Lemma 15.1.

$$\kappa(\ell, f_e) \geq \kappa(-\log f_e, f_e), \qquad \forall \ell \tag{15.3}$$

Proof. We have by partial integration

$$\int \ell''(x) f_e(x) dx = -\int \ell'(x) f_e'(x) dx$$

Cauchy's inequality now gives

$$\left[\int \ell''(x) f_e(x) dx\right]^2 = \left[\int \ell'(x) \frac{f_e'(x)}{f_e(x)} \cdot f_e(x) dx\right]^2$$

$$\leq \int [\ell'(x)]^2 f_e(x) dx \cdot \int \left[\frac{f_e'(x)}{f_e(x)}\right]^2 f_e(x) dx$$

with equality when

$$\ell'(x) = C_1 \frac{f_e'(x)}{f_e(x)}$$

which proves that

$$\ell(x) = C_1 \log f_e(x) + C_2$$

gives the minimum in (15.3). □

The lemma tells us that the best choice is

$$\ell_{\text{opt}}(\varepsilon) = -\log f_e(\varepsilon) \tag{15.4a}$$

which can be seen as a restatement of the fact that the maximum likelihood method is asymptotically efficient. The lemma deals with the case of a stationary innovations sequence $\{e_0(t)\}$. If the distribution of $e_0(t)$ depends on t, $f_e, (x, t)$, then the optimal norm is also time varying.

$$\ell_{\text{opt}}(\varepsilon, t) = -\log f_e(\varepsilon, t) \tag{15.4b}$$

This follows from the fact that (15.4b) gives the MLE. It can also be established directly; see Problem 15T.2. If the innovations are Gaussian, with known variances, then (15.4b) tells us to use a quadratic norm, scaled by the inverse innovation variances.

An annoying aspect of these results is that the PDF f_e, may not be known. There are two remedies for this: to simultaneously estimate f_e or to select an ℓ that is insensitive to the different ℓ that possibly could be at hand.

Adapting the Norm

In the first case we include additional parameters α: $\ell(\varepsilon, \alpha)$ so as to allow for an adjustment of the norm. Then we know from (8.59) that the norm will adjust itself so that it becomes close to the optimal choice (15.4). If a reasonably good estimate of f_e in (15.4) will do, the adaptive norm approach will be a good solution. This might, however, not be the case, as we shall now demonstrate.

Sensitivity of the Optimal Norm

The optimal variance scaling $\kappa(\ell, f)$ in (15.2) could be quite sensitive with respect to the PDF f. That is, as a function of f, the scalar $\kappa(-\log f_e, f)$ could have a very sharp minimum at $f = f_e$. This is illustrated in the following example.

Example 15.1 Sensitivity of Optimal Norm

Let the nominal PDF f_e, be a normal with variance 1.

$$f_e(x) = \frac{1}{\sqrt{2\pi}} e^{-x^2/2} = \varphi(x)$$

Then $-\log f_e(x) = \frac{1}{2}x^2$ (disregarding a constant term) and

$$\kappa(-\log f_e, f_e) = \frac{\int x^2 \varphi(x) dx}{[\int \varphi(x) dx]^2} = 1 \tag{15.5}$$

Suppose now that the prediction errors with a very small probability can assume a certain large value. This could, for example, correspond to a certain failure in the measurement or data transmission equipment. Such data are called *outliers*. We thus assume that ε is almost normal, but with probability $\frac{1}{2}10^{-3}$ it may assume the value 100, and with probability $\frac{1}{2}10^{-3}$ the value -100. The actual f then is

$$f(x) = (1 - 10^{-3})\varphi(x) + 10^{-3}\left[\tfrac{1}{2}\delta(x - 100) + \tfrac{1}{2}\delta(x + 100)\right] \tag{15.6}$$

This gives

$$\kappa(-\log f_e, f) = (1 - 10^{-3}) + 10^4 \cdot 10^{-3} = 10.999 \tag{15.7}$$

The variance thus becomes 11 times larger, even though the change of probabilities in absolute terms was very small. □

Robust Norms

It is obvious that such a sensitivity to the true PDF f_e, is not acceptable in practical use. Adapting the norm is not the solution in most cases, since a finite number of data may not render an accurate enough estimate of the best norm. Instead we must look for norms that are robust with respect to unknown variations in the PDF. This is a well-developed topic in statistics; see, for example, the monograph by Huber (1981). A useful formalization is to seek the norm ℓ that minimizes the largest variance scaling that may result in a certain class of PDFs:

$$\ell_{\text{opt}} = \arg\min_{\ell} \max_{f \in \mathcal{F}} \kappa(\ell, f) \tag{15.8}$$

This norm gives, in Huber's terminology, the *minimax M-estimate*. The problem (15.8) with $\kappa(\ell, f)$ given by (15.2) is a variational problem whose solution depends only on the family of f functions. It is thus decoupled from its statistical context, and the discussion of (15.8) in, for example, Chapter 4 in Huber (1981)applies equally well to our system identification framework.

Typical families $\mathcal{F}$ are environments of the normal distribution, much in the spirit of our example. Solutions to such problems have the characteristic feature that $\ell'(x)$ behaves like x for small x, then saturates, and may even tend to zero as x increases ("redescending" ℓ'). Some typical curves are shown in Figure 15.1.

Let us return to our example to check how such ℓ may handle outliers.

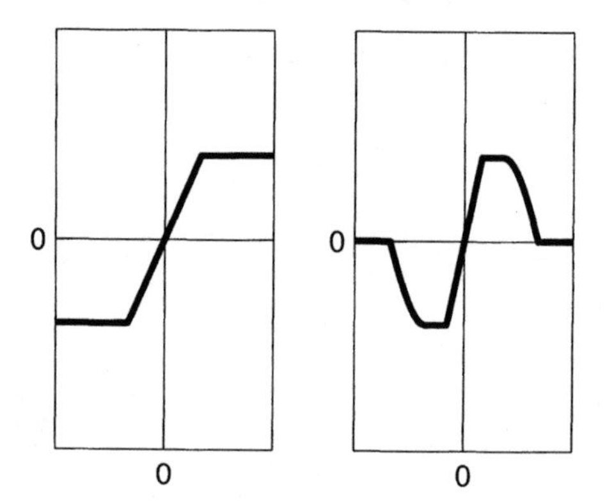

Figure 15.1 Some typical robust choises of $\ell'(x)$

Example 15.1 (continued)

Let $\ell_*(x)$ be such that

$$\ell_*'(x) = \begin{cases} x, & |x| < 4 \\ 4, & x \geq 4 \\ -4, & x \leq -4 \end{cases}$$

Then with f as in (15.6)

$$\kappa(\ell_*, f) = \frac{0.999 \int_{|x|\leq 4} x^2 \varphi(x)dx + 0.999 \int_{|x|>4} 16 \cdot \varphi(x)dx + 0.001 \cdot 16}{\left[\int_{|x|\leq 4} \varphi(x)dx\right]^2}$$

$$\approx 1.015$$

The decrease in variance compared to (15.7) is drastic. Let us also check what we loose in optimality if the true PDF indeed was normal:

$$\kappa(\ell_*, f_e) = \frac{\int_{|x|\leq 4} x^2 \varphi(x)dx + \int_{|x|>4} 16 \cdot \varphi(x)dx}{\left[\int_{|x|\leq 4} \varphi(x)dx\right]^2} \approx 1.0001$$

This price of increased variance for the nominal case is thus worth paying to obtain resilience against small variations in the PDF. □

As a recommended choice of robust norm, we may suggest the following, see Ruppert (1985)and Hempel et.al. (1986), p 105.

$$\ell'_R(x) = \begin{cases} x & |x| < \rho \cdot \hat{\sigma} \\ \rho \cdot \hat{\sigma} & x > \rho \cdot \hat{\sigma} \\ -\rho \cdot \hat{\sigma} & x < -\rho \cdot \hat{\sigma} \end{cases} \tag{15.9}$$

Here $\hat{\sigma}$ is the estimated standard deviation of the prediction errors, while ρ is a scalar in the range $1 \leq \rho \leq 1.8$. The estimate $\hat{\sigma}$ in turn should be robust so that it is not disturbed by outliers. A recommended estimate is to take

$$\hat{\sigma} = \frac{\text{MAD}}{0.7} \tag{15.10}$$

Here MAD $=$ the median of $\{|\varepsilon(t) - \tilde{\varepsilon}|\}$ with $\tilde{\varepsilon}$ as the median of $\{\varepsilon(t)\}$

Influence Function

A basic idea behind robust norms is to limit the influence of single observations on the resulting estimate. It is reasonable to ask how the estimate would change had a certain observation been lacking. For the least-squares estimate, an exact answer can be given. It follows from (11.9) that

$$\begin{aligned} \hat{\theta}_N - \hat{\theta}_{N,t} &= \overline{R}^{-1}(N)\varphi(t)[y(t) - \hat{\theta}^T_{N,t}\varphi(t)] \\ &= \overline{R}_t^{-1}(N)\varphi(t)[y(t) - \hat{\theta}^T_N\varphi(t)] \end{aligned} \tag{15.11}$$

where $\hat{\theta}_N$ is the actual estimate and the $\hat{\theta}_{N,t}$ is the estimate with the measurement $(y(t), \varphi(t))$ removed. Also,

$$\overline{R}(N) = \sum_{k=1}^{N} \varphi(t)\varphi^T(k), \qquad \overline{R}_t(N) = \overline{R}(N) - \varphi(t)\varphi^T(t)$$

The influence of measurement $(y(t), \varphi(t))$ can thus be evaluated by

$$\overline{R}_t^{-1}(N)\varphi(t)\varepsilon(t, \hat{\theta}_N)$$

This is, somewhat simplified, the idea behind Hampel's (1974)*influence function.* Generalized to arbitrary norms and model structures, the influence of measurement t can approximately be evaluated by

$$S(t) = \overline{R}_t^{-1}(N)\psi(t, \hat{\theta}_N)\ell'(\varepsilon(t, \hat{\theta}_N))$$

$$\overline{R}_t(N) = \sum_{\substack{k=1 \\ k \neq t}}^{N} \psi(k, \hat{\theta}_N)\ell''(\varepsilon(k, \hat{\theta}_N))\psi^T(k, \hat{\theta}_N) \tag{15.12}$$

Compare (11.52). The objective with robust norms ℓ as in Figure 15.1 could then be expressed as minimizing

$$\max_t |S(t)|$$

More pragmatically, it makes good sense to critically evaluate $S(t)$ in (15.12) after the parameter fit has been completed so as to reveal which observations have considerably influenced the estimate. Such observations had better be reliable or their influence should be reduced.

Detecting Outliers

Outliers that are as drastic as in Example 15.1 and data points that have a large influence on the estimate will often be detected by eye inspection of the data record. It is good practice, even when robust norms are used, to display the data before they are used for identification. Outliers are most easily detected in plots of the residuals $\varepsilon(t, \hat{\theta}_N)$. See Example 14.1 and Section 16.5.

Multivariable Case (*)

The covariance matrix for multivariable systems is given by (9.47). Choices of the function $\ell(\varepsilon)$ from $\mathbf{R}^p$ to $\mathbf{R}$ are quite analogous to the scalar case discussed previously. The multivariable case introduces a new issue, though: How should the different components of the vector ε be weighted together? We shall illustrate this question by considering the family (7.28) and (7.29) of quadratic criteria.

The covariance matrix associated with the quadratic norm Λ^{-1} in (7.27) is, according to (9.47) (see also Problem 9E.4),

$$P_\theta(\Lambda) = \left[\overline{E}\psi(t, \theta_0)\Lambda^{-1}\psi^T(t, \theta_0)\right]^{-1}\left[\overline{E}\psi(t, \theta_0)\Lambda^{-1}\Lambda_0\Lambda^{-1}\psi^T(t, \theta_0)\right]$$
$$\times \left[\overline{E}\psi(t, \theta_0)\Lambda^{-1}\psi^T(t, \theta_0)\right]^{-1}$$

Here $\Lambda_0 = Ee_0(t)e_0^T(t)$ is the true innovations covariance matrix. It is straightforward to establish (cf. Problem 9E.4) that

$$P_\theta(\Lambda) \geq P_\theta(\Lambda_0) \qquad \forall\Lambda$$

so that the best norm requires knowledge of the true covariance. Since anyway the minimization of (7.29) is carried out iteratively, the following variant to implement $\Lambda = \Lambda_0$ suggests itself:

$$\hat{\Lambda}_N^{(i)} = \frac{1}{N} \sum_{t=1}^{N} \varepsilon(t, \hat{\theta}_N^{(i)}) \varepsilon^T(t, \hat{\theta}_N^{(i)})$$
$$\hat{\theta}_N^{(i+1)} = \arg\min_{\theta} \frac{1}{N} \sum_{t=1}^{N} \varepsilon^T(t, \theta) \left[\hat{\Lambda}_N^{(i)}\right]^{-1} \varepsilon(t, \theta) \tag{15.13}$$

where superscript (i) denotes the ith iterate.

It is interesting to note, though, that the criterion

$$\hat{\theta}_N = \arg\min_{\theta} \det \left[\frac{1}{N} \sum_{t=1}^{N} \varepsilon(t, \theta) \varepsilon^T(t, \theta) \right] \tag{15.14}$$

gives the same asymptotic covariance matrix for $\hat{\theta}_N$ as the quadratic norm (7.27), with Λ being the true innovations covariance. See Problem 9E.5 and also (7.92) to (7.96).

15.3 VARIANCE-OPTIMAL INSTRUMENTS

Consider now the IV method (7.129). Under assumptions (9.80) and (9.81) the asymptotic covariance matrix P_θ of the estimates is given by (9.83). Clearly, the choice of instruments $\zeta(t, \theta_0)$ and the choice of prefilter $L(q)$ may have a considerable effect on P_θ. Now, what might the optimal choices be?

A Lower Bound

Suppose that the true system is given by

$$y(t) = G_0(q)u(t) + H_0(q)e_0(t) \tag{15.15}$$

and that the transfer function G_0 is to be estimated, while H_0 is assumed known. Let

$$G_0(q) = \frac{B_0(q)}{A_0(q)} \tag{15.16}$$

and let the model be parametrized as

$$G(q, \theta) = \frac{B(q)}{A(q)} \tag{15.17}$$

with appropriate model orders. The limit of the Cramér-Rao bound for this estimation problem is given by (9.31) (normality assumed here):

$$P_{\text{CR}} = \lambda_0[\overline{E}\psi(t,\theta_0)\psi^T(t,\theta_0)]^{-1} \tag{15.18}$$

$$\begin{aligned} \psi^T(t,\theta_0) &= \left[\frac{d}{d\theta}\hat{y}(t|\theta)\right]^T\Bigg|_{\theta=\theta_0} \\ &= \frac{1}{H_0(q)A_0(q)}[-G_0(q)u(t-1)\dots \\ &\quad -G_0(q)u(t-n_a) \quad u(t-1)\dots u(t-n_b)] \end{aligned} \tag{15.19}$$

[cf. (10.55); here we have used A for what is called F in the model (4.33)].

Optimal Instruments

Equation (15.18) gives a lower bound for any unbiased method aiming at estimating θ in (15.17). It thus applies also to the IV method, so (15.18) is a lower bound for (9.83). However, this lower bound is achieved for

$$\begin{aligned} L^{\text{opt}}(q) &= \frac{1}{H_0(q)A_0(q)} \\ \zeta^{\text{opt}}(t) &= \psi(t,\theta_0) \end{aligned} \tag{15.20}$$

To see this, we note that

$$\varphi_F(t) = L^{\text{opt}}(q)\varphi(t) = \psi(t,\theta_0) + \tilde{\varphi}_e(t)$$

where $\tilde{\varphi}_e(t)$ depends on $\{e_0(t)\}$ only, while $\psi(t,\theta_0)$ depends on $\{u(t)\}$ only. Hence

$$\overline{E}\psi(t,\theta_0)\varphi_F^T(t) = \overline{E}\psi(t,\theta_0)\psi^T(t,\theta_0)$$

and (9.83) simplifies to (15.18), when the system operates in open loop. The optimal design variables for the IV method are thus given by (15.20).

Adaptive IV Methods

While (15.20) gives direct advice about the best IV method, the annoying aspect is that the optimal prefilter and instruments depend on unknown properties of the true system. This could be handled by letting the instruments $\zeta(t,\theta)$ depend on θ in a proper way and by simultaneously estimating the noise properties. This leads to algorithms that are closely related to and of about the same complexity as the corresponding prediction-error method. An alternative is to approximately realize the optimal choices in a multistep algorithm.

Multistep Algorithm

The choices of instruments and prefilters in the IV method primarily affect the asymptotic variance, while the consistency properties are generically secured. This suggests that minor deviations from the optimal values according to (15.20) will only cause second-order effects in the resulting accuracy. It could thus be sufficient to use consistent, but not necessarily efficient, estimates of the dynamics and of the noise when forming the instruments and prefilter. To retain the appealing simplicity of the IV method, we should work with linear regression structures for these steps. We thus suggest the following four-step IV estimator for a system that operates in open loop.

Step 1: Write the model structure (15.17) as a linear regression

$$\hat{y}(t|\theta) = \varphi^T(t)\theta \tag{15.21}$$

Estimate θ by the LS method (7.34). Denote the estimate by $\hat{\theta}_N^{(1)}$ and the corresponding transfer function by $\hat{G}_N^{(1)}(q)$

Step 2: Generate the instruments as in (7.122) and (7.123):

$$x^{(1)}(t) = \hat{G}_N^{(1)}(q)u(t) \tag{15.22}$$

$$\zeta^{(1)}(t) = \left[-x^{(1)}(t-1)\ldots - x^{(1)}(t-n_a) \quad u(t-1)\ldots u(t-n_b)\right]^T \tag{15.23}$$

and determine the IV estimate (7.118) of θ in (15.21) using these instruments. Denote the estimate $\hat{\theta}_N^{(2)}$ and the corresponding transfer-function estimate

$$\hat{G}_N^{(2)}(q) = \frac{\hat{B}_N^{(2)}(q)}{\hat{A}_N^{(2)}(q)}$$

Step 3: Let

$$\hat{w}_N^{(2)}(t) = \hat{A}_N^{(2)}(q)y(t) - \hat{B}_N^{(2)}(q)u(t)$$

and postulate an AR model of order $n_a + n_b$ (order chosen to balance the computational efforts in each step) for $\hat{w}_N^{(2)}(t)$:

$$L(q)\hat{w}_N^{(2)}(t) = e(t)$$

Estimate $L(q)$ using the LS method and denote the result by $\hat{L}_N(q)$.

Step 4: Let $x^{(2)}(t)$ be defined analogously to (15.22), and let

$$\zeta^{(2)}(t) = \hat{L}_N(q)[-x^{(2)}(t-1)\ldots -x^{(2)}(t-n_a) \quad u(t-1)\ldots u(t-n_b)]^T \tag{15.24}$$

Using these instruments and a prefilter $\hat{L}_N(q)$ in (7.129) with $\alpha(x) = x$, determine the IV estimate of θ in (15.21), giving the final estimate

$$\hat{\theta}_N = \left[\sum_{t=1}^{N} \zeta^{(2)}(t)\varphi_F^T(t)\right]^{-1} \sum_{t=1}^{N} \zeta^{(2)}(t)y_F(t) \tag{15.25}$$

$$\varphi_F(t) = \hat{L}_N(q)\varphi(t), \qquad y_F(t) = \hat{L}_N(q)y(t)$$

This algorithm is a special case of a multistep procedure discussed in Section 6.4 of Söderström and Stoica (1983). They show that the asymptotic covariance matrix of $\hat{\theta}_N$ indeed is the Cramér-Rao bound (15.18), provided (in our case) the true H_0 is an autoregression of order n_b.

Example 15.2 Four-Step IV Algorithm

The system

$$y(t) = \frac{1.0q^{-2} + 0.5q^{-3}}{1 - 1.5q^{-1} + 0.7q^{-2}} u(t) + e(t)$$

was simulated over 400 samples with $\{e(t)\}$ white Gaussian noise of variance 1 and $\{u(t)\}$ as a white binary ± 1 signal. A second-order ARX-model structure with two delays was used. The four-step IV method gave the following transfer-function estimates:

$$\hat{G}_{400}^{(1)}(q) = \frac{1.0597q^{-2} + 1.1546q^{-3}}{1 - 1.0255q^{-1} + 0.2965q^{-2}}$$

$$\hat{G}_{400}^{(2)}(q) = \frac{0.9778q^{-2} + 0.2750q^{-3}}{1 - 1.6072q^{-1} + 0.7803q^{-2}}$$

$$\hat{G}_{400}(q) = \frac{0.9688q^{-2} + 0.5216q^{-3}}{1 - 1.5038q^{-1} + 0.7023q^{-2}}$$

□

The example indicates that the extra work of steps 3 and 4 is worthwhile. An additional advantage is that these steps provide a noise characteristics estimate, necessary for the computation of the asymptotic covariance (9.83). In fact, as remarked previously, the particular choices of ζ and L give the following estimated covariance matrix of $\hat{\theta}_N$:

$$\frac{1}{N} P_\theta = \hat{\lambda}_N \left[\sum_{t=1}^{N} \zeta^{(2)}(t) \left[\zeta^{(2)}(t)\right]^T \right]^{-1} \tag{15.26a}$$

$$\hat{\lambda}_N = \frac{1}{N} \sum_{t=1}^{N} \left[y_F(t) - \varphi_F^T(t)\hat{\theta}_N \right]^2 \tag{15.26b}$$

15.4 SUMMARY

It is an unavoidable consequence of the structure of this book that useful results and advice on various identification methods are scattered over several chapters. Therefore, we give here a concrete user-oriented summary of suggested parametric identification procedures.

We consider prediction-error methods (PEM) to be the basic approach to system identification. They have three important advantages:

1. Applicability to general model structures
2. Optimal asymptotic accuracy when the true system can be represented within the model structure
3. Reasonable approximation properties when the true system cannot be represented within the model structure

For a given model structure $\hat{y}(t|\theta)$, the PEM can be summarized as follows: Select a prefilter $L(q)$, guided by the discussion of Section 14.4. Form the criterion

$$
\begin{aligned}
V_N(\theta, Z^N) &= \frac{1}{N}\sum_{t=1}^{N} \ell(\varepsilon_F(t,\theta)) \\
\varepsilon_F(t,\theta) &= L(q)[y(t) - \hat{y}(t|\theta)] \\
&\ell(\cdot) \text{ given by (15.9)}
\end{aligned}
$$

For a linear black-box model, form an initial estimate $\hat{\theta}_N^{(1)}$ by the procedure (10.79). Then minimize V_N iteratively using the damped Gauss-Newton method (10.40), (10.41), and (10.46) [(10.47) when necessary]. The asymptotic properties of the resulting estimate $\hat{\theta}_N$ are then given by (8.103) and (9.90).

However, it is still true that other methods may be preferable in certain cases. Especially for a linear system with several outputs, that requires a model structure with many parameters, the *subspace methods* form a valuable alternative. They have the advantage of allowing an estimate, using efficient and numerically robust calculations without iterative search.

The main advantage of the *IV method* is its simplicity. It is often worth-while to use the four-step procedure (15.21)–(15.26), as well as the subspace method, for a first quick estimate of the system transfer function. This may then be refined by PEM, if necessary.

We may note that it is not necessary to be able to tell which of the approaches is "best." Experience says that each may have its advantages. It is good practice to have them all in one's toolbox, compute models with the different methods, and subject them to validation according to the next chapter.

15.5 BIBLIOGRAPHY

Robust norms have been extensively discussed in the statistical literature. Huber (1981)gives a comprehensive account of robust estimates. Hempel et.al. (1986)discusses the use of the influence function. Krasker and Welsch (1982)describe how to "robustify" linear regressions also with respect to the magnitude of the regressors. Polyak and Tsypkin (1980)have advocated the use of robust norms for on-line applications. Tsypkin (1984)contains a comprehensive treatment of this application.

Optimal instruments and their approximate implementation have been extensively studied by Söderström and Stoica (1981, 1983)and Stoica and Söderström (1983). Several mixed IV-PEM schemes have been studied also by Young and Jakeman (1979).

15.6 PROBLEMS

15G.1 Consider an identification criterion

$$V_N(\theta, Z^N) = \frac{1}{N}\sum_{t=1}^{N}\alpha_t \ell(\varepsilon(t,\theta), t)$$

where the functions $\ell(\cdot, t)$ are given and the positive scalars α_t, are to be selected. Suppose that $S \in \mathcal{M}$ and that the true innovations $\{e_0(t)\}$ are white, zero mean, but not necessarily stationary. Show that the choice of $\{\alpha_t\}$ that minimizes the variance of the parameter estimate is

$$\alpha_t^{\text{opt}} = \left[\frac{E[\ell'_\varepsilon(e_0(t), t)]^2}{[E\ell''_{\varepsilon\varepsilon}(e_0(t), t)]^2}\right]^{-1} \qquad \text{(times arbitrary scaling)}$$

Conclude that, if $\ell(x, t) = x^2$ all t, then the optimal weights α_t are the inverse variances of the innovations $e_0(t)$, regardless of their distribution. [Compare (II.65).] Conclude also that, if $e_0(t)$ is Gaussian with variance λ_t and $\ell(x) = |x|$, then $\alpha_t^{\text{opt}} = 1/\sqrt{\lambda_t}$.

15E.1 Consider the output error model structure

$$\hat{y}(t|\theta) = G(q, \theta)u(t)$$

Suppose that the true system is given by

$$y(t) = G_0(q)u(t) + H_0(q)e_0(t)$$

where $e_0(t)$ is white noise. Suppose also that $G_0(q) = G(q, \theta_0)$. Apply a prediction-error method with prefilter $L(q)$ to estimate θ and compute the variance of $\hat{\theta}_N$. Show that the variance is minimized for $L(q) = H_0^{-1}(q)$.

15T.1 Prove that (15.18) gives a lower bound for (9.83) by direct algebraic methods (reference: Söderström and Stoica, 1983, p. 97).

15T.2 In case a time-varying norm is allowed, (15.1) takes the form

$$\kappa(\ell(\varepsilon, \cdot)) = \frac{\overline{E}[\ell'_\varepsilon(e_0(t), t)]^2}{[\overline{E}\ell''_{\varepsilon\varepsilon}(e_0(t), t)]^2} \tag{15.27}$$

(see Problem 9T.1). Recall that

$$\overline{E}g(e(t), t) = \lim_{N\to\infty}\frac{1}{N}\sum_{t=1}^{N}Eg(e(t), t)$$

Establish that (15.4b) minimizes (15.27) by applying Schwarz's inequality to the expression corresponding to (15.27) with finite N values replacing the limit $\overline{E}$.

15T.3 Suppose that the innovations have a time-varying distribution, but that the norm $\ell(\varepsilon)$ is constrained to be time invariant. Show that (15.1) is then minimized by

$$\ell(\varepsilon) = -\log\left(\lim_{N\to\infty}\frac{1}{N}\sum_{t=1}^{N}f_e(\varepsilon, t)\right)$$

16

MODEL STRUCTURE SELECTION AND MODEL VALIDATION

The choice of an appropriate model structure $\mathcal{M}$ is most crucial for a successful identification application. This choice must be based both on an understanding of the identification procedure and on insights and knowledge about the system to be identified. In Chapters 4 and 5 we provided lists of typical model structures to be used for identification. In this chapter we shall complement these lists by discussing how to arrive at a suitable structure, guided by system knowledge and the collected data set.

Once a model structure has been chosen, the identification procedure provides us with a particular model in this structure. This model may be the best available one, but the crucial question is whether it is good enough for the intended purpose. Testing if a given model is appropriate is known as *model validation*. Such techniques, which are closely related to the choice of model structure, will also be described in this chapter.

16.1 GENERAL ASPECTS OF THE CHOICE OF MODEL STRUCTURE

The route to a particular model structure involves, at least, three steps:

1. To choose the *type* of model set. (16.1)
   This involves, for example, the selection between nonlinear and linear models, between input-output, black-box and physically parametrized state-space models, and so on.
2. To choose the *size* of the models set. (16.2a)
   This involves issues like selecting the order of a state-space model, the degrees of the polynomials in a model like (4.33) or the number of "neurons" in a neural network. It also contains the problem of which variables to include in the model description. We thus have to select $\mathcal{M}$ from a given, increasing chain of structures

$$\mathcal{M}_1 \subset \mathcal{M}_2 \subset \mathcal{M}_3 \dots \tag{16.2b}$$

[Recall the definition of $\mathcal{M}_1 \subset \mathcal{M}_2$ in (4.127).] This problem (16.2) will be called the *order selection problem.*

3. To choose the model parametrization. (16.3)
When a model set $\mathcal{M}^*$ has been decided on (like a state-space model of a given order), it remains to parametrize it, that is, to find a suitable model structure $\mathcal{M}$ whose range equals $\mathcal{M}^*$ (see Section 4.5).

In this section we shall discuss basic guidelines for these three steps. We said in Chapter 12 that the goal of the user is to "obtain a good model at a low price." The choice of model structure certainly has a considerable effect on both the quality of the resulting model and the price for it.

Quality of the Model

The *quality* of the resulting model can, for example, be measured by a mean-square error criterion $J(\mathcal{D})$ as in (12.26), where the design variables $\mathcal{D}$ include the model structure $\mathcal{M}$. (For nonlinear systems and models we could, at least conceptually, give analogous formalizations.) In Chapter 12 we found it convenient to split up the mean-square error into a *bias* contribution and a *variance* contribution:

$$J(\mathcal{D}) = J_B(\mathcal{D}) + J_P(\mathcal{D}) \tag{16.4}$$

[see (12.26)]. We would thus select $\mathcal{M}$ so that both bias and variance are kept small. These are usually, however, conflicting requirements. To reduce bias one basically has to employ larger and more flexible model structures, requiring more parameters. Since the variance typically increases with the number of estimated parameters [see (9.92)], the best model structure is thus a trade-off between:

- *Flexibility:* Employing model structures that offer good capabilities of describing different possible systems. Flexibility can be obtained either by using many parameters or by placing them in "strategic positions." (16.5)
- *Parsimony:* Not to use unnecessarily many parameters: to be "parsimonious" with the model parametrization. (16.6)

This trade-off can be formalized objectively as a minimization of (16.4) with respect to the model structures.

Price of the Model

The price of the model is associated with the effort to calculate it, that is, to perform the minimization in (7.155) or to solve the equation (7.156). This work is highly dependent on the model structure, which influences:

- *The algorithm complexity:* We saw in Chapter 10 that solving for $\hat{\theta}_N$ involves evaluation of the prediction errors $\varepsilon(t, \theta)$ and their gradients $\psi(t, \theta)$ for a number of θ. The work associated with these evaluations depends critically on $\mathcal{M}$. (16.7)
- *The properties of the criterion function:* The amount of work to solve for $\hat{\theta}_N$ also depends on how many evaluations of the criterion function and its gradient are necessary. This is determined by the "shape" of the criterion function

(7.155) or (7.156): nonunique minimum, possible undesired local solutions, and so on. The "shape" in turn is a result of the choice of $\ell(\cdot)$ and of how the $\varepsilon(t, \theta)$ depend on θ (i.e., the model structure). (16.8)

There is also a price associated with the use of the model. A high-order complex model is more difficult to use for simulation and control design. If it is only marginally better [in the sense of (16.4)] than a simpler model, it may not be worth the higher price. Consequently, also

- *The intended use of the model* (16.9)

will affect the choice of model structure.

General Considerations

The final choice of model structure will be a compromise between the listed aspects (16.5) to (16.9). The techniques and considerations that are used when evaluating these aspects can be split into different categories:

- *A priori considerations:* Certain aspects are independent of the data set Z^N and can be evaluated a priori, before the data have been measured. We shall discuss these in Section 16.2.
- *Techniques based on preliminary data analysis:* With the data available, certain testing and evaluation of Z^N can be carried out that give insights into possible and suitable model structures. These techniques do not necessarily require the computation of a complete model. Section 16.3 contains a discussion of such preliminary data analysis.
- *Comparing different model structures:* Before a final model structure is chosen, it is advisable to shop around in different model structures and compare quality and prices of the models offered there. This will require the computation and comparison of several models; such procedures are described in Section 16.4.
- *Validation of a given model:* Regardless of how a given model is obtained, we can always use Z^N to evaluate whether it seems likely that it will serve its purpose. If a certain model is accepted, we have also implicitly approved the choice of the underlying model structure. Such model-validation techniques are described in Sections 16.5 and 16.6.

16.2 A PRIORI CONSIDERATIONS

Type of Model

The choice of which type of model to use is quite subjective and involves several issues that are independent of the data set Z^N. It is usually the result of a compromise between the aspects listed previously, combined with more irrational factors like the availability of computer programs and familiarity with certain models. Let us briefly comment on the rational issues involved in the choice.

The compromise between parsimony and flexibility is at the heart of the identification problem. How shall we obtain a good fit to data with few parameters? The answer usually is to use a priori knowledge about the system, intuition, and ingenuity. These facts stress that identification can hardly be brought into a fully automated procedure. The problem of minimizing (16.4) thus favors *physically parametrized models*. It will depend on our insight and understanding of the process whether it is feasible to build a well-founded physically parametrized model structure. This is of course an application-dependent problem.

For a physical system, a priori information can typically best be incorporated into a continuous-time model such as (4.62). This means that the computation of $\varepsilon(t, \theta)$ and the minimization of (7.155) become a laborious task both regarding the programming effort and the computation time required. Aspects of algorithmic complexity as well as the shape of the criterion function therefore favor *black-box models*. By this we mean a model like (4.33) that adapts its parameters to data, without imposing any physical interpretation of their values.

A general advice is to "try simple things first." One should go into sophisticated model structures only if simpler ones do not pass the model-validation tests. Especially linear regression models like (4.12) lead to simple and robust minimization schemes (the least-squares method; see Section 7.3). They are therefore often a good first choice for an identification problem.

One should note that using physical a priori knowledge does not necessarily mean that fancy continuous-time model structures have to be constructed. Some thinking about the nature of the relationships between the measured signals can give good hints for model structures. This was illustrated in Example 5.1, where a "semiphysical" model structure of linear regression type was obtained from fairly simple a priori considerations. In general, one should contemplate whether *nonlinear transformations of data* [such as (5.20) or a logarithmic transformation] will make it easier for the transformed data to fit a linear model. Note in particular that nonlinear effects in actuators and sensors may be known and may be used for helpful redefinitions of the input-output signals. See Kashyap and Rao (1976), Box and Cox (1964), Daniel and Wood (1980), or Carroll and Ruppert (1988)for further discussion on data transformations.

Model Order

Solving problem (16.2) usually requires help from the data. However, physical insight and the intended model application will often tell which range of model orders should be considered. Also, even when the data have not been evaluated, knowing N and the data quality will indicate how many parameters it is feasible to estimate. With few data points, it is not reasonable to try to determine a model in a complex model structure.

A related problem is how many different time scales it is feasible to let one and the same model handle. Problem 13G.1 indicated that, for numerical reasons, it may be difficult to adequately describe more than two to three decades of the frequency range within one model. Considerations on sampling rates, proper excitation, and data record lengths strongly suggest that one should not aim at covering more than three decades of time constants in one experiment. If the system is stiff so that it contains widely separated time constants of interest, the conclusion thus is to build two (or more) models, each covering a proper part of the frequency range and each sampled with a corresponding, suitable sampling interval. For a high-frequency model, the low-frequency dynamics for all practical purposes look like integrators (the number being equal to the pole excess at low frequencies); think of a Bode diagram representation. These should thus be postulated for the high-frequency model. Correspondingly, the high-frequency dynamics look like static (instantaneous) relationships to the low-frequency model. Thus introduce a no-delay term $b_0 u(t)$ in this model.

Model Parametrization

The issue of model parametrization is basically numerical. We seek model parametrizations that are well conditioned so that a round-off or other numerical error in one parameter has a small influence on the input-output behavior of the model. This is a problem that has been widely recognized in the digital filtering area, but less so in the identification literature. In fact, the standard input-output model structures like (4.7) to (4.33) could be quite sensitive to numerical errors. See Problem 16E.1 for an example. (Compare also with Problem 13G.1.) The choice of parametrization of a linear model essentially amounts to picking a certain state-space representation. The difference equation models correspond to the observability canonical form of Example 4.2. Other choices of state variables, such as wave-digital filters or ladder/lattice filters (cf. Section 10.1) give better conditioned parametrizations. See Mullis and Roberts (1976)or Oppenheim and Schafer (1975)for a discussion of this issue. Middleton and Goodwin (1990)has advocated parametrizations in terms of

$$\delta = 1 - q^{-1}$$

rather than q^{-1} to cope with this problem. See also Gevers and Li (1993)for a thorough discussion of parameterization issues.

16.3 MODEL STRUCTURE SELECTION BASED ON PRELIMINARY DATA ANALYSIS

By preliminary data analysis, we mean calculations that do not involve the determination of a complete model of the system. Such analysis could prove helpful for finding suitable model structures.

Estimating the Type of Model

Generally, data-aided model structure selection appears to be an underdeveloped field. An exception is the order determination in linear structures, to be discussed shortly. It is conceivable that various nonparametric techniques could be helpful to find out suitable nonlinear transformations of data, as well as to indicate what type of dependences between measured variables should be considered. In the statistical literature, such procedures are discussed (e.g., in Daniel and Wood, 1980, and Parzen, 1985), but they have not really found their way into system identification applications yet.

A particular problem forms an exception; to test for nonlinear effects. That is, is it likely that the data can be explained by a linear relationship or will a nonlinear model structure be required? Such tests are based on the relationships between higher (than second) order correlations and spectra that follow from linear descriptions. See Billings and Voon (1983), Rajbman (1981), Haber (1985), Tong (1990), and Varlaki, Terdik, and Lototsky (1985)for further details.

Order Estimation

The order of a linear system can be estimated in many different ways. Methods that are based on preliminary data analysis fall into the following categories.

1. Examining the spectral analysis estimate of the transfer function
2. Testing ranks in sample covariance matrices
3. Correlating variables
4. Examining the information matrix

We shall give a brief account of each of these approaches.

1. Spectral analysis estimate: A nonparametric estimate of the transfer function $\hat{G}_N(e^{i\omega})$ as in (6.82) will give valuable information about resonance peaks and the high-frequency roll-off and phase shift. All this gives a hint as to what model orders will be required to give an adequate description of the (interesting part of the) dynamics. Note, though, that discrete-time Bode plots show some artifacts in their interpretation in terms of poles and zeros, compared to continuous-time Bode plots. Thus, use the observations with some care.

2. Testing ranks in covariance matrices: Suppose that the true system is described by

$$\begin{aligned} y(t) + a_1 y(t-1) + \cdots + a_n y(t-n) \\ = b_1 u(t-1) + \cdots + b_n u(t-n) + v_0(t) \end{aligned} \tag{16.10}$$

for some noise sequence $\{v_0(t)\}$. Suppose also that n is the smallest number for which this holds ("n is the true order"). As usual, let

$$\varphi_s(t) = [-y(t-1)\ldots -y(t-s) \quad u(t-1)\ldots u(t-s)]^T \tag{16.11}$$

Suppose first that $v_0(t) \equiv 0$. Then (16.10) implies that the matrix

$$R^s(N) = \frac{1}{N}\sum_{t=1}^{N} \varphi_s(t)\varphi_s^T(t) \tag{16.12}$$

will be nonsingular for $s \leq n$ (provided $\{u(t)\}$ is persistently exciting) and singular for $s \geq n+1$. $\kappa(s) = \det R^s(N)$ could thus be used as a test quantity for the model order. This was first suggested by Woodside (1971). The relationship between singularity of (16.12) and the corresponding model order, however, goes back to Lee (1964)and the realization algorithm by Ho and Kalman (1966). (See also Lemma 4A.1.)

In case a noise $\{v_0(t)\}$ is present in (16.10), (16.12) can still be used, with a suitable threshold, provided the signal-to-noise ratio is high. If this is not the case, Woodside (1971)suggested the use of the "enhanced" matrix

$$\hat{R}^s(N) = R^s(N) - \hat{\sigma}^2 R_v \tag{16.13}$$

where $\hat{\sigma}^2 R_v$ is the estimated influence of $v_0(t)$ on $R_s(N)$.

A better alternative, when the influence of $v_0(t)$ is not negligible, is to use other correlation vectors. If $\{v_0(t)\}$ and $\{u(t)\}$ are uncorrelated, we could use

$$\zeta_s(t) = [u(t-1) \quad u(t-2) \ldots u(t-2s)]^T \tag{16.14}$$

and find that

$$\overline{R}_\zeta^S(N) = \overline{E}\varphi_s(t)\zeta_s^T(t) \tag{16.15}$$

is nonsingular for $s \leq n$ and singular for $s \geq n+1$ [cf. the discussion on consistency of the IV method, (8.98)]. Replacing $\overline{E}$ by sample mean then gives a usable test quantity. If $\{v_0(t)\}$ is known to be a moving average of order r, so that $y(t-r-1)$ and $v_0(t)$ are uncorrelated, we could also use

$$\zeta_s(t) = \varphi_s(t-r) \tag{16.16}$$

or any combination of such correlators with (16.14). This order-determination test has been discussed by Wellstead (1978)and Wellstead and Rojas (1982), and was apparently first described for multivariable structures in Tse and Weinert (1975).

3. Correlating variables: The order-determination problem (16.2b) is whether to include one more variable in a model structure or not. This variable could be $y(t-n-1)$ in (16.10) (a true order-determination problem) or a measured possible disturbance variable $w(t)$. In any case, the question is whether this new variable has anything to contribute when explaining the output variable $y(t)$. This is measured by the correlation between $y(t)$ and $w(t)$. However, to discount the possible relationship between $y(t)$ and $w(t)$, already accounted for by the smaller model structure, the correlation should be measured between $w(t)$ and what remains to be explained [i.e., the residuals $\varepsilon(t, \hat{\theta}_N) = y(t) - \hat{y}(t|\hat{\theta}_N)$]. This is known as *canonical correlation* or *partial correlation* in regression analysis (see Draper and Smith, 1981). See also the discussion in Section 16.6.

We may also note that the determination of the state-space model order, i.e., determining how many of the singular values in (10.127) are significant (or step 2 of (7.66), is a test of the same kind.

4. The information matrix: It follows from Theorem 4.1 that, if the model orders are overestimated in certain model structures, global and local identifiability will be lost. This means that $\psi(t, \theta)$ will not have full rank at $\theta = \theta^*$ (the limit value), and hence the information matrix (7.89) will be singular. Since the Gauss-Newton search algorithm uses the inverse of the information matrix, a natural test quantity for whether the model order is too high will be the conditioning number of this matrix. See Young, Jakeman, and McMurtrie (1980), Mehra (1974), Söderström (1975a), and Stoica and Söderström (1982c).

A related situation occurs when the IV method is used. Then the matrix

$$R_{\zeta}(N) = \frac{1}{N}\sum_{t=1}^{N} \zeta(t)\varphi^T(t)$$

in (7.118) will be singular when the orders are overestimated, as we found in (8.98) and under paragraph 2. Testing the conditioning of this matrix is thus naturally incorporated in the IV approach.

Multivariable Case: Model Parametrization (*)

The black-box multivariable parametrization problem is to select the multi-index $\overline{\nu}_n$ in (4A.16) to (4A.18). Some different methods for this have been discussed in the literature. The observability indexes $\{\sigma_i\}$, defined in the proof of Lemma 4A.2, form one possible choice for $\overline{\nu}_n$. The indexes $\{\sigma_i\}$ are defined by the rank structure of $\mathcal{H}_{np+p}$, which in turn, in the noise-free case, is related to the rank structure of $R^s(N)$ in (16.12). Guidorzi (1975)has suggested the use of $R^s(N)$ to determine the observability indexes and, in the noise-corrupted one, the analog of the "enhanced" matrix (16.13). Tse and Weinert (1975)use instead an estimate of the matrix (16.15) and (16.16) (in the input-free case) for the same purpose.

An alternate route has been considered by van Overbeek and Ljung (1982). They use the overlapping model structure (4A.33) and switch, during the criterion minimization, from one parametrization to another when the information matrix is ill-conditioned. They also link the conditioning of this matrix to the conditioning of the state covariance matrix.

Other non-canonical parameterizations, like a tridiagonal form and a full parameterization, have been discussed in McKelvey and Helmersson (1996)and McKelvey (1994). The use of balanced realization parameterizations is described in Ober (1987)and Hanzon and Ober (1997).

16.4 COMPARING MODEL STRUCTURES

A most natural approach to search for a suitable model structure is simply to test a number of different ones and to compare the resulting models. In this section we shall discuss what to compare and how to evaluate the comparisons. The model to be evaluated will generically be denoted by $m = \mathcal{M}(\hat{\theta}_N)$. It is estimated within the model structure $\mathcal{M}$, which is supposed to have $d_{\mathcal{M}} = \dim \theta$ free parameters. By the *Estimation Data* we mean the data that were used to estimate m, while *Validation Data* will denote any data set available that has not been used to build any of the models we would like to evaluate.

What to Compare?

There are of course a number of ways to evaluate a model. We shall here describe evaluations and comparisons that are based on data sets from the system. Generally speaking, the tests should bring out the relevant features for the intended model application, so it is desirable that these data sets have been collected under conditions that are close to the intended operating conditions. The model tests are then basically tests of *how well the model is capable of reproducing these data.*

We shall generally work with *k-step ahead model predictions* $\hat{y}_k(t|\mathcal{m})$ as the basis of the comparisons. By that we mean that $\hat{y}_k(t|\mathcal{m})$ is computed from past data

$$u(t-1), \ldots, u(1),\; y(t-k), \ldots, y(1) \tag{16.17}$$

using the model $\mathcal{m}$. The case when k equals ∞ corresponds to the use of past inputs only, i.e., a pure *simulation*. We use the notation $\hat{y}_\infty(t|\mathcal{m}) = \hat{y}_s(t|\mathcal{m})$ for this case. Similarly we introduce $\hat{y}_1(t|\mathcal{m}) = \hat{y}_p(t|\mathcal{m})$ for the standard one-step ahead predictor. For a linear model $y = \hat{G}u + \hat{H}e$ we thus have, according to Chapter 3,

$$\hat{y}_s(t|\mathcal{m}) = \hat{G}(q)u(t) \tag{16.18a}$$

$$\hat{y}_p(t|\mathcal{m}) = \hat{H}^{-1}(q)\hat{G}(q)u(t) + \left(1 - \hat{H}^{-1}(q)\right)y(t) \tag{16.18b}$$

$$\hat{y}_k(t|\mathcal{m}) = \hat{W}_k(q)\hat{G}(q)u(t) + \left(1 - \hat{W}_k(q)\right)y(t) \tag{16.18c}$$

with $\hat{W}_k$ determined as in (3.29). For an output error model, $H(q) = 1$, there is clearly no difference between the expressions in (16.18). Otherwise, note the considerable conceptual difference between $\hat{y}_s$ and $\hat{y}_p$. The latter has $y(t-1)$ and earlier y-values available and can therefore give fits that "look good," even though the model may be bad.

Example 16.1 A Trivial Model

Consider the model

$$\mathcal{m}: \quad \hat{y}(t|\theta) = y(t-1)$$

It will predict the next output to be the previous one. For a data record that is sampled fast (like the one in Figure 14.1a), $\hat{y}_p(t|\mathcal{m})$ will be practically indistinguishable from $y(t)$. On the other hand, $\hat{y}_s(t|\mathcal{m}) \equiv 0$, so the model is useless for simulation. □

For the general model (5.66) the simulated output is defined recursively as

$$\hat{y}_s(t|\mathcal{m}) = g(t, Z_s^{t-1}, \hat{\theta}_N) \tag{16.19}$$

$$Z_s^{t-1} = \left\{\hat{y}_s(t-1|\mathcal{m}), u(t-1), \hat{y}_s(t-2|\mathcal{m}), u(t-2), \ldots, \hat{y}_s(1|\mathcal{m}), u(1)\right\}$$

For control applications, the predicted output over a time span that corresponds to the dominating time constant will be an adequate variable to look at. The simulated output may be more revealing, since it is a more demanding task to reproduce the output from input only. For an unstable model, we clearly have to use predictions.

Now, the models can either be evaluated by visual inspection of plots of $y(t)$ and $\hat{y}_k(t|m)$, or by the numerical value

$$J_k(m) = \frac{1}{N}\sum_{t=1}^{N}\left|y(t) - \hat{y}_k(t|m)\right|^2 \tag{16.20}$$

We will also use the notation $J_p = J_1$ and $J_s = J_\infty$. It is useful to give some normalized measure of this fit. Assume that y has been detrended to zero mean and define

$$R^2 = 1 - \frac{J_k(m)}{\frac{1}{N}\sum_{t=1}^{N}|y(t)|^2} \tag{16.21}$$

Then R is that part of the output variation that is explained by the model, and is often expressed in %. See also (II.38).

The quality measure $J_k(m)$ will depend on the actual data record for which the comparison is made. It is therefore natural also to consider the expected value of this measure, where expectation is taken with respect to the data, regarding the model as a fixed, deterministic quantity:

$$\overline{J}_k(m) = EJ_k(m) \tag{16.22}$$

This gives a quality measure for the given model. Now, $m = \mathcal{M}(\hat{\theta}_N)$ is itself a random variable, being estimated from noisy data. The expectation of the model fit with respect to $\hat{\theta}_N$ gives a *quality measure for the model structure* $\mathcal{M}$:

$$\overline{J}_k(\mathcal{M}) = E\overline{J}_k(\mathcal{M}(\hat{\theta}_N)) \tag{16.23}$$

It should be noted that for linear regression models, the measure $J_p(m)$ can be computed for many models simultaneously. The requirement is only that the models are obtained by deleting trailing regressors. This follows from (10.11), which shows that the norm of the k:th row of R_2 gives the increase $J_p(m_1) - J_p(m_2)$ when the k:th parameter is removed from the model structure.

Comparing Models on Fresh Data Sets: Cross-Validation

It is not so surprising that a model will be able to reproduce the estimation data. The real test is whether it will be capable of also describing fresh data sets from the process. A suggestive and attractive way of comparing two different models m_1 and m_2 is to evaluate their performance on validation data, e.g., by computing $J_k(m_i)$ in (16.20). We would then favor that model that shows the better performance. Such procedures are known as *cross-validation* and several variants have been developed. See, for example, Stone (1974)and Snee (1977). An attractive feature of cross-validation procedures is their pragmatic character: the comparison makes sense without any probabilistic arguments and without any assumptions about the true system. Their only disadvantage is that we have to save a fresh data set for the validation, and therefore cannot use all our information to build the models.

For linear regressions, we can use $J_p(m)$ in the following way: Let $\hat{y}_p(t|m_t)$ be computed for a model m_t which is estimated from *all data except the observation* $(y(t), \varphi(t))$, within the model structure $\mathcal{M}$. Form J by summing over all the corresponding squared errors. Then J is a measure of the predictive power of this model structure in a cross-validation sense, yet no data have been "wasted" in the estimation phase. The procedure is known as *PRESS* (Prediction sum of squares); see Allen (1971)and Draper and Smith (1981), Section 6.8. For dynamic systems that do not have predictors with finite impulse responses, this is less easy to accomplish.

Comparing Models on Second-hand Data Sets: Evaluating the Expected Fit

The proper quality measure for the model m is the expected criterion $\overline{J}_k$ in (16.22). If the model is evaluated on validation data, the observation J_k is a reasonable and unbiased estimate of $\overline{J}_k$. This is why model evaluation on validation data is to be preferred.

If we use estimation data for the comparisons, then J_k is no longer an unbiased estimate of $\overline{J}_k$. In this section we shall discuss the nature of this discrepancy in case the comparison criterion coincides with the estimation criterion. This means that the value J_p will equal the value of the identification criterion:

$$J_p(m) = \frac{1}{N}\sum_{t=1}^{N}\left|y(t) - \hat{y}(t|\hat{\theta}_N)\right|^2 = \min_{\theta}\frac{1}{N}\sum_{t=1}^{N}\left|y(t) - \hat{y}(t|\theta)\right|^2 \tag{16.24}$$

A Pragmatic Preview. The model obtained in the larger model structure will automatically yield a smaller value of the criterion of fit, since it is the minimizing value obtained by minimization over a larger set. As the model structure increases, as in (16.2b), the minimal value of the criterion will thus behave as depicted in Figure 16.1; it is a monotonically decreasing function of the model structure flexibility. To begin with, the value V_N decreases since the model picks up more of the relevant features of the data. But even after a model structure has been reached that allows a correct description of the system, the value V continues to decrease, now because the additional (unnecessary) parameters adjust themselves to features of the particular realization of the noise. This is known as *overfit* and this extra improved fit is of course of no value to us, since we are going to apply the model to data with different noise realizations. It is reasonable that the decrease from overfit should be less significant than the decrease that results when more relevant features are included in the model. We will thus be looking for the "knee" in the curve of Figure 16.1. Indeed, it is good practice to plot this curve to get a subjective opinion on whether the improved fit is significant and worthwhile.

A Formal Result. For the case that the the comparison criterion coincides with the estimation criterion, we have the following result:

Theorem 16.1. Let

$$\hat{\theta}_N = \arg\min_{\theta} V_N(\theta, Z^N) = \arg\min_{\theta}\frac{1}{N}\sum_{t=1}^{N}\ell\left(\varepsilon(t,\theta),\theta,t\right)$$

$$\overline{V}(\theta) = \overline{E}\ell\left(\varepsilon(t,\theta),\theta,t\right) = \lim_{N\to\infty} EV_N(\theta, Z^N)$$

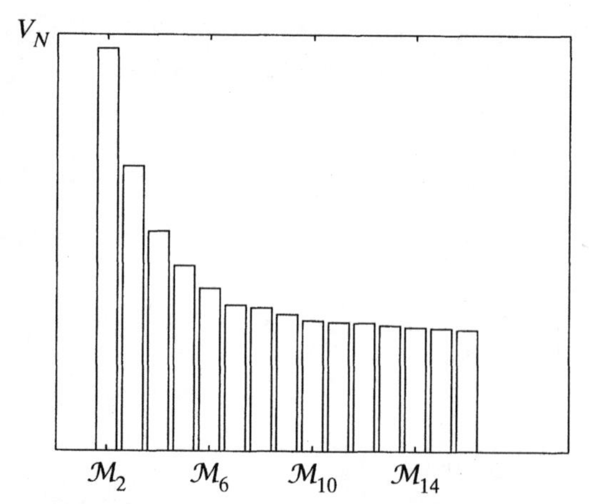

Figure 16.1 The minimal value of the loss function as a function of the size of the model structures (16.2b). $V_N = \min V_N(\theta)$.

Let θ^* be the minimizing argument of $\overline{V}(\theta)$ and suppose

$$EN(\hat{\theta}_N - \theta^*)(\hat{\theta}_N - \theta^*)^T \to P_\theta \text{ as } N \to \infty$$

Then, asymptotically, as $N \to \infty$

$$E\overline{V}(\hat{\theta}_N) \approx EV_N(\hat{\theta}_N, Z^N) + \frac{1}{N}\text{tr}\overline{V}''(\theta^*)P_\theta \tag{16.25}$$

with expectation over the random variable $\hat{\theta}_N$.

Proof. Expand $\overline{V}(\theta)$ around θ^*:

$$\overline{V}(\hat{\theta}_N) = \overline{V}(\theta^*) + \tfrac{1}{2}(\hat{\theta}_N - \theta^*)^T\overline{V}''(\zeta_N)(\hat{\theta}_N - \theta^*) \tag{16.26}$$

Similarly, since $V_N'(\hat{\theta}_N, Z^N) = 0$,

$$V_N(\hat{\theta}_N, Z^N) = V_N(\theta^*, Z^N) - \tfrac{1}{2}(\hat{\theta}_N - \theta^*)^T V_N''(\overline{\zeta}_N, Z^N)(\hat{\theta}_N - \theta^*) \tag{16.27}$$

Take expectation of these two expressions and use the following asymptotic relationships

$$\begin{aligned} &E\tfrac{1}{2}(\hat{\theta}_N - \theta^*)^T\overline{V}''(\zeta_N)(\hat{\theta}_N - \theta^*) \\ &\quad = \tfrac{1}{2}E \text{ tr}\left\{\overline{V}''(\zeta_N)(\hat{\theta}_N - \theta^*)(\hat{\theta}_N - \theta^*)^T\right\} \approx \tfrac{1}{2}\text{tr}\overline{V}''(\theta^*)P_N \end{aligned} \tag{16.28}$$

where $P_N = (1/N)P_\theta$ is the asymptotic covariance matrix of $\hat{\theta}_N$. Note that P_N decays as $1/N$. Also,

$$E\tfrac{1}{2}(\hat{\theta}_N - \theta^*)^T V_N''(\overline{\zeta}_N, Z^N)(\hat{\theta}_N - \theta^*) \approx \tfrac{1}{2}\text{tr}\overline{V}''(\theta^*)P_N \tag{16.29}$$

and

$$EV_N(\theta^*, Z^N) \approx \overline{V}(\theta^*)$$

This gives, from (16.26) and (16.27), respectively,

$$E\overline{V}(\hat{\theta}_N) \approx \overline{V}(\theta^*) + \tfrac{1}{2}\mathrm{tr}\overline{V}''(\theta^*)P_N \tag{16.30a}$$

$$EV_N(\hat{\theta}_N, Z^N) \approx \overline{V}(\theta^*) - \tfrac{1}{2}\mathrm{tr}\overline{V}''(\theta^*)P_N \tag{16.30b}$$

which concludes the proof. □

Note the important difference between $E\overline{V}(\hat{\theta}_N)$ and $EV_N(\hat{\theta}_N)$. If we generate many estimation and validation data sets in a Monte-Carlo manner, the second measure would be the averages of the fits of the models as they are fitted to estimation data, while the first one is the average as the estimated models are evaluated on validation data. Clearly, it is this value that is the one to consider for validation purposes.

Akaike's Final Prediction-Error Criterion (FPE). Let us now specialize to $\ell(\varepsilon(t,\theta),\theta,t) = \varepsilon^2(t,\theta)$. Then we have

$$\overline{J}_p(\mathcal{M}) = E\overline{V}(\hat{\theta}_N)$$

Let us also assume that

- The true system is described by $\theta^* = \theta_0$, (i.e., $\mathcal{S} \in \mathcal{M}$)
- The parameters are identifiable so that $\overline{V}''(\theta_0)$ is invertible (16.31)
- The validation data have the same second order properties as the estimation data

Then we can use Theorem 16.1 to determine a suitable estimate of $\overline{J}_p$ that can be formed from estimation data only: We know from (9.17) that (note that $\overline{V}'' = 2\overline{E}\psi\psi^T$ in our case)

$$P_\theta = 2\lambda_0\left[\overline{V}''(\theta_0)\right]^{-1}, \text{ where } \lambda_0 = Ee_0^2(t) = \overline{V}(\theta_0) \tag{16.32}$$

The last of the assumptions (16.31) means that this function $\overline{V}(\theta)$ is the same as the one in (16.25). We can consequently use this expression for P_θ in (16.25) which gives

$$\begin{aligned}\overline{J}_p(\mathcal{M}) &= E\overline{V}(\hat{\theta}_N) \approx V_N(\hat{\theta}_N, Z^N) + \frac{2\lambda_0}{N}\mathrm{tr}\left[\overline{V}''(\theta_0)[\overline{V}''(\theta_0)]^{-1}\right] \\ &= V_N(\hat{\theta}_N, Z^N) + \lambda_0\frac{2d_{\mathcal{M}}}{N}\end{aligned}$$

In the first step, we replaced $EV_N(\hat{\theta}_N, Z^N)$ with the only observation we have of it, viz. $V_N(\hat{\theta}_N, Z^N)$. In the last step we used that

$$\mathrm{tr}\left\{\overline{V}''(\theta_0)\left[\overline{V}''(\theta_0)\right]^{-1}\right\} = \dim\theta^{\mathcal{M}} = d_{\mathcal{M}}$$

The expression

$$\overline{J}_p(\mathcal{M}) = E\overline{V}_N(\hat{\theta}_N) \approx V_N(\hat{\theta}_N, Z^N) + \lambda_0 \frac{2d_{\mathcal{M}}}{N} \tag{16.33}$$

shows the fundamental *cost of parameters.* The more parameters are used by the model structure, the smaller the first term will be. However, each parameter carries a variance penalty that will contribute with $2\lambda_0/N$ to the expected mean square error fit. This is true regardless of the importance of the parameter for the fit, i.e., how much it reduces the first term. Any parameter that improves the fit of V_N by less than $2\lambda_0/N$ will thus be harmful in this respect.

Now, λ_0 is not known, but can easily be estimated. According to (16.30) (ignoring the first expectation):

$$V_N(\hat{\theta}_N, Z^N) \approx \overline{V}(\theta_0) - \tfrac{1}{2}\text{tr}\overline{V}''(\theta_0)P_N \approx \lambda_0 - \frac{d_{\mathcal{M}} \cdot \lambda_0}{N}$$

A suitable estimate for λ_0 is thus obtained as

$$\hat{\lambda}_N = \frac{V_N(\hat{\theta}_N, Z^N)}{1 - (d_{\mathcal{M}}/N)}$$

which, inserted into (16.33), gives

$$\begin{aligned} \overline{J}_p(\mathcal{M}) &\approx \frac{1 + (d_{\mathcal{M}}/N)}{1 - (d_{\mathcal{M}}/N)} V_N(\hat{\theta}_N, Z^N) \\ &= \frac{1 + (d_{\mathcal{M}}/N)}{1 - (d_{\mathcal{M}}/N)} \frac{1}{N} \sum_{t=1}^{N} \varepsilon^2(t, \hat{\theta}_N) \end{aligned} \tag{16.34}$$

This criterion was first described by Akaike (1969)as a final prediction-error (FPE) criterion. It shows how to modify the loss function to get a reasonable estimate of the validation and comparison criterion $\overline{J}$ from estimation data only. The observed fit must be compensated for using the number of estimated parameters to give a fair picture of the model quality.

A Note on Regularization. Suppose that we are using the regularized criterion (7.107):

$$W_N(\theta, Z^N) = \frac{1}{N} \sum_{t=1}^{N} \ell(\varepsilon(t, \theta)) + \delta|\theta - \theta^{\#}|^2 = V_N(\theta, Z^N) + \delta|\theta - \theta^{\#}|^2 \tag{16.35}$$

Assume also that the true system is in the model set, that (16.31) holds and that $\theta^{\#} = \theta_0$. The latter assumption is of course not quite realistic, and we shall comment on it below. The calculations in Chapter 9 can directly be applied to (16.35), to give

$$P_\theta = \left[\overline{W}''(\theta_0)\right]^{-1} Q \left[\overline{W}''(\theta_0)\right]^{-1}, \quad \overline{W}''(\theta_0) = \overline{V}''(\theta_0) + \delta I, \quad Q = \lambda_0 \overline{V}''(\theta_0)$$

See (9.9)–(9.11). Inserted into the equation leading to (16.33), the trace-term becomes ($H = \overline{V}''(\theta_0)$)

$$\text{tr} H(H + \delta I)^{-1} H (H + \delta I)^{-1} = \sum_{k=1}^{d_{\mathcal{M}}} \frac{\sigma_i^2}{(\sigma_i + \delta)^2}$$

where σ_i are the eigenvalues (singular values) of $\overline{V}''(\theta_0)$. This follows since H and $H + \delta I$ can be simultaneously diagonalized. This means that instead of (16.33), the regularized criterion leads to

$$\overline{J}_p(\mathcal{M}) = E\overline{V}_N(\hat{\theta}_N) \approx V_N(\hat{\theta}_N, Z^N) + \frac{2\lambda_0}{N} \sum_{k=1}^{d_{\mathcal{M}}} \frac{\sigma_i^2}{(\sigma_i + \delta)^2} \tag{16.36}$$

We note that with $\delta = 0$ the sum will equal $d_{\mathcal{M}}$, so the special case (16.33) is then re-obtained. Typically, the singular values of $\overline{V}''(\theta_0)$ are widely spread so that either $\sigma_i \ll \delta$ or $\sigma_i \gg \delta$, which means that the sum really just counts the number of singular values that are larger then δ. We could think of this as *the efficient number of parameters* used by the model structure with the regularized criterion. The other parameters can be thought of as locked in by the regularization. Comparing with (16.33) we may regard the regularization parameter δ as a knob by which we control the number of free parameters in the model structure, without having to decide which ones to set free. The criterion will then let $\mathcal{M}$ use those that have the largest influence on the fit.

The expression (16.36) has been derived under the (unrealistic) assumption $\theta^{\#} = \theta_0$. In the common case that $\theta^{\#} = 0$, the regularization will cause a small bias in the parameters, due to the pull towards the origin. This is similar to the bias introduced by using too few parameters, so the bias-variance trade-off in terms of the regularization parameter δ is still analogous to the one obtained by explicitly controlling the number of parameters.

Model Structure Selection Criteria: AIC, BIC, and MDL

Suppose that the prediction-error criterion is chosen as the normalized log-likelihood function [see (7.84)]:

$$\begin{aligned} V_N(\theta, Z^N) &= -\frac{1}{N} \text{(log likelihood for the estimation problem)} \\ &= -\frac{1}{N} L_N(\theta, Z^N) \end{aligned} \tag{16.37}$$

Then provided that (16.31) holds we know from the asymptotic distribution result of Section 9.3 that

$$P_N = \left[\overline{V}''(\theta_0)\right]^{-1} \cdot \frac{1}{N} = \left[EL_N''(\theta_0)\right]^{-1} \tag{16.38}$$

[see (7.80), (7.89), and (9.29)]. This inserted into (16.25) gives the criterion

$$\overline{E}\,\overline{V}_N(\hat{\theta}_N) = -\frac{1}{N}L_N(\hat{\theta}_N, Z^N) + \frac{d_{\mathcal{M}}}{N} \tag{16.39}$$

since

$$\operatorname{tr}\left\{\overline{V}''(\theta_0) \cdot \left[\overline{V}''(\theta_0)\right]^{-1}\right\} = \dim \theta^{\mathcal{M}} = d_{\mathcal{M}}$$

This expression is the Akaike AIC criterion (7.106). We can thus phrase the joint problem of model structure determination and parameter estimation as

$$\{\hat{\theta}_N^{\mathcal{M}}, \hat{\mathcal{M}}\} = \arg \min_{\mathcal{M} \in M} \min_{\theta^{\mathcal{M}} \in D_{\mathcal{M}}} \frac{1}{N}\left[-L_N(\theta^{\mathcal{M}}, Z^N) + d_{\mathcal{M}}\right] \tag{16.40}$$

where L_N is the log-likelihood function and superscript $\mathcal{M}$ denotes that $\theta^{\mathcal{M}}$ is associated with the model structure $\mathcal{M}$.

Example 16.2 AIC for Gaussian Innovations

Suppose that the process innovations are Gaussian with unknown variance λ. Then

$$L_N(\theta, Z^N) = -\frac{1}{2}\sum_{t=1}^{N} \frac{\varepsilon^2(t, \theta')}{\lambda} - \frac{N}{2}\log \lambda - \frac{N}{2}\log 2\pi$$

where $\theta = [\theta', \lambda]$ [see (7.87) and (7.17)]. For the inner minimization (within a given model structure) in (16.40), we have (see Problem 7E.7)

$$\hat{\theta}_N = [\hat{\theta}_N', \hat{\lambda}_N]$$

$$\hat{\lambda}_N = \frac{1}{N}\sum_{t=1}^{N} \varepsilon^2(t, \hat{\theta}_N')$$

$$\hat{\theta}_N' = \arg \min_{\theta \in D_{\mathcal{M}}} \sum_{t=1}^{N} \varepsilon^2(t, \theta)$$

Hence

$$L_N(\hat{\theta}_N, Z^N) = -\frac{N}{2} - \frac{N}{2}\log \hat{\lambda}_N - \frac{N}{2}\log 2\pi$$

and the outer minimization (w.r.t $\mathcal{M}$) in (16.40) takes the form

$$\widehat{\mathcal{M}} = \arg \min_{\mathcal{M} \in M} \left[\frac{1}{2} + \frac{1}{2}\log 2\pi + \frac{1}{2}\log\left[\frac{1}{N}\sum_{t=1}^{N} \varepsilon^2(t, \hat{\theta}_N^{\mathcal{M}})\right] + \frac{d_{\mathcal{M}}}{N}\right]$$

The term to minimize is

$$\log\left[\frac{1}{N}\sum_{t=1}^{N}\varepsilon^2(t,\hat{\theta}_N^{\mathcal{M}})\right]+\frac{2d_{\mathcal{M}}}{N}$$

$$\approx \log\left[\left(1+\frac{2d_{\mathcal{M}}}{N}\right)\frac{1}{N}\sum_{t=1}^{N}\varepsilon^2(t,\hat{\theta}_N^{\mathcal{M}})\right] \tag{16.41}$$

where the last approximation follows when $\dim \theta^{\mathcal{M}} \ll N$. Note the similarity with (16.34). □

The criteria that we have discussed here may, from a more pragmatic point of view, be seen as joint criteria for the determination of model structure *and* parameter values within the structure. Conceptually, they could be written

$$W_N^0(\theta,\mathcal{M},Z^N) = V_N(\theta,Z^N)(1+U_N(\mathcal{M})) \tag{16.42}$$

where V_N is the prediction-error criterion (7.155) within certain model structures $\mathcal{M}$, and $U_N(\mathcal{M})$ is a function that measures some "complexity" of the model structure. In the cases so far, this measure has been related to the dimensionality of θ:

$$U_N(\mathcal{M}) = \frac{2\dim\theta}{N} \tag{16.43}$$

These criteria have been directed to find system descriptions that give the smallest mean-square error. A model that apparently gives a smaller mean-square (prediction) error fit will be chosen even if it is quite complex. In practice, one may want to add extra penalty in (16.43) for model complexity, reflecting the cost of using it: "If I am going to accept a more complex model (according to my own complexity measure) it has to prove to be significantly better!"

What is meant by a complex model and what penalty should be associated with it are usually subjective issues. An interesting approach to this problem is taken by Rissanen (1978). He asserts that the ultimate goal of identification is to achieve the shortest possible description of data. This leads to a criterion of the type (16.42) with

$$U_N(\mathcal{M}) = \dim\theta\cdot\frac{\log N}{N} \tag{16.44}$$

called the MDL (minimum description length) criterion. This has also been termed "BIC" by Akaike.

Statistical Hypothesis Tests (*)

The selection between two model structures $\mathcal{M}_0$ and $\mathcal{M}_1$, subject to (16.2b), can also be approached by the theory of statistical tests. The idea is to pose the hypothesis

$$H_0: \quad \text{the data have been generated by } \mathcal{M}_0(\hat{\theta}_N^{(0)}) \tag{16.45}$$

This hypothesis is to be tested against the alternative

$$H_1: \quad \text{the data have been generated by } \mathcal{M}_1(\hat{\theta}_N^{(1)}) \tag{16.46}$$

If now $\mathcal{M}_1$ is "larger" than $\mathcal{M}_0$, we should introduce a prejudice against $\mathcal{M}_1$. This means that we would prefer the set $\mathcal{M}_0$ unless there is "convincing evidence" that the hypothesis H_1 is true. In statistics this is expressed as H_0 is the "null hypothesis."

We should now like to make a decision between H_0 and H_1 such that the risk (the probability) of rejecting H_0 when it is true is less than a certain number α. (The smaller α is chosen the more prejudice is inflicted against $\mathcal{M}_1$.) At the same time, we would like to maximize the probability that H_0 is rejected when H_1 is true. The latter quantity is known as the "power" of the test.

Let $\theta_*^{(k)}$ denote the limiting estimate in the model structure $\mathcal{M}_k$. If this value gives a correct description of the system ($S \in \mathcal{M}_k$), it can be shown under general conditions that

$$N \cdot \frac{V_N(\theta_*^{(k)}, Z^N) - V_N(\hat{\theta}_N^{(k)}, Z^N)}{V_N(\theta_*^{(k)}, Z^N)} \in As\chi^2(d(k)) \tag{16.47}$$

That is, the random variable on the left side converges in distribution to the χ^2 distribution with $d(k)(= \dim \theta^{(k)})$ degrees freedom. This is proved in Lemma II.4 for the case of linear regressions and by Åström and Bohlin (1965)for ARMAX models. Notice that (16.47) is consistent with the expression (16.30b) in case $S \in \mathcal{M}_k$.

Under the null hypothesis (16.45), it follows from (16.47) that

$$N \cdot \frac{V_N(\hat{\theta}_N^{(0)}, Z_N) - V_N(\hat{\theta}_N^{(1)}, Z_N)}{V_N(\hat{\theta}_N^{(1)}, Z^N)} \in As\chi^2(d(1) - d(0)) \tag{16.48}$$

The null hypothesis can thus be tested at any desired confidence level α using this expression.

In case V_N is chosen as the log-likelihood function, this test also becomes the likelihood ratio (LR) test (see, e.g., Kendall and Stuart, 1961), which has maximum power. Bohlin (1978)has derived a maximum power test that does not require the computation of the model in the large set.

Using (16.48) amounts to rejecting H_0 (and hence choosing the structure $\mathcal{M}_1$) if

$$V_N(\hat{\theta}_N^{(0)}, Z^N) - V_N(\hat{\theta}_N^{(1)}, Z^N) \geq V_N(\hat{\theta}_N^{(1)}, Z^N) \cdot \frac{1}{N} \cdot k_d(\alpha) \tag{16.49}$$

where $k_d(\alpha)$ is the α level for the χ^2 distribution with $d(1)-d(0)$ degrees of freedom. From a user's point of view, (16.49) thus coincides with the use of (16.41) (AIC) or (16.34) (FPE), for $d \ll N$, with α such that

$$k_d(\alpha) = 2(d(1) - d(0)) \tag{16.50}$$

This is satisfied for α somewhere between 7% and 1% depending on the difference $d(1) - d(0)$ (within reasonable ranges). There is consequently a clear relationship between FPE, AIC, and hypothesis testing in practical use. See Söderström (1977)for further details.

For small N, a more accurate expression than (16.48) is

$$\frac{V_N(\hat{\theta}_N^{(0)}, Z^N) - V_N(\hat{\theta}_N^{(1)}, Z^N)}{V_N(\hat{\theta}_N^{(1)}, Z^N)} \cdot \frac{N - d(1)}{d(1) - d(0)}$$

$$\in AsF\,(N - d(1), d(1) - d(0)) \tag{16.51}$$

(cf. Lemma II.4). Based on this asymptotic F-distribution, *F-tests* can be applied.

16.5 MODEL VALIDATION

The parameter estimation procedure picks out the "best" model within the chosen model structure. The crucial question then is whether this "best" model is "good enough." This is the problem of model validation. The question has several aspects:

1. Does the model agree sufficiently well with the observed data?
2. Is the model good enough for my purpose?
3. Does the model describe the "true system"?

Generally (Bohlin, 1991), the method to answer these questions is to confront the model $\mathcal{M}(\hat{\theta}_N)$ with as much information about the true system as is practical. This includes a priori knowledge, experiment data, and experience of using the model. In an identification application the most natural entity with which to confront the model is the data themselves. Model-validation techniques thus tend to focus on question 1. We shall in this section list a number of tools that are useful for discarding models, as well as for developing confidence in them. A particularly useful technique, *residual analysis*, is treated separately in the following section.

Validation with Respect to the Purpose of the Modeling

While question 3 is intriguing, it is also, philosophically, impossible to answer. What matters in practice is question 2. There is always a certain purpose with the modeling. It might be that the model is required for regulator design, prediction, or simulation. The ultimate validation then is to test whether the problem that motivated the modeling exercise can be solved using the obtained model. If a regulator based on the model gives satisfactory control, then the model was a "valid" one, regardless of the formal aspects on this concept that can be raised. Often it will be impossible, costly, or dangerous to test all possible models with respect to their intended use. Instead, one has to develop confidence in the model in other ways.

Feasibility of Physical Parameters

For a model structure that is parametrized in terms of physical parameters, a natural and important validation is to confront the estimated values and their estimated variances with what is reasonable from prior knowledge. It is also good practice to evaluate the sensitivity of the input-output behavior with respect to these parameters to check their practical identifiability (this should also be reflected by the estimated variances).

Consistency of Model Input-Output Behavior

For black-box models, we focus our interest on their input-output properties. For linear models, we would normally display these as Bode diagrams. For non-linear models, they would be inspected by simulation (see the following). It is always good practice to evaluate and compare different linear models in Bode plots, possibly with the estimated variance translated to confidence intervals of $\hat{G}$ (and $\hat{H}$), see (9.59). Comparisons between spectral analysis estimates (6.46) and Bode plots derived from parametric models are especially useful, since they are formed from quite different underlying assumptions (i.e., model structures; see Problem 7G.2).

Generally, when the true system does not belong to the model set, we obtain an approximation whose character will depend on the experimental conditions, the prefilters used, the criterion, and the model structure (Chapter 8). Thus, comparing Bode plots of models obtained by prediction error methods in different structures, as well as with different prefilters, by the subspace method and by spectral analysis will give a good feel for whether the essential features of the dynamics have been captured.

Model Reduction

One procedure that tests if the model is a simple and appropriate system description is to apply some model-reduction technique to it. If the model order can be reduced without affecting the input-output properties very much, then the original model was "unnecessarily complex." Söderström (1975b)has developed this idea for pole-zero cancellations.

Parameter Confidence Intervals

Another procedure that checks whether the current model contains too many parameters is to compare the estimate with the corresponding estimated standard deviation (see Section 9.6). If the confidence interval contains zero, we could consider whether this parameter should be removed. This is usually of interest only when the corresponding parameter reflects a physical structure, such as model order or time delay. If the estimated standard deviations are all large, the information matrix is close to singular. This also is an indication of too large model orders (see Section 16.3).

Simulation and Prediction

In Section 16.4 we used the models' ability to reproduce input-output data in terms of simulations and predictions as a main tool for comparisons. Such plots, and the numerical fits associated with them, are of course most useful and intuitively appealing also for evaluating a given model. We see exactly what features the model is capable of reproducing and what features it has not captured. The discrepancies can be due to noise or model errors, and we will see the combined effects of these sources. If we have an independent estimate of the noise level, e.g., from (7.154), we will also be able to tell from $J_k(m)$ in (16.20) what the size of the model error is.

16.6 RESIDUAL ANALYSIS

The "leftovers" from the modeling process—the part of the data that the model could not reproduce—are *the residuals*

$$\varepsilon(t) = \varepsilon(t, \hat{\theta}_N) = y(t) - \hat{y}(t|\hat{\theta}_N) \tag{16.52}$$

It is clear that these bear information about the quality of the model. In this section we shall discuss informal and formal methods to draw conclusions about the model validity from analysis of the residuals.

Pragmatic Viewpoints

The bottom line really is that we have a data set Z^N, be it estimation or validation data, and a nominal model $\mathcal{m}$. We want to know the quality of the model, which in a sense is a statement about how it will be able to reproduce new data sets. A simple and pragmatic starting point is to compute basic statistics for the residuals from the model:

$$S_1 = \max_t |\varepsilon(t)|, \quad S_2^2 = \frac{1}{N}\sum_{t=1}^{N} \varepsilon^2(t) \tag{16.53}$$

The intuitive use of these statistics would then be like this: "This model has never produced a larger residual than S_1 (or an average error of S_2) for all data we have seen. It's likely that such a bound will hold also for future data." Indeed, the rationale of the different identification criteria could be said to allow for as strong such statements as possible. Now, this use of the statistics (16.53) has an implicit invariance assumption: The residuals do not depend on something that is likely to change. Of special importance is, of course, that they do not depend on the particular input used in Z^N. If they did, the value of (16.53) would be limited, since the model should work for a range of possible inputs. To check this, it is reasonable to study the covariance between residuals and past inputs:

$$\hat{R}_{\varepsilon u}^N(\tau) = \frac{1}{N}\sum_{t=1}^{N} \varepsilon(t)u(t - \tau) \tag{16.54}$$

If these numbers are small (and we shall shortly quantify what that should mean) we have some reason to believe that the measures (16.53) could have relevance also when the model is applied to other inputs.

Another way to express the importance of $\hat{R}_{\varepsilon u}^N$ being small is as follows: If there are traces of past inputs in the residuals, then there is a part of $y(t)$ that originates from the past input and that has not been properly picked up by the model $\mathcal{m}$. Hence, the model could be improved.

Similarly, if we find correlation among the residuals themselves, i.e., if the numbers

$$\hat{R}_{\varepsilon}^N(\tau) = \frac{1}{N}\sum_{t=1}^{N} \varepsilon(t)\varepsilon(t - \tau) \tag{16.55}$$

are not small for $\tau \neq 0$, then part of $\varepsilon(t)$ could have been predicted from past data. This means that $y(t)$ could have been better predicted, which again is a sign of deficiency of the model.

From a more formal point of view we have motivated the estimation criterion as a maximum likelihood method, assuming that the data have been generated as in (7.82):

$$y(t) = g(t, Z^N; \hat{\theta}_N) + \varepsilon(t) \tag{16.56}$$

where the $\varepsilon(t)$ have the properties given by (7.81) to be independent of each other and past data. The model validation question related to data, then, is "Is it likely that the data record Z^N actually has been generated by (16.56)?" This question is equivalent to "Is it likely that

$$\varepsilon(t) = y(t) - g(t, Z^N; \hat{\theta}_N) \tag{16.57}$$

is a sequence of independent random variables with PDF $f_e(x, t; \hat{\theta}_N)$?" Clearly (16.55) and (16.54) form the basis to part of the answer.

Whiteness Test

The numbers $\hat{R}_\varepsilon^N(\tau)$ carry information about whether the residuals can be regarded as white. To get an idea of how large these numbers may be if indeed $\varepsilon(t)$ are white, we reason as follows: Suppose $\{\varepsilon(t)\}$ is a white noise sequence with zero mean and variance λ. Then it follows from Lemma 9A.1 that

$$\frac{1}{\sqrt{N}} \sum_{t=1}^{N} \begin{bmatrix} \varepsilon(t-1) \\ \vdots \\ \varepsilon(t-M) \end{bmatrix} \varepsilon(t) \in AsN(0, \lambda^2 \cdot I)$$

The k:th row of this vector is $\sqrt{N}\hat{R}_\varepsilon^N(k)$. Under the assumption that the ε are white, this consequently means that

$$\frac{N}{\lambda^2} \sum_{\tau=1}^{M} \left(\hat{R}_\varepsilon^N(\tau)\right)^2$$

should be asymptotically $\chi^2(M)$-distributed (cf. Appendix II). Replacing the unknown λ by the obvious estimate does not change this, asymptotically. (Rightly, the distribution becomes an F-distribution, see (II.79), but we anyway assume N to be large.) The test for whiteness will thus be if

$$\zeta_{N,M} = \frac{N}{\left(\hat{R}_\varepsilon^N(0)\right)^2} \sum_{\tau=1}^{M} \left(\hat{R}_\varepsilon^N(\tau)\right)^2 \tag{16.58}$$

will pass a test of being $\chi^2(M)$ distributed, i.e., by checking if $\zeta_{N,M} < \chi_\alpha^2(M)$, the α level of the $\chi^2(M)$-distribution.

In addition to this whiteness test, more tests can be performed, like the number of sign changes of $\varepsilon(t)$, and histogram test for the distribution of ε. See, e.g., Draper and Smith (1981), Chapter 3, for details of such tests.

Independence between Residuals and Past Inputs

To investigate which requirements should be associated with (16.54), we define

$$r_M^N = \frac{1}{\sqrt{N}} \sum_{t=1}^{N} \varepsilon(t)\varphi(t), \quad \varphi(t) = \begin{bmatrix} u(t - M_1) \\ \vdots \\ u(t - M_2) \end{bmatrix}, \tag{16.59}$$

$$M = M_2 - M_1 + 1 = \dim \varphi$$

Note that the k:th component of r is equal to $\sqrt{N}\hat{R}_{\varepsilon u}^N(k + M_1 - 1)$, given by (16.54). If the ε are independent of φ and can be written as

$$\varepsilon(t) = \sum_{k=0}^{\infty} f_k e(t - k), \quad f_0 = 1, \quad e(t) \text{ white noise with } Ee^2(t) = \lambda \tag{16.60}$$

then it follows from (9.38)–(9.40) that

$$r_M^N \in AsN(0, \lambda P), \quad P = \overline{E}\tilde{\varphi}(t)\tilde{\varphi}^T(t), \quad \tilde{\varphi}(t) = \sum_{k=1}^{\infty} f_k \varphi(t + k) \tag{16.61}$$

It can be shown that the (k, ℓ) element of P can also be expressed as

$$\sum_{\tau=-\infty}^{\infty} R_\varepsilon(\tau) R_u(\tau - (k - \ell)), \; R_\varepsilon(\tau) = E\varepsilon(t)\varepsilon(t - \tau), \; R_u(\tau) = \overline{E}u(t)u(t - \tau) \tag{16.62}$$

Now, (16.61) implies that

$$\zeta_{N,M}^u = \frac{1}{\lambda} r_M^N P^{-1} [r_M^N]^T \in As\chi^2(M) \tag{16.63}$$

if ε is independent of the inputs. Thus, $\zeta_{N,M}^u$ is the right quantity to subject to a $\chi^2(M)$ test. Note that we need to estimate a model (16.60) to be able to form this quantity. If ε is assumed to be white noise, or has passed the test (16.58), the calculation of $\zeta_{N,M}^u$ is simplified.

A simple use of (16.54) is to consider just one given τ as follows: From the calculations above, it follows that

$$\sqrt{N}\hat{R}_{\varepsilon u}^N(\tau) \in AsN(0, P_1), \quad P_1 = \sum_{k=-\infty}^{\infty} R_\varepsilon(k) R_u(k) \tag{16.64}$$

If N_α denotes the α level of the $N(0, 1)$ distribution, we could thus check if

$$\left|\hat{R}_{\varepsilon u}^N(\tau)\right| \leq \sqrt{\frac{P_1}{N}} N_\alpha \tag{16.65}$$

If not, the hypothesis that $\varepsilon(t)$ and $u(t-\tau)$, are independent should be rejected. An appealing way to carry out the test is to plot $\hat{R}^N_{\varepsilon u}(\tau)$ as a function of τ. Since P_1 in (16.55) does not depend on τ, the confidence limits will be horizontal lines. Such a plot gives valuable insight in the correctness of the model structure. If, for example, a time delay of two samples has been assumed in the model, but the true delay is one sample, then a clear correlation between $u(t-1)$ and $\varepsilon(t)$ will show up. When examining plots of $\hat{R}^N_{\varepsilon u}(\tau)$, note the following points:

1. Correlation between $u(t-\tau)$ and $\varepsilon(t)$ for negative τ is an indication of output feedback in the input, not that the model structure is deficient.
2. The least-squares method constructs $\hat{\theta}_N$ so that $\varepsilon(t, \hat{\theta}_N)$ is uncorrelated with the regressors. We thus have $\hat{R}^N_{\varepsilon u}(\tau) = 0$ for $\tau = 1, \ldots, n_b$ automatically for the model structure (4.7), when the analysis is carried out on estimation data.

This means that some care should be exercised when the numbers M_1 and M_2 in $\varphi(t)$ are selected. If estimation data are used, together with an ARX model of order n_a, n_b, we should thus have $M_1 > n_b$. Similarly, it is natural to take $M_1 \geq 0$ if only causal dependence from past inputs shall be tested. Note that the levels of the tests are somewhat effected by whether estimation or validation data are used. See Söderström and Stoica (1990)for an analysis of this.

Independence between u and ε can also be measured in other terms. Unmodeled nonlinear effects can, for example, be seen in scatter plots of the pairs $(\varepsilon(t), u(t-\tau))$, or by correlating non-linear transformations of ε and u. See, for example, Draper and Smith (1981), Chapter 3, Cook and Weisberg (1982), and Anscombe and Tukey (1963)for general aspects on residual testing, and Billings and Tao (1991), Lee, White, and Granger (1993), and Luukkonen, Saikkonen, and Teräsvirta (1988)for tests that specifically aim at detecting nonlinearities.

Tests for Dynamical Systems

Testing the correlation between past inputs and the residuals is natural to evaluate if the model has picked up the essential part of the (linear) dynamics from u to y. For a dynamical model, the results of the tests can however be visualized more effectively if we view them as estimates of the residual dynamics or *Model Error Model*:

$$\varepsilon(t) = G_\varepsilon(q)u(t) \tag{16.66}$$

In fact, if the input is white, then $\hat{R}^N_{\varepsilon u}(\tau)$ are approximately the components of the estimate $\hat{\theta}_N$ obtained in the FIR-model

$$\varepsilon(t) = \theta^T\varphi(t)$$

with φ given by (16.59), i.e., the impulse response of (16.66). For the case of non-white input, it will be easier to evaluate a plot of the impulse response estimate of $\hat{G}_\varepsilon$ than of the correlation estimates $\hat{R}^N_{\varepsilon u}$, since the latter also are affected by the internal correlation of u.

Even more effective from a control point of view will be to display the frequency function of the estimate $\hat{G}_\varepsilon(e^{i\omega})$, along with estimated confidence regions. This gives a picture of what frequency ranges the model has not captured in the input-output behavior. Depending on the intended use, the model could then be accepted, even when (16.65) is violated, provided the errors occur in ''harmless'' frequency ranges.

Example 16.3 Residual Analysis for Dynamic Systems

The system

$$y(t) - 1.2y(t-1) - 0.15y(t-1) + 0.35y(t-3)$$
$$= u(t-1) + 0.5u(t-2) + e(t) - e(t-1) + 0.4e(t-2)$$

was simulated over 500 samples with an input consisting of sinusoids between 0.3 and 0.6 rad/sec, and white Gaussian noise e with variance 1. A second order ARX model $\mathcal{m}$ was identified from these data. A validation data set was generated using a random binary input with a resonance peak around 0.3 rad/sec. Figure 16.2a shows the result of conventional residual analysis when $\mathcal{m}$ was confronted with these data. Figure 16.2b shows the impulse and frequency responses of the model-error model (16.66), estimated as a 10th order ARX model. It is clear that the frequency function plot of the error model gives much more precise information about the model's qualities, from a control point of view. In Figure 16.3, the amplitude Bode plots of the model and the true system are compared, and we see that the model error information from validation data is quite reliable. □

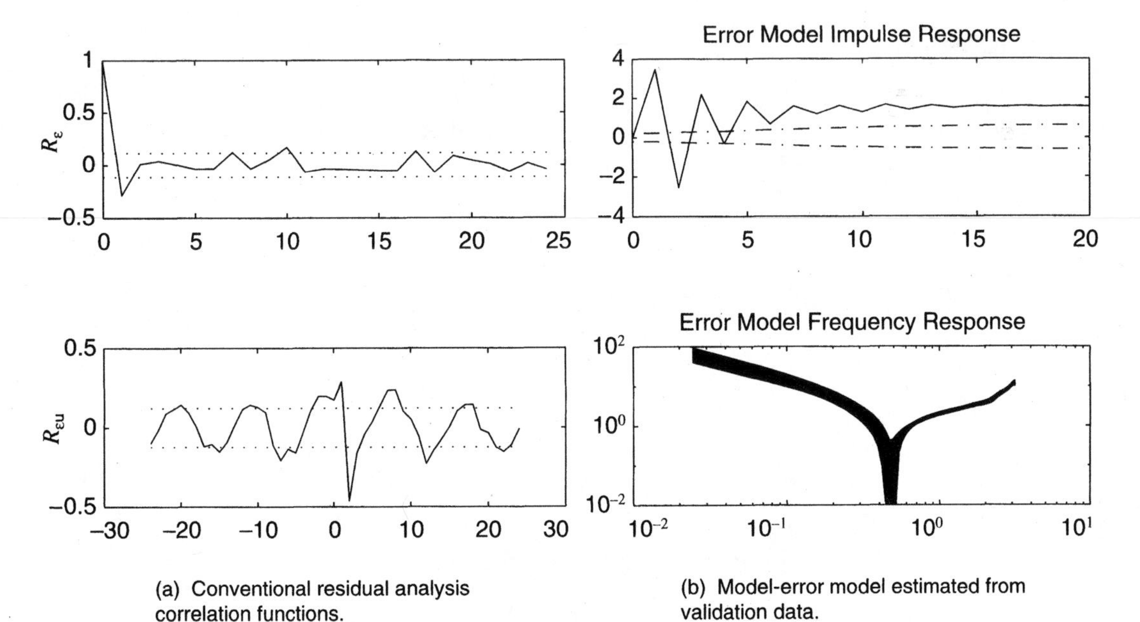

(a) Conventional residual analysis correlation functions.

(b) Model-error model estimated from validation data.

Figure 16.2 Model validation of the second order ARX model using validation data. Dash-dotted lines denote confidence intervals. For the frequency domain plot, the confidence interval is marked as a shaded region.

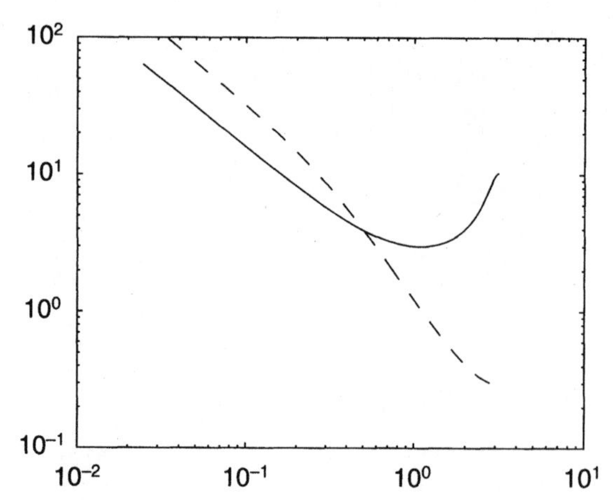

Figure 16.3 Amplitude Bode plots of the model m (solid line) and the true system (dashed line).

Critical Data Evaluation

The residuals $\varepsilon(t, \hat{\theta}_N)$ will also tell us, inserted into the influence function (15.12), which data points have had a large impact on the estimate. The reliability of these points should be critically evaluated as part of the model-validation procedure. It is always good practice to plot $\varepsilon(t, \hat{\theta}_N)$ to inspect the data for outliers and "bad data." Compare Example 14.1.

16.7 SUMMARY

The "true system" is an esoteric entity that cannot be attained in practical modeling. We have to be content with partial descriptions that are purposeful for our applications. Sometimes this means that we may have to work with several models of the same system that are to be used for different operating points, for different time scale problems, and so on.

In this chapter we have described various methods by which suitable model structures can be found and by which we can reject or develop confidence in particular models. Among *a priori* considerations, we may single out the principle "Try simple things first." This usually means that one should start by testing simple linear regressions, such as ARX models in the linear case, and variants with non-linear data transformations based on physical insight, whenever appropriate.

For validation of models, we have described an arsenal of methods of different characters. It is wise to include several of these in one's toolbox. We may point to the following:

- Comparing linear models obtained under various conditions in various model structures (including spectral analysis estimates) in Bode plots
- Comparing measured and simulated outputs for models obtained in different structures
- Testing residuals for independence of past inputs and possibly for whiteness

- Monitoring parameter estimate confidence intervals for trailing and leading zeros in transfer-function polynomials, as well as for possible loss of local identifiability

Finally, the subjective ingredient in model validation should be stressed. The techniques presented here should be viewed as advisers to the user. It is the user that makes the decision. In the words of Draper and Smith (1981, p. 273), "The screening of variables should never be left to the sole discretion of any statistical procedure."

16.8 BIBLIOGRAPHY

General treatments of the model structure determination problem for dynamical systems are given by Bohlin (1991)and Söderström (1987). General discussions on this problem can also be found in Hoaglin, Mosteller, and Tukey (1983). Experimental comparisons between different techniques are carried out in van den Boom and van den Enden (1974), Unbehauen and Göhring (1974), and Freeman (1985). Kashyap and Rao (1976)contains detailed discussions on the problem.

General hypothesis testing, power of tests, and the like, can be studied in Kendall and Stuart (1961). Bohlin (1978)developed a general maximum power test for dynamical systems. Akaike's AIC has been widely applied. See Akaike (1974a, 1981). Cross-validation and its relationship to AIC have been studied by Stone (1974, 1977a and b). Rissanen's criterion (16.44) has been studied also by Schwarz (1978)and another variant by Hannan and Quinn (1979). Rissanen has developed his MDL criterion in a very interesting framework of information, coding, prediction, and estimation. See, for example, Rissanen (1984, 1985, 1986, 1989).

The model order determination problem in time-series modeling has received considerable attention. See, for example, Hannan (1980b), Shibata (1976, 1980), and Fine and Hwang (1979).

Tests for residual independence are treated in many textbooks (e.g., Draper and Smith, 1981; Jenkins and Watts, 1968; Box and Jenkins, 1970). Ljung and van Overbeek (1978)describe how to validate models based on how the resulting model depends on the experimental condition.

Regularization is treated, e.g., in Wahba (1987). It can also be shown that the number of iterations used when minimizing the criterion plays a role similar to regularization: First the parameters that are very important for the fit will get close to their minimizing values, while the less important parameters will move slower (unless we have a pure quadratic criterion and apply a Newton method). Increasing the number of iterations will thus be the same as decreasing the regularization parameter. "Too many" iterations (when the model really is overparameterized) will thus lead to overfit. In the neural network literature this has been called *overtraining*. See, e.g., Wahba (1987)and Sjöberg and Ljung (1995).

From a quite different perspective, in the 1990's there has been a substantial interest in model validation for control applications. This literature typically addresses the question of whether a given bound on the model-error model (16.66), and on the system disturbance v, together with a nominal model for control design, is consistent with a given validation data set. See, e.g., Skelton (1989), Kosut, Lau, and Boyd (1992), Smith and Doyle (1992), Poolla et.al. (1994), Rangan and Poolla (1996), and Smith and Dullerud (1996). These approaches typically deal with a *worst case* scenario for the noise sequence, i.e., no averaging properties of the noise are taken into account. This also gives rise to so-called *hard bounds* on the uncertainty. See Wahlberg and Ljung (1992), Mäkilä, Partington, and Gustafsson (1995), Mäkilä (1992), and Tse, Dahleh, and Tsitsiklis (1993). Another focus is to characterize the model-error model in suitable terms: Goodwin, Givers, and Ninness (1992), Ninness and Goodwin (1995), and Ljung and Guo (1997).

16.9 PROBLEMS

16G.1 *Mallow's C_p-criterion:* Mallows (1973)has suggested the following criterion for selecting model structures:

$$C_p = \frac{\sum_{t=1}^{N} \varepsilon^2(t, \hat{\theta}_N)}{\hat{s}_N^2} - (N - 2p)$$

Here p is the number of estimated parameters, and $\hat{s}_N^2$ is an estimate of the innovations variance, normally taken as the normalized sum of prediction errors for the largest model structure considered. C_p is to be minimized w.r.t p. Discuss the relationship between C_p and AIC.

16E.1 Consider the discrete-time system

$$S: \quad (1 - 0.95q^{-1})^5 y(t) = u(t-1)$$

and a model parametrized in terms of the ARX structure

$$\mathcal{M}: \quad y(t) + a_1 y(t-1) + \cdots + a_5 y(t-5) = u(t-1)$$

The true value of a_5 is thus $(-0.95)^5 = -0.77378$. Suppose that we obtain a model where $a_1, \ldots, a_4$ are exactly correct, but where $a_5 = -0.77379$. Where are the poles of this model? Suggest an alternative model structure for identifying S that is less sensitive to numerical errors.

16E.2 Consider the ARMAX model

$$A(q)y(t) = B(q)u(t) + C(q)e(t)$$

Discuss how to test the hypothesis that $A(q) = C(q)$, that is, we have white measurement errors.

16E.3 Consider a model structure

$$A(q)y(t) = q^{-k}B(q)u(t) + e(t)$$

with a time delay of k units. Discuss how to determine a good value of k based on several different techniques described in this chapter.

16E.4 Show that R defined by (16.21) is the correlation coefficient between $\hat{y}(t)$ and $y(t)$ in case the model has been estimated from the data by minimization of $J_1(\mathcal{M})$ in a linear regression model structure.

16T.1 Let $\{\varepsilon(t)\}$ be a process independent of $u(t)$ and such that

$$e(t) = H^{-1}(q)\varepsilon(t)$$

is white noise with variance λ. Let

$$\zeta(t) = \begin{bmatrix} u(t-s-1) \\ \vdots \\ u(t-s-M) \end{bmatrix}$$

Let

$$\hat{R}_u = \frac{1}{N}\sum_{t=1}^{N} \zeta(t)\zeta^T(t), \qquad \hat{r} = \frac{1}{\sqrt{N}}\sum_{t=1}^{N} \left[H^{-1}(q)\varepsilon(t)\right]\zeta(t)$$

Show that

$$\frac{1}{\lambda}\hat{r}^T \hat{R}_u^{-1}\hat{r} \in As\chi^2(M)$$

17

SYSTEM IDENTIFICATION IN PRACTICE

In this book we have dealt with the theory of system identification. Performing the identification task in practice enhances the "art side" of the topic. Experience, intuition, and insights will then play important roles. In this final chapter we shall discuss the system identification techniques as a toolbox for investigating, understanding, and mastering real-life systems. We shall first, in Section 17.1, describe the system identification tool in the hand of the user: interactive computing. Section 17.2 discusses the practical side of identification; how to approach the task. Then, in Section 17.3 we discuss a few applications to real data sets. Finally, in Section 17.4 we try to answer the ultimate question: What does system identification have to offer for real problems in engineering and applied science?

17.1 THE TOOL: INTERACTIVE SOFTWARE

The work to produce a model by identification is characterized by the following sequence:

1. Specify a model structure.
2. The computer delivers the best model in this structure.
3. Evaluate the properties of this model.
4. Test a new structure, go to step 1.

See Figure 17.1, which is a more elaborate version of Figure 1.10. The first thing that requires help is to compute the model and to evaluate its properties. There are now many commercially available program packages for identification that supply such help. They typically contain the following routines:

A *Handling of data, plotting, and the like*
Filtering of data, removal of drift, choice of data segments, and so on.

B *Nonparametric identification methods*
Estimation of covariances, Fourier transforms, correlation and spectral analysis, and so on.

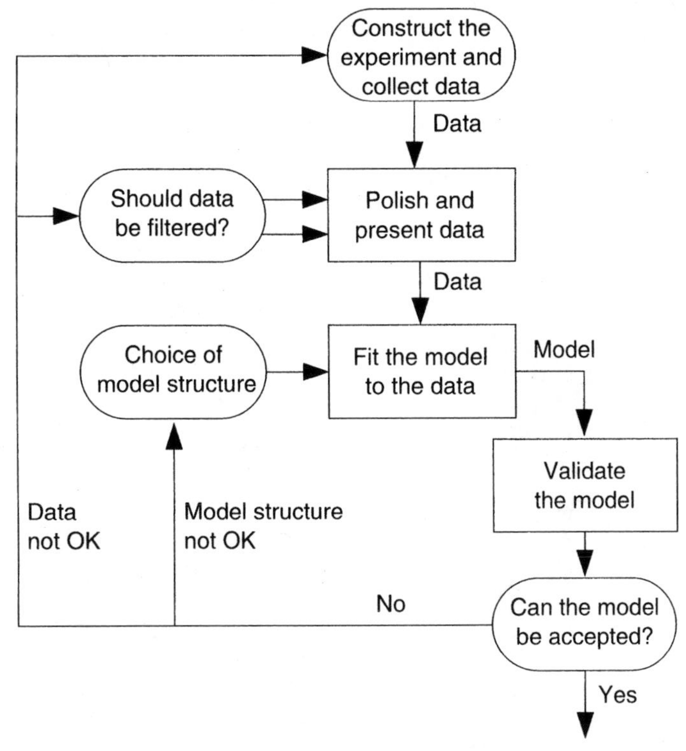

Figure 17.1 Identification cycle. Rectangles: the computer's main responsibility. Ovals: the user's main responsibility.

C *Parametric estimation methods*
Calculation of parametric estimates in different model structures.

D *Presentation of models*
Simulation of models, estimation and plotting of poles and zeros, computation of frequency functions and plotting in Bode diagrams, and so on.

E *Model validation*
Computation and analysis of residuals ($\varepsilon(t, \hat{\theta}_N)$); comparison between different models' properties, and the like.

The existing program packages differ mainly by various user interfaces and by different options regarding the choice of model structure according to item C.

One of the most used packages is MathWork's SYSTEM IDENTIFICATION TOOLBOX (SITB), Ljung (1995), which is used together with MATLAB. The command structure is given by MATLAB's programming environment with the work-space concept and MACRO possibilities in the form of m-files. SITB gives the possibility to use all model structures of the black-box type described in Section 4.2 with an arbitrary number of inputs. ARX-models and state-space models with an arbitrary number of inputs and outputs are also covered. Moreover, the user can define arbitrary tailor-made linear state-space models in discrete and continuous time as in (4.62), (4.84), and (4.91). A Graphical User Interface helps the user both to keep track of identified models and to guide him or her to available techniques.

Other packages of general type include PIM, Landau (1990), the Identification module in Matrix-X, Van Overschee et.al. (1994), Adapt$_X$, Larimore (1997), and the Frequency-Domain Identification Toolbox in MATLAB, Kollàr (1994). A tool for specifically dealing with gray box identification, IDKIT, is described in Graebe (1990).

17.2 THE PRACTICAL SIDE OF SYSTEM IDENTIFICATION

It follows from our discussion that the most essential element in the process of identification—once the data have been recorded—is to try out various model structures, compute the best model in the structures, using the techniques of Chapter 16, and then validate this model. Typically this has to be repeated with quite a few different structures before a satisfactory model can be found.

The difficulties of this process should not be underestimated, and it will require substantial experience to master it. Here follows a procedure that could prove useful to try out. This is adapted from Ljung (1995).

Step 1: Looking at the Data Plot the data. Look at them carefully. Try to see the dynamics with your own eyes. Can you see the effects in the outputs of the changes in the input? Can nonlinear effects be seen, like different responses at different levels, or different responses to a step up and a step down? Are there portions of the data that appear to be "messy" or carry no information? Use this insight to select portions of the data for estimation and validation purposes.

Do physical levels play a role in the model? If not, detrend the data by removing their mean values. The models will then describe how changes in the input give changes in output, but not explain the actual levels of the signals. This is the normal situation. The default situation, with good data, is to detrend by removing means, and then select the first two thirds or so of the data record for estimation purposes, and use the remaining data for validation. (All of this corresponds to the "Data Quickstart" in the MATLAB Identification Toolbox.)

Step 2: Getting a Feel for the Difficulties Compute and display the spectral analysis frequency response estimate, the correlation analysis impulse response estimate, as well as a fourth order ARX model with a delay estimated from the correlation analysis, and a default order state-space model computed by a subspace method. (All of this corresponds to the "Estimate Quickstart" in the MATLAB Identification Toolbox.) Look at the agreement between the

- Spectral Analysis estimate and the ARX and state-space models' frequency functions.
- Correlation Analysis estimate and the ARX and state-space models' transient responses.
- Measured Validation Data output and the ARX and state-space models' simulated outputs. We call this the *Model Output Plot.*

If these agreements are reasonable, the problem is not so difficult, and a relatively simple linear model will do a good job. Some fine tuning of model orders and noise models may have to be made, and we can proceed to Step 4. Otherwise go to Step 3.

Step 3: Examining the Difficulties There may be several reasons why the comparisons in Step 2 did not look good. This step discusses the most common ones, and how they can be handled:

- **Model Unstable:** The ARX or state-space model may turn out to be unstable, but could still be useful for control purposes. Then change to a 5- or 10-step ahead prediction instead of simulation when the agreement between measured and model outputs is considered. See (16.20).
- **Feedback in Data:** If there is feedback from the output to the input, due to some regulator, then the spectral and correlation analysis estimates, as well as the state-space model, are not reliable. Discrepancies between these estimates and the ARX model can therefore be disregarded in this case. In residual analysis of the parametric models, feedback in data can also be visible as correlation between residuals and input for negative lags.
- **Noise Model:** If the state-space model is clearly better than the ARX model at reproducing the measured output, this is an indication that the disturbances have a substantial influence, and it will be necessary to carefully model them.
- **Model Order:** If a fourth order model does not give a good Model Output plot, try eighth order. If the fit clearly improves, it follows that higher order models will be required, but that linear models could be sufficient.
- **Additional Inputs:** If the Model Output fit has not significantly improved by the tests so far, think over the physics of the application. Are there more signals that have been, or could be, measured that might influence the output? If so, include these among the inputs and try again a fourth order ARX model from all the inputs. (Note that the inputs need not at all be control signals; anything measurable, including disturbances, should be treated as inputs).
- **Nonlinear Effects:** If the fit between measured and model output is still bad, consider again the physics of the application. Are there nonlinear effects in the system? In that case, form the nonlinearities from the measured data. This could be as simple as forming the product of voltage and current measurements, if it is the electrical power that is the driving stimulus in, say, a heating process, and temperature is the output. This is of course application dependent. It does not cost very much work, however, to form a number of additional inputs by reasonable nonlinear transformations of the measured signals, and just test whether inclusion of them improves the fit.
- **General Nonlinear Mappings:** In some applications physical insight may be lacking, so it is difficult to come up with structured non-linearities on physical grounds. In such cases, nonlinear, black box models could be a solution. See Sections 5.4–5.6.
- **Still Problems?** If none of these tests leads to a model that is able to reproduce the validation data reasonably well, the conclusion might be that a sufficiently good model cannot be produced from the data. There may be many reasons for this. The most important one is that the data simply do not contain sufficient information, e.g., due to bad signal to noise ratios, large and non-stationary disturbances, varying system properties, etc.

Otherwise, use the insights on which inputs to use and which model orders to expect and proceed to Step 4.

Step 4: Fine Tuning Orders and Noise Structures For real data there is no such thing as a "correct model structure." However, different structures can give quite different model quality. The only way to find this out is to try out a number of different structures and compare the properties of the obtained models. There are a few things to look for in these comparisons:

- **Fit Between Simulated and Measured Output.** Look at the fit between the model's simulated output and the measured one for the validation data. Formally, pick that model, for which this number is the lowest. In practice, it is better to be more pragmatic, and also take into account the model complexity, and whether the important features of the output response are captured.
- **Residual Analysis Test.** For a good model, the cross correlation function between residuals and input does not go significantly outside the confidence region. See Section 16.6. A clear peak at lag k shows that the effect from input $u(t-k)$ on $y(t)$ is not properly described. A rule of thumb is that a slowly varying cross correlation function outside the confidence region is an indication of too few poles, while sharper peaks indicate too few zeros or wrong delays.

  For models that are to be used for control design, it is quite valuable to display the result of residual analysis in the frequency domain as in Example 16.3.
- **Pole Zero Cancellations.** If the pole-zero plot (including confidence intervals) indicates pole-zero cancellations in the dynamics, this suggests that lower order models can be used. In particular, if it turns out that the order of ARX models has to be increased to get a good fit, but that pole-zero cancellations are indicated, then the extra poles are just introduced to describe the noise. Then try ARMAX, OE, or BJ model structures with an A- or F-polynomial of an order equal to that of the number of non-cancelled poles.

What Model Structures Should be Tested? Well, any amount of time can be spent on checking out a very large number of structures. It often takes just a few seconds to compute and evaluate a model in a certain structure, so one should have a generous attitude to the testing. However, experience shows that when the basic properties of the system's behavior have been picked up, it is not much use to fine tune orders in absurdum just to improve the fit by fractions of percents. For ARX models and state-space models estimated by subspace methods there are also efficient algorithms for handling many model structures in parallel.

Multivariable Systems. Multivariable systems are often more challenging to model. In particular, systems with several outputs could be difficult. A basic reason for the difficulties is that the couplings between several inputs and outputs leads to more complex models: The structures involved are richer and more parameters will be required to obtain a good fit.

Generally speaking, it is preferable to work with state-space models in the multivariable case, since the model structure complexity is easier to deal with. It is essentially just a matter of choosing the model order.

Working with Subsets of the Input-Output Channels. In the process of identifying good models of a system it is often useful to select subsets of the input and output channels. Partial models of the system's behavior will then be constructed. It might not, for example, be clear if all measured inputs have a significant influence on the outputs. That is most easily tested by removing an input channel from the data, building a model for how the output(s) depend on the remaining input channels, and checking if there is a significant deterioration in the model output's fit to the measured one. See also the discussion under Step 3 above. Generally speaking, the fit gets better when more inputs are included and worse when more outputs are included. To understand the latter fact, it should be realized that a model that has to explain the behavior of several outputs has a tougher job than one that simply must account for a single output. If there are difficulties to obtain good models for a multi-output system, it might thus be wise to model one output at a time, to find out which are the difficult ones to handle. Models that just are to be used for simulations could very well be built up from single-output models, for one output at a time. However, models for prediction and control will be able to produce better results if constructed for all outputs simultaneously. This follows from the fact that knowing the set of all previous output channels gives a better basis for prediction than just knowing the past outputs in one channel.

Step 5: Accepting the Model The final step is to accept, at least for the time being, the model to be used for its intended application. Note the following, though: *No matter how good an estimated model looks on the computer screen, it has only picked up a simple reflection of reality. Surprisingly often, however, this is sufficient for rational decision making.*

17.3 SOME APPLICATIONS

The Hairdryer; A Laboratory Scale Application

Consider as a real, but laboratory scale process, Feedback's Process Trainer PT326, depicted in Figure 17.2. Its function is like a hairdryer: air is fanned through a tube and heated at the inlet. The input u is the power of the heating device, which is just a mesh of resistor wires. The output is the outlet air temperature. It should be said that the process is well behaved: it has reasonably simple dynamics with quite small disturbances. It also allows measurements with good signal-to-noise ratio.

Transient Response. The step response of the process is given in Figure 17.3. It reveals that the dynamics is simple, with no oscillatory poles, the dominating time constant is around 0.4 seconds, and there is a pure time delay of about 0.14 seconds.

Experiment Design. To collect data for further analysis, a few decisions have to be taken. Following the discussion of Section 13.7, we select a sampling interval of 0.08 s, since Figure 17.3 clearly shows that the dominating time constant is not much less than 0.4 s. A shorter sampling interval would also mean several delays between the (sampled) input and output sequences. The input was chosen to be a binary random signal shifting between 35 and 65 W. The probability of shifting the input at each sample was set to 0.2. A record of 1000 samples was collected, and the data set is shown in Figure 17.4. As a first step, the sample means of the input and

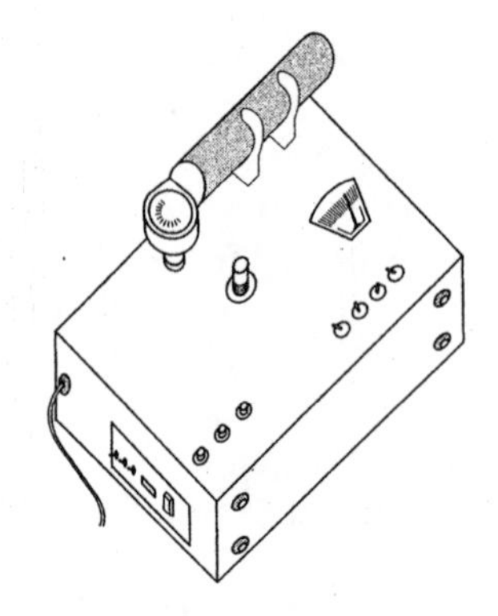

Figure 17.2 The hairdryer process.

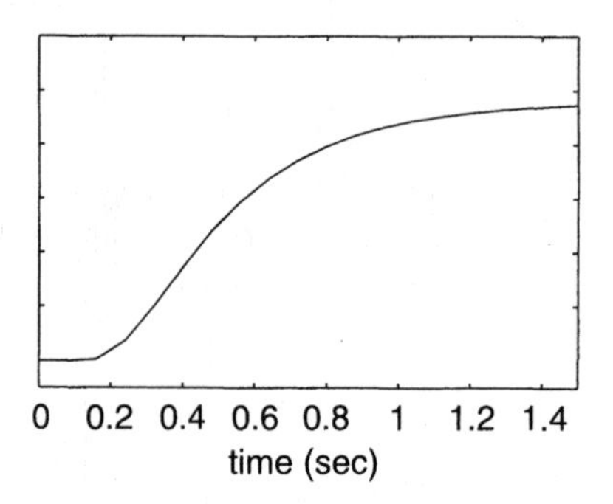

Figure 17.3 The step response from the process.

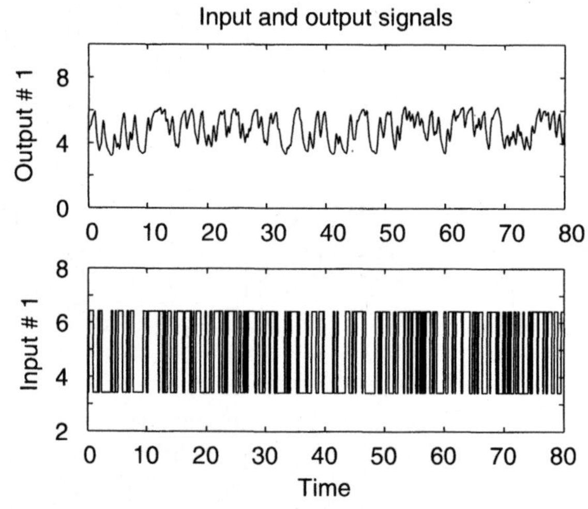

Figure 17.4 The data set from the process trainer. This is the same set as **`dryer2`** supplied with the System Identification Toolbox.

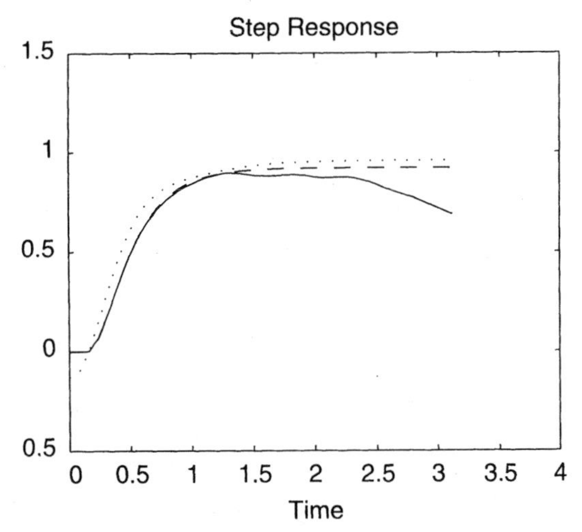

Figure 17.5 The step responses according to the correlation analysis estimate (solid), the fourth order ARX model (dashed), and the 3rd order state-space model (dotted).

output sequences were removed (see the discussion of Section 14.1). Then the data set was split into two halves, the first to be used for estimation, and the second one for validation.

Preliminary Models. Following Step 2 in Section 17.2 we estimate the step/impulse response by correlation analysis, as described in Section 6.1, compute the spectral analysis estimate of the frequency function, as well as a fourth order ARX model and a state-space model (using a subspace method, according to (7.66)). The order of this model is selected automatically, and turned out to be 3 in this case. The results of these calculations are shown in Figures 17.5, 17.6, and 17.7. These

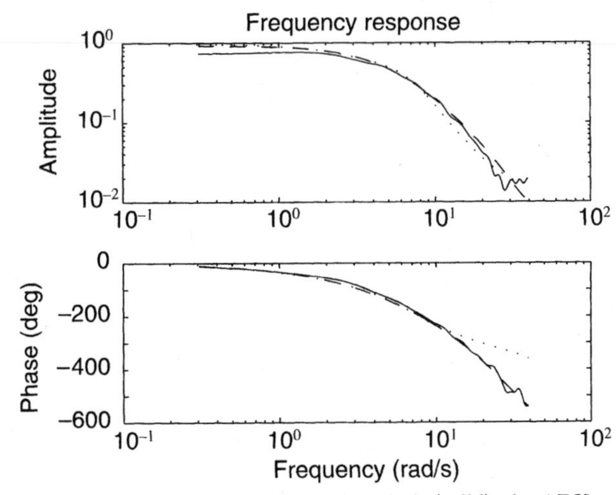

Figure 17.6 The Bode plots from spectral analysis (solid), the ARX model (dashed) and the state-space model (dotted).

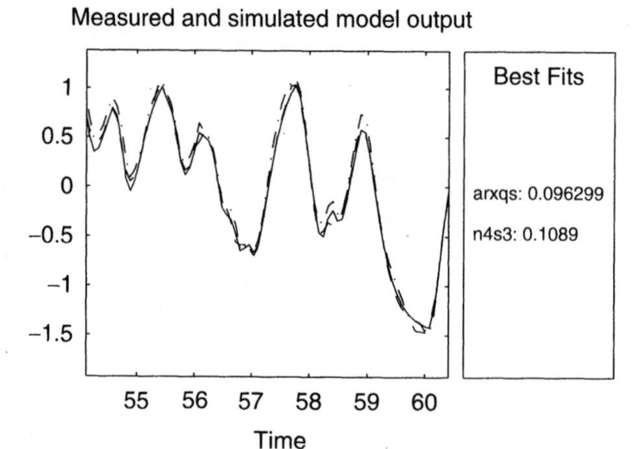

Figure 17.7 The measured output (dash-dotted) for the validation data set, together with the output from the ARX model (arxqs, solid), and the state-space model (n4s3, dashed) when simulated with the input sequence from the validation data set. The figures shown are the RMS-values of the difference between measured and simulated output.

plots show that we have good agreement between the models computed in different ways. This is a clear indication that the models have picked up essential features of the true process. Moreover, the comparisons in Figure 17.7 show that even these "immediate" models are able to reproduce the input-output behavior quite well. All this indicates, according to Step 4 of Section 17.2, that a linear model will do fine, and some further work to fine-tune orders and delays is all that remains.

Further Models. To look into suitable orders and delays, we compute, simultaneously, 1000 ARX-models of the type

$$y(t)+a_1y(t-1)+\ldots+a_{n_a}y(t-n_a)=b_1u(t-n_k)+\ldots+b_{n_b}u(t-n_k-n_b+1)$$

consisting of all combinations n_a, n_b, and n_k in the range 1 to 10. The different ARX models will be referred to as ARX(n_a, n_b, n_k). The prediction errors of each model are then computed for the validation data, and their sum of squares is computed. The result is shown in Figure 17.8, where the fit for the models is depicted as a function of the number of parameters used. Only the fit for best model with a given number of parameters is shown. The overall best fit is obtained for a model with 15 parameters, which turns out to be $n_a = 6$, $n_b = 9$ and $n_k = 2$. The figure also shows that almost as good a fit is obtained also for models with much less parameters, like 4. In this case the best orders turn out to be $n_a = 2, n_b = 2$, and $n_k = 3$. These models are added to the model output comparisons in Figure 17.9. We see that the higher order ARX model is able to reproduce the validation data best, but that the differences between the models really are minor. We compute also a state-space model of order 6 as well as an ARMAX-model with $n_a = 3, n_b = 3, n_c = 2$, and

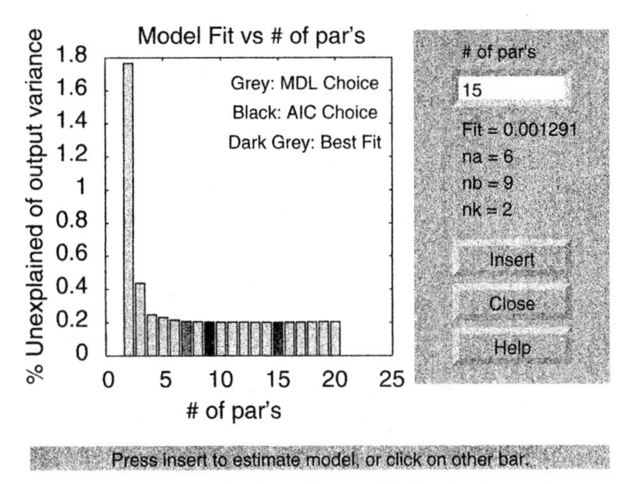

Figure 17.8 The best fit to validation data for ARX-models as a function of the number of used parameters.

$n_k = 2$. These models are also shown in the same plot. The residuals of the models (computed from the validation data) are analyzed in Figure 17.10. It shows that the ARMAX(3,3,2,2) model and the ARX(9,6,2) model both give residuals that pass whiteness and independence tests, while the model ARX(2,2,3) shows statistically significant correlation between past inputs and the residuals.

Final Choice of Model. Based on this analysis we conclude that there are many linear models that give a good fit to the system. The ARX(9,6,2) model shows the best fit to validation data, but is at the same time only marginally better than the simpler, third order ARMAX(3,3,2,2) model. Both also pass the residual analysis

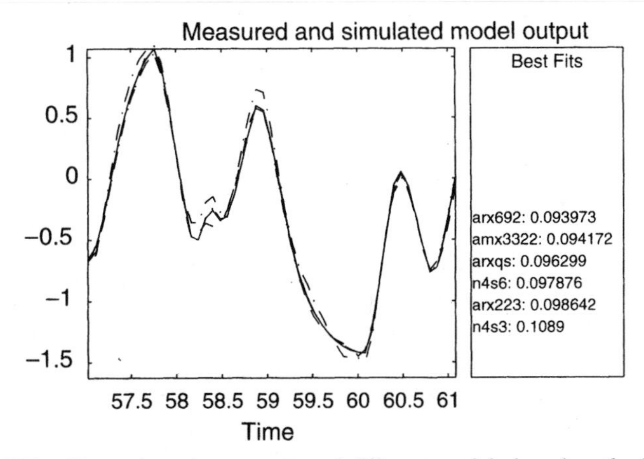

Figure 17.9 Comparisons between several different models, based on the fit between measured and simulated output.

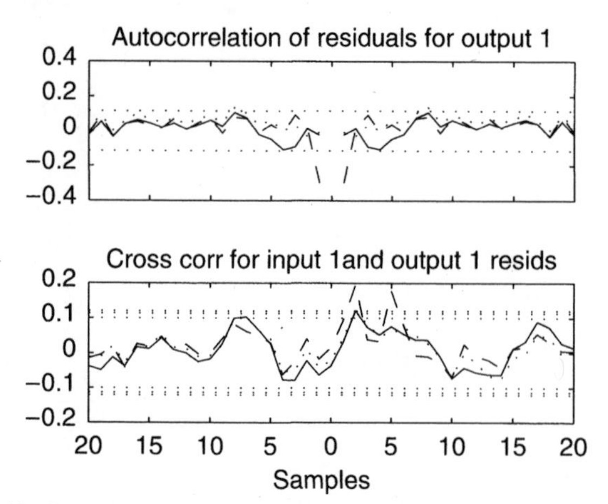

Figure 17.10 Results from the residual analysis of three different models: solid line ARX(9,6,2), dashed line ARX(2,2,3). Dotted lines ARMAX(3,3,2,2). The horizontal lines mark the confidence regions.

tests. It seems reasonable to pick this simple model as the final choice. The numerical value is

$$
\begin{aligned}
y(t) &- 1.4898y(t-1) + 0.7025y(t-2) - 0.1123y(t-3) = 0.0039u(t-2) \\
&+ 0.0621u(t-3) + 0.0284u(t-4) + e(t) - 0.5474e(t-1) \\
&+ 0.2236e(t-2)
\end{aligned}
$$

The estimated standard deviations of the 8 parameters are

$$
[\,0.0574 \quad 0.0849 \quad 0.0333 \quad 0.0015 \quad 0.0023 \quad 0.0055 \quad 0.0710 \quad 0.0523\,]
$$

We see that the coefficient for $u(t-2)$ is on the borderline from being significantly different from zero. This is the reason why models with delay n_k=3 also work well. However, the small effect from the term $u(t-2)$ does give an improved fit.

The estimated standard deviation of the noise source $e(t)$ is 0.0388.

A Fighter Aircraft

Consider the aircraft Example 1.2 with data shown in Figure 1.6. Note that these data were collected under closed loop operation.

To develop models of the aircraft's pitch channel from these data, we proceed as follows. The data set is first detrended, so that the means of each signal is removed. Then the data is split into one set consisting of the first 90 samples, to be used for estimation, and a validation data set consisting of the remaining 90 samples. As a main tool to screen models we computed the RMS fit between the measured output and the 10-step ahead predicted output according to the different models. In these

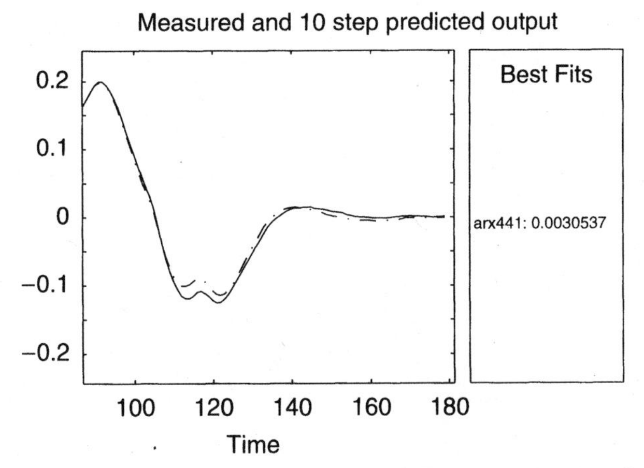

Figure 17.11 Measured output (dash-dotted line) and 10-step ahead predicted output (solid line) for aircraft validation data, using an ARX model with $n_a = 4, n_b^k = 4, n_k^k = 1, k = 1, 2, 3$.

calculation the whole data set was used—in order to let transients die out—but the fit was computed only for the validation part of the data. The reason for using 10 step ahead predictions rather than simulations is that the pitch cannel of the aircraft is unstable, and so will most of the estimated models also be. A simulation comparison may therefore be misleading.

A typical starting ARX model, using 4 past outputs and 4 past values of each of the 3 inputs, gave a fit according to Figure 17.11. We see that we get a good fit, so it seems reasonable that we can do a good modeling job with fairly simple models. As a next step we calculate 1000 ARX models corresponding to orders in inputs and outputs and delays ranging between 1 and 10. (In this case all 3 input orders were kept the same.) The best 1-step ahead prediction fit to the validation data turned out to be for a model

$$\begin{aligned} y(t) + a_1 y(t-1) + \ldots + a_{n_a} y(t-n_a) \\ = b_1^{(1)} u_1(t-1) + b_1^{(2)} u_2(t-1) + b_1^{(3)} u_3(t-1) + e(t) \end{aligned} \tag{17.1}$$

with $n_a = 8$. See Figure 17.14. Note in particular that models that use many parameters are considerably much worse for the validation data. Models of the kind (17.1) with other values of n_a were also estimated, as well as ARMAX models and state-space models using the N4SID method. A comparison plot based for several such models is shown in Figure 17.12. The best 10-step ahead prediction fit is obtained for the ARX model with $n_a = 4$. (Note, though, that the best 1-step ahead prediction is obtained for $n_a = 8$, as was said above.) The comparison for that model is shown in Figure 17.13. The result of residual analysis for this model on validation data is shown in Figure 17.15. We see that this simple model with 7 parameters is capable of reproducing new measurements quite well, at the same time it is not falsified by residual analysis.

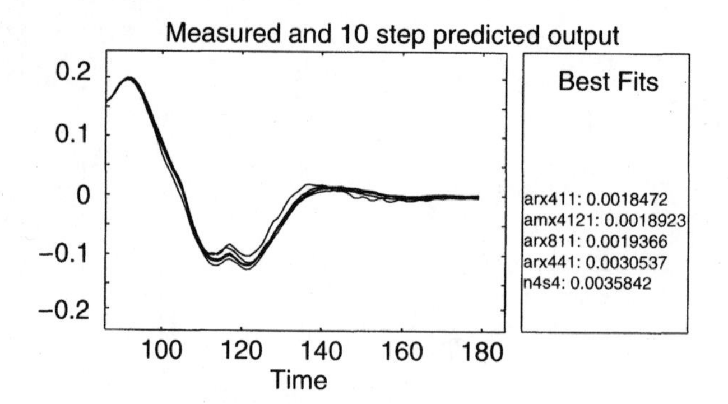

Figure 17.12 As Figure 17.11 but for several different models.

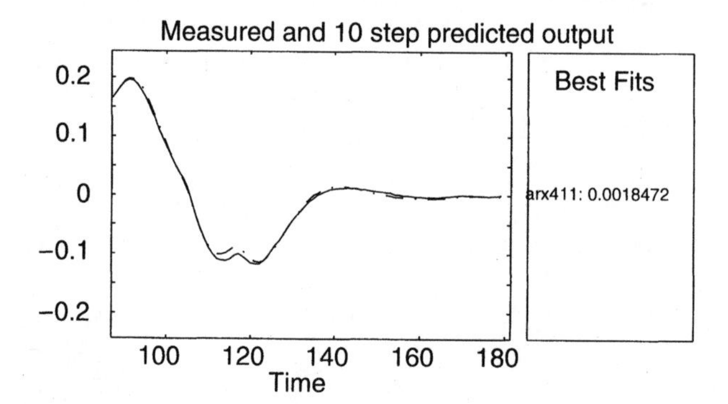

Figure 17.13 As Figure 17.11 but for the best ARX model.

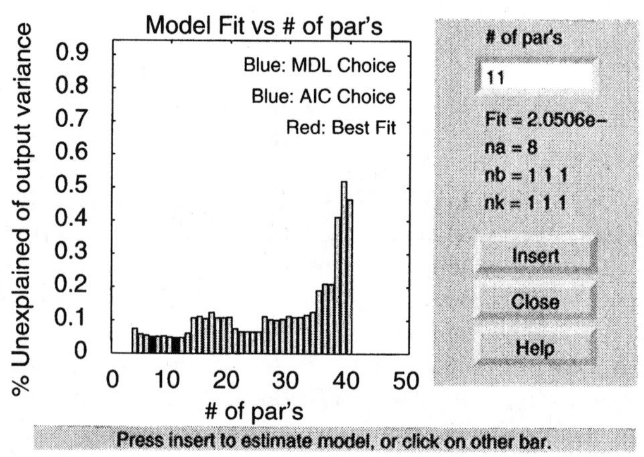

Figure 17.14 Comparisons of the 1-step ahead prediction error for 1000 ARX-models for the aircraft data.

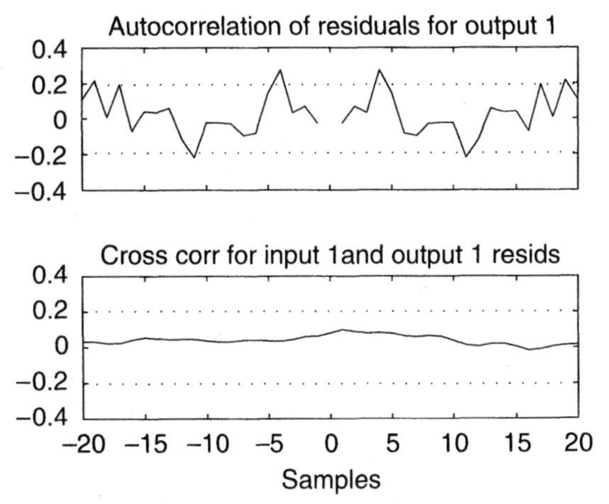

Figure 17.15 Residual analysis the best ARX model for the validation aircraft data.

Buffer Vessel Dynamics

This example concerns a typical problem in process industry. It is taken from the pulp factory in Skutskär, Sweden. Wood chips are cooked in the digester and the resulting pulp travels through several vessels where it is washed, bleached etc.

The pulp spends about 48 hours total in the process, and knowing the residence time in the different vessels is important in order to associate various portions of the pulp with the different chemical actions that have taken place in the vessel at different times. Figure 17.16 shows data from one buffer vessel. We denote the measurements as follows:

$y(t)$: The κ-number of the pulp flowing out

$u(t)$: The κ-number of the pulp flowing in

$f(t)$: The output flow

$h(t)$: The level of the vessel

The problem is to determine the residence time in the buffer vessel. (The κ-number is a quality property that in this context can be seen as a marker allowing us to trace the pulp.)

To estimate the residence time of the vessel it is natural to estimate the dynamics from u to y. That should show how long time it takes for a change in the input to have an effect on the output.

We can visually inspect the input-output data and see that the delay seems to be at least an hour or two. The sampling rate may therefore be too fast and we resample the data (decimate it) by a factor of 3, thus giving a sampling interval of 12

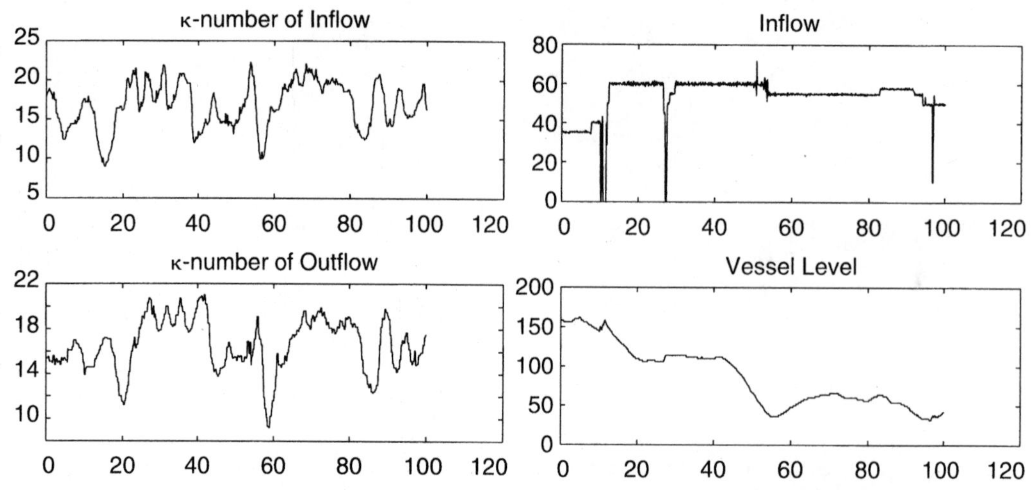

Figure 17.16 From the pulp factory at Skutskär, Sweden. The plots show the κ-number of the pulp flowing into a buffer vessel. The κ-number of the pulp coming out from the buffer vessel. Flow out from the buffer vessel. Level in the buffer vessel. The sampling interval is 4 minutes, and the time scale shown in hours.

minutes. We proceed as before, remove the means from the κ-number signals, split into estimation and validation data and estimate simple ARX-models. This turns out to give quite bad results.

According to the recipe of Section 17.2 we should then contemplate if there are more input signals that may affect the process. Yes, clearly the flow and level of the vessel should have something to do with the dynamics, so we include these two inputs. The best model output comparison was achieved for an ARX model with 4 parameters associated with the output and each of the inputs, a delay of 12 from u and a delay of 1 from f and h. This comparison is shown in Figure 17.17. This does not look good.

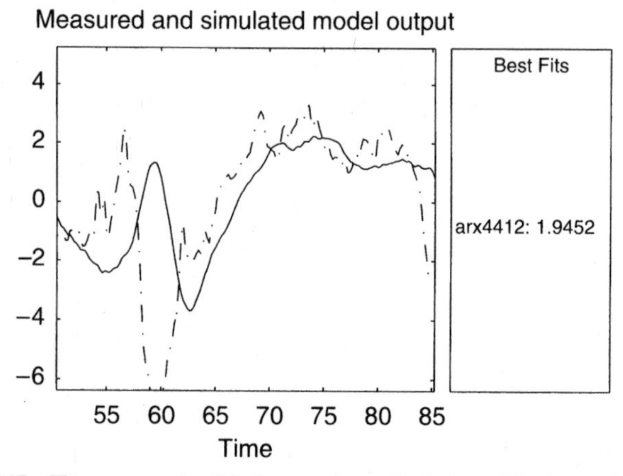

Figure 17.17 The measured validation output y (dash-dotted line) together with the best linear simulated model output for the system from u, f, h to y.

Some reflection shows that this process indeed must be non-linear (or time-varying): the flow and the vessel level definitely affect the dynamics. For example, if the flow was a plug flow (no mixing in the vessel) the vessel would have a dynamics of a pure delay equal to vessel volume divided by flow. This ratio, which has dimension time, is really the natural time scale of the process, in the sense that the delay would be constant in this time scale for a plug flow, even if vessel flow and level vary.

Let us thus resample the date accordingly, i.e. so that a new sample is taken (by interpolation from the original measurement) equidistantly in terms of integrated flow divided by volume. In MATLAB terms this will be

```
z = [y,u]; pf = f./h;
t =1:length(z)
newt = interp1(cumsum(pf+0.00001),t,[pf(1):sum(pf)]' );
newz = interp1(t,z, newt);
y1=newz(:,1); u1=newz(:,2)
```

(The small added number to **pf** is in order to overcome those time points where the flow is zero.) The resampled data are shown in Figure 17.18. We now apply the same procedure to the resampled data u_1 and y_1. The best ARX model fit was obtained for

$$y_1(t) + a_1 y_1(t-1) + \ldots + a_4 y_1(t-1) = b_1 u_1(t-9) + e(t)$$

Slightly better fit was obtained for an output-error model (4.25) with the same orders ($n_b = 1, n_f = 4, n_k = 9$). The comparison is shown in Figure 17.19. This "looks good."

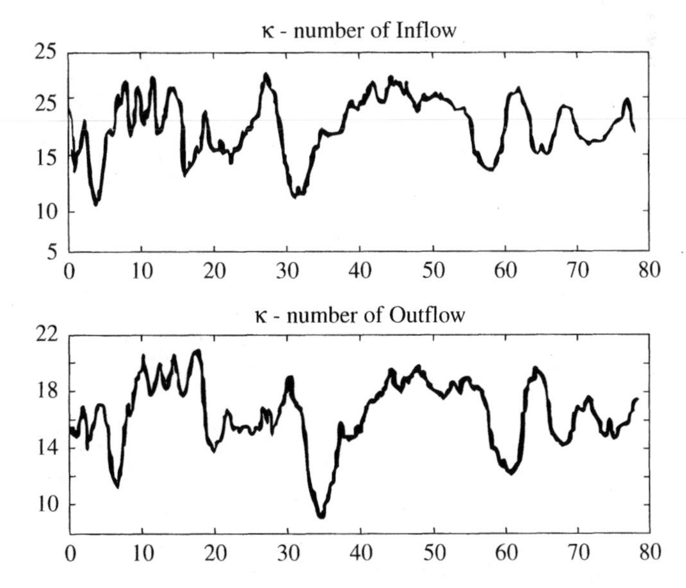

Figure 17.18 The input and output κ-numbers resampled according to the text.

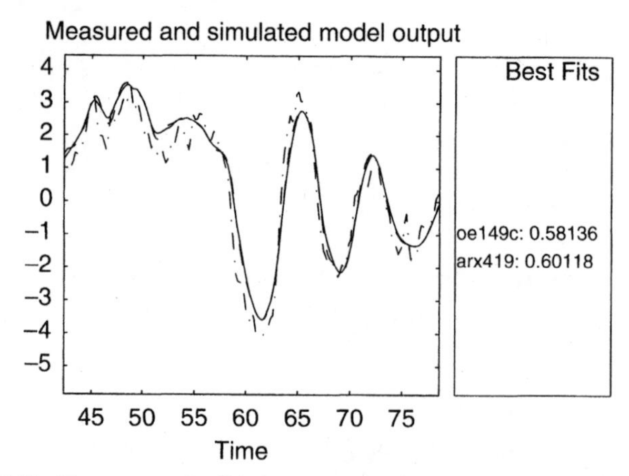

Figure 17.19 The measured validation output y_1 (dash-dotted line) together with simulated model outputs from resampled u_1. An ARX(419) model is shown as well as an OE model of the same orders.

The impulse responses of these models are shown in Figure 17.20. We see a delay of about 1.75 hours and then a time constant of about 2 hours. The vessel thus gives a pure delay as well as some mixing of the contents. The two impulse responses are in good agreement, if we take into account their uncertainties. See Andersson and Pucar (1995)for a more comprehensive treatment of the data in this example.

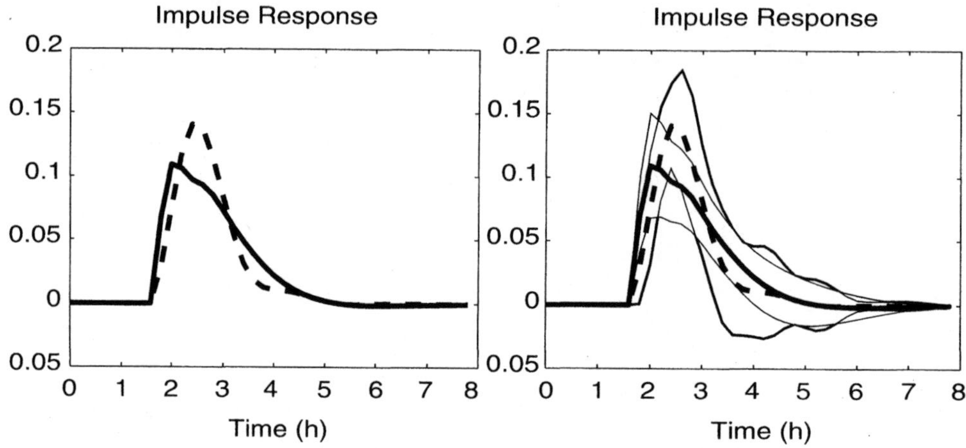

Figure 17.20 The impulse response of the ARX (solid) and OE (dashed) models. The right figure shows also the corresponding estimated 99% confidence intervals.

17.4 WHAT DOES SYSTEM IDENTIFICATION HAVE TO OFFER?

System identification techniques form a versatile tool for many problems in science and engineering. The techniques are, as such, application independent. The value of the tool has been evidenced by numerous applications in diverse fields. For example, the proceedings from the IFAC symposium series in System Identification contain thousands of successful applications from a wide selection of areas. Still, there are some limitations associated with the techniques, and we shall in this final section give some comments on this.

Adaptive and Robust Designs: Have They Made Modeling Obsolete?

As we discussed in Section 1.2, models of dynamical systems are instrumental for many purposes: prediction, control, simulation, filter design, reconstruction of measurements, and so on. It is sometimes claimed that the need for a model can be circumvented by more elaborate solutions: adaptive mechanisms where the decision parameters are directly adjusted or robust designs that are insensitive to the correctness of the underlying model. One should note, though, that adaptive schemes typically can be interpreted as the recursive identification algorithms described in Chapter 11 applied to a specific model structure (e.g., the model parameterized in terms of the corresponding optimal regulator); see Chapter 7 in Ljung and Söderström (1983). The model-building feature is thus very much present also in adaptive mechanisms.

Robust design is based on a nominal model and is determined so that good operation is secured even if the actual system deviates from the nominal model. Usually, a neighborhood around the nominal model can be specified within which performance degradation is acceptable. It is then a very useful fact that models obtained by system identification can be delivered with a quality tag: estimated deviations form a true description in the parameter domain or in the frequency domain. Such models are thus suited for robust design.

Limitations: Data Quality

It is obvious that the limitation of the use of system identification techniques is linked to the availability of good data and good model structures. Without a reasonable data record not much can be done, and there are several reasons why such a record cannot be obtained in certain applications. A first and quite obvious reason is that the time scale of the process is so slow that any informative data records by necessity will be short. Ecological and economical systems may clearly suffer from this problem. Another reason is that the input may not be open to manipulations, either by its nature or due to safety and production requirements. The signal-to-noise ratio could then be bad, and identifiability (informative data sets) perhaps cannot be guaranteed. Bad signal-to-noise ratios can, in theory, be compensated for by longer data records. Even if the plant, as such, admits long experimentation time, it may not always be a feasible way out, due to time variations in the process, drift, slow disturbances, and so on.

Finally, even when we are allowed to manipulate the inputs, can measure for long periods, and have good signal-to-noise ratios, it may still be difficult to obtain a good data record. The prime reason for this is the presence of unmeasurable disturbances that do not fit well into the standard picture of "stationary stochastic processes". We have discussed how to cope with such slow disturbances in Section

14.1 and how to handle occasional "bursts" by robust norms (Section 15.2), and such measures may often be successful. The fact remains though: data quality must be a prime concern in system identification applications. This also determines the cost of the exercise.

Limitations: Model Structures

It is trivial that a bad model structure cannot offer a good model, regardless of the amount and quality of the available data. For example, the ARX model structure in Figure 17.17 can never provide a good description of the buffer vessel dynamics even if fitted to data collected over several years. The crucial nonlinear mechanisms must be built in, and this requires physical insight.

The first problem thus is whether the process (around its operation point of interest) admits a standard, linear, ready-made ("black-box") model description, or whether a tailor-made model set must be constructed. In the first case, our chances of success are good; in the second, we have to resort to some physical insight before a model can be estimated or hoping that the nonlinear dynamics can be picked up by a nonlinear black-box structure. This problem clearly is application dependent and therefore not so much discussed in the identification literature. It cannot, however, be sufficiently stressed that the key to success lies here: Thinking, intuition, and insights cannot be made obsolete by automated model construction.

Appendix I

SOME CONCEPTS FROM PROBABILITY THEORY

In this appendix we list some basic concepts and notions from probability theory that are used in the book. See a textbook like Papoulis (1965)or Chung (1979)for a proper treatment.

A *random variable* e describes the possible numerical outcomes of experiments whose results cannot be exactly predicted beforehand. The probability of the numerical values falling in certain ranges is then expressed by the *probability density function* (PDF) $f_e(x)$:

$$P(a \leq e < b) = \int_a^b f_e(x)dx \tag{I.1}$$

If e may assume a certain value with nonzero probability, we can think of f_e containing a δ-function component of that value. A formal treatment then replaces (I.1) by a Stieltjes integral, but this is not essential to our needs.

For a random vector

$$e = \begin{bmatrix} e_1(t) \\ \vdots \\ e_n(t) \end{bmatrix}$$

a corresponding PDF $f_e(x) = f_e(x_1, \ldots, x_n)$ from $\mathbf{R}^n$ to $\mathbf{R}$ is defined and

$$P(e \in B) = \int_{x \in B} f_e(x)dx \tag{I.2}$$

B here is a subset of $\mathbf{R}^n$ and $P(A)$ means "the probability of the event A." $f_e(x)$ is also known as the *joint PDF* for $e_1, \ldots, e_n$. The *expectation* or *mean value* of e is defined as

$$Ee = \int_{\mathbf{R}^N} x f_e(x)dx \tag{I.3}$$

while the *covariance matrix* is

$$\text{Cov}\, e = E(e - m)(e - m)^T; \qquad m = Ee \tag{I.4}$$

The random vector e is said to have *Gaussian* or *normal* distribution if

$$f_e(x) = \frac{1}{(2\pi)^{n/2}} \frac{1}{(\det P)^{1/2}} \exp\left[-\tfrac{1}{2}(x - m)^T P^{-1}(x - m)\right] \tag{I.5}$$

The mean is then m and the covariance matrix is P. This will be written

$$e \in N(m, P) \tag{I.6}$$

With two random variables y and z we may define the joint PDF as $f(x_y, x_z)$. The probabilities associated with outcomes of y only, disregarding z, are then given by the PDF for y, $f_y(x_y)$:

$$P(a \leq y < b) = \int_a^b f_y(x_y) dx_y$$

Since

$$\begin{aligned} P(a \leq y < b) &= P(a \leq y < b \text{ and } -\infty \leq z \leq +\infty) \\ &= \int_{x_y=a}^{b} \int_{x_z=-\infty}^{\infty} f(x_y, x_z) dx_z dx_y \end{aligned}$$

we find that

$$f_y(x_y) = \int_{x_z=-\infty}^{\infty} f(x_y, x_z) dx_z \tag{I.7}$$

We can now introduce the *conditional PDF of z given y* as

$$f_{z|y}(x_z|x_y) = \frac{f(x_y, x_z)}{f_y(x_y)} \tag{I.8}$$

We then have

$$P(a \leq z < b|y = x_y) = \int_a^b f_{z|y}(x_z|x_y) dx_z \tag{I.9}$$

Here $P(A|B)$ is the *conditional probability of the event A given the event B*. Intuitively, we can think of (I.9) as the probability that z will assume a value between a and b if we already know that the outcome of y was x_y. Note that formal definitions of these concepts require more attention, and the reader should consult a textbook on probability for that. The expression (I.8) can be seen as a version of Bayes's rule:

$$P(A|B) = \frac{P(A \text{ and } B)}{P(B)} \tag{I.10}$$

Two random variables y and z are *independent* if

$$f(x_y, x_z) = f_y(x_y) \cdot f_z(x_z) \tag{I.11}$$

Then

$$f_{z|y}(x_z|x_y) = f_z(z_z) \tag{I.12}$$

We occasionally deal with *complex-valued random variables* in this book. If e may assume complex values, we define the covariance matrix as

$$\text{Cov } e = E(e - m)(\overline{e} - \overline{m})^T \tag{I.13}$$

where the overbar means complex conjugate. Notice that this concept does not give full information about the covariation between the real and imaginary parts of e. For a complex-valued random vector e, the notation

$$e \in N_c(m, P) \tag{I.14}$$

will mean that

1. The real and imaginary parts of e are jointly normal.
2. $Ee = m$ (a complex number).
3. $\text{Cov } e = P$ [defined as in (I.13)].
4. $\text{Re } e$ and $\text{Im } e$ are independent.
5. $\text{Cov Re } e = \text{Cov Im } e = \frac{1}{2}P$.

Let $y(t)$, $t = 1, 2, \ldots$, be a sequence of random vectors (a discrete time stochastic process). The outcome or *realization* of this sequence will then be a sequence of vectors. Suppose that the event that this sequence converges to a limit y^* (that may depend on the realization) as t tends to infinity has probability 1. Then we say that $\{y(t)\}$ *converges* to y^* (a random vector) *with probability 1* (w.p. 1) (or "almost surely," a.s., "almost everywhere," a.e.):

$$y(t) \to y^*, \qquad \text{w.p. 1 as } t \to \infty \tag{I.15}$$

Often in our applications y^* will in fact not depend on the realization.

If the associated sequence of PDFs, $f_{y(t)}(x)$, converges (weakly) to a PDF f^*,

$$f_{y(t)}(x) \to f^*(x) \tag{I.16}$$

we say that $\{y(t)\}$ *converges in distribution* to the PDF f^*. In the special case when f^* is the Gaussian distribution (I.5), we say that $\{y(t)\}$ is *asymptotically normal* with mean m and covariance P, and denote it as

$$y(t) \in AsN(m, P) \tag{I.17}$$

Useful theorems for proving results like (I.17) are given in Problem II.3 and in Lemmas 9A.1, 9A.2. They are usually known as *central limit theorems* (CLTs). To prove convergence with probability 1, the following version of *Borel-Cantelli's lemma* is a good tool:

$$\text{“If } \sum_{k=1}^{\infty} P(|y(k)| > \varepsilon) < \infty, \text{ for all } \varepsilon > 0, \text{ then} \quad y(t) \to 0 \text{ w.p. 1 as } t \to \infty\text{”} \tag{I.18}$$

(see Chung, 1974, for a proof).

To estimate probabilities of this kind, *Chebyshev's inequality* is useful:

$$P(|y| > \varepsilon) \leq \frac{1}{\varepsilon^2} \cdot E|y|^2 \tag{I.19}$$

Appendix II

SOME STATISTICAL TECHNIQUES FOR LINEAR REGRESSIONS

The purpose of this appendix is twofold: First, to provide a refresher of basic statistical techniques so as to form a proper background for Part II of this book; second, methods, algorithms, theoretical analysis, and statistical properties for linear regression estimates are all archetypal for the more complicated structures we discuss in Part II. This appendix can therefore also be read as a preview of ideas and analysis, maximally stripped from technical complications. The appendix has a format so that it can be read independently of the rest of the book (and vice versa).

II.1 LINEAR REGRESSIONS AND THE LEAST SQUARES ESTIMATE

Linear regressions are among the most common models in statistics, and the least-squares technique with its root in Gauss's (1809) work is certainly classical. Treatments of these techniques are given in many textbooks, and we may mention Rao (1973)(Chapter 4), Draper and Smith (1981), and Daniel and Wood (1980)as suitable references for further study.

The Regression Concept

The statistical theory of regression is concerned with the prediction of a variable y, on the basis of information provided by other measured variables $\varphi_1, \ldots \varphi_d$. The dependent variable y could, for example, be the yield of a certain crop, while the independent variables φ_i (the *regressors*) give information about rainfall, sunshine, soil quality, and the like. There are abundant examples of this situation across all fields of science and society. The dynamical systems that we consider in Part I clearly form another application of the regression concept, y being the output of a system (at a given time) and φ_i containing information about past behavior. Let us denote

$$\varphi = \begin{bmatrix} \varphi_1 \\ \varphi_2 \\ \vdots \\ \varphi_d \end{bmatrix}$$

The problem is to find a function of the regressors $g(\varphi)$ such that the difference

$$y - g(\varphi)$$

becomes small (i.e., so that $\hat{y} = g(\varphi)$ is a good prediction of y). If y and φ are described within a stochastic framework, one could, for example, aim at minimizing

$$E\,[y - g(\varphi)]^2 \tag{II.1}$$

It is well known that the function g that minimizes (II.1) is the conditional expectation of y, given $\varphi_1 \ldots \varphi_d$:

$$g(\varphi) = E[y|\varphi] \tag{II.2}$$

This is also known as the *regression function* or the regression of y on φ.

Another approach would be to look for the function $\hat{g}(\varphi)$ that has maximal correlation with y. The answer is essentially the regression function. See Problem II.2.

Linear Regressions

With unknown properties of the variables y and φ, it is not possible to determine the regression function $g(\varphi)$ a priori. It has to be estimated from data and must therefore be suitably parametrized. The special case where this parametrization is constrained to be linear has been studied extensively. We are then trying to fit y to a linear combination of the φ_i:

$$g(\varphi) = \theta_1\varphi_1 + \theta_2\varphi_2 + \ldots + \theta_d\varphi_d \tag{II.3}$$

With the vector

$$\theta = \begin{bmatrix} \theta_1 \\ \theta_2 \\ \vdots \\ \theta_d \end{bmatrix}$$

(II.3) can be written

$$g(\varphi) = \varphi^T\theta \tag{II.4}$$

Remark: Of course, "affine" functions

$$g(\varphi) = \theta_{d+1} + \varphi^T\theta \tag{II.5}$$

could also be considered. By extending the regressors by the constant $\varphi_{d+1} = 1$ and the parameter vector θ accordingly, the case (II.5) is however subsumed in (II.4).

Least-squares Estimate

Typically, we are not supplied with exact a priori information about the relationship between $y(t)$ and $\varphi(t)$. What we have instead are "historic data," a collection of previous observations of related values of y and φ. It is convenient to enumerate these values using an argument t:

$$y(t), \varphi(t), \qquad t = 1, \dots, N \tag{II.6}$$

With the historic data we could replace the variance (II.1) by the sample variance

$$\frac{1}{N} \sum_{t=1}^{N} [y(t) - g(\varphi(t))]^2$$

In the linear case (II.4), we thus have

$$V_N(\theta) = \frac{1}{N} \sum_{t=1}^{N} \left[y(t) - \varphi^T(t)\theta\right]^2 \tag{II.7}$$

instead of (II.1), and a suitable θ to choose is the minimizing argument of (II.7):

$$\hat{\theta}_N = \arg\min V_N(\theta) \tag{II.8}$$

This is the least-squares estimate (LSE). Based on the previous observations, we would thus use

$$\varphi^T \hat{\theta}_N$$

as a predictor function.

Notice that this method of selecting θ makes sense whether or not we have imposed a stochastic framework for the problem. The parameter $\hat{\theta}_N$ is simply the value that gives the best performing predictor when applied to historic data. This "pragmatic" interpretation of the LSE was given also by its inventor, K. F. Gauss:

> *In conclusion, the principle that the sum of the squares of the differences between the observed and the computed quantities must be minimum may, in the following manner, be considered independently of the calculus of probabilities (Gauss, 1809).*

The unique feature of (II.7) is that it is a quadratic function of θ. Therefore, it can be minimized analytically (see Problem 7D.2). We find that all $\hat{\theta}_N$ that satisfy

$$\left[\frac{1}{N} \sum_{t=1}^{N} \varphi(t)\varphi^T(t)\right] \hat{\theta}_N = \frac{1}{N} \sum_{t=1}^{N} \varphi(t) y(t) \tag{II.9}$$

yield the global minimum of $V_N(\theta)$. This set of linear equations is known as the *normal equations*. If the matrix on the left is invertible, we have the LSE

$$\hat{\theta}_N = \left[\frac{1}{N} \sum_{t=1}^{N} \varphi(t)\varphi^T(t)\right]^{-1} \frac{1}{N} \sum_{t=1}^{N} \varphi(t) y(t) \tag{II.10}$$

Matrix Formulation

For some calculations, the expressions (II.6) to (II.10) can be written more conveniently in matrix form. Define the $N \times 1$ column vector

$$Y_N = \begin{bmatrix} y(1) \\ \vdots \\ y(N) \end{bmatrix} \tag{II.11}$$

and the $N \times d$ matrix

$$\Phi_N = \begin{bmatrix} \varphi^T(1) \\ \vdots \\ \varphi^T(N) \end{bmatrix} \tag{II.12}$$

Then the criterion (II.7) can be written

$$V_N(\theta) = \frac{1}{N}|Y_N - \Phi_N\theta|^2 = \frac{1}{N}(Y_N - \Phi_N\theta)^T(Y_N - \Phi_N\theta) \tag{II.13}$$

The normal equations take the form

$$[\Phi_N^T\Phi_N]\hat{\theta}_N = \Phi_N^T Y_N \tag{II.14}$$

and the estimate

$$\hat{\theta}_N = [\Phi_N^T\Phi_N]^{-1}\Phi_N^T Y_N \tag{II.15}$$

We may in (II.15) recognize the (Moore-Penrose) pseudoinverse of Φ_N:

$$\Phi_N^\dagger = [\Phi_N^T\Phi_N]^{-1}\Phi_N^T \tag{II.16}$$

Equation (II.15) thus gives the pseudoinverse solution to the overdetermined $(N > d)$ system of linear equations

$$Y_N = \Phi_N\theta \tag{II.17}$$

Geometric Interpretation

The least-squares solution can be given a geometric interpretation that may be helpful when determining certain properties. Let

$$\Phi_N = [\phi_1 \dots \phi_d]$$

and consider Y_N and $\phi_1 \dots \phi_d$ as vectors in the vector space $\mathbf{R}^N$. The problem expressed in (II.17) is to find a linear combination of the vectors ϕ_i, $i = 1, \dots, d$, that approximates Y_N as well as possible. Let D_d be the d-dimensional subspace that is spanned by the ϕ_i. If Y_N happens to belong to this subspace, we can describe it as a unique linear combination of ϕ_i. Otherwise, the best approximation of Y_N in the subspace D_d is the vector in D_d that has the smallest distance to Y_N, which is well known to be the orthogonal projection of Y_N on D_d. See Figure II.1.

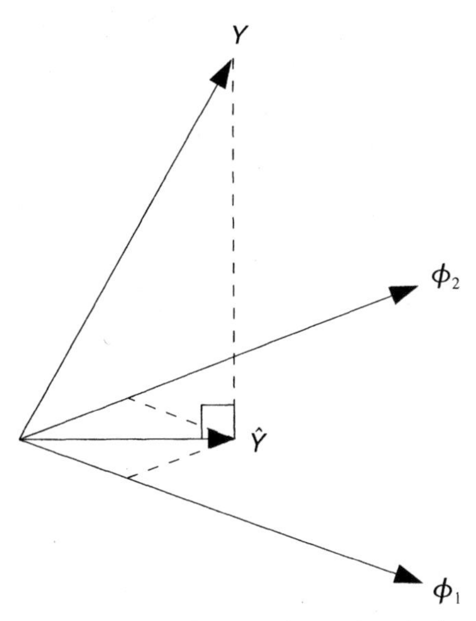

Figure II.1 The least-squares solution as orthogonal projection. ($d = 2$ and $N = 3$.

Let this projection be denoted by $\hat{Y}_N$. Since it is the orthogonal projection, we have

$$(Y_N - \hat{Y}_N) \perp \phi_i$$

That is,

$$(Y_N - \hat{Y}_N)^T \phi_i = 0, \qquad i = 1 \dots d$$

and since $\hat{Y}_N \in D_d$, we have for some coordinates $\hat{\theta}_i$

$$\hat{Y}_N = \sum_{j=1}^{d} \hat{\theta}_j \phi_j$$

This gives

$$Y_N^T \phi_i = \sum_{j=1}^{d} \hat{\theta}_j \phi_j^T \phi_i, \qquad i = 1 \dots d \tag{II.18}$$

which in matrix form is (II.14).

Weighted Least Squares

In (II.7) the different observations are given equal weight in the criterion. Sometimes there is occasion to consider a weighted criterion

$$V_N(\theta) = \sum_{t=1}^{N} \alpha_t \left[y(t) - \varphi^T(t)\theta \right]^2 \tag{II.19}$$

The reason for this could be twofold.

1. The observations y could be of varying *reliability*. Some observation could, for example, be subject to more disturbances and should therefore be down-weighted. (II.20)
2. The observations could be of varying *relevance*. It is perhaps not believed that a linear model holds over all ranges of φ. An observation, corresponding to φ in such a questionable region, even if accurate, should therefore carry less weight. (II.21)

With the diagonal matrix

$$Q_N = \begin{bmatrix} \alpha_1 & & 0 \\ & \ddots & \\ 0 & & \alpha_N \end{bmatrix} \tag{II.22}$$

the criterion (II.19) can be written

$$V_N(\theta) = (Y_N - \Phi_N\theta)^T Q_N (Y_N - \Phi_N\theta) \triangleq |Y_N - \Phi_N\theta|^2_{Q_N} \tag{II.23}$$

It is immediate to verify that the minimizing element is given by

$$\begin{aligned} \hat{\theta}_N &= [\Phi_N^T Q_N \Phi_N]^{-1} \Phi_N^T Q_N Y_N \\ &= \left[\sum_{t=1}^{N} \alpha_t \varphi(t)\varphi^T(t)\right]^{-1} \sum_{t=1}^{N} \alpha_t \varphi(t) y(t) \end{aligned} \tag{II.24}$$

There could also be reason to use the criterion (II.23) for a general, symmetric positive definite Q_N. The former part of (II.24) then still holds. To interpret what is going on in terms of the original measurements, it is convenient to factorize Q_N:

$$Q_N = L_N^T D_N L_N \tag{II.25}$$

with L_N as a lower triangular matrix with 1's along the diagonal:

$$L_N = \begin{bmatrix} 1 & 0 & 0 & \dots 0 \\ \ell_{21} & 1 & 0 & \dots 0 \\ \ell_{31} & \ell_{32} & 1 & 0 \dots 0 \\ \vdots & \vdots & & \vdots \\ \ell_{N1} & \ell_{N2} & \dots & 1 \end{bmatrix} \tag{II.26}$$

and D_N a diagonal matrix as in (II.22). Then (II.23) takes the form

$$V_N(\theta) = |\tilde{Y}_N - \tilde{\Phi}_N\theta|^2_{D_N} \tag{II.27}$$

$$\tilde{Y}_N = L_N Y_N, \qquad \tilde{\Phi}_N = L_N \Phi_N \tag{II.28}$$

The elements of these matrices are

$$\tilde{y}(t) = \sum_{k=0}^{t-1} \ell_{t(t-k)} y(t-k) \tag{II.29a}$$

$$\tilde{\varphi}(t) = \sum_{k=0}^{t-1} \ell_{t(t-k)} \varphi(t-k) \tag{II.29b}$$

We thus have

$$V_N(\theta) = \sum_{t=1}^{N} \alpha_t \left[\tilde{y}(t) - \tilde{\varphi}^T(t)\theta\right]^2 \tag{II.30a}$$

$$\hat{\theta}_N = \left[\sum_{t=1}^{N} \alpha_t \tilde{\varphi}(t)\tilde{\varphi}^T(t)\right]^{-1} \sum_{t=1}^{N} \alpha_t \tilde{\varphi}(t)\tilde{y}(t) \tag{II.30b}$$

The effect of the general norm Q_N in (II.23) is consequently that the original observations have been filtered by the filter (II.25) to (II.29).

Residuals and Prediction Errors

The difference

$$\varepsilon(t, \theta) = y(t) - \varphi^T(t)\theta \tag{II.31}$$

is the error associated with the value θ. We shall call this error the *prediction error* corresponding to θ. The vector of prediction errors is

$$E_N(\theta) = \begin{bmatrix} \varepsilon(1, \theta) \\ \vdots \\ \varepsilon(N, \theta) \end{bmatrix} \tag{II.32}$$

and the criteria (II.7) and (II.23) are just different quadratic norms of this vector. Norms Q_N that are not diagonal correspond to sums of squares of filtered prediction errors analogously to (II.29).

We shall call

$$\hat{\varepsilon}_N(t) = \varepsilon(t, \hat{\theta}_N)$$

the *residuals* ("leftovers") associated with the model $\hat{\theta}_N$.

Consider now for simplicity the case $Q_N = I$. Denote the residual vector

$$\hat{E}_N = E_N(\hat{\theta}_N) \tag{II.33}$$

and the predicted output

$$\hat{Y}_N = \Phi_N \hat{\theta}_N \tag{II.34}$$

From the geometrical interpretation we know that $\hat{E}_N$ and $\hat{Y}_N$ are orthogonal. Hence

$$|Y_N|^2 = (\hat{Y}_N + \hat{E}_N)^T(\hat{Y}_N + \hat{E}_N) = |\hat{Y}_N|^2 + |\hat{E}_N|^2 \tag{II.35}$$

which also can be written

$$\sum_{t=1}^{N} y^2(t) = \sum_{t=1}^{N} \hat{y}_N^2(t) + \sum_{t=1}^{N} \hat{\varepsilon}_N^2(t) \tag{II.36}$$

which shows how the sum of squared observation splits into predictions

$$\hat{y}_N(t) = \varphi^T(t)\hat{\theta}_N$$

and residuals

$$\hat{\varepsilon}_N(t) = \varepsilon(t, \hat{\theta}_N) \tag{II.37}$$

The ideal situation is when the predicted outputs $\hat{y}_N$ are capable of explaining a major part of the actual output. The ratio

$$R_y^2 = \frac{\sum_{t=1}^{N} \hat{y}_N^2(t)}{\sum_{t=1}^{N} y^2(t)} = 1 - \frac{\sum_{t=1}^{N} \varepsilon_N^2(t)}{\sum_{t=1}^{N} y^2(t)} \tag{II.38}$$

measures the proportion of the total variation of y that is explained by the regression. It is known as the *multiple correlation coefficient* (squared) and is often expressed in percent. Sometimes the mean value of y is subtracted from $\hat{y}$ and y before calculating R_y^2.

Quality of the Parameter Estimate

To investigate what properties the estimate $\hat{\theta}_N$ may have, let us assume that the actual measurements $y(t), t = 1, \ldots, N$, can be described by

$$y(t) = \varphi^T(t)\theta_0 + w_0(t) \tag{II.39}$$

where $\{w_0(t)\}$ is some disturbance or error sequences of yet unspecified nature. If this sequence has some "nice" (to be specified later) properties, it is natural to call θ_0 "the true parameter."

If we denote

$$W_N = \begin{bmatrix} w_0(1) \\ \vdots \\ w_0(N) \end{bmatrix} \tag{II.40}$$

we may write (II.39) as

$$Y_N = \Phi_N\theta_0 + W_N \tag{II.41}$$

Inserting this into (II.24) gives

$$\hat{\theta}_N = [\Phi_N^T Q_N \Phi_N]^{-1} \Phi_N^T Q_N [\Phi_N \theta_0 + W_N] = \theta_0 + \tilde{\theta}_N \tag{II.42}$$

where

$$\tilde{\theta}_N = [\Phi_N^T Q_N \Phi_N]^{-1} \Phi_N^T Q_N W_N$$

which in the case $Q_N = 1/N \cdot I$ also reads

$$\tilde{\theta}_N = \left[\frac{1}{N}\sum_{t=1}^{N} \varphi(t)\varphi^T(t)\right]^{-1} \left[\frac{1}{N}\sum_{t=1}^{N} \varphi(t) w_0(t)\right] \tag{II.43}$$

This expression for the parameter error is of purely algebraic nature and holds for all sequences $\{w_0(t)\}$. If φ and w are quasistationary, we see that, as N tends to infinity, $\tilde{\theta}_N$ tends to

$$\tilde{\theta} = \left[R_\varphi(0)\right]^{-1} R_{\varphi w}(0) \tag{II.44}$$

with the notation (2.62). If $R_\varphi(0)$ is invertible [which corresponds to an assumption that the sequence $\varphi(t)$ has full rank] and $R_{\varphi w}(0)$ is zero (which corresponds to a certain "independence" between the regressors and the disturbance), then $\hat{\theta}_N$ will tend to the true value θ_0 when more observations become available.

To be able to tell more about the properties of $\hat{\theta}_N$, it is natural to create a probabilistic framework for the disturbance sequence. This will be done in the next section.

II.2 STATISTICAL PROPERTIES OF THE LEAST-SQUARES ESTIMATE

A Probabilistic Setup

To achieve further results on the properties of the LSE, we shall introduce more specific assumptions about the generation of the observations y. Typical assumptions follow:

- The sequence of regressors $\{\varphi(t)\}$ is a deterministic sequence. (II.45)
- $y(t) = \varphi^T(t)\theta_0 + w_0(t)$ where $w_0(t)$ is a sequence of independent random variables with zero mean values and variances λ_0. (II.46)

Let it immediately be said that assumption (II.45) is too restrictive for most applications to system identification, since then the regressors typically contain past outputs. This means that the analysis will not be applicable to system identification methods. However, (II.45) greatly simplifies the analysis, at the same time as the results are archetypal for what holds also in the general case. The contents of this section can therefore be read as a simple preview of the material of Chapters 8 and 9.

In this section we shall relax assumption (II.46) somewhat. We shall allow more general disturbances $\{w_0(t)\}$ and also, occasionally, that the true regression function is not compatible with the given linear model. We thus have the description

$$y(t) = g_t(\varphi(t)) + w_0(t) \tag{II.47a}$$

where

$$\varphi(t) \text{ is deterministic} \tag{II.47b}$$

$$Ew_0(t) = 0 \tag{II.47c}$$

$$Ew_0(t)w_0(s) = r_{ts} \tag{II.47d}$$

In matrix form, with (II.11), (II.12), (II.40), and

$$G_N(\Phi_N) = \begin{bmatrix} g_1(\varphi(1)) \\ \vdots \\ g_N(\varphi(N)) \end{bmatrix} \tag{II.48}$$

we have

$$Y_N = G_N(\Phi_N) + W_N \tag{II.49a}$$

$$EW_N = 0 \tag{II.49b}$$

$$EW_N W_N^T = R_N \tag{II.49c}$$

We shall frequently specialize to

$$G_N(\Phi_N) = \Phi_N \theta_0 \tag{II.50}$$

and

$$R_N = \lambda_0 \cdot I \tag{II.51}$$

Convergence and Consistency

Consider the special case (II.10) (i.e., $Q_N = (1/N)I$). Then if (II.50) holds we can write, as in (II.43),

$$\hat{\theta}_N - \theta_0 = \left[\frac{1}{N}\sum_{t=1}^{N} \varphi(t)\varphi^T(t)\right]^{-1} \left[\frac{1}{N}\sum_{t=1}^{N} \varphi(t)w_0(t)\right] \tag{II.52}$$

The first sum consists of deterministic variables. Suppose that it converges to an invertible matrix $R_\varphi(0)$:

$$\frac{1}{N}\sum_{t=1}^{N} \varphi(t)\varphi^T(t) \to R_\varphi(0) \quad \text{as } N \to \infty, \qquad R_\varphi(0) \text{ invertible} \tag{II.53}$$

The second sum consists of random variables with zero mean values under assumption (II.47b, c, d). Various "strong laws of large numbers" describe when such sums converge to zero with probability 1. This will depend on the properties of $\{w_0(t)\}$. See, for example, Chung (1974)or Hall and Heyde (1980)for thorough treatments of such results. Our Theorem 2.3 shows that

$$\frac{1}{N}\sum_{t=1}^{N}\varphi(t)w_0(t) \rightarrow 0, \qquad \text{w.p. 1 as } N \rightarrow \infty \tag{II.54}$$

if $\{w_0(t)\}$ can be described as filtered white noise as in (2.88) and $\{\varphi(t)\}$ is a bounded sequence. When (II.53) and (II.54) hold, we have

$$\hat{\theta}_N \rightarrow \theta_0, \qquad \text{w.p. 1 as } N \rightarrow \infty \tag{II.55}$$

This means that $\hat{\theta}_N$ is a *strongly consistent* estimate of θ_0.

Bias and Variance

Consider the general weighted LSE

$$\hat{\theta}_N = [\Phi_N^T Q_N \Phi_N]^{-1}\Phi_N^T Q_N Y_N \tag{II.56}$$

Since Φ_N and Q_N are deterministic, it is easy to calculate the expectation of (II.56).

$$E\hat{\theta}_N = \theta^* = [\Phi_N^T Q_N \Phi_N]^{-1}\Phi_N^T Q_N G_N(\Phi_N) \tag{II.57}$$

When (II.50) holds, we have

$$E\hat{\theta}_N = \theta_0 \tag{II.58}$$

which means that the estimate $\hat{\theta}_N$ is *unbiased* in case a true description (II.50) is available.

For the parameter difference, we obtain from (II.56), (II.57), and (II.49a)

$$\tilde{\theta}_N \triangleq \hat{\theta}_N - E\hat{\theta}_N = [\Phi_N^T Q_N \Phi_N]^{-1}\Phi_N^T Q_N W_N \tag{II.59}$$

which gives the covariance matrix

$$\operatorname{Cov}\hat{\theta}_N = P_N = E\tilde{\theta}_N\tilde{\theta}_N^T = [\Phi_N^T Q_N \Phi_N]^{-1}\Phi_N^T Q_N R_N Q_N \Phi_N [\Phi_N^T Q_N \Phi_N]^{-1} \tag{II.60}$$

Notice that this holds regardless of the form of $G_N(\Phi_N)$.

For the nonweighted LSE $[Q_N = (1/N)\cdot I]$ and independent disturbances, case (II.51), we find that

$$P_N = \lambda_0[\Phi_N^T\Phi_N]^{-1} = \lambda_0\left[\sum_{t=1}^{N}\varphi(t)\varphi^T(t)\right]^{-1} \tag{II.61}$$

The covariance matrix of $\hat{\theta}_N$ is thus determined by the residuals' variance λ_0 and properties of the regressors. When the $\varphi(t)$ are open to manipulation during the data collection, it is an important *experiment design* issue to make the inverse in (II.61) "small," subject to constraints that may be at hand.

Note that computation of P_N requires knowledge of λ_0. Since this may not be known to the user, it is important to estimate it from data.

Estimating the Noise Variance

We have the following result:

Lemma II.1. Let the criterion be given by (II.7) and suppose that (II.49) to (II.51) hold. Then

$$\hat{\lambda}_N = \frac{N}{N-d} V_N(\hat{\theta}_N) = \frac{1}{N-d} \sum_{t=1}^{N} \left[y(t) - \varphi^T(t)\hat{\theta}_N \right]^2 \tag{II.62}$$

is an unbiased estimate of λ_0. (Recall that $d = \dim \theta$.)

Proof. We have

$$\begin{aligned} Y_N - \Phi_N \hat{\theta}_N &= Y_N - \Phi_N [\Phi_N^T \Phi_N]^{-1} \Phi_N^T Y_N \\ &= \left[I - \Phi_N [\Phi_N^T \Phi_N]^{-1} \Phi_N^T \right] [\Phi_N \theta_0 + W_N] \\ &= \left[I - \Phi_N [\Phi_N^T \Phi_N]^{-1} \Phi_N^T \right] W_N \triangleq F_N W_N \end{aligned} \tag{II.63}$$

Note that $F_N^T F_N = F_N$. Hence

$$\begin{aligned} E|Y_N - \Phi_N \hat{\theta}_N|^2 &= E W_N^T F_N^T F_N W_N = E W_N^T F_N W_N \\ &= E \text{ tr } F_N W_N W_N^T = \lambda_0 \text{ tr } F_N \end{aligned}$$

Now

$$\begin{aligned} \text{tr } F_N &= \text{tr} \left[I - \Phi_N [\Phi_N^T \Phi_N]^{-1} \Phi_N^T \right] = \text{tr } I - \text{tr } \Phi_N [\Phi_N^T \Phi_N]^{-1} \Phi_N^T \\ &= \text{tr } I - \text{tr} [\Phi_N^T \Phi_N]^{-1} \Phi_N^T \Phi_N = N - d \end{aligned} \tag{II.64}$$

Recall that F_N is an $N \times N$ matrix and $\Phi_N^T \Phi_N$ is a $d \times d$ matrix. Here we made use of $\text{tr } AB = \text{tr } BA$.

Consequently,

$$E\hat{\lambda}_N = E \frac{1}{N-d} |Y_N - \Phi_N \hat{\theta}_N|^2 = \lambda_0$$

and the lemma is proved. □

Notice that both assumptions (II.50) and (II.51) are important for this result. It is not feasible to estimate a general covariance matrix R_N from data. Also, if the true description is of the general form (II.47a), the estimate $\hat{\lambda}_N$ will contain a contribution

$$\frac{1}{N-d}\sum_{t=1}^{N}\left[g_t(\varphi(t)) - \varphi^T(t)\hat{\theta}_N\right]^2$$

that will not tend to zero. This makes $\hat{\lambda}_N$ systematically larger than λ_0, and the noise level is overestimated. This points to a very typical dilemma in many applications: to distinguish noise effects from bias effects.

Minimizing the Variance

From (II.60) we know that the variance of $\hat{\theta}_N$ depends on the choice of norm Q_N. One might ask what the best choice of Q_N is in the sense that it gives the smallest covariance matrix. This question is answered by the following lemma:

Lemma II.2. Let R_N be a positive definite matrix and define

$$P_N(Q) = [\Phi_N^T Q \Phi_N]^{-1} \Phi_N^T Q R_N Q \Phi_N [\Phi_N^T Q \Phi_N]^{-1}$$

Then for all symmetric, positive semidefinite Q,

$$P_N(R_N^{-1}) \le P_N(Q)$$

Proof. The matrix

$$\begin{bmatrix} \Phi_N^T \\ \Phi_N^T Q R_N \end{bmatrix} R_N^{-1} \begin{bmatrix} \Phi_N^T \\ \Phi_N^T Q R_N \end{bmatrix}^T = \begin{bmatrix} \Phi_N^T R_N^{-1} \Phi_N & \Phi_N^T Q \Phi_N \\ \Phi_N^T Q \Phi_N & \Phi_N^T Q R_N Q \Phi_N \end{bmatrix}$$

is positive semidefinite by construction. Hence, according to Problem 7D.8,

$$\Phi_N^T R_N^{-1} \Phi_N \ge \Phi_N^T Q \Phi_N [\Phi_N^T Q R_N Q \Phi_N]^{-1} \Phi_N^T Q \Phi_N$$

Inverting both sides proves the lemma. □

The estimate (II.24) obtained for $Q = R_N^{-1}$ is also known as the *Markov estimate* or the best linear unbiased estimate (BLUE). Notice that it requires knowledge of the covariance matrix R_N, which might not be a realistic assumption.

In case the noise terms in (II.47) are independent with different variances,

$$E w_0^2(t) = \lambda_t$$

the lemma tells us that the variance of the estimate is minimized when the criterion (II.19) is used with

$$\alpha_t = \frac{1}{\lambda_t} \tag{II.65}$$

That is, the observations should be weighted with their inverse variances. This choice of weights is thus optimal in the sense of (II.20). Notice, though, that when (II.50) does not hold, the weights α_t will also affect the bias of $\hat{\theta}_N$, and this effect may be in conflict with (II.65).

Distribution of the Estimates

The estimates $\hat{\theta}_N$ of θ_0 and $\hat{\lambda}_N$ of λ_0 are random variables, since they are constructed from the random variables $\{y(t)\}$. It is thus of interest to determine their distribution. In order to do that we shall introduce the following additional assumption:

$$\text{The vector } W_N \text{ of disturbance terms has a Gaussian distribution.} \tag{II.66}$$

This assumption implies that Y_N will also have a Gaussian distribution with mean value $G_N(\Phi_N)$ and variance R_N:

$$Y_N \in N(G_N(\Phi_N), R_N) \tag{II.67}$$

Since $\hat{\theta}_N$ in (II.56) is a linear combination of Y_N, $\hat{\theta}_N$ will also be Gaussian:

$$\hat{\theta}_N \in N(\theta^*, P_N) \tag{II.68}$$

where θ^* and P_N are given by (II.57) and (II.60), respectively. This answers the question of the distribution of $\hat{\theta}_N$ under assumption (II.66).

Even when the observations are not normally distributed, it is often the case that the distribution of $\hat{\theta}_N$ approaches the normal distribution as N increases to infinity. This follows from application of central limit theorems (CLTs) to the sum of random variables that constitutes the estimates. See Problem II.3.

The distribution of $\hat{\lambda}_N$ in (II.62) is somewhat more technical to determine. Let us again consider the special case (II.50) and (II.51) and let $V_N(\theta)$ be defined by (II.7). Then

$$N \cdot V_N(\theta_0) = \sum_{t=1}^{N} \left[y(t) - \varphi^T(t)\theta_0\right]^2 = \sum_{t=1}^{N} w_0^2(t) \tag{II.69}$$

Now, under (II.51) and (II.66), $\{w_0(t)\}_1^N$ is a sequence of independent Gaussian random variables with variances λ_0. Hence

$$\frac{N}{\lambda_0} V_N(\theta_0) \in \chi^2(N) \tag{II.70}$$

That is, the left side is χ^2-distributed with N degrees of freedom, by definition of the χ^2-distribution. When θ_0 is replaced by $\hat{\theta}_N$, we have a related result:

Lemma II.3. Assume that (II.49) to (II.51) and (II.66) hold. Then

$$\frac{N}{\lambda_0} \cdot V_N(\hat{\theta}_N) = \frac{1}{\lambda_0} \sum_{t=1}^{N} \left[y(t) - \varphi^T(t)\hat{\theta}_N\right]^2 \in \chi^2(N-d) \tag{II.71}$$

Proof. Compare with the proof of Lemma II.1. Let

$$F_N = I - \Phi_N[\Phi_N^T\Phi_N]^{-1}\Phi_N^T$$

which is a symmetric $N \times N$ matrix. Then, as in (II.63), $F_N = F_N F_N$, which implies that all eigenvalues of F_N are either 0 or 1. Since tr $F_N = N - d$ by (II.64), we find that there are $N - d$ eigenvalues that are 1 and d that are 0. Since F_N is symmetric, it can be diagonalized by an orthogonal matrix U_N:

$$U_N F_N U_N^T = D_N$$

where D_N consists of $N - d$ ones and d zeros along the diagonal. As in the proof of Lemma II.1,

$$\frac{N}{\lambda_0}V_N(\hat{\theta}_N) = \frac{1}{\lambda_0}|Y_N - \Phi_N\hat{\theta}_N|^2 = \frac{1}{\lambda_0}|F_N W_N|^2 = \frac{1}{\lambda_0}W_N^T F_N^T F_N W_N$$

$$= \frac{1}{\lambda_0}W_N^T F_N W_N = \frac{1}{\lambda_0}W_N^T U_N^T D_N U_N W_N = \frac{1}{\lambda_0}\sum_{t=1}^{N-d}\overline{w}_N^2(t)$$

where $\overline{w}_N(t)$ are the components of the vector $U_N W_N$. But since the components of W_N are independent and normal, with variance λ_0, so are those of $U_N W_N$, the matrix U_N being orthogonal. This proves the lemma. □

A consequence of the lemma is that the estimate $\hat{\lambda}_N$ of λ_0 obeys

$$(N - d)\frac{\hat{\lambda}_N}{\lambda_0} \in \chi^2(N - d) \tag{II.72}$$

which implies that

$$E(\hat{\lambda}_N - \lambda_0)^2 = \frac{2\lambda_0^2}{N - d} \tag{II.73}$$

Confidence Intervals

In the case (II.50) where a true parameter θ_0 exists, the distribution result (II.68) tells us how the deviation between $\hat{\theta}_N$ and θ_0 is distributed:

$$\hat{\theta}_N - \theta_0 \in N(0, P_N) \tag{II.74}$$

For the ith component, we thus have

$$\hat{\theta}_N^{(i)} - \theta_0^{(i)} \in N(0, P_N^{(ii)})$$

or

$$\frac{\hat{\theta}_N^{(i)} - \theta_0^{(i)}}{\sqrt{P_N^{(ii)}}} \in N(0, 1) \tag{II.75}$$

Here $P_N^{(ii)}$ indicates the ith diagonal element of P_N. Hence the probability that $\theta_0^{(i)}$ deviates from $\hat{\theta}_N^{(i)}$ with more than $\alpha \cdot \sqrt{P_N^{(ii)}}$ is the $(1 - \alpha)$-level of the normal distribution, which is available in standard statistical tables.

In fact, (II.68) tells us more than how each component of $\hat{\theta}_N$ is distributed. Since P_N is the covariance matrix of the joint distribution of the vector $\hat{\theta}_N$, we also have useful information about the covariance and correlation between the different components of $\hat{\theta}_N$. This is most easily utilized as follows. From (II.74) we have

$$(\hat{\theta}_N - \theta_0)^T P_N^{-1}(\hat{\theta}_N - \theta_0) \in \chi^2(d) \tag{II.76}$$

by a direct application of the definition of the χ^2-distribution. The probability that

$$|\hat{\theta}_N - \theta_0|^2_{P_N^{-1}} = (\hat{\theta}_N - \theta_0)^T P_N^{-1}(\hat{\theta}_N - \theta_0) \geq \alpha \tag{II.77}$$

is thus $\chi^2_\alpha(d)$, the α-level of the $\chi^2(d)$-distribution. The expressions (II.77) define ellipsoids in $\mathbf{R}^d$, whose shape is determined by P_N. See Figure II.2.

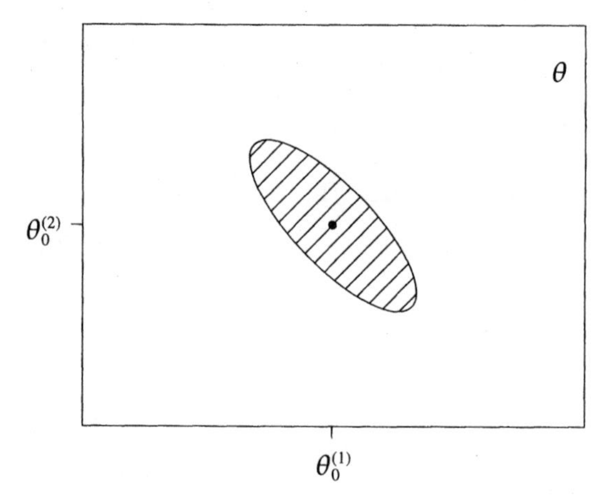

Figure II.2 Shaded area: $|\hat{\theta}_N - \theta_0|^2_{P^{-1}} \leq$ constant.

In case $Q_N = I$ and $R_N = I$, the expression for P_N is (II.61)

$$P_N = \lambda_0[\Phi_N^T \Phi_N]^{-1} \tag{II.78}$$

While the matrix $[\Phi_N^T \Phi_N]^{-1}$ is known to the user, λ_0 is typically not, which impairs the use of the results (II.75) and (II.76). An immediate approach would be to replace λ_0 by the estimate $\hat{\lambda}_N$. In view of (II.73), this is a good approximation for large N, and the use of the preceding confidence limits is still reasonable. With Lemma II.2, a more exact result can, however, be achieved. We have

$$\frac{1}{\hat{\lambda}_N} \cdot (\hat{\theta}_N - \theta_0)^T [\Phi_N^T \Phi_N](\hat{\theta}_N - \theta_0) = \frac{(\hat{\theta}_N - \theta_0)^T P_N^{-1}(\hat{\theta}_N - \theta_0)}{\hat{\lambda}_N/\lambda_0}$$
$$\in F(d, N-d) \tag{II.79}$$

where the last step follows from the definition of the F-distribution as the distribution of the ratio of two χ^2-distributed variables. The leftmost expression is available to the user, and thus the probability that θ_0 deviates from $\hat{\theta}_N$ in such a way that

$$\frac{1}{\hat{\lambda}_N}|\hat{\theta}_N - \theta_0|^2_{[\Phi_N^T\Phi_N]} \geq \alpha \tag{II.80}$$

can be computed as the α-level of the $F(d, N-d)$ distribution. Note that, as $N \to \infty$, $F(d, N-d)$ approaches $\chi^2(d)$. Hence using $\hat{\lambda}_N$ for λ_0 in P_N in (II.77) is often reasonable. Similarly, using $\hat{\lambda}_N$ for λ_0 in (II.75) replaces the normal distribution by Student's t-distribution.

II.3 SOME FURTHER TOPICS IN LEAST-SQUARES ESTIMATION

Selecting Regressors

A major problem when applying regression analysis to practical problems is to settle for a good set of regressor variables. This means that we have to determine which variables $\varphi_i(t)$ may influence the output $y(t)$ in (II.3). Clearly, this is a very application dependent task and requires a good understanding of the process to be described. The choice of φ_i may, however, also be supported by some formal procedures that are described in the statistical literature.

Selecting the regressors in (II.3) corresponds to the choice of model structure in system identification setup. This problem is discussed in more detail in Chapter 16, and we shall only provide a short preview here.

A basic tool is to investigate the sequence of residuals $\{\hat{\varepsilon}(t)\}_1^N$, defined by (II.37). If (II.49) to (II.51) hold and $\hat{\theta}_N$ gives a correct description of the process, then $\hat{\varepsilon}_N(t)$ would equal $w_0(t)$ in (II.46) and would thus be a sequence of random variables. Such a hypothesis can be tested in various ways. Also, the residual sequence should be uncorrelated with all potential regressors. If it is not, the regressor in question has something to offer for the prediction of $y(t)$ and it should thus be included in the regressor set. This may be a useful way of conducting the search for informative regressors. Daniel and Wood (1980)contains practical advice in this respect.

Another way of determining whether a regressor should be included in the regressor set is to check whether it leads to a significant reduction of the criterion function $V_N(\hat{\theta}_N)$. It should be clear that the minimal value of the criterion function will automatically decrease when a new regressor is added, whether or not it is actually correlated with the output. This follows since the minimization of V_N is performed over a larger set. As a simple special case, think of the process (II.49) to (II.51) with $\theta_0 = 0$. The criterion function

$$V_N(0) = \frac{1}{N}\sum_{t=1}^{N} y^2(t) = \frac{1}{N}\sum_{t=1}^{N} w_0^2(t)$$

then has mean value λ_0. If, however, we minimize over a d-dimensional regressor space θ, we obtain

$$V_N(\hat{\theta}_N) = \frac{1}{N}\sum_{t=1}^{N} \hat{\varepsilon}_N^2(t)$$

which according to Lemma II.1 has mean value $[(N-d)/N]\lambda_0$. We thus see an average decrease of the criterion function of $(d/N)\lambda_0$, despite the fact that there was nothing in $y(t)$ to explain by the regressors! The improved fit is spurious and can be seen as an overfit to the particular realization of $\{w_0(t)\}_1^N$.

Consequently, the observed decrease in V_N when new regressors are added must be matched against this overfit. Several ways to do this are discussed in Chapter 16. For the present setup we have the following formal result:

Lemma II.4. Suppose that the data can be described by

$$y(t) = \varphi^T(t)\theta_0 + e_0(t) \tag{II.81}$$

where $\{e_0(t)\}$ is white Gaussian noise with variance λ_0. Let

$$V_N^{(1)} = \min_{\theta} \sum_{t=1}^{N} [y(t) - \varphi^T(t)\theta]^2 \tag{II.82}$$

Consider another regression with some added regressors

$$y(t) = \varphi^T(t)\theta + \zeta^T(t)\eta$$

and let

$$V_N^{(2)} = \min_{\theta,\eta} \sum_{t=1}^{N} [y(t) - \varphi^T(t)\theta - \zeta^T(t)\eta]^2 \tag{II.83}$$

Let

$$d = \dim\theta, \qquad r = \dim\eta$$

Then

(a) $\dfrac{V_N^{(2)}}{\lambda_0} \in \chi^2(N-d-r)$

(b) $\dfrac{V_N^{(1)} - V_N^{(2)}}{\lambda_0} \in \chi^2(r)$

(c) $V_N^{(1)} - V_N^{(2)}$ and $V_N^{(2)}$ are independent

(d) $$t(d,r,N) = \frac{N-d-r}{V_N^{(2)}} \cdot \frac{V_N^{(1)} - V_N^{(2)}}{r} \in F(r, N-d-r) \tag{II.84}$$

Proof. (a) is a restatement of Lemma II.2 and (d) is a consequence of (a), (b), and (c) by the definition of the F-distribution. The proofs of (b) and (c) are analogous to the proof of Lemma II.2. □

The lemma implies that, if (II.81) holds so that the inclusion of η is "unnecessary," then the normalized decrease in the criterion is distributed according to (II.84). If the observed decrease is significantly larger [i.e., if $t(d, r, N) > F_\alpha(r, N-d-r)$], then the conclusion should be that the inclusion of η is useful and that, consequently, (II.81) does not hold. Recall that the F-distribution can often be approximated by the χ^2-distribution for large N.

Multiple Regressions

Sometimes there is occasion to study the simultaneous prediction of several, say p, variables. This means that our variable $y(t)$ will be a p-dimensional column vector. In systems applications this corresponds to multivariable systems. Much of what has been said here about linear regression will hold also for the multivariate case, but some algebraic expressions take a slightly different form.

It is convenient to distinguish between two cases:

The same set of regressors is used for each component of $y(t)$: Denote the regressors by the r-dimensional column vector $\varphi(t)$. Then write the regression as

$$y(t) = \boldsymbol{\theta}^T \varphi(t) \tag{II.85}$$

where $\boldsymbol{\theta}$ now is an $r \times p$-dimensional matrix with its ith column containing the coefficients associated with the ith component of y. The number of estimated parameters thus is $d = rp$. The LS criterion becomes

$$V_N(\boldsymbol{\theta}) = \frac{1}{N}\sum_{t=1}^{N} |y(t) - \boldsymbol{\theta}^T \varphi(t)|^2 \tag{II.86}$$

which is minimized by

$$\hat{\boldsymbol{\theta}}_N = \left[\frac{1}{N}\sum_{t=1}^{N} \varphi(t)\varphi^T(t)\right]^{-1} \frac{1}{N}\sum_{t=1}^{N} \varphi(t)y^T(t) \tag{II.87}$$

(see Problem 7D.2).

Different regressor sets for the different components of y: In case the different outputs are associated with different regressor sets, one must introduce a $d \times p$ matrix $\boldsymbol{\varphi}(t)$ whose ith column contains the regressors associated with output i and very possibly several zeros. Then the regression is written

$$y(t) = \boldsymbol{\varphi}^T(t)\theta \tag{II.88}$$

with θ as a d-dimensional column vector. The LS criterion becomes

$$\begin{aligned} V_N(\theta) &= \frac{1}{N}\sum_{t=1}^{N} |y(t) - \boldsymbol{\varphi}^T(t)\theta|_{\Lambda^{-1}} \\ &= \frac{1}{N}\sum_{t=1}^{N} [y(t) - \boldsymbol{\varphi}^T(t)\theta]^T \Lambda^{-1} [y(t) - \boldsymbol{\varphi}^T(t)\theta] \end{aligned} \tag{II.89}$$

Here we allowed a general quadratic norm to give different weights to the different components of $y(t)$. The element that minimizes (II.89) is

$$\hat{\theta}_N = \left[\frac{1}{N}\sum_{t=1}^{N}\varphi(t)\Lambda^{-1}\varphi^T(t)\right]^{-1}\left[\frac{1}{N}\sum_{t=1}^{N}\varphi(t)\Lambda^{-1}y(t)\right] \tag{II.90}$$

(see Problem 7D.2).

Comparing (II.90) with (II.87), we see the advantage with the special structure (II.85): To determine the $r \times p$ estimate $\hat{\boldsymbol{\theta}}_N$ in (II.87), it is sufficient to invert an $r \times r$ matrix. In (II.90), $\hat{\theta}_N$ is a pr vector and the matrix inversion involves a $pr \times pr$ matrix.

Correlation Interpretation and the Instrumental-variable Method

A useful interpretation of the LS method is as follows. Given the description

$$y(t) = \varphi^T(t)\theta_0 + w_0(t)$$

multiply both sides by $\varphi(t)$ and sum over $t = 1, \dots, N$. This leads to

$$\frac{1}{N}\sum_{t=1}^{N}\varphi(t)y(t) = \left[\frac{1}{N}\sum_{t=1}^{N}\varphi(t)\varphi^T(t)\right]\theta_0 + \frac{1}{N}\sum_{t=1}^{N}\varphi(t)w_0(t) \tag{II.91}$$

Provided that the disturbance $\{w_0(t)\}$ and the regression vector $\varphi(t)$ are uncorrelated, which means that the rightmost term is small, we find that the LSE

$$\hat{\theta}_N = \left[\frac{1}{N}\sum_{t=1}^{N}\varphi(t)\varphi^T(t)\right]^{-1}\frac{1}{N}\sum_{t=1}^{N}\varphi(t)y(t) \tag{II.92}$$

is a reasonable estimate of θ_0. We have, thus "correlated out" θ_0 from the noise using the regressor sequence.

In some cases we may expect correlation between the noise and the regressors. A natural extension of the preceding correlation idea would then be to use a d-dimensional vector sequence $\{\zeta(t)\}$ that is uncorrelated with the noise, but correlated with the regressor. Multiplying by $\zeta(t)$ and summing over t leads to

$$\frac{1}{N}\sum_{t=1}^{N}\zeta(t)y(t) = \frac{1}{N}\left[\sum_{t=1}^{N}\zeta(t)\varphi^T(t)\right]\theta_0 + \frac{1}{N}\sum_{t=1}^{N}\zeta(t)w_0(t) \tag{II.93}$$

If $\{\zeta(t)\}$ has the two above-mentioned properties, we find that

$$\hat{\theta}_N = \frac{1}{N}\left[\sum_{t=1}^{N}\zeta(t)\varphi^T(t)\right]^{-1}\frac{1}{N}\sum_{t=1}^{N}\zeta(t)y(t) \tag{II.94}$$

is a reasonable estimate of θ_0. This estimate was introduced by Reiersøl (1941)and is discussed in some detail in Kendall and Stuart (1961). It is known as the *instrumental-variable estimate* (IVE), and $\zeta(t)$ is called instrumental variables or the instruments. It remains to be discussed how to choose $\zeta(t)$, and that is done in Section 7.6 in connection with applications to dynamical systems.

Nonlinear Regressions

The characterizing feature of the linear regression model (II.3) is that the regression function is linearly parametrized in θ. Often one has to consider more general parametrizations of the regression function:

$$g(\varphi, \theta) \tag{II.95}$$

We thus obtain a *nonlinear regression*

$$y(t) = g(\varphi(t), \theta). \tag{II.96}$$

The weighted least-squares criterion

$$V_N(\theta) = \frac{1}{N}\sum_{t=1}^{N} \alpha_t[y(t) - g(\varphi(t), \theta)]^2 \tag{II.97}$$

can still be used as a measure of fit, and the estimate becomes

$$\hat{\theta}_N = \arg\min_{\theta} V_N(\theta)$$

just as in (II.7) and (II.8). The important difference is that it may not be possible to find explicit expressions for $\hat{\theta}_N$ as in (II.10), but one has to resort to numerical, iterative techniques.

To make things still more general, it is not necessary to use a quadratic measure of fit in (II.97). Let $\ell(\varepsilon)$ be a function that assumes positive values suitable to measure the "size" of ε and take

$$V_N(\theta) = \frac{1}{N}\sum_{t=1}^{N} \ell(y(t) - g(\varphi(t), \theta))$$

$$\hat{\theta}_N = \arg\min_{\theta} V_N(\theta) \tag{II.98}$$

The problem (II.96) and (II.98) clearly is more general and more difficult than (II.7) and (II.8), but it certainly is in the same spirit. Gauss's pragmatic interpretation of the LS criterion as a measure of fit of historic data applies equally well to (II.98). This way of selecting an estimate makes sense also without a probabilistic framework.

II.4 PROBLEMS

II.1 Show that

$$\bar{g}(\varphi) = E[y|\varphi]$$

minimizes

$$E[y - g(\varphi)]^2$$

with respect to g [see (II.1) and (II.2)]. *Hint:* Add and subtract $\bar{g}(\varphi)$ in the latter expression.

II.2 Let $\rho(a, b)$ be the correlation between the random variables a and b:

$$\rho(a, b) = \frac{\text{Cov}\,(a, b)}{(\text{Var}\,(a) \cdot \text{Var}\,(b))^{1/2}}$$

Let $\bar{g}(\varphi)$ be defined as in Problem II.1. Show that

$$\rho(y, \bar{g}(\varphi)) \geq |\rho(y, g(\varphi))|$$

for any function $g(\varphi)$ [Rao (1973), Section 4g.1].

II.3 Lyaponov's central limit theorem states: "Let

$$Z_N = \sum_{k=1}^{N} \alpha(k, N) w(k)$$

where $\{w(k)\}$ is a sequence of independent random variables with

$$Ew(k) = 0$$

$$Ew^2(k) = \lambda_k$$

$$E|w^3(k)| = \gamma_k$$

Assume that

$$\lim_{N\to\infty} \sum_{k=1}^{N} \alpha^2(k, N)\lambda_k = \lambda$$

$$\lim_{N\to\infty} \sum_{k=1}^{N} \alpha^3(k, N)\gamma_k = 0$$

Then

$$Z_N \in AsN(0, \lambda)\text{''}$$

Use this result to prove that the estimate (II.10) is asymptotically normal under suitable assumptions on $\{w_0(t)\}$.

References

Abraham, B. and Chuang, A. (1993). Estimation of time series models in presence of outliers. *Journal of Time Series Analysis*, 14(3):221–234.

Ahlfors, L. V. (1979). *Complex Analysis*. McGraw-Hill, New York.

Akaike, H. (1967). Some Problems in the Application of the Cross-Spectral Method. In Harris, B., editor, *Spectral Analysis of Time Series*, pages 81–107. John Wiley & Sons.

Akaike, H. (1968). On the use of a linear model for the identification of feedback systems. *Ann. Inst. Stat. Math.*, 20:425–439.

Akaike, H. (1969). Fitting autoregressive models for prediction. *Ann. Inst. Stat. Math.*, 21:243–347.

Akaike, H. (1972). Information theory and an extension of the maximum likelihood principle. *Proc. 2nd Int. Symp. Information Theory, Supp. to Problems of Control and Information Theory*, 267–281.

Akaike, H. (1973). Maximum likelihood identification of Gaussian autoregressive moving average models. *Biometrika*, 60:255–265.

Akaike, H. (1974a). A new look at the statistical model identification. *IEEE Trans. Autom. Control*, AC–19:716–723.

Akaike, H. (1974b). Stochastic theory of minimal realization. *IEEE Trans. Automatic Control*, AC–19:667–674.

Akaike, H. (1976). Canonical correlation. In Mehra, R. and Lainiotis, D., editors, *Advances in System Identification*. Academic Press, New York.

Akaike, H. (1981). Modern development of statistical methods. In Eykhoff, P., ed.; *Trends and Progress in System Identification*, Pergamon Press, Elmsford, N.Y.

Allen, D. M. (1971). The relationship between variable selection and data augmentation and a method for prediction. *Technometrics*, 13:469–475. discussion 477–481.

Anderson, B. D. O. (1985). Identification of scalar errors-in-variables models with dynamics. *Automatica*, 21:709–716.

Anderson, B. D. O. and Gevers, M. R. (1982). Identifiability of linear stochastic systems operating under linear feedback. *Automatica*, 18:195–213.

Anderson, B. D. O. and Moore, J. B. (1979). *Optimal Filtering*. Prentice-Hall, Upper Saddle River, N.J.

Anderson, B. D. O., Moore, J. B., and Hawkes, R. M. (1978). Model approximation via prediction error identification. *Automatica*, 14:615–622.

Anderson, T. W. (1959). On asymptotic distributions of estimated parameters of stochastic difference equations. *Ann. Math. Stat.*, 30:676–687.

Anderson, T. W. (1971). *The Statistical Analysis of Time Series*. Wiley, New York.

Anderson, T. W. (1975). Maximum likelihood estimation of parameters of autoregressive processes with moving average residuals and other covariance matrices with linear structure. *Ann. Stat.*, 3:1283–1304.

Anderson, T. W. and Taylor, J. B. (1979). Strong consistency of least-squares estimates in dynamic models. *Ann. Stat.*, 7:484–489.

Andersson, P. (1985). Adaptive forgetting in recursive identification through multiple models. *Int. J. Control*, 42:1175–1194.

Andersson, T. and Pucar, P. (1995). Estimation of residence time in continuous flow systems with dynamics. *Journal of Process Control*, 5:9–17.

Anscombe, F. J. and Tukey, J. W. (1963). The examination and analysis of residuals. *Technometrics*, 5:141–160.

Ansley, C. F. and Kohn, R. (1983). Exact likelihood of vector autoregressive-moving average processes. *Biometrica*, 70(1):275–278.

Aoki, M. (1987). *State Space Modeling of Time Series*. Springer Verlag, Berlin.

Åström, K. J. (1968). Lectures on the identification problem—the least squares method. Technical Report 6806, Division of Automatic Control, Lund Institute of Technology, Lund, Sweden.

Åström, K. J. (1969). On the choice of sampling rates in parametric identification of time series. *Inf. Sci.*, 1:273–278.

Åström, K. J. (1970). *Introduction to Stochastic Control Theory*. Academic Press, New York.

Åström, K. J. (1972). Unpublished notes.

Åström, K. J. (1980). Maximum likelihood and prediction error methods. *Automatica*, 16:551–574.

Åström, K. J. and Bohlin, T. (1965). Numerical identification of linear dynamic systems from normal operating records. In *IFAC Symposium on Self-Adaptive Systems*, Teddington, England.

Åström, K. J. and Eykhoff, P. (1971). System identification—a survey. *Automatica*, 7:123–167.

Åström, K. J. and Källström, C. G. (1976). Identification of ship steering dynamics. *Automatica*, 12:9–22.

Åström, K. J. and Söderström, T. (1974). Uniqueness of the maximum likelihood estimates of the parameters of an ARMA model. *IEEE Trans. Automatic Control*, AC–19:769–773.

Åström, K. J. and Wittenmark, B. (1971). Problems of identification and control. *J. Math. Analysis and Applications*, 34:90–113.

Åström, K. J. and Wittenmark, B. (1984). *Computer Controlled Systems* (3rd ed. 1997). Prentice-Hall, Upper Saddle River, N.J.

Åström, K. J. and Wittenmark, B. (1989). *Adaptive Control*. Addison-Wesley, Reading, MA.

Åström, K. J., Jury, E. I., and Agniel, R. G. (1970). A numerical method for the evaluation of complex integrals. *IEEE Trans. Automatic Control*, AC–15:468–471.

Bai, E. W. and Sastry, S. S. (1985). Persistency of excitation, sufficient richness and parameter convergence in discrete time adaptive control. *System and Control Letters*, 6:153–163.

Banks, H. T., Crowley, J. M., and Kunisch, K. (1983). Cubic spline approximation techniques for parameter estimation in distributed systems. *IEEE Trans. Automatic Control*, AC–28:773–786.

Barnett, S. (1975). *Introduction to Mathematical Control Theory*. Oxford University Press, New York.

Barron, A. (1993). Universal approximation bounds for superpositions of a sigmoidal function. *IEEE Trans. Inf. Theory*, IT–39:930–945.

Basseville, M. and Nikiforov, I. V. (1993). *Detection of Abrupt Changes*. Prentice-Hall, Upper Saddle River, N.J.

Beck, M. B. and Van Straten, G., eds. (1983). *Uncertainty and Forecasting of Water Quality*. Springer-Verlag, New York.

Beghelli, S. and Guidorzi, R. P. (1983). Transformations between input-output multistructural models—properties and applications. *Int. J. Control*, 37:1385–1400.

Belanger, P. R. (1985). Model-based tuning of controllers, in *On-line Process Simulation Techniques in Industrial Control* (E. J. Kompass and T. J. Williams, eds.) Proceedings of the 11th Annual Advanced Control Conference, Purdue University, Lafayette, Ind., 11–27.

Bellman, R. and Åström, K. J. (1970). On structural identifiability. *Math. Biosci.*, 7:329–339.

Bendat, J. S. and Piersol, A. G. (1980). *Engineering Applications of Correlation and Spectral Analysis*. Wiley, New York.

Benveniste, A., Métivier, M., and Priouret, P. (1990). *Adaptive Algorithms and Stochastic Approximations*. Springer Verlag, Berlin.

Benveniste, A., Goursat, M., and Ruget, G. (1980). Robust identification of a nonminimum phase system: Blind adjustment of a linear equalizer in data communication. *IEEE Trans. Automatic Control*, AC–25:385–399.

Benveniste, A. and Ruget, G. (1982). A measure of the tracking capability of recursive stochastic algorithms with constant gains. *IEEE Trans. Automatic Control*, AC–27:639–649.

Berk, K. N. (1974). Consistent autoregressive spectral estimates. *Ann. Stat.*, 2:489–502.

Bertsekas, D. P. (1982). *Constrained Optimization and Lagrange Multiplier Methods*. Academic Press, New York.

Bethoux, G. (1976). Approche unitaire des méthodes d'identification et de commande adaptive des procédes dynamiques. Thése de 3éme cycle, Université de Grenoble.

Bierman, G. J. (1977). *Factorization Methods for Discrete Sequential Estimation*. Academic Press, New York.

Billings, S. A. (1980). Identification of nonlinear systems—A survey. *IEE Proc. D*, 127:272–285.

Billings, S. A. and Voon, W. S. F. (1983). Structure detection and model validity test in the identification of nonlinear systems. *IEE Proc. D.*, 130(4):193–199.

Billings, S. A. and Voon, W. S. F. (1984). Least-squares parameter estimation algorithms for nonlinear systems. *Intern. J. Systems Science*, 15:601–615.

Billings, S. and Tao, Q. (1991). Model validity tests for non-linear signal processing applications. *Int. J. Control*, 54(1):157–194.

Billingsley, P. (1965). *Ergodic Theory and Information*. Wiley, New York.

Björck, A. (1967). Solving least squares problems by Gram-Schmidt orthogonalization. *BIT*, 7:1–27.

Blackman, R. B. and Tukey, J. W. (1958). The measurement of power spectra from the point of view of communications engineering. *Bell Syst. Tech. J.*, 37(I):183–282, (II):485–569.

Bohlin, T. (1970). Information pattern for linear discrete time models with stochastic coefficients. *IEEE Trans. Automatic Control*, AC–15:104–106.

Bohlin, T. (1971). On the problem of ambiguities in maximum likelihood identification. *Automatica*, 7:199–210

Bohlin, T. (1976). Four cases of identification of changing systems. In Mehra, R. K. and Lainiotis, D. G., editors, *System Identification: Advances and Case Studies.*, pages 441–518. Academic Press, New York.

Bohlin, T. (1978). Maximum-power validation without higher-order fitting. *Automatica*, 14:137–146.

Bohlin, T. (1991). *Interactive System Identification: Prospects and Pitfalls*. Springer-Verlag, Berlin.

Boothby, W. M. (1975). *An Introduction to Differentiable Manifolds and Riemannian Geometry*. Academic Press, New York.

Box, G. E. P. and Cox, D. R. (1964). An analysis of transformation. *J. Roy. Statist. Soc.*, 26(B):211–252.

Box, G. E. P. and Jenkins, G. M. (1970). *Time Series Analysis, Forecasting and Control* (3rd ed. 1994). Holden-Day, San Francisco.

Brillinger, D. R. (1981). *Time Series: Data Analysis and Theory*. Holden-Day, San Francisco.

Brillinger, D. R. (1983). The finite Fourier transform of a stationary process. In *Handbook of Statistics*, 3(Brillinger, D. R. and Krishnaiah, P. R., eds.):21–37, North-Holland, Amsterdam.

Brillinger, D. R. and Krishnaiah, P. R., eds. (1983). *Handbook of Statistics 3: Time Series in the Frequency Domain*. North-Holland, Amsterdam.

Brockett, R. W. (1970). *Finite Dimensional Linear System*. Wiley, New York.

Broersen, P. M. T. (1997). On orders of long AR models for ARMA estimation. In Prochazka, A., Uhlir, J., and Sovka, P., editors, *Proc. ECSAP-97, 1st European Conference on Signal Analysis and Prediction*, pages 83–86, Prague.

Brown, B. M. (1971). Martingale central limit theorems. *Ann. Math. Stat.*, 42:59–66.

Brown, M. and Harris, C. (1994). *Neurofuzzy Adaptive Modelling and Control*. Prentice-Hall, Upper Saddle River, N.J.

Burg, J. P. (1967). Maximum entropy spectral analysis. Paper presented at the 37th Annual International Meeting, Soc. of Explor. Geophys., Oklahoma City.

Byrnes, C. I. (1976). A brief tutorial on calculus on manifolds, with emphasis on applications to identification and control. In Bucy, R. S. and Moura, J. M. F., editors, *Nonlinear Stochastic Problems*, NATO ASI Series no C104. Reidel, Boston.

Cadzow, J. A. (1980). High performance spectral estimation—A new ARMA method, *IEEE Trans. Acoustics, Speech and Signal Proc.*, ASSP–28:524–529.

Caines, P. E. (1976a). Prediction error identification methods for stationary stochastic processes. *IEEE Trans. Automatic Control*, AC–21:500–505.

Caines, P. E. (1976b). On the asymptotic normality of instrumental variable and least squares estimators. *IEEE Trans. Automatic Control*, AC–21:598–600.

Caines, P. E. (1978). Stationary linear and nonlinear system identification and predictor set completeness. *IEEE Trans. Automatic Control*, AC–23:583–594.

Caines, P. E. (1988). *Linear Stochastic Systems*. John Wiley, New York.

Caines, P. E. and Chan, C. W. (1975). Feedback between stationary stochastic processes. *IEEE Trans. Autom. Control*, AC–20:498–508.

Carayannis, G., Manolakis, D., and Kalouptsidis, N. (1983). A fast sequential algorithm for least squares filtering and prediction. *IEEE Trans. Acoustics, Speech and Signal Processing*, ASSP–31:1394–1402.

Carroll, R. J. and Ruppert, D. (1988). *Transformation and Weighting in Regression*. Chapman and Hill, New York.

Cellier, F. E. (1990). *Continuous System Modeling*. Springer-Verlag, Berlin and New York.

Chen, H. F. and Guo, L. (1991). *Identification and Stochastic Adaptive Control*. Birkhäuser, Boston.

Chou, C. T. and Verhaegen, M. (1997). Subspace algorithms for the identification of multivariable errors-in-variables models. *Automatica*, 33(10):1857–1869.

Chung, K. L. (1974). *A Course in Probability Theory*, 2nd ed. Academic Press, New York.

Chung, K. L. (1979). *Elementary Probability Theory with Stochastic Processes*, 3rd ed. Springer-Verlag, New York.

Cioffi, J. and Kailath, T. (1984). Fast recursive least-squares transversal filters for adaptive filtering. *IEEE Trans. Acoustics, Speech and Signal Processing*, ASSP–32:304.

Clark, J. M. C. (1976). The consistent selection of parametrization in systems identification. In *Proc. Joint Automatic Control Conference (JACC)*, pages 576–580, Purdue University, Lafayette, Ind.

Clarke, D. W. (1967). Generalized least squares estimation of parameters of a dynamic model. In *First IFAC Symposium on Identification in Automatic Control Systems*, Prague.

Cochrane, D. and Orcutt, G. H. (1949). Application of least squares regression to relationships containing autocorrelated error terms. *J. Amer. Statist. Assoc.*, 44:32–61.

Cook, D. R. and Weisberg, S. (1982). *Residuals and Influence in Regression*. Chapman and Hall, New York.

Cooley, J. W. and Tukey, J. W. (1965). An algorithm for the machine calculation of complex Fourier series, *Math. Comput.*, 19:297–301.

Correa, G. O. and Glover, K. (1984). Pseudo-canonical forms, identifiable parameterizations and simple parameter estimation for linear multivariable systems. *Automatica*, 20:429–452.

Cramér, H. (1946). *Mathematical Methods of Statistics*. Princeton University Press, Princeton, N.J.

Cybenko, G. (1980). The numerical stability of the Levinson-Durbin algorithm for Toeplitz systems of equations, *SIAM J. Science*, 1(3):303–319.

Cybenko, G. (1984). The numerical stability of the lattice algorithm for least squares linear prediction problems. *BIT*, 24:441–455.

Cybenko, G. (1989). Approximation by superpositions of a sigmoidal functions. *Mathematics of Control, Signals and Systems*, 2:303–314.

Daniel, C. and Wood, F. S. (1980). *Fitting Equation to Data*, 2nd ed. Wiley, New York.

Daniell, P. J. (1946). Discussion of paper by M. S. Bartlett. *J. Roy. Statist. Soc.*, suppl. 8:27.

Davies, W. D. T. (1970). *System Identification for Self-Adaptive Control.* Wiley Interscience, New York.

Davis, M. H. A. and Vinter, R. B. (1985). *Stochastic Modelling and Control.* Chapman and Hall, London.

De Boor, C. D. (1978). *Practical Guide to Splines*. Springer Verlag, New York.

Delchamps, D. F. and Byrnes, C. I. (1982). Critical point behavior of objective functions defined in spaces of multivariable systems. *Proc. 21st IEEE Conference on Decision and Control*, Orlando, Fla., 919–924.

Deller, J. (1990). Set membership identification in digital signal processing. *Acoust. Speech and Signal Process. Mag.*, 6(4):4–20.

Dempster, A. P., Laird, N. M., and Rubin, D. B. (1977). Maximum likelihood from incomplete data via the EM algorithm. *J. Royal Stat. Soc.*, B–39(1):1–38.

Denery, D. G. (1971). Simplification in computation of sensitivity function for constant coefficient linear systems. *IEEE Trans. Automatic Control*, AC–16:348.

Dennis Jr., J. E. and Schnabel, R. B. (1983). *Numerical Methods for Unconstrained Optimization and Nonlinear Equations*. Prentice-Hall, Upper Saddle River, N.J.

Dennis Jr., J. E., Gay, D. M., and Welsch, R. E. (1981). An adaptive nonlinear least-squares algorithm, *ACM T. Math. Software*, 7:348–368, 369–383.

Diananda, P. H. (1953). Some probability limit theorems with statistical applications. *Proc. Cambridge Philos. Soc.*, 49:239–246.

Dickinson, B. W., Kailath, T., and Morf, M. (1974). Canonical matrix fraction and state-space descriptions for deterministic and stochastic linear systems. *IEEE Trans. Automatic Control*, AC–19:656–667.

Draper, N. R. and Smith, H. (1981). *Applied Regression Analysis*, 2nd ed. Wiley, New York.

Dudley, D. G. (1983). Parametric identification of transient electro-magnetic systems. *Wave Motion*, 5:369–384.

Dugard, L. and Landau, I. D. (1980). Recursive output error identification algorithms. *Automatica*, 16:443–462.

Dunsmuir, W. and Hannan, E. J. (1976). Vector linear time series models. *Advances in Applied Prob.*, 8:339–364. See also Deistler, M., Dunsmuir, W., and Hannan, E. J., vector linear time series models, corrections and extensions, in *Advances in Applied Prob.* 10:360–372, 1978.

Durbin, J. (1959). Efficient estimators of parameters in moving average models. *Biometrica*, 46:306–316.

Durbin, J. (1960). The fitting of time-series models, *Rev. Inst. Int. Stat.*, 28(3):233–243.

Dvoretzky, A. (1972). Asymptotic normality for sums of dependent random variables. *Proc. 6th Berkeley Symp. on Math Statist.*, Berkeley, Calif., 513–535.

Dzhaparidze, K. O. and Yaglom, A. M. (1983). Spectrum parameter estimation in time series analysis. In *Developments in Statistics* (Krishnaiah, P. R., ed.). Academic Press, New York.

Eaton, J. H. (1967). *Identification for Control Purposes*. IEEE winter meeting, New York.

Efron, B. and Tibshirani, B. (1993). *An Introduction to the Bootstrap*. Chapman and Hall, New York.

Eykhoff, P. (1974). *System Identification*. Wiley, New York.

Eykhoff, P., editor (1981). *Trends and Progress in System Identification*. Pergamon Press, Elmsford, N.Y.

Eykhoff, P. and Parks, P., editors (1990). *Special Issue on Identification and System Parameter Estimation*. Automatica, Vol 26, No 1.

Farlow, S. J. (1984). *Self-organizing Methods in Modelling—GMDH Type Algorithms*, Marcel Dekker, New York.

Fedorov, V. V. (1972). *Theory of Optimal Experiments*. Academic Press, New York.

Fine, T. L. and Hwang, W. G. (1979). Consistent estimation of system order. *IEEE Trans Automatic Control*, AC–24:387–402.

Finigan, B. M. and Rowe, I. H. (1974). Strongly consistent parameter estimation by the introduction of strong instrumental variables. *IEEE Trans. Automatic Control*, AC–19:825–930.

Fisher, R. A. (1912). On an absolute criterion for fitting frequency curves. *Mess. Math.* 41:155.

Fnaiech F. and Ljung, L. (1987). Recursive identification of bilinear systems, *Int. J. Control*, 45(2):453–470.

Fogel, E. and Huang, Y. F. (1982). On the value of information in system identification-bounded noise case. *Automatica*, 18:224–238.

Forssell, U. and Ljung, L. (1998a). Closed loop identification revisited. To appear in *Automatica* 1999.

Forssell, U. and Ljung, L. (1998b). Identification for control: Some results on optimal experiment design. In *Proceedings of the 37th IEEE Conference on Decision and Control*, Tampa, FL.

Forssell, U. and Ljung, L. (1998c). Identification of unstable systems using output error and Box-Jenkins model structures. *IEEE Trans. Automatic Control*. To appear.

Forssell, U. and Ljung, L. (1998d). A projection method for closed-loop identification. *IEEE Trans. Automatic Control*. To appear.

Fortesque, T. R., Kershenbaum, L. S., and Ydstie, B. E. (1981). Implementation of self-tuning regulators with variable forgetting factors. *Automatica*, 17:831–835.

Frederick, K. and Close, C. M. (1978). *Modelling and Analysis of Dynamic Systems*, Houghton Mifflin, Boston.

Freeman, T. G. (1985). Selecting the best linear transfer function model. *Automatica*, 21(4):361–370.

Friedlander, B. (1982). Lattice filters for adaptive processing. *Proc. IEEE*, 70:829–867.

Furth, B. P. (1973). New estimator for the identification of dynamic processes. Technical report, Institute Boris Kidric Vinca, Belgrade, Yugoslavia.

Gabr, M. M. and Subba Rao, T. (1984). On the identification of bilinear systems from operating records. *Int. J. Control*, 40:121–128.

Gauss, K. F. (1809). *Theoria Motus Corporum Celestium, English Translation: Theory of the Motion of Heavenly Bodies*. Dover (1963), New York.

Gauthier, A. and Landau, I. D. (1978). On the recursive identification or multi-input, multi-output systems. *Automatica*, 14:609–614.

Gecklini, N. C. and Yavuz, D. (1978). Some novel windows and concise tutorial comparison of window families. *IEEE Trans. Acoustics, Speech, and Signal Processing*, ASSP–26:501–507.

Gertler, J. and Cs. Bànyàsz (1974). A recursive (on-line) maximum likelihood identification method. *IEEE Trans. Automatic Control*, AC–19:816–820.

Gevers, M. (1993). Towards a joint design of identification and control? In Trentelman, H. L. and Willems, J. C., editors, *Essays on control: Perspectives in the theory and its applications*, ECC '93 Groningen.

Gevers, M. and Li, G. (1993). *Parametrizations in Control, Estimation and Filtering Problems: Accuracy Problems*. Springer Verlag, Berlin.

Gevers, M. and Ljung, L. (1986). Optimal experiment designs with respect to the intended model application. *Automatica*, 22:543–555.

Gevers, M. and Wertz, V. (1984). Uniquely identifiable state-space and ARMA parametrizations for multivariable systems. *Automatica*, 20:333–347.

Girosi, F., Jones, M., and Poggio, T. (1995). Regularization theory and neural network architectures. *Neural Computation*, 7:219–269.

Glover, K. and Willems, J. C. (1974). Parametrizations of linear dynamical systems—canonical forms and identifiability. *IEEE Trans. Automatic Control*, AC–19:640–646.

Godfrey, K. R. (1980). Correlation methods. *Automatica*, 16:527–534.

Godfrey, K. R. (1983). *Compartmental Models and Their Applications*. Academic Press, New York.

Godfrey, K., editor (1993). *Perturbation Signals for System Identification*. Prentice Hall International, New York.

Godfrey, K. R. and Distefano III, J. J. (1985). Identifiability of model parameters. *Proc. 7th IFAC Symp. on Identification Systems and Parameter Estimation*, York, U.K., 89–114.

Golub, G. H. and van Loan, C. F. (1996). *Matrix Computations*. Johns Hopkins University Press, Balitmore, 3rd edition.

Golub, G. H. and Pereyra, V. (1973). The differentiation of pseudo-inverse and non-linear least squares problems whose variables separate. *SIAM J. Numer. Anal.*, 10(2):413–432.

Goodwin, G. C. (1985). Some observations on robust estimation and control. *Proc. 7th IFAC Symp. on System Identification*, York, U.K., 853–860.

Goodwin, G. C. (1987). Experiment design for system identification. In *Encyclopedia of Systems and Control* (M. Singh, ed.). Pergamon Press, Oxford.

Goodwin, G. C., Evans, R. J., Lozano Leal, R., and Feik, R. (1986). Sinusoidal disturbance rejection with application to helicopter flight data. *IEEE Trans. Acoustics, Speech and Signal Proc.*, ASSP–34:479–484.

Goodwin, G., Gevers, M., and Ninness, G. M. (1992). Quantifying the error in estimated transfer functions with application to model order selection. *IEEE Trans. Automatic Control*, AC–37:913–928.

Goodwin, G. C. and Feuer, A. (1998). Estimation with missing data. *Mathematical Modeling of Systems*. To appear.

Goodwin, G. C. and Payne, R. L. (1977). *Dynamic System Identification: Experiment Design and Data Analysis*. Academic Press, New York.

Goodwin, G. C. and Sin, K. S. (1984). *Adaptive Filtering, Prediction and Control*. Prentice-Hall, Upper Saddle River, N.J.

Graebe, S. (1990). *Theory and Implementation of Gray Box Identification*. PhD thesis, Automatic Control, Royal Institute of Technology, Stockhom, Sweden. TRITA-REG 90/3 and 90/6.

Granger, C. W. J. and Newbold, P. (1977). *Forcasting Economic Time Series*. Academic Press, New York.

Graupe, D., Jain, V. K., and Salahi, J. (1980). A comparative analysis of various least-squares identification algorithms. *Automatica*, 16:663–681.

Grenander, U. and Rosenblatt, M. (1957). *Statistical Analysis of Stationary Time Series*. Chelsea, New York (2nd ed., 1984).

Griffiths, L. J. (1977). A continuously adaptive filter implemented as a lattice structure. *Proc. 1977 IEEE Intern. Conf. Acoustics, Speech and Signal Processing*, 683–686.

Grübel, G. (1985). Uncertainty and control - some activities at DFVLR. In Ackermann, J., editor, *Lecture Notes in Control and Information Sciences*, volume 70, pages 1–47. Springer Verlag, New York.

Gu, G. and Khargonekar, P. (1992). A class of algorithms for identification in H_∞. *Automatica*, 28:299–312.

Guidorzi, R. (1975). Canonical structures in the identification of multivariable systems. *Automatica*, 11:361–374.

Guidorzi, R. (1981). Invariants and canonical forms for systems structural and parametric identification. *Automatica*, 17:117–133.

Gunnarsson, S. and Ljung, L. (1989). Frequency domain tracking characteristics of adaptive algorithms. *IEEE Trans. on Acoustics, Speech and Signal Processing*, ASSP–37:1072–1084.

Guo, L. and Ljung, L. (1995). Performance analysis of general tracking algorithms. *IEEE Trans. Automatic Control*, AC–40:1388–1402.

Gupta, N. K. and Mehra, R. K. (1974). Computational aspects of maximum likelihood estimation and reduction in sensitivity function calculations. *IEEE Trans. Automatic Control*, AC–19(6):774–783.

Gustafsson, F. (1996). The marginalized likelihood ratio test for detecting abrupt changes. *IEEE Trans. Autom. Control*, AC–41:66–78.

Gustavsson, I., Ljung, L., and Söderström, T. (1977). Identification of processes in closed loop—identifiability and accuracy aspects. *Automatica*, 13:59 – 77.

Gustavsson, I., Ljung, L., and Söderström, T. (1981). Choice and effect of different feedback configurations. In Eykhoff, P., editor, *Trends and Progress in System Identification*, pages 367–388. Pergamon Press, Oxford.

Haber, R. (1985). Nonlinearity tests for dynamic processes. *Proc. 7th IFAC Symp. on System Identification*, York, U.K., 409–413.

Haber, R. and Unbehauen, H. (1990). Structure identification of nonlinear dynamic systems - a survey on input/output approaches. *Automatica*, 26:177–202.

Hagberg, B. and Schöön, N. H. (1974). Kinetic aspects of the acid sulfite cooking process. Part 3: Mathematical simulation of cooks. *Svensk Papperstidning*, 72:557–562.

Hägglund, T. (1984). Adaptive control of systems subject to large parameter changes. *Proc. 9th IFAC World Congress*, Budapest Hungary, 993–998.

Hall, P. and Heyde, C. C. (1980). *Martingale Limit Theory and Its Applications*. Academic Press, New York.

Hampel, F. R. (1974). The influence curve and its role in robust estimation. *J. Am. Stat. Assoc.*, 69:383–394.

Hampel, F. R., Ronchetti, E. M., Rousseeuw, P. J., and Stahel, W. (1986). *Robust Statistics, the Approach Based on Influence Functions*. Wiley, New York.

Hannan, E. J. (1969). The identification of vector mixed autoregressive moving average systems. *Biometrika*, 56:223–225.

Hannan, E. J. (1970). *Multiple Time Series*. Wiley, New York.

Hannan, E. J. (1971a). The identification problem for multiple equation systems with moving average errors. *Econometrica*, 39:751–766.

Hannan, E. J. (1971b). Nonlinear time series regression. *J. Appl. Prob.*, 8:767–780.

Hannan, E. J. (1973). The asymptotic theory of linear time series models. *J. App. Prob.*, 10:135–145.

Hannan, E. J. (1976). The identification and parametrization of ARMAX and state space forms. *Econometrica*, 44:713–723.

Hannan, E. J. (1979). The statistical theory of linear systems. In *Developments in Statistics*, vol. 2(Krishnaiah, P., ed.). Academic Press, New York.

Hannan, E. J. (1980a). Recursive estimation based on ARMA models. *Ann. Statis.*, 8:762–777.

Hannan, E. J. (1980b). The estimation of the order of an ARMA process. *Ann. Statis.*, 8:1071–1081.

Hannan, E. J. and Deistler, M. (1988). *The Statistical Theory of Linear Systems*. John Wiley, New York.

Hannan, E. J. and Kavalieris, L. (1984). Multivariate linear time series models. *Adv. Appl. Prob.* 16:492–561.

Hannan, E. J., Krishnaiah, P. R., and Rao, M. M., editors (1985). *Handbook of Statistics 5: Time Series in the Time Domain*. North Holland, Amsterdam.

Hannan, E. J. and Quinn, B. G. (1979). The determination of the order of an autoregression. *J. Royal Stat. Soc. Ser. B.*, 41:713–723.

Hansen, F. R., Franklin, G. F., and Kosut, R. L. (1989). Closed-loop identification via the fractional representation: Experiment design. In *Proceeding American Control Conference*, pages 386–391.

Hanzon, B. and Ober, R. J. (1997). Overlapping block-balanced canonical forms for various classes of linear systems. *SIAM J. Control and Optim.*, 35(1):228–242.

Hastings-James, R. and Sage, M. W. (1969). Recursive generalized least squares procedure for on-line identification of process parameters. *IEE Proc.*, 116:2057–2062.

Haykin, S. (1986). *Adaptive Filter Theory*. Prentice-Hall, Upper Saddle River, N.J.

Haykin, S. (1994). *Neural Networks: A Comprehensive Foundation*. 2nd ed. 1999, Prentice-Hall, Upper Saddle River, N.J.

Hazewinkel, M. and Kalman, R. E. (1976). On invariants, canonical forms and moduli for linear constant, finite-dimensional dynamical systems. In *Lecture Notes, Econo-Math. Systems Theory*, volume 133, pages 48–60. Springer Verlag, New York.

Hellendorn, H. and Driankov, D., editors (1997). *Fuzzy Model Identification—Selected Approaches*. Springer Verlag, Berlin.

Helmicki, A. J., Jacobson, C. A., and Nett, C. N. (1991). Control-oriented system identification: a worst case/deterministic approach in H_∞. *IEEE Trans. Automatic Control*, AC–36:1163–1176.

Hill, S. D. (1985). Reduced gradient computation in prediction error identification. *IEEE Trans. Automatic Control*, AC–30:776–778.

Hjalmarsson, H., Gevers, M., and De Bruyne, F. (1996). For model-based control design, closed loop identification gives better performance. *Automatica*, 32:1659–1674.

Hjalmarsson, H. and Ljung, L. (1992). Estimating model variance in the case of undermodeling. *IEEE Trans. Automatic Control*, AC–37:1004–1008.

Ho, B. and Kalman, R. E. (1966). Efficient construction of linear state variable models from input/output functions. *Regelungstechnik*, 14:545–548.

Ho, Y. C. (1963). On the stochastic approximation method and optimal filtering theory. *J. Math. Analysis and Applications*, 6:152–154.

Hoaglin, D. C., Mosteller, F., and Tukey, J. W., eds. (1983). *Understanding Robust and Exploratory Data Analysis*. Wiley, New York.

Holmberg, A. and Ranta, J. (1982). Procedures for parameter and state estimation of microbial growth processes. *Automatica*, 18:181–193.

Honig, M. L. and Messerschmitt, D. G. (1984). *Adaptive Filters: Structures, Algorithms and Applications*. Kluwer, Boston.

Householder, A. S. (1964). *The Theory of Matrices in Numerical Analysis*, Blaisdell, New York.

Hsia, T. C. (1977). *Identification: Least Squares Methods*. Lexington Books, Lexington, Mass.

Huber, P. J. (1981). *Robust Statistics*. Wiley, New York.

Ibragimov, I. A. and Linnik, Yu. V. (1971). *Independent and Stationary Sequences of Random Variables*. Amsterdam: Groningen, Wolters, Noordhoff.

Isaksson, A. (1993). Identification of ARX models subject to missing data. *IEEE Trans. Automatic Control*, 38(5):813–819.

Isermann, R. (1980). Practical aspects of process identification, *Automatica*, 16:575–587.

Isermann, R., editor (1981). *Special Issue on System Identification*. Automatica, Vol 17, No 1.

Ivakhenko, A. G. (1968). The group method of data handling—A rival of stochastic approximation. *Sov. Autom. Control*, 3:43.

Jakeman, A. and Young, P. C. (1979). Refined instrumental variable methods of recursive time-series analysis. Part II. Multivariable systems. *Intern. J. Control*, 29:621–644.

James, H. M., Nichols, N. B., and Phillips, R. S. (1947). *Theory of Servomechanisms*. McGraw-Hill, New York.

Jang, J.-S. R. and Sun, C.-T. (1995). Neuro-fuzzy modeling and control. *Proc. of the IEEE*, 83(3):378–406.

Jansson, M. and Wahlberg, B. (1996). A linear regression approach to state-space subspace system identification. *Signal Processing (EURASIP), Special Issue on Subspace Methods, Part II: System Identification*, 52(2):103–129.

Jazwinski, A. H. (1970). *Stochastic Processes and Filtering Theory*. Academic Press, New York.

Jenkins, G. M. and Watts, D. G. (1968). *Spectral Analysis and Its Applications*. Holden-Day, San Francisco.

Jennrich, R. I. (1969). Asymptotic properties of nonlinear least squares estimators. *Ann. Math. Statis.*, 40(2):633–643.

Johansen, T. A. (1996). Identification of non-linear systems using empirical data and prior knowledge—an optimization approach. *Automatica*, 32(3):337–356.

Johansen, T. A. and Foss, B. A. (1995). Identification of nonlinear-system structure and parameters using regime decomposition. *Automatica*, 31(2):321–326.

Johansson, R. (1993). *System Modeling and Identification*. Prentice-Hall, Upper Saddle River, N.J.

Juditsky, A., Hjalmarsson, H., Benveniste, A., Delyon, B., Ljung, L., Sjöberg, J., and Zhang, Q. (1995). Nonlinear black-box modeling in system identification: Mathematical foundations. *Automatica*, 31(12):1724–1750.

Kabaila, P. V. (1983). On output-error methods for system identification. *IEEE Trans. Autom. Control*, AC–28:12–23.

Kabaila, P. V. and Goodwin, G. C. (1980). On the estimation of the parameters of an optimal interpolator when the class of interpolators is restricted. *SIAM J. Control and Optimization*, 18(2):121–144.

Kailath, T. (1974). A view on three decades of linear filtering theory. *IEEE Trans. Information Theory*, IT–20:146–181.

Kailath, T. (1980). *Linear Systems*. Prentice-Hall, Upper Saddle River, N.J.

Kailath, T., Mehra, R. K., and Mayne, D. Q., eds. (1974). *IEEE Trans. Automatic Control*, vol. AC–19(6), Special Issue on System Identification.

Kalafatis, A., Wang, L., and Cluett, W. R. (1997). Identification of Wiener-type nonlinear systems in a noisy environment. *Int. J. Control*, 66(6):923–941.

Kalman, R. E. (1960). On the general theory of control systems. *Proc. First IFAC Congress, Moscow*. Butterworths, London, 1:481–492.

Kalman, R. E. (1974). Algebraic geometric description of the class of linear systems of constant dimensions. In *8th Ann. Princeton Conf. Information Sciences and Systems*, pages 189–191, Princeton, N.J.

Kalman, R. E. (1991). Identification from real data. In Hazewinkel, M. and Kan, A. H. G. R., editors, *Current Delevopments in the Interface: Economics, Econometrics, Mathematics*, pages 161–196. D. Riedel, Dordrecht,The Netherlands.

Kalman, R. E. and Bucy, R. S. (1961). New results in linear filtering and prediction theory. *Trans. ASME, J. Basic Engineering (ser. D)*, 83:95–108.

Kashyap, R. L. (1970). Maximum likelihood identification of stochastic linear system. *IEEE Trans. Automatic Control*, AC–15:25.

Kashyap, R. L. and Nasburg, R. E. (1974). Parameter estimation in multivariable stochastic difference equations. *IEEE Trans. Automatic Control*, AC–19:784–797.

Kashyap, R. L. and Rao, A. R. (1976). *Dynamic Stochastic Models from Empirical Data*. Academic Press, New York.

Kay, S. M. (1988). *Modern Spectral Estimation, Theory and Application*. Prentice-Hall, Upper Saddle River, N.J.

Kendall, M. G. and Stuart, A. (1961). *Advanced Theory of Statistics*. Griffin, London.

Khasminski, R. Z. (1966). On stochastic processes defined by differential equations with small parameter. *Theory of Probability and Its Applications*, 11:211–288.

King, A., Desai, U., and Skelton, R. (1988). A generalized approach to q-markov covariance equivalent realizations of discrete systems. *Automatica*, 24(4):507–515.

Knudsen, T. (1994). A new method for estimating ARMAX models. In *Proc. 10th IFAC Symposium on System Identification*, pages 611–617, Copenhagen, Denmark.

Kohn, R. and Ansley, C. F. (1986). Estimation, prediction, and interpolation for ARMA models with missing data. *J. Amer. Stat. Assoc.*, 81:751–761.

Kollàr, I. (1994). *Frequency Domain System Identification Toolbox*. The MathWorks, Inc., Natick, MA.

Kolmogorov, A. N. (1941). Interpolation und Extrapolation von stationären zufälligen Folgen. *Bull. Acad. Sci. de l'U.R.S.S.*, 5:3–14.

Korenberg, M. J. (1991). Parallel cascade identification and kernel estimation for nonlinear systems. *Ann. Biomed. Eng.*, 19:429–455.

Kosut, R., Goodwin, G. C., and Polis, M. P., editors (1992). *Special Issue on System Identification for Robust Control Design*, volume AC–37(7). IEEE Trans. Automatic Control.

Kosut, R., Lau, M. K., and Boyd, S. P. (1992). Set-membership identification of systems with parametric and nonparametric uncertainty. *IEEE Trans. Automatic Control*, AC–37:929–941.

Krasker, W. S. and Welsch, R. E. (1982). Efficient bounded influence regression estimation. *J. Am. Stat. Assoc.*, 77:595–605.

Kubrusly, C. S. (1977). Distributed parameter system identification - a survey. *Int. J. Control*, 26:509–535.

Kulhavy, R. and Kraus, J. (1996). On duality of regularized exponential and linear forgetting. *Automatica*, 32(10):1403 – 1415.

Kullback, S. (1959). *Information Theory and Statistics*. Wiley, New York.

Kullback, S. and Leibler, R. A. (1951). On information and sufficiency. *Ann. Math. Statis.*, 22:79–86.

Kung, S. Y. (1978). A new identification and model reduction algorithm via singular value decomposition. In *In Proc. 12th Asilomar Conf. on Circuits, Systems and Computers*, pages 705–714, Pacific Grove, CA.

Kung, S. Y. (1993). *Digital Neural Networks*. Prentice-Hall, Upper Saddle River, N.J.

Kushner, H. J. and Clark, D. S. (1978). *Stochastic Approximation Methods for Constrained and Unconstrained Systems*. Springer-Verlag, New York.

Kushner, H. J. and Huang, H. (1979). Rates of convergence for stochastic approximation type algorithms. *SIAM J. Control and Optimization*, 17:607–617.

Kushner, H. J. and Huang, H. (1981). Asymptotic properties of stochastic approximations with constant coefficients. *SIAM J. Control and Optimization*, 19:87–105.

Lai, T. L. and Wei, C. Z. (1982). Asymptotic properties of projections with applications to stochastic regression problems. *J. Multivariate Analysis*, 12:346–370.

Landau, I. D. (1976). Unbiased recursive identification using model reference techniques. *IEEE Trans. Automatic Control*, AC–21:194–202. See also I. D. Landau, an addendum to "Unbiased recursive identification using model reference adaptive techniques," *IEEE Trans. Automatic Control*, AC–23:97–99.

Landau, I. D. (1979). *Adaptive Control. The Model Reference Approach*. Marcel Dekker, New York.

Landau, I. D. (1990). *System Identification and Control Design Using P.I.M. + Software*. Prentice-Hall, Upper Saddle River, N.J.

Larimore, W. E. (1983). System identification, reduced order filtering and modelling via canonical variate analysis. In *Proc 1983 American Control Conference*, San Francisco.

Larimore, W. E. (1990). Canonical variate analysis in identification, filtering and adaptive control. In *Proc. 29th IEEE Conference on Decision and Control*, pages 596–604, Honolulu, Hawaii.

Larimore, W. E. (1997). *Adapt$_X$ Automated System Identification Software, Users Manual*. Adaptics, Inc., 1717 Briar Ridge Road, McLean, VA.

Lawson, C. L. and Hanson, R. J. (1974). *Solving Least Squares Problems*. Prentice-Hall, Upper Saddle River, N.J.

Leden, B., Hamza, M. H., and Sheirah, M. A. (1976). Different methods for estimating thermal diffusivity of heat process. *Automatica*, 12:445–456.

Lee, D. T., Morf, M., and Friedlander, B. (1981). Recursive least squares ladder estimation algorithms. *IEEE Trans. Acoustics, Speech and Signal Processing*, ASSP–29:627–641.

Lee, R. C. K. (1964). *Optimal Estimation, Identification and Control*. MIT Press, Cambridge, Mass.

Lee, T.-H., White, H., and Granger, C. (1993). Testing for neglected non-linearity in time series models. *Journal of Econometrics*, 56:269–290.

Lee, W. S., Anderson, B. D. O., Mareels, I. M. Y., and Kosut, R. L. (1995). On some key issues in the windsurfer approach to adaptive robust control. *Automatica*, 31(11):1619–1636.

Leondes, C. L., ed. (1987). *Advances in Control*, vol. 26. Academic Press, New York.

Leontaritis, I. J. and Billings, S. A. (1985). Input-output parametric models for nonlinear systems. *Intern. J. Control*, 41:303–344.

Levenberg, K. (1944). A method for the solution of certain nonlinear problems in least squares. *Quart. Appl. Math.*, 2:164–168.

Levi, E. C. (1959). Complex curve fitting. *IRE Trans. Automatic Control*, AC–4:37–43.

Levinson, N. (1947). The Wiener rms (root mean square) error criterion-filter design and prediction. *J. Math. Phys.*, 25:261–278.

Lin, D. W. (1984). On digital implementation of the fast Kalman algorithm. *IEEE Trans. Acoustics, Speech and Signal Processing*, ASSP–32:998–1005.

Lindgren, B. L. (1976). *Statistical Theory*, 3rd Ed. Macmillan, New York.

Lindskog, P. and Ljung, L. (1995). Tools for semiphysical modeling. *Int. J. Adaptive Control and Signal Processing*, 9(6):509–523.

Little, R. J. A. and Rubin, D. B. (1987). *Statistical Analysis with Missing Data*. Wiley, New York.

Liu, K. and Skelton, R. E. (1992). Identification of linear systems from their pulse responses. In *Proc. Amer. Contr. Conf.*, pages 1243–1247.

Ljung, L. (1971). Characterization of the concept of "persistently exciting" in the frequency domain. Report 7119, Division of Automatic Control, Lund Institute of Technology, Lund, Sweden.

Ljung, L. (1974). On convergence for prediction error identification methods. Technical Report 7405, Division of Automatic Control, Lund Institute of Technology, Lund, Sweden.

Ljung, L. (1976a). On consistency and identifiability. *Mathematical Programming Study No. 5*, North-Holland, Amsterdam, 169–190.

Ljung, L. (1976b). Consistency of the least-squares identification method. *IEEE Trans. Automatic Control*, AC–21:779–781.

Ljung, L. (1976c). On the consistency of prediction error identification methods. In *System Identification, Advances and Case Studies* (R. K. Mehra and D. G. Lainiotis, eds.), Academic Press, New York, 121–164.

Ljung, L. (1977a). On positive real transfer functions and the convergence of some recursive schemes. *IEEE Trans. Automatic Control*, AC–22(4):539–551.

Ljung, L. (1977b). Analysis of recursive stochastic algorithms. *IEEE Trans. Automatic Control*, AC–22(4):551–575.

Ljung, L. (1978a). Convergence analysis of parametric identification methods. *IEEE Trans. Automatic Control*, AC–23:770–783.

Ljung, L. (1978b). Strong convergence of a stochastic approximation algorithm. *Ann. Statis.*, 6:680–696.

Ljung, L. (1979a). Asymptotic behavior of the extended Kalman filter as a parameter estimator for linear systems. *IEEE Trans. Automatic Control*, AC–24:36–50.

Ljung, L. (1979b). Convergence of recursive estimators. In *5th IFAC Symp. on Identification and System Parameter Estimation*, pages 131–144, Darmstadt, FRG. Pergamon Press, Elmsford, N.Y.

Ljung, L. (1981). Analysis of a general recursive prediction error identification algorithm. *Automatica*, 27:89–100.

Ljung, L. (1984). Analysis of stochastic gradient algorithms for linear regression problems. *IEEE Trans. Information Theory*, IT–30(2):151–160.

Ljung, L. (1985a). On the estimation of transfer functions. *Automatica*, 21:677–696.

Ljung, L. (1985b). Asymptotic variance expressions for identified black-box transfer function models. *IEEE Trans. Automatic Control*. AC–30:834–844.

Ljung, L. (1985c). Asymptotic variances of transfer function estimates obtained by the instrumental variable method. *Proc. 7th IFAC Symp. on Identification and System Parameter Estimation*, York, U.K., 1341–1344.

Ljung, L. (1985d). A nonprobabilistic framework for signal spectra. *Proc. 24th IEEE Conf. on Decision and Control*, Fort Lauderdale, Fla., 1056–1060.

Ljung, L. (1986). Parametric methods for identification of transfer functions of linear systems. In Leondes, C. L., editor, *Advances in Control*, vol. 26(2). Academic Press, New York.

Ljung, L. (1995). *The System Identification Toolbox: The Manual*. The MathWorks Inc. 1st edition 1986, 4th edition 1995, Natick, MA.

Ljung, L. and Caines, P. E. (1979). Asymptotic normality of prediction error estimation for approximate system models. *Stochastics*, 3:29–46.

Ljung, L. and Glad, T. (1994a). *Modeling of Dynamic Systems*. Prentice-Hall, Upper Saddle River, N.J.

Ljung, L. and Glad, T. (1994b). On global identifiability for arbitrary model parametrizations. *Automatica*, 30(2):265–276.

Ljung, L. and Glover, K. (1981). Frequency domain versus time domain methods in system identification. *Automatica*, 17(1):71–86.

Ljung, L. and Gunnarsson, S. (1990). Adaptation and tracking in system identification—a survey. *Automatica*, 26(1):7–21.

Ljung, L. and Guo, L. (1997). The role of model validation for assessing the size of the unmodeled dynamics. *IEEE Trans. Automatic Control*, AC–42:1230–1240.

Ljung, L., Morf, M., and Falconer, D. (1978). Fast calculations of gain matrices for recursive estimation schemes. *Int. J. Control*, 27:1–19.

Ljung, L. and van Overbeek, A. J. M. (1978). Validation of approximate models obtained from prediction error identification, *Proc. 7th IFAC Congress*, Helsinki, paper 45 A.3, 1899–1980.

Ljung, L., Pflug, G., and Walk, H. (1992). *Stochastic Approximation and Optimization of Random Systems*. Birkhäuser, Berlin.

Ljung, L. and Rissanen, J. (1976). On canonical forms, parameter identifiability and the concept of complexity. In *Proc. 4th IFAC Symposium on Identification and System Parameter Estimation*, pages 58–69. Tblisi, USSR. Paper 13.6.

Ljung, L. and Söderström, T. (1983). *Theory and Practice of Recursive Identification*. MIT press, Cambridge, Mass.

Ljung, L., Söderström, T., and Gustavsson, I. (1975). Counterexamples to general convergence of a commonly used recursive identification method. *IEEE Trans. Automatic Control*, AC–20(5):643–652.

Ljung, L. and Wahlberg, B. (1992). Asymptotic properties of the least-squares method for estimating transfer functions and disturbance spectra. *Adv. Appl. Prob.*, 24:412–440.

Ljung, L. and Yuan, Z. D. (1985). Asymptotic properties of black-box identification. *IEEE Trans. Automatic Control*, AC–30:514–530.

Ljung, S. and Ljung, L. (1985). Error propagation properties of recursive least-squares adaptation algorithms. *Automatica*, 21(2):157–167.

Luenberger, D. G. (1967). Canonical forms for linear multivariable systems. *IEEE Trans. Automatic Control*, AC–12:290.

Luenberger, D. G. (1971). An introduction to observers. *IEEE Trans. Automatic Control*, AC–16:596–603.

Luenberger, D. G. (1973). *Introduction to Linear and Nonlinear Programming*. Addison-Wesley, Reading, Mass.

Luukkonen, R., Saikkonen, P., and Teräsvirta, T. (1988). Testing linearity in univariate times-series models. *Scand. J. Statist.*, 15:161–175.

Maciejowski, J. M. (1985). Balanced realizations in system identification. In *Proc. 7th IFAC Symposium on Identification and Parameter Estimation*, York, U.K.

Makhoul, J. (1977). Stable and efficient lattice methods for linear prediction. *IEEE Trans. Acoustics, Speech and Signal Processing*, ASSP–25:423–428.

Makhoul, J. and Wolf, J. (1972). Linear prediction and the spectral analysis of speech, NTIS, No. AD–749066. *BBN Report No. 2304*. Bolt, Baraneck and Newman, Inc., Cambridge, Mass.

Mäkilä, P. M. (1992). Worst-case input-output identification. *Int. J. Control*, 56:673–689.

Mäkilä, P. M., Partington, J. R., and Gustafsson, T. (1995). Worst-case control-relevant identification. *Automatica*, 31(12):1799–1819.

Malinvaud, E. (1980). *Statistical Methods of Econometrics*. North-Holland, Amsterdam.

Mallows, C. L. (1973). Some comments on C_p, *Technometrics*, 15:661–676.

Mann, H. B. and Wald, A. (1943). On the statistical treatment of linear stochastic difference equations. *Econometrica*, 11:173–220.

Markel, J. D. and Gray Jr., A. H. (1976). *Linear Prediction of Speech*. Springer-Verlag, New York.

Marple, S. L. (1987). *Digital Spectral Analysis with Applications*. Prentice-Hall, Upper Saddle River, N.J.

Marquardt, D. W. (1963). An algorithm for least squares estimation of nonlinear parameters. *J. SIAM*, 11:431–441.

Mayne, D. Q. (1967). A method for estimating discrete time transfer functions. In *Advances in Computer Control*, Second UKAC Control Convention, University of Bristol.

Mayne, D. Q. and Firoozan, F. (1982). Linear identification of ARMA processes. *Automatica*, 18:461–466.

McKelvey, T. (1994). Fully parametrized state-space models in system identification. In *Proc. 10th IFAC Symposium on system Identification*, volume 2, pages 373–378, Copenhagen, Denmark.

McKelvey, T. (1996). Periodic excitation for identification of dynamic errors-in-variables systems operating in closed loop. In Gertler, J. J., Cruz, J. B., and Peshkin, M., editors, *Proc. 13th IFAC World Congress*, volume J, pages 155–160, San Francisco, CA.

McKelvey, T., Akcay, H., and Ljung, L. (1996). Subspace-based identification of infinite-dimensional multivariable systems from frequency-response data. *IEEE Trans. Automatic Control*, AC–41:960–979.

McKelvey, T. and Helmersson, A. (1996). State-space parametrizations of multivariable linear systems using tridiagonal matrix forms. In *IEEE Proceedings of the 35th Conference on Decision and Control*, pages 3654–3659, Kobe, Japan.

McKelvey, T. and Ljung, L. (1997). Frequency domain maximum likelihood identification. In *Proc. 10th IFAC Symposium on System Identification*, 4:1741–1746, Fokuoka, Japan.

Mehra, R. K. (1974). Optimal input signals for parameter estimation in dynamic systems—A survey and new results. *IEEE Trans. Automatic Control*, AC–19:753–768.

Mehra, R. K. (1979). Nonlinear system identification. *Proc. 5th IFAC Symp. Identification and System Parameter Estimation*, Darmstadt, FRG, (Pergamon Press, New York) paper S–4, 77–85.

Mehra, R. K. (1981). Choice of input signals. In *Trends and Progress in Systems Identification* (P. Eykhoff, ed.). Pergamon Press, Elmsford, N.Y.

Mehra, R. K. and Lainiotis, D. G., editors (1976). *System Identification - Advances and Case Studies*. Academic Press, New York.

Mehra, R. K. and Tyler, J. S. (1973). Case studies in aircraft parameter identification. In *3rd IFAC Symposium on Identification and System Parameter Estimation*, pages 117–144, Amsterdam. North-Holland.

Mendel, J. M. (1973). *Discrete Techniques of Parameter Estimation. The Equation Error Formulation*. Marcel Dekker, New York.

Mendel, J. (1983). *Optimal Seismic Deconvolution: An Estimation-based Approach*. Academic Press, New York.

Middleton, R. H. and Goodwin, G. C. (1990). *Digital Control and Estimation: A Unified Approach*. Prentice-Hall, Upper Saddle River, N.J.

Milanese, M. and Tempo, R. (1985). Optimal algorithms theory for robust estimation and prediction. *IEEE Trans. Autom. Control*, AC–30:730–738.

Milanese, M., Tempo, R., and Vicino, A. (1986). Strongly optimal algorithms and optimal information in estimation problems. *J. of Complexity*, 2:78–94.

Milanese, M. and Vicino, A. (1991). Optimal estimation theory for dynamic systems with set membership uncertainty: an overview. *Automatica*, 27:997–1009.

Mohler, R. R. (1973). *Bilinear Control Processes*. Academic Press, New York.

Moonen, M., DeMoor, B., Vandenbergh, L., and Vanderwalle, J. (1989). A QSVD approach to on- and off-line state space identification. *Int. J. Control*, 49:219–232.

Moore, J. B. (1983). Persistence of excitation in extended least squares. *IEEE Trans. Automatic Control*, AC–28:60–68.

Moore, J. B. and Ledwich, G. (1980). Multivariable adaptive parameter and state estimators with convergence analysis. *J. Australian Math. Soc.*, 21:176–197.

Moore, J. B. and Weiss, H. (1979). Recursive prediction error methods for adaptive estimation. *IEEE Trans. Systems, Man and Cybernetics*, SMC–9:197–205.

Morf, M. and Kailath, T. (1975). Square-root algorithms for least squares estimation. *IEEE Trans. Automatic Control*, AC–20:487–497.

Morf, M., Dickinson, B., Kailath, T., and Vieira, A. (1977). Efficient solution of covariance equations for linear prediction. *IEEE Trans. Acoustics, Speech and Signal Processing*, ASSP–25:429–433.

Mueller, M. S. (1981). Least squares algorithms for adaptive equalizers. *Bell Syst. Tech. J.*, 60:1905.

Mullis, C. T. and Roberts, R. A. (1976). Synthesis of minimum round-off noise in fixed point digital filters. *IEEE Trans. Circuits and Systems*, CAS–23:551–562.

Nadaraya, E. (1964). On estimating regression. *Theory of Prob. and Applic.*, 9:141–142.

Narendra, K. S. and Gallman, P. G. (1966). An iterative method for identification of nonlinear systems using a Hammerstein model. *IEEE Trans. Automatic Control*, AC–11:546.

Nerrand, O., Roussel-Ragot, P., Personnaz, L., and Drefys, G. (1993). Neural networks and nonlinear adaptive filtering; unifying concepts and new algorithms. *Neural Comput.*, 5:165–199.

Neuman, C. D. and Sood, A. K. (1972). Sensitivity functions for multi-input linear time-invariant systems. Part 2: Minimal order models. *Intern. J. Control*, 15:451.

Nguyen, V. V. and Wood, E. G. (1982). Review and unification of linear identifiability. *SIAM Rev.*, 24:34–51.

Nicholson, H., editor (1981). *Modelling of Dynamical Systems, Vol 2.* IEE Control Engineering Series, N 13. Peter Peregrinus Ltd, Stevenage, UK.

Niederlinski, A. (1984). Convergence of least squares dynamic system identification with finite-accuracy data. *Intern. J. Control*, 15:479–486.

Ninness, B. M. (1993). *Stochastic and Deterministic Modeling*. PhD thesis, Dept. of Electrical Engineering, University of Newcastle, NSW, Australia.

Ninness, B. M. and Goodwin, G. C. (1995). Estimation of model quality. *Automatica*, 31(12):1771–1797.

Ober, R. J. (1987). Balanced realizations: Canonical form, parameterization, model reduction. *Int. J. Control*, 46(2):643–670.

Oppenheim, A. V. and Schafer, R. W. (1975). *Digital Signal Processing*. Prentice-Hall, Upper Saddle River, N.J.

Oppenheim, A. V. and Willsky, A. S. (1983). *Signals and Systems*. Prentice-Hall, Upper Saddle River, N.J.

Orey, S. (1958). A central limit theorem for m-dependent random variables. *Duke Math. J.*, 25:543–546.

Paley, R. E. A. C. and Wiener, N. (1934). Fourier transforms in the complex domain. *Am. Math. Soc. Colleq.*, Publ. Col. 19, New York.

Pandya, R. N. (1974). A class of bootstrap estimators and their relationship to the generalized two stage least squares estimators. *IEEE Trans. Automatic Control*, AC–19:831–835.

Panuska, V. (1968). A stochastic approximation method for identification of linear systems using adaptive filtering. In *Joint Automatic Control Conference*, pages 1014–1021, Ann Arbor, Mich.

Papoulis, A. (1965). *Probability, Random Variables and Stochastic Processes*. McGraw-Hill, New York.

Papoulis, A. (1973). Minimum-bias windows for high-resolution spectral estimates. *IEEE Trans. Inform. Theory*, IT–19:9–12.

Parks, T. W. and Burrus, C. S. (1987). *Digital Filter Design*. Wiley Interscience, New York.

Parzen, E. (1985). Time series model identification and quantile spectral analysis. *Proc. 7th IFAC Symposium on System Identification*, York, U.K., Pergamon Press, Elmsford, N.Y., 731–736.

Payne, R. L., Goodwin, G. C., and Zarrop, M. (1975). Frequency domain approach for designing sampling rates for system identification. *Automatica*, 11:189–191.

Peterka, V. (1981a). Bayesian approach to system identification. In *Trends and Progress in System Identification* (P. Eykhoff, ed.). Pergamon Press, Elmsford, N.Y.

Peterka, V. (1981b). Bayesian system identification. *Automatica*, 17:41–53.

Peternell, K., Scherrer, W., and Deistler, M. (1996). Statistical analysis of novel identification methods. *Signal Processing*, 52:161–177.

Pintelon, R., Guillaume, P., Rolain, Y., Schoukens, J., and Van hamme, H. (1994). Parametric identification of transfer functions in the frequency domain – a survey. *IEEE Trans. Automatic Control*, AC-39(11):2245–2260.

Plackett, R. L. (1950). Some theorems in least squares. *Biometrika*, 37:149.

Poggio, T. and Girosi, F. (1990). Networks for approximation and learning. *Proc. of the IEEE*, 78:1481–1497.

Polis, M. P. and Goodson, R. E. (1976). Parameter identification in distributed systems - a synthesizing overview. *Proc. IEEE*, 64:45–61.

Politis, D. B. (1998). Computer-intensive methods in statistical analysis. *IEEE Signal Processing Magazine*, 15(1):39–55.

Polyak, B. T. and Juditsky, A. B. (1992). Acceleration of stochastic approximation by averaging. *SIAM J. Control Optim.*, 30:838–855.

Polyak, B. T. and Tsypkin, Ya. Z. (1980). Robust identification. *Automatica*, 16:53–63.

Poolla, K., Khargonekar, P., Tikku, A., Krause, J., and Nagpal, K. (1994). A time domain approach to model validation. *IEEE Trans. Automatic Control*, AC–39:951–959.

Popper, K. R. (1934). *The Logic of Scientific Discovery*. Basic Books, New York.

Potter, J. E. (1963). New statistical formulas. Memo 40, Instrumentation Laboratory, Massachusetts Institute of Technology, Cambridge, Mass.

Powell, M. J. D. (1964). An efficient method for finding the minimum of a function of several variables without calculating derivatives. *Comput. J.*, 7:155–162.

Priestley, M. B. (1981). *Spectral Analysis and Time Series*. Academic Press, New York.

Qureshi, Z. H., Ng, T. S., and Goodwin, G. C. (1980). Optimum experimental design for identification of distributed parameter systems. *Int. J. Control*, 31:21–29.

Rabiner, L. R. and Schafer, R. W. (1978). *Digital Processing of Speech Signals*. Prentice-Hall, Upper Saddle River, N.J.

Rajbman, N. S. (1976). The application of identification methods in the USSR—A survey. *Automatica*, 12:73–95.

Rajbman, N. S. (1981). *Dispersional Identification* (in Russian). Nauka, Moscow.

Rake, H. (1980). Step response and frequency response methods. *Automatica*, 16:519–526.

Rangan, S. and Poolla, K. (1996). Time-domain validation for sample-data uncertainty models. *IEEE Trans. Automatic Control*, AC–41:980–991.

Rao, C. R. (1973). *Linear Statistical Inference and Its Applications*. Wiley, New York.

Reiersøl, O. (1941). Confluence analysis by means of lag moments and other methods of confluence analysis. *Econometrica*, 9:1–23.

Rissanen, J. (1974). Basis of invariants and canonical forms for linear dynamic systems. *Automatica*, 10:175–182.

Rissanen, J. (1978). Modelling by shortest data description. *Automatica*, 14:465–471.

Rissanen, J. (1984). Universal coding, information, prediction and estimation. *IEEE Trans. Inform. Theory*, IT–30:629–636.

Rissanen, J. (1985). Minimum description length principles. In *Encyclopedia of Statistical Sciences*, vol. V, (S. Kotz and N. L. Johnson, eds.), Wiley, New York.

Rissanen, J. (1986). Prediction minimum description length principles. *Ann. Statist.* 14:1080–1100.

Rissanen, J. (1989). *Stochastic Complexity in Statistical Inquiry*. World Scientific, Singapore.

Rissanen, J. and Barbosa, L. (1969). Properties of infinite covariance matrices and stability of optimum predictors. *Inform. Sci.*, 1:221–236.

Rissanen, J. and Caines, P. E. (1979). Strong consistency of maximum likelihood estimators for ARMA process. *Ann. Statist.*, 7:297–315.

Rivera, D. E., Pollard, J. F., and Garcia, C. E. (1992). Control-relevant prefiltering: a systematic design approach and case study. *IEEE Trans. Automatic Control*, AC–37:964–974.

Robinson, E. A. (1967). *Multichannel Time Series Analysis with Digital Computer Programs*. Holden-Day, San Francisco.

Robinson, E. A. and Treitel, S. (1980). *Geophysical Signal Analysis*. Prentice-Hall, Upper Saddle River, N.J.

Rootzén, H. and Sternby, J. (1984). Consistency in least-squares estimation—A Bayesian approach. *Automatica*, 20:471–777.

Rosén, B. (1967). On the central limit theorem for sums of dependent random variables. *Z. Wahrsch verw. Geb.*, 7:48–82.

Rozanov, Yu. A. (1967). *Stationary Random Processes*. Holden-Day, San Francisco.

Rugh, W. J. (1981). *Nonlinear Systems Theory. The Volterra/Wiener Approach*. Johns Hopkins University Press, Baltimore.

Rumelhart, D., Hinton, G., and Cybenko, G. (1986). Learning representations by back-propagating errors. *Nature*, 323:533–536.

Ruppert, D. (1985). M-estimators. In Kotz, S., Johnson, N. L., and Reed, C. B., editors, *Encyclopedia of Statistical Sciences, volume 5*. John Wiley, New York. 443–449.

Saarinen, S., Bramley, R., and Cybenko, G. (1993). Ill-conditioning in neural network training problems. *IEEE J. Sci. Comput.*, 14:693–714.

Sage, A. P. and Melsa, J. L. (1971). *System Identification*. Academic Press, New York.

Samson, C. (1982). A unified treatment of fast algorithms for identification. *Int. J. of Control*, 35:909–934.

Samson, C. and Reddy, V. K. (1983). Fixed point error analysis of the normalized ladder algorithm. *IEEE Trans. Acoustics, Speech and Signal Processing*, ASSP–31:1177–1191.

Scales, L. E. (1985). *Introduction to Non-linear Optimization*. Computer Science Series. MacMillan.

Schoukens, J., Guillaume, P., and Pintelon, R. (1993). Design of broadband excitation signals. In Godfrey, K., editor, *Perturbation Signals for System Identification*, chapter 3. Prentice-Hall International, Hemel Hampstead, UK.

Schoukens, J. and Pintelon, R. (1991). *Identification of Linear Systems: A Practical Guideline to Accurate Modeling*. Pergamon Press, London (U.K.).

Schoukens, J., Pintelon, R., and Van hamme, H. (1994). Identification of linear dynamic systems using piecewise constant excitations: Use, misuse and alternatives. *Automatica*, 30(7):1153–1169.

Schoukens, J., Pintelon, R., Vandersteen, G., and Guillaume, P. (1997). Frequency-domain system identification using non-parametric noise models estimated from a small number of data sets. *Automatica*, 33(6):1073–1087.

Schrama, R. J. P. (1991). An open-loop solution to the approximate closed-loop identification problem. In *Proc. 9th IFAC Symposium on Identification and System Parameter Estimation*, pages 1602–1607, Budapest.

Schrama, R. J. P. (1992). Accurate models for control design: the necessity of an iterative scheme. *IEEE Trans. Automatic Control*, 37:991–994.

Schroeder, M. (1970). Synthesis of low-peak factor signals and binary sequences with low autocorrelation. *IEEE Trans. Inform. Theory*, IT–16:85–89.

Schumaker, L. L. (1981). *Spline functions: Basic Theory*. Wiley, Chichester.

Schuster, A. (1894). On interference phenomena. *Phil. Mag.*, 37:509–545.

Schwarz, G. (1978). Estimating the dimension of a model. *Ann. Statist.*, 6:461–464.

Schwarze, G. (1964). Algoritmische Bestimmung der Ordnung und Zeitkonstanten bei P-, I- und D-gliedern mit zwei underschiedlichen Zeitkonstanten und Verzögerung bis 6. Ordnung. *Messen., Steuern., Regeln.*, 7:10–19.

Schweppe, F. C. (1973). *Uncertain Dynamic Systems*. Prentice-Hall, Upper Saddle River, N.J.

Shibata, R. (1976). Selection of an autoregressive model by Akaike's information criteria. *Biometrica*, 63:117–126.

Shibata, R. (1980). Asymptotically efficient selection of the order of the model for estimating parameters of a linear process. *Ann. Statist.*, 8:147–164.

Sin, K. S. and Goodwin, G. C. (1980). Checkable conditions for identifiability of linear systems operating in closed loop. *IEEE Trans. Autom. Control*, AC–25:722–729.

Sinha, S. and Caines, P. E. (1977). On the use of shift register sequences as instrumental variables for the recursive identification of multivariable linear systems. *Intern. J. Systems Sciences*, 4:131–138.

Sjöberg, J. and Ljung, L. (1995). Overtraining, regularization and searching for a minimum, with application to neural networks. *Int. J. Control*, 62:1391–1407.

Sjöberg, J., McKelvey, T., and Ljung, L. (1993). On the use of regularization in system identification. In *Preprints 12th IFAC World Congress*, volume 7, pages 381–386, Sydney.

Sjöberg, J., Zhang, Q., Ljung, L., Benveniste, A., Delyon, B., Glorennec, P., Hjalmarsson, H., and Juditsky, A. (1995). Nonlinear black-box modeling in system identification: A unified overview. *Automatica*, 31(12):1691–1724.

Skelton, R. E. (1989). Model error concepts in control design. *Int. J. Control*, 49:1725–1753.

Skeppstedt, A., Ljung, L., and Millnert, M. (1992). Construction of composite models from observed data. *Int. J. Control*, 55(1):141–152.

Slutsky, E. (1929). Sur l'extension de la théorie de periodogrammes aux suites des quantiés dépendentes. *Comptes Rendues*, 189:722–733.

Smith, R. and Doyle, J. C. (1992). Model validation: a connection between robust control and identification. *IEEE Trans. Automatic Control*, AC–37:942–952.

Smith, R. and Dullerud, G. (1996). Continuous-time control model validation using finite experimental data. *IEEE Trans. Automatic Control*, AC–41:1094–1105.

Snee, R. D. (1977). Validation of regression models. Methods and examples. *Technometrics*, 19:415–428.

Söderström, T. (1973). An on-line algorithm for approximative maximum likelihood identification of linear systems. Technical Report 7308, Department of Automatic Control, Lund Institute of Technology, Lund, Sweden.

Söderström, T. (1974). Convergence properties of the generalized least squares identification method. *Automatica*, 10:617–626.

Söderström, T. (1975a). Comments on order assumption and singularity of information matrix for pulse transfer function models. *IEEE Trans. Automatic Control*, AC–20:445–447.

Söderström, T. (1975b). Tests of pole-zero cancellation in estimated models. *Automatica*, 11:537–541.

Söderström, T. (1975c). On the uniqueness of maximum likelihood identification, *Automatica*, 11:193–197.

Söderström, T. (1977). On model structure testing in system identification. *Intern. J. Control*, 26:1–18.

Söderström, T. (1981). Identification of stochastic linear systems in presence of input noise. *Automatica*, 17:713–725.

Söderström, T. (1987). Model structure determination. In *Encyclopedia of Systems and Control* (M. Singh, ed.), Pergamon Press, Elmsford, N.Y.

Söderström, T. and Åström, K. J., editors (1995). *Special Issue on Trends in System Identification*. Automatica, Vol 31, No 12.

Söderström, T., Gustavsson, I., and Ljung, L. (1975). Identifiability conditions for linear systems operating in closed loop, *Intern. J. Control*, 21(2):56–60.

Söderström, T., Ljung, L., and Gustavsson, I. (1976). Identifiability conditions for linear multivariable systems operating under feedback. *IEEE Trans. Automatic Control*, AC–21(6):837–840.

Söderström, T. and Stoica, P. (1981). Comparison of some instrumental variable methods. Consistency and accuracy aspects. *Automatica*, 17:101–115.

Söderström, T. and Stoica, P. (1983). *Instrumental Variable Methods for System Identification*. Springer-Verlag New York. Lecture Notes in Control and Information Sciences.

Söderström, T. and Stoica, P. (1989). *System Identification*. Prentice-Hall Int., London.

Söderström, T. and Stoica, P. (1990). On covariance function tests used in system identification. *Automatica*, 26(1):125–133.

Söderström, T., Stoica, P., and Trulsson, E. (1987). Instrumental variable methods for closed loop systems. *Proc 10th IFAC Congress*, Munich.

Solbrand, G., Ahlén, A., and Ljung, L. (1985). Recursive methods for off-line identification. *Int. J. Control*. 41:177–191.

Solo, V. (1978). *Time series recursion and stochastic approximation*. PhD thesis, Australian National University, Canberra.

Solo, V. (1979). The convergence of AML. *IEEE Trans. Automatic Control*, AC–24:958–963.

Solo, V. (1981). The second order properties of a time series recursion. *Ann. Statist.*, 9:307–317.

Solo, V. and Kong, X. (1995). *Adaptive Signal Processing Algorithms*. Prentice-Hall, Upper Saddle River, N.J.

Sontag, E. (1993). Neural networks for control. In *Essays on Control: Perspectives in the Theory and its Applications*, volume 14, pages 339–380. Birkhäuser.

Spriet, J. A. and Vansteenkiste, G. C. (1982). *Computer Aided Modelling and Simulation*. Academic Press, New York.

Staley, R. M. and Yue, P. C. (1970). On system parameter identifiability. *Inform. Sci.*, 2:127–138.

Steiglitz, K. and McBride, L. E. (1965). A technique for the identification of linear systems. *IEEE Trans. Automatic Control* AC–10:461–464.

Stoica, P. (1976). The repeated least squares identification method. *Journal A*, 17:151–156.

Stoica, P. (1981). On multivariable persistently exciting signals. *Bul. Inst. Politechnic*, "Gh. Gheorghiu-dej" Bucuresti, seria Electrotechnica, XLIII:59–64.

Stoica, P., Cedervall, M., Sorelius, J., and Söderström, T. (1997). An instrumental variable solution to an extended Frisch problem with application to blind channel estimation. In Huffel, S. V., editor, *Recent Advances in Total Least Squares Techniques and Error-in-Variables Modeling*. SIAM, Philadelphia, USA.

Stoica, P. and Moses, R. L. (1997). *Introduction to Spectral Analysis*. Prentice-Hall, Upper Saddle River, N.J.

Stoica, P. and Söderström, T. (1981a). The Steiglitz-McBride identification algorithms revisited—convergence analysis and accuracy aspects. *IEEE Trans. on Automatic Control*, AC–26:712–717.

Stoica, P. and Söderström, T. (1981b). Asymptotic behavior of some bootstrap estimators. *Intern. J. Control*, 33:433–454.

Stoica, P. and Söderström, T. (1982a). A useful parametrization for optimal experimental design. *IEEE Trans. Automatic Control*, AC–27:986–989.

Stoica, P. and Söderström, T. (1982b). Instrumental variable methods for identification of Hammerstein systems, *Intern. J. Control*, 35(3):459–476.

Stoica, P. and Söderström, T. (1982c). On nonsingular information matrices and local identifiability. *Intern. J. Control*, 36:323–329.

Stoica, P. and Söderström, T. (1983). Optimal instrumental variable estimation and approximate implementation. *IEEE Trans. Automatic Control*, AC–28:757–772.

Stoica, P., Friedlander, B., and Söderström, T. (1986). Instrumental variable methods for ARMA models. In *Control and Dynamic Systems-Advances in Theory and Applications*. vol. 26 (C. T. Leondes, ed.), Academic Press, New York.

Stoica, P., Holst, J., and Söderström, T. (1982). Eigenvalue location of certain matrices arising in convergence analysis problems. *Automatica*, 18:487–491.

Stoica, P., Söderström, T., and Friedlander, B., (1985). Optimal instrumental variable estimates of the AR-parameters of an ARMA-process. *IEEE Trans. Automatic Control*, AC–30:1066–1074.

Stoica, P., Söderström, T., Ahlén, A., and Solbrand, G. (1984). On the asymptotic accuracy of pseudo-linear regression algorithm. *Intern. J. Control*, 39:115–126.

Stoica, P., Söderström, T., Ahlén, A., and Solbrand, G. (1985). On the convergence of pseudo-linear regression algorithms. *Intern. J. Control*, 41:1429–1444.

Stone, M. (1974). Cross-validity choice and assessment of statistical predictors. *J. Royal Stat. Soc.*, ser. B. 36:111–147.

Stone, M. (1977a). Asymptotics for and against cross-validation. *Biometrika*, 64:29–35.

Stone, M. (1977b). An asymptotic equivalence of choice of model by cross-validation and Akaike's criterion. *J. Royal Stat. Soc.*, ser. B, 39:44–47.

Takagi, T. and Sugeno, M. (1985). Fuzzy identification of systems and its applications to modeling and control. *IEEE Trans. Systems, Man, and Cybernetics*, SCM–15(1):116–132.

Talmon, J. L. and van den Boom, A. J. W. (1973). On the estimation of transfer function parameters of process and noise dynamics using a single-stage estimator. *Proc. 3rd IFAC Symposium on Identification and System Parameter Estimation*, The Hague (North-Holland, Amsterdam), 929–938.

Tikhonov, A. N. and Arsenin, V. Y. (1977). *Solutions of ill-posed problems*. Winston, Washington, D.C.

Tong, H. (1990). *Nonlinear Time-series Analysis*. Oxford University Press, Oxford.

Traub, J. F. and Wozniakowski, H. (1980). *A General Theory of Optimal Algorithms*, Academic Press, New York.

Tse, D., Dahleh, M., and Tsitsiklis, J. (1993). Optimal asymptotic identification under bounded disturbances. *IEEE Trans Automatic Control*, AC–38:1176–1190.

Tse, E. and Anton, J. (1972). On the identifiability of parameters. *IEEE Trans. Automatic Control*, AC–17:637–646.

Tse, E. and Weinert, H. L. (1975). Structure determination and parameter identification for multivariable stochastic linear systems. *IEEE Trans. Automatic Control*, AC–20(5):603–613.

Tsypkin, Ya. Z. (1984). *On the Foundations of Information and Identification Theory* (in Russian). Izd., Nauka, Moscow.

Unbehauen, H. and Göhring, B. (1974). Test for determining model order in parameter estimation. *Automatica*, 10:233–244.

Unbehauen, H., Göhring, B., and Bauer, B. (1974). *Parameterschätzverfahren zur Systemidentification*. R. Oldenburg Verlag, Munich.

Unbehauen, H. and Rao, G. P. (1987). *Identification of Continuous-Time Systems*. North-Holland, Amsterdam.

van den Boom, A. J. and van den Enden, A. E. M. (1974). The determination of the orders of process and noise dynamics. *Automatica*, 10:244–256.

Van den Hof, P. M. J., Heuberger, P. S., and Bokor, J. (1995). System identification with generalized orthonormal basis functions. *Automatica*, 31(12):1821–1834.

Van den Hof, P. M. J. and Schrama, R. J. P. (1993). An indirect method for transfer function estimation from closed loop data. *Automatica*, 29(6):1523–1527.

Van den Hof, P. M. J. and Schrama, R. J. P. (1995). Identification and control - closed-loop issues. *Automatica*, 31(12):1751–1770.

Van den Hof, P. M. J., Schrama, R. J. P., de Callafon, R. A., and Bosgra, O. H. (1995). Identification of normalised coprime plant factors from closed-loop experiment data. *Europ. J. Control*, 1:62–74.

Van der Ouderaa, E., Schoukens, J., and Renneboog, J. (1988). Peak factor minimization using a time-frequency domain swapping algorithm. *IEEE Trans. Instrum. and Meas.*, 37:342–351.

van Overbeek, A. J. M. and Ljung, L. (1982). On-line structure selection for multivariable state space models. *Automatica*, 18(5):529–543.

Van Overschee, P. and DeMoor, B. (1994). N4SID: Subspace algorithms for the identification of combined deterministic-stochastic systems. *Automatica*, 30:75–93.

Van Overschee, P. and DeMoor, B. (1996). *Subspace Identification of Linear Systems: Theory, Implementation, Applications*. Kluwer Academic Publishers.

Van Overschee, P. and DeMoor, B. (1997). Closed-loop subspace system identification. *Submitted to Automatica.*

Van Overschee, P., DeMoor, B., Aling, H., Kosut, R. L., and Boyd, S. (1994). *Xmath interactive system identification module*. Integrated Systems, Inc., Santa Clara, CA.

van Zee, G. A. and Bosgra, O. H. (1982). Gradient computation in prediction error identification of linear discrete-time systems. *IEEE Trans. Automatic Control*, AC–27:738–739.

Vandersteen, G., Van hamme, H., and Pintelon, R. (1996). General framework for asymptotic properties of generalized weighted nonlinear least squares estimators with deterministic and stochastic weighting. *IEEE Trans. Automatic Control*, AC–41(10):1501–1507.

Varlaki, P., Terdik, G., and Lototsky, V. A. (1985). Tests for linearity and bilinearity of dynamic systems. *Proc. 7th IFAC Symp. on System Identification*, York, U.K. Pergamon Press, New York, 731–736.

Verhaegen, M. (1991). A novel non-iterative MIMO state space model identification technique. In *Proc. 9th IFAC/IFORS Symposium on Identification and System Parameter Estimation*, pages 1453–1458, Budapest, Hungary.

Verhaegen, M. (1993). Application of a subspace model identification technique to identify LTI systems operating in closed-loop. *Automatica*, 29(4):1027–1040.

Verhaegen, M. (1994). Identification of the deterministic part of MIMO state space models, given in innovations form from input-output data. *Automatica*, 30(1):61–74.

Viberg, M. (1995). On subspace-based methods for the identification of linear time-invariant systems. *Automatica*, 31(12):1835–1852.

Viberg, M., Ljung, L., Ottersten, B., and Wahlberg, B. (1993). Performance of subspace based state-space, system identification methods. In *Proc. 12th IFAC Congress, Session Th-M-5*, vol. 7:369, Sydney, Australia.

Viberg, M., Wahlberg, B., and Ottersten, B. (1997). Analysis of state space system identification methods based on instrumental variables and subspace fitting. *Automatica*, 33(9):1603–1616.

Vidyasagar, M. (1985). *Control System Synthesis: A Factorization Approach*. MIT Press, Cambridge, MA.

Wahba, G. (1980). Parameter estimation in linear dynamic systems. *IEEE Trans. Automatic Control*, AC–25:235–238.

Wahba, G. (1987). Inverse and ill-posed problems. In Engel, H. and Groetsch, C., editors, *Three topics in ill-posed problems*. Academic Press, New York.

Wahba, G. (1990). *Spline Models for Observational Data. SIAM*. 3600 University City Center, Philadelphia, PA 19104–2688.

Wahlberg, B. (1986). On model reduction in system identification. *Proc. Americ. Control Conf.*, Seattle, Wash.

Wahlberg, B. (1989). Model reduction of high order estimated models: The asymptotic ML approach. *Int. J. Control*, 49(1):169–192.

Wahlberg, B. (1991). System identification using Laguerre models. *IEEE Trans. Automatic Control*, AC–39:1276–1282.

Wahlberg, B. and Ljung, L. (1986). Design variables for bias distribution in transfer function estimation. *IEEE Trans. Automatic Control*, AC–31:134–144.

Wahlberg, B. and Ljung, L. (1992). Hard frequency domain model error bounds from least-squares like identification techniques. *IEEE Trans. Automatic Control*, AC–37:900–912.

Wald, A. (1949). Note on the consistency of the maximum likelihood estimate. *Anti. Math. Statis.*, 20:595–601.

Walker, A. M. (1961). Large-sample estimation of parameters for moving-average models. *Biometrika*, 48:343–357.

Walker, A. M. (1964). Asymptotic properties of least-squares estimates of parameters of the spectrum of a stationary non-deterministic time-series. *J. Aust. Math. Soc.*, 4:363–384.

Walker, G. (1931). On periodicity in series of related terms. *Proc. Roy Soc. London*, ser. A, 131:518–532.

Walter, E. (1982). *Identifiability of State Space Models*. Springer-Verlag, New York.

Walter, E. and Piet-Lahanier, H. (1990). Estimation of parameter bounds from bounded error data: a survey. *Math. Comput Simul.*, 32:449–468.

Watson, G. (1969). Smooth regression analysis. *Sankhya, Ser. A*, 26:359–372.

Wei, W. W.-S. (1990). *Time Series Analysis: Univariate and Multivariate Methods*. Addison-Wesley, Reading, MA.

Welch, P. D. (1967). The use of the fast Fourier transform for the estimation of power spectra: A method based on time averaging over short modified periodograms. *IEEE Trans. Audio Electroacoust.*, AU–15:70–73.

Wellstead, P. E. (1978). An instrumental product moment test for model order estimation. *Automatica*, 14:89–91.

Wellstead, P. E. (1979). *Introduction to Physical System Modelling*. Academic Press, New York.

Wellstead, P. E. (1981). Non-parametric methods of system identification. *Automatica*, 17:55–69.

Wellstead, P. E. and Rojas, R. A. (1982). Instrumental product moment model-order testing: Extensions and applications. *Intern. J. Control*, 35(6):1013–1027.

Werbos, P. (1974). *Beyond regression: new tools for prediction and analysis in the behavioral sciences*. PhD thesis, Harvard University.

White, S. (1975). An adaptive recursive digital filter. *Proc. 9th Annual Asilomar Conference on Circuits, Systems and Computers*, 21–25.

Whittle, P. (1951). *Hypothesis Testing in Time Series Analysis*, thesis, Uppsala University, Almqvist and Wiksell, Uppsala. Hafner, New York.

Whittle, P. (1963). *Prediction and Regulation by Linear Least Square Methods*, English University Press, London.

Widrow, B. and Stearns, S. (1985). *Adaptive Signal Processing*. Prentice-Hall, Upper Saddle River, N.J.

Wiener, N. (1930). Generalized harmonic analysis. *Acta. Math.*, 55:117–258.

Wiener, N. (1949). *The Extrapolation, Interpolation and Smoothing of Stationary Time Series with Engineering Applications*. Wiley, New York.

Wiggins, R. A. and Robinson, E. A. (1965). Recursive solution to the multichannel filtering problem, *J. Geophys. Res.*, 70:1885–1891.

Wigren, T. (1993). Recursive prediction error identification using the nonlinear Wiener model. *Automatica*, 29(4):1011–1025.

Willems, J. C. (1985). Modelling, approximation and complexity of linear systems. *Proc. of MTNS–85*, Stockholm, Sweden (C. I. Byrnes, A. Lindquist, eds.), North-Holland, Amsterdam, 277–285.

Willems, J. C. (1987). From time series to linear systems—part III. Approximate modelling. *Automatica*, 23(1):87–115.

Williams, P. M. (1995). Bayesian regularization and pruning using a Laplace operator. *Neural Computation*, 7:117–143.

Withers, C. S. (1981). Central limit theorems for dependent variables I. *Z. Wahrsch. verw Gebiete*, 57:509–534.

Wold, H. (1938). *A Study in the Analysis of Stationary Time Series*. Almqvist and Wiksell, Uppsala, Sweden.

Wong, K. Y. and Polak, E. (1967). Identification of linear discrete time systems using the instrumental variable approach. *IEEE Trans. Automatic Control*, AC–12:707–718.

Woodside, C. M. (1971). Estimation of the order of linear systems. *Automatica*, 7:727–733.

Youla, D. C. (1961). On the factorization of rational matrices. *IEEE Trans. Inform. Theory*, IT–7:172–189.

Young, P. C. (1965). On a weighted steepest descent method of process parameter estimation. Report, Cambridge University, Engineering Laboratory, Cambridge, England.

Young, P. C. (1968). The use of linear regression and related procedures for the identification of dynamic processes. In *Proc. 7th IEEE Symposium on Adaptive Processes*, pages 501–505, San Antonio, TX.

Young, P. C. (1976). Some observations on instrumental variable methods of time series analysis. *Intern. J. Control*, 23:593–612.

Young, P. C. (1984). *Recursive Estimation and Time-Series Analysis*. Springer Verlag, Berlin.

Young, P. C. and Jakeman, A. J. (1979). Refined instrumental variable methods of recursive time-series analysis. Part I: Single input, single output systems. *Intern. J. Control*, 29:1–30.

Young, P. C., Jakeman, A. J., and McMurtrie, R. (1980). An instrumental variable method for model order identification. *Automatica*, 16:281–294.

Yuan, Z. D. and Ljung, L. (1985). Unprejudiced optimal open loop input design for identification of transfer functions. *Automatica*, 21:697–708.

Yule, G. U. (1927). On a method for investigating periodicities in disturbed series with special reference to Wolfer's sunspot numbers. *Philos. Trans. Roy. Soc. London*, ser. A, 226:267–298.

Zadeh, L. (1975). Fuzzy logic and approximate reasoning. *Synthese*, 30:407–428.

Zang, Z., Bitmead, R. R., and Gevers, M. (1995). Iterative weighted least-squares identification and weighted LQG control design. *Automatica*, 31(11):1577–1594.

Zarrop, M. B. (1979). *Optimal Experimental Design for Dynamic System Identification*. Springer-Verlag, New York.

Zhang, Q. and Benveniste, A. (1992). Wavelet networks. *IEEE Trans. Neural Networks*, 3:889–898.

Zhu, Y. C. (1989). Asymptotic properties of prediction error models. *Int. J. Adaptive Control and Signal Processing*, 3:357–373.

Zhu, Y. C. and Backx, A. C. M. P. (1993). *Identification of Multivariable Industrial Process for Simulation Diagnosis and Control*. Springer Verlag, London.

Ziegler, J. G. and Nichols, N. B. (1942). Optimum settings for automatic controllers. *Trans. ASME*, 64:759–768.

Zoubir, A. M. and Boashash, B. (1998). The bootstrap and its application in signal processing. *IEEE Signal Processing Magazine*, 15(1):56–76.

Subject Index

A

B

C

D

E

F

G

H

I

J

K

L

M

N

O

P

Q

R

S

T

Reference Index

A

B

C

D

E

F

G

H

I

J

K

L

M

N

O

P

Q

R

S

T

U

V

W

Y

Z